D1552369

Wetland and Stream Rapid Assessments

Wetland and Stream Rapid Assessments

Development, Validation, and Application

Edited by

John Dorney
Moffatt and Nichol, Raleigh, NC, United States

Rick Savage
Carolina Wetlands Association, Raleigh, NC, United States

Ralph W. Tiner
Institute for Wetland & Environmental Education & Research, Leverett, MA, United States

Paul Adamus
Oregon State University, Corvallis, OR, United States

Academic Press is an imprint of Elsevier
125 London Wall, London EC2Y 5AS, United Kingdom
525 B Street, Suite 1650, San Diego, CA 92101, United States
50 Hampshire Street, 5th Floor, Cambridge, MA 02139, United States
The Boulevard, Langford Lane, Kidlington, Oxford OX5 1GB, United Kingdom

Cover credit: Marc Adamus Wilderness Photography https://www.marcadamus.com/

Library of Congress Cataloging-in-Publication Data
A catalog record for this book is available from the Library of Congress

British Library Cataloguing-in-Publication Data
A catalogue record for this book is available from the British Library

ISBN 978-0-12-805091-0

For information on all Academic Press publications
visit our website at https://www.elsevier.com/books-and-journals

Publisher: Candice Janco
Acquisition Editor: Louisa Hutchins
Editorial Project Manager: Emily Thomson
Production Project Manager: Omer Mukthar
Cover Designer: Victoria Pearson Esser

Typeset by SPi Global, India

Contents

Section 4.2
Stream Condition Methods

Section 4.3
Wetland Assessment Methods

Section 4.4
Implementing National-Scale and Regional-Scale Wetland Assessments

Section 4.5
Other Methods

Contributors

Numbers in parentheses indicate the pages on which the authors' contributions begin.

Francesco Accatino (189), Department of Biology, Western University (Ontario), London, ON, Canada

Paul Adamus (1, 209, 217, 219, 229, 233, 239, 241, 245, 249, 259, 261, 343, 453), Oregon State University, Corvallis, OR, United States

David A. Aldred (189), Department of Biology, Western University (Ontario), London, ON, Canada

Kory Angstadt (135, 433), College of William and Mary, Virginia Institute of Marine Science, Gloucester Point, VA, United States

Virginia Baker (251), Division of Water Resources, North Carolina Department of Water Quality, Raleigh, NC, United States

Florence Baptist (495), Biotope, Mèze, France

Mark A. Biddle (117), Delaware Department of Natural Resources and Environmental Control, Division of Watershed Stewardship, Wetlands Monitoring and Assessment Program, Dover, DE, United States

Donna M. Bilkovic (135), College of William and Mary, Virginia Institute of Marine Science, Gloucester Point, VA, United States

Juan Bravo (539), Wetlands Project, Heredia, Costa Rica

Robert Brooks (433), Department of Geography, Pennsylvania State University, University Park, PA, United States

Pierre Caessteker (495), Agence Française pour la Biodiversité, Direction de la Recherche, de l'Expertise et du Développement des Compétences, Vincennes, France

Beverley Clarkson (511), Landcare Research, Hamilton, New Zealand

Jean-Christophe Clément (495), Centre, Alpin de Recherche sur les Réseaux Trophiques et les Écosystèmes Limniques-UMR INRA-USMB, Le Bourget du Lac, France

Josh Collins (353, 443), San Francisco Estuary Institute and Aquatic Science Center, Richmond, CA, United States

Travis W. Crayosky (293), Stantec, Williamsburg, VA, United States

Irena F. Creed (189), University of Saskatchewan, Saskatoon, SK; Department of Biology, Western University (Ontario), London, ON, Canada

Kari F. Cretini (459), Cherokee Nation Technologies, Wetland and Aquatic Research Center, Lafayette, LA, United States

Dane Cunningham (305), AllStar Ecology LLC, Fairmont, WV, United States

Dave Davis (135), Virginia Department of Environmental Quality, Richmond, VA, United States

Douglas A. DeBerry (293), Environmental Science and Policy, Integrated Science Center, College of William and Mary; VHB, Inc., Williamsburg, VA, United States

John Dorney (1, 209, 261, 273, 325, 335, 491), Moffatt and Nichol, Raleigh, NC, United States

Chad Fizzell (79), Michigan Department of Environmental Quality, Lansing, MI, United States

Maxime Fossey (495), Muséum national d'Histoire naturelle, UMS Patrimoine Naturel, Paris, France

Juliette Gaillard (495), Muséum national d'Histoire naturelle, UMS Patrimoine Naturel, Paris, France

Anne Garwood (363), Michigan Department of Environmental Quality, Lansing, MI, United States

Stéphanie Gaucherand (495), Institut national de recherche en sciences et technologies pour l'environnement et l'agriculture, Unité de Recherche sur les Ecosystèmes Montagnards, Saint Martin d'Hères, France

Guillaume Gayet (495), Muséum national d'Histoire naturelle, UMS Patrimoine Naturel, Paris, France

Marcela Gutiérrez (539), National University of Costa Rica, Heredia, Costa Rica

Keto Gyekis (363), Michigan Department of Environmental Quality, Lansing, MI, United States

Marcy Hamilton (79), Southwest Michigan Planning Commission, Benton Harbor, MI, United States

Patricia Harris (453), Marine Biologist (retired), Juneau, AK, United States

Kirk J. Havens (135, 433), College of William and Mary, Virginia Institute of Marine Science, Gloucester Point, VA, United States

Michelle Henicheck (135), Virginia Department of Environmental Quality, Richmond, VA, United States

Ainsley Henry (529), University of the West Indies (Mona), Kingston, Jamaica

Erica Hernandez (371), University of Florida, Gainesville, FL, United States

Carl Hershner (135), College of William and Mary, Virginia Institute of Marine Science, Gloucester Point, VA, United States

Adam Hogg (173), Ontario Ministry of Natural Resources and Forestry, Peterborough, ON, Canada

Thomas Hruby (423), Washington State Department of Ecology, Lacey, WA, United States

Hannah Ingram (433), Department of Geography, Pennsylvania State University, University Park, PA, United States

Francis Isselin-Nondedeu (495), Ecole Polytechnique de l'Université François Rabelais, UMR 7324, CNRS CITERES, Tours, France

Amy Jacobs (433), The Nature Conservancy, Denton, MD, United States

Jeremy Jones (79), Michigan Department of Environmental Quality, Lansing, MI, United States

Susan Jones (363), Michigan Department of Environmental Quality, Lansing, MI, United States

Steven M. Kloiber (91), Minnesota Department of Natural Resources, St. Paul, MN, United States

Donovan C. Kotze (545), University of KwaZulu-Natal, Pietermartizburg, South Africa

Ken W. Krauss (459), U.S. Geological Survey, Wetland and Aquatic Research Center, Lafayette, LA, United States

José C. Leal (539), National System of Conservation Areas, San José, Costa Rica

Joanna Lemly (105), Colorado Natural Heritage Program, Fort Collins, CO, United States

Todd Losee (363), Niswander Environmental, Brighton, MI, United States

Douglas M. Macfarlane (545), Eco-Pulse Environmental Consulting Services, Hilton, South Africa

John J. Mack (401), Permit and Resource Management Department of Sonoma County, Santa Rosa, CA, United States

Jan M. Mackinnon (49), Georgia Department of Natural Resources, Coastal Resources Division, Brunswick, GA, United States

Robb D. Macleod (91), Ducks Unlimited, Memphis, TN, United States

Maryann M. McGraw (387), New Mexico Environment Department, Surface Water Quality Bureau, Santa Fe, NM, United States

Matt Meersman (79), Friends of the St. Joseph River, South Bend, IN, United States

Mick Micacchion (401), Midwest Biodiversity Institute, Hilliard, OH, United States

Elizabeth R. Milford (387), Natural Heritage New Mexico Division, Museum of Southwestern Biology, Albuquerque, NM, United States

Paul Minkin (381), New England District, US Army Corps of Engineers, Concord, MA, United States

Miriam Miranda (539), Wetlands Project, Heredia, Costa Rica

Esteban H. Muldavin (387), Natural Heritage New Mexico Division, Museum of Southwestern Biology, Albuquerque, NM, United States

Breda Munoz (251), RTI International, Durham, NC, United States

Tracie-Lynn Nadeau (281), United States Environmental Protection Agency, Region 10 (Pacific Northwest), Portland, OR, United States

Michael Nassry (433), Department of Geography, Pennsylvania State University, University Park, PA, United States

Dean J. Ollis (545), Freshwater Research Centre, Cape Town, South Africa

Kyle Paffett (475), Northern Arizona University, Flagstaff, AZ, United States

LeiLani Paugh (325, 335), North Carolina Department of Transportation, Raleigh, NC, United States

Sarai C. Piazza (459), U.S. Geological Survey, Wetland and Aquatic Research Center, Baton Rouge, LA, United States

Claire Poinsot (495), Biotope, Mèze, France

Brian Potter (173), Ontario Ministry of Natural Resources and Forestry, Peterborough, ON, Canada

Fabien Quétier (495), Biotope, Mèze, France

Bettina Rayfield (317), Virginia Department of Environmental Quality, Richmond, VA, United States

Kelly C. Reiss (371), American Public University, Charles Town, WV; University of Florida, Gainesville, FL, United States

Jeanne Richardson (317), U.S. Army Corps of Engineers, Norfolk District, Lynchburg, VA, United States

Tamia Rudnicky (135), College of William and Mary, Virginia Institute of Marine Science, Gloucester Point, VA, United States

Periann Russell (273), Division of Mitigation Services, NC Department of Environmental Quality, Raleigh, NC, United States

Erica Sachs-Lambert (381), US Environmental Protection Agency, Boston, MA, United States

Rick Savage (251), Carolina Wetland Association, Raleigh, NC, United States

Dan Schatt (135), College of William and Mary, Virginia Institute of Marine Science, Gloucester Point, VA, United States

Jacqueline N. Serran (189), Department of Biology, Western University (Ontario), London, ON, Canada

Gary P. Shaffer (459), Department of Biological Sciences, Southeastern Louisiana University, Hammond, LA, United States

Subodh Sharma (521), Aquatic Ecology Centre, Kathmandu University, Kathmandu, Nepal

Leigh Anne Sharp (459), Coastal Protection and Restoration Authority of Louisiana, Lafayette Field Office, Lafayette, LA, United States

M. Siobhan Fennessy (401, 443), Department of Biology, Kenyon College, Gambier, OH, United States

Sandy Smith (325, 335), Axiom Environmental, Raleigh, NC, United States

Brian Sorrell (511), Aarhus University, Aarhus, Denmark

Abraham E. Springer (475), Northern Arizona University, Flagstaff, AZ, United States

David Stanhope (135, 433), College of William and Mary, Virginia Institute of Marine Science, Gloucester Point, VA, United States

Stephen Stanley (151), Washington Department of Ecology, Shorelands and Environmental Assistance Program, Lacey, WA, United States

Eric D. Stein (353), Southern California Coastal Water Research Project, Costa Mesa, CA, United States

Lawrence E. Stevens (475), Museum of Northern Arizona, Flagstaff, AZ, United States

Gregory D. Steyer (459), U.S. Geological Survey, Southeast Region, Baton Rouge, LA, United States

Jeremy Sueltenfuss (105), Colorado State University, Fort Collins, CO, United States

Ralph W. Tiner (9, 21,163), Institute for Wetland & Environmental Education & Research, Leverett, MA, United States

Linda Vance (67), Montana Natural Heritage Program, University of Montana, Missoula, MT, United States

Regina Varrin (413), Ontario Ministry of Natural Resources and Forestry, Peterborough, ON, Canada

Walter Veselka (305), AllStar Ecology LLC, Fairmont, WV, United States

Jenneke M. Visser (459), School of Geosciences, University of Louisiana at Lafayette, Lafayette, LA, United States

Gang Wang (91), Ducks Unlimited, Memphis, TN, United States

Ryan Ward (305), AllStar Ecology LLC, Fairmont, WV, United States

Denice Wardrop (433), Department of Geography, Pennsylvania State University, University Park, PA, United States

Dale Webber (529), School for Graduate Studies and Research, University of the West Indies, Kingston, Jamaica

Mona Webber (529), Centre for Marine Sciences, Department of Life Sciences, University of the West Indies, Kingston, Jamaica

William B. Wood (459), Coastal Protection and Restoration Authority of Louisiana, Lafayette Field Office, Lafayette, LA, United States

Rebecca Zeran (413), Ontario Ministry of Natural Resources and Forestry, Peterborough, ON, Canada

Foreword

It is estimated that wetlands comprise only around 5% of the global land surface area, with streams and rivers adding an additional 0.5%. Nevertheless, thousands of research studies over the past fifty years have demonstrated that wetlands and streams are among the most vital ecosystems on the planet as they provide unique functions, which in turn provide values and services to society, that are much greater than their areas suggest. For example, scientific findings confirm that wetlands store water and attenuate flooding and transform elements like nitrate–nitrogen back into the atmosphere as N_2 gas, thus reducing pollution downstream. Both streams and wetlands also provide vital habitat for thousands of species and function as refuge for rare flora and fauna. These scientific studies provided important information, as millions of hectares of wetlands were being drained and converted to agriculture, forestry, and urban development, with the US destroying >50% in the lower 48 states by the end of the 20th century. Concurrently, thousands of streams and rivers were being channelized, polluted, and dammed, resulting in significant losses in water quality, habitat and fisheries. Scientific studies were also undertaken to assess wetland and stream structure and function as well as quantify the effects of drainage and development on these ecosystems. However, scientific studies take considerable time, effort, expertise, and significant funding to collect and process the detailed functional and structural data needed to quantify ecosystem processes and functions and, in turn, each system's value. For example, over the last four decades my graduate students and colleagues in the Duke University Wetland Center needed millions of dollars of funding to quantify wetlands functions such as carbon and phosphorus cycling, primary productivity, habitat suitability, and hydrologic flow in the Everglades of Florida, the Pocosins of North Carolina, and the Mesopotamian marshes of Iraq. We tried with only moderate success to take the detailed scientific data we collected on key functions like carbon sequestration, groundwater recharge, nutrient transformations, and decomposition and translate these into values and services that society would recognize and support. Unfortunately, wetland science, while invaluable, often needs a more transferable way to communicate the functions and values of wetlands and streams to gain public interest.

Fortunately, fabulous popular books like John and Mildred Teal's *Life and Death of the Salt Marsh* (1969) spurred public interest in wetlands, and water quality, and by the 1970s the value of wetlands and streams was becoming recognized by the public, policy makers and land managers. As a result, Congress passed the Clean Water Act of 1972 (amended 1977 and 1987), which provided some wetlands protection under section 404. This act, along with several key Supreme Court decisions, provided some restrictions and guidelines for determining isolated wetlands (SWANCC, 2001) and determined that wetlands without a hydrological or ecological connection to other navigable waters do not fall within the jurisdiction of the Clean Water Act (Rapanos, 2006). In the 1990s the EPA and the U.S. Army Corps of Engineers promulgated additional mitigation rules to reduce the impacts of development on wetlands and streams by requiring replacement of ecosystem functions on the landscape.

Detailed and expensive scientific studies had proven the value of these wetland ecosystems. However, it was also evident that not all wetlands and streams were functioning at the same level, and thus the value and services they provided to society were not of equal value. The scientific and management community had to come up with a rapid and less expensive way to accurately assess the functions and values of tens of thousands wetlands and streams on the landscape.

Fittingly, a diversity of multiple Rapid Assessment Methods (RAMs) were being developed in many U.S. states and in countries around the world. These methods required less time in the field and less expertise than more quantitative scientific methods, leading to cost reductions and the ability to rapidly assess more sites. Most approaches developed a single (often numerical) score method that could be used by many wetland and stream managers for assessing the overall condition, quality, or health of a wetland or stream. RAMs were to characterize individual stream or wetland sites and, importantly, compare them with reference sites to determine a loss of function. Accordingly, RAMs took on an important role in wetland and stream monitoring as they provided key information to decision makers about watershed impacts or the conservation potential of wetlands and streams. They also provided a rapid method for evaluating project impacts, designing appropriate mitigation, and/or developing strategies that promote conservation. However, in the past there has been no easy way to compare and assess the multitude of RAMS being developed and used across the country or to inform managers, policy makers, and scientists with the strengths and weakness of each approach.

Fortunately, we now have the comprehensive volume entitled *Wetland and Stream Rapid Assessments: Development, Validation and Application,* written and edited by renowned scientist, managers and practioners who have specialized in the development and use of RAMs for specific regions of the United States as well as a few international countries. As noted by the editors, this book was largely inspired by the need for rapid systematic tools to help evaluate development impacts on wetlands and streams, and to achieve appropriate mitigation quotients based on both the quality and quantity of changes in the resource. The depth of coverage in this book will help the user community make better use of wetland and stream RAM methods as they gain a deeper understanding of how RAMs have been developed, tested, and applied. The book provides the reader with a multitude of approaches that at first glance may appear different, but importantly most use some type of scaling procedure that allows for the amalgamation of diverse ecosystem variables (hydrologic flow, litter depth, habitat, etc.) selected as indicators metrics into scores or ratings for functions, values and in many cases health. It should also be noted that the framework of Dr. Mark Brinson's 1993 hydrogeomorphic (HGM) assessment system underlies many of the RAM approaches used for wetlands throughout the country.

Finally, the book presents not only the foundation for RAM development but also individual state-by-state case studies and historical perspective on the development of RAM as it has evolved over the past decades. This book is more than a typical recipe book, as the authors share their trials and errors as well as make suggestions for the future improvement needs in RAM methodology. The list of authors is a who's who of scientists and managers who have not only developed and tested RAMs but have also provided wonderful regional case studies to provide the reader with a diversity of approaches. This book should occupy a critical place on the shelves of all wetland and stream scientists and managers as well as policy makers and practitioners who are interested in evaluating development impacts to wetlands and streams as well as developing appropriate mitigation requirements and selection of areas for conservation.

REFERENCES

Rapanos, 2006. Rapanos v. United States, 547 U.S. 715.

SWANCC, 2001. Solid Waste Agency of Northern Cook County versus Army Corps of Engineers, 531 U.S. 159.

Curtis J. Richardson

Duke University Wetland Center, Nicholas School of the Environment, Durham, NC, United States

Introduction

John Dorney*, Paul Adamus†, Ralph W. Tiner‡, Mary E. Kentula§ and Rick Savage¶

**Moffatt and Nichol, Raleigh, NC, United States, †Oregon State University, Corvallis, OR, United States, ‡Institute for Wetland & Environmental Education & Research, Leverett, MA, United States, §US Environmental Protection Agency, Office of Research and Development, Corvallis, OR, United States, ¶Carolina Wetland Association, Raleigh, NC, United States*

BACKGROUND

The scientific study of wetlands and streams goes back many decades (e.g., Warming, 1895; Warming et al., 1909; MacMillan, 1898; Cowles, 1899; Gates, 1926). It is probably an inherent characteristic of humans to be aware of the quality of wetlands and streams as well as their presence and extent. However, formal and standardized rapid assessment methods (called RAMs in this book) date from the 1950s in the United States and Canada for wetlands and from the 1980s for streams. RAMs were originally developed for evaluating fish and wildlife habitats. They have evolved to include many other attributes of wetlands and streams as resource management has become more urgent. The development and expansion of RAMs based on geographic information systems (GIS) have also been notable over the last several decades. RAMs have continued to evolve in response to increased experience in their use among practitioners. While wetlands and streams continue to be the main foci of RAM development and use, efforts have begun to develop RAMs for other types of ecosystems. At least in North America, the initial focus on RAMs for wetlands and streams reflects the legal needs of programs that singled out these two ecosystems for enhanced protection, restoration, and/or prioritization. There is also growing interest in developing or modifying RAMs to cover additional—and in some cases, broader—geographic areas.

This book aims to describe processes used to develop, calibrate, and test both GIS-based and field-based wetland and stream RAMs, recognizing that these processes as well as many of the RAMs themselves will continue to evolve and be updated. Examples of a variety of RAMs are provided to illustrate alternative approaches. We emphasize RAM development in North America but also provide case histories from some other continents. We have chosen not to recommend any particular wetland or stream RAM, allowing readers to choose which best fits their own intended purposes and geography as well as meets their criteria for speed, required skill level, comprehensiveness, accuracy, repeatability, sensitivity, documentation, and other factors that may be considered important in a particular local or regional context.

Our hope is that this book will also be helpful for potential RAM developers as they create, revise, and/or employ RAMs that are robust and useful for a wide range of purposes. Our goal for this book was to produce an essential resource in response to the fact that RAMs have become critical tools in implementing environmental management measures across the world in a wide variety of settings. Because this book was developed in the 2016–2017 time frame, readers should contact the authors of any particular RAM to learn of revisions and updates as well as to determine how that RAM is currently being used.

DEFINITIONS

The definition of various terms is vital to correctly to understand this book and the RAMs it describes. Unfortunately, different RAM authors have used some of these definitions for key terms in multiple ways. This situation is to be expected given the fact that most RAMs have been developed independently of each other, in different regions, and, in some cases, for different purposes. We provide the following definitions to provide consistency in the use of terms in this book and to hopefully clarify the discussion of RAMs here and in the future. We provided these definitions to the chapter authors and suggested (but did not require) their consistent use. However, if case history authors preferred a different definition, they were urged to note that difference in their chapter.

RAMs are standardized procedures that generate a score, index, or rating for the ecological status of a specified site (individual wetland, stream, watershed, etc.) and/or its individual ecosystem services (functions, values), or other attributes (e.g., condition, sensitivity), based mainly on ground-level observations and/or by using aerial imagery/GIS. In general, there are two types of RAMs—landscape-level and field-based. Landscape-level RAMs are desktop approaches that utilize existing mapped data, the interpretation of aerial imagery, and GIS technology to produce assessments of wetland functions and/or condition. Field-based RAMs may use aerial imagery and maps to provide a context for a particular site, but their main focus is on features that are identified on site through ground surveys. The *observational* nature of most RAM procedures contrasts with procedures based mainly on ground-level *measurements* such as pounds of nutrients removed, mayfly counts, or acre-feet of water storage. Field-based RAMs commonly require just one field visit lasting less than 1 day. RAM scores are either on a categorical scale (for instance, high, medium, or low), or ordinal scale (for instance, 0–100). Both scales are often relative to a qualitative or quantitative description of a set of other wetlands/streams in a defined region or to a theoretical description reference condition for the wetland type being examined.

Variables are characteristics of a wetland, stream, or other landscape that can be observed or measured.

Indicators comprise the subset of variables used to calculate a score or rating for a site because they are believed most relevant to predicting or representing a specified ecosystem service or other attribute of the site (e.g., water storage).

Metrics are of two basic types, as described below. The decision to use either type is made on a case-by-case basis by developers of the RAM, depending on the intended purpose of the metric.

> *Indicator metrics* are derived from stand-alone variables or indicators such as presence of invasive species or soil texture. They can then be used in the subsequent calculation of scores/ratings in combination with synthesis metrics.
> *Synthesis metrics* are derived by combining multiple variables or indicators. They may be used as indicators in subsequent calculations of scores/ratings (e.g., perimeter-area ratio) or may themselves comprise the endpoint rating or score (e.g., index of biotic integrity, functional capacity index).

Condition—A measure of the integrity, health, or quality of a wetland, stream, or other ecosystem. This is generally synonymous with the term "status" of the wetland or stream. Condition is often reported within a range from least disturbed to highly altered, degraded, or disturbed, along with intermediate categories.

Functions—The natural processes performed by wetlands, streams, and other ecosystems. Hydrologic and water quality functions are usually defined by transfer rates or net fluxes (input minus output) of water, gases, sediment, and other substances. Habitat functions are usually defined in terms of the suitability of a wetland or stream to support particular species, species groups, or biodiversity generally. RAMs typically assess the relative levels of only a subset of all possible functions that ecosystems perform, and represent those levels using scores or ratings rather than by requiring direct measurement.

Values—The benefits that the functions of an ecosystem provide to humans, including consumable products as well as aesthetics and an appreciation for the existence of a pristine ecosystem (e.g., wilderness). RAMs represent these as scores or ratings rather than in monetary units, and focus on sustainable values such as hay or timber production rather than exploitive values such as most types of mining and developable real estate. In some RAMs, functions and values are treated equivalently.

Verification—Comparison of the results of a RAM to the expected results from a suite of study sites. In other words, does the RAM produce results that a designated group of people (subject experts, the developers of the method, and/or unaffiliated individuals) expect based on their experience and/or judgment? To preserve objectivity, the designated group commonly provides rankings or ratings of the sites before seeing the score-based rankings and/or ratings generated for the same sites by the RAM. Verification testing may also include testing a RAM's sensitivity and/or repeatability among independent users relative to other methods or relative to prespecified standards.

Validation—Similar to verification but actual measurements (rather than simply group opinions) are compared with the results of a RAM across a broad range of sites, usually based on long-term monitoring data. The measurements are believed to more directly and accurately represent what the RAM assesses, partly because they may be collected during repeated site visits over an extended period and/or use equipment or procedures that are more sophisticated or time- or cost-intensive (e.g., water level gauges, DNA analysis, radioisotope tracers) than the observational data collected as part of a RAM.

Calibration—Mathematical conversion ("normalization") of the score or scores from a specific site to a number (e.g., percentile) that represents the site's ranking relative to a larger set of sites within a specified area (e.g., a state or river basin). This adjustment can be based on the theoretical or actual maximum of the scores or measured condition of all other sites, or on the score or measured condition of one or more "reference" sites believed to represent a specified desirable state (e.g., presumed "least altered").

OVERVIEW OF THE STRUCTURE OF THE BOOK

This book starts with a summary of the history of wetland and stream RAMs (Chapter 1.0). Next, we discuss landscape-level methods that have evolved rapidly since the introduction of GIS and we provide numerous examples of these from

North America (Section 2). Then, Section 3 describes a general process for development, testing, and use of field-based methods (Chapters 3.1–3.12). Field-based case studies then follow in Section 4 starting with stream methods (first identification/flow duration methods and then condition-based methods) followed by wetland RAMs, then RAMs for other aquatic ecosystems, and, finally, a discussion of several large-scale methods. Section 5 addresses several RAMs from outside North America and provides examples from around the world in a wide variety of settings. Finally, Section 6 presents our conclusions about RAMs as well as our speculation as to the future direction of RAMs.

ACKNOWLEDGMENTS

The editors are grateful for the critical contributions of the numerous chapter authors who took their valuable time to summarize their work for this book. This book is not a comprehensive tally of all the methods in existence when the book was written (2016–2017) because that would be impossible. Rather, the chapters represent methods of which the editors were aware and which we believe provide valuable examples for the development or refinement of other RAMs. We sincerely appreciate the hard work and dedication of the authors who agreed to donate their valuable time to contribute a chapter to this book. We are especially grateful for the editing and encouragement of Mary E. Kentula, who was unable to be a formal editor but who continually encouraged us to move forward to this final product. Finally, we are extremely grateful for the help of Emily Thomson (and her team) with Elsevier Scientific, whose patience and encouragement over many months were critical to developing this final product.

REFERENCES

Cowles, H.C., 1899. The ecological relations of the vegetation on the sand dunes of Lake Michigan. Part I. Geographical relations of the dune floras. Bot. Gaz. 27 (2), 95–117.

Gates, F.C., 1926. Plant succession about Lake Douglas, Cheboygan County, MI. Bot. Gaz. 82, 170–182.

MacMillan, C., 1898. The occurrence of Sphagnum atolls in Central Minnesota. Minn. Bot. Stud. 9, 12.

Warming, E., 1895. Plantesamfund—grundtræk af den økologiske plantegeografi. P.G. Philipsens Forlag, Kjøbenhavn.

Warming, E., Vahl, M., Groom, P., Balfour, I.H., 1909. Oecology of Plants: An Introduction to the Study of Plant Communities. Oxford at the Clarendon Press, Oxford.

Chapter 1.0

History of Wetland and Stream RAMs

Paul Adamus* and John Dorney[†]
*Oregon State University, Corvallis, OR, United States, [†]Moffatt and Nichol, Raleigh, NC, United States

Chapter Outline

RAPID ASSESSMENT METHODS FOR WETLANDS

Since the beginnings of civilization, humans have sought to identify and communicate which natural places might be most important for providing food, shelter, water, or spiritual sustenance. In many communities, the favored places have often been in or around wetlands. During the latter part of the 20th century, a growing awareness of the capacity of wetlands to fill these human needs led to the enactment of laws that partially restrict wetland loss from human activities (e.g., drainage and filling) in a few western nations. Because not all wetlands were deemed essential or desirable to protect, and because of lack of consistency in making those decisions, a need became apparent for prioritizing individual wetlands using standardized criteria.

The earliest wetland rapid assessment methods (RAMs) grew from decades of experimentation with land classification criteria (Wathern et al., 1986). The land classification efforts had initially focused on production functions of the landscape (e.g., suitability for agriculture, forestry) but expanded into a concern for reducing hazards of development as might be predicted by measures of landscape sensitivity and vulnerability (e.g., McHarg and Mumford, 1969). Attention also focused on the capacity of natural systems to support ecological attributes such as biodiversity, naturalness (Siipi, 2004), and various concepts of "environmental quality" (e.g., Dee et al., 1973). The first efforts to develop criteria and methods specifically for wetlands focused mainly on the widely recognized importance of wetlands, in general, as waterbird habitat (e.g., Allan, 1956). For example, in 1961 the state of Maine initiated an inventory that not only mapped wetlands visible in aerial photographs but also implemented a RAM that focused on the relative importance of wetlands as waterbird habitat. These early efforts were based on observations of vegetation and other features made by a biologist during a single visit and recorded on a standardized data form (McCall, 1972). Maine's effort was part of a national effort coordinated by the US Fish and Wildlife Service (USFWS) that sought to identify "wetlands of importance" to waterfowl (Shaw and Fredine, 1956).

By the mid-1970s, a multidisciplinary research group at the University of Massachusetts had called attention to the fact that some wetlands might support other functions and values besides waterbird habitat. They developed simple models, applicable only to the northeastern United States, for scoring visual-cultural values (Smardon and Fabos, 1975), wildlife habitat (Golet, 1976), and groundwater recharge potential of individual wetlands (Larson, 1975). Parallel with a growing body of research that documented the existence of multiple functions of wetlands considered important to society, an increasing number of scientists began calling the attention of planners and bureaucrats to those functions. Literature syntheses and published discussions resulting from national symposia during that period helped articulate this collective knowledge (e.g., Greeson et al., 1979) and a growing number of states passed legislation that outlined a public interest in protecting a variety of wetland functions.

About the same time, in response to requirements of the new National Environmental Policy Act (NEPA), scientists needing to rate natural places according to their wildlife habitat importance devised the habitat evaluation procedures (U.S. Fish and Wildlife Service, 1980). Although focusing only on biological resources and not applicable exclusively

Wetland and Stream Rapid Assessments. https://doi.org/10.1016/B978-0-12-805091-0.00001-3

to wetlands, it was perhaps the first RAM applied at hundreds of locations nationwide and was used on a routine basis for assessing habitat functions. Focusing specifically on wetlands, in 1979 scientists and planners within a section of the U.S. Army Corps of Engineers prepared a RAM, intended for national use, for assessing eight broadly defined wetland functions or values (Reppert et al., 1979). The RAM provided no models or standardized guidance for rolling up scores of individual variables into ratings for a wetland or its eight potential functions. It was never formally adopted by the Corps, and apparently was seldom used in the years following its publication.

In 1980 the Federal Highway Administration (FHWA) sought to sponsor the development of a more structured, comprehensive, and well-documented RAM applicable to all wetland types present in the conterminous United States. The resulting RAM (Adamus, 1983) was not only considered the most comprehensive RAM available at that time (Stuber and Sather, 1984), but it added impetus to the growing political recognition of the multiplicity of economically important functions that wetlands provide. As Hollis and Bedding (1994) commented: "The general shift in thinking about wetlands has its roots in a report published in 1983 by Paul Adamus, a consultant working for the Federal Highway Administration in the U.S." Soon after the method's publication, 15 federal agencies and two NGOs sponsored a 3-day peer review by 40 recognized wetland scientists and resource managers. They concluded that the FHWA method was the best existing framework for wetland evaluation (Stuber and Sather, 1984). During the next 5 years, additional feedback was obtained from dozens of individuals, many who attended training sessions sponsored by the Corps of Engineers and who applied the RAM to wetlands in their regions of the United States. New comments were also provided by scientists recruited to provide peer review in a series of regional workshops (e.g., Kusler and Riexinger, 1986). From that feedback, a version revised by its author and renamed "Wetland Evaluation Technique" (WET) by the U.S. Army Corps of Engineers was published (Adamus et al., 1987) and distributed nationally while finding a growing number of users despite never being required by federal agencies for use in evaluating wetlands.

WET's attempt to be comprehensive may have been partly responsible for some users finding it too cumbersome for routine use, and frequency of its use eventually declined. Others considered that its synthesis of a huge amount of knowledge from diverse disciplines was a strong asset (Brinson, 1995). It was the first published RAM to be accompanied by free software that aided, but was not essential to, the scoring calculations. Additionally, it was the first RAM to clearly segregate, in the scoring/rating process, a wetland's functions and the values or benefits potentially resulting from those functions. It also was the first RAM to group the input data into three levels of increasing effort and precision: office, rapid field, and intensive field—which foreshadowed the levels 1 (landscape), 2 (rapid field), and 3 (intensive) scheme described by Brooks et al. (1996) and adopted by the US Environmental Protection Agency (EPA). But perhaps its most lasting contribution was that, for the first time, it provided (and where possible, documented) a relatively complete list of easily observable indicators ("predictors") for all the major functions and values of wetlands. In doing so, WET helped spawn a number of other RAMs (some described in this book's case histories) that focused on wetlands in selected parts of North America and used the same or similar indicators. Reviews of many of those methods can be found in Adamus (1992) and Bartoldus (1998).

Partly because those regional spinoffs of WET were being developed with little coordination at a national level, the Corps of Engineers published guidance for developing RAMs at finer-than-national scales. The guidance was termed the hydrogeomorphic (HGM) approach (Smith et al., 1995, 2013). It was drafted in response to a growing recognition that RAM accuracy might benefit from creating separate RAMs for each wetland type or at least for each region (as recommended by Adamus, 1984). The HGM approach was not an operational RAM, per se, but rather a conceptual framework to be used as a template for developing RAMs specific to particular regions and wetland types, using HGM classes defined by Brinson (1993). More than 30 such HGM-based RAMs have been published since the HGM initiative began. Many were funded by state governments and the USEPA, but apparently few came to be used routinely (Cole and Kooser, 2002), largely because no agencies required their use. Chapter 4.3.2 explains similarities and differences between the HGM approach and a more recent approach, WESP (Wetland Ecosystem Services Protocol, Adamus, 2016), which is noted here because it has been used as a template for regional RAMs developed and field-calibrated for agencies in Oregon, Alberta, four Atlantic Canada provinces, and parts of Alaska (Chapter 4.3.2).

While most RAMs focus on wetland functions and related attributes, a more recent development has been the creation and application of RAMs intended to assess wetland integrity or "condition," as defined and described further in Section 3.0. In most applications, these RAMs complement rather than substitute for those that assess wetland functions and values. A US Bureau of Land Management team led by Pritchard (1994) designed one of the first such RAMs and extensively applied it in the western United States. Like the FHWA method and some other RAMs used for function assessment, many condition assessment RAMs allow for increasing levels of data collection intensity with—it is hoped—increasing accuracy. As described by Brooks et al. (1996), "level 1" mainly uses existing spatial data and GIS to generate scores or ratings with relatively few data inputs but relatively low accuracy, whereas optional levels 2 and

3 refine those outputs using more intensive field observations or measurements. A recent attempt at creating a national-scale "level 2" RAM for condition assessment was USA-RAM (Collins and Fennessy, 2011), described in Chapter 4.4.2. Several other RAMs developed for assessing wetland condition, not functions or ecosystem services, were reviewed by Fennessy et al. (2004).

Increasingly, RAMs are using GIS and aerial imagery to provide preliminary estimates of the functions, values, and/or condition of individual wetlands or, more often, wetlands across entire landscapes (Lyon and McCarthy, 1995; Lyon et al., 2001). In addition, several RAMs now require GIS-derived data to complement ground-level observations. The increased use of GIS-derived data in RAMs has largely been the result of growing recognition in the early 1990s of landscape ecology as an important and distinct discipline, accompanied by increased availability of spatial data sets covering entire regions of the country, and increased use of computers and GIS for spatial analysis. Tiner (1996, 1997, 2002, 2003, 2005; Tiner et al., 1999) and Sutter and Wuenscher (1996) were among the first to describe systematic procedures for using GIS and aerial imagery for rapidly assessing wetland functions and/or condition. Tiner's procedure—termed "NWI-Plus" (NWI+)—has been employed in several watersheds and regions throughout the United States. Chapter 2.2.1 discusses this in more detail.

RAMs FOR STREAM AND RIPARIAN AREAS

Stream and riparian RAMs date from the late 1990s to the present. Work is ongoing in many states and Corps Districts to develop, test, refine, and implement these. In general, stream and riparian RAMs have taken one of two basic forms: (1) assessment of stream flow duration (delineating where streams begin and classifying different stream segments as ephemeral, intermittent, or perennial), and (2) assessment of the overall condition of streams or their riparian areas. Development of these two forms of RAMs has been a fairly recent phenomenon compared to the longer history of wetland RAMs described above. Similar to wetland RAMs, most stream and riparian methods in the United States have been developed to address needs under the CWA 404/401 regulatory program[1] or as tools to help implement requirements for riparian buffers in local and state regulatory programs. Descriptions of the RAMs below are mostly from Somerville (2010).

Stream Flow Duration Methods

The earliest formal stream identification and flow duration method apparently is the NC method (NC Stream Identification Method), which was initially developed in 1999 for use by the state of North Carolina for the Neuse River Riparian Protection Rules (15A NCAC 2B 0.0242, and HB 1257; NC Division of Water Quality, 2010) but has since been modified several times and is now in its fourth version (September 2010). Its use has expanded to the adjacent states of South Carolina, Georgia, Virginia, and Tennessee with Virginia subsequently developing a modification of the NC method. The method is described in more detail in Chapter 4.1.2. Although the NC method has also been used in India (Kumar et al., 2014), few RAMs have been developed and applied extensively to stream and riparian areas in other countries.

Use of the NC method then spread to Virginia (the Perennial Stream Identification Protocol) where the NC method was modified to identify perennial streams that are the focus of protected stream buffers in the Chesapeake Bay Rules by local governments in the area (Fairfax County Public Works and Environmental Services, 2003). Recently, a more comprehensive method has been developed for the Chesapeake Bay and is described in more detail in Chapter 4.1.3.

More recently (2009), in Oregon an Interim Streamflow Duration Assessment Method was developed initially based on the framework of the NC method, which evolved to a more streamlined, statistically based method (model) following a three-state, multiyear validation study supporting applicability of the revised model across the Pacific Northwest (Nadeau et al., 2015). This method is described in more detail in Chapter 4.1.1 of this book. The Rapanos decision of the US Supreme Court (2006) has been the main impetus for the development of stream flow duration RAMs because that decision requires agencies to distinguish between streams of differing flow regimes for determining jurisdiction under the US Clean Water Act.

Stream Condition and Function Assessment Methods

Similarly, RAMs to rapidly measure the overall condition or function of streams or their riparian areas are a fairly recent development, starting in the 1990s. A recent impetus for many of these methods in the United States appears to be the Joint Mitigation Rule of 2008 (U.S. Army Corps of Engineers and U.S. Environmental Protection Agency, 2008), which made

1. The 404/401 regulatory program is authorized under the Clean Water Act in the United States and requires approval for certain wetland and stream alterations from the US Army Corps of Engineers (404 Permit) and the corresponding state government (401 Certification).

clear that compensatory mitigation is required for all aquatic resources, not just wetlands. However, stream condition RAMs published by the US Bureau of Land Management (Prichard et al., 1993) and the states of Virginia, Ohio, and West Virginia (described elsewhere in this book) preceded that actual rule. In 2000, the West Virginia Stream and Wetland Valuation Metric was published as a method developed by the state of West Virginia to assess the quality of both streams and wetlands in the state (Barbour et al., 2000). This RAM is described in more detail in Chapter 4.2.1. In 2002, the state of Ohio developed the Field Evaluation Manual for Ohio's Primary Headwater Habitat Streams (Ohio EPA, 2002). In 2007, Virginia released the Unified Stream Methodology (USM), which was a collaborative effort between the U.S. Army Corps of Engineers Norfolk District and the Virginia DEQ (U.S. Army Corps of Engineers and Norfolk District and Virginia Department of Environmental Quality, 2007). This method is described in more detail in Chapter 4.2.2. In 2010, the Functional Assessment Approach for High Gradient Streams was developed by the Interagency Review Team (which is a multi-agency team established to provide oversight to the compensatory mitigation process in a particular Corps of Engineers District) led by the Huntington District of the U.S. Army Corps of Engineers in West Virginia (West Virginia Interagency Review Team, 2010). Similar to the WV Stream Valuation Metric described earlier, it too is widely used in the CWA 404/401 regulatory program in West Virginia, notably in reviews of surface coal mining permits. In 2011, the NC Stream Assessment Method (NC SAM) was developed by an interagency team of federal and state agency staff starting in 2003 with the final method adopted in 2011 (N.C. Stream Functional Assessment Team, 2015; Dorney et al., 2014). It is intended to be used by the CWA 404/401 Regulatory program; that method is described in more detail in Chapter 4.3.2.

In the western United States, the Stream Functional Assessment Method (SFAM)—developed by an interagency team led by the Oregon Department of State Lands, the EPA, and the U.S. Army Corps of Engineers working with partner organizations—is expected to become available in 2018. Intended for initial application in Oregon, SFAM is anticipated to be relatively straightforward to adapt for use in other states of the Pacific Northwest. It was developed considering stream processes similar to those described by Harman et al. (2012) in their Stream Pyramid and is intended mainly to address requirements under the 2008 Joint Mitigation Rule of the EPA and U.S. Army Corps of Engineers (Tracie Nadeau, EPA, personal communication, Dec. 28, 2015). In the meantime, in the state of Washington, agencies and consultants are using a riparian RAM described by Hruby (2009).

CONCLUSIONS

Wetland RAMs have been available since the 1970s and have grown in sophistication over time. There has been a general trend toward regionalizing RAMs to better address the wetlands in particular states and provinces. Some of these RAMs are now required for a variety of regulatory and nonregulatory purposes, and some use both field data and spatial data extracted using GIS. In contrast, stream and riparian RAMs are much less common and have been widely used only since the 1990s. No RAMs for assessing streams or riparian areas are currently required at a national scale but in the United States, several states have developed stream RAMs or adapted ones from other states. Most have a regulatory focus and are field-based. A next logical step would be to use an assortment of existing stream RAMs to guide the development of stream/riparian RAMs that are applicable at a national scale in the United States or other countries.

REFERENCES

Adamus, P.R., 1983. A method for wetland functional assessment. Vol. II. Methodology. Report No. FHWA-IP-82-24, Federal Highway Administration, Washington, DC.

Adamus, P.R., 1984. In: Sather, J.H., Stuber, P.R. (Eds.), Responding comments. Proceedings of the National Wetland Values Assessment Workshop. FWS-OBS-84/12. US Fish & Wildlife Service, Washington, DC. p. 85.

Adamus, P.R., 1992. Data sources and evaluation methods for addressing wetland issues. In: Statewide Wetlands Strategies. World Wildlife Fund and Island Press, Washington, DC, pp. 171–224.

Adamus, P.R., 2016. Manual for the Wetland Ecosystem Services Protocol (WESP). Version 1.3. people.oregonstate.edu/~adamusp/WESP.

Adamus, P.R., Clairain, E.J., Smith, R.D., Young, R.E., 1987. Wetland Evaluation Technique (WET). Volume II. Methodology. US Army Corps of Engineers Waterways Experiment Station, Vicksburg, MS.

Allan, P.F., 1956. A system for evaluating coastal marshes as duck winter range. J. Wildl. Manag. 20 (3), 247–252.

Barbour, M.T., J. Burton, and J. Gerritsen. 2000. A stream condition index for West Virginia wadeable streams. March 28, 2000 (revised July 21, 2000). EPA 68-C7-0014. U.S. Environmental Protection Agency, Region 3 Environmental Services Division and U.S. Environmental Protection Agency, Office of Science and Technology, Office of Water.

Bartoldus, C., 1998. A Comprehensive Review of Wetland Assessment Procedures: A Guide for Wetland Practitioners. Environmental Concern, Inc., Edgewater, MD.

Brinson, M.M., 1993. A hydrogeomorphic classification for wetlands. Tech. Rep. WRP-DE-4, U.S. Army Corps of Engineers Waterways Exp. Stn., Vicksburg, MS.

Brinson, M.M., 1995. The HGM approach explained. Natl Wetl. Newsl., 7–15 November–December 1995.

Brooks, R.P., Cole, C.A., Wardrop, D.H., Bishchel-Machung, L., Prosser, D.J., Campbell, D.A., Gaudette, M.T., 1996. Wetlands, Wildlife, and Watershed Assessment Techniques for Evaluation and Restoration (W3ATER), vol. 1, 2A, and 2B. Penn State Coop. Wetlands Ctr., University Park, PA. Report No. 96-2, 782 pp.

Cole, C.A., Kooser, J.G., 2002. HGM: hidden, gone, missing? Natl. Wetl. Newsl. 24 (2), 1.

Collins, J., Fennessy, M.S., 2011. USA RAM v. 1.1. In: National Wetland Condition Assessment: Field Operations Manual for Wetlands. EPA-843-R-10-001, U.S. Environmental Protection Agency, Washington, DC.

Dee, N., Baker, J., Drobny, N., Duke, K., Whitman, I., Fahringer, D., 1973. An environmental evaluation system for water resource planning. Water Resour. Res. 9 (3), 523–535.

Dorney, J.R., Paugh, L., Smith, S., Lekson, D., Tugwell, R., Allen, B., Cusack, M., 2014. Development and testing of rapid wetland and stream functional assessment methods in North Carolina. Natl Wetl. Newsl. 36 (4), 31–35.

Fairfax County Public Works and Environmental Services, 2003. The perennial stream identification protocol. Available at http://www.fairfaxcounty.gov/dpwes/watersheds/ps_protocols.pdf. Accessed 31 December 2015.

Fennessy, M.S., Jacobs, A.D., Kentula, M.E., 2004. Review of rapid methods for assessing wetland condition. EPA/620/R-04/009, US Environmental Protection Agency, Washington, DC.

Golet, F.C., 1976. Wildlife wetland evaluation model. In: Larson, J.S. (Ed.), Models for Assessment of Freshwater Wetlands. Completion Report 76-5, Water Resource Research Center, University of Massachusetts, Amherst, MA.

Greeson, P.E., Clark, J.R., Clark, J.E. (Eds.), 1979. Wetland Functions and Values: The State of Our Understanding. Proceedings of a National Symposium on Wetlands. American Water Resources Association, Minneapolis, MN.

Harman, W., Starr, R., Carter, M., Tweedy, K., Clemmons, M., Suggs, K., Miller, C., 2012. A function-based framework for stream assessment and restoration projects. US Environmental Protection Agency, Office of Wetlands, Oceans, and Watersheds, Washington, DC. EPA 843-K-12-006, http://www.fws.gov/chesapeakebay/StreamReports/Stream%20Functions%20Framework/Final%20Stream%20Functions%20Pyramid%20Doc_9-12-12.pdf. Accessed 31 December 2015.

Hollis, T., Bedding, J., 1994. Can we stop the wetlands from drying up? New Scientist, 30–35 (2 July 1994).

Hruby, T., 2009. Developing rapid methods for analyzing upland riparian functions and values. Environ. Manag. 43 (6), 1219–1243.

Kumar, R.S., Rao, V.V., Sasikala, C., Nagulu, V., 2014. Wetland birds of Srikakulam District, Andhra Pradesh, India. Int. J.Fauna Biol. Stud. 1 (6), 42–49.

Kusler, J. A. and P. Riexinger (eds). 1986. Proceedings of the National Wetland Assessment Symposium. Association of State Wetland Managers, Albany, NY.

Larson, J.S. (Ed.), 1975. Models for Evaluation of Freshwater Wetlands. Publication 32, Massachusetts Water Resources Research Center, Univ. of Massachusetts, Amherst, MA.

Lyon, J.G., McCarthy, J., 1995. Wetland and Environmental Applications of GIS. CRC Press, Boca Raton, FL.

Lyon, J.G., Lopez, R.D., Lyon, L.K., Lopez, D.K., 2001. Wetland Landscape Characterization: GIS, Remote Sensing and Image Analysis. CRC Press, Boca Raton, FL.

McCall, C.A., 1972. Manual for Maine Wetlands Inventory. Maine Department of Inland Fisheries and Game, Augusta, ME.

McHarg, I.L., Mumford, L., 1969. Design With Nature. American Museum of Natural History, New York.

N.C. Division of Water Quality, 2010. Methodology for the Identification of Intermittent and Perennial Streams and Their Origins. Version 4.11., 1 September 2010, North Carolina Department of Environment and Natural Resources, Raleigh, NC. Available athttp://portal.ncdenr.org/c/document_library/get_file?uuid=0ddc6ea1-d736-4b55-8e50-169a4476de96&groupId=38364. Accessed 31 December 2015.

N.C. Stream Functional Assessment Team. 2015. N.C. Stream Assessment Method (NC SAM) User Manual. N.C. Department of Transportation, US Army Corps of Engineers, NC Department of Environment and Natural Resources, US Environmental Protection Agency, and US Fish and Wildlife Service. Raleigh, NC. Available at https://ribits.usace.army.mil/ribits_apex/f?p=107:27:1277742720629::NO:RP:P27_BUTTON_KEY:20, under Wilmington District, Assessment Tools. Accessed 31 December 2015.

Nadeau, T.-L., Leibowitz, S.G., Wigington Jr., R.J., Eversole, J.L., Fritx, K.M., Coulombe, R.A., Comeleo, R.L., Blocksom, K.A., 2015. Validation of rapid assessment methods to determine streamflow duration classes in the Pacific Northwest, USA. Environ. Manag. 56 (1), 34–53.

Ohio EPA, 2002. Field Evaluation Manual for Ohio's Primary Headwater Habitat Streams. Version 1.0, July 2002, Ohio Environmental Protection Agency, Division of Surface Water, Columbus, OH.

Prichard, D., H. Barrett, J. Cagney, R. Clark, J. Fogg, K. Gebhardt, P. Hansen, B. Mitchell, and D. Tippy. 1993. Riparian area management: process for assessing proper functioning condition. TR 1737-9. Bureau of Land Management, BLM/SC/ST-93/003+1737, Service Center, CO. 60 pp.

Pritchard, D., (work group leader), 1994. Process for assessing proper functioning condition for lentic riparian-wetland areas. TR 1737-11, Bureau of Land Management, US Department of the Interior, Denver, CO.

Reppert, R.T., Sigleo, W., Stackhiv, E., Messman, L., Meyers, C., 1979. Wetland values: concepts and methods of wetland evaluation. IWR Research Report 79-R-1, U.S. Army Engineers, Fort Belvoir, VA.

Shaw, S.P., Fredine, C.G., 1956. Wetlands of the United States, Their Extent, and Their Value for Waterfowl and Other Wildlife. Circular 39. US Department of Interior, Fish and Wildlife Service, Washington, DC.

Siipi, H., 2004. Naturalness in biological conservation. J. Agric. Environ. Ethics 17 (6), 457–477.

Smardon, R.C., Fabos, J.C., 1975. Assessing visual-cultural values of inland wetlands in Massachusetts. In: Larson, J.S. (Ed.), Models for Assessment of Freshwater Wetlands. Completion Report 76-5, Water Resource Research Center, University of Massachusetts, Amherst, MA.

Smith, R.D., Ammann, A., Bartoldus, C., Brinson, M.M., 1995. An approach for assessing wetland functions using hydrogeomorphic classification, reference wetlands, and functional indices. Tech. Rept. WRP-DE-9, Waterways Exp. Stn., US Army Corps of Engineers, Vicksburg, MS.

Smith, R.D., Noble, C.V., Berkowitz, J.F., 2013. Hydrogeomorphic (HGM) approach to assessing wetland functions: guidelines for developing guidebooks (Version 2). No. ERDC/EL-TR-13-11, Engineer Research & Development Center, Environmental Lab, Vicksburg, MS.

Somerville, D.E. 2010. Stream assessment and mitigation protocols: a review of commonalities and differences. May 4, 2010. Prepared for the U.S. Environmental Protection Agency, Office of Wetlands, Oceans, and Watersheds (Contract No. GS-00F-0032M). Washington, DC. Document No. EPA 843-12-003.

Stuber, P.R., Sather, J.H., 1984. Research gaps in assessing wetland functions. Natl Wetl. Newsl. March–April 1984.

Sutter, L.A., Wuenscher, J.R., 1996. NC-CREWS: a wetland functional assessment procedure for the North Carolina coastal area. Draft. Division of Coastal Management, North Carolina Department of Environment and Natural Resources, Raleigh, NC. 61 pp.

Tiner, R.W., 1996. A landscape and landform classification for Northeast wetlands (operational draft). U.S. Fish and Wildlife Service, Ecological Services, Hadley, MA.

Tiner, R.W., 1997. Piloting a more descriptive NWI. Natl Wetl. Newsl. 19 (5), 14–16.

Tiner, R.W., 2002. Enhancing wetlands inventory data for watershed-based wetland characterizations and preliminary assessments of wetland functions. R.W. Tiner (compiler), In: Watershed-based Wetland Planning and Evaluation: A Collection of Papers From the Wetland Millennium Event. (August 6–12, 2000; Quebec City, Quebec, Canada), Association of State Wetland Managers, Berne, NY, pp. 17–39.

Tiner, R.W., 2003. Correlating Enhanced National Wetlands Inventory Data With Wetland Functions for Watershed Assessments: A Rationale for Northeastern U.S. Wetlands. U.S. Fish and Wildlife Service, National Wetlands Inventory Program, Northeast Region, Hadley, MA (26 pp).

Tiner, R.W., 2005. Assessing cumulative loss of wetland functions in the Nanticoke River watershed using enhanced National Wetlands Inventory data. Wetlands 25 (2), 405–419.

Tiner, R.W., Schaller, S., Petersen, D., Snider, K., Ruhlman, K., Swords, J., 1999. Wetland characterization study and preliminary assessment of wetland functions for the Casco Bay watershed, southern Maine. U.S. Fish and Wildlife Service, Ecological Services, Northeast Region, Hadley, MA. Prepared for the Maine State Planning Office, Augusta, ME. (51 pp).

U.S. Army Corps of Engineers, Norfolk District and Virginia Department of Environmental Quality. 2007. Unified Stream Methodology for use in Virginia. Norfolk, VA. 37 pp. Available at http://www.deq.virginia.gov/Portals/0/DEQ/Water/WetlandsStreams/USMFinal_01-18-07.pdf. Accessed 31 December 2015.

U.S. Army Corps of Engineers and U.S. Environmental Protection Agency. 2008. Compensatory mitigation for losses of aquatic resources. Fed. Regist. 73(70):19594–19705; final rule. April 10, 2008. 33 CFR Parts 235 and 332; 40 CFR Part 230. Washington, DC. Available at http://www.epa.gov/sites/production/files/2015-03/documents/2008_04_10_wetlands_wetlands_mitigation_final_rule_4_10_08.pdf. Accessed 31 December 2015.

U.S. Fish and Wildlife Service, 1980. Habitat Evaluation Procedures (HEP) Manual (102ESM). U.S. Fish and Wildlife Service, Department of the Interior, Washington, DC.

US Supreme Court, 2006. Rapanos et ux., et al. v. United States. Opinion announcement. June 19, 2006. Available at http://www.aswm.org/pdf_lib/rapanos-decision-SCOTUS.pdf. Accessed 31 December 2015.

Wathern, P., Young, S.N., Brown, I.W., Roberts, D.A., 1986. Ecological evaluation techniques. Landsc. Plan. 12, 403–420.

West Virginia Interagency Review Team, 2010. West Virginia stream and wetland valuation metric, Version 1.1. March 2010, USACE Huntington District, USACE Pittsburgh District, USEPA, USFWS, USDA NRCS, West Virginia Department of Environmental Protection, and West Virginia Division of Natural Resources.

FURTHER READING

Ohio, E.P.A., 2006. Methods for Assessing Habitat in Flowing Waters: Using the Qualitative Habitat Evaluation Index (QHEI). OEPA Technical Bulletin EAS/2006-06-1, Ohio Environmental Protection Agency, Division of Surface Water, Ecological Assessment Section, Columbus, OH (26 pp).

Rankin, E., 1989. The Qualitative Habitat Evaluation Index (QHEI): Rational, Methods, and Applications. Ohio Environmental Protection Agency, Division of Surface Water, Columbus, OH.

U.S. Army Corps of Engineers, 2012. Regional supplement to the Corps of Engineers wetland delineation manual: Eastern Mountains and Piedmont Region. Version 2.0. ERDC/EL TR-12-9, US Army Corps of Engineers, Engineering Research and Development Center, Vicksburg, MS.

Section 2

Landscape-Level Approaches

Chapter 2.1

Introduction to Landscape-Level Wetland Assessment

Ralph W. Tiner
Institute for Wetland & Environmental Education & Research, Leverett, MA, United States

Chapter Outline

BACKGROUND

Prior to the 1970s, while some groups (e.g., botanists, waterfowl hunters, and fishermen) and wildlife agencies recognized the value of certain wetlands to wildlife, wetlands in the United States were largely viewed as wastelands whose best use would be attained through "reclamation"—conversion to productive farmland by drainage, to real estate for residential or commercial development, or to landfills to dispose of society's wastes (Waring, 1867; Smith, 1907; Elliott, 1912; Vileisis, 1997; Tiner, 2013). Since then, wetlands have gained increasing recognition as valuable natural resources—serving not only as vital fish and wildlife habitats but providing valued environmental services such as temporarily storing water to help reduce flood damage, improving water quality, sequestering carbon, and stabilizing shorelines (e.g., Kusler and Montanari, 1978; Sather and Smith, 1984). The general values of wetlands to society are understood by many Americans today due to the efforts of scientists, environmental organizations, concerned citizens, and natural resources and permitting agencies.

In the United States, individuals and companies seeking to alter wetlands (e.g., drain, fill, or excavate) are usually required to get a permit from the federal government (in accordance with the Clean Water Act and Rivers and Harbors Act) as well as, in many areas, state permits (state wetland protection laws) and even local permission from planning boards or conservation commissions (local wetland ordinances). Given widespread interest in wetland conservation and expanding governmental jurisdiction over wetlands across the country, standardized techniques were first developed to classify and map wetlands, then to delineate wetlands on the ground for regulatory purposes. In the late 1970s, federal and state agencies began devising rapid assessment methods (RAMs) to assess wetland functions to better understand and evaluate the impact of alterations on the ecology and environmental services of wetlands and to assist in the design of appropriate mitigation (see Chapter 1.0). Two types of approaches have been developed for this purpose: landscape-level (described in this section of the book) and field-level (Section 3.0). These approaches are not competing. The choice of one or the other, or both, depends on several factors: (1) available time and budget, (2) availability of relevant spatial data at appropriate scales, (3) technical capacity (particularly geographic information systems (GIS)), (4) the desired extent of geographic coverage, (5) legal and physical accessibility of wetlands, (6) desired levels of accuracy and repeatability, and (7) the intended uses of the output (e.g., how much and what kind of information does one need to plan for wetland conservation or to make an informed decision regarding wetland impacts and to analyze alternatives).

Wetland and Stream Rapid Assessments. https://doi.org/10.1016/B978-0-12-805091-0.00016-5

Wetland and stream assessment can be focused on the health or quality of the wetland—the so-called "wetland condition"—or on the functions that wetlands perform. Wetland condition (or wetland status[1]) ranges from pristine or relatively pristine to highly altered, degraded, or disturbed. To determine wetland condition, one must examine factors operating both inside and outside the wetland. To evaluate functions, one would consider the wetland's properties that influence its ability to perform the suite of functions attributed to wetlands, then rate them in some fashion (e.g., high, medium, low, or no) or measure performance (e.g., how much water is stored and for how long, or how many amphibian species utilize the wetland as a breeding habitat). Both wetland condition and functions can be evaluated remotely through landscape-level approaches or on the ground through field observations, data collection (e.g., measurement), and comparison to "reference wetlands" that may represent the highest quality or least impacted wetland, or the best of the type for the geographic area of interest (Smith et al., 1995; Brinson and Rheinhardt, 1996; Gaucherand et al., 2015).

While individual wetlands perform various functions, it is the collection of wetlands on the landscape that is of utmost importance in providing ecosystem services. Turner et al. (2000) underscored the need for landscape-level assessment: "The full range of public and private instrumental and non-instrumental values all depend on protection of the processes that support the functioning of larger-scale ecological systems. Thus when a wetland, for example, is disturbed or degraded, we need to look at the impacts of the disturbance across the larger level of the landscape." Assessing the cumulative impact of wetland losses requires a landscape-level analysis (e.g., Gosselink and Lee, 1989; Bedford and Preston, 1988; Preston and Bedford, 1988).

WHAT IS A LANDSCAPE-LEVEL WETLAND ASSESSMENT?

Landscape-level approaches provide the most rapid assessment of wetland condition or wetland function as they are based on analysis of existing information from maps and aerial imagery coupled with a basic understanding of wetlands (expert judgment) that is used to develop models of wetland condition or wetland function. It is the first of a three-tiered approach to wetland assessment (Table 2.1.1).

Landscape-level assessments are designed to evaluate wetland condition or functions for large geographic areas, although the analysis can be done for individual sites (off-site evaluation) before conducting site-specific investigations. Once the spatial data have been assembled from existing sources, landscape-level approaches are capable of assessing larger numbers of wetlands in a shorter period of time than is the case with field-level assessments. Moreover, because wetlands do not need to be visited, issues with physical and legal access are avoided. Landscape-level assessments are clearly preliminary in nature, a first approximation that serves as a starting point to inform natural resource managers and others on the condition of wetlands across a broad area of the landscape or on the functions that wetlands are likely to perform for such an area. Besides presenting the large-scale perspective on the status of wetlands in a watershed or other

TABLE 2.1.1 Three-Tiered Approaches to Assessment of Wetland Condition

Tiers	Definitions
Level 1	Landscape assessments rely entirely on GIS data, utilizing landscape disturbance indices to assess wetland condition. This approach involves characterizing the lands that surround wetlands through the use of landscape metrics (e.g., percent forest cover and land use category). Assessment results can provide a coarse gauge of wetland condition within a watershed.
Level 2	Rapid assessments use relatively simple metrics to assess wetland condition. They are customarily based on the readily observable hydrogeomorphic and plant community attributes of wetlands. They also can employ the use of a "stressor checklist." Rapid assessment methods typically produce a single score that describes where a wetland generally falls along a gradient of human disturbance and with respect to ecological integrity.
Level 3	Intensive site assessments provide a more thorough and rigorous measure of wetland condition by gathering direct and detailed measurements of biological taxa and/or hydrogeomorphic functions. Two examples of the type of indicators that might be used in Level 3 assessment are plant composition/structure and soil organic matter content. Normally, Level 3 assessments take years to complete and involve repeated site visits.

Source: U.S. Environmental Protection Agency; https://www.epa.gov/sites/production/files/2015-09/documents/monitoring_and_assessment_cef.pdf.

1. Various researchers have referred to these types of assessment as wetland condition assessment (e.g., U.S. EPA, 2002; Tibbets et al., 2016), wetland functional health (e.g., Patience and Klemas, 1993), or wetland status assessment (Paul Adamus, pers. comm. Aug. 4, 2017). For this chapter, I will use the term "condition."

area of interest, where historical data are available, landscape-level assessments can be used to identify what wetland functions have been lost or diminished in a watershed and to help locate areas where wetland creation or restoration can be initiated strategically to address watershed problems such as flood damage, water quality degradation, and loss of wildlife habitat.

GUIDING PRINCIPLE

The location of a wetland in and within an ecoregion (e.g., mountain, hill, valley, or variations in climate or altitude) and in a watershed (e.g., headwater or downstream), its position on the landscape (e.g., along a waterbody, at the toe of a slope, on a broad flat, or surrounded by upland), its form (e.g., depression, flat, or slope), its vegetation type (e.g., predominant life form—tree, shrub, emergent, or aquatic plant), and the surrounding landscape influence its ability to perform the variety of functions attributed to wetlands. Its condition is largely dependent on disturbances (stressors) within and outside the wetland and on its landscape setting (developed (e.g., urban, suburban, or agricultural land) vs undeveloped (e.g., forest or prairie)). Many of these features can be identified on maps and aerial imagery. The more two wetlands share characteristics, the more likely they are to be similar in condition and function.

SOURCE DATA

Assessments are based on interpretation of thematic maps, geospatial data, and aerial imagery, with or without field review. Typical source data include wetland inventory data (e.g., National Wetlands Inventory (NWI) geospatial data for the United States), river/stream network data (e.g., National Hydrography Data for the United States), topographic data, and digital aerial imagery. These sources each have limitations due to scale, mapping objectives, ability to recognize the target features, and production date. While all wetlands and all streams are not shown on any of the geospatial data products, they represent the best available information for landscape-level assessments. Digital imagery is used to supplement the mapped data, primarily to update the results (i.e., maps are dated—based on the year of the imagery used to prepare them) and to improve the classifications and delineations (e.g., identify wetlands and streams that were not depicted on the maps or changes in wetland type, or expand the wetland classification to add other attributes to the digital database for analysis). The use of GIS technology also permits integration with other geospatial data sets. For example, if locations of habitat for rare, endangered, and threatened species are available in a geospatial database, they can be added to the assessment procedure to identify "red flag" wetlands that may be considered to be off-limits for development. Other "red flag" wetlands could be highlighted by using other information. By knowing the location of areas experiencing major flood damage, one could designate all floodplain wetlands upstream of such areas as "red flag" wetlands. Wetlands upstream of public water supply reservoirs could also be similarly labeled. This designation process as performed by natural resource managers, often with public input, is where "values" come into play. Wetland size may be an important consideration for valuing many wetlands, although there are numerous exceptions (e.g., vernal pools and prairie potholes being two notable exceptions; see Adamus, 2013 for others).

The minimum requirements for landscape-level assessments are wetland inventory maps and aerial photographs; for GIS analyses these data must be available in digital formats. In cases where wetland inventories have not been completed, aerial imagery will be the foundation for the entire project. The investigator must then have the ability to interpret and classify wetlands and other features sufficiently to obtain variables necessary for predicting wetland functions and condition.

FEATURES TO IDENTIFY

As mentioned above, landscape analyses rely on interpretation of information from existing maps (geospatial databases for GIS analyses) or from aerial imagery. Features that may be mapped or interpretable from imagery that is commonly available (e.g., Google Earth) are listed in Table 2.1.2.

For condition assessments, one would examine a wetland and its surrounding area to determine the level of disturbance. Many disturbances are readily interpreted from aerial imagery. Within a wetland one would look for alterations such as ditches, excavations, stressed vegetation (e.g., dead or downed trees or chlorosis—yellowing of leaves), the presence of invasive species (if interpretable, e.g., monotypic stands of *Phragmites australis*, *Lythrum salicaria*, and *Phalaris arundinacea*; Johnston, 2015), and road/railroad crossings that fragment wetlands. Outside, one would consider the area immediately surrounding the wetland (e.g., naturally vegetated buffer, cropland, lawn, or impervious surface), the general landscape setting (natural habitat to highly disturbed, e.g., urban) as an indicator of the quality of runoff, and other external disturbances (stressors) that influence the ecological integrity of the wetland. It must be emphasized that not all

TABLE 2.1.2 Some Features That May Be Interpretable on Aerial Imagery That Can Be Used to Assess Wetland Condition and Function

Vegetation-related features	Life-form, dominant species (for monotypic stands), diversity of cover types, interspersion, stressed vegetation (e.g., dead trees), downed trees, vegetated buffer (cover type and width), and certain invasive species (photointerpretable)
Hydrologic features	Presence of water (inundation or saturation at the surface) and seasonal patterns, presence of an inlet, presence of an outlet, changes in open water areas within wetlands, entrenched/channelized streams, and evidence of restricted hydrology by roads, railroads, or other fragmentation
Physical features	Position/location of wetland on the landscape (e.g., geographically isolated, streamside, or headwater), landscape setting of surrounding area (land cover and/or land use), landform, shoreline edge (linear v. irregular; width), association with water (contiguous or connected), association with other wetlands (connectivity), overwash deposits, and salt deposits (crusts)
Human impacts	Dikes (impoundments), levees, drainage ditches, fill, and fragmentation by development
Other features	Beaver dams/lodges, muskrat eat-outs/mounds, accumulation of detritus (e.g., along upper portion of salt marsh), animal trails in marshes, eutrophication, trash, overgrazing, and soil pugging by livestock

Multiple images are required to identify seasonal trends in vegetation and hydrology. For condition assessments, interpretation of features in lands immediately surrounding the wetland and in the contributing watershed are necessary.

disturbances are readily determined through image analysis; these include tile drainage and some historic disturbances (e.-g., land leveling, bedding, or shallow ditches under forest canopy). Using maps and digital imagery, it should be fairly easy to determine the condition of the wetland simply as pristine (or relatively pristine) or disturbed (or stressed). The real challenge is determining critical levels or thresholds of disturbance for rating as good, fair, poor, and very poor. Ultimately, deciding on the range of conditions that defines those ratings depends on professional judgment.[2] Judgment is also necessary to construct models to predict different levels of expected performance for a variety of wetland functions based on wetland properties.

OVERVIEW OF LANDSCAPE-LEVEL WETLAND ASSESSMENT IN NORTH AMERICA

The earliest uses of maps and aerial photographs for wetland assessment in North America were found in methods focused on identifying significant wildlife habitats (e.g., Golet, 1972, 1976, 1978; Schamberger et al., 1978). Later, as other functions became the subject of attention, multifunction methods developed for local, regional, provincial, and national applications also included review of this material as a screening tool before conducting site inspections (e.g., Larson, 1976; Ecologistics Limited, 1981; Adamus and Stockwell, 1983; Euler et al., 1983; Hollands and McGee, 1986; Adamus et al., 1987). Consulting imagery and maps remains a first step in field-based methods—to collect information on the location, type, and configuration of wetland or wetland mosaic and the wetland's contributing area (e.g., "office procedures" in Adamus, 2016; see Schempf (1992) and Section 4.0 for examples).

Offsite or desktop assessment was also an important component of methods specifically designed for use by local governments: the Connecticut and New Hampshire methods (Ammann et al., 1986; Ammann and Lindley Stone, 1991).[3] These state methods were designed, in large part, to educate the public on wetland functions and values and to provide communities, conservation groups, and wetland scientists with a practical method for evaluating 12 wetland functions and values: ecological integrity, wetland-dependent wildlife habitat, fish and aquatic life habitat, scenic quality, educational potential, wetland-based recreation, flood storage, groundwater recharge, sediment trapping, nutrient trapping/retention/transformation, shoreline anchoring, and noteworthiness (Stone et al., 2015).

As GIS technology advanced and hardcopy maps were replaced by geospatial data, integration of geospatial databases made it possible for landscape-level wetland assessments to be performed over large geographic areas. The United States

2. Consequently some researchers have tended to focus on departure from reference rather than labeling as good, fair, etc. while others have devised numerical rating schemes for field-based assessments (see chapters in Sections 4.0 and 5.0 dealing with field-based rapid assessment methods (RAMs) for details).

3. The New Hampshire method was updated recently to make use of digital information; an online mapper—NH Wetlands Mapper—was created in 2013 to facilitate use of the method (Stone et al., 2015).

Environmental Protection Agency (U.S. EPA) developed the synoptic method for wetland assessment (Leibowitz et al., 1992) for geographic prioritization of ecological restoration. It has been used for evaluating cumulative impacts to wetlands (Abbruzzese and Leibowitz, 1997); for prioritizing land units for protection and restoration (Hyman and Leibowitz, 2000) and sites for restoring wetlands for flood storage and sediment reduction (McAllister et al., 2000; Vellidis et al., 2003); and for identifying the vulnerability of wetland biodiversity (Schweiger et al., 2002).

Sutter and Wuenscher (1996) developed a GIS-based method for use in North Carolina to predict the ecological significance of wetlands at the watershed level. It evolved into the NC CREWS system (North Carolina Department of Environmental Quality, 2017), which is applicable to the North Carolina coastal plain where most of the state's wetlands are found. The ecological significance of wetlands is evaluated for four functions: water quality, hydrologic, habitat, and risk to the watershed. Roise et al. (2004) used this method in multiobjective programming to test its applicability for evaluating construction costs and wetland impacts for transportation corridors. However, the system is not widely used in North Carolina for planning or permitting purposes (John Dorney, personal communication, July 26, 2017).

When Brinson (1993) created his hydrogeomorphic (HGM) classification, the United States Fish and Wildlife Service Northeast Region's National Wetlands Inventory (NWI) program embraced the concept and developed HGM descriptors for landscape position, landform, water-flow path, and waterbody type (LLWW descriptors) that could be added to existing or ongoing NWI geospatial databases (Tiner, 1995, 1997, 2000). The main reason for this was to provide better characterizations of wetlands sufficient to predict wetland functions at the watershed and regional levels in the northeastern United States. The combination of NWI types (Cowardin et al., 1979) and LLWW descriptors (Tiner, 1995, 1997, 2000, 2003a, 2011a, 2014) provided sufficient attributes for categorizing wetlands into like groups for predicting their functions through an approach called "Watershed-Based Preliminary Assessment of Wetland Functions" (Tiner, 2003b, 2010; see Chapter 2.2.1 for details). The relationships between wetland attributes and wetland functions were established by a team of wetland specialists and should be reviewed and modified, as necessary, when applying the approach in any area (e.g., Tiner, 2003b, 2011b). In their national standard for wetland classification, the Federal Geographic Data Committee (FGDC) recommended including the LLWW descriptors to provide a more useful wetland database (FGDC Wetlands Subcommittee, 2013). The FWS and several states (e.g., Connecticut, Delaware, Georgia, Minnesota, Montana, and New Mexico) have used this approach for predicting functions for more than 20 years (see Chapter 2.2.1 for overview). Today it is part of ongoing wetland mapping efforts in these and other states (e.g., Colorado). Michigan officials have adapted this approach to produce information vital to watershed planning and have incorporated the procedures into their statewide wetland monitoring and assessment strategy (Michigan Department of Environmental Quality, 2015; see example in Chapter 2.2.4).

The use of HGM features was also a key component of a landscape-level approach used to assess wetlands in Alaska's Kenai Peninsula (http://www.homerswcd.org/user-files/pdfs/KPWetlandsGuideForEveryone.pdf) and in the Matanuska-Susitna Borough, the fastest growing area in the state (Matanuska-Susitna Borough Planning Department, 2014). These approaches were derived from earlier work in Cook Inlet by Michael Gracz of the Kenai Watershed Forum that includes a web mapper and references to LLWW types (http://cookinletwetlands.info/).

Landscape-level approaches have also been used to predict wetland condition. Early examples of these applications of GIS and aerial imagery include Brooks et al. (2004), Brown and Vivas (2005), Mack (2006), Reiss (2006), Hychka et al. (2007), Mita et al. (2007), Wardrop et al. (2007), Weller et al. (2007), Stein et al. (2009), and Vance (2009). These studies examined the landscape at various distances from the wetland: within a 100–3000 m buffer zone surrounding the wetland, a 300 m radius from the center of the wetland, and a 1 km radius from the center of the wetland. Brown and Vivas (2005) found that the 100 m buffer was the best as there was no difference in their Land Development Intensity scores when larger areas (i.e., 200 and 500 m) were examined. Zampella et al. (2008) evaluated trends in the integrity of the landscape, aquatic systems, and wetland-drainage areas for describing the ecological integrity of the New Jersey Pinelands using a variety of sources and 10 m × 10 m cells for analysis. Rooney et al. (2012) demonstrated a strong relationship between wetland condition and the surrounding landscape. Tibbets et al. (2015, 2016) combined landscape-level assessment and field investigations to produce a wetland profile and condition assessment for Wyoming's Laramie Plains Wetland Complex and the Upper Green River Basin. Likewise, Hart et al. (2015) used a combination of GIS analyses and field observations to evaluate the wetland condition in Montana's Blackfoot and Swan subbasins.

CONSIDERATIONS IN THE APPLICATION OF LANDSCAPE-LEVEL APPROACHES BEYOND NORTH AMERICA

Landscape-level wetland assessments can be used anywhere, provided that certain requirements are met. One needs an inventory of wetlands in the area of interest and the maps and data should ideally be available in digital form for GIS analysis (e.g., ArcGIS Pro, or desktop versions such as ArcGIS Desktop, Global Mapper, or TatuckGIS). GlobWetland (http://www.globwetland.org/), a collaboration project between the Ramsar Secretariat and the European Space Agency, is making wetland data acquired by Landsat satellites available for certain regions of the world (e.g., Africa; http://globwetland-africa.org/). The information on wetland status, trends, assessment, and monitoring is designed for use in wetland conservation and management. If wetland maps are not available or more detailed data are desired, a wetland inventory can be prepared by examining aerial imagery (e.g., Google Earth or other high-resolution digital imagery; Tiner et al. 2015); classifying and delineating wetlands sufficiently to provide information on wetland vegetation, expected hydrology (water regime—frequency and duration of inundation and/or saturation), and hydrogeomorphology; and cre ating a geospatial database. Whenever possible, wetlands subjected to overbank flooding from rivers, streams, and estuaries should be distinguished from wetlands whose surface water mainly comes from precipitation, high groundwater levels, or local surface runoff.

Although digital data are required for GIS applications and analyzing wetlands across large geographic areas, landscape-level assessments can be done on a wetland-by-wetland basis by manual interpretation (i.e., examining existing maps and aerial imagery) for relatively small geographic areas (e.g., wetlands appearing on a single large-scale map) or for a group of selected wetlands of interest.

The number and name of functions to be assessed may be determined by the project sponsor based on scientific information on the functions in the region; values can also be predicted. Knowledge of the region's wetlands is essential in formulating relationships between wetland characteristics and functions. Consequently, a group of local/regional wetland experts needs to be assembled to develop correlations or conceptual models that would use the variables in the database to predict different levels of performance for wetland functions of interest. The correlations presented in Chapter 2.2.1 and those from similar projects and field-based RAMs could serve as a starting point, especially for temperate zone wetlands. Field review (e.g., locating onsite indicators that reflect the performance of wetland functions) is recommended, at least initially when developing the correlations, to review the accuracy of the predictions and modify the relationships, as necessary. Following similar steps, the condition of wetlands can be determined by examining the wetland for alterations and externally for disturbances (stressors) that may affect the condition of the wetland. The condition of watersheds and subbasins can be accessed as well, following an approach proposed by Tiner (2004) and used by Virginia (Ciminelli and Scrivani 2007) and adapted for use in Montana (Vance 2005) or by other approaches referenced in the previous subsection.

GENERAL LIMITATIONS OF LANDSCAPE-LEVEL APPROACHES

Given the occurrence of many wetlands on seasonally saturated or temporarily flooded soils, wetlands pose a serious challenge for mapping through remote sensing (Tiner, 2015). There are also limitations of other source data, especially streams (e.g., Colson et al., 2008; Vanderhoof et al., 2017). Consequently, existing maps and databases do not show all wetlands and streams. The results of landscape-level assessments are dependent upon the data sources, the wetland classifications, existing knowledge of the relationships between wetland characteristics and wetland functions and condition, and how that information is used to predict wetland functions or condition. The results of landscape-level wetland assessments are therefore preliminary determinations to be verified through field investigations. On-the-ground observations will provide additional insight into the relationships between functions or condition and the wetland and landscape variables used to make predictions.

A comparison of classification approaches used in landscape-level assessment showed that the more localized information included in the classification, the more accurate the predictions of wetland functions (Gracz and Glaser, 2017). The Cook Inlet classification system incorporated seven geomorphic components and six hydrologic components to classify oligotrophic peatlands in Cook Inlet, Alaska. Not surprisingly, it was found to be a better predictor of wetland functions for these wetlands than either the NWI classification (Cowardin et al., 1979), the LLWW system (Tiner, 2003a), or a combination of NWI and LLWW, although the latter combination was a much better predictor than relying solely on the attributes from one or the other classification schemes.[4] This underscores the need to conduct field investigations in the area of interest and to incorporate those findings into the model or relationships used to predict wetland functions.

SUMMARY

Landscape-level assessments involve applying our current knowledge of wetlands to interpretation of existing maps, geospatial data, and aerial imagery for predicting wetland functions or condition. As such, they can provide a broad overview of

4. The LLWW system was developed explicitly to be used in combination with the NWI types (Cowardin et al., 1979) for functional assessment.

wetland condition and functions for geographic areas of varying size. The results of landscape-level assessments are particularly useful for watershed planning, initial screening of sites for acquisition or options for development (e.g., transportation and utility corridors) to be verified by field inspections, and for public education and policy analysis. The classification component of landscape-level analysis is valuable for designing research studies and RAMs. Landscape-level assessments do not replace the need for field examinations, but they do provide the opportunity to access areas inaccessible by foot and can evaluate more wetlands than can be visited on the ground. They can be census surveys covering all mapped wetlands in a particular area or can be used to produce estimates of wetland condition or functions for a given region by employing a probabilistic sampling design to evaluate a random sample of wetlands (e.g., U.S. EPA, 2006; Stevens and Jensen, 2007). Landscape-level assessments are usually the first step in doing RAMs as they provide the preliminary (offsite) assessment before collecting data in the field. Concurrent or follow-up field checking of classifications as well as searching for field indicators of wetland function are recommended to produce the most accurate predictions of functions. The results of RAMs and research studies can also help improve the ability of landscape-level assessments to accurately predict wetland functions and condition.

REFERENCES

Abbruzzese, B., Leibowitz, S.G., 1997. A synoptic approach for assessing cumulative impacts to wetlands. Environ. Manag. 21, 457–475.

Adamus, P.R., 2013. Wetland functions: not only about size. Natl. Wetl. Newsl. 2013, 18–19.

Adamus, P.R., 2016. Manual for the Wetland Ecosystem Services Protocol (WESP). http://people.oregonstate.edu/~adamusp/WESP/Manual_WESP1.3_13Oct2016_Adamus.pdf.

Adamus, P.R., Stockwell, L.T., 1983. A method of wetland functional assessment: Volumes I and II. Offices of Research and Development, Federal Highway Administration, U.S. Department of Transportation, Washington, DC. Report No. FHWA-1P-82-23 and FHWA-1P-82-24.

Adamus, P.R., Clairain Jr., E.J., Smith, R.D., Young, R.E., 1987. Wetland evaluation technique (WET). Volume II: Methodology. Federal Highway Administration, Office of Implementation, McLean, VA. FHWA-IP-88-029.

Ammann, A.P., and A. Lindley Stone, 1991. Method for the comparative evaluation of nontidal wetlands in New Hampshire. Published by the New Hampshire Department of Environmental Services, NHDES-WRD-1991-3.

Ammann, A.P., Franzen, R.W., Johnson, J.L., 1986. Method for the Evaluation of Inland Wetlands in Connecticut. Connecticut Department of Environmental Protection. Bulletin No. 9.

Bedford, B.L., Preston, E.M., 1988. Developing the scientific basis for assessing cumulative effects of wetland loss and degradation on landscape functions: status, perspectives, and prospects. Environ. Manag. 12, 751–771.

Brinson, M.M., 1993. A hydrogeomorphic classification for wetlands. U.S. Army Corps of Engineers, Washington, DC. Wetlands research program tech. rep. WRP-DE-4.

Brinson, M.M., Rheinhardt, R., 1996. The role of reference wetlands in functional assessment and mitigation. Ecol. Appl. 6, 69–76.

Brooks, R.P., Wardrop, D.H., Bishop, J.A., 2004. Assessing wetland condition on a watershed basis in the mid-Atlantic region using synoptic land-cover maps. Environ. Monit. Assess. 94, 9–22.

Brown, M.T., Vivas, M.B., 2005. Landscape development intensity index. Environ. Monit. Assess. 101, 289–309.

Ciminelli, J., Scrivani, J., 2007. Virginia conservation lands needs assessment Virginia watershed integrity model. Virginia Department of Conservation and Recreation, Division of Natural Heritage; Virginia Department of Forestry; Virginia Commonwealth University Center for Environmental Studies, and Virginia DEQ Coastal Zone Management Program, Richmond, VA.http://www.dcr.virginia.gov/natural-heritage/document/watershedintegritymodel.pdf.

Colson, T., Gregory, J., Dorney, J., Russell, P., 2008. Topographic and soil maps do not accurately depict headwater stream networks. Natl. Wetl. Newsl. 30 (3), 25–28.

Cowardin, L.M., Carter, V., Golet, F.C., LaRoe, E.T., 1979. Classification of wetlands and deepwater habitats of the United States. U.S. Fish and Wildlife Service, Washington, DC. FWS/OBS-79/31.

Ecologistics Limited. 1981. A wetland evaluation system for southern Ontario. Prepared for the Canada/Ontario Steering Committee on Wetland Evaluation and the Canadian Wildlife Service, Environment Canada.

Elliott, C.G., 1912. Engineering for Land Drainage: A Manual for the Reclamation of Lands Injured by Water. John Wiley & Sons, New York, NY.

Euler, D.L., Carreiro, J.F.T., McCullough, G.B., Snell, E.A., Glooschenko, V., Spurr, R.H., 1983. An Evaluation System for Wetlands of Ontario South of the Precambrian Shield. Ontario Ministry of Natural Resources and Canadian Wildlife Service, Ontario, Canada.

FGDC Wetlands Subcommittee, 2013. Wetlands classification standard. Federal Geographic Data Committee, Washington, DC. FGDC-STD-004-2013, http://www.fgdc.gov/standards/projects/FGDC-standards-projects/wetlands/nvcs-2013.

Gaucherand, S., Schwoertzig, E., Clement, J.-C., Johnson, B., Quétier, F., 2015. The cultural dimensions of freshwater wetland assessments: lessons learned from the application of US rapid assessment methods in France. Environ. Manag. 56, 245–259.

Golet, F.C., 1972. Classification and Evaluation of Freshwater Wetlands as Wildlife Habitat in the Glaciated Northeast. University of Massachusetts, Amherst, MA (Ph. D. dissertation).

Golet, F.C., 1976. Wildlife wetland evaluation model. In: Larson, J.S. (Ed.), Models for Evaluation of Freshwater Wetlands. Water Resources Research Center, University of Massachusetts, Amherst, MA, pp. 13–34. Publication No. 32.

Golet, F.C., 1978. Rating the wildlife value of northeastern freshwater wetlands. In: Greeson, P.E., Clark, J.R., Clark, J.E. (Eds.), Wetland Functions and Values: The State of Our Understanding. American Water Resources Association, Minneapolis, MN, pp. 63–73.

Gosselink, J.G., Lee, L.C., 1989. Cumulative impact assessment in bottomland hardwood forests. Wetlands 9, 169–174.

Gracz, M., Glaser, P.H., 2017. Evaluation of a wetland classification system devised for management in a region with a high cover of peatlands: an example from the Cook Inlet Basin, Alaska. Wetl. Ecol. Manag. 25, 87–104.

Hart, M., L. Vance, K. Newlon, J. Chutz, and J. Hahn. 2015. Estimating wetland condition locally: an intensification study in the Blackfoot and Swan River watersheds. Report to the U.S. Environmental Protection Agency. Montana Natural Heritage Program. Helena, MT.

Hollands, G.G., McGee, D.W., 1986. In: Kusler, J.A., Riexinger, P. (Eds.), A method for assessing the functions of wetlands.Proceedings of the National Wetland Assessment Symposium. Association of State Wetland Managers (ASWM), Chester, VT, pp. 108–118. ASWM Technical Report 1.

Hychka, K.C., Wardrop, D.H., Brooks, R.P., 2007. Enhancing a landscape assessment with intensive data: a case study in the Upper Juniata watershed. Wetlands 27, 446–461.

Hyman, J.B., Leibowitz, S.G., 2000. A general framework for prioritizing land units for ecological protection and restoration. Environ. Manag. 25, 23–35.

Johnston, C.A., 2015. Mapping invasive wetland plants. Chapter 23, In: Tiner, R.W., Lang, M.W., Klemas, V.V. (Eds.), Remote Sensing of Wetlands: Applications and Advances. CRC Press, Boca Raton, FL, pp. 491–510.

Kusler, J.A. and J.H. Montanari. 1978. National Wetland Protection Symposium. U.S. Fish and Wildlife Service, Office of Biological Services, Washington, DC. FWS/OBS-78-97.

Larson, J.S., 1976. Models for Assessment of Freshwater Wetlands. Water Resources Research Center, University of Massachusetts, Amherst, MA. Publication No. 32.

Leibowitz, S.G., Abbruzzese, B., Adamus, P.R., Hughes, L.E., Irish, J.T., 1992. A synoptic approach to cumulative impact assessment: a proposed methodology. U.S. Environmental Protection Agency. EPA/600/R-92/167.

Mack, J.J., 2006. Landscape as a predictor of wetland condition: an evaluation of the Landscape Development Index (LDI) with a large reference wetland dataset from Ohio. Environ. Monit. Assess. 120, 221–241.

Matanuska-Susitna Borough Planning Department, 2014. Matanuska-Susitna Wetland Functions and Values Landscape-Level Assessment Methodology and Mapping, first ed. Matanuska-Susitna Borough, Palmer, AK.

McAllister, L.S., Peniston, B.E., Leibowitz, S.G., Abbrussese, B., Hyman, J.B., 2000. A synoptic assessment for prioritizing wetland restoration efforts to optimize flood attenuation. Wetlands 20, 70–83.

Michigan Department of Environmental Quality, 2015. Michigan Wetland Monitoring and Assessment Strategy. Water Resources Division, Lansing, MI. https://www.michigan.gov/documents/deq/wrd-wetlands-strategy_555457_7.pdf.

Mita, D., DeKeyser, E., Kirby, D., Easson, G., 2007. Developing a wetland condition prediction model using landscape structure variability. Wetlands 27, 1124–1133.

North Carolina Department of Environmental Quality, 2017. NC CREWS, NC Division of Coastal Management. Available at https://deq.nc.gov/about/divisions/coastal-management/coastal-management-data/setback-factor-maps-1998-shoreline/nc-crews-wetlands-functional-assessment. Accessed 26 July 2017.

Patience, N., Klemas, V.V., 1993. Wetland functional health assessment using remote sensing and other techniques: literature search. NOAA Technical Memorandum NMFSSEFSC-319.

Preston, E., Bedford, B.L., 1988. Evaluating cumulative effects on wetland functions: a conceptual overview and generic framework. Environ. Manag. 12, 565–583.

Reiss, K.C., 2006. Florida wetland condition index for depressional forested wetlands. Ecol. Indic. 6, 337–352.

Roise, J.P., Shear, T.H., Bianco, J.V., 2004. Sensitivity analysis of transportation corridor location in wetland areas: a multiobjective programming and GIS approach. Wetl. Ecol. Manag. 12, 519–529.

Rooney, R.C., Bayley, S.E., Creed, I.F., Wilson, M.J., 2012. The accuracy of land cover-based assessments is influenced by landscape extent. Landsc. Ecol. 27, 1321–1335.

Sather, J.H., Smith, R.D., 1984. An overview of major wetland functions and values. U.S. Fish and Wildlife Service, Office of Biological Services, Washington, DC. FWS/OBS-84/18.

Schamberger, M.L., Short, C., Farmer, A., 1978. Evaluation wetlands as wildlife habitat. In: Greeson, P.E., Clark, J.R., Clark, J.E. (Eds.), Wetland Functions and Values: The State of Our Understanding. American Water Resources Association, Minneapolis, MN, pp. 74–83.

Schempf, J.H., 1992. Wetland classification, inventory, and assessment methods: an Alaska guide to their fish and wildlife application. Alaska Department of Fish and Game, Habitat Division, Juneau, AK. Habitat Division Technical Report 93-2.

Schweiger, E.W., Leibowitz, S., Hyman, J., Foster, W.E., Downing, M.C., 2002. Synoptic assessment of wetland function: a planning tool for protection of wetland species biodiversity. Biodivers. Conserv. 11, 379–406.

Smith, J.B. 1907. The New Jersey salt marsh and its improvement. NJ Agricultural Experiment Station Bulletin 207.

Smith, R.D., Ammann, A., Bartoldus, C., Brinson, M.M., 1995. An approach for assessing wetland functions using hydrogeomorphic classification, reference wetlands, and functional indices. U.S. Army Corps of Engineers, Washington, DC. Wetland research program tech. rep. WRP-DE-9.

Stein, E.D., Fetscher, A.E., Clark, R.P., Wiskind, A., Grenier, J.L., Sutula, M., Collins, J.N., Grosso, C., 2009. Validation of a wetland rapid assessment method: use of EPA's level 1-2-3 framework for method testing and refinement. Wetlands 29, 648–665.

Stevens, D.L., Jensen, S.F., 2007. Sample design, execution, and analysis for wetland assessment. Wetlands 27, 515–523.

Stone, A.L., Mitchell, F., Van de Poll, R., Rendall, N., 2015. Method for Inventorying and Evaluating Freshwater Wetlands in New Hampshire (NH Method). University of New Hampshire Cooperative Extension, Durham, NH.https://nhmethod.org/.

Sutter, L.A., Wuenscher, J.R., 1996. NC-CREWS: A Wetland Functional Assessment Procedure for the North Carolina Coastal Area. Draft, Division of Coastal Management, North Carolina Department of Environment and Natural Resources, Raleigh, NC.

Tibbets, T.M., H.E. Copeland, L. Washkoviak, S. Patla, and G. Jones. 2015. Wetland profile and condition assessment of the Upper Green River Basin, Wyoming. Report to the U.S. Environmental Protection Agency. The Nature Conservancy–Wyoming Chapter, Lander, WY.

Tibbets, T.M., L. Washkoviak, S.A. Tessmann, G. Jones and H.E. Copeland. 2016. Wetland profile and condition assessment of the Laramie Plains Wetland Complex, Wyoming. Report to the U.S. Environmental Protection Agency. The Nature Conservancy–Wyoming Chapter, Lander, WY.

Tiner, R.W., 1995. A Landscape and Landform Classification for Northeast Wetlands (Operational Draft). U.S. Fish and Wildlife Service, National Wetlands Inventory Program, Hadley, MA.

Tiner, R.W., 1997. Keys to Landscape Position and Landform Descriptors for U.S. Wetlands (Operational Draft). U.S. Fish and Wildlife Service, National Wetlands Inventory Project, Northeast Region, Hadley, MA.

Tiner, R.W., 2000. Keys to Waterbody Type and Hydrogeomorphic-Type Descriptors for U.S. Waters and Wetlands (Operational Draft). U.S. Fish and Wildlife Service, National Wetlands Inventory Program, Hadley, MA.

Tiner, R.W., 2003a. Dichotomous Keys and Mapping Codes for Wetland Landscape Position, Landform, Water Flow Path, and Waterbody Type Descriptors. U.S. Fish and Wildlife Service, National Wetlands Inventory Program, Northeast Region, Hadley, MA.

Tiner, R.W., 2003b. Correlating Enhanced National Wetlands Inventory Data with Functions for Watershed Assessments: A Rationale for Northeastern U.S. Wetlands. U.S. Fish and Wildlife Service, National Wetlands Inventory Program, Hadley, MA.https://www.fws.gov/northeast/ecologicalservices/pdf/wetlands/CorrelatingEnhancedNWIDataWetlandFunctionsWatershedAssessments.pdf.

Tiner, R.W., 2004. Remotely-sensed indicators for monitoring the general condition of natural habitat in watersheds: an application for Delaware's Nanticoke River watershed. Ecol. Indic. 4, 227–243.

Tiner, R.W., 2010. NWIPlus: geospatial database for watershed-level functional assessment. Natl. Wetl. Newsl. 32 (3), 4–7. 23.

Tiner, R.W., 2011a. Dichotomous Keys and Mapping Codes for Wetland Landscape Position, Landform, Water Flow Path, and Waterbody Type Descriptors. Version 2.0. U.S. Fish and Wildlife Service, Northeast Region, Hadley, MA.

Tiner, R.W. 2011b. Predicting Wetland Functions at the Landscape Level for Coastal Georgia Using NWIPlus Data. U.S. Fish and Wildlife Service, National Wetlands Inventory Program; In cooperation with the Georgia Department of Natural Resources, Coastal Resources Division, Region 5, Hadley, MA; Brunswick, GA and Atkins North America, Raleigh, NC. http://coastalgadnr.org/sites/uploads/crd/CORRELATION%20REPORT%20Georgia_FINAL_September-20-2011.pdf

Tiner, R.W., 2013. Tidal Wetlands Primer: An Introduction to Their Ecology, Natural History, Status, and Conservation. University of Massachusetts Press, Amherst, MA.

Tiner, R.W., 2014. Dichotomous Keys and Mapping Codes for Wetland Landscape Position, Landform, Water Flow Path, and Waterbody Type Descriptors. Version 3.0. U.S. Fish and Wildlife Service, Northeast Region, Hadley, MA. https://www.fws.gov/wetlands/Documents/Dichotomous-Keys-and-Mapping-Codes-for-Wetland-Landscape-Position-Landform-Water-Flow-Path-and-Waterbody-Type-Version-3.pdf.

Tiner, R.W., 2015. Introduction to wetland mapping and its challenges. Chapter 3, In: Tiner, R.W., Lang, M.W., Klemas, V.V. (Eds.), Remote Sensing of Wetlands: Applications and Advances. CRC Press, Boca Raton, FL, pp. 43–65.

Tiner, R.W., Lang, M.W., Klemas, V.V. (Eds.), 2015. Remote Sensing of Wetlands: Applications and Advances. CRC Press, Boca Raton, FL.

Turner, R.K., van den Bergh, J.C.J.M., Söderqvist, T., Barendregt, A., van der Stratten, J., Maltby, E., van Ierland, E.C., 2000. Ecological-economic analysis of wetlands: scientific integration for management and policy. Ecol. Econ. 35, 7–23.

U.S. EPA, 2002. Methods for evaluating wetland condition: Study design for monitoring wetlands. Office of Water, U.S. Environmental Protection Agency, Washington, DC. EPA-822-R-02-015.

U.S. EPA, 2006. Application of Elements of a State Water Monitoring and Assessment Program for Wetlands. Wetlands Division, Office of Wetlands, Oceans, and Watersheds, U.S. Environmental Protection Agency, Washington, DC.

Vance, L.K. 2005. Watershed assessment of the Cottonwood and Whitewater watersheds. Prepared for the Malta Field Office, Bureau of Land Management; Montana Natural Heritage Program, Malta, Montana; Helena, MT. 57 pp. plus appendices. https://archive.org/details/836FFF8A-3313-460E-8CDE-B7176B2CA4A8

Vance, L.K. 2009. Assessing wetland condition with GIS: a landscape integrity model for Montana. A Report to the Montana Department of Environmental Quality and the Environmental Protection Agency. Montana Natural Heritage Program, Helena, MT.

Vanderhoof, M.K., Distler, H.E., Lang, M.W., Alexander, L.C., 2017. The influence of data characteristics on detecting wetland/stream surface-water connections in the Delmarva Peninsula, Maryland and Delaware. Wetl. Ecol. Manag. https://doi.org/10.1007/s11273-017-9554-y.

Vellidis, G., Leibowitz, S.G., Ainslie, W.B., Pruitt, B.A., 2003. Prioritizing wetland restoration for sediment yield reduction: a conceptual model. Environ. Manag. 31, 301–312.

Vileisis, A., 1997. Discovering the Unknown Landscape: A History of America's Wetlands. Island Press, Washington, DC.

Wardrop, D.H., Kentula, M.E., Stevens, D.L., Jensen, S.F., Brooks, R.P., 2007. Assessment of wetland condition: an example from the Upper Juniata watershed in western Pennsylvania. Wetlands 27, 416–431.

Waring, G.E., 1867. Drainage for Profit, and Drainage for Health. Orange Judd and Company, New York.

Weller, D.E., Snyder, M.N., Whigham, D.F., Jacobs, A.D., Jordon, T.E., 2007. Landscape indicators of wetland condition in the Nanticoke River Watershed, Maryland and Delaware, USA. Wetlands 27, 498–514.

Zampella, R.A., Procopio III, N.A., Du Brul, M.U., Bunnell, J.F., 2008. An Ecological-Integrity Assessment of the New Jersey Pinelands. A Comprehensive Assessment of the Landscape and Aquatic and Wetland Systems of the Region. New Jersey Pinelands Commission, New Lisbon, NJ.

FURTHER READING

Adamus, P.R., 1983. FHWA assessment method, volume 2 of method for wetland functional assessment. U.S. Department of Transportation, Federal Highway Administration, Washington, DC. FHWA-IP-82-24.

Johnson, J.B., 2005. Hydrogeomorphic wetland profiling: an approach to landscape and cumulative impacts analysis. U.S. Environmental Protection Agency, Washington, DC. EPA/620/R-05/001.

Tiner, R.W., 2005. Assessing cumulative loss of wetland functions in the Nanticoke River watershed using enhanced National Wetlands Inventory data. Wetlands 25 (2), 405–419.

Section 2.2

Case Studies—Landscape-Level Approaches

Chapter 2.2.1

A Landscape-Level Wetland Functional Assessment Tool: Building a Framework for Watershed-Based Assessments in the United States

Ralph W. Tiner
Institute for Wetland & Environmental Education & Research, Leverett, MA, United States

Chapter Outline

INTRODUCTION

Watershed-based strategic planning for wetlands has become a major theme of government agencies at the federal and state levels. In fact, the Environmental Protection Agency (EPA) has encouraged and is actively working with states to develop statewide wetland conservation plans (http://www.epa.gov/owow/wetlands/initiative/swcp.html). An inventory and assessment of wetland resources are vital components of such plans.

The U.S. Fish and Wildlife Service (FWS) established the National Wetlands Inventory (NWI) program in 1974 to conduct an inventory of the nation's wetlands and provide this information to decision-makers so that they could make more informed decisions on the fate of wetlands. The 1977 amendments to the Federal Clean Water Act state that the FWS is to conduct its NWI and provide such information to states to aid in formulating plans that will help protect the nation's water quality. While the NWI initially produced hardcopy maps, in the mid-1990s production of such maps was replaced by digital data for display through an online mapping tool (https://www.fws.gov/wetlands/data/mapper.html) and for downloading for geographic information system (GIS) analyses (Tiner, 2009). NWI digital data could be integrated with other digital data sources (e.g., soils and hydrology) and even superimposed on digital images of topographic maps (digital raster graphics (DRGs), or orthophotoquads) or on digital aerial imagery (e.g., to aid in detecting wetland changes). Recognizing the power of GIS and the value of NWI data, the FWS sought ways to expand the application of NWI data for watershed-based or landscape-level functional assessments of wetlands, among other things. The Northeast Region of the FWS took the lead in this development, aided by various state agencies and other partners, especially Virginia Tech's Conservation Management Institute.

NWI data are based on the Cowardin et al. (1979) classification system with wetlands and deepwater habitats classified by ecological system (marine, estuarine, palustrine, riverine, and lacustrine), subsystems, class (vegetated and nonvegetated types), water regime (degree of flooding or saturation), and other modifiers that describe human impacts and other

Wetland and Stream Rapid Assessments. https://doi.org/10.1016/B978-0-12-805091-0.00017-7

features of interest. The classification system was designed for descriptive purposes (i.e., grouping wetlands with similar properties) and, most importantly, to produce categories for mapping. Cowardin and others recognized that their classification did not cover all the possibilities for describing wetlands and explicitly stated, "below the level of class, the system is open-ended and incomplete." When the Cowardin system was developed, the main concerns were where the wetlands were and what kinds were there—questions answered by a traditional wetlands inventory. With increased regulations from the late 1970s into the 1990s, other questions focusing on wetland functions (e.g., what functions does this wetland perform, what is the impact of the wetland alteration on functions, and how do we mitigate for losses of wetland functions?) and wetland restoration also became important to resource managers. The Cowardin et al. classification did not contain all the features necessary for this type of assessment because physical and topographic features could be viewed on the maps. Brinson (1993) noted that information on more physical attributes—so-called "hydrogeomorphic properties"—was needed to address many wetland functions. Rather than develop attributes to expand the Cowardin et al. system, Brinson designed a separate classification system—the hydrogeomorphic (HGM) classification to cover these features (Brinson, 1993). His system used several key terms from the FWS system but defined them differently (i.e., riverine and lacustrine), making it difficult to simply add these features to the existing NWI database. Brinson did not, however, intend that his HGM classification be used for inventories, as evidenced by his claim that "the report describes a general approach to classification and not a specific one to be used in practice" (p. 2). HGM is more an approach to separating wetlands into groupings for better understanding functions and comparing individual wetlands for impact analysis and mitigation. Brinson's work did provide the impetus to expand the characteristics that could be classified by the NWI and others interested in broadening the functionality of NWI digital data. Many of these features were landscape-level properties that could be interpreted from a traditional NWI map. Given that the maps are now digital products that facilitate analysis of large datasets, Tiner (1995) realized the value of adding hydrogeomorphic-type descriptors to the NWI database and did so to create what is now called an NWI+ or NWIPlus database (Tiner, 2010a).

EXPANDING THE NWI DATABASE

In 1995, the first set of hydrogeomorphic-type descriptors called "**LLWW** descriptors" was created to provide additional information on wetland characteristics, namely their **L**andscape position, **L**andform, **W**ater flow path, and **W**aterbody type (Tiner, 1995). They were developed to be added to the NWI database on a project-specific basis. These descriptors were initially created to help evaluate the likely functions of potential wetland restoration sites for watershed-based restoration planning in Massachusetts (i.e., to target restoration efforts to address problems in watersheds such as increase flood storage upstream of areas experiencing flooding problems) and to estimate the effect of wetland trends (gains and losses) on wetland functions (Tiner, 1997a). The LLWW descriptors were refined over the years (Tiner, 1997b,c, 2000, 2002a,b, 2003a, 2011a, 2014) through application in various locations of the Northeast and elsewhere. Currently, they are published as "Dichotomous Keys and Mapping Codes for Wetland Landscape Position, Landform, Water Flow Path, and Waterbody Type Descriptors Version 3.0" (Tiner, 2014). These descriptors provide a practical application of the hydrogeomorphic approach to NWI data for producing landscape-level assessments of wetland functions. In drafting national wetland mapping standards, the Wetlands Subcommittee of the Federal Geographic Data Committee (FGDC) recommended the use of LLWW descriptors for wetland inventories (FGDC Wetlands Subcommittee, 2009). They recognized that adding these attributes to the existing NWI data makes the database a much more useful and powerful analytical tool.

LLWW Descriptors

A set of dichotomous keys is used to identify the LLWW descriptors (see Tiner, 2014 for details). *Landscape position* reflects the position of a wetland on the land, emphasizing its relationship to a waterbody, if present (Fig. 2.2.1.1). Five basic landscape positions are recognized: Marine, Estuarine, Lotic (along rivers and streams and subject to overbank flooding—Lotic River and Lotic Stream), Lentic (lakes and reservoirs), and Terrene (surrounded by upland or hydrologically decoupled from a river, stream, lake, or reservoir, with respect to surface water). Terrene wetlands may be further separated into three general types: Terrene Headwater (those that are the source of a stream or river), Terrene Riparian (those on the 100-year floodplain but not frequently overflowed by river or stream), and Terrene Nonriparian (including but not necessarily limited to geographically isolated wetlands). *Landform* represents the physical shape or form of a wetland. Several landform types are recognized including Fringe, Island, Flat, Basin, Floodplain, Peatland, and Slope. *Water flow path* characterizes the directional flow of water associated with a wetland. The six basic paths are (1) Throughflow, (2) Outflow, (3) Inflow, (4) Bidirectional Nontidal, (5) Bidirectional Tidal, and (6) Vertical Flow, with

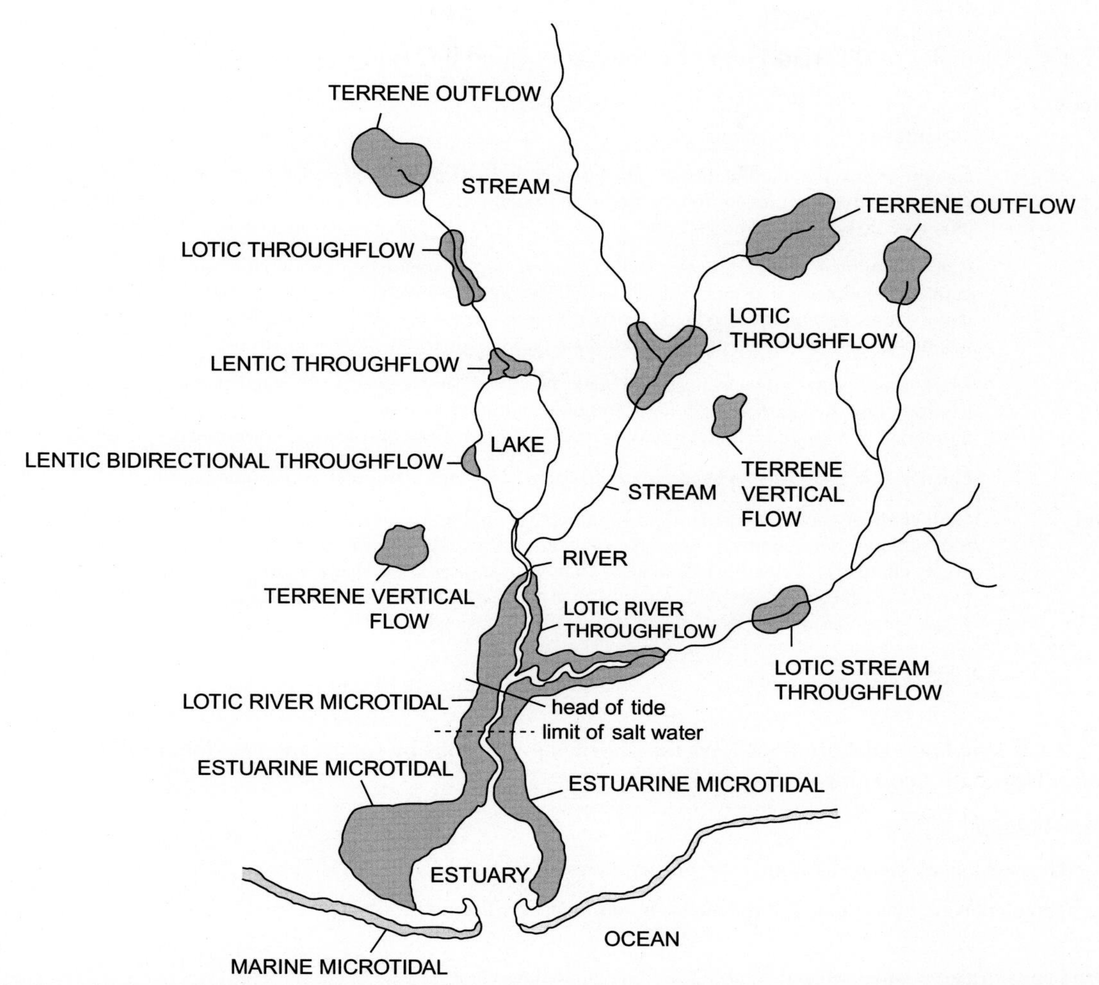

FIG. 2.2.1.1 Generalized classification of wetlands by landscape position and water flow path. Terrene Headwater wetlands are the sources of streams in this diagram while the geographically isolated wetlands shown are Terrene Vertical Flow wetlands. *(From Tiner, R.W. 2014. Dichotomous Keys and Mapping Codes for Wetland Landscape Position, Landform, Water Flow Path, and Waterbody Type Descriptors. Version 3.0. U.S. Fish and Wildlife Service, Northeast Region, Hadley, MA. https://www.fws.gov/wetlands/Documents/Dichotomous-Keys-and-Mapping-Codes-for-Wetland-Landscape-Position-Landform-Water-Flow-Path-and-Waterbody-Type-Version-3.pdf).)*

variations possible. For example, if water flow is intermittent, this term can be added to the first three water flow paths (e.g., Throughflow Intermittent, Outflow Intermittent, and Inflow Intermittent). Bidirectional Nontidal is used to indicate water levels affected by lakes and reservoirs. While water rises and falls in all wetlands, Vertical Flow defines the rise and fall of water tables in wetlands with no apparent surface-water inflow or outflow (i.e., geographically isolated), although if groundwater hydrology relations are known, additional modifiers can be added to vertical flow to describe the groundwater relationship (i.e., Vertical Flow Throughflow, Vertical Flow Outflow, or Vertical Flow Inflow with a "groundwater-connected" modifier). *Waterbody type* descriptors provide additional information on the type of lake, pond, river, stream, or estuary (see Table 2.2.1.1 for some examples of the diversity of types). Simplified keys are provided in Table 2.2.1.2; the more detailed technical keys contain many other descriptors to better characterize wetlands. One of particular importance is "headwater," which is used to identify wetlands that are sources of streams or along first-order perennial streams and any intermittent ones upstream.

GIS technology is used to help assign these new properties to wetlands and deepwater habitats in the NWI database. Using ArcGIS with personal geodatabases, NWI data are combined with other digital data for analysis. These data sources include the United States Geological Survey (USGS) digital raster graphics (DRGs) and 1:24,000 USGS national hydrography data (NHD; https://nhd.usgs.gov/). More detailed stream data may be available from USGS or from other government agencies. Using these sources and the dichotomous keys, LLWW descriptors can be added to all wetlands and deepwater habitats in the NWI database to create what is called an "*NWI+*" or "*NWIPlus*" database.

TABLE 2.2.1.1 Examples of Different Types of Waterbodies Based on Tiner (2003a).

Waterbody Type	Variants
Lake	Natural, excavated, dammed river valley (reservoir, hydropower), other dammed (former natural, artificial), seiche-influenced, river-fed, stream-fed, barrier beach lagoon, semienclosed embayment, open embayment, and main body, plus specific types listed under Pond.
Pond	Natural, artificial, beaver, alligator, marsh, swamp, vernal, prairie pothole, sandhill, sinkhole/karst, Grady, interdunal, farm-cropland, farm-livestock, golf, industrial, sewage/wastewater treatment, stormwater, aquaculture-catfish, aquaculture-shrimp, aquaculture-crayfish, cranberry, irrigation, aesthetic-business, acid-mine, arctic polygonal, bog, woodland, borrow pit, Carolina bay, tundra, coastal plain, tidal, and in-stream.
Estuary	Fjord, island-protected rocky headland, rocky headland bay, tectonic, river-dominated, drowned river valley, bar-built (coastal pond, hypersaline lagoon), island protected, and shoreline bay, plus descriptors for tidal range (macrotidal, mesotidal, and microtidal) and hydrologic circulation patterns (salt-wedge, homogeneous, and partially mixed)
Ocean	Open ocean, reef-protected waters, atoll lagoon, fjord, and semiprotected oceanic bay
River and stream	High gradient, moderate gradient, low gradient, intermittent gradient, tidal gradient, ditch, channelized, canal, restored, lock and dammed, run-of-the-river, other dammed, natural, single thread (one channel), multiple thread (many channels), braided (wide, shallow channels), anastomosed (deep, narrow channels), connecting channel (joins two lakes), riffle, pool, waterfall, deep (>2 m at low water), shallow (<2 m at low water), and channel patterns (straight, slightly meandering, moderate meandering, and high meandering)

TABLE 2.2.1.2 Simplified Dichotomous Keys for Classifying Wetlands by Landscape Position, Landform, and Water Flow Path According to Tiner (2014)

Landscape Position	
1. Wetland borders a river, stream, lake, reservoir, estuary, or ocean	2
1. Wetland does not border these waters; it is completely surrounded by upland (dryland)	Terrene Non-riparian
2. Wetland lies along an ocean shore and is subject to frequent tidal flooding	Marine
2. Wetland does not lie along an ocean shore	3
3. Wetland lies in an estuary and is subject to frequent tidal flooding	Estuarine
3. Wetland does not lie in an estuary	4
4. Wetland forms the shore of a lake or reservoir or lies within its basin	Lentic
4. Wetland lies along a river or stream, or if along an estuary or estuarine wetland, it is a freshwater type that is only flooded by tides during severe storms	5
5. Wetland is the source of a river or stream; minor (typically intermittent) streams may flow into it from adjacent hillsides	Terrene Headwater
5. Wetland is not the source of a river or stream	6
6. Wetland is a freshwater type bordering an estuarine wetland and subject only to tidal flooding during severe storms	Terrene Non-riparian
6. Wetland occurs in or along a river or stream	7
7. Wetland is in a river or stream or on its active floodplain where it is frequently flooded by overflow	8
7. Wetland is along a river or stream but it is not frequently flooded by the river or stream	9
8. Wetland is on the active floodplain of a stream	Lotic Stream[a]
8. Wetland is on the active floodplain of a river	Lotic River
9. Wetland is on the 100-year floodplain	Terrene Riparian
9. Wetland is above the 100-year floodplain	Terrene Non-riparian

TABLE 2.2.1.2 Simplified Dichotomous Keys for Classifying Wetlands by Landscape Position, Landform, and Water Flow Path According to—cont'd

Landform	
1. Wetland is formed by the accumulation of peat (i.e., bog or fen)	Peatland
1. Wetland does not accumulate peat at the surface	2
2. Wetland occurs on a slope >2%	Slope
2. Wetland does not occur on a slope >2%	3
3. Wetland forms an island completely surrounded by water (excluding wetlands where extensive grid ditching might create this condition at a small scale)	Island
3. Wetland does not form on an island	4
4. Wetland occurs in the shallow water zone of a permanent nontidal waterbody,[b] the intertidal zone of an estuary (regularly and irregularly flooded zones), or the regularly flooded (daily tidal inundation) zone of freshwater tidal wetlands	Fringe[c]
4. Wetland does not occur in these waters or intertidal zones	5
5. Wetland forms a nonvegetated bank or is within the banks of a river or stream	Fringe
5. Wetland is not a nonvegetated river or stream bank or within the banks	6
6. Wetland occurs on an alluvial plain	Floodplain[d]
6. Wetland does not occur on an alluvial plain	7
7. Wetland occurs on a nearly level landform	Flat
7. Wetland occurs in a distinct depression	8
8. Wetland is a shallow open waterbody less than 20 acres in size	Pond
8. Wetland is a vegetated wetland and not solely open water	Basin
Water Flow Path[e]	
1. Wetland is typically surrounded by upland (nonhydric soil); receives precipitation and runoff from adjacent areas with no apparent outflow	Vertical Flow[f]
1. Wetland is not geographically isolated but is connected to other wetlands or waters	2
2. Wetland is a sink receiving water from a river, stream, or other surface water source, lacking surface water outflow	Inflow
2. Wetland is not a sink; surface water flows through or out of the wetland	3
3. Wetland is subjected to frequent tidal flooding (consider applying tidal range descriptor)	Bidirectional-Tidal
3. Wetland is not tidally influenced or if flooded infrequently by storm tides, it is a freshwater wetland	4
4. Water flows out of the wetland but does not flow into this wetland from another wetland or waterbody (except perhaps during severe storms)	Outflow[g]
4. Water flows in and out of the wetland	5
5. Water flows through the wetland, often coming from upstream or uphill sources	Throughflow[g]
5. Wetland is along a lake or reservoir and its water levels are subjected to the rise and fall of this waterbody; no stream flow through the wetland	Bidirectional-Nontidal

[a] *The active floodplain is flooded every other year on average. Lotic wetlands are separated into Lotic River and Lotic Stream types based on watercourse width—approximately 25 ft (e.g., a polygon = Lotic River versus linear = Lotic Stream at a scale of 1:24,000) and then are classified by flow durations: perennial, intermittent, ephemeral, and tidal.*
[b] *If the wetland is a shallow waterbody less than 20 acres in size, it should be considered a Pond.*
[c] *Tidally restricted wetlands behind causeways, dikes, and similar structures are classified as Basins.*
[d] *Basin and Flat sublandforms can be identified within this landform when desirable.*
[e] *Surface water connections are emphasized because they are more readily identified than groundwater linkages. Note: The "Paludified" water flow path is not included in this simple key; paludified wetlands are common in northern climates where bogs are one of the most common wetland types in the region.*
[f] *Wetland is "geographically isolated;"; hydrological relationship to other wetlands and watercourses may be more complex than can be determined by simple visual assessment of surface water conditions.*
[g] *If stream data are separated into intermittent and perennial flows, consider separating Outflow and Throughflow into perennial (OU and TH) and intermittent (OI and TI) water flow paths.*

PREDICTING WETLAND FUNCTIONS FROM NWI+ DATA

Once the NWI database was expanded to include LLWW descriptors, wetlands of potential significance for various functions can be identified and added to the NWI+ database. The functional assessment approach—"*Watershed-based Preliminary Assessment of Wetland Functions*" (W-PAWF)—applies general knowledge about wetlands and their functions to develop a watershed overview that highlights possible "wetlands of significance" in terms of performance of numerous functions. "Significance" means that the wetland likely performs a given function at a level greater than wetlands not designated as having high or moderate potential. To accomplish this objective, the relationships between wetlands and various functions are simplified into a set of practical criteria or observable characteristics.

Predictions can be made for 11 wetland functions from the NWI+ data: surface-water detention, streamflow maintenance, nutrient transformation, carbon sequestration, coastal storm surge detention, sediment and other particulate retention, shoreline stabilization, provision of fish and aquatic invertebrate habitat, provision of waterfowl and waterbird habitat, provision of other wildlife habitat, and provision of habitat for unique, uncommon, and highly diverse plant communities (Table 2.2.1.3). The models or criteria used for these preliminary predictions were the result of the Northeast Region's NWI Program working with wetland specialists from several states (Delaware, Georgia, Maine, Maryland, New Mexico, New York, and Wisconsin) and scientists from the Smithsonian Institution and universities. Documentation supporting the models is provided in Tiner (2003b, 2011b) while minor changes have been made for specific projects that

TABLE 2.2.1.3 Examples of Correlations Between Wetland Functions and Properties in the Enhanced NWI Database for the Northeastern United States

Function/Level of Function	Wetland Types
Surface water detention	
High	Estuarine fringe, basin, and island; Lentic fringe, basin and island; Lentic flats (in reservoirs/dammed areas only); Lotic basin, floodplain, fringe, and island; Marine fringe and island; Pond throughflow, bidirectional-nontidal, and bidirectional-tidal; Terrene fringe-pond throughflow, basin-pond throughflow, basin-pond bidirectional-tidal, basin throughflow, and basin throughflow-intermittent
Moderate	Lotic flat; Lentic flat (not listed as high); Terrene interfluve, basin (not rated as high), floodplain, and -pond (not listed as high); Ponds (other than those listed as high excluding wastewater, sewage treatment, and dredged material ponds and impounded ponds with vertical flow)
Coastal storm surge detention	
High	Estuarine basin, fringe, and island; Marine fringe; Lotic River (tidal) floodplain, fringe, and island (excluding diked or restricted tidal wetlands)
Moderate	Other tidal wetlands (including tidally restricted Lotic and Terrene ones); Terrene and Lotic wetlands (not slope landform) contiguous to an estuarine or marine wetland
Streamflow maintenance	
High	Headwater (not ditched)
Moderate	Headwater-ditched; Lotic (low gradient) floodplain; Lotic stream basin; Pond throughflow and outflow; Terrene-pond throughflow, -pond outflow, and outflow (associated with streams only, not rivers); Lentic wetlands associated with throughflow lakes
Nutrient transformation	
High	Palustrine vegetated seasonally flooded and wetter (including tidal regimes) plus saturated types not on coastal or glaciolacustrine plains; Estuarine vegetated wetlands
Moderate	Palustrine vegetated temporarily flooded plus saturated types on coastal or glaciolacustrine plains and temporarily flooded-tidal types
Carbon sequestration	
High	Estuarine and tidal freshwater vegetated wetlands; Nontidal vegetated wetlands with seasonally flooded or wetter; Permanently saturated freshwater wetlands (bogs and fens)

TABLE 2.2.1.3 Examples of Correlations Between Wetland Functions and Properties in the Enhanced NWI Database for the Northeastern United States—cont'd

Function/Level of Function	Wetland Types
Moderate	Temporarily flooded or seasonally saturated vegetated wetlands; Estuarine aquatic beds and intertidal mud or organic flats; Mixed unconsolidated shore/vegetated wetlands; Ponds (not commercial, industrial, residential, sewage treatment, golf, and mining types)
Sediment and other particulate retention	
High	Estuarine vegetated; Lentic vegetated; Lotic basin, floodplain, fringe (vegetated), and island (vegetated); Pond throughflow and bidirectional-tidal; Terrene basin throughflow and throughflow-intermittent; Terrene any landform-pond throughflow, bidirectional-tidal, and bidirectional-nontidal; Terrene interfluve basin throughflow and throughflow-intermittent
Moderate	Estuarine nonvegetated (excluding rocky shore): Marine unconsolidated shore; Lotic flat (not shrub or forested bog); Lotic fringe (nonvegetated) and island (nonvegetated); Ponds (not rated as high and not commercial, industrial, residential, sewage treatment, golf, and mining); Other Terrene basins (not listed as High but not shrub or forested bog); Terrene any landform-pond (not shrub or forested bog); Terrene floodplain
Shoreline stabilization	
High	Estuarine vegetated, Estuarine rocky shore (not island); Marine rocky shore (not island); Lotic vegetated (not island); Lentic vegetated (not island)
Moderate	Terrene any landform-pond and vegetated; Terrene outflow-headwater and vegetated
Provision of fish and shellfish habitat	
High	Estuarine emergent (excluding *Phragmites*-dominated irregularly flooded), unconsolidated shore, reef, aquatic bed, rocky shore (vegetated with macroalgae); Lacustrine aquatic bed, vegetated mixtures of unconsolidated bottom with vegetation; Lentic vegetated that are permanently flooded; Marine aquatic bed, rocky shore, unconsolidated shore, and reef; Palustrine semipermanently flooded and adjacent to a waterbody (pond, lake, estuary, perennial river, and perennial stream), Palustrine vegetated permanently flooded, Palustrine aquatic bed, mixtures of unconsolidated bottom with vegetation, Palustrine emergent regularly flooded, seasonally flooded-tidal, semipermanently flooded-tidal (except *Phragmites* marshes); Ponds associated with Palustrine vegetated semipermanently flooded; Riverine tidal emergent and unconsolidated shore (except temporarily flooded-tidal)
Moderate	Lentic and Palustrine emergent seasonally flooded-saturated; Lotic and Palustrine emergent seasonally flooded-saturated; Palustrine *Phragmites* semipermanently flooded and adjacent to a waterbody (lake, estuary, perennial river, perennial stream, ponds except commercial, industrial, residential, sewage treatment, golf, and mining types); Estuarine Phragmites along the estuarine shoreline (not "interior marshes"); Estuarine vegetated mixes where emergent is subordinate (e.g., FO/EM and SS/EM); Lotic tidal and forested/emergent seasonally flooded-tidal and wetter; Ponds (all except commercial, industrial, residential, sewage treatment, golf, dredged disposal, and mining); Terrene fringe-pond (not listed as high)
Stream shading	Lotic Stream (not intermittent) and Palustrine forested or scrub-shrub (not shrub bog)
Provision of waterfowl and waterbird habitat	
High	Estuarine emergent (not *Phragmites*), unconsolidated shore, reef, aquatic bed, rocky shore; Lacustrine semipermanently flooded; Lacustrine vegetated permanently flooded (including mixes with nonvegetated); Lacustrine aquatic bed, unconsolidated shore (seasonally flooded or wetter), Lacustrine vegetated permanently flooded; Marine aquatic bed, rocky shore, unconsolidated shore, and reef; Palustrine semipermanently flooded (excluding *Phragmites*-dominated) and adjacent to a waterbody (pond, lake, estuary, perennial river and perennial stream), Palustrine vegetated (including mixtures with nonvegetated) permanently flooded, Palustrine aquatic bed, Palustrine seasonally flooded-saturated diked/impounded or beaver-influenced; Lotic and Palustrine emergent (including mixtures; excluding *Phragmites*-dominated) seasonally flooded-saturated; Lentic and Palustrine emergent (including mixtures; excluding *Phragmites*-dominated) seasonally flooded-saturated, Terrene headwater and Palustrine emergent (including mixtures; excluding *Phragmites*-dominated) seasonally flooded-saturated, Palustrine emergent (except *Phragmites*) regularly flooded, seasonally flooded-tidal, semipermanently flooded-tidal, and subtidal); Ponds associated with Palustrine vegetated semipermanently flooded, Palustrine emergent (and mixes) seasonally flooded-tidal and semipermanently flooded-tidal; Riverine tidal emergent and unconsolidated shore (except temporarily flooded-tidal)

Continued

TABLE 2.2.1.3 Examples of Correlations Between Wetland Functions and Properties in the Enhanced NWI Database for the Northeastern United States—cont'd

Function/Level of Function	Wetland Types
Moderate	Estuarine Phragmites-dominated regularly flooded; Estuarine emergent (*Phragmites*) irregularly flooded and adjacent to estuarine water (not "interior marshes"); Palustrine emergent (*Phragmites*) seasonally flooded-saturated and wetter (including seasonally flooded-tidal) and adjacent to a waterbody (pond, lake, perennial river or stream, estuary); Lacustrine littoral unconsolidated bottom (not designated as high); Ponds (not listed as high or classified as commercial, industrial, residential, sewage treatment, golf, and mining types); Palustrine emergent (and mixes) seasonally flooded-saturated and adjacent to a waterbody (lake, estuary, perennial river, perennial stream, ponds except commercial, industrial, residential, sewage treatment, golf, and mining types)
Wood duck	Lotic perennial (not high gradient) or tidal, basin and floodplain-basin with forested or scrub-shrub cover (and mixtures with other vegetation where trees or shrubs predominate); Lotic floodplain-basin and Palustrine unconsolidated bottom/forested, Palustrine forested or scrub-shrub (and mixes) seasonally flooded-tidal and wetter and contiguous with open water
Provision of other wildlife habitat	
High	Any vegetated wetland complex 20 acres or greater; vegetated wetlands 10–20 acres in size with two or more vegetation classes (excluding *Phragmites*); narrow wetlands connecting large wetlands; small isolated wetlands (basins or ponds) in a dense cluster in a forest matrix (for the Northeast; possible vernal pools)
Moderate	Other vegetated wetlands
Provision of habitat for unique, uncommon, or highly diverse plant communities (formerly "Conservation of biodiversity" focusing on regionally uncommon types)	
Examples from the Northeast United States	Palustrine scrub-shrub seasonally flooded-tidal and semipermanently flooded-tidal; Tidal fresh marshes; Low salt marsh; Oligohaline marsh; Atlantic white cedar swamps; Rich fens; Bald cypress swamps; Eelgrass beds; Lotic fringe; Woodland vernal pools; Bogs (where uncommon)

This table is included in this chapter for readers to gain a general sense of the relationships between database variables and specific functions; "carbon sequestration" was added to the functions around 2010. (Revised slightly from Tiner, 2003b; note that these correlations are typically reviewed and modified as necessary relative to the area of application during the data analysis phase for special projects and are included in project reports.)

are reflected in tables published in project reports. Models for other functions (e.g., provision of amphibian habitat or groundwater recharge) as well as values could be developed where supported by research data and expert knowledge, depending on user needs.

NWI+ DATABASE

The enhanced NWI database—*NWI+ data*—includes the standard NWI classification (Cowardin et al., 1979 and amendments), LLWW classifications (various dates based on project initiation: Tiner, 1997b,c, 2000, 2002a,b, 2003a, 2011a, 2014), and predicted functions for each mapped wetland. Depending on project objectives and funding, other data may be included such as "p-wet areas" (potential wetlands based on USDA soil mapping—undeveloped hydric soil map units that were not mapped as wetland by NWI) and potential wetland restoration sites. NWI+ data can be analyzed through GIS techniques to produce maps and statistics for each watershed or area of interest. Wetlands having properties listed in Table 2.2.1.3 are designated as potentially significant wetlands for particular functions and then can be checked for accuracy through field observations of indicators of wetland functions or the use of field-level rapid assessment methods (RAMs; Sections 4 and 5).

NWI+ PRODUCTS

The outputs of these analyses are the NWI+ geospatial database and, in many cases, technical reports summarizing findings through narrative, tabular data and figures. Thematic maps were initially printed and available as a PDF (Fig. 2.2.1.2), but were later replaced by an online NWI+ mapper created by Virginia Tech that could display the data in full color on topographic maps or aerial images (Fig. 2.2.1.3). Due to changing agency priorities, the NWI+ mapper has not been supported

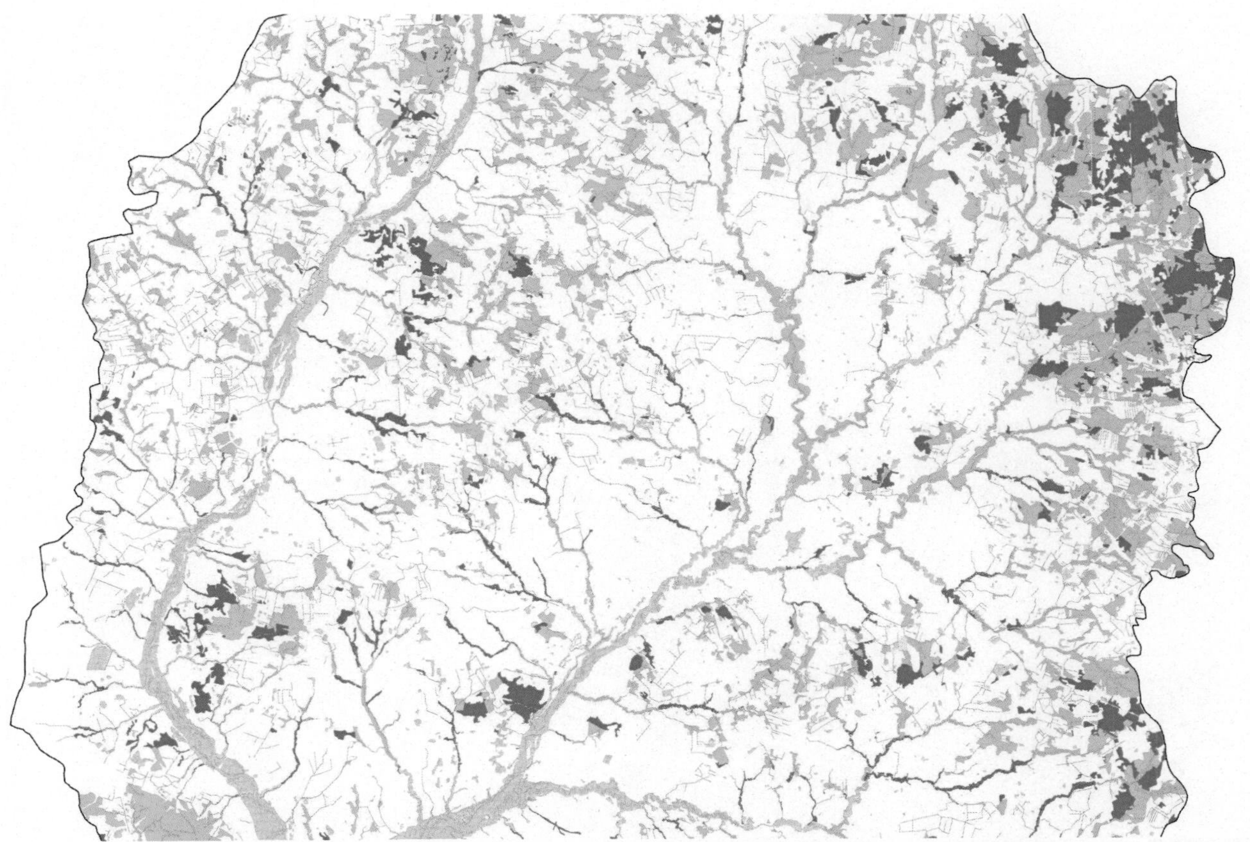

FIG. 2.2.1.2 Prior to development of the NWI+ mapper, hardcopy maps were produced and included with the technical reports. This figure is a clip from the streamflow maintenance function map for part of the Nanticoke watershed, Maryland/Delaware. Red areas indicate wetlands with potential for high function, orange areas represent wetlands with predicted moderate function, and other wetlands are gray (Tiner and Bergquist, 2003).

by the FWS since 2015, so changes in GIS and browser technology left the mapper only accessible through Internet Explorer. Currently, Virginia Tech is graciously supporting the mapper with data displayed in shades of black and white (Fig. 2.2.1.4; http://nwiplus.cmi.vt.edu/nwiplusmapper/). The future of the existing NWI+ databases and mapper remains uncertain. Meanwhile, state-supported NWI+ data should be available through the sponsor agency.

USES OF THE NWI+ DATA

NWI+ data provides valuable information for watershed-based wetland characterization reports and landscape-level assessment of wetland functions. The LLWW descriptors increase the number of variables used to describe individual wetlands beyond those employed by most wetland classification systems. For example, palustrine wetlands, the dominant wetland type in much of the United States, can be subdivided by their landscape position (i.e., association with a waterbody), landform (e.g., basin versus flat versus slope), and water flow path (e.g., vertical flow versus throughflow versus outflow). The expanded classification also allows for improved description of ponds by separating natural ponds (different types) from created ponds and by associating ponds with adjacent land uses (e.g., farm ponds from industrial ponds from stormwater ponds) (Table 2.2.1.1).

This expanded and more refined classification approach allows wetlands to be further divided into similar groupings. This is valuable for selecting "reference wetlands" for various studies (e.g., functions, condition assessment, and wetland plant communities). Researchers in New York, Maryland, and Delaware have used the enhanced NWI classification to identify reference wetlands while developing hydrogeomorphic profiles and for assessing plant communities and water quality relations. Johnson (2005) adapted some of the LLWW descriptors to create a wetland classification system for Summit County, Colorado. Water flow paths were used to separate wetlands into different types for use in developing hydrogeomorphic profiles for the county.

FIG. 2.2.1.3 Creation of the NWI+ mapper allowed users to display results in colors on topographic maps, aerial imagery, or other maps. This example was clipped from the original NWI+ mapper showing wetlands classified by landscape position in Bristol, Pennsylvania. Coding: blue – Lotic; orange – Lentic; brown – Terrene; blue-green – Pond; gray – Deepwater Habitat (Tiner et al., 2015b).

LLWW descriptors, when applied to wetlands lost and gained, provide wetland managers with information on the likely functions that were lost or gained, respectively. This greatly strengthens the findings from wetland status and trend analysis studies and aids in projecting the effect of the losses and gains on wetland functions (cumulative impacts). Tiner and Bergquist (2003) and Tiner (2005) demonstrated this in an analysis of historic changes in wetlands of the Nanticoke River watershed. The results of landscape-level functional assessments can be used to develop conservation strategies for watersheds (e.g., what types of wetlands have suffered the greatest losses and may be priorities for wetland restoration). The state of Michigan includes this type of landscape-level functional assessment in its statewide wetland monitoring and assessment plan (Michigan Department of Environmental Quality, 2015). Also when applied to potential restoration sites, the LLWW descriptors can help provide an estimate of the likely functions to be gained through such efforts (Tiner et al., 2000b,c, 2001b; Tiner, 2001).

The watershed-based wetland characterization and functional assessment reports can serve as basic tools to educate public agencies, private organizations, the general public, elected officials, and others on the diversity of wetlands, differences that affect wetland functions, and where wetlands of potential significance for certain functions are located (see Michigan case study, Chapter 2.2.3). They provide a presumption of "value" that local governments can refer to when reviewing projects that may adversely affect these resources (see Lake County, Illinois, example, Chapter 2.2.13).

Finally, even in the absence of digital geospatial data, the relationships between the enhanced NWI attributes and wetland functions provide a type of map-based "rapid" assessment of wetland functions. Using these correlations plus information from field-based RAMs, people can interpret vital properties and predict wetland functions for individual wetlands in areas of interest from existing wetland maps. This provides a type of rapid functional assessment tool that can be used for desktop assessment prior to field inspection.

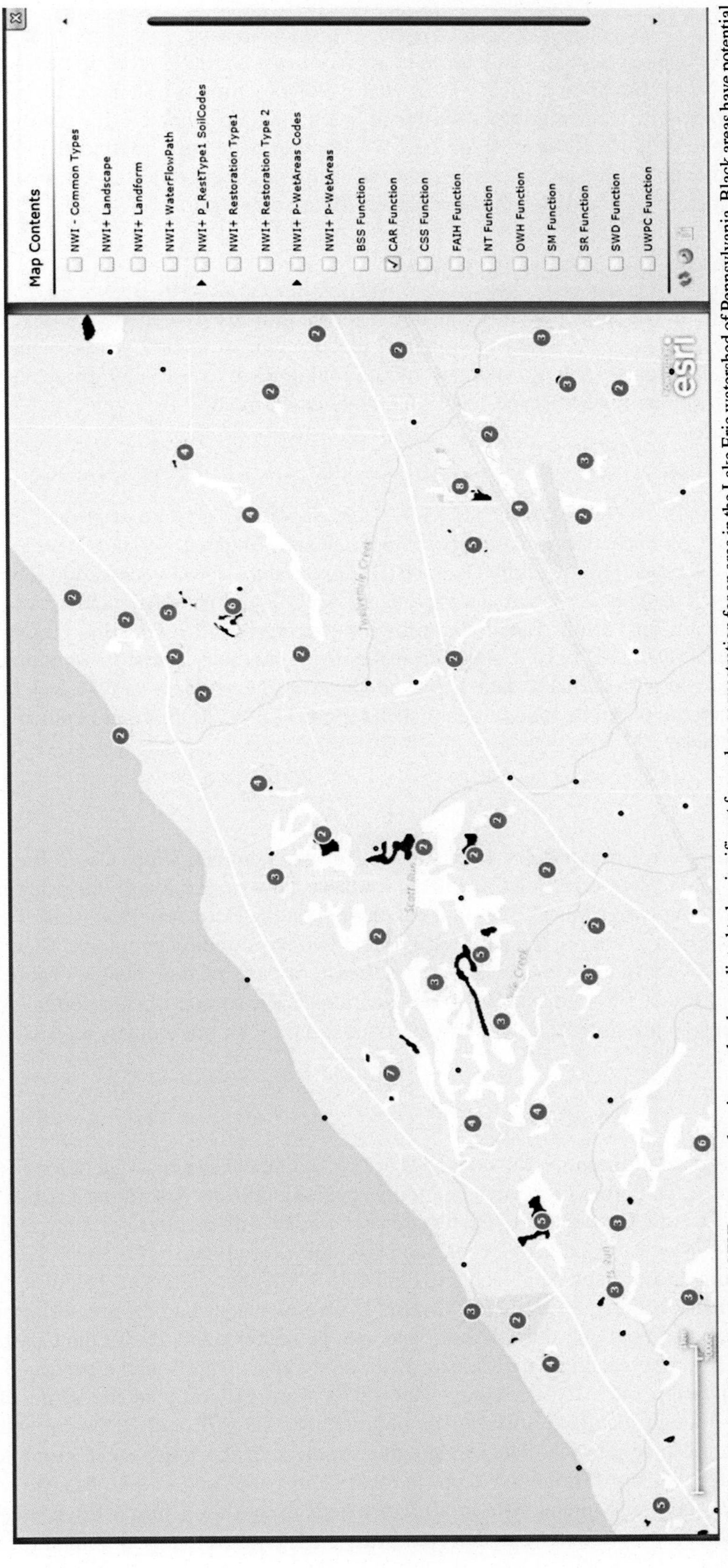

FIG. 2.2.1.4 Clip from the current NWI+ mapper showing wetlands predicted to be significant for carbon sequestration for an area in the Lake Erie watershed of Pennsylvania. Black areas have potential for storing carbon at a high level and medium gray areas at moderate levels. Other wetlands are not shown but their locations may be represented by red or black dots. *(From http://nwiplus.cmi.vt.edu/nwiplusmapper/ (accessed 19.07.17).)*

APPLICATIONS TO DATE

The LLWW descriptors have been applied to many areas in the Northeast United States, first as demonstration projects in Massachusetts, Maine, and New York (Tiner, 1997a; Tiner et al., 1999a,b), then as a standard for updated NWI mapping in this region and as pilot projects in other regions. As a result, such data are available for many areas of the Northeast, including numerous National Wildlife Refuges (Fig. 2.2.1.5). Reports have been produced for many areas as of mid-2017 (Table 2.2.1.4). For other areas, the data are preliminary and have not been subject to a more thorough quality review, but in the future their findings may be summarized in first approximation reports.

CASE STUDY EXAMPLES OF APPLICATIONS

The purpose of this subsection is to demonstrate how existing NWI data was used for landscape-level assessment of wetland functions in six watersheds and one state. From this review, readers should quickly see how such enhancement greatly expands the utility of NWI data for describing wetlands and arranging them into similar groups that allows for improved wetland characterization as well as for landscape-level functional assessment.

Basic Procedures

LLWW descriptors (Tiner et al., 1999a,b; Tiner, 2003a, 2011a,b; depending on date of project initiation) were added to existing NWI data to produce an expanded wetland geospatial database. Mapped wetlands were reexamined using digital geospatial data for streams (National Hydrography Data, NHD), topography (DRGs), elevation (Digital Elevation Models, DEMs), and digital imagery. The data were then analyzed using ArcGIS and personal databases to identify wetlands that were predicted to be significant for various functions following peer-reviewed correlation reports developed with input from regional experts (Tiner, 2003b, 2011a,b). Using GIS technology, summary statistics were prepared along with maps for the watersheds in the study or with thematic data displayed via an online mapper—NWI+ web mapper—for the state of Connecticut. Details of the methods can be found in published reports (Tiner, 2005; Tiner et al., 1999a,b, 2000a,b,c, 2004, 2013; Tiner and Stewart, 2004).

Watershed Examples

Six watersheds in the northeastern United States were selected for this demonstration: Casco Bay watershed in southern Maine, three watersheds (Croton, Delaware, and Catskill) in southern New York that provide drinking water for New York City residents, and the remaining two (Coastal Bays and Nanticoke) on the Delmarva Peninsula. These watersheds, which range in size from 480 to 2400 km^2, occur in three ecoregions and two physiographic regions. Therefore, they reflect some of the landscape and forest diversity associated with the northeastern parts of the country (Table 2.2.1.5). Three of the watersheds contain estuaries: Casco Bay, Coastal Bays, and Nanticoke. The former is a macrotidal system (mean tide range about 9.0 ft) while the latter two are microtidal systems (0.5 and 2.3 ft mean tide ranges, respectively).

Watershed Results

Traditional NWI Findings

Standard NWI mapping produced information on wetland types by ecological system, vegetation life form, substrate (non-vegetated types), water regime, and other properties including special modifiers describing various alterations (Cowardin et al., 1979; Table 2.2.1.6). Overall, the percent of each watershed varied with its physiographic region. The watersheds in the Appalachian Highlands—New England Province (Casco Bay and Croton) had 7.3% and 6.5% of their watershed area represented by wetlands while those in the Appalachian Highlands—Allegheny Plateau (Delaware and Catskill) had only one percent of their watersheds occupied by wetlands. Coastal Plain watersheds had the greatest proportion of their watersheds occupied by wetlands with about one-third of their land area in wetlands (30.5% for the Coastal Bays and 33.7% for the Nanticoke River). The low flat terrain of the Coastal Plain study areas clearly had a profound influence on wetland formation as annual precipitation rates for the study watersheds were generally in the same range (1016–1270 mm; http://www.ocs.orat.edu), although localized areas in the mountainous Catskill and Delaware watersheds may have up to 1524 mm or more of annual precipitation. The topographic gradient of the latter areas promoted runoff and did not provide much opportunity for wetland formation except in local depressions and in lake basins.

Palustrine wetlands were the predominant type in all watersheds except the Coastal Bays, where estuarine wetlands prevailed. Differences within the systems were detected with emergent wetlands dominating estuarine wetlands in

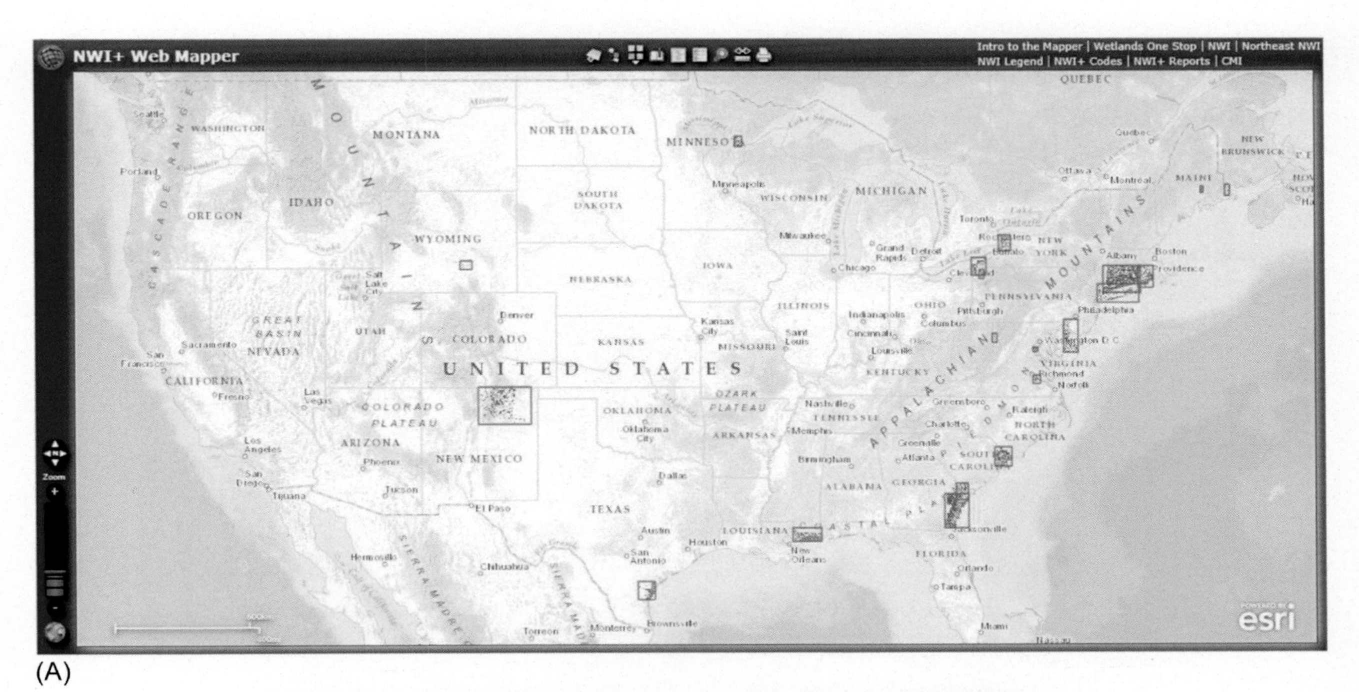

(A)

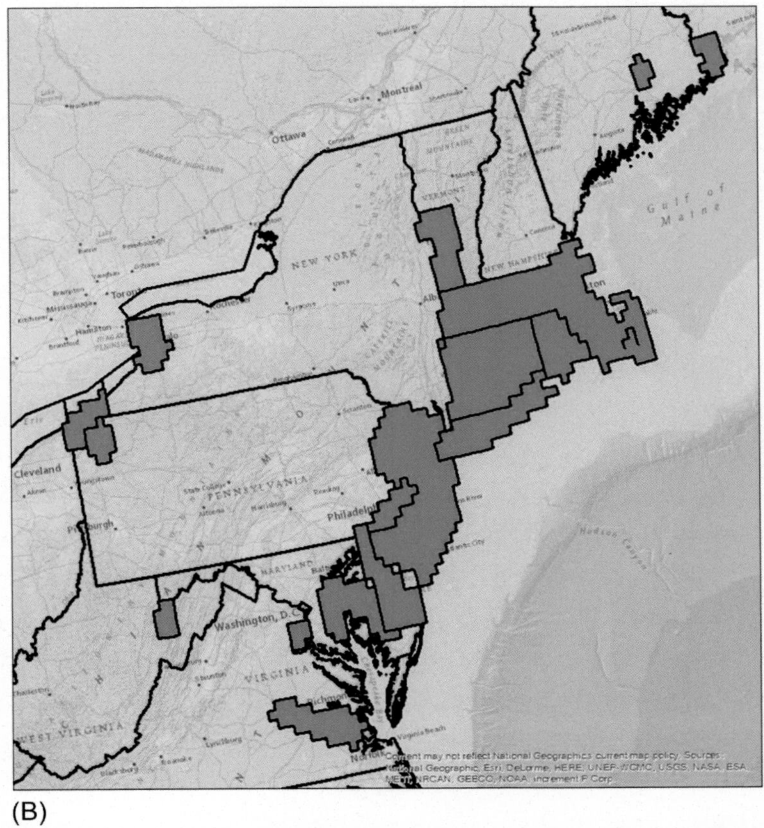

(B)

FIG. 2.2.1.5 Areas where NWI+ data have been created: (A) Sites where data were available online via the NWI+ web mapper as of July 2017 and (B) areas in the Northeast where LLWW descriptors were recently applied and preliminary landscape-level functional assessments performed (figure prepared by Ingrid Mans); all work was completed by Virginia Tech's Conservation Management Institute by the end of 2015. Smaller plots represent lands associated with National Wildlife Refuges. Note: Does not show all areas done prior to 2012 that are listed in Table 2.2.1.4; digital data were not posted online for many of these areas as hardcopy maps were included as figures in the reports. *(From (A) http://nwiplus.cmi.vt.edu/nwiplusmapper/)*

TABLE 2.2.1.4 Published Reports That Include Landscape-Level Assessments

Type Report	Subject Area	Citation
Watershed	Casco Bay, ME	Tiner et al., 1999a
	Boyds Corners and West Branch Sub-basins, NY	Tiner et al., 1999b
	Nanticoke River and Coastal Bays, MD	Tiner et al., 2000a
	Mill and Manhan Rivers, MA	Tiner et al., 2000b
	Shawsheen River, MA	Tiner et al., 2000c
	Nanticoke River, DE	Tiner et al., 2001a
	Upper Ipswich River, MA	Tiner et al., 2001b
	Upper Blackstone River, MA	Tiner, 2001
	Neversink and Cannonsville Reservoirs, NY	Tiner et al., 2002
	Croton, NY	Tiner et al., 2004
	Delaware River and Catskills, NY	Tiner and Stewart, 2004
	Nanticoke River, MD/DE	Tiner, 2005
	Cumberland Bay, NY	Tiner, 2006a
	Upper Wappinger Creek, NY	Tiner, 2006b
	Upper Tioughnioga River, NY	Tiner, 2006c
	Sodus Bay to Wolcott Creek, NY	Tiner, 2006d
	Sodus Creek, NY	Tiner, 2006e
	Post Creek to Sing Sing Creek, NY	Tiner, 2006f
	Sucker Brook to Grass River, NY	Tiner, 2006g
	Peconic River	Tiner, 2006h
	Salmon River to South Sandy Creek, NY	Tiner, 2006i
	Catherine Creek, NY	Tiner, 2006j
	Hudson River-Snook Kill to Fish Creek, NY	Tiner, 2006k
	Hackensack River, NJ/NY	Tiner and Bergquist, 2007
	Lake Erie, PA	Tiner et al., 2014a
Focus Area	Pennsylvania Coastal Zone	Tiner et al., 2002
	Boston Harbor Islands National Recreation Area, MA	Tiner et al., 2003
	Cape Cod and the Islands, MA	Tiner, 2010b
	Long Island, NY	Tiner et al., 2015a
	Delaware River Coastal Zone of Pennsylvania	Tiner et al., 2015b
	Southwestern Vermont and Neighboring NY	Tiner and Herman, 2015
State	Delaware	Tiner et al., 2011
	Connecticut	Tiner et al., 2013
	Rhode Island	Tiner et al., 2014b

Most of these reports can be accessed at https://www.fws.gov/northeast/ecologicalservices/wetlandspubs.html, or by email request from the author; U.S. Fish and Wildlife Service (Region 5, Ecological Services, Hadley, MA), Virginia Tech's Conservation Management Institute (Blacksburg, VA), or the sponsoring agency should have geospatial data for these projects.

TABLE 2.2.1.5 General Characteristics of Study Watersheds

Watershed	State	Bailey	Physiographic		Percent	Percent	Percent
		Ecoregion	Region	Area (est.)	Wetland	Upland	Open Water
Casco Bay	ME	EBDF	AH-NE	2400 km^2	7.3	84.4	8.3[a]
Croton	NY/CT	EBDF	AH-NE	1000 km^2	6.5	87.6	5.9
Catskill	NY	ANEF	AH-AP	1500 km^2	1.0	96.4	2.6
Delaware	NY	ANEF	AH-AP	2000 km^2	0.8	96.9	2.3[b]
Coastal Bays	MD	OCPF	CP	480 km^2	30.5	69.4	0.1[a]
Nanticoke	DE/MD	OCPF	CP	840 km^2	33.7	65.8	0.5[a]

[a]Excludes marine and estuarine waters.
[b]Based on reservoir full condition.
Codes: EBDF, Eastern Broadleaf Deciduous Forest; ANEF, Adirondack-New England Mixed Forest-Coniferous Forest-Alpine Meadow; OCPF, Outer Coastal Plain Mixed Forest; AH-NE, Appalachian Highlands-New England Province; AH-AP, Appalachian Highlands-Allegheny Plateau (Bailey, 1995).

TABLE 2.2.1.6 Summary of Wetlands by NWI Type for Study Watersheds (Sources: Tiner et al., 1999a, 2000a, 2004; Tiner and Stewart, 2004; Tiner, 2005)

Watershed	Wetland	Wetland		% of Watershed's
(State)	System	Class	Area (ha)	Wetlands
Casco Bay	Marine	Aquatic Bed	643	3.4
(ME)		Unconsolidated Shore	1058	5.6
		Rocky Shore	170	0.9
		Subtotal	1871	9.9
	Estuarine	Aquatic Bed	76	0.4
		Emergent	605	3.2
		Unconsolidated Shore	1966	10.4
		Subtotal	2647	14.0
	Lacustrine	Unconsolidated Shore	5	<0.1
	Palustrine	Emergent	1871	9.9
		Forested	8051	42.6
		Scrub-Shrub	3742	19.8
		Unconsolidated Bottom	718	3.8
		Subtotal	14,382	76.1
	All combined		18,905	
Croton	Palustrine	Aquatic Bed	6	0.1
(NY)		Emergent	434	6.9
		Forested	4529	71.9
		Scrub-Shrub	488	7.7
		Unconsolidated Bottom	844	13.4
	All combined		6301	
Catskill	Palustrine	Emergent	378	24.2

Continued

TABLE 2.2.1.6 Summary of Wetlands by NWI Type for Study Watersheds (Sources:Tiner et al., 1999a, 2000a, 2004;;)—cont'd

Watershed	Wetland	Wetland		% of Watershed's
(State)	System	Class	Area (ha)	Wetlands
(NY)		Forested	528	33.9
		Scrub-Shrub	280	18.0
		Unconsolidated Bottom	298	19.1
		Subtotal	1484	95.2
	Riverine	Unconsolidated Shore	75	4.8
	All combined		1559	
Delaware	Palustrine	Emergent	708	33.8
(NY)		Forested	374	17.9
		Scrub-Shrub	361	17.3
		Unconsolidated Bottom	590	28.2
		Subtotal	2033	97.2
	Riverine	Unconsolidated Shore	59	2.8
	All combined		2092	
Coastal Bays	Estuarine	Emergent	6680	45.9
(MD)		Forested	22	0.2
		Scrub-Shrub	208	1.4
		Unconsolidated Shore	439	3.0
		Subtotal	7349	50.5
	Palustrine	Emergent	593	4.1
		Farmed	19	0.1
		Forested	5312	36.5
		Scrub-Shrub	1017	7.0
		Unconsolidated Bottom	249	1.7
		Subtotal	7190	49.5
	All combined		14,539	
Nanticoke	Estuarine	Emergent	6463	11.2
(MD/DE)		Scrub-Shrub	56	0.1
		Forested	98	0.2
		Unconsolidated Shore	232	0.4
		Subtotal	6849	11.9
	Palustrine	Emergent	2289	4.0
		Farmed	1428	2.5
		Forested	40,208	69.9
		Scrub-Shrub	6075	10.6
		Unconsolidated Bottom	488	0.8
		Subtotal	50,488	87.8

TABLE 2.2.1.6 Summary of Wetlands by NWI Type for Study Watersheds (Sources:Tiner et al., 1999a, 2000a, 2004;;)—cont'd

Watershed	Wetland	Wetland		% of Watershed's
(State)	System	Class	Area (ha)	Wetlands
	Riverine	Emergent (tidal)	134	0.2
		Unconsolidated Shore (tidal)	19	<0.1
		Subtotal	153	0.3
	All combined		57,490	

microtidal estuaries on the Coastal Plain and unconsolidated shores (tidal flats) being the chief type in the macrotidal estuary of the Casco Bay watershed. It is no surprise that forested wetlands were the dominant freshwater type in most watersheds because the natural vegetation of the eastern United States is forest. Emergent wetlands were the main freshwater type in the portion of the Delaware watershed that provides drinking water to New York City while the neighboring Catskill watershed also had a high percentage of emergent wetlands. Both these watersheds had a more evenly mixed distribution of freshwater wetland types than the other watersheds in this comparison. The occurrence of numerous lakes increased the likelihood for marsh formation along shorelines. Although water regime data are not presented in the tables, the seasonally flooded/saturated type tends to dominate the northern watersheds while temporarily flooded and seasonally saturated types are the more common water regimes for palustrine wetlands in the coastal plain watersheds.

Enhanced NWI Findings

Enhanced wetland classification provided a *better characterization of wetlands* by adding information on the proximity of wetlands to waterbodies, connectivity, landform, and water flow patterns as well as more specific classification of waterbodies (Tiner, 2003a,b). In particular, more characteristics were assigned to the palustrine wetlands that dominate all but the Coastal Bays watershed. Table 2.2.1.7 summarizes wetland types by landscape position, landform, and water flow path for the subject areas. In northern watersheds, more than half the freshwater wetland area was located along rivers and streams, with the more mountainous areas having 67%–77% of the freshwater wetland area in these lotic positions. In contrast, the southern coastal plain watersheds had most of their freshwater wetland area situated in headwater positions (terrene) serving as sources of streams—75%–81%. Less than 20% of their wetland area was found in locations subject to stream or river overflow. The Casco Bay watershed had its wetland area more evenly divided among the lotic and terrene landscape positions (55% in the former and 40% in the latter). Lakeside wetlands (lentic) are insignificant on the coastal plain watersheds but occupy from 4% to 9% of the wetland area in watersheds in the glaciated Northeast. The Nanticoke and Coastal Bays watersheds had 9% and 5% of their freshwater wetland area located along estuaries, respectively. These were tidal freshwater wetlands that are likely to be converted to brackish marshes with rising sea levels, along with nontidal wetlands bordering the estuarine wetlands (see Tiner, 2005 for more details).

The added information on wetland classification allows for the *prediction of wetland functions* (Fig. 2.2.1.6). Comparisons of wetland functions between watersheds can now be made; these are given for the subject watersheds in Table 2.2.1.8. Using the geospatial database, similar data can be generated for smaller watersheds (subbasins). As expected, 90% or more of the wetland area provided functions generally attributed to wetlands such as surface water detention, nutrient transformation, sediment and particulate retention, and provision of habitat for various wildlife. Provision of habitat for fish, aquatic invertebrates, waterfowl, and waterbirds was performed by less wetland area than for other wildlife but was higher than one would expect. Their totals are higher than actual habitat due to the inclusion of forested and shrub wetlands bordering rivers and streams that moderate temperatures for aquatic ecosystems and provide habitat for wood ducks. The smallest amount of wetland was predicted as significant for "conservation of biodiversity" in watersheds where this function was evaluated.[1] This is not surprising because it is the most restrictive of the functions intended to highlight

1. The "conservation of biodiversity" function was simply an attempt to identify wetlands with vegetation that was markedly different than other wetlands based on their classification by the NWI; it was not based on an examination of rare, threatened, or endangered species habitat data. In more recent assessments, this function was changed to "Provision of Habitat for Unique, Uncommon, or Highly Diverse Plant Communities" as the reference to wetlands identified as significant for "conservation of biodiversity" might have led readers to believe that other wetlands were not important for biodiversity.

TABLE 2.2.1.7 Summary of Enhanced NWI Characteristics (Landscape Position, Landform, and Water Flow Path) for Wetlands in the Subject Watersheds

Watershed	Landscape Position—% of Wetland Area	Landform—% of Wetland Area	Water Flow Path—% of Wetland Area	Pond % of Wetland Area
Casco Bay	Marine—9.9	Basin—58.7	Inflow—0.4	4.3
	Estuarine—14.2[a]	Flat—4.4	Vertical Flow—13.2	
	Terrene—30.6	Fringe—25.0	Outflow—16.3	
	Lotic River—7.7	Floodplain—6.0	Throughflow—41.6	
	Lotic Stream—34.0	Island—<0.1	Bidirectional-nontidal—3.6	
	Lentic—3.6	Slope—5.8	Bidirectional-tidal—24.8	
Croton[b]	Lotic River—1.5	Basin—40.7	Vertical Flow—4.2	10.4
	Lotic Stream—75.5	Flat—1.4	Throughflow—83.6	
	Lentic—8.1	Fringe—2.4	Outflow—10.5	
	Terrene—14.9	Floodplain—55.4	Inflow—0.3	
		Island—<0.1	Bidirectional-nontidal—1.4	
		Slope—<0.1		
Delaware[b]	Lotic River—12.5	Basin—26.6	Vertical Flow—7.8	24.5
	Lotic Stream—63.3	Flat—7.0	Throughflow—79.1	
	Lentic—3.8	Fringe—6.7	Outflow—12.1	
	Terrene—20.4	Floodplain—57.6	Inflow—0.1	
		Island—0.4	Bidirectional-nontidal—0.9	
		Slope—1.8		
Catskill[b]	Lotic River—13.5	Basin—36.0	Vertical Flow—7.5	16.3
	Lotic Stream—53.6	Flat—6.1	Throughflow—73.9	
	Lentic—8.7	Fringe—9.1	Outflow—15.9	
	Terrene—24.2	Floodplain—47.4	Inflow—0.1	
		Island—0.2	Bidirectional-nontidal—2.5	
		Slope—1.0		
Coastal Bays	Estuarine—52.7[a]	Fringe—48.0	Vertical Flow—2.4	1.7
	Lotic Stream—10.0	Basin—2.1	Throughflow—6.4	
	Lentic—<0.1	Floodplain—9.9	Outflow—33.8	
	Terrene—37.3	Flat—10.8	Bidirectional-tidal—57.3	
		Island—4.7	Bidirectional-nontidal—< 0.1	
		Interfluve—24.6		
Nanticoke	Estuarine—16.1[a]	Fringe—16.6	Vertical Flow—3.6	0.9
	Lotic River—2.1	Basin—0.5	Throughflow—10.4	
	Lotic Stream—9.9	Floodplain—10.6	Outflow—67.6	

TABLE 2.2.1.7 Summary of Enhanced NWI Characteristics (Landscape Position, Landform, and Water Flow Path) for Wetlands in the Subject Watersheds—cont'd

Watershed	Landscape Position—% of Wetland Area	Landform—% of Wetland Area	Water Flow Path—% of Wetland Area	Pond % of Wetland Area
	Lentic—0.2	Flat—1.1	Bidirectional-tidal—18.3	
	Terrene—71.7	Island—0.2	Bidirectional-nontidal—0.2	
		Interfluve—71.0		

[a]*Difference in estuarine proportion versus NWI summary (Table 2.2.1.2) is due to classification of tidal fresh wetlands contiguous to estuarine waters and wetlands as part of the estuary because they were not located along a freshwater tidal river.*
[b]*Data did not include the exposed portions of the New York City reservoirs during drawdown as they were considered part of the reservoir (normal pool). Nonvegetated (open water) portions of ponds were treated as waterbodies, except for the earliest of the studies (Casco Bay) where all portions of ponds were treated as wetlands.* Note: *Vertical Flow water flow was classified as "isolated" in these projects; the landform "interfluve" has been eliminated and is now identified as either basin or flat with most of them on the coastal plain being flats that are seasonally saturated.*

wetlands different from the rest, largely from a vegetative standpoint. Wetlands in watersheds that did not include tidal wetlands, of course, did not have any wetlands performing the coastal storm surge detention function.

Comparison Between Watersheds

No two watersheds are alike, although those in the same physiographic region have certain features in common. As noted above, wetlands accounted for a much higher percentage of the watersheds on the Coastal Plain than in the Adirondack—New England Highlands or the Eastern Broadleaf Deciduous Forest. While estuarine wetlands represented a significant amount of the wetland area in the Coastal Plain watersheds (Coastal Bays and Nanticoke), palustrine wetlands still accounted for 15%–25% of the land surface area of these watersheds. Moreover, the area of today's wetlands does not include the tremendous area of former wetlands that has been effectively drained in these watersheds. Many of the remaining interfluve wetlands are sources of low-gradient streams. In contrast, wetlands in glaciated, hilly to mountainous areas tend to be more frequently associated with streams and rivers. Due to the presence of natural lakes in these regions, lentic wetlands are also more extensive in glaciated regions than in the coastal plain where such wetlands only exist along artificial lakes created by damming river and stream valleys. Based on the study, geographically isolated wetlands (vertical flow water flow path) appeared to be more common in watersheds in glaciated regions than in coastal plain landscapes. For this region of the country, throughflow and outflow wetlands predominate outside the tidal zone. For wetlands on the coastal plain watersheds in the study area, the chief water flow path was outflow as most wetlands appeared as sources of streamflow and were not located on floodplains. In contrast, more wetland area was associated with floodplains (throughflow water flow path) in the hilly to mountainous watersheds. Although not presented in this chapter, additional information is available from the digital wetland database for further analysis (e.g., the percentage of the area or number of forested wetlands that are associated with rivers, streams, or are geographically isolated). Differences in the percent of wetlands that were predicted to perform various functions can be seen in Table 2.2.1.8 for each watershed with areal extent figures (i.e., acreage) given in the published reports.

Statewide Assessment Example—Connecticut

Following the basic procedures outlined above and detailed in the technical report (Tiner et al., 2013), a statewide assessment of wetland functions was performed. Statistics were summarized in tables like those shown for the watersheds above for the entire state plus major watersheds, with results also presented in a series of graphics as depicted in Figs. 2.2.1.7–2.2.1.9. The geospatial data were posted via an online mapper—NWI+ mapper—that was developed and is maintained by Virginia Tech in Blacksburg, Virginia (http://nwiplus.cmi.vt.edu/nwiplusmapper/; Note: This mapper is only functional when accessed with Internet Explorer as it has not been upgraded with changes in technology).

LIMITATIONS OF THIS APPROACH

Numerous wetlands were predicted to perform a given function at a significant level presumably important to a watershed's ability to provide that function; nearly all wetlands provide at least one function at a significant level. "Significance" is a

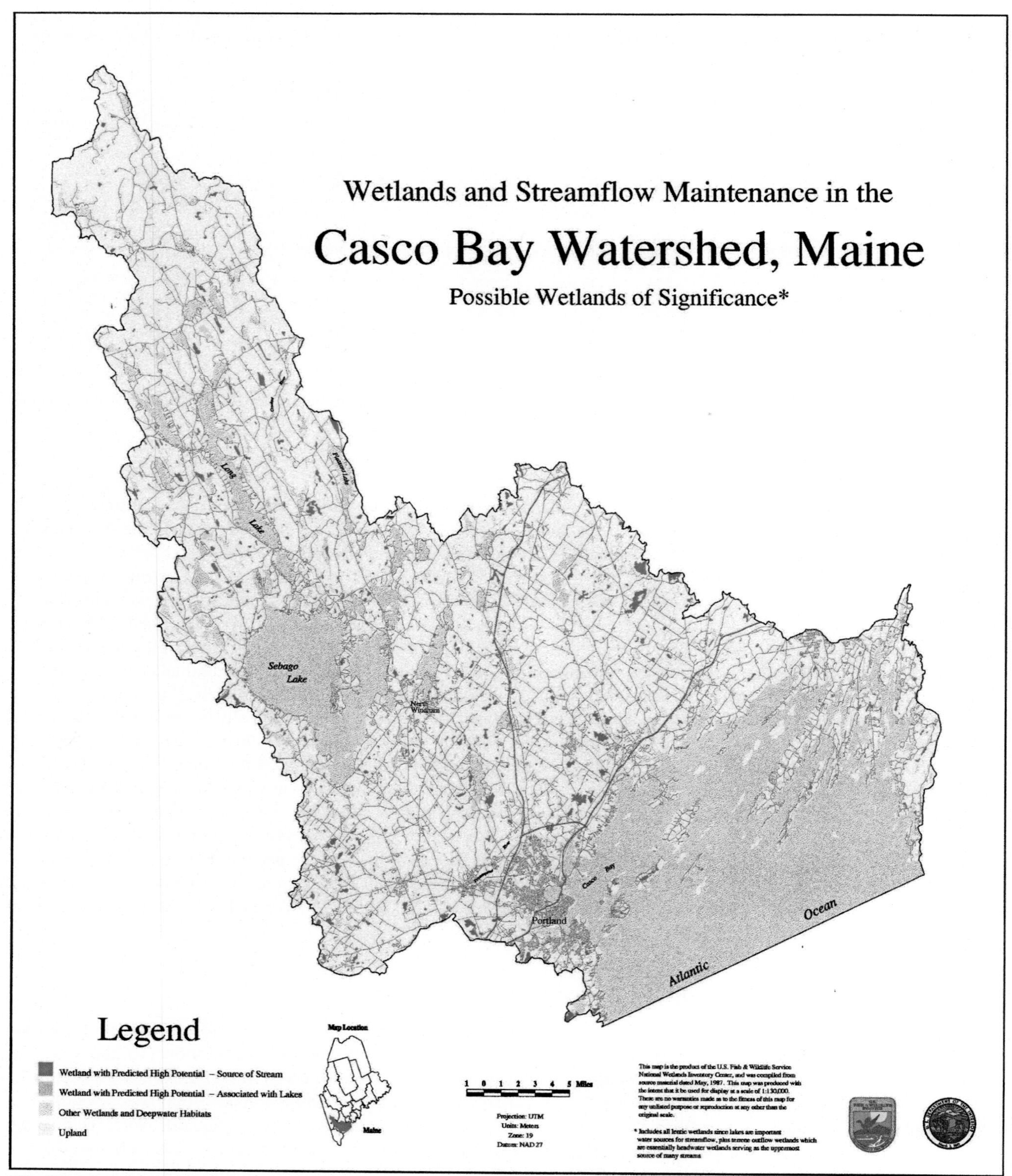

FIG. 2.2.1.6 Map highlighting wetlands predicted to be significant for streamflow maintenance for the Casco Bay Watershed, Maine (Tiner et al., 1999a).

TABLE 2.2.1.8 Summary of Wetland Functional Assessments Derived From Enhanced NWI Data

Study Watershed	Level	% of Wetland Area Predicted to Perform Functions at Significant Levels									
		SWD	SFM	NT	SPR	SLS	CSD	FSH +SS	WWH +WD	OHW	CBD
Casco Bay[a]	High	63.0	19.9[b]	56.9	42.8	52.6	24.8	24.4	24.4 +36.8	46.2	22.4
	Moderate	30.3	–	nd	49.6	nd	–	nd +23.3	4.3	29.4	na
		(93.3)	(19.9)	(56.9)	(92.4)	(52.6)	(24.8)	(47.7)	(65.5)	(75.6)	(22.4)
Croton	High	80.7	37.9	87.4	74.6	75.4	na	1.3	5.3+60.8	57.5	ne
	Moderate	16.4	49.3	1.1	17.0	2.5	na	12.1 +62.2	10.0	28.1	ne
		(97.1)	(87.2)	(88.5)	(91.6)	(77.9)		(75.6)	(76.1)	(85.6)	
Delaware	High	56.1	61.4	48.3	54.0	56.8	na	1.3	17.2+9.4	19.0	ne
	Moderate	39.7	17.0	24.2	36.1	2.6	na	36.4 +14.0	22.9	50.6	ne
		(95.8)	(78.4)	(72.5)	(90.1)	(59.4)		(51.7)	(49.5)	(69.6)	
Catskill	High	58.7	48.1	56.3	55.5	58.4	na	1.1	15.9 +10.4	27.0	ne
	Moderate	34.9	17.4	22.3	33.5	3.4	na	29.1 +18.5	17.8	42.7	ne
		(93.6)	(65.5)	(78.6)	(89.0)	(61.8)		(48.7)	(44.1)	(69.7)	
Nanticoke	High	27.6	16.7	25.2	27.2	27.5	18.1	12.4	12.8+9.5	91.5	25.2
	Moderate	69.2	57.9	71.0	3.3	–	–	1.0+9.3	0.8	4.7	na
		(96.8)	(74.6)	(96.2)	(30.5)	(27.5)	(18.1)	(22.7)	(23.1)	(96.2)	(25.2)
Coastal Bays[a]	High	62.2	31.5	59.3	57.8	57.4	51.8	48.9	49.1+3.2	85.5	67.9
	Moderate	27.3	8.4	34.8	4.3	–	3.8[c]	1.7+8.7	1.7	9.7	na
		(89.5)	(39.9)	(94.1)	(62.1)	(57.4)	(55.6)	(59.3)	(54.0)	(95.2)	(67.9)

[a]*Estimated from data in original report.*
[b]*For this watershed only wetlands that were the actual sources of streams were highlighted as significant for streamflow maintenance; other studies also included wetlands along first-order streams, so the percent of wetlands contributing to streamflow for Casco Bay is underrepresented.*
[c]*Includes nontidal wetlands contiguous with tidal marshes.*
Abbreviations for functions: SWD, surface water detention; SFM, streamflow maintenance; NT, nutrient transformation; SPR, sediment and other particulate retention; SLS, shoreline stabilization; CSD, coastal storm surge detention; FSH, provision of fish and shellfish habitat; WWH, provision of waterfowl and waterbird habitat; OWH, provision of other wildlife habitat; CBD, conservation of biodiversity; SS, stream shading; WD, wood duck potential; na, not applicable; nd, not determined; ne, not evaluated.

relative term and is used in these analyses to identify wetlands that are likely to perform a given function at a high or moderate level. Wetlands not highlighted may perform the function at a low level or may not perform the function at all.

Source data is a primary limiting factor for landscape-level functional assessment. All wetland and stream mapping have limitations for many reasons, such as scale, image quality, date of the survey, and the difficulty of photo-interpreting certain wetland types (especially evergreen forested wetlands and drier-end wetlands; Tiner, 1990; Tiner, 2017) and narrow or intermittent streams, especially those flowing through dense evergreen forests or beneath built-up lands. Consequently, many small streams were not identifiable on the source data (Colson et al., 2008 and Vanderhoof et al., 2018) or imagery used for these projects. This would affect their LLWW classification. Also joining different geospatial data sources is challenging and oftentimes inexact because they were interpreted from different imagery and aligned to different products (i.e., aerial imagery or maps).

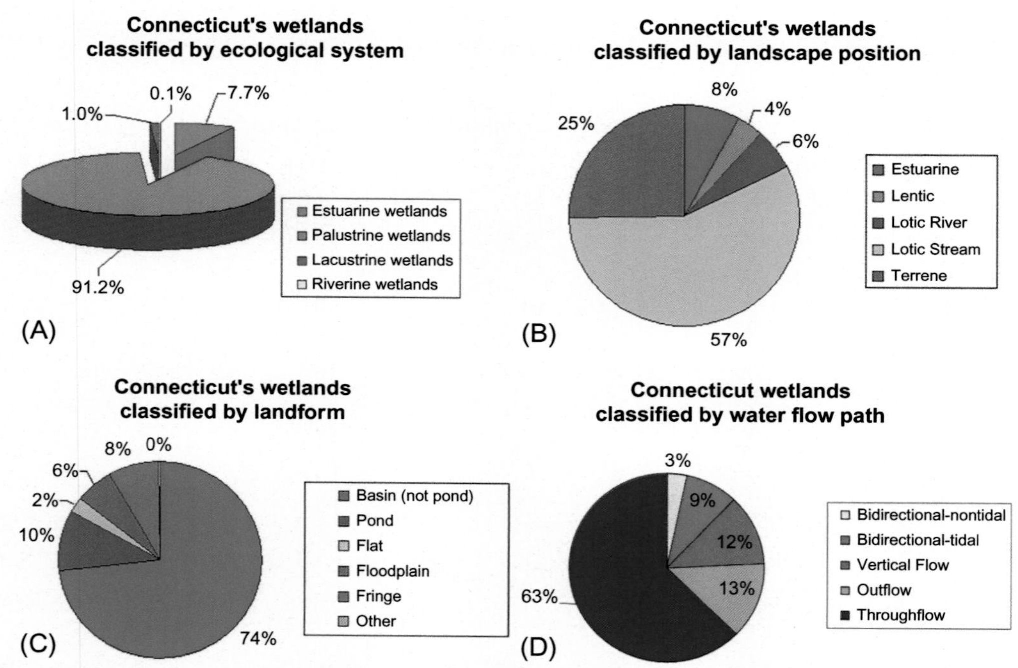

FIG. 2.2.1.7 Percent of Connecticut's wetland area classified by features: (A) ecological system, (B) landscape position, (C) landform, and (D) water flow path. Note: In "D" the original document listed "isolated" as a water flow path; this has been changed to "vertical flow" to be consistent with current terminology (Tiner et al., 2013).

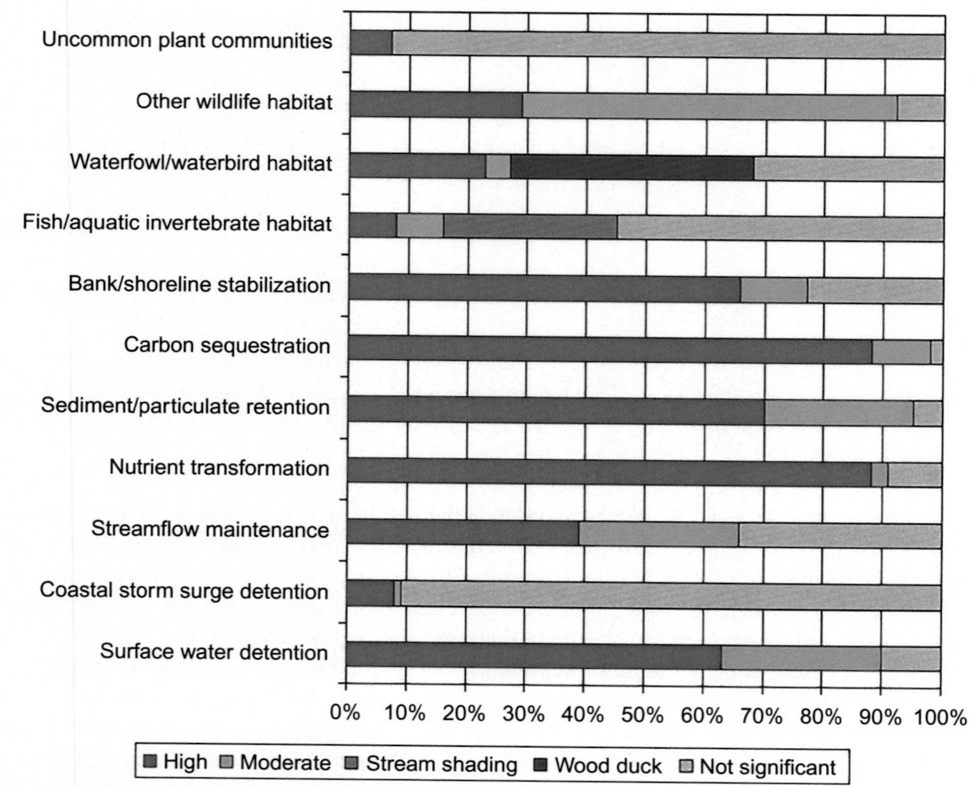

FIG. 2.2.1.8 Percentage of Connecticut's wetland area that support various functions (Tiner, 2013).

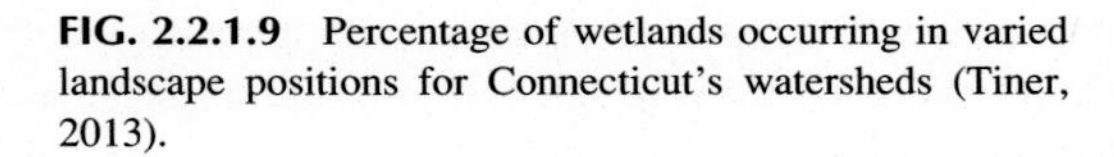

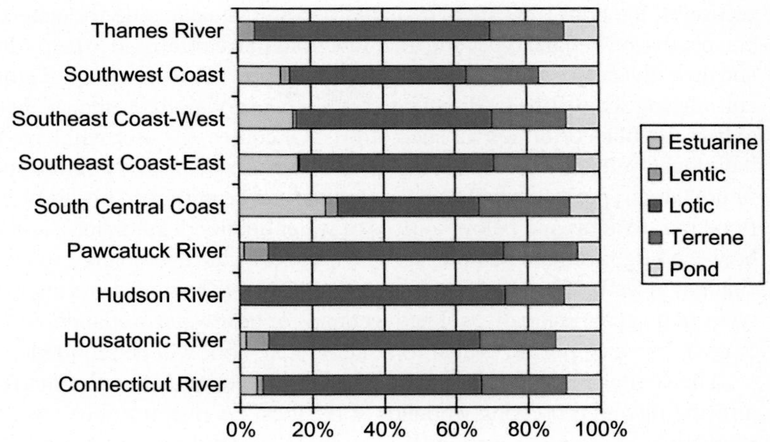

FIG. 2.2.1.9 Percentage of wetlands occurring in varied landscape positions for Connecticut's watersheds (Tiner, 2013).

Besides the inherent limitations of wetland mapping through remote sensing (Tiner, 2015) and of other source data (e.g., streams), the results of these assessments are dependent upon the classifications and the relationships determined to predict wetland functions from the variables in the database. For example, wetlands designated with the water flow path—vertical flow—lack a stream connection, according to the image analyst, but upon field inspection may have a narrow drainage outlet that would connect these wetlands to other wetlands or waterbodies. This would change the classification and the functions predicted. Published reports specify limitations and recommended uses of the data. Field checking would help improve the results.

The approach focuses on wetland functions rather than on wetland condition, recognizing that wetlands even in an altered condition can provide significant environmental services. Additional analyses of landscape context (area surrounding the wetland and waterbodies) and various stressors (disturbances) could be done to provide some perspective on condition. Examples of these analyses were done for the Nanticoke River and Coastal Bays watersheds in Maryland and Delaware (Tiner et al., 2000a,b,c; Tiner and Bergquist, 2003; Tiner, 2005) and the Hackensack River watershed in New Jersey and New York (Tiner and Bergquist, 2007).

When the landscape-level assessment technique is applied in a new area, users should first review the correlations to determine their applicability for that region. While most of the correlations should be more or less universal (e.g., headwater wetlands are vital for streamflow maintenance, floodplain wetlands store water temporarily, and the organic soils of bogs and fens are evidence of carbon sequestration), there are some, particularly the habitat functions, that may need to be tailored to specific locales.

Overall, the landscape-level approach is a first cut at predicting wetland functions and producing a preliminary assessment of those functions based on interpretations of existing information and the collective professional judgment (expert opinion) of many wetland scientists, mostly from the eastern United States. It does not obviate the need for more detailed assessments of the various functions as well as an assessment of wetland condition and opportunities to provide more benefits given the state of the contributing watershed and adjacent land use activities. This preliminary assessment should be viewed as a starting point for more rigorous assessments because it attempts to highlight wetlands that may likely provide significant functions based on generally accepted principles and the source information used for this analysis. Fieldwork can refine the results by validating or adjusting wetland classifications or interpretations and also by contributing knowledge to the relationship between wetland classification attributes and functions through the observation of field indicators of performance (e.g., direct or indirect evidence of inundation). The method also does not utilize size as a factor for rating wetlands or their functions and does not combine function ratings to generate a score for individual wetlands. Any scoring or comparisons between wetlands for minimizing impacts from proposed projects (e.g., road alignments) would be left to the agencies responsible for managing the resource. Consequently, when using the results of this type of assessment for comparing the functions between two or more wetlands, the user will need to consider additional features such as size, condition of the wetland buffer, landscape setting, or context when ranking or prioritizing wetlands of interest.

USES OF THIS ASSESSMENT METHOD

This type of assessment is most useful for regional or watershed planning purposes, for a cursory screening of sites for acquisition, and to aid in developing landscape-level wetland conservation and protection strategies (see other case studies in this section of the book). As such, it should be an invaluable aide for watershed associations, providing them with a basin-wide perspective of wetland types and their functions that can be used to help educate stakeholders on the values that wetlands provide and their diversity across the watershed as well as to help conserve and protect wetlands in the watershed. When such groups are considering acquisition by direct purchase or conservation easements, the information produced by this assessment can be used to help establish priorities for such efforts. Of course, the user will have to establish his or her own rating system for ranking individual wetlands by considering factors such as size, scarcity of the type, uniqueness, historic losses, current and anticipated future threats, problems in the watershed that can be mitigated in part by wetland restoration (e.g., by restoring wetlands above flood hazard areas and below sources of water quality degradation), and maintaining natural corridors for wildlife movement between large wetland complexes. The approach can also be used to evaluate cumulative impacts of various alterations and changes in wetlands on key functions or to consider the national and regional scale impacts of policy changes on certain wetland types (e.g., geographically isolated wetlands or headwater wetlands, or determining significant nexus to waters of the United States). For site-specific evaluations, additional work will be required.

The Center for Watershed Protection has developed the "Wetlands-At-Risk Protection Tool" (WARPT) as a guide to prioritizing and protecting wetlands at the local level (http://www.wetlandprotection.org/). The plan is a five-step process to identify and protect wetlands: (1) update wetland maps, (2) estimate wetland loss, (3) identify priority wetlands, (4) estimate wetland values, and (5) protect wetlands. The first component of Step 3 is a desktop assessment of wetlands fol lowing the approach mentioned in this chapter—the "Watershed-based Preliminary Assessment of Wetland Functions." WARPT is designed for use by local governments and watershed associations. Frederick County (Maryland) has used the assessment in greenway infrastructure planning.

The work done to date has utilized digital databases, but it is also possible to use the correlations to make preliminary interpretations of wetland functions for individual wetlands from standard NWI data. One would simply review the NWI classification, add the appropriate LLWW classifications, and determine what functions are likely to be performed by the wetland(s) in question from the correlation reports/tables.

FUTURE APPLICATIONS

Many states are taking the lead in updating NWI and advancing this approach, including Colorado, Georgia, Michigan, Minnesota, Montana, and New Mexico (see other chapters in this section of the book for examples of their applications and adaptations). Local governments have applied or are also applying this approach to gain insight into possible functions for wetlands under their jurisdiction (e.g., New York City water supply watersheds and Lake County, Illinois north of Chicago). While the initial objective was to use these data for watershed assessments, such data have been used to generate preliminary wetland functional assessments for political units (county or state) or other geographic areas of interest (e.g., Cape Cod, National Wildlife Refuges, or National Parks and Recreation Areas). In formulating state wetland protection strategies required by the EPA, the availability of an enhanced NWI database and functional correlations for a region would make it possible, for the first time, to produce a statewide perspective on wetlands and their functions (e.g., Connecticut).

The enhancement of NWI data through adding LLWW descriptors creates a more powerful tool for natural resource planning and management of wetlands, as evidenced by the adoption of the technique for state-funded wetland inventory initiatives. It provides an analytical tool that can produce a preliminary landscape-level assessment of wetland functions for large geographic areas at a relatively low cost compared with field-based evaluations. While GIS users can make use of the existing data, providing such information through an upgraded web mapper would provide access to a larger audience.

CONCLUSION

While the existing NWI database is useful for showing the distribution of wetlands by major type, expanding these data by adding LLWW descriptors creates a powerful database for natural resource planners and decision-makers. Wetland functions can be predicted across geographic areas of varying sizes based on the relationships developed by numerous applications across the country. Using GIS technology, maps highlighting individual functions can be produced so that the resource managers as well as the general public can readily see that all wetlands do not perform all functions at high levels and which wetlands are predicted to be significant performers of various functions. The characterizations and preliminary functional assessments coupled with a user-created rating system can produce a decision support tool for evaluating wetland impacts from proposed construction projects (e.g., alternative routes for corridor projects such as road alignments and pipelines), for prioritizing wetlands for conservation and protection, and for planning a green infrastructure network.

ACKNOWLEDGMENTS

Many people and agencies have supported the development of NWI+ data and have been acknowledged in the various reports published to date. Key players in developing and applying the tool were Jason Herman and Kevin McGuckin (Conservation Management Institute, Virginia Tech), David Foulis, Glenn Smith, John Swords, Herb Bergquist, Gabe DeAlessio, and Matt Starr (U.S. Fish and Wildlife Service), Todd Nuerminger and Irene Huber (University of Massachusetts), Andy Robertson and John Anderson (St. Mary's University of Minnesota), and Rainor Gresham, Benjamin Cogdell, and John Hefner (Atkins North America). The following individuals and agencies were instrumental in supporting development of the tool during the two decades of its development: Christy Foote-Smith (Massachusetts Executive Office of Environmental Affairs), Jackie Sartoris and Liz Hertz (Maine State Planning Office), Bill Jenkins (Maryland Department of Natural Resources), Mark Biddle and Amy Jacobs (Delaware Department of Natural Resources and Environmental Control), Robert Gilmore (Connecticut Department of Environmental Protection), Laurie Machung (New York City Department of Environmental Protection), Patricia Riexinger (New York State Department of Environmental Conservation), David Stout, Bill Wilen, and Clifford Day (Fish and Wildlife Service), Peter Veneman (University of Massachusetts), Maryann McGraw (New Mexico Environment Department), and Jan Mackinnon (Georgia Department of Natural Resources, Coastal Resources Division). Other persons working for state governments (including contributors to other chapters in this section of the book) have initiated expanded wetland classification in their ongoing wetland inventories.

REFERENCES

Bailey, R.G., 1995. Descriptions of the Ecoregions of the United States, second ed. U.S.D.A. Forest Service, Washington, DC Misc. Publ. 1391.

Brinson, M.M., 1993. A Hydrogeomorphic Classification for Wetlands. U.S. Army Corps of Engineers, Washington, DC Wetlands Research Program Tech. Rep. WRP-DE-4.

Colson, T., Gregory, J., Dorney, J., Russell, P., 2008. Topographic and soil maps do not accurately depict headwater stream networks. Natl. Wetl. Newsl. 30 (3), 25–28.

Cowardin, L.M., Carter, V., Golet, F.C., LaRoe, E.T., 1979. Classification of Wetlands and Deepwater Habitats of the United States. U.S. Fish and Wildlife Service, Washington, DC. FWS/OBS-79/31.

FGDC Wetlands Subcommittee, 2009. Wetlands Mapping Standard. Federal Geographic Data Committee, Washington, DC. FGDC-STD-015-2009. https://www.fgdc.gov/standards/projects/FGDC-standards-projects/wetlands-mapping/2009-08FGDCWetlandsMappingStandard_final.pdf.

Johnson, J.B., 2005. Hydrogeomorphic Wetland Profiling: An Approach to Landscape and Cumulative Impacts Analysis. U.S. Environmental Protection Agency, Washington, D.C. EPA/620/R-05/001.

Michigan Department of Environmental Quality, 2015. Michigan Wetland Monitoring and Assessment Strategy. Water Resources Division, Lansing, MI. https://www.michigan.gov/documents/deq/wrd-wetlands-strategy_555457_7.pdf.

Tiner, R.W., 1990. Use of high-altitude aerial photography for inventorying forested wetlands in the United States. For. Ecol. Manag. 33/34, 593–604.

Tiner, R.W., 1995. A Landscape and Landform Classification for Northeast Wetlands (Operational Draft). U.S. Fish and Wildlife Service, National Wetlands Inventory Program, Hadley, MA.

Tiner, R.W., 1997a. Piloting a more descriptive NWI. Natl. Wetl. Newsl. 19, 14–16.

Tiner, R.W., 1997b. Keys to Landscape Position and Landform Descriptors for Northeast Wetlands. U.S. Fish and Wildlife Service, National Wetlands Inventory Program, Hadley, MA.

Tiner, R.W., 1997c. Keys to Landscape Position and Landform Descriptors for U.S. Wetlands (Operational Draft). U.S. Fish and Wildlife Service, National Wetlands Inventory Program, Hadley, MA.

Tiner, R.W., 2000. Keys to Waterbody Type and Hydrogeomorphic-Type Descriptors for U.S. Waters and Wetlands (Operational Draft). U.S. Fish and Wildlife Service, National Wetlands Inventory Program, Hadley, MA.

Tiner, R.W., 2001. Wetlands, Potential Wetland Restoration Sites, and Functional Deficits for the Upper Blackstone River Watershed. Department of Plant and Soil Sciences, University of Massachusetts, Amherst. Prepared for the Worcester County Soil and Water Conservation District in support of the Massachusetts Wetlands Restoration & Banking Program, Executive Office of Environmental Affairs, Boston, MA.

Tiner, R.W., 2002a. Keys to Waterbody Type and Hydrogeomorphic-Type Descriptors for U.S. Waters and Wetlands (Operational Draft). U.S. Fish and Wildlife Service, National Wetlands Inventory Program, Hadley, MA.

Tiner, R.W., 2002b. Preliminary Functional Assessment Correlations for the Northeast. U.S. Fish and Wildlife Service, National Wetlands Inventory Program, Hadley, MA.

Tiner, R.W., 2003a. Dichotomous Keys and Mapping Codes for Wetland Landscape Position, Landform, Water Flow Path, and Waterbody Type Descriptors. U.S. Fish and Wildlife Service, National Wetlands Inventory Program, Hadley, MA.

Tiner, R.W., 2003b. Correlating Enhanced National Wetlands Inventory Data with Functions for Watershed Assessments: A Rationale for Northeastern U.S. Wetlands. U.S. Fish and Wildlife Service, National Wetlands Inventory Program, Hadley, MA.https://www.fws.gov/northeast/ecologicalservices/pdf/wetlands/CorrelatingEnhancedNWIDataWetlandFunctionsWatershedAssessments.pdf.

Tiner, R.W., 2005. Assessing cumulative loss of wetland functions in the Nanticoke River watershed using enhanced National Wetlands Inventory data. Wetlands 25 (2), 405–419.

Tiner, R.W., 2006a. Wetlands of the Cumberland Bay Watershed, Clinton County, New York. U.S. Fish and Wildlife Service, National Wetlands Inventory, Northeast Region; Prepared for the New York State Department of Environmental Conservation, Division of Fish, Wildlife, and Marine Resources, Bureau of Habitat, Hadley, MA; Albany, NY.

Tiner, R.W., 2006b. Wetlands of the Upper Wappinger Creek Watershed, Dutchess County, New York. U.S. Fish and Wildlife Service, National Wetlands Inventory, Northeast Region; Prepared for the New York State Department of Environmental Conservation, Division of Fish, Wildlife, and Marine Resources, Bureau of Habitat, Hadley, MA; Albany, NY.

Tiner, R., 2006c. Wetlands of the Upper Tioughnioga River Watershed, Cortland and Onondaga Counties, New York. U.S. Fish and Wildlife Service, National Wetlands Inventory, Northeast Region; Prepared for the New York State Department of Environmental Conservation, Division of Fish, Wildlife, and Marine Resources, Bureau of Habitat, Hadley, MA; Albany, NY.

Tiner, R., 2006d. Wetlands of the Sodus Bay to Wolcott Creek Watershed, Wayne County, New York. U.S. Fish and Wildlife Service, National Wetlands Inventory, Northeast Region; Prepared for the New York State Department of Environmental Conservation, Division of Fish, Wildlife, and Marine Resources, Bureau of Habitat, Hadley, MA; Albany, NY.

Tiner, R., 2006e. Wetlands of the Sodus Creek Watershed, Wayne County, New York. U.S. Fish and Wildlife Service, National Wetlands Inventory, Northeast Region; Prepared for the New York State Department of Environmental Conservation, Division of Fish, Wildlife, and Marine Resources, Bureau of Habitat, Hadley, MA; Albany, NY.

Tiner, R., 2006f. Wetlands of the Post Creek to Sing Sing Creek Watershed, Chemung and Steuben Counties, New York. U.S. Fish and Wildlife Service, National Wetlands Inventory, Northeast Region; Prepared for the New York State Department of Environmental Conservation, Division of Fish, Wildlife, and Marine Resources, Bureau of Habitat, Hadley, MA; Albany, NY.

Tiner, R., 2006g. Wetlands of the Sucker Brook to Grass River Watershed, St Lawrence County, New York. U.S. Fish and Wildlife Service, National Wetlands Inventory, Northeast Region; Prepared for the New York State Department of Environmental Conservation, Division of Fish, Wildlife, and Marine Resources, Bureau of Habitat, Hadley, MA; Albany, NY.

Tiner, R., 2006h. Wetlands of the Peconic River Watershed, Suffolk County, New York. U.S. Fish and Wildlife Service, National Wetlands Inventory, Northeast Region; Prepared for the New York State Department of Environmental Conservation, Division of Fish, Wildlife, and Marine Resources, Bureau of Habitat, Hadley, MA; Albany, NY.

Tiner, R., 2006i. Wetlands of the Salmon River to South Sandy Creek Watershed, Oswego County, New York. U.S. Fish and Wildlife Service, National Wetlands Inventory, Northeast Region; Prepared for the New York State Department of Environmental Conservation, Division of Fish, Wildlife, and Marine Resources, Bureau of Habitat, Hadley, MA; Albany, NY.

Tiner, R., 2006j. Wetlands of the Catherine Creek Watershed, Schuyler and Chemung Counties, New York. U.S. Fish and Wildlife Service, National Wetlands Inventory, Northeast Region; Prepared for the New York State Department of Environmental Conservation, Division of Fish, Wildlife, and Marine Resources, Bureau of Habitat, Hadley, MA; Albany, NY.

Tiner, R., 2006k. Wetlands of the Hudson River-Snook Kill to Fish Creek Watershed, Saratoga County, New York. U.S. Fish and Wildlife Service, National Wetlands Inventory, Northeast Region; Prepared for the New York State Department of Environmental Conservation, Division of Fish, Wildlife, and Marine Resources, Bureau of Habitat, Hadley, MA; Albany, NY.

Tiner, R.W. (Ed.), 2009. Status Report for the National Wetlands Inventory Program-2009. U.S. Fish and Wildlife Service, Division of Habitat and Resource Conservation, Branch of Resource and Mapping Support, Arlington, VA. https://www.fws.gov/wetlands/Documents/Status-Report-for-the-National-Wetlands-Inventory-Program-2009.pdf.

Tiner, R.W., 2010a. NWIPlus: geospatial database for watershed-level functional assessment. Natl. Wetl. Newsl. 32 (3), 4–7. 23.

Tiner, R.W., 2010b. Wetlands of Cape Cod and the Islands, Massachusetts: Results of the National Wetlands Inventory and Landscape-level Functional Assessment. U.S. Fish and Wildlife Service, Northeast Region, Hadley, MA.

Tiner, R.W., 2011a. Dichotomous Keys and Mapping Codes for Wetland Landscape Position, Landform, Water Flow Path, and Waterbody Type Descriptors. Version 2.0. U.S. Fish and Wildlife Service, Northeast Region, Hadley, MA.

Tiner, R.W., 2011b. Predicting Wetland Functions at the Landscape Level for Coastal Georgia Using NWIPlus Data. U.S. Fish and Wildlife Service, National Wetlands Inventory Program; In cooperation with the Georgia Department of Natural Resources, Coastal Resources Division, Region 5, Hadley, MA; Brunswick, GA and Atkins North America, Raleigh, NC. http://coastalgadnr.org/sites/uploads/crd/CORRELATION%20REPORT%20Georgia_FINAL_September-20-2011.pdf.

Tiner, R.W., 2013. Tidal Wetlands Primer: An Introduction to Their Ecology, Natural History, Status, and Conservation. University of Massachusetts Press, Amherst, MA.

Tiner, R.W., 2014. Dichotomous Keys and Mapping Codes for Wetland Landscape Position, Landform, Water Flow Path, and Waterbody Type Descriptors. Version 3.0. U.S. Fish and Wildlife Service, Northeast Region, Hadley, MA. https://www.fws.gov/wetlands/Documents/Dichotomous-Keys-and-Mapping-Codes-for-Wetland-Landscape-Position-Landform-Water-Flow-Path-and-Waterbody-Type-Version-3.pdf.

Tiner, R.W., 2015. Introduction to wetland mapping and its challenges. Chapter 3. In: Tiner, R.W., Lang, M.W., Klemas, V.V. (Eds.), Remote Sensing of Wetlands: Applications and Advances. CRC Press, Boca Raton, FL, pp. 43–65.

Tiner, R.W., 2017. Wetland Indicators: A Guide to Wetland Formation, Identification, Delineation, Classification, and Mapping. CRC Press, Boca Raton, FL.

Tiner, R.W., Bergquist, H.C., 2003. Historical Analysis of Wetlands and Their Functions for the Nanticoke River Watershed: A Comparison between Pre-Settlement and 1998 Conditions. U.S. Fish and Wildlife Service, National Wetlands Inventory Program, Northeast Region, Hadley, MA.

Tiner, R.W., Bergquist, H.C., 2007. The Hackensack River Watershed, New Jersey/New York Wetland Characterization, Preliminary Assessment of Wetland Functions, and Remotely-Sensed Assessment of Natural Habitat Integrity. U.S. Fish and Wildlife Service, National Wetlands Inventory, Ecological Services, Region 5, Hadley, MA.

Tiner, R.W., Bergquist, H.C., McClain, B.J., 2002. Wetland Characterization and Preliminary Assessment of Wetland Functions for the Neversink Reservoir and Cannonsville Reservoir Watersheds. U.S. Fish and Wildlife Service, National Wetlands Inventory, Ecological Services, Region 5, Hadley, MA.

Tiner, R.W., DeAlessio, G., 2002. Wetlands of Pennsylvania's Coastal Zone: Wetland Status, Preliminary Functional Assessment, and Recent Trends (1986–1999). U.S. Fish and Wildlife Service, National Wetlands Inventory Program, Northeast Region, Hadley, MA.

Tiner, R.W., Herman, J., 2015. Wetlands of Southwestern Vermont and Neighboring New York: Inventory, Characterization, and Preliminary Landscape-level Functional Assessment. U.S. Fish and Wildlife Service, Northeast Region, Hadley, MA.

Tiner, R.W., Stewart, J., 2004. Wetland Characterization and Preliminary Assessment of Wetland Functions for the Delaware and Catskill Watersheds of the New York City Water Supply System. U.S. Fish and Wildlife Service, National Wetlands Inventory Program, Ecological Services, Northeast Region, Hadley, MA.

Tiner, R., Schaller, S., Peterson, D., Snider, K., Ruhlman, K., Swords, J., 1999a. Wetland Characterization Study and Preliminary Assessment of Wetland Functions for the Casco Bay Watershed, Southern Maine. U.S. Fish and Wildlife Service, National Wetlands Inventory Program; Prepared for the Maine State Planning Office, Northeast Region, Hadley, MA; Augusta, ME.

Tiner, R.W., Schaller, S., Starr, M., 1999b. Wetland Characterization and Preliminary Assessment of Wetland Functions for the Boyds Corners and West Branch Sub-basins of the Croton Watershed, New York. Prepared for New York City Dept. of Environmental Protection, Division of Drinking Water Quality Control, Natural Resources Section, Valhalla, NY.

Tiner, R., Starr, M., Bergquist, H., Swords, J., 2000a. Watershed-Based Wetland Characterization for Maryland's Nanticoke River and Coastal Bays Watersheds: A Preliminary Assessment Report. U.S. Fish and Wildlife Service, National Wetlands Inventory Program, Northeast Region, Hadley, MA.

Tiner, R.W., Huber, I., Starr, M., Veneman, P., 2000b. Wetlands and Potential Wetland Restoration Sites for the Shawsheen Watershed. University of Massachusetts, Department of Plant and Soil Sciences; and U.S. Fish and Wildlife Service, Ecological Services; A Cooperative Report produced for the Massachusetts Wetlands Restoration & Banking Program, Executive Office of Environmental Affairs, Amherst, MA; Northeast Region, Hadley, MA; Boston, MA.

Tiner, R.W., Swords, J., Huber, I., Nuerminger, T., Starr, M., 2000c. Wetlands and Potential Wetland Restoration Sites for the Mill Rivers and Manhan River Watersheds. U.S. Fish and Wildlife Service, Ecological Services; National Wetlands Inventory Report prepared for the U.S. Army Corps of Engineers, Northeast Region, Hadley, MA; New England District, Concord, MA.

Tiner, R.W., Bergquist, H.C., Swords, J.Q., McClain, B.J., 2001a. Watershed-Based Wetland Characterization for Delaware's Nanticoke River Watershed: A Preliminary Assessment Report. U.S. Fish and Wildlife Service, National Wetlands Inventory Program, Northeast Region, Hadley, MA.

Tiner, R.W., Huber, I., Starr, M., 2001b. Wetlands and Potential Wetland Restoration Sites for the Upper Ipswich Watershed. U.S. Fish and Wildlife Service, Ecological Services, Northeast Region; in cooperation with the University of Massachusetts, Natural Resources Assessment Group, Department of Plant and Soil Sciences; National Wetlands Inventory Report prepared for the Massachusetts Wetlands Restoration & Banking Program, Executive Office of Environmental Affairs, Hadley, MA; Amherst, MA; Boston, MA.

Tiner, R.W., Swords, J.Q., Bergquist, H.C., 2003. Wetlands of the Boston Harbor Islands National Recreation Area. U.S. Fish and Wildlife Service, National Wetlands Inventory Program, Hadley, MA. NWI Technical Report.

Tiner, R.W., Polzen, C.W., McClain, B.J., 2004. Wetland Characterization and Preliminary Assessment of Wetland Functions for the Croton watershed of the New York City Water Supply System. U.S. Fish and Wildlife Service, National Wetlands Inventory Program, Ecological Services, Northeast Region, Hadley, MA.

Tiner, R.W., Biddle, M.A., Jacobs, A.D., Rogerson, A.B., McGuckin, K.G., 2011. Delaware Wetlands: Status and Changes from 1992 to 2007. Cooperative National Wetlands Inventory Publication. U.S. Fish and Wildlife Service, Northeast Region; and the Delaware Department of Natural Resources and Environmental Control, Hadley, MA; Dover, DE.

Tiner, R.W., Herman, J., Roghair, L., 2013. Connecticut Wetlands: Characterization and Landscape-level Functional Assessment. Prepared for the Connecticut Department of Environmental Protection; U.S. Fish and Wildlife Service, Northeast Region, Hartford, CT; Hadley, MA, p. 45. plus appendices.

Tiner, R.W., Diggs, B., Mans, I., Herman, J., 2014a. Wetlands of Pennsylvania's Lake Erie Watershed: Status, Characterization, Landscape-level Functional Assessment, and Potential Restoration Sites. Prepared for the Pennsylvania Department of Environmental Protection, Coastal Zone Management Program; U.S. Fish and Wildlife Service, Harrisburg, PA; Northeast Region, Hadley, MA.

Tiner, R.W., McGuckin, K., Herman, J., 2014b. Rhode Island Wetlands: Updated Inventory, Characterization, and Landscape-level Functional Assessment. U.S. Fish and Wildlife Service, Northeast Region, Hadley, MA.

Tiner, R.W., McGuckin, K., Herman, J., 2015a. Wetland Characterization and Landscape-level Functional Assessment for Long Island. New York. U.S, Fish and Wildlife Service, Northeast Region, Hadley, MA.

Tiner, R.W., Olson, E., Cross, D., Herman, J., 2015b. Wetlands of Pennsylvania's Delaware Estuary Coastal Zone and Vicinity: Characterization and Landscape-level Functional Assessment. Prepared for the Pennsylvania Department of Environmental Protection, Coastal Zone Management Program; U.S. Fish and Wildlife Service, Northeast Region, Harrisburg, PA; Hadley, MA.

Vanderhoof, M.K., Distler, H.E., Lang, M.W., Alexander, L.C., 2018. The influence of data characteristics on detecting wetland/stream surface-water connections in the Delmarva Peninsula, Maryland and Delaware. Wetl. Ecol. Manag. 26 (1), 63–86. https://doi.org/10.1007/s11273-017-9554-y.

FURTHER READING

Ciminelli, J., Scrivani, J., 2007. Virginia Conservation Lands Needs Assessment Virginia Watershed Integrity Model. Virginia Department of Conservation and Recreation, Division of Natural Heritage; Virginia Department of Forestry; Virginia Commonwealth University Center for Environmental Studies, and Virginia DEQ Coastal Zone Management Program, Richmond, VA.http://www.dcr.virginia.gov/natural-heritage/document/watershedintegritymodel.pdf.

Tiner, R.W., 2004. Remotely-sensed indicators for monitoring the general condition of natural habitat in watersheds: an application for Delaware's Nanticoke River watershed. Ecol. Indic. 4, 227–243.

Tiner, R.W., Lang, M.W., Klemas, V.V. (Eds.), 2015. Remote Sensing of Wetlands: Applications and Advances. CRC Press, Boca Raton, FL.

Turner, R.K., van den Bergh, J.C.J.M., Söderqvist, T., Barendregt, A., van der Stratten, J., Maltby, E., van Ierland, E.C., 2000. Ecological-economic analysis of wetlands: scientific integration for management and policy. Ecol. Econ. 35, 7–23.

Chapter 2.2.2

Georgia Coastal Wetlands Landscape-Level Assessment

Jan M. Mackinnon
Georgia Department of Natural Resources, Coastal Resources Division, Brunswick, GA, United States

Chapter Outline

INTRODUCTION

Coastal Georgia was among the first areas in the nation to be inventoried by the National Wetlands Inventory Program of the Fish and Wildlife Service (FWS). The early wetland mapping work was done in cooperation with the Georgia Department of Natural Resources (DNR) in the late 1970s, and though it was not widely distributed, it was instrumental in formulating NWI procedures. In the late 1980s the NWI was again conducted for coastal Georgia, and the results of the inventory were published in the form of hardcopy 1:24,000 scale maps that were later digitized to create a data layer for geographic information systems (GIS) applications.

Remote-sensing technology has advanced considerably since the early mapping was conducted (Tiner et al., 2015). More accurate aerial imagery is increasingly available and geospatial technology has evolved to make desktop interpretation of digital imagery possible. These advances allow production of a more comprehensive inventory with both improved detection (i.e., more wetlands identified) and better classification detail. The NWI also created additional descriptors for landscape position, landform, water flow path, and waterbody type (known as LLWW descriptors) to expand wetland classification. The enhanced classification (NWI and LLWW) produced an enhanced NWI database, referred to as NWI+ data, that allows for more detailed classification of types and can be used to perform a preliminary assessment of functions for wetlands in the region.

Need for More Detailed Classification

As with most locations in the United States, many changes have occurred since the 1980s inventory, and the original mapping was no longer relevant for most of the coastal counties, especially in areas where development activity and natural coastal geophysical processes have taken place. Many agencies and organizations in coastal Georgia depend on the NWI dataset for planning and regulatory springboards. Due to the dependency on the NWI dataset for these purposes, it was determined that updates were much needed in order to provide a better characterization of wetlands for many purposes.

Wetland and Stream Rapid Assessments. https://doi.org/10.1016/B978-0-12-805091-0.00018-9

Funding Source

The Coastal Resources Division (CRD) of DNR utilized a grant from the U.S. Environmental Protection Agency (EPA) and worked in cooperation with the FWS to update and enhance the NWI.

Study Area

The study area is comprised of Georgia's six coastal counties, each with direct access to the Atlantic Ocean. They are, from north to south: Chatham, Bryan, Liberty, McIntosh, Glynn, and Camden counties. These counties cover a land area of approximately 3159 square miles and represent about 5.5% of the State of Georgia. County acreages used in this study are based on the 2006 United States Census Bureau, Geography Division, TIGER/Line Shapefiles.

METHODS

The CRD of the Georgia Department of Natural Resources, with assistance from a support contractor, Atkins North America, Inc., updated the NWI for coastal Georgia with strict adherence to the Wetland Mapping Standard of the Federal Geographic Data Committee (FDGC) Wetland Subcommittee (2009) and following the Data Collection Requirements and Procedures for Mapping Wetland, Deepwater, and Related Habitats of the United States (Dahl et al., 2009). Both documents were available in final draft format at the beginning of the project, making it possible to apply the new mapping standards to the coastal Georgia NWI updates. The FWS actively participated in the updating process by providing quality control review of the draft wetland delineations to assure that the revised NWI was consistent with the NWI nationally and suitable for inclusion as part of the wetland data layer of the National Spatial Data Infrastructure.

Source Data and Technical Requirements

An on-screen or "heads up" digitization process was employed using ArcMap 9 software for identifying, classifying, and delineating wetlands. Wetlands were interpreted from U.S. Geological Survey (USGS) high-resolution (0.5 m) color orthoimagery acquired in 2006. The orthoimagery also served as the base photography for displaying the NWI update. For locations along the southern coast where USGS imagery was unavailable, Florida Bureau of Survey and Mapping LABINS high-resolution color infrared imagery from 2004 served as the base photography. Along the western portion of the study area, where neither of these data sets were available, 2007 National Agriculture Imagery Program (NAIP) imagery was utilized as the base photography. Care was taken to place wetland boundaries of well-defined features within 20 ft of the boundary position on the imagery, as practicable, to ensure that National Map Accuracy Standards were met. The imagery was routinely interpreted at a scale of approximately 1:7000, but was viewed at much larger scales as interpretation questions arose. Regular utilization of collateral data was an important part of the wetland identification and classification process. Digital georeferenced collateral information was layered in GIS for contemporaneous viewing during the interpretation process. Wetlands were classified in accordance with the Cowardin et al. (1979) to system, subsystem, class, subclass level with water regime, and special modifiers. The minimum size of wetland regularly mapped and classified was between 0.25 and 0.5 acres.

During the interpretation process, natural resource professionals routinely reviewed collateral digital data sets, as available, including the 2007 NAIP imagery, USGS Orthophoto Quadrangle color-infrared imagery with one-meter resolution (taken in 1999), USGS 1:24,000 topographic quadrangles, the USGS National Hydrography Dataset (NHD) (published in 2009) depicting streams, the Natural Resources Conservation Service (NRCS) soil survey geographic data (SSURGO), LiDAR elevation data for Glynn County, the previous NWI representing 1983 conditions, and the DNR Wildlife Resources Division National Vegetation Classification System data for Glynn County and portions of other counties.

Wetland interpretations, delineations, and classifications were reviewed at least two times prior to submittal. In addition, the FWS Wetland Verification Tool was then applied to the dataset as a final quality control check for incorrect wetland codes, adjacent wetlands, sliver wetlands, sliver uplands, and lake and pond sizes.

Draft data files were submitted to the FWS Southeast Regional NWI coordinator for review and evaluation on a regular basis as sections of the coast were completed. After each review, editorial comments were discussed with the NWI coordinator to ensure that they were interpreted correctly and to incorporate suggestions for improvement into the ongoing database development. Final editorial changes were then incorporated into the database.

After the incorporation of all editorial comments and suggestions, work areas were edge-matched, topology was rechecked, and the FWS verification tool was reapplied as a final check. The updated NWI data were submitted to CRD as a single, high-quality seamless ESRI ArcGIS 9.2 File Geodatabase in Albers Equal Area Conic, NAD 83, meters projection. The updated NWI for coastal Georgia can be viewed online at the FWS NWI Wetland Mapper Site.

NWI Wetland Classification

For the NWI, wetlands were classified following the Service's official wetland classification: Classification of Wetlands and Deepwater Habitats of the United States (Cowardin et al., 1979). This classification system has also been adopted as the federal wetland classification standard by the Federal Geographic Data Committee (FGDC Wetlands Subcommittee, 2009). The following discussion represents a simplified overview of the Service's wetland classification system.

The Service's wetland classification system is hierarchical, proceeding from general to specific. In this approach, wetlands are first defined at a rather broad level–the *system*. The term *system* represents "a complex of wetlands and deepwater habitats that share the influence of similar hydrologic, geomorphologic, chemical, or biological factors." Five systems are defined: marine, estuarine, riverine, lacustrine, and palustrine. The marine system generally consists of the open ocean and its associated high-energy coastline while the estuarine system encompasses salt and brackish marshes, nonvegetated tidal shores, and brackish waters of coastal rivers and embayments. Freshwater wetlands and deepwater habitats fall into one of the other three systems: riverine (rivers and streams), lacustrine (lakes, reservoirs, and large ponds), or palustrine (e.g., marshes, bogs, swamps, and small shallow ponds). Thus, at the most general level, wetlands can be defined as either marine, estuarine, riverine, lacustrine, or palustrine.

Each system, with the exception of the palustrine, is further subdivided into *subsystems*. The marine and estuarine systems both have the same two subsystems, which are defined by tidal water levels: (1) subtidal—continuously submerged areas, and (2) intertidal—areas alternately flooded by tides and exposed to air. Similarly, the lacustrine system is separated into two systems based on water depth: (1) littoral—wetlands extending from the lake shore to a depth of 6.6 ft (2 m) below low water or to the extent of nonpersistent emergents (e.g., arrowheads, pickerelweed, or spatterdock) if they grow beyond that depth, and (2) limnetic—deepwater habitats lying beyond the 6.6 ft (2 m) mark at low water. By contrast, the riverine system is further defined by four subsystems that represent different reaches of a flowing freshwater or lotic system: (1) tidal—water levels subject to tidal fluctuations for at least part of the growing season, (2) lower perennial—permanent, flowing waters with a well-developed floodplain, (3) upper perennial—permanent, flowing water with very little or no floodplain development, and (4) intermittent—channel containing nontidal flowing water for only part of the year.

The next level, *class*, describes the general appearance of the wetland or deepwater habitat in terms of the dominant vegetative life form or the nature and composition of the substrate, where vegetative cover is <30%. Of the 11 classes, five refer to areas where vegetation covers 30% or more of the surface: aquatic bed, moss-lichen wetland, emergent wetland, scrub-shrub wetland, and forested wetland. The remaining six classes represent areas generally lacking vegetation, where the composition of the substrate and degree of flooding distinguish classes: rock vottom, unconsolidated bottom, reef (sedentary invertebrate colony), streambed, rocky shore, and unconsolidated shore. Permanently flooded nonvegetated areas are classified as either rock bottom or unconsolidated bottom while exposed areas are typed as streambed, rocky shore, or unconsolidated shore. Reefs formed by the colonization of invertebrates (oysters in Georgia) are found in both permanently flooded and exposed areas.

Each class is further divided into *subclasses* to better define the type of substrate in nonvegetated areas (e.g., bedrock, rubble, cobble-gravel, mud, sand, and organic) or the type of dominant vegetation (e.g., persistent or nonpersistent emergents, moss, lichen, or broad-leaved deciduous, needle-leaved deciduous, broad-leaved evergreen, needle-leaved evergreen, and dead woody plants). Below the subclass level, *dominance level* can be applied to specify the predominant plant or animal in the wetland community.

To allow better description of a given wetland or deepwater habitat in regard to hydrologic, chemical, and soil characteristics and to human impacts, the classification system contains four types of specific modifiers: (1) water regime, (2) water chemistry, (3) soil, and (4) special. These modifiers may be applied to class and lower levels of the classification hierarchy.

Water regime modifiers describe flooding or soil saturation conditions and are divided into two main groups: tidal and nontidal. Tidal water regimes are used where water level fluctuations are largely driven by oceanic tides. Tidal regimes can be subdivided into two general categories, one for salt and brackish water tidal areas and another for freshwater tidal areas. This distinction is needed because of the special importance of seasonal river overflow and groundwater inflows in freshwater tidal areas. By contrast, nontidal modifiers define conditions where surface water runoff, ground-water discharge, and/or wind effects (i.e., lake seiches) cause water level changes.

Water chemistry modifiers are divided into two categories that describe the water's salinity or hydrogen ion concentration (pH): (1) salinity modifiers and (2) pH modifiers. Like water regimes, salinity modifiers have been further subdivided into two groups: halinity modifiers for tidal areas and salinity modifiers for nontidal areas. Estuarine and marine waters are dominated by sodium chloride, which is gradually diluted by fresh water as it moves upstream in coastal rivers. On the other hand, the salinity of inland waters is dominated by four major cations (i.e., calcium, magnesium, sodium, and potassium) and three major anions (i.e., carbonate, sulfate, and chloride). Interactions between precipitation, surface runoff, groundwater flow, evaporation, and sometimes plant evapotranspiration form inland salts that are most common in arid and semiarid regions of the country. The other set of water chemistry modifiers are pH modifiers for identifying acid (pH <5.5), circumneutral (5.5–7.4), and alkaline (pH >7.4) waters. Some studies have shown a good correlation between plant distribution and pH levels (Sjors, 1950; Jeglum, 1971). Moreover, pH can be used to distinguish between mineral-rich (e.g., fens) and mineral-poor wetlands (e.g., bogs).

The third group of modifiers, *soil modifiers*, are presented because the nature of the soil, which exerts strong influence on plant growth and reproduction as well as on the animals living in it. Two soil modifiers are given: (1) mineral and (2) organic. In general, if a soil has 20% or more organic matter by weight in the upper 16 in., it is considered an organic soil, whereas if it has less than this amount, it is a mineral soil.

The final set of modifiers, *special modifiers*, was established to describe the activities of people or beavers affecting wetlands and deepwater habitats. These modifiers include: excavated, impounded (i.e., to obstruct outflow of water), diked (i.e., to obstruct inflow of water), partly drained, farmed, and artificial (i.e., materials deposited to create or modify a wetland or deepwater habitat).

Expanded Classification for Developing Functional Correlations or Condition Assessment

A set of abiotic attributes was developed by FWS to increase the information contained in the NWI database and to create what is known as the NWI+ database. The four groups of attributes describe landscape position (relationship of a wetland to a waterbody if present), landform (physical shape of the wetland), water flow path, and waterbody type (different types of estuaries, rivers, lakes, and ponds). Collectively, the attributes are known as LLWW descriptors, which represent the first letter of each descriptor. Dichotomous keys have been developed to interpret these attributes (Tiner, 2003a, 2014). Other modifiers are also included in these keys to further describe wetland characteristics. LLWW descriptors can be added to the NWI database by interpreting topography from digital raster graphics (DRGs) or digital elevation model data (DEMs), stream courses from the NHD and/or aerial imagery, and waterbody types from aerial imagery. The interpretations can be done by employing some automated GIS routines coupled with manual review and interpretation by wetland specialists.

The NWI+ database adds value and increases the functionality of the original NWI database. Besides providing more features that can be used to predict wetland functions from the NWI database, NWI+ makes it possible to better characterize the nation's wetlands. For example, all of the palustrine wetlands, which account for 95% of the wetlands in the conterminous United States, can now be linked to rivers, streams, lakes, and ponds, where appropriate, so that the acreage of floodplain wetlands, lakeside wetlands, and geographically isolated wetlands can be reported. The Wetlands Subcommittee of the FGDC recognized the value added by the LLWW descriptors and recommended that they be included in wetland mapping to increase the functionality of wetland inventory databases (FGDC Wetlands Subcommittee, 2009).

For this project, LLWW descriptors were applied to all wetlands in the NWI digital database in accordance with the definitions and dichotomous keys developed by the FWS (Tiner, 2003a). For consistency and accuracy, the LLWW descriptors were added to the NWI database by the wetland scientists who updated the NWI and were familiar with the study area. NWI data were viewed with online USGS topographic maps (DRGs) to identify wetlands along streams and general slope characteristics. Aerial imagery was used to determine waterbody types (e.g., ponds).

Six wetland landscape positions (including two lotic types) describing the relationship between a wetland and an adjacent waterbody were identified:

(1) Marine—on the shores of the open ocean and its embayments.

(2) Estuarine—associated with tidal brackish waters (estuaries).

(3) Lotic (river or stream; see below)—along freshwater rivers and streams and periodically flooded, at least during high discharge periods (including freshwater tidal reaches of coastal rivers).

(4) Lentic—in lakes, reservoirs, and their basins where water levels are significantly affected by the presence of these waterbodies.

(5) Terrene—isolated or headwater wetlands, fragments of former isolated or headwater wetlands that are now connected to downslope wetlands via drainage ditches, and wetlands on broad, flat terrain cut through by streams but where overbank flooding does not occur (e.g., hydrologically decoupled from streams).

Lotic wetlands were further separated by river and stream sections based on watercourse width (i.e., polygon = river; linear = stream at a scale of 1:24,000) and then divided into one of five gradients:

(1) High (e.g., shallow mountain streams on steep slopes—not present in the study area).
(2) Middle (e.g., streams with moderate slopes—not present in the study area).
(3) Low (e.g., mainstem rivers with considerable floodplain development and slow-moving streams).
(4) Intermittent (i.e., periodic flows).
(5) Tidal (i.e., under the influence of tides).

Eight landforms describing the physical form of a wetland or the predominant land mass upon which it occurs (e.g., floodplain), were identified: basin, flat, floodplain, fringe, island, slope, and interfluves.

Additional modifiers were assigned to indicate water flow paths associated with wetlands: bidirectional tidal, bidirectional nontidal, throughflow, inflow, outflow, or isolated. Surface water connections were emphasized because they are more readily observable than groundwater linkages. Bidirectional-tidal flow paths were assigned to all intertidal wetlands. Throughflow wetlands were identified as having either a watercourse or another type of wetland above and below them. Most lotic wetlands were observed to be throughflow types. Inflow pathways were determined where watercourses could be observed entering the wetland but no surface water outlet could be seen. Outflow wetlands were identified as those appearing to have water leaving them and moving downstream via a watercourse or a slope wetland. Isolated wetlands[1] were observed to be closed ("geographically isolated") depressions or flats where water appeared to come from direct precipitation, localized surface water runoff, and/or groundwater discharge. From the surface water perspective, these wetlands appear to be isolated from other wetlands because they lack an apparent surface water connection; however it should be recognized that they may be hydrologically linked to other wetlands and waterbodies via groundwater or by intermittent overflow during extreme precipitation events while others may be connected by small streams that were not mapped on the collateral data sources.

Other descriptors applied to mapped wetlands include headwater, drainage divide, and partly drained. Headwater wetlands appear to be sources of streams or wetlands along first-order (perennial) streams. Wetlands described as drainage-divide wetlands have outflow in two directions to separate creeks (drainage systems). Partly drained wetlands were typically ditched wetlands. For open water habitats, additional descriptors following Tiner (2003a) were applied, including water flow path and pond, estuary, and lake types.

Because ponds were separated from wetlands for the LLWW classification, wetland acreage totals are different for NWI and LLWW. NWI routinely classifies open water areas <20 acres in size as palustrine unconsolidated bottom wetlands. These areas were not reclassified as lacustrine in the NWI database, so deepwater habitat acreage of lacustrine waters and acreage of palustrine unconsolidated bottoms based on NWI will be different than LLWW totals for lakes and ponds. Ponds were separated into three categories: natural, dammed/impounded, and excavated.

Preliminary Assessment of Wetland Functions

After creating the NWI+ database (the enhanced NWI database), analyses were performed to produce a preliminary assessment of wetland functions for the study area. Both wetlands and ponds were evaluated for performance of 11 functions:

(1) Surface water detention.
(2) Coastal storm surge detention.
(3) Streamflow maintenance.
(4) Nutrient transformation.
(5) Carbon sequestration.
(6) Retention of sediment and other particulates.
(7) Bank and shoreline stabilization.
(8) Provision of fish and aquatic invertebrate habitat.
(9) Provision of waterfowl and waterbird habitat.
(10) Provision of other wildlife habitat.
(11) Provision of habitat for unique, uncommon, or highly diverse plant communities.

1. In the current version of the dichotomous keys (Tiner, 2014), the isolated water flow path was changed to vertical flow, which is more representative of the flow of water (up and down).

The preliminary assessment of wetland functions for coastal Georgia was accomplished under the guidance of Ralph Tiner (FWS, Hadley, MA). This study employed a landscape-level functional assessment approach that may be called "Watershed-based Preliminary Assessment of Wetland Functions" (W-PAWF). W-PAWF applies general knowledge about wetlands and their functions to develop a watershed or area-wide overview that highlights possible wetlands of significance in terms of performance of various functions. The rationale for correlating wetland characteristics with wetland functions in the northeastern United States is described in Tiner (2003b). The procedure begins with the identification of wetland attributes or characteristics from the suite of characteristics described by the NWI, with the addition of LLWW modifiers that contribute to the performance of each wetland function. Then, using GIS technology, wetlands are selected that exhibit those particular characteristics. The information resulting from the selection process can be portrayed graphically on maps or in tabular form.

In order to develop region-specific information for the six-county study area, the relationships (formerly called correlations) developed for use in the northeastern United States were introduced to and reviewed by a group of Georgia scientists from federal, state, and local agencies, nonprofit organizations, and academic institutions at an August 31, 2010, workshop on Little St. Simons Island. The peer group provided comments that were used to reevaluate the relationships and tailor them to coastal Georgia. In cases where there were differences in opinions, the points were considered and decisions were made by consensus between the DNR-CRD, Atkins North America, and Ralph Tiner. A detailed rationale for the selection of Georgia-specific characteristics and their relationship to wetland functions is found in Tiner (2011). Using the sets of characteristics important to each of the 11 functions developed from the workshop, ArcView 10 software was utilized to select wetlands from the NWI+ database that exhibited those characteristics.

Data Analysis and Compilation

GIS was used to analyze the data and produce wetland statistics for the overall study area and for each of the six coastal counties. Tables were prepared to summarize the results of the NWI update (i.e., the extent of different wetland types by NWI classification) and to correlate wetland characteristics with wetland functions to identify wetlands of significance for 11 functions. After running the analyses, a series of maps was generated to display the variety of wetland types and to highlight wetlands that may perform various functions at significant levels. Statistics were mostly generated from Microsoft's Excel program whereas thematic maps were generated by ArcView software.

RESULTS

The project produced an enhanced digital database with NWI classifications, LLWW classification, predictions for 11 functions in the subject area, and two reports—one providing the documentation and rationale for predicting wetlands of significance for each function (Tiner, 2011) and the final project report presenting the results of the enhanced inventory (Georgia Department of Natural Resources, 2012). The latter report includes a series of thematic maps—a set for the entire region (wetlands by NWI types, landscape position, landform, and water flow path plus one for each of 11 functions) and a set of function maps for each county. The reports and maps are available online at: http://coastalgadnr.org/Wetlands.

NWI Types

Wetlands of the six-county region total 804,227 acres (Table 2.2.2.1) and cover nearly 40% of the study area. Palustrine wetlands (freshwater) are most abundant, occupying 432,419 acres and comprising about 54% of the region's wetlands (Fig. 2.2.2.1). Nearly 79% of palustrine wetlands are forested. Palustrine emergent and scrub-shrub wetlands account for only 12% and 7% of freshwater wetlands, respectively. Estuarine wetlands are second in abundance, occupying 368,484 acres, or about 46% of the area's wetlands. Emergent wetlands are the most common estuarine type (95%). Marine wetlands inventoried total 3084 acres, comprised exclusively of unconsolidated shore (marine beaches and flats). Marine wetlands make up $<1\%$ of the coastal wetlands total. Only 151 acres of lacustrine and 90 acres of riverine wetlands were inventoried.

LLWW Types

Wetlands in the estuarine landscape position account for less than half of the wetlands (46%) in the region while wetlands in the marine landscape position represent $<1\%$ of the total (Table 2.2.2.2). By definition, all estuarine and marine wetlands

TABLE 2.2.2.1 Wetlands of Coastal Georgia Classified by NWI Types (Cowardin et al., 1979)

System	Class	Acreage
Marine	Unconsolidated shore	3084
Total marine wetlands		*3084*
Estuarine	Emergent	304,920
	Emergent/forested	2
	Emergent/scrub-shrub	107
	Emergent/unconsolidated shore	46,206
	(Subtotal emergent)	(351,236)
	Forested, broad-leaved deciduous	13
	Forested, broad-leaved evergreen	1832
	Forested, needle-leaved evergreen	206
	Forested/emergent	2
	(Subtotal forested)	(2053)
	Scrub-shrub, broad-leaved deciduous	533
	Scrub-shrub, broad-leaved evergreen	3464
	Scrub-shrub, needle-leaved evergreen	383
	Scrub-shrub/emergent	115
	(Subtotal scrub-shrub)	(4495)
	Unconsolidated shore	10,509
	Unconsolidated shore/emergent	190
	(Subtotal nonvegetated)	(10,700)
Total estuarine wetlands		*368,484*
Lacustrine	Aquatic bed	108
	Emergent	10
	Unconsolidated shore	32
Total lacustrine wetlands		*151*
Palustrine	Aquatic bed	826
	Aquatic bed/unconsolidated bottom	6
	(Subtotal aquatic bed)	(832)
	Emergent	50,147
	Emergent/aquatic bed	178
	Emergent/forested	1638
	Emergent/scrub-shrub	548
	(Subtotal emergent)	(52,511)
	Forested, broad-leaved deciduous	202,949
	Forested, broad-leaved evergreen	30,450
	Forested, needle-leaved deciduous	83,007
	Forested, needle-leaved evergreen	21,739

Continued

TABLE 2.2.2.1 Wetlands of Coastal Georgia Classified by NWI Types —cont'd

System	Class	Acreage
	Forested/emergent	434
	Forested/scrub-shrub	1075
	(Subtotal forested)	(339,743)
	Scrub-shrub, broad-leaved deciduous	21,750
	Scrub-shrub, broad-leaved evergreen	5670
	Scrub-shrub, needle-leaved deciduous	1113
	Scrub-shrub, needle-leaved evergreen	1453
	Scrub-shrub/emergent	393
	Scrub-shrub/forested	520
	(Subtotal scrub-shrub)	(30,899)
	Unconsolidated bottom	8242
	Unconsolidated shore	192
	Unconsolidated shore/emergent	1
	(Subtotal nonvegetated)	(8434)
Total palustrine wetlands		*432,419*
Riverine	Unconsolidated shore	90
Total riverine wetlands		*90*
Grand total (all wetlands)		**804,227**

have bidirectional-tidal water flows. Almost 31% of the area's wetlands are in the lotic landscape position (i.e., associated with rivers and streams; Fig. 2.2.2.2). Most of the region's lotic wetlands exhibited throughflow water pathways (63% of the lotic acreage) or bidirectional-tidal (freshwater tidal) water pathways. <1% of the wetlands are lentic types (along lakes and deep ponds classified as palustrine unconsolidated bottoms by NWI). The water flow path of 78% of the lentic wetlands is classified as isolated whereas about 22% of the lentic wetlands have an obvious stream running from them. A total of 22% of wetlands are located in the terrene landscape position, mainly in headwater locations or in isolated depressions. Most (83%) of the region's terrene wetlands are outflow types (typically the source of a stream; Fig. 2.2.2.3). The remainder are either wetlands that receive surface or groundwater, which flows through the wetland and into another wetland or stream, or are geographically isolated wetlands (surrounded by upland and lacking a detectable surface water connection to other wetlands or waters).

All marine wetlands and 93% of the estuarine wetlands are classified as fringe landform types with open access to bays, sounds, or the Atlantic Ocean (Fig. 2.2.2.4). Estuarine wetlands classified as basin landforms are usually the result of partial hydrologic blockage by roads or railroad crossings. Most lotic wetlands (88%) are basin types (subject to prolonged seasonal flooding), while nearly all remaining lotic wetlands are classified as flats (subject to short-term flooding). Of the terrene wetlands, 61% are classified as basins (depressions) and 39% as flats. Terrene basins are seasonally flooded or wetter while terrene flats are temporarily flooded or seasonally saturated. Lentic wetlands are by definition fringe landforms.

Ponds occupy 9266 acres or 1% of the region's wetlands. A total of 4416 ponds were inventoried with nearly all (95%) identified as excavated (Table 2.2.2.3). The average size of ponds in coastal Georgia is about 2.1 acres. Nearly three quarters (74%) of ponds appear to be hydrologically isolated while most of the remainder have outflow or throughflow water pathways.

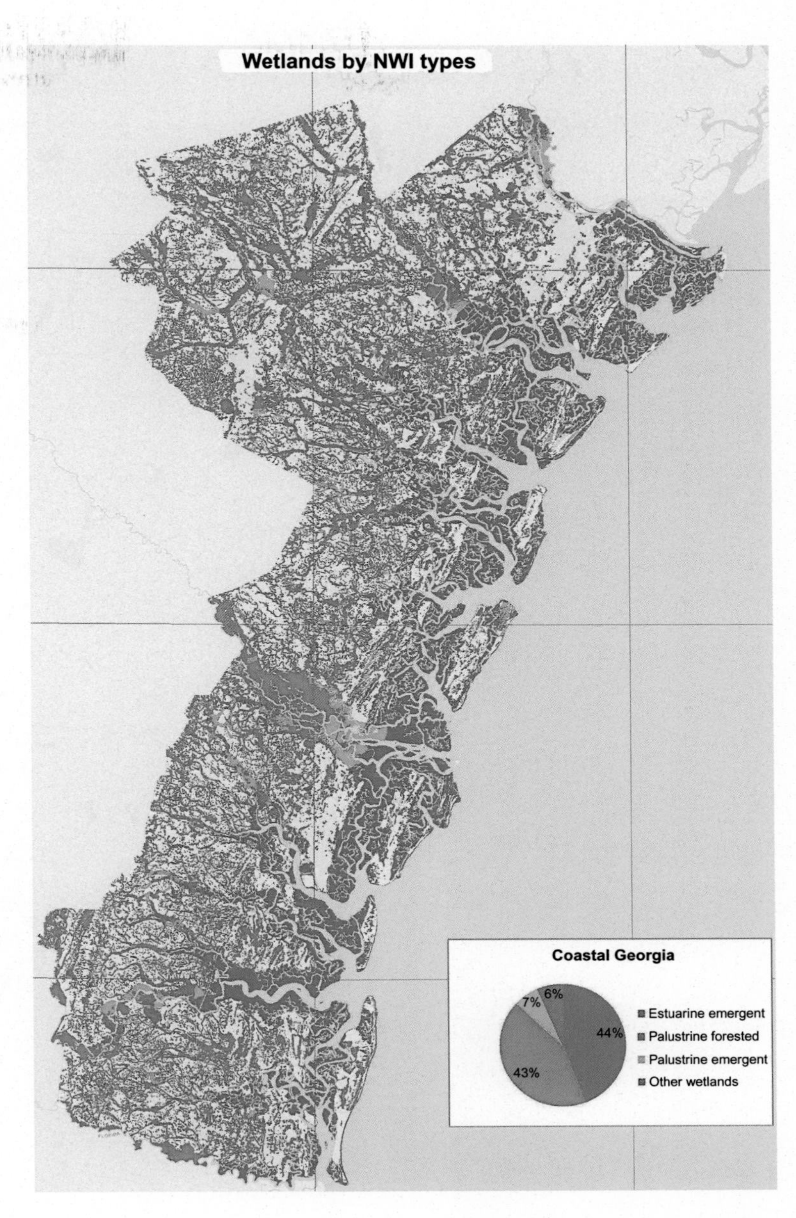

FIG. 2.2.2.1 Distribution of wetlands by NWI type. *(Source: Georgia Department of Natural Resources, Coastal Resources Division.)*

Wetland Functions

In coastal Georgia, 60% of the wetlands (including ponds) are predicted to perform eight of the 11 functions at high-to-moderate levels (Table 2.2.2.4). As much as 97% of the wetlands is deemed important for nutrient transformation, carbon sequestration, habitat for wildlife other than waterfowl and other waterbirds, and retention of sediment and other particulates. >75% of the wetlands are predicted to contribute to bank and shoreline stabilization. >60% of the wetlands are

TABLE 2.2.2.2 Wetlands Classified by Landscape Position, Landform, and Water Flow Path for Coastal Georgia

Landscape Position	Landform	Water Flow Path	Acreage
Marine	Fringe	Bidirectional tidal	3084
Total marine			*3084*
Estuarine	Fringe	Bidirectional tidal	341,187
	Basin	Bidirectional tidal	27,334
Total estuarine			*368,521*
Lentic	Fringe	Isolated	355
		Outflow	99
Total lentic			*454*
Lotic river	Fringe	Bidirectional tidal	73
	(Subtotal fringe)		(73)
	Floodplain basin	Bidirectional tidal	87,044
		Throughflow	11,940
	(Subtotal basin)		(98,983)
	Floodplain flat	Bidirectional tidal	4479
		Throughflow	2598
	(Subtotal flat)		(7077)
Total lotic river			*106,134*
Lotic stream	Basin	Outflow	57
		Throughflow	109,543
	(Subtotal basin)		(109,600)
	Flat	Throughflow	29,744
	(Subtotal flat)		(29,744)
Total lotic stream			*139,344*
Terrene	Basin	Isolated	22,975
		Outflow	85,596
	(Subtotal basin)		(108,571)
	Flat	Isolated	6550
		Outflow	62,266
		Throughflow	13
	(Subtotal flat)		(68,828)
	Island	Isolated	26
	(Subtotal island)		(26)
Total terrene			*177,425*
Grand total			**794,961**

Note: Ponds were treated as waterbody type (see Table 2.2.2.3) for summary.

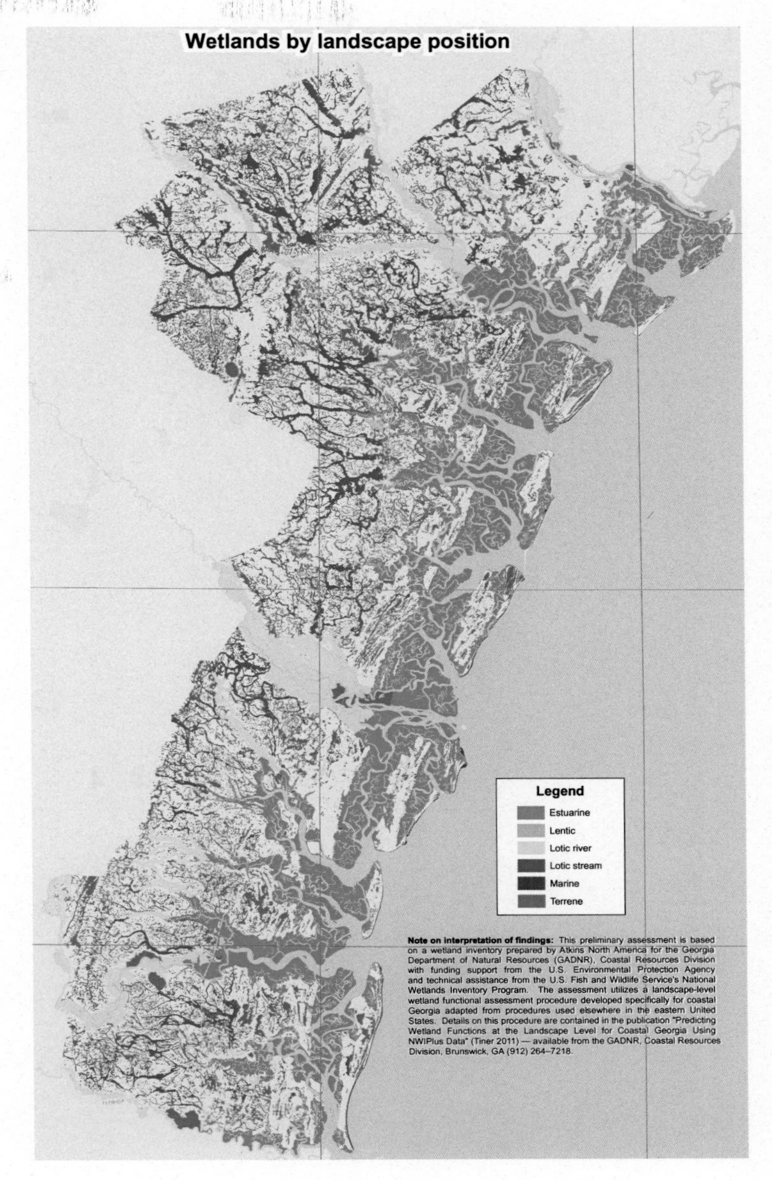

FIG. 2.2.2.2 Distribution of wetlands by landscape position. *(Source: Georgia Department of Natural Resources, Coastal Resources Division.)*

predicted to provide coastal storm surge detention (Fig. 2.2.2.5), fish and aquatic invertebrate habitat, and waterfowl and waterbird habitat. A total of 41% of wetlands provide surface water detention. Relatively few wetlands (23%) are located in landscape positions where they could contribute to maintaining streamflow. Only 4% of the wetlands are recognized as uncommon or highly diverse plant communities that contribute significantly to the area's biodiversity. These plant communities included the following types: Palustrine tidal emergent wetlands (regularly flooded, seasonally flooded tidal, and semipermanently flooded tidal water regimes), Palustrine tidal scrub-shrub wetlands (regularly flooded, seasonally flooded tidal, and semipermanently flooded tidal water regimes), freshwater vegetated wetlands on barrier islands (semipermanently flooded, semipermanently flooded tidal, and permanently flooded), Carolina bay wetlands (relatively intact), and Palustrine vegetated wetlands that are permanently flooded. (*Note*: Because this assessment was based on remotely sensed information and largely on observable life-form differences in plant communities and water regimes, it did not attempt to identify wetlands that do or may support rare or endangered species. Such wetlands would have to be identified through other means, such as Georgia's Natural Heritage Program.)

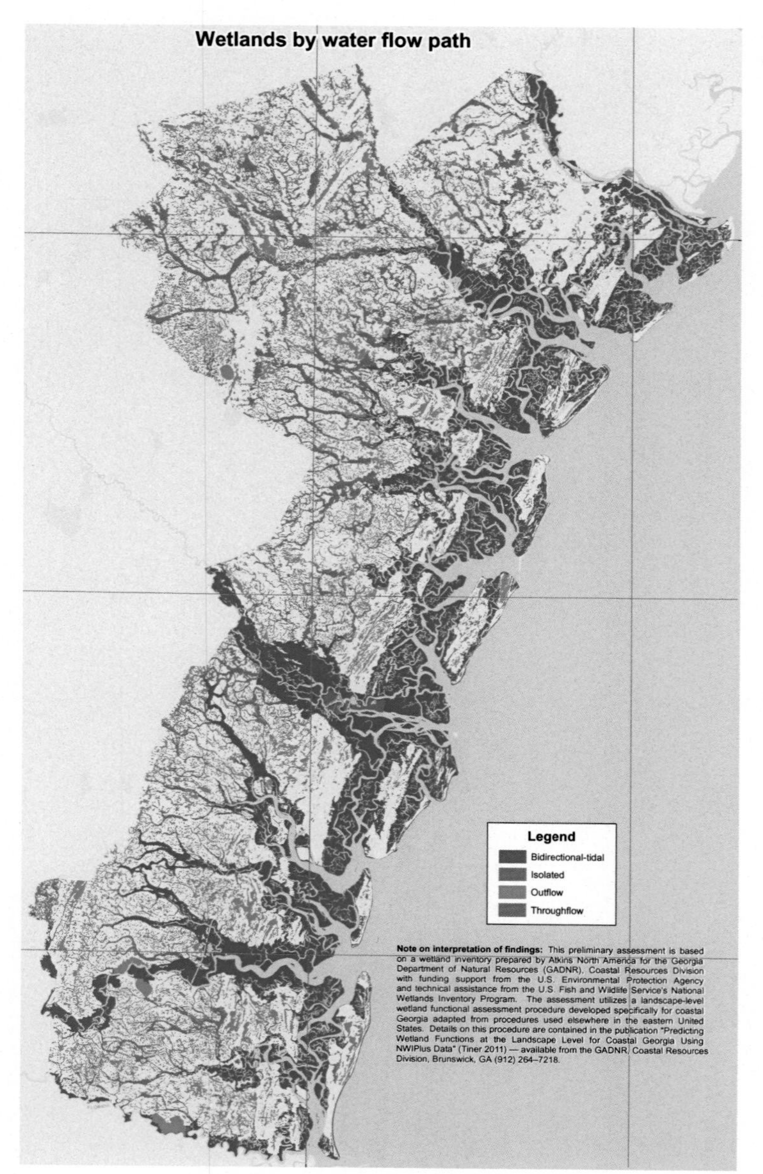

FIG. 2.2.2.3 Wetlands classified by water flow path. *(Source: Georgia Department of Natural Resources, Coastal Resources Division.)*

CURRENT USE OF THE ASSESSMENT

In coastal Georgia, NWI remains the most detailed source of wetland data. Agencies and organizations are using the NWI+ to better characterize wetlands for educational purposes and to provide enhanced data for restoration planning as well as for use as a tool to evaluate functional loss assessments in wetland trend studies. CRD has used the NWI+ data as a platform to launch a finer-scale marsh classification. In 2015, CRD funded a study to produce high-resolution mapping of vegetation, elevation, salinity, and bathymetry to advance coastal habitat management in Georgia. This study produced vegetation mapping that incorporates elevation data corrections based on doctoral research conducted by Dr. Christine Hladik on Sapelo Island, Georgia. The data layers were produced by the University of Georgia Skidaway Institute of Oceanography and Georgia Southern University. Once completed, CRD combined all applicable datasets into a web application called the Georgia Wetland Restoration Access Portal (G-WRAP). The portal is accessible by all users and is available at http://geospatial.gatech.edu/G-WRAP/. Wetland classification was only pursued for estuarine wetlands due to a lack of funding.

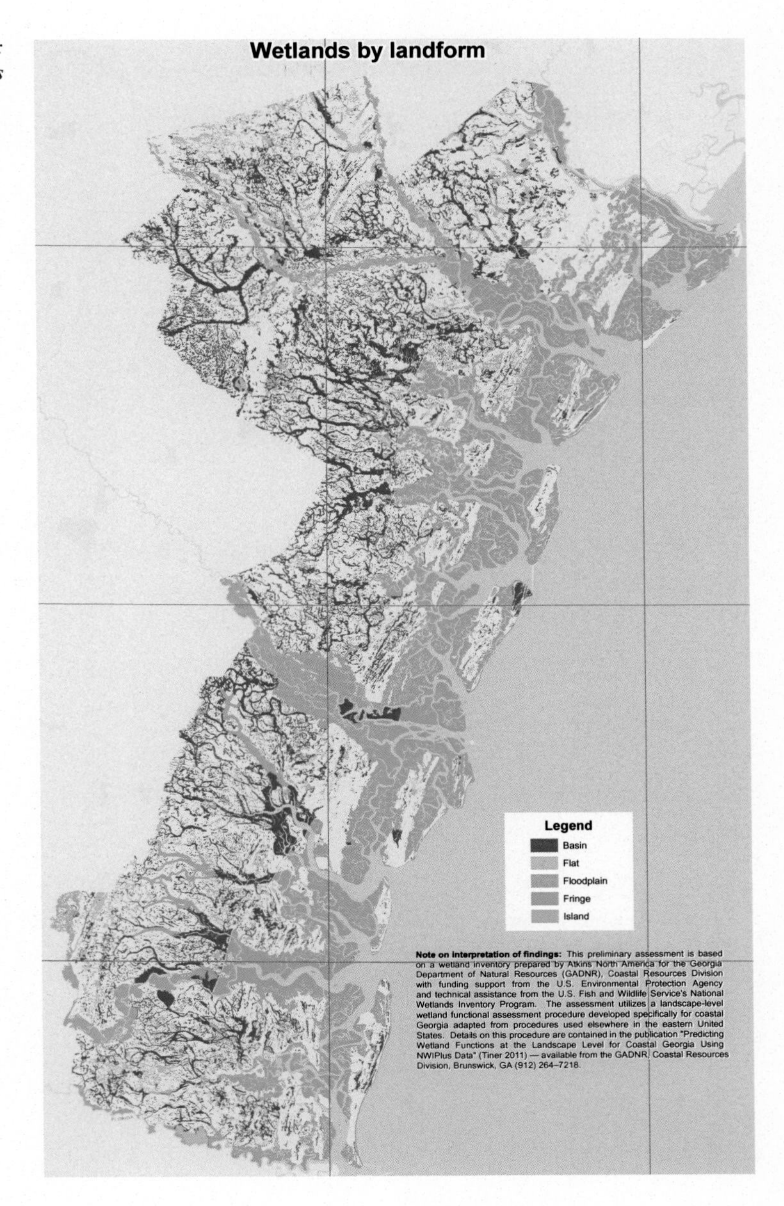

FIG. 2.2.2.4 Wetland landforms across the study area. *(Source: Georgia Department of Natural Resources, Coastal Resources Division.)*

TABLE 2.2.2.3 Pond Acreage for Coastal Georgia

Type of Pond	Water Flow Path	Number of Ponds	Acreage
Natural	Isolated	32	38
	Mesotidal	43	126
	Outflow	44	274
	Throughflow	19	172
Total natural ponds		*138*	*610*
Impounded	Isolated	31	107
	Mesotidal	5	15
	Outflow	37	172
	Throughflow	7	49

Continued

TABLE 2.2.2.3 Pond Acreage for Coastal Georgia—cont'd

Type of Pond	Water Flow Path	Number of Ponds	Acreage
Total impounded ponds		*80*	*343*
Excavated	Isolated	3191	5767
	Mesotidal	29	91
	Outflow	702	1787
	Throughflow	276	667
Total excavated ponds		*4198*	*8313*
Grand total		**4416**	**9266**

TABLE 2.2.2.4 Wetlands of Potential Significance for Various Functions for Coastal Georgia

Function	Significance	Acreage	% of All Wetlands
Surface water detention	High	122,923	15
	Moderate	206,768	26
	Total	*329,691*	*41*
Coastal storm surge detention	High	462,862	58
	Moderate	20,059	2
	Total	*482,921*	*60*
Streamflow maintenance	High	57,965	7
	Moderate	126,006	16
	Total	*183,971*	*23*
Nutrient transformation	High	680,893	85
	Moderate	101,185	13
	Total	*782,078*	*97*
Carbon sequestration	High	679,414	84
	Moderate	119,280	15
	Total	*798,694*	*99*
Retention of sediments	High	567,281	71
	Moderate	157,944	20
	Total	*725,225*	*90*
Shoreline stabilization	High	605,410	75
	Moderate	16,598	2
	Total	*622,008*	*77*
Fish and shellfish habitat	High	470,370	58
	Moderate	38,883	5
	Total	*509,253*	*63*
Waterfowl and waterbird habitat	High	456,882	57
	Moderate	43,552	5
	Total	*500,434*	*62*

TABLE 2.2.2.4 Wetlands of Potential Significance for Various Functions for Coastal Georgia—cont'd

Function	Significance	Acreage	% of All Wetlands
Other wildlife habitat	High	738,574	92
	Moderate	42,566	5
	Total	*781,140*	*97*
Unique, diverse communities	Palustrine vegetated (H WR)	78	–
	Selected PEM (N,R,T WR)	21,462	3
	Selected PSS (N,R,T WR)	8843	1
	Barrier island (F,T,H WR)	1307	–
	Carolina bays (relatively intact)	919	–
	Total	*32,609*	*4*

Note: Results include ponds.

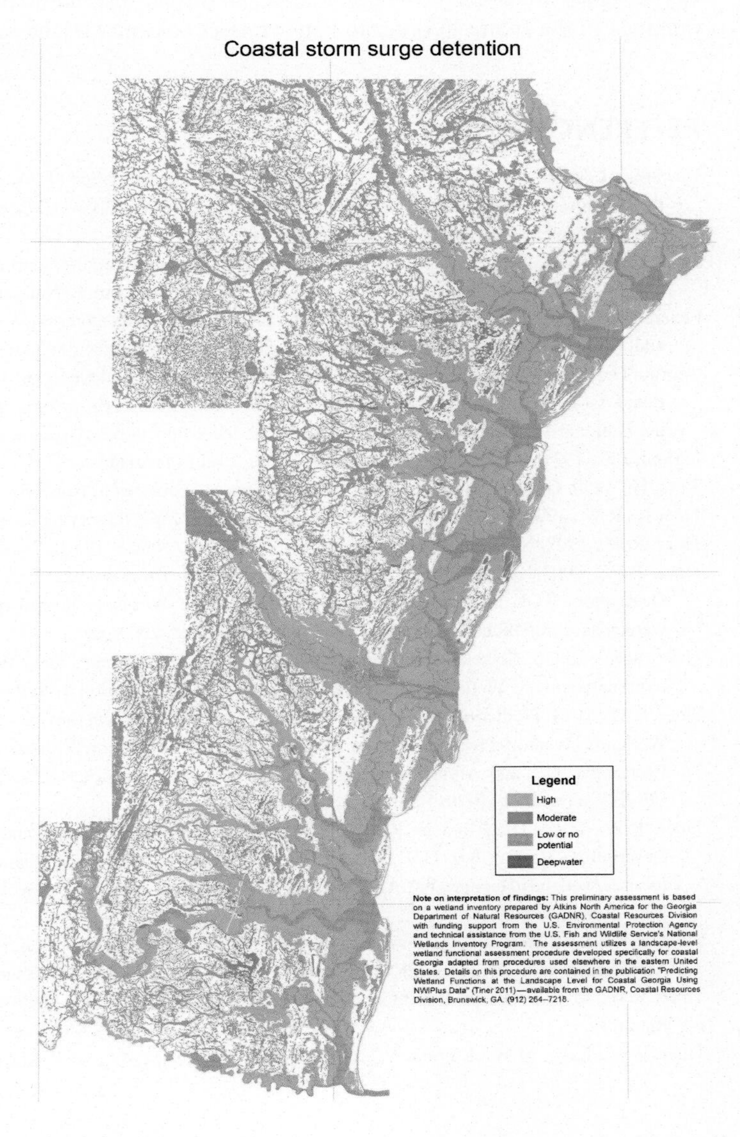

FIG. 2.2.2.5 Wetlands predicted to be significant for coastal storm surge detention. (Source: *Georgia Department of Natural Resources, Coastal Resources Division.*)

LESSONS LEARNED

Because the NWI data were derived from 2006 imagery, they do not reflect changes in some wetlands that have occurred in the past 6 years. These changes may be due to permitted alterations by federal, state, and local governments or to natural processes including erosion, accretion, and sea level rise. Despite this, the 2006 database should reasonably reflect contemporary conditions because wetlands in this area are well regulated.

It is important to recognize the limitations of any wetland mapping effort derived mainly through photo-interpretation techniques (see Tiner, 1990, 1997, 2015, 2017 for details). NWI data, or any other wetland data derived from these techniques, do not include all wetlands. Some wetlands are simply too small to map given the imagery used while others avoid detection due to evergreen tree cover, dry surface conditions, or other factors. For this inventory and assessment, the minimum size of the wetland-targeted mapping unit was one-half acre, but many wetlands (especially ponds) smaller than this were mapped. Wetland units may contain small areas that are different from the mapped type (i.e., inclusions) due to scale and map complexity issues. For example, a 10 acre forested wetland may include small areas of emergent wetlands or small upland islands not discernable from aerial photography due to canopy cover or minimum mapping units. Drier-end wetlands such as temporarily flooded palustrine wetlands are often difficult to separate from nonwetlands through photo interpretation. Finally, despite our best attempts at quality control, some errors of interpretation and classification are likely to occur due to the sheer number of polygons in the wetland database (>52,000).

Even with limitations, however, the NWI+ dataset remains the most comprehensive coverage of wetland data in Georgia. Multiple agencies (DNR, U.S. Army Corps of Engineers, and others) as well as organizations such as the Nature Conservancy are using the Georgia NWI+ to further restoration planning. CRD hopes to update the functional assessment rankings in the future in order to better reflect coastal wetland status.

REFERENCES

Cowardin, L.M., Carter, V., Golet, F.C., LaRoe, E.T., 1979. Classification of Wetlands and Deepwater Habitats of the United States. U.S. Department of the Interior, Fish and Wildlife Service, Washington, DC (FWS/OBS-79/31), https://www.fws.gov/wetlands/Documents/Classification-of-Wetlands-and-Deepwater-Habitats-of-the-United-States.pdf.

Dahl, T.E., Dick, J., Swords, J., Wilen, B.O., 2009. Data Collection Requirements and Procedures for Mapping Wetland, Deepwater and Related Habitats of the United States. Division of Habitat and Resource Conservation, National Standards and Support Team, Madison, WI. 85 p.

Federal Geographic Data Committee, 2009. Federal Geographic Data Committee Wetlands Inventory Mapping Standard. FGDC Wetland Subcommittee and Wetland Mapping Standard Workgroup. 50 p, http://www.fgdc.gov/participation/working-groups-subcommittees/wsc/.

Georgia Department of Natural Resources, 2012. Wetlands of Coastal Georgia: Results of the National Wetlands Inventory and Landscape-level Functional Assessment. Coastal Resources Division, Brunswick, GA. https://www.fws.gov/wetlands/documents/Wetlands-of-Coastal-Georgia-Results-of-the-National-Wetlands-Inventory-and-Landscape-level-Functional-Assessment.pdf.

Jeglum, J.K., 1971. Plant indicators of pH and water level in peatlands at Candle Lake, Saskatchewan. Can. J. Bot. 49, 1661–1676.

Sjors, H., 1950. On the relation between vegetation and electrolytes in North Swedish mire waters. Oikos 2, 241–258.

Tiner Jr., R.W., 1990. Use of high-altitude aerial photography for inventorying forested wetlands in the United States. For. Ecol. Manag. 33 (34), 593–604.

Tiner, R.W., 1997. NWI maps: what they tell us. Natl. Wetl. Newsl. 19 (2), 7–12.

Tiner, R.W., 2003a. Dichotomous Keys and Mapping Codes for Wetland Landscape Position, Landform, Water Flow Path, and Waterbody Type Descriptors. U.S. Fish and Wildlife Service, National Wetlands Inventory Program, Northeast Region, Hadley, MA. http://www.aswm.org/wetlandsonestop/dichotomous_keys_and_mapping_coded_2003.pdf.

Tiner, R.W., 2003b. Correlating Enhanced National Wetlands Inventory Data With Wetland Functions for Watershed Assessments: A Rationale for Northeastern U.S. Wetlands. U.S. Fish and Wildlife Service, National Wetlands Inventory Program, Northeast Region, Hadley, MA.

Tiner, R.W., 2011. Predicting Wetland Functions at the Landscape Level for Coastal Georgia Using NWI+ Data. U.S. Fish and Wildlife Service, National Wetlands Inventory Program, Region 5, Hadley, MA. In Cooperation With the Georgia Department of Natural Resources, Coastal Resources Division, Brunswick, GA and Atkins North America, Raleigh, NC, http://coastalgadnr.org/sites/uploads/crd/CORRELATION%20REPORT%20Georgia_FINAL_September-20-2011.pdf.

Tiner, R.W., 2014. Dichotomous Keys and Mapping Codes for Wetland Landscape Position, Landform, Water Flow Path, and Waterbody Type Descriptors: Version 3.0. U.S. Fish and Wildlife Service, National Wetlands Inventory Program, Northeast Region, Hadley, MA. https://www.fws.gov/wetlands/Documents/Dichotomous-Keys-and-Mapping-Codes-for-Wetland-Landscape-Position-Landform-Water-Flow-Path-and-Waterbody-Type-Version-3.pdf.

Tiner, R.W., 2015. Introduction to wetland mapping and its challenges. In: Tiner, R.W., Lang, M.D., Klemas, V.V. (Eds.), Remote Sensing of Wetlands: Applications and Advances. CRC Press, Boca Raton, FL, pp. 43–65 (Chapter 3).

Tiner, R.W., 2017. Wetland Indicators: A Guide to Wetland Formation, Identification, Delineation, Classification, and Mapping. CRC Press, Boca Raton, FL.

Tiner, R.W., Lang, M.W., Klemas, V.V. (Eds.), 2015. Remote Sensing of Wetlands: Applications and Advances. CRC Press, Boca Raton, FL.

FURTHER READING

Meyer, J.L., Kaplan, L.A., Newbold, D., Strayer, D.L., Woltemade, C.J., Zedler, J.B., Beilfuss, R., Carpenter, Q., Semlitsch, R., Watzin, M.C., Zedler, P.H., 2003. Where Rivers are Born: The Scientific Imperative for Defending Small Streams and Wetlands. American Rivers and Sierra Club, Washington, DC.

Mitsch, W.J., Gosselink, J.G., 2008. Wetlands, fourth ed. John Wiley & Sons, Inc., Hoboken, NJ.

Sandifer, P.A., Miglarese, J.V., Calder, D.R., 1980. Ecological Characterization of the Sea Island Coastal Region of South Carolina and Georgia. Vol. III: Biological Features of the Characterization Area. U.S. Department of the Interior, Fish and Wildlife Service, Washington, DC (FWS/OBS-79/45).

Tiner, R.W., 2005. In Search of Swampland: A Wetland Sourcebook and Field Guide, second ed. (Revised and Expanded) Rutgers University Press, New Brunswick, NJ.

Tiner, R.W., 2010. Wetlands of Cape Cod and the Islands, Massachusetts: Results of the National Wetlands Inventory and Landscape-level Functional Assessment (National Wetlands Inventory Report), U.S. Fish and Wildlife Service, Northeast Region, Hadley, MA.

Tiner, R.W., 2013. Tidal Wetlands Primer: An Introduction to Their Ecology, Natural History, and Conservation. The University of Massachusetts Press, Amherst, MA.

Chapter 2.2.3

Assessing Streamflow Maintenance Functions in Wetlands of the Blackfoot River Subbasin in Montana, United States

Linda Vance
Montana Natural Heritage Program, University of Montana, Missoula, MT, United States

Chapter Outline

INTRODUCTION

The Montana Natural Heritage Program (MTNHP) took on the task of mapping the state's wetlands and riparian areas in 2007. This effort was critical to MTNHP's mission of being the state's authoritative source for unbiased information on species and their habitats, and to its programmatic goals of documenting the extent, distribution, and condition of wetlands statewide. However, because we mapped to National Wetlands Inventory (NWI) standards, which are based on the Cowardin classification system (Cowardin et al., 1979), we soon found that our mapping could not, by itself, provide all the answers we needed, especially about wetland functions. To address this gap, we were early adopters of Landform, Landscape Position, Water Flow Path and Water Body Type (LLWW) descriptors (Tiner, 2003), adapting them to conditions in the western United States and, ultimately, developing a geographic information systems (GIS) toolbox that would automate >90% of the attribution.

The integration of LLWW descriptors into our mapping has greatly streamlined our landscape-level wetland assessment efforts. Coupled with GIS-based disturbance evaluations, the LLWW descriptors help us characterize wetland function while identifying the degree to which those functions may be compromised. Although this approach does not provide the level of precision and accuracy that can be attained through field assessments, it does allow us to evaluate suites of functions across broad landscapes, something that would be prohibitively expensive (in both time and money) with field campaigns. Here, we report on one such effort that was recently completed for the Blackfoot River 8-digit Hydrologic Unit (HUC) in Montana. This subbasin is one of several headwater sources for the Columbia River system as well as the location of the Blackfoot River, immortalized in Norman Maclean's *A River Runs Through It* (Maclean, 1976). Despite its iconic status, the river and the subbasin that holds it both suffered extensive impacts from human land use in the 20th century, especially mining, forest harvesting, cattle grazing, and associated road building (MTDOJ, 2011).

BACKGROUND

Within stream and river networks, headwater streams are the most abundant in both length and number, typically contributing more than two-thirds of total stream length in a river drainage (Freeman et al., 2007). Many of these streams originate in high-elevation wetlands whose soils store early season snowmelt, recharging groundwater and/or discharging surface water to the

Wetland and Stream Rapid Assessments. https://doi.org/10.1016/B978-0-12-805091-0.00019-0

streams. As such, these headwater wetlands provide critical functions for the health of aquatic systems, including water storage, maintenance of surface/groundwater connections and biochemical processes, support for hydrodynamic balance, and habitat for diverse assemblages of wetland-dependent native species (Meyer et al., 2007). Despite their significance, the extent, distribution, characteristics, and functions of headwater wetlands in Montana were not systematically examined until 2015 when the MTNHP completed a series of landscape-level, rapid, and intensive assessments in the Upper Missouri Headwaters (Vance et al., 2015). At that time, our automated LLWW attribution method was not fully developed, and so headwater wetlands were only characterized in terms of their Cowardin types. However, we recognized the potential for combining the broad categorization of headwater wetlands with LLWW descriptors in subsequent assessments. In the particular case of the Blackfoot River, we were especially interested in whether we could use our GIS-based tools to characterize wetlands associated with streamflow maintenance functions, and carry out landscape-level assessments to determine whether these wetlands were at risk from current and ongoing human activities. We felt that such a case study would illustrate both the advantages of our methodologies and the areas that needed refinement.[1]

STUDY AREA

The Blackfoot River subbasin (HUC 17010203) covers 598,285 ha (1,478,394 acres) in western Montana (Fig. 2.2.3.1). The area is a complex mix of publicly and privately owned lands; the Forest Service, the state of Montana, and the Bureau of Land Management manage the largest public holdings (Hart et al., 2015). Public and private lands still support the full suite of native wildlife species, including gray wolf, bull trout, trumpeter swan, and grizzly bear. With some of the highest wetland densities in Montana, the Blackfoot supports a diversity of wetland systems, including forested wetlands, fens, and potholes. Timber harvest and livestock grazing are the historically predominant land uses, with recreation, wildlife habitat, and rural residential development becoming increasingly important.

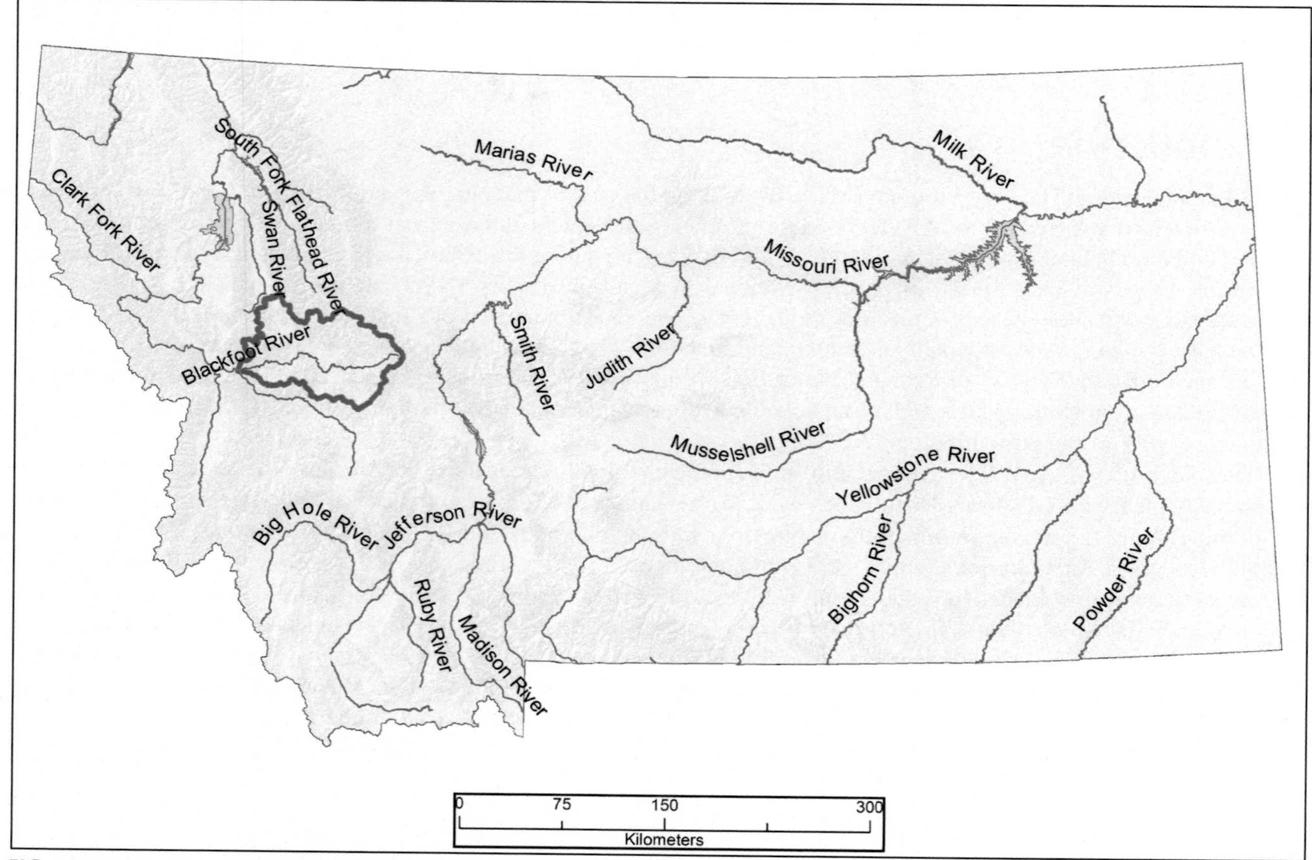

FIG. 2.2.3.1 Blackfoot River subbasin, Montana.

1. This project was funded by an Environmental Protection Agency Wetland Program Development Grant, CD-96815301.

The topography of the project area, for the most part, has been shaped by the disintegration of alpine glaciers, which shaped the hummocky moraines of the Blackfoot Valley. Soils are fine-textured and have low permeability, with many wetlands perched above the water table and receiving most hydrologic inputs from precipitation and runoff. Climate is characterized by long, cold winters and moist springs, with peak precipitation occurring in May and June and the highest temperatures coming in July and August. The major river in the project area is the Blackfoot River, originating along the Continental Divide at Rogers Pass and flowing southwesterly to its confluence with the Clark Fork River just east of Missoula, Montana. Upper portions of the river are on the Montana Department of Environmental Quality's Section 303(d) list of impaired waters due to heavy metals and sedimentation; the middle and lower reaches are considered to be impaired for temperature, nutrients, and sedimentation (MTDEQ, 2016). Low flows and high air temperatures in summer impact water temperature in most years by July, often resulting in angling closures throughout the entire river length (MTFWP, 2015). Several nonprofit groups, including Trout Unlimited and the Blackfoot Challenge, have undertaken wetland and riparian restoration projects in an attempt to address temperature, water quality, and habitat limitations (Blackfoot Challenge, 2014). However, these efforts have largely been driven by opportunity rather than by landscape-level analysis, which has only recently become possible with the availability of mapping and GIS tools.

METHODS

Source Data

MTNHP completed NWI mapping of the Blackfoot River between 2011 and 2012, using heads-up photointerpretation to digitize and classify 1-m aerial imagery from 2005 and 2009 provided by the National Agriculture Imagery Program. Additional base data layers required for the assignment of LLWW descriptors included the 1:24,000 National Hydrography Dataset (NHD) and a 10 m Digital Elevation Model (DEM), which was used to create a slope layer calculated in percent. These two datasets, projected into Montana State Plane 2500 coordinates, were downloaded from the Montana Geographic Information Clearing House (http://geoinfo.msl.mt.gov/Home/msdi). Three additional data products were used: a valley bottom shapefile created by MTNHP through heads-up digitizing for large river valleys across the state, which supported identification of wetlands as FL; a topographic position index layer (Jenness et al., 2010); and a binary classification database, also developed by MTNHP, that assigns each wetland polygon into headwater or nonheadwater categories. This binary database, adapted from earlier work (Vance et al., 2015), uses a DEM to classify wetlands in individual subbasins. After calculating a midpoint of elevations for a given basin, wetlands are given a headwater status if they occur in the upper 50% of elevation bands within the subbasin or they are found on steep slopes or ridgelines (as defined in the topographic position index layer) in the lower 50% of elevation bands.

Wetland Classification

We adapted the LLWW classification scheme developed by Tiner (2003) to reflect western conditions. In a GIS, wetlands polygons were assigned one of three landscape positions: (1) Lentic (LE), (2) Lotic (LO), or (3) Terrene (TE). Wetlands without a distinct waterbody type were further classified by landform: (1) Floodplain (FP), (2) Basin (BA), (3) Island (IL), (4) Fringe (FR), (5) Slope (SL), and (6) Flat (FL). If a Slope or Flat wetland occurred in a valley bottom position, we added "va" to the landform. Waterbody type was assigned to lakes, rivers, streams, and Palustrine wetlands with aquatic beds or unconsolidated bottoms/shores as follows: (1) Lake (LK), (2) River (RV), (3) Stream (ST), or (4) Pond (PD).[2] Streams and Rivers were further attributed by waterbody flow (Perennial = 1, Ephemeral = 3, and Intermittent = 4). Water flow paths were assigned as follows: (1) Inflow (IN), (2) Inflow Artificial (IA), (3) Outflow (OU), (4) Bidirectional (BI), (5) Throughflow (TH), (6) Vertical Flow (VR), or (7) Vertical Flow Complex. Wetlands were given special modifiers in the LLWW process based on whether or not they had the Cowardin modifiers: b (beaver), d (drained), farmed (f), impounded (h), or excavated (x), or had been designated as a headwater (hw) in our headwaters database.

Functional Correlations

Correlations between streamflow maintenance functions and wetland classification were made based on a search of the literature (Arnold et al., 2000; Galat et al., 1998; Hauer et al., 2002; Lane and D'Amico, 2010; Tiner et al., 2013). Because we define headwater wetlands as occurring anywhere in a headwater area, rather than requiring they be specifically linked

2. Palustrine emergent, Palustrine forested, and Palustrine shrub-scrub wetlands were not assigned a waterbody type.

to a stream or river feature in the NWI, we did not consider all headwater wetlands to have high streamflow maintenance function, as has been suggested by others using a more restricted definition of "headwater." Instead, we assigned ratings of "high" and "moderate" as follows:

High	Headwater Lotic wetlands; headwater Lentic wetlands with a TH or OUT water flow path
Moderate	Nonheadwater Lotic wetlands; nonheadwater Lentic wetlands with a TH or OU water flow path; and headwater Terrene wetlands with a B, C, or F water regime and a TH or OU water flow path. All wetlands with a d, f, or h modifier were excluded

Landscape Condition Assessment

To evaluate the landscape around wetlands identified as important to streamflow maintenance, we used the Montana Human Disturbance Index (HDI; MTNHP, 2014). This index, based on the Montana Land Cover Land Use data layer (MSDI, 2014) is a distance-decay model incorporating six disturbance categories: development, transportation, agriculture, resource extraction/energy development, introduced vegetation, and forestry practices. As an index, this dataset has no units, but rather "scores" ranging from 0 to 4314, with a value of 0 indicating no human disturbance and higher values indicating increasing levels. We buffered each wetland polygon in the High Streamflow Maintenance or Moderate Streamflow Maintenance category by 300m, and used "Zonal Statistics by Table" in ArcGIS 10.3 to calculate the mean HDI score for the area. Because ArcGIS Spatial Analyst does not handle overlapping polygons well, we did this iteratively for groups of buffered wetlands, then used the "Append" tool to rejoin the separate groups into a single shapefile or table. The tabular HDI scores were then joined to the original shapefiles using the common field "Unique ID," and the attribute tables were exported as dBASE IV (dbf) files for analysis in Microsoft Excel.

We also conducted an optimized hotspot analysis in ArcGIS 10.3 to identify clusters of wetlands with high HDI scores. The hotspot analysis tool examines the data to determine whether observed spatial clustering of high or low values for a particular attribute is statistically different than would be expected with a random distribution of those values. Results for values higher than expected (CI ≥95%) were displayed on separate maps (high streamflow maintenance and moderate streamflow maintenance) so that spatial patterns could be visualized.

RESULTS

Using the criteria we selected, >60% of the 27,685 wetland[3] polygons in the Blackfoot River HUC were considered to have high (16.8%) or moderate (43.7%) streamflow maintenance function. By areal extent, the percentages differed slightly. High streamflow maintenance polygons accounted for 2839.6 of the 24,630.4 ha of mapped wetlands in the HUC, or 11.5%. However, moderate streamflow maintenance polygons accounted for 15,438.7 ha, or 62.7% of the mapped wetland extent. Overall, the high percentage of wetlands captured by our criteria reflects the importance of wetlands for this function (Johnston et al., 1990). Indeed, given the topography and nature of the watershed, and its overall landscape position on the west slope of the Continental Divide, it is likely that this approach underestimates the percentage of wetlands contributing to this function, especially insofar as our method does not assign any functional significance to geographically isolated wetlands that may have an important groundwater and shallow surface water linkage with streams (McGuire and McDonnell, 2010; Evenson et al., 2015). As we noted earlier, the entire HUC can be regarded as a headwater area for larger downstream rivers.

One notable feature in our dataset was the importance of beaver-modified polygons in areas of high or moderate streamflow maintenance: 308 and 590, respectively, contrasted with only 12 beaver-modified polygons in areas not meeting our criteria. These numbers alone do not fully capture the influence of beavers. In our mapping work, we only assign the "b" modifier to specific polygons where active or recent beaver dams are visible in the imagery. On the ground, of course, that specific polygon is likely to be bordered in all directions by wetlands that are strongly affected by flooding and damming, but which do not, themselves, have beaver-created structures. In our datasets, for example, an additional 211 polygons in the high streamflow maintenance category—337 ha—are within five meters of a beaver-modified polygon. In the case of polygons with a moderate stream maintenance classification, 360 polygons (1531 ha) are similarly adjacent to beaver-modified polygons.

Although some lentic wetlands in headwater areas met our criteria for inclusion in the high streamflow maintenance category, by far the most common LLWW descriptors involved lotic landscape positions (Table 2.2.3.1). Within these,

3. MTNHP also maps riparian polygons, but these were excluded from the analysis.

TABLE 2.2.3.1 LLWW Attributes of Wetlands Categorized as "High" for Stream Maintenance

LLWW_Code	Number	Hectares
LOSLvaTHhw	2235	1298.3
LOFLvaTHhw	1176	1012.0
LOPDTHhw	266	146.7
LOFRTHhw	468	130.2
LOLKTHhw	4	73.5
LOFPTHhw	69	57.3
LOPDTHbvhw	257	27.8
LOFLvaTHbvhw	14	27.6
LOBATHhw	43	24.3
LOILTHhw	67	23.0
LEBATHhw	12	8.4
LOSLvaTHbvhw	16	3.8
LOFRTHbvhw	10	2.4
LEBAOUhw	1	1.8
LOILTHbvhw	8	1.2
LEPDTHbvhw	3	0.6
LOPDINxhw	4	0.4
LEFRTHhw	4	0.2

slope and flat landforms dominated, reflecting the biophysical characteristics of the landscape, where headwater streams flow from steep, V-shaped corridors into broader and flatter subalpine meadows before dropping gradually to the river's mainstem. By Cowardin class, the wetlands typical of streamside habitats—palustrine emergent (PEMA and PEMC), palustrine shrub scrub (PSSA and PSSC), and palustrine forested wetlands—dominated the distribution in terms of both area and numbers, with linear riverine upper perennial (R3USA) and the small open pools (PABF) also plentiful, albeit covering less area (Table 2.2.3.2).

The numbers and extent of LLWW codes differed for wetlands rated as "moderate" on stream maintenance function, largely because of the selection criteria; again, though, lotic landscape positions dominated. Here, however, lakes constituted a significant percentage of the total area while slope wetlands were uncommon, reflecting the flatter topography outside areas identified as headwater (Table 2.2.3.3). The distribution and relative weight of Cowardin types is similar, however, again reflecting the wetlands most characteristic of streamside areas (Table 2.2.3.4).

We calculated ownership percentages for selected wetlands using cadastral records. As might be expected, the majority of high streamflow maintenance wetlands found in headwater areas are in public land ownership, primarily the Forest Service, which manages three National Forests in this area: the Helena National Forest, the Lewis and Clark National Forest, and the Lolo National Forest (Table 2.2.3.5). In contrast, the majority of moderate streamflow maintenance wetlands, many concentrated along the corridors of the Blackfoot River mainstem and its larger tributaries, are under private ownership (Table 2.2.3.6).

In rugged rural landscapes like those found in the western part of the Blackfoot River HUC, land ownership is not a predictor of landscape condition. National forests are often crisscrossed with roads from timber harvests, recreation, and fire management, and in the case of this subbasin, large-scale mining activities from the last century. Private lands, except in towns, tend to be either recreational lands or large ranches that, along with adjacent public land leases, are primarily used for summer grazing. In the middle and eastern parts of the subbasin, however, there is more irrigated agriculture, heavier grazing, and more private and recreational development. In the following section, we examine the results of our landscape condition assessment.

TABLE 2.2.3.2 Cowardin Attributes of Wetlands Categorized as "High" for Stream Maintenance

Cowardin Code	Number	Hectares
PEMA	1545	894.1
PSSA	919	552.3
PFOA	371	401.1
PEMC	410	290.3
PSSC	227	234.4
L1UBH	3	63.8
PABF	210	61.8
R3USA	305	56.0
PEMB	57	39.5
PEMF	60	38.0
PABG	9	35.4
PSSCb	37	32.2
PUBH	23	29.5
PABFb	258	28.1
PFOC	53	24.0

TABLE 2.2.3.3 LLWW Attributes of Wetlands Categorized as "Moderate" for Stream Maintenance

LLWW Code	Number	Hectares
LOFLvaTH	5167	8784.3
LOFPTH	1802	1991.5
LOLKTH	21	1948.3
LOSLvaTH	1482	998.9
LOFRTH	1243	582.0
LOPDTH	992	277.5
LEBATH	109	249.8
LEFPTH	29	169.3
LOILTH	441	102.6
LOBATH	101	95.8
LOPDTHbv	553	66.8
LOFLvaTHbv	12	64.7
LEFRTH	18	57.7
TESLOUhw	25	13.4
LOPDINx	28	8.5

TABLE 2.2.3.4 Cowardin Attributes of Wetlands Categorized as "Moderate" for Stream Maintenance

Attribute	Number	Hectares
PEMA	2973	5397.4
L1UBH	20	1945.7
PFOA	1006	1894.1
PSSA	2347	1863.1
PEMC	1286	1556.5
PSSC	1034	1438.9
PABF	982	257.9
PEMF	205	226.6
R3USA	669	198.2
PEMB	70	164.8
R3USC	597	105.3
PFOC	76	97.2
PSSCb	14	66.7
PABFb	550	64.5
PSSB	17	57.9

TABLE 2.2.3.5 Ownership of Wetlands Categorized as "High" for Stream Maintenance

Owner	Percent by Number	Percent by Area
Private	23.1%	23.8%
Bureau of Land Management	5.0%	5.3%
Forest Service	66.7%	64.9%
State of Montana	0.3%	0.2%
Montana Fish, Wildlife, and Parks	2.0%	4.3%

TABLE 2.2.3.6 Ownership of Wetlands Categorized as "Moderate" for Stream Maintenance

Owner	Percent by Number	Percent by Area
Private	72.3%	78.8%
Bureau of Land Management	2.1%	1.8%
Fish and Wildlife Service	1.7%	1.1%
Forest Service	11.9%	7.9%
State of Montana	4.6%	0.2%
Montana Fish, Wildlife, and Parks	6.4%	6.1%
Other Montana State Agencies	<0.5%	0.5%

Landscape Condition Assessment

In our previous landscape-level analysis of wetlands in the Blackfoot watershed, we found moderate levels of disturbance (Hart et al., 2015) with HDI for a random selection of 434 wetlands organized by broad wetland type (emergent, forested/shrub, and pond) at 574.39, 588.84, and 629.92, respectively. In the present analysis, wetlands with high streamflow maintenance function had mean HDI scores well below those found through the random selection (Table 2.2.3.7). Wetlands with moderate function had much higher mean HDI scores (Table 2.2.3.8), likely reflecting their occurrences in the more heavily populated portions of the subbasin.

In the earlier analysis, we established narrative thresholds for HDI scores. Scores of 0–150 were characterized as representing conditions "at or near reference;" scores of 151–1025 were considered to be "slight departure from reference;" scores of 1026–1900 were classified as "moderate departure from reference;" and scores of 1901–4314 were recorded as "severe departure from reference." While mean HDI scores for wetlands in both functional rankings in the current study would indicate that most wetlands fell into the "slight departure from reference" category, the mean scores are misleading. Of the high streamflow maintenance wetlands, 60% can be considered to be at or near reference (Fig. 2.2.3.2) while only 15.6% of the moderate streamflow maintenance wetlands fall into that category (Fig. 2.2.3.3).

Hotspot analysis on HDI scores for each wetland functional group helped illustrate where clusters of high scores occurred. Fig. 2.2.3.4 illustrates the overall absence of clustering for disturbed high streamflow maintenance wetlands, except for an area of historic mine activity and current cleanup operations in the western part of the HUC. Fig. 2.2.3.5 reveals several clusters of disturbance, mostly around population centers and concentrations of agricultural activities. As a whole, then, these are likely to be the wetlands at the greatest risk of degradation in the future.

LESSONS LEARNED

Montana is rich in geospatial data, with a spatial data infrastructure (http://geoinfo.msl.mt.gov/Home/msdi), dedicated theme stewards, and a central data repository where both downloadable geodatabases and map services are readily accessed. Consequently, MTNHP—itself the theme steward for both the Wetlands and Land Cover Land Use themes—has long made use of these resources to analyze wetland distribution and condition at the landscape level (Vance, 2009; Newlon et al., 2013). In particular, landscape-level assessments allow us to overcome a major impediment to

TABLE 2.2.3.7 HDI Scores for High Streamflow Maintenance Wetlands, by Type

Wetland Type	Number	Mean HDI
Freshwater Emergent	2073	192.10
Freshwater Forested/Shrub	1624	262.82
Freshwater Pond	565	212.11
Lake	4	0.26
Riverine	391	100.03

TABLE 2.2.3.8 HDI Scores for Moderate Streamflow Maintenance Wetlands, by Type

Wetland Type	Number	Mean HDI
Freshwater Emergent	4559	722.12
Freshwater Forested/Shrub	4514	698.58
Freshwater Pond	1727	674.51
Lake	21	516.85
Riverine	1267	732.00

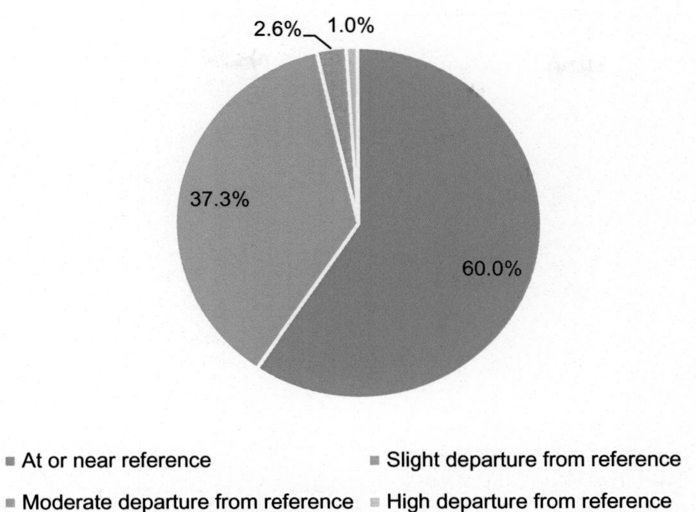

FIG. 2.2.3.2 High streamflow maintenance wetlands, by disturbance category.

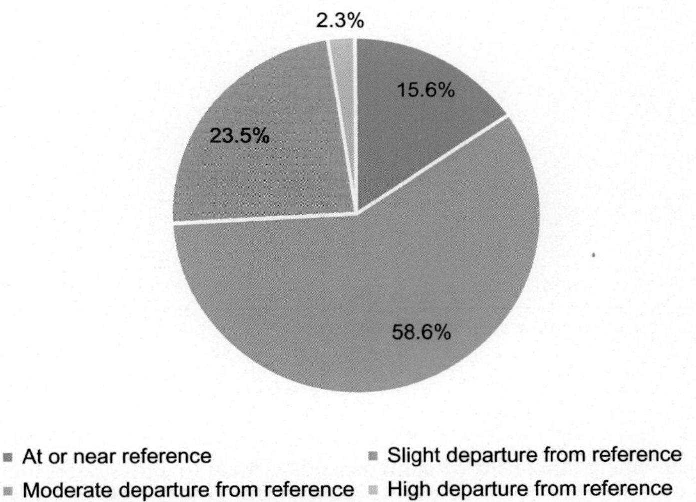

FIG. 2.2.3.3 Moderate streamflow maintenance wetlands, by disturbance category.

field-based assessments: a general unwillingness on the part of landowners to allow assessments on their property. Our subbasin assessments typically follow a nested pattern where we use landscape-scale methods to assess 1000 wetlands, rapid field assessments for 100 of those, and intensive field assessments for 30. In general, while landscape-level and rapid-assessment scores for surveyed wetlands show fairly good correspondence, at the broader scale, landscape-level assess ments tend to show higher overall disturbance for wetlands in a given subbasin (e.g., Hart et al., 2015). We attribute this in large part to the access issue—because of landowner refusals, our Level 2 assessments tend to be concentrated in publicly owned areas and frequently include sites accessible only by foot trails. Thus, landscape-level assessments, in general, may provide a more accurate view of wetland condition.

Although we have used LLWW descriptors in the past to suggest areas that are especially strong for specific wetland functions (Vance et al., 2006), this is the first case study since completing our automated GIS assignment procedures. For us, this is part of an overall direction we have taken to develop a Montana NWI Plus-Plus (MTNWI++) product, which, in addition to Cowardin classification, integrates both LLWW descriptors and a suite of additional attributes that support desktop-based functional and ecological assessments. This case study was undertaken to provide one example of how value-added attributes can be used to answer specific questions, beyond those that can be pursued with the Cowardin

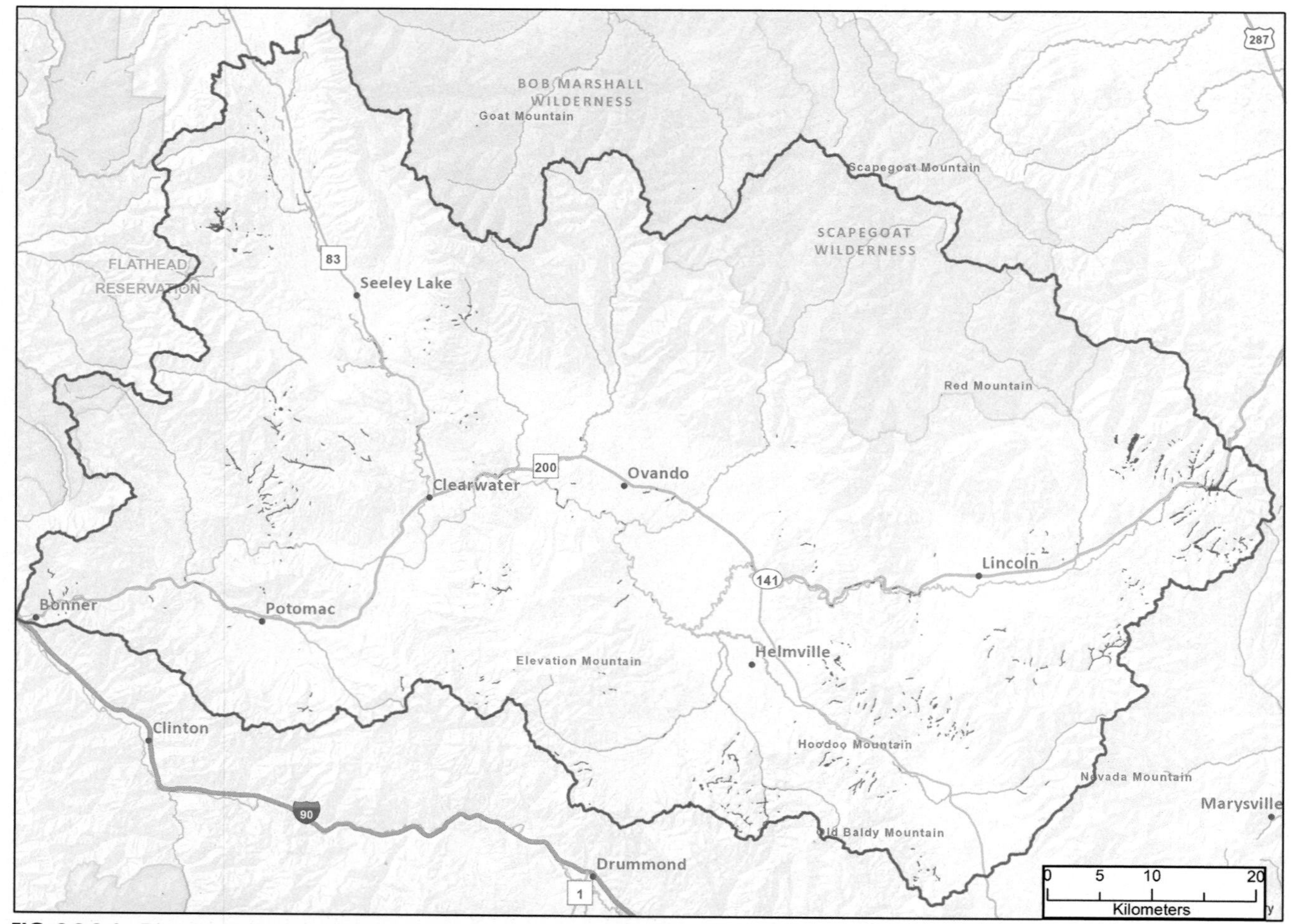

FIG. 2.2.3.4 Disturbance clusters, high streamflow maintenance wetlands.

classification alone. In a state where streamflow maintenance is critical to agriculture, tourism, and habitat maintenance, this was an appropriate and timely way to showcase the possibilities of this kind of assessment.

At the same time, we recognize two central limitations to landscape-level assessments. First, they are only as good as their data. In this case, for example, our HDI layer dates from 2014. Our current Land Cover Land Use (LCLU) layer was released in 2016, and incorporates a number of changes, especially in human disturbances. Fortunately, the particular subbasin we are focusing on here is one that does not undergo rapid change, and so the earlier layer probably captures current conditions fairly well; however, that would not be the case everywhere in the state. Nevertheless, even the most accurate parts of the LCLU layer may suffer from problems of scale. Unlike wetlands, which are mapped at a scale of 1:5000, the LCLU layer is intended for use at a 1:100,000 scale. Even the data layers that go into the LLWW attributes themselves are lacking in some areas. The assignment of landscape position depends on a slope raster and a headwaters model, which themselves depend on a Digital Elevation Model. In Montana, our best statewide digital elevation product is a 10 m product, which means that neither the slope raster nor the headwaters model is highly accurate at the wetland scale. The second limitation to landscape-level assessment is the uncertainty involved in basing characterizations of condition on disturbance. In general, landscape-level assessment presumes that impacts on the landscape will translate to a loss of condition in wetlands. Although we see fairly good correlation between landscape-level and rapid assessments, the area in which the correlation is weakest is generally vegetation (Hart et al., 2015; Newlon et al., 2013), which seems more resistant to disturbance at the landscape scale (Vance, 2009). For this reason, we are trying to deemphasize vertical integration of the three assessment levels in our future work, and instead promote landscape-level work as more of a characterization than an assessment.

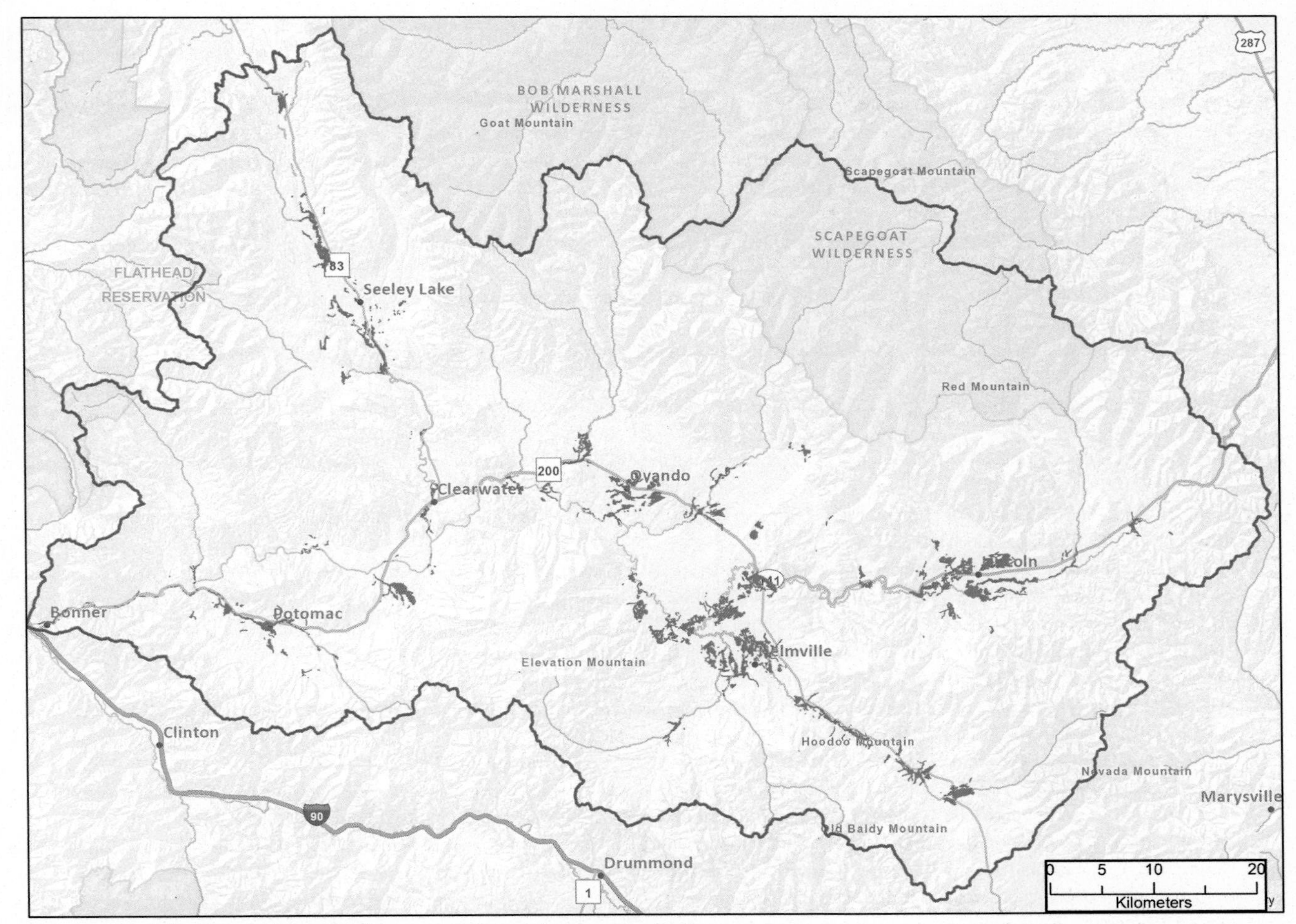

FIG. 2.2.3.5 Disturbance clusters, moderate streamflow maintenance wetlands.

CONCLUSION

The current case study illustrates an approach to evaluating and characterizing specific wetland functions at the subbasin scale using LLWW descriptors and available GIS data. It demonstrates how problems associated with large study areas, difficult access (both physical and legal), and complex questions can be addressed in desktop analysis. While we encourage users and others interested in this approach to keep in mind the limitations associated with landscape-level assessment, we believe that work of this nature offers important baseline information that can be built upon to develop more accurate field-based studies.

REFERENCES

Arnold, J.G., Muttiah, R.S., Srinivasan, R., Allen, P.M., 2000. Regional estimation of base flow and groundwater recharge in the Upper Mississippi River basin. J. Hydrol. 227, 21–40.

Blackfoot Challenge, 2014. Conservation partners tour the upper Blackfoot watershed. Available at http://biz170.inmotionhosting.com/~blackf22/Clone//conservation-partners-tour-the-upper-blackfoot-watershed/.

Cowardin, L.M., Carter, V., Golet, F.C., LaRoe, E.T., 1979. Classification of Wetlands and Deepwater Habitats of the United States. U.S. Fish and Wildlife Service, Washington, DC. FWS/OBS-79/31.

Evenson, G.R., Golden, H.E., Lane, C.R., D'Amico, E., 2015. Geographically isolated wetlands and watershed hydrology: a modified model analysis. J. Hydrol. 539, 240–256.

Freeman, M.C., Pringle, C.M., Jackson, C.R., 2007. Hydrologic connectivity and the contribution of stream headwaters to ecological integrity at regional scales. J. Am. Water Resour. Assoc. 43, 5–14.

Galat, D.L., Fredrickson, L.H., Humburg, D.D., Bataille, K.J., Bodie, J.R., Dohrenwend, J., Gelwicks, G.T., Havel, J.E., Helmers, D.L., Hooker, J.B., Jones, J.R., Knowlton, M.F., Kubisiak, J., Mazourek, J., McColpin, A.C., Renken, R.B., Semlitsch, R.D., 1998. Flooding to restore connectivity to regulated, large-river wetlands. Bioscience 48 (9), 721–733.

Hart, M., Vance, L., Newlon, K., Chutz, J., Hahn, J., 2015. Estimating Wetland Condition Locally: An Intensification Study in the Blackfoot and Swan River Watersheds. Report to the U.S. Environmental Protection Agency. Montana Natural Heritage Program, Helena, MT.

Hauer, F.R., Cook, B.J., Gilbert, M.C., Clairain Jr., E.J., Smith, R.D., 2002. A Regional Guidebook for Applying the Hydrogeomorphic Approach to Assessing Wetland Functions of Riverine Floodplains in the Northern Rocky Mountains. U.S. Army Engineer Research and Development Center, Vicksburg, MS. ERDC/EL TR-02-21.

Jenness, J., Brost, B., Beier, P., 2010. Land Facet Corridor Designer: Extension for ArcGIS. Available at http://www.jennessent.com/arcgis/land_facets.htm.

Johnston, C., Detenbeck, N., Niemi, G., 1990. The cumulative effect of wetlands on stream water quality and quantity. A landscape approach. Biogeochemistry 10, 105–141.

Lane, C.R., D'Amico, E., 2010. Calculating the ecosystem service of water storage in isolated wetlands using LIDAR in north Central Florida, USA. Wetlands 30, 967–977.

Maclean, N., 1976. A river runs through it. In: A River Runs Through It and Other Stories, 25th Anniversary Edition. University Press of Chicago, Chicago, IL.

McGuire, K.J., McDonnell, J.J., 2010. Hydrological connectivity of hillslopes and streams: characteristic time scales and nonlinearities. Water Resour. Res. 46, W10543,

Meyer, J.L., Strayer, D.L., Wallace, J.B., Eggert, S.L., Helfman, G.S., Leonard, N.E., 2007. The contribution of headwater streams to biodiversity in river networks. J. Am. Water Resour. Assoc. 43, 86–103.

MTDEQ, 2016. Montana's Clean Water Act Information Center. Draft 2016 Water Quality Integrated Report. Montana Department of Environmental Quality, Helena, MT. Available at http://deq.mt.gov/Water/WQPB/cwaic/reports.

MTDOJ, 2011. Draft Conceptual Restoration Plan for the Upper Blackfoot Mining Complex. Montana Department of Justice, Helena, MT.

MTFWP, 2015. Warm Water Prompts "Hoot-Owl" Fishing Restrictions on Blackfoot, Bitterroot and Clark Fork Rivers plus Flint and Silver Bow Creek. Montana Fish Wildlife and Parks, Helena, MT. Available at http://fwp.mt.gov/news/newsReleases/fishing/nr_4051.html.

MTNHP, 2014. Human Disturbance Index Dataset. Available at https://mslservices.mt.gov/Geographic_Information/Data/DataList/datalist_Details?did=%7B639e7c86-8224-11e4-b116-123b93f75cba%7D.

MSDI, 2014. Montana Land Cover and Land Use Dataset. Available at http://geoinfo.msl.mt.gov/Home/msdi/land_use_land_cover.

Newlon, K.R., Ramstead, K.M., Hahn, J., 2013. Southeast Montana Wetland Assessment: Developing and Refining Montana's Wetland Assessment and Monitoring Strategy. Report to the U.S. Environmental Protection Agency. Montana Natural Heritage Program, Helena, MT.

Tiner, R.W., 2003. Dichotomous Keys and Mapping Codes for Wetland Landscape Position, Landform, Water Flow Path, and Waterbody Type Descriptors. U.S. Fish and Wildlife Service, National Wetlands Inventory Program, Northeast Region, Hadley, MA. Available at https://www.fws.gov/northeast/ecologicalservices/pdf/wetlands/dichotomouskeys0903.pdfNote. Updated keys available at: https://www.fws.gov/wetlands/Documents/Dichotomous-Keys-and-Mapping-Codes-for-Wetland-Landscape-Position-Landform-Water-Flow-Path-and-Waterbody-Type-Version-3.pdf.

Tiner, R.W., Herman, J., Roghair, L., 2013. Connecticut Wetlands: Characterization and Landscape-Level Functional Assessment. Prepared for the Connecticut Department of Environmental Protection, Hartford, CT. U.S. Fish and Wildlife Service, Northeast Region, Hadley, MA.https://www.fws.gov/northeast/ecologicalservices/pdf/wetlands/CT_Wetland_CharacterizationFunctional_Assessment_FINALREPORT_Nov_2013.pdf.

Vance, L.K., Kudray, G.M., Cooper, S.V., 2006. Crosswalking National Wetland Inventory Attributes to Hydrogeomorphic Functions and Vegetation Communities: A Pilot Study in the Gallatin Valley, Montana. Report to the Montana Department of Environmental Quality and the U.S. Environmental Protection Agency. Montana Natural Heritage Program, Helena, MT.

Vance, L.K., 2009. Assessing Wetland Condition with GIS: A Landscape Integrity Model for Montana. A Report to the Montana Department of Environmental Quality and the Environmental Protection Agency. Montana Natural Heritage Program, Helena, MT.

Vance, L.K., Tobalske, C., Chutz, J., Zaret, K., 2015. Headwater Wetlands in the Missouri River Basin of Southwestern Montana: Characterization and Description of their Extent, Distribution and Condition. Montana Natural Heritage Program, Helena, MT.

Chapter 2.2.4

Landscape-Level Wetland Functional Assessment for the St. Joseph River Watershed, Southwest Michigan, United States

Chad Fizzell*, Jeremy Jones*, Matt Meersman[†] and Marcy Hamilton[‡]

**Michigan Department of Environmental Quality, Lansing, MI, United States, [†]Friends of the St. Joseph River, South Bend, IN, United States, [‡]Southwest Michigan Planning Commission, Benton Harbor, MI, United States*

Chapter Outline

INTRODUCTION

The Michigan Department of Environmental Quality (MDEQ) has been working with federal and local agencies, universities, and watershed associations to develop protocols and sample designs for three levels of monitoring the state's wetland resources (landscape assessment, rapid wetland assessment, and intensive site assessment). Their Wetland Assessment and Monitoring Strategy (MDEQ, 2015) therefore includes an objective for completing landscape-level functional assessments for all watersheds within 5 years. Objective 4 specifies applying "landscape level wetland assessment methods to support the protection, management, and restoration of wetlands on a watershed scale." MDEQ has been doing this type of assessment for about 10 years. The first application of this method was in the St. Joseph River Watershed (SJRW) (http://www.swmpc.org/downloads/pprw_WetlandFunctionAssmnt.pdf).

With the support of a 319 grant from MDEQ, the Van Buren Conservation District initiated the Paw Paw and Black Rivers Wetland Protection and Restoration Project to address this need locally. The goals of the project included restoring and protecting wetlands in high-priority areas as demonstration projects, and implementing a replicable wetland education strategy for landowners and municipalities. In order to accomplish these goals, a wetland prioritization process was developed utilizing geographic information systems (GIS). This chapter describes the baseline data, the methodology utilized, and the results of the prioritization process.

STUDY AREA

The SJRW is the third-largest river basin in Michigan (Fig. 2.2.4.1).

Beginning at Baw Beese Lake in Michigan's Hillsdale County, the SJRW spans the Michigan-Indiana border and empties into Lake Michigan in the city of St. Joseph. The watershed drains 4685 $mile^2$ from 15 counties, and the main stem is 210 miles long. Major tributaries include the Prairie, Pigeon, Fawn, Portage, Coldwater, Elkhart, Dowagiac, and Paw Paw Rivers as well as Nottawa Creek.

Wetland and Stream Rapid Assessments. https://doi.org/10.1016/B978-0-12-805091-0.00020-7

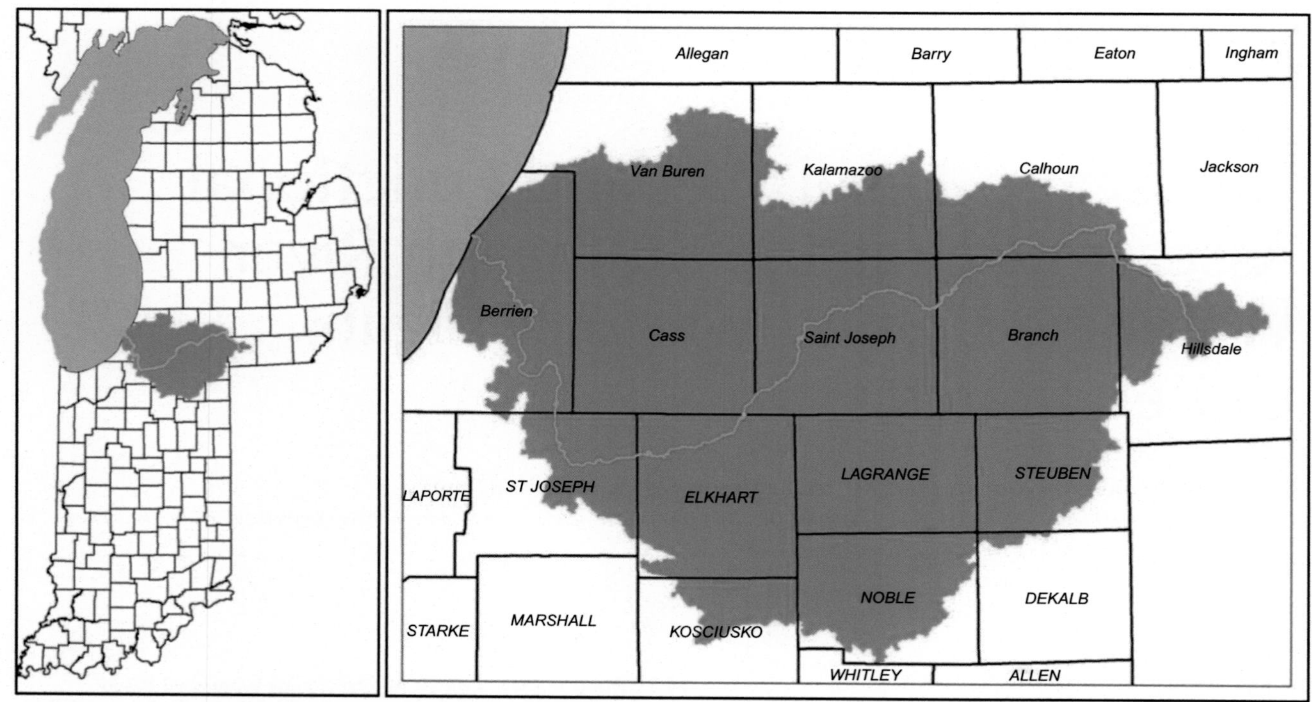

FIG. 2.2.4.1 Location of the St. Joseph River Watershed in Michigan.

DEVELOPING A WATERSHED MANAGEMENT PLAN

In the fall of 2002, the Friends of the St. Joseph River group was awarded a grant to develop a watershed management plan for the entire SJRW. This plan would cross state lines, as a major percentage of the watershed falls within Indiana. The major impairments to the water quality within the watershed were pathogens, sediment loading, and nutrient loading. Given that the SJRW as a whole had lost approximately 53% (Fig. 2.2.4.2) of its wetlands since pre-European settlement, the degradations in water quality were almost entirely due to historic wetland loss and fragmentation. Wetland loss underlies most of the water quality issues in Michigan, representing the systemic cause of excessive sedimentation and nutrient loading, habitat fragmentation, and reduced streamflow.

When implementation funds were being sought after the planning period, the watershed management group focused on prioritizing wetlands for preservation and restoration based on the significance of the functions they serve or once served in the case of restorable historic wetlands. This "Wetland Partnership Project" identified four major project objectives to accomplish the stated goals of the planning effort:

1. Develop and coordinate a bistate wetland partnership to direct wetland protection and restoration efforts.
2. Conduct functional assessments of all current and historic wetlands.
3. Prioritize counties, townships, and smaller areas to focus on wetland protection and restoration efforts.
4. Develop an educational strategy for municipal officials and landowners, including model planning/zoning language and maps as well as outreach materials for wetland protection and restoration.

These concepts boil down to a three-step strategy for watershed improvement: (1) enhance existing and historic wetland data, (2) prioritize by geography to identify priority areas for protection and restoration, and (3) utilize the results to assist in the planning effort.

Baseline Data

The baseline data for this project included a Landscape-Level Wetland Functional Assessment (LLWFA) and parcel/ownership information provided by Allegan, Berrien, and Van Buren Counties. Data sources used for this project are outlined in Table 2.2.4.1.

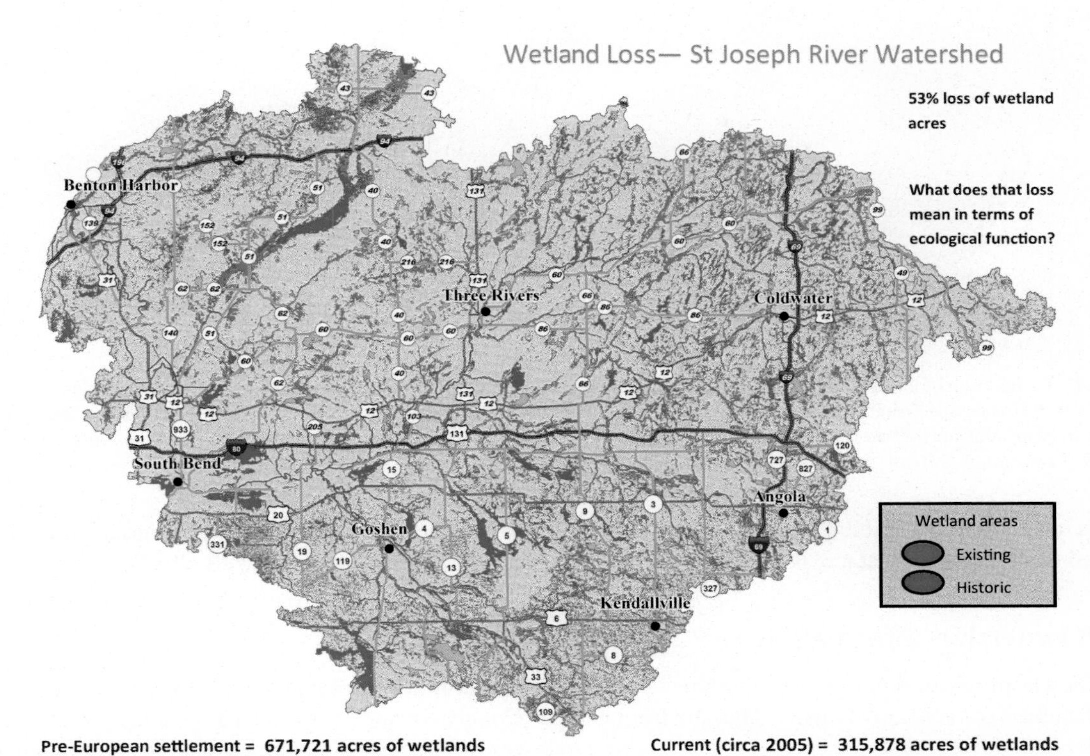

FIG. 2.2.4.2 Wetland loss in the St. Joseph River Watershed.

TABLE 2.2.4.1 Data Sources Used in Compiling the LLWFA

Layer Name	Data Source	Description
National Wetlands Inventory	Fish and Wildlife Service, National Wetlands Inventory	2005 National Wetland Inventory competed by Ducks Unlimited (GLARO)
National Hydrography Dataset-High Resolution	Geological Survey and EPA	Based upon Digital Line Graph (DLG) hydrography at 1:24,000 scale
Digital Raster Graphic (DRG) Topography and DEM	Geological Survey	Scanned USGS Topo quads
SSURGO Soil Surveys	Natural Resource Conservation Service	Digitized from Paper Soil Surveys at 1:24,000
NAPP 1998 Digital Orthophoto Mosaics	Geological Survey	Color Infrared Aerial Imagery
NAIP 2005 and 2010 Digital Orthophoto Mosaics	Natural Resource Conservation Service (NRCS)	Natural Color Aerial Imagery
CGI Framework Data	Michigan Center for Geographic Information	Includes roads, political boundaries, hydrography, census figures, etc.
Michigan Natural Features Inventory Land Cover 1800	Michigan Natural Features Inventory (MNFI)	Land Cover data derived from GLO Surveys from early to mid 1800s

Landscape-Level Wetland Functional Assessment

The LLWFA performed by the MDEQ classifies current and historic wetlands (Fig. 2.2.4.3), which are likely to perform certain functions at a level significantly greater than others, as "moderate" or "high" for each assessed function. The functions utilized in this project were primarily related to water quality (floodwater storage, streamflow maintenance, sediment retention, nutrient transformation, and shoreline stabilization), but wildlife-related functions (fish, waterfowl, forest bird,

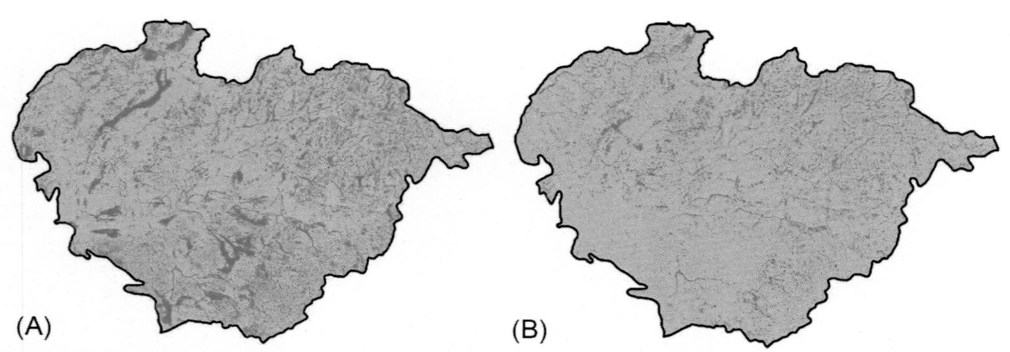

FIG. 2.2.4.3 Comparison of historic and contemporary wetlands for the St. Joseph River Watershed: (A) presettlement wetlands and (B) present-day wetlands. LLWW (Landscape Position (L), Landform (L), Waterflow Path (W), Waterbody Type (W)) descriptors are applied to these wetlands to help predict wetland functions for the two time periods and estimate the effect of cumulative losses on functions. *(Source: http://fotsjr.org/Resources/Documents/SJRW_Presentation_Zbiciak_3-15-10.pdf (accessed 10.08.17.)*

shore bird, and amphibian habitat) were also considered for use by other groups. Detailed information about the methodology used to develop this dataset is available in the LLWFA methodology report (Fizzell, 2011).

Parcel/Ownership Information

Parcel data was provided by each county in the watershed. The data from Berrien and Van Buren Counties contained the name and address of each parcel owner. The data from Allegan County contained only the name of the parcel owner. Parcels from each county that intersected current or historic wetlands were selected and exported as a new dataset. Parcels with the exact same owner name were automatically dissolved. Each landowner name in the resulting dataset was reviewed, utilizing good judgment and local knowledge to identify separate records that should be treated as one. For example, "Timothy S. and Linda A. Johnson" was merged with "Tim and Linda Johnson," and local knowledge was used to connect "Timothy Hood" with "Hood Dairy Farm."

THE THREE-STEP PROCESS FOR DEVELOPING WATERSHED-BASED WETLAND CONSERVATION

To meet the ultimate objective of developing a watershed-based strategy for wetland protection and restoration, a three-step process was initiated. The first step involved expanding existing wetland inventory data to include more variables for predicting wetland functions. The second step was prioritizing wetlands for protection or restoration. The final step required use of the data to improve the status of wetlands across the watershed through protection and restoration initiatives. These steps are discussed below.

Step 1—Enhancement

The enhancement step in the context of the St. Joseph project involves the development of the LLWFA for the watershed. Enhancement of the current National Wetlands Inventory (NWI) is the first step in this process by adding hydrogeomorphic-type attributes for landscape position, landform, water flow path, and waterbody type (LLWW descriptors based on Tiner 2002, 2003). Each NWI polygon is coded with the LLWW descriptors: Landscape Position (L), Landform (L), Waterflow Path (W), and in the case of ponds, rivers, and lakes, a Waterbody Type (W) (see LLWW descriptor assignment based on Tiner 2002, 2003). Some of this information can be garnered using automation; however, all polygons are interpreted on aerial imagery to verify this information and to make necessary changes to the codes. In the case of the SJRW, 68,294 wetlands were coded with the LLWW descriptors across Michigan and Indiana.

The same enhancement process was also performed on the presettlement wetland inventory that was developed by the MDEQ Wetlands, Lakes, and Streams Unit. This presettlement wetland inventory was created using hydric soils data and a historic land cover dataset created by the Michigan Natural Features Inventory (Fig. 2.2.4.3A). With a presettlement inventory, the enhancement process could now be completed for the Michigan portion of the watershed. MDEQ employees had to determine a way to create the same layer for Indiana or risk only being able to compare historic data to present day for half the watershed. Using hydric soils data for Indiana as the base (as was done for Michigan), all that was left was to determine what type of historic land cover might have existed before settlement in Indiana. Using the Natural Resource

Conservation Service's soil series descriptions, the MDEQ was able to assign an estimated historic land cover for the presettlement wetlands in Indiana. The two state datasets were then combined and coded with LLWW descriptors. Ponds were not coded in the historic wetland inventory due to the coarseness of the data. The historic data was coded using topographic maps and three-dimensional elevation models to help interpreters determine historic water courses. In the SJRW, 34,594 wetlands were coded with the LLWW descriptors for the historic wetland datasets.

Once presettlement and current wetlands were classified by hydrogeomorphic descriptors, the functional significance of each wetland area was predicted by correlating wetland characteristics with specific functions. Functions were organized into two major groups: water quality functions and habitat functions. Water quality functions included flood water storage, streamflow maintenance, nutrient transformation, sediment and other particulate retention, shoreline stabilization, stream shading, groundwater influence, carbon sequestration, and pathogen retention. Predicted habitat functions included habitats for fish, waterfowl/waterbirds, shorebirds, interior forest birds, and amphibians plus conservation for rare and imperiled wetlands and species. Step 2 addresses the landscape-level functional assessment procedures.

Step 2—Prioritization

The second step in the process is intended to narrow down the universe of priority wetland areas from an entire major watershed—where pollution and wetland impacts are diffuse and scattered—to specific target areas where limited watershed funding can achieve the greatest net result. Identifying the highest areas of loss for a given wetland function is a relatively simple step from a technical perspective, but the implications of prioritizing the subwatersheds or townships with the most loss allows work to begin in areas where it is needed most. An example of how this can be done for a watershed is outlined below.

Wetland protection and restoration priorities were determined by calculating functional units, analyzing functional loss, and scoring wetland areas. Detailed information about the functional unit datasets and the GIS processes used to create them is available in the LLWFA Version 1: methodology report (Fizzell, 2011).

Calculating Functional Units

Calculating the number of functional units for current and historic wetland polygons is necessary to determine functional loss trends and prioritize wetlands for protection or restoration. Functional units are calculated by multiplying the size (acres) of each current and historic wetland polygon by its significance value (1—moderate, 2—high) for each function. For example, a 2.5-acre wetland polygon designated as moderate for floodwater storage and high for nutrient transformation would have 2.5 floodwater storage functional units and 5 nutrient transformation functional units (Fig. 2.2.4.4).

Functional Status and Trends

Comparing the total number of current wetland functional units for a particular function within a watershed to the number of historical functional units for that same function determines the predicted change in functional capacity. A functional status and trends report for the Paw Paw and Black River Watersheds and for two selected subwatersheds is provided in Table 2.2.4.2. Analyzing differences between acreage loss and functional loss can help in developing implementation strategies. For example, outreach efforts in the South Branch of the Paw Paw River, where there is significantly more floodwater storage loss than wetland acreage loss, might focus on wetland restoration while outreach efforts in the South Branch of the Black River, where there has been relatively little floodwater storage loss, might focus on wetland protection.

Scoring Wetland Areas

In order to identify landowners and municipalities with the most significant wetland resources, current and lost wetlands were carved into wetland pieces based on property ownership or municipal boundaries using a GIS process called a union. The functional units for each wetland piece are calculated by multiplying its size (acres) by the significance value (1—moderate, 2—high) for each function. The wetland pieces are then merged together into wetland areas based on common ownership or jurisdiction using a GIS process called a dissolve. The functional units of each wetland piece are combined or summed during the dissolve process to determine the functional unit score of the wetland area for each function.

Ranking wetland loss by geography and function takes into account the practicality of completing a wetland-related project and eases potential complications by targeting large areas with a single landowner, or a township with high loss of sediment retention wetlands (Figs. 2.2.4.5 and 2.2.4.6; Table 2.2.4.2). In an area as large as the St. Joseph with a multitude of political jurisdictions and myriad interested stakeholders, this sort of targeting is essential for successful implementation on a limited budget. The key to any wetland inventory effort should be to protect resources on the ground and to utilize the data in real-world situations. This includes protection of existing wetlands and identification of opportunities for

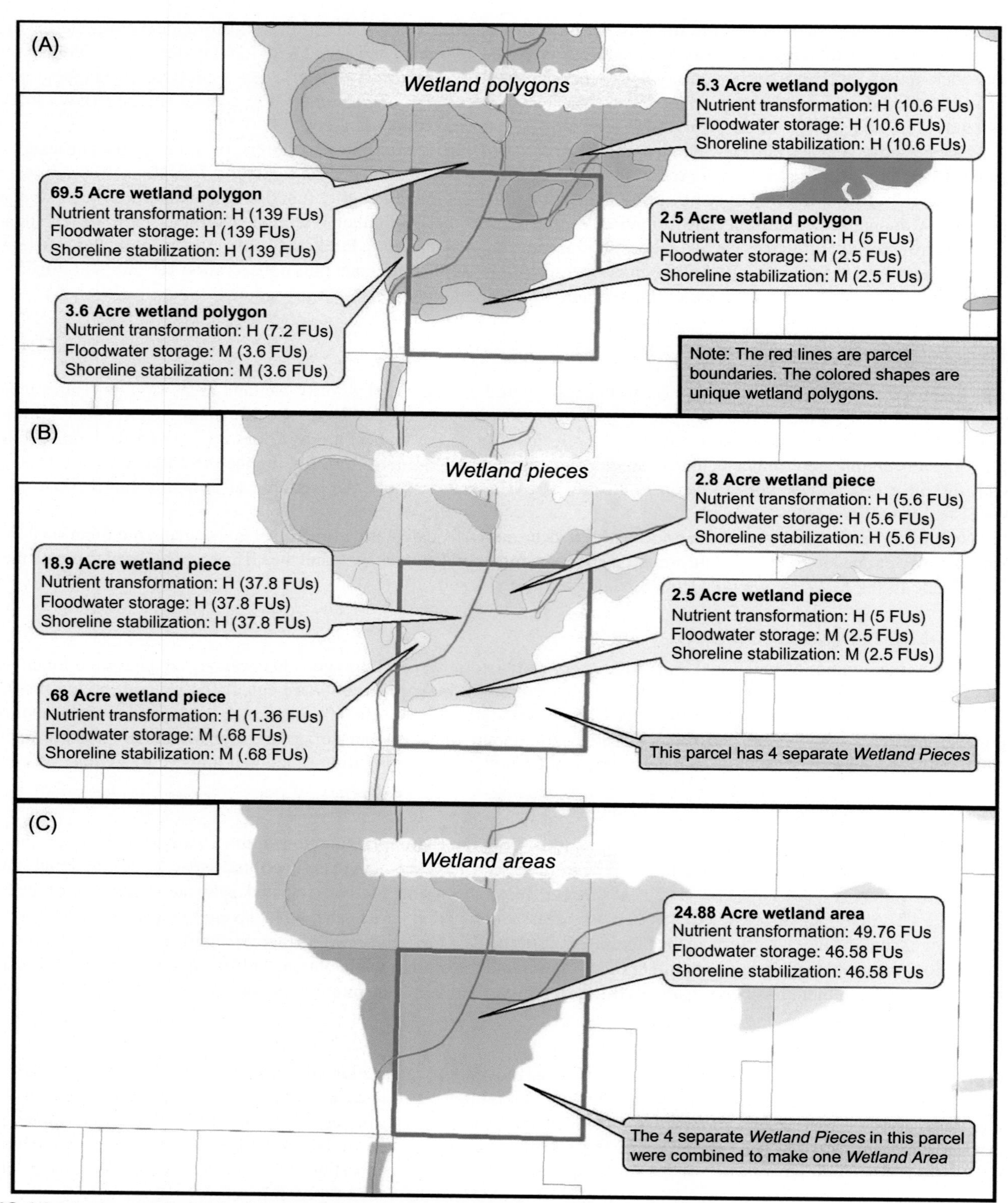

FIG. 2.2.4.4 An example showing the procedure for calculating functional units. The Original NWI Wetland Polygons (A) are broken up by parcel boundaries (B), then adjacent parcels with identical ownership are merged into Wetland Areas (C)

TABLE 2.2.4.2 The Predicted Cumulative Impact of Historic Wetland Losses on Wetland Functions for the Paw Paw and Black River Watersheds (Similar Tables Were Generated for Townships)

Landscape Level Wetland Functional Assessment Status and Trends Report				
Paw Paw and Black River Watersheds				
	Watershed (HUC 10)	Watershed (HUC 10)	Subwatershed (HUC 12)	Subwatershed (HUC 12)
Name and/or ID	Paw Paw River (0405000124/5)	Black River (0405000202)	South Branch Paw Paw River (40500012405)	South Branch Black River (40500020210)
Total acres	285,799	183,419	26,625	12,268
Existing wetland acres	37,391	28,169	3212	1298
Historic wetland acres	64,792	58,929	5679	2959
Wetland loss	42%	52%	43%	56%
Predicted percent change in functional capacity[a]				
Function				
Water quality combined	*−45%*	*−53%*	*−58%*	*−42%*
Floodwater storage	−52%	−61%	−64%	−46%
Streamflow maintenance	−38%	−58%	−44%	−57%
Nutrient transformation	−44%	−45%	−51%	−44%
Sediment retention	−53%	−43%	−70%	−17%
Shoreline stabilization	−40%	−54%	−55%	−41%
Habitat combined	*−50%*	*−54%*	*−51%*	*−45%*
Fish	−51%	−72%	−64%	−54%
Waterfowl	−17%	41%	−28%	23%
Shore bird	−44%	−41%	−45%	−55%
Forest bird	−45%	−54%	−51%	−52%
Amphibian	−72%	−69%	−57%	−36%

[a]Functional capacity estimates the ability of the wetlands in each geographic area to perform the listed functions. Increases in predicted functional capacity change can be attributed to mapping differences in the two wetland layers and may not represent the current conditions on the ground. The LLWFA is preliminary, based on wetland characteristics interpreted through remote sensing and the professional judgment of various specialists who developed correlations between wetlands and their functions. The LLWFA does not consider the condition of adjacent uplands and it does not eliminate the need for more detailed assessment of wetlands in the field.

wetland restoration/creation/enhancement. Prioritizing by watershed based on percentage of wetland loss is one of many methods for utilizing the enhanced NWI data to pinpoint wetland preservation and restoration efforts to ensure conservation dollars are spent effectively. Prioritizations based on municipality, lost wetland function, and wetland type are also possible and expand the usefulness of the highly scalable dataset.

Step 3—Utilization

The true power of the watershed approach comes in the utilization of all the wetland information that has been compiled up to this point in the process. Targeting outreach materials to interested landowners, public officials, municipalities, zoning boards and the like allowed the Friends of the St. Joseph River to reach their audience without over broadcasting their

Step 2. Prioritize

Rank by geography and function

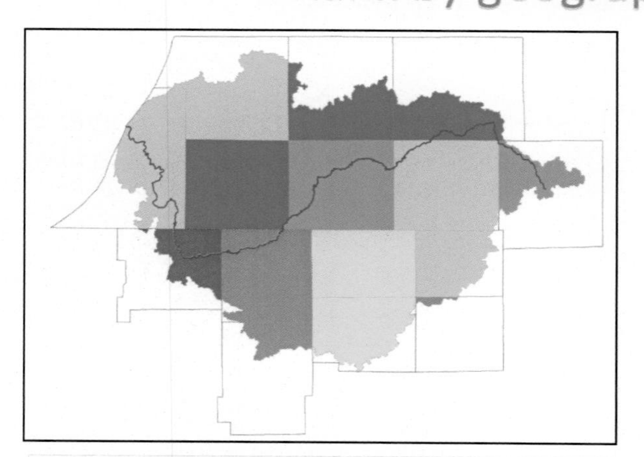

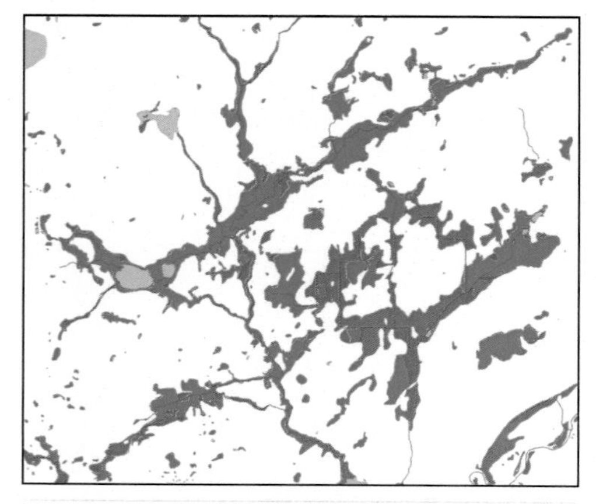

High **remaining** sediment retention capacity

Rank	Municipality Name	Wetland Acres	% Wetlands Loss	Sediment Retention Functional Units
1	Almena Township	5400.3	32.7%	6096.9
2	Waverly Township	4190.6	47.8%	5005.4
3	Hartford Township	3566.1	44.2%	4377.5
4	Lawrence Township	2885.6	36.1%	3996.8
5	Paw Paw Township	3263.0	49.7%	3121.3
6	Benton Township	1614.4	42.6%	2633.7
7	Antwerp Township	1904.3	24.6%	2319.1
8	Hamilton Township	1889.9	6.3%	2229.3
9	Watervliet Township	1423.9	31.5%	1807.7
10	Keeler Township	1548.8	33.3%	1708.2
11	Bainbridge Township	1237.4	27.5%	1679.7
12	Hagar Township	1000.0	28.1%	1667.1
13	Coloma Township	941.2	28.2%	1438.1
14	Bangor Township	1104.3	60.5%	1059.1
15	Bloomingdale Township	816.6	65.7%	821.6
16	Decatur Township	866.2	78.0%	742.4

"Top 25" wetland restoration owners

	Landowner Name	Restoration Acres	Sediment Retention Functional Units
1	State of michigan	4,411	3,848
2	Geerlings Hillside Farms	777	993
3	Scenic View Dairy	567	763
4	Blue Goose Farms Inc	384	748
5	Ghidotti Bert	612	281
6	Onesimus LLC	176	258
7	Copeland Paule	112	214
8	Borden Processing Inc	154	195
9	Roedger Bros Real Estate LLC	97	188
10	Stokes Roger	299	177
11	Arnold Gene & Shirley	115	156
12	Jorgensen Donald O	99	127
13	Busy Bee Farms	339	126
14	Reimink Edward & Cynthia	77	122
15	Scholten Cathryn	72	122
16	Tate Billy	60	121
17	Leduc Bros	72	107
18	Priebe Karen Murphy LLC	96	106

FIG. 2.2.4.5 Example of ranking of wetlands by townships based on remaining acreage supporting the sediment retention function and list of top 25 landowners possessing the most acreage of those wetlands.

message to stakeholders with minimal interest in wetlands protection. It also allowed the wise and efficient use of the limited financial resources available to the group. By subdividing the information by property owner (parcel), the group was able to send direct mailings to the major landowners in the watershed with the most restorable acreage suited to perform the targeted wetland functions that the group was looking to replace on the landscape.

The Friends of the St. Joseph River also targeted municipalities, harbor authorities, drain officials, and conservation groups with direct mailings. Stakeholders were invited to informational sessions where the results of the analysis could be explained in layman's terms, and the benefits of wetland restoration and preservation activities could be explained. As a result of these efforts, several local governmental entities have spent considerable amounts of local taxpayer funds on wetland projects outside their jurisdiction to address local water quality problems, after finally understanding the concept that pollution doesn't recognize political boundaries.

The Friends of the St. Joseph River also spent considerable time producing educational materials exploring the different aspects of this approach (see video on wetland assessment; http://fotsjr.org/WetlandPartnershipOutreach). Brochures were developed to explain the complex LLWFA process in layman's terms. Template landowner letters, attached to maps and figures directly targeted at the landowner, were created to expedite the outreach process and to begin educating the land owner about the value and function of their particular wetland area. Information was also provided on the different resto ration/preservation opportunities available to the landowner through the federal government and the state, informing them of the financial incentives that were possible when participating in these programs.

Explaining to municipalities the benefits of advanced planning and zoning considerations as it applies to their current and former natural resources helps these entities make wise land-use decisions when a project is faced with wetland/upland alternatives. In certain cases, giving justification to a local zoning board that the upland parcel may be more suitable for a given project, given the potential of a site to be restored in the future, even if that site is currently in agriculture, preserves a local entity's opportunity to complete a restoration project when a need arises. This sort of advanced planning allows

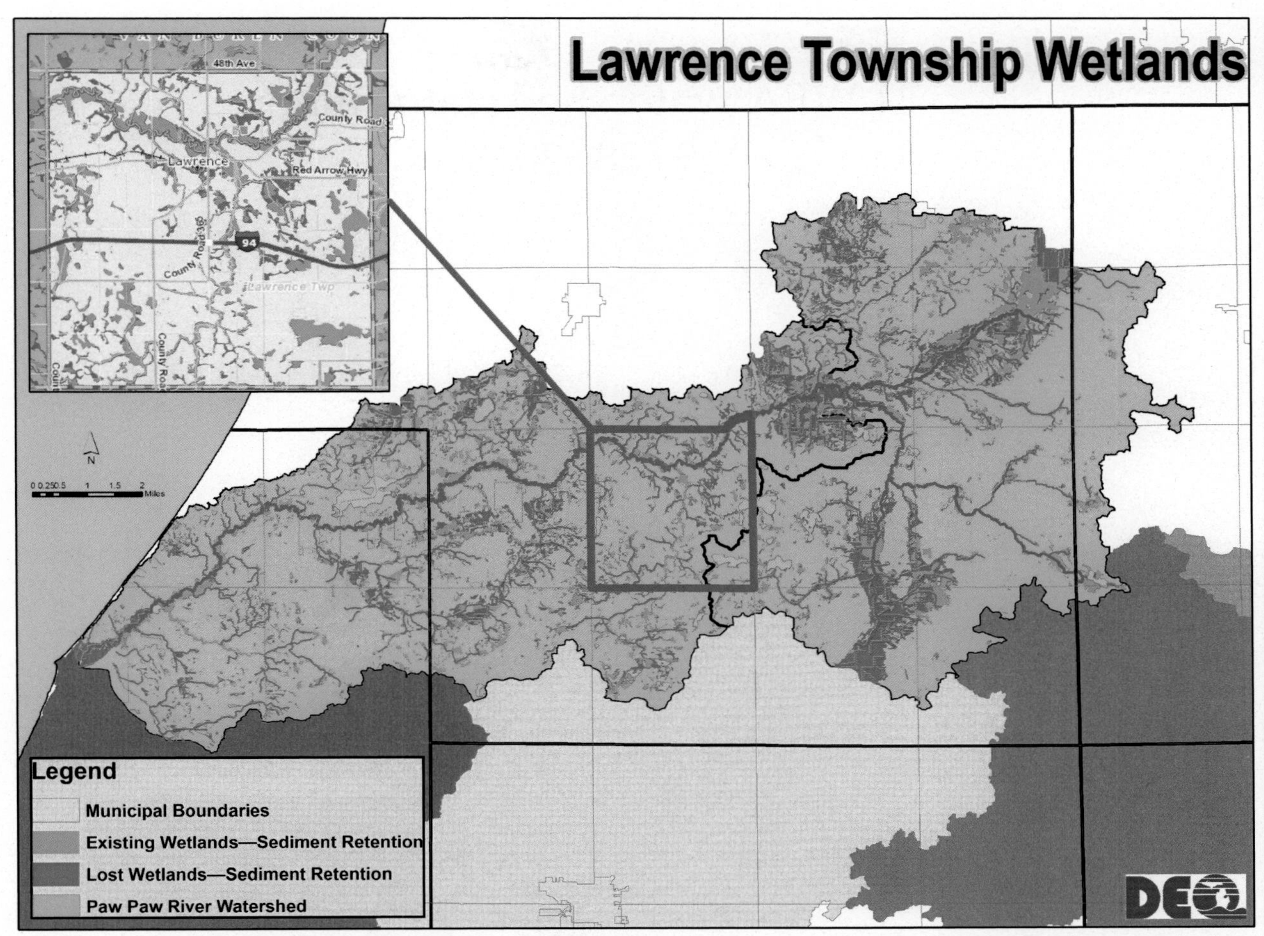

FIG. 2.2.4.6 Example of township map showing current wetlands and lost wetlands that were rated as significant for the sediment retention function.

municipalities to consider a parcel's highest and best use, not only taking into account the normal bounds of increasing the local tax base but also preserving the community's ability to manage its natural resources wisely and effectively and to avoid costly water quality problems before they arise.

CONCLUSION

The Friends of the St. Joseph River partnership continues to work together to increase wetland protection and restoration in the SJRW. As a direct result of this project, the Friends of the St. Joseph River and its stakeholders gained an increased understanding of existing and lost wetland functions in the watershed. With the completion and sharing of the LLWFA data, citizens all over the watershed became aware of the critical role their wetlands play in the local ecosystem. The outreach to citizens, local government, and other interested stakeholders drove the success of this project overall. More than 2000 letters were sent to high-priority landowners in the watershed, explaining the project and inviting them to landowner events. During the project, more than 200 landowners attended landowner events and about 200 local officials were reached through municipal presentations. The fact that MDEQ and the Friends of the St. Joseph River are still discussing this project several years after its completion demonstrates that the value of utilizing this type of landscape-level assessment in watershed planning efforts. The information generated by this assessment is driving increased consideration of wetland function during watershed planning. Watershed groups, lake associations, and others continue to ask the Friends of the St. Joseph River for the wetland function data for specific areas to support grant projects or to help with advocacy efforts to protect wetlands. The information generated for the watershed will continue to have an impact long into the future, and the project overall still stands as a great example of collaboration between state and local partners.

The success of this project has led to similar initiatives across the state. Fig. 2.2.4.7 shows the status of watershed-based wetland functional assessments statewide. MDEQ's goal is to have these watershed assessments completed for the entire state in the next 5 years.

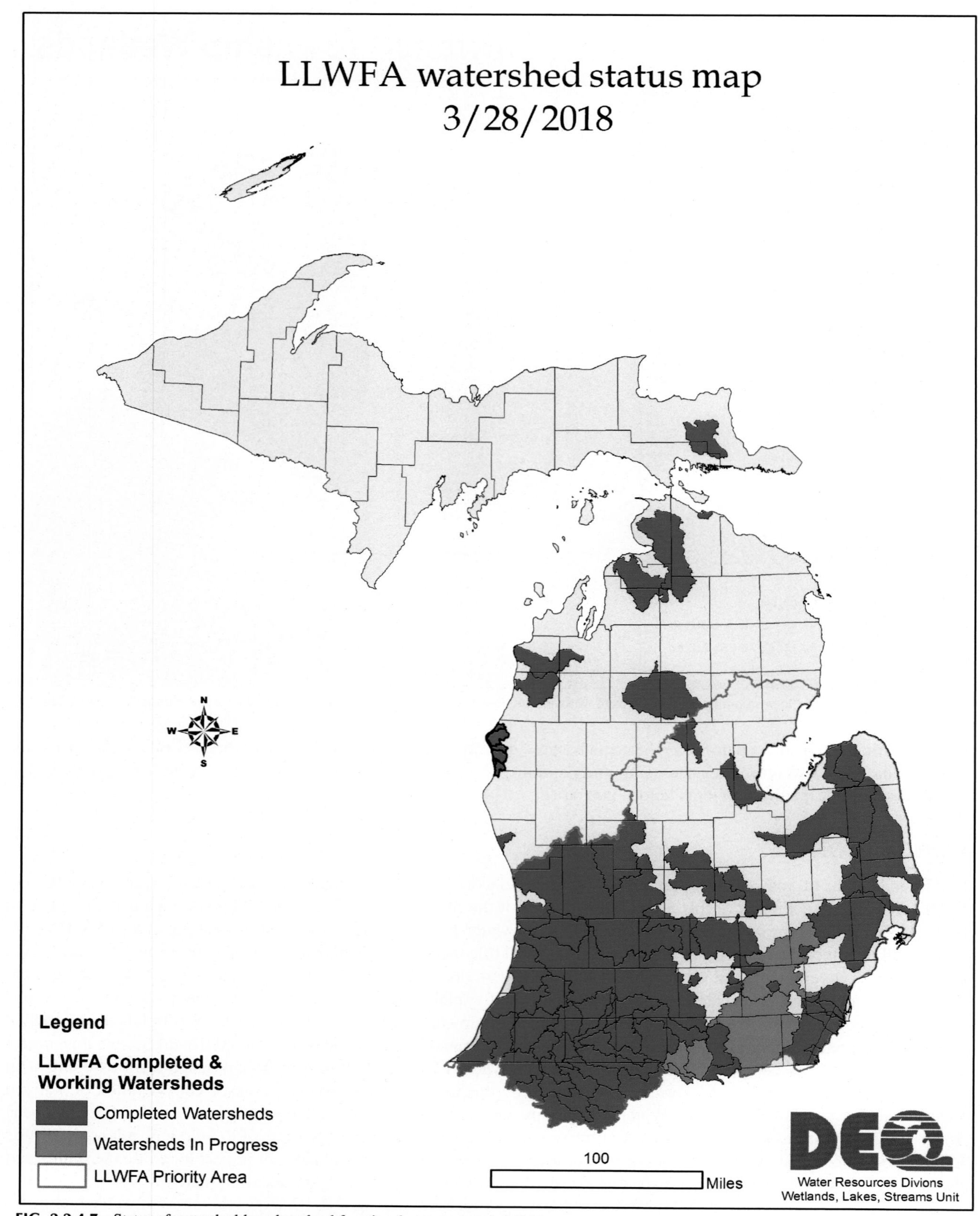

FIG. 2.2.4.7 Status of watershed-based wetland functional assessments in Michigan.

REFERENCES

Fizzell, C.J., 2011. Landscape Level Wetland Functional Assessment (LLWFA) Version 1: Methodology Report. Michigan Department of Environmental Quality. Water Resources Division, Lansing, MI.

Michigan Department of Environmental Quality, 2015. State of Michigan Wetland Monitoring and Assessment Strategy. Water Quality Division, Lansing, MI. https://www.michigan.gov/documents/deq/wrd-wetlands-strategy_555457_7.pdf.

Tiner, R.W., 2002. In: Tiner, R. (Ed.), Enhancing wetlands inventory data for watershed-based wetland characterizations and preliminary assessment of wetland functions. Watershed-Based Wetland Planning and Evaluation: A Collection of Papers From the Wetland Millennium Event August 6–12, 2000; Quebec City, Quebec, Canada. Association of State Wetland Managers, Inc., Berne, NY, pp. 17–39.

Tiner, R.W., 2003. Keys to Waterbody Type and Hydrogeomorphic-type Wetland Descriptors for U.S. Waters and Wetlands. U.S. Fish and Wildlife Service, Northeast Region, Hadley, MA.

Chapter 2.2.5

An Automated Procedure for Extending the NWI Classification System for Wetland Functional Assessment in Minnesota, United States

Steven M. Kloiber*, Robb D. Macleod† and Gang Wang†

**Minnesota Department of Natural Resources, St. Paul, MN, United States, †Ducks Unlimited, Memphis, TN, United States*

Chapter Outline

INTRODUCTION

Wetlands perform a variety of different ecological functions and many of these functions are beneficial to society. Among the more commonly cited beneficial functions are flood reduction, water quality improvement, wildlife habitat, shoreline erosion protection, and recreational opportunity (Zedler and Kercher, 2005). These functions depend on complex relationships between the wetland and other ecosystems in a watershed. In recognition of these benefits, it has been the policy of the U.S. government to achieve a goal of no overall net loss of values and functions (Environmental Protection Agency (EPA) and U.S. Army Corps of Engineers (USACE), 1990). Achieving this goal requires the ability to accurately and efficiently assess wetland functions.

One such assessment approach is the hydrogeomorphic (HGM) method proposed by Brinson (1993). The general premise of the HGM approach is that hydrologic and geomorphic conditions govern many of the functional aspects of wetland ecosystems and that by classifying wetlands according to their hydrologic and geomorphic characteristics, we can assess these functions. This classification approach has three main components: geomorphic setting, water source, and hydrodynamics. The geomorphic setting, or landscape position, includes the following types: depressions, sloped wetlands, extensive peatlands, riverine wetlands, and fringe wetlands (tidal and lacustrine). This approach also relies on establishing regionally specific reference wetlands, conducting detailed monitoring and assessment of functions for these wetlands, and extrapolating to other wetlands of similar HGM class. These HGM classes have been shown to correlate to wetland function (Cole et al., 1997; Whigham et al., 2007).

Wetland functional assessment is frequently used within wetland regulatory processes to assess what functions might be lost due to a permitted wetland impact and to ensure that these functions are replaced through wetland mitigation. Because of this regulatory aspect, many efforts at wetland functional assessment are targeted at the site scale rather than the landscape or watershed scale. Yet there is increasing recognition of the need for better understanding of wetland function across

Wetland and Stream Rapid Assessments. https://doi.org/10.1016/B978-0-12-805091-0.00021-9

broad geographic areas in support of strategic planning efforts for wetland conservation and restoration. Functional assessments at these broader scales require the use of geospatial data and GIS tools.

The National Wetlands Inventory (NWI) is the most comprehensive geospatial data for wetlands in the United States, but the primary classification system used for the NWI largely relies on plant community (Cowardin et al. 1979), which somewhat limits its utility for functional assessment (Brinson, 1993). To address this shortcoming, Tiner (2002, 2003, 2011, 2014) developed a classification system based on the HGM approach that could be applied to the NWI dataset to enhance the ability to use these data for functional assessment. This classification system includes descriptors for landscape position (relation to a waterbody), landform (physical shape), water flow path (hydrologic connectivity), and waterbody type. The system is known as the LLWW system. In most early efforts, these descriptors were added to the NWI database manually using GIS and interpreting geospatial data for wetlands, topography, watercourses, and aerial imagery; however, there have been recent efforts to automate the process (Dvorett et al., 2012).

Wetland Assessment Efforts in Minnesota

Wetland assessment efforts in Minnesota can be classified according to a "three-tier framework" developed by the EPA, as described by Fennessy et al. (2007). Level I is a landscape-scale assessment using readily available geospatial data. Level II is a localized, field-scale assessment using rapid methods based on simple observational metrics. Level III is a more intensive site assessment based on quantitative biological and physicochemical measurements. To date, most efforts to assess wetland function and condition in Minnesota have generally been of Level II or Level III in this system.

In the mid-1990s, an interagency group in Minnesota began developing a rapid functional assessment method for wetlands called the Minnesota Routine Wetland Assessment Method (MnRAM). The purpose of this method was to provide a practical assessment tool to provide information on wetland function to local authorities charged with making wetland management decisions under the Minnesota Wetland Conservation Act of 1991 (BWSR, 2010). This method relies primarily on easily observable field characteristics and some simple desktop measurements to assess wetland functions. While it is a rapid method, it is not a wetland condition assessment method like that of the California Rapid Assessment Method or the Ohio Rapid Assessment Method (Fennessy et al., 2007).

At approximately the same time the state was developing MnRAM, the Minnesota Pollution Control Agency (MPCA) was developing methods for assessing wetland condition. Much of this effort was focused on developing quantitative indices of biological integrity (IBIs) for depressional wetlands, which is a Level III assessment method (Gernes and Norris, 2006). Genet (2015) reported on depressional wetland condition for the Mixed Wood Plains and Temperate Prairie ecoregions of the state. This analysis was based on a probabilistic sample of 100 depressional wetlands.

Additionally, the MPCA has developed a rapid floristic quality assessment for wetland condition (Bourdaghs, 2012). The floristic quality assessment (FQA) is a vegetation-based ecological condition assessment approach. The FQA combines onsite plant community data with a coefficient of conservatism based on the plant species' fidelity to the specific habitat and tolerance of disturbance.

Conducting a comprehensive wetland assessment for Minnesota's estimated 10.6 million acres of wetlands is a significant logistical challenge. The field-based methods described above can only realistically assess a tiny fraction of the wetlands in the state. The LLWW system will be incorporated into the statewide NWI data for Minnesota; therefore, it fills an important need for broad landscape-scale wetland assessments.

The NWI dataset for Minnesota is being updated through a statewide effort conducted in geographic phases. The updated NWI data include both a Cowardin classification (Cowardin et al., 1979) and a simplified HGM classification based on an adaptation of the LLWW system (Tiner, 2002). For southern Minnesota, which includes the study area for this investigation, the NWI update was conducted using manual photo interpretation of high-resolution (0.5 m), multispectral spring imagery and 3 m resolution digital elevation models derived from LiDAR (Rokus, 2015). Various other ancillary datasets, including SSURGO soils data from the Natural Resources Conservation Service (NRCS) and summer aerial imagery from the U.S. Department of Agriculture (USDA) National Agricultural Imagery Program, were also used. The data inputs for the NWI update are similar to those used for other geographic phases of the project and are described elsewhere (Kloiber et al., 2015). The HGM classifications in southern Minnesota were also assigned using manual photo interpretation, whereas for other geographic regions, the simplified HGM classes were assigned using an automated procedure.

Efforts are currently underway to develop predictive relationships for wetland function, particularly hydrologic function, using the updated and enhanced NWI data. The approach will use methods similar to those described by Dvorett et al. (2010), Hruby et al. (1999), and Tiner (2005). These predictive relationships will be developed through a combination of statistical analyses, modeling, and literature review. The procedures developed from this effort will serve as guidance to natural resource managers for applications including flood analysis, water quality improvement, and wildlife habitat suitability assessment.

Objective

Herein, we describe an automated GIS method for assigning HGM classifications to an existing NWI dataset. These results are compared to HGM classifications assigned using a more traditional photo-interpretation method. An automated GIS HGM classification entails using various hydrological and topographic data along with the NWI to automatically assign HGM codes to the NWI data.

METHODS

Study Area

The study area is the headwater watershed of the Des Moines River, located in southwest Minnesota (Fig. 2.2.5.1). The watershed is 801,772 acres (324,465 ha) and lies within the Loess Prairies and Des Moines Lobe portions of the Western Corn Belt Plains Ecoregion (Omernik, 1987). Approximately 96% of the watershed is under private ownership and is dominated by agricultural land use (NRCS n.d.). Wetlands comprise 9% of the watershed (71,984 acres or 29,131 ha) and are dominated by emergent wetlands (60%), followed by lakes (27%), woody wetlands-forest and shrub (6%), ponds (4%), and riverine (3%).

Classification System

The classification system used for this effort was a simplified HGM classification system based on the LLWW system developed by Tiner (2002, 2003). The system applied here groups the waterbody types for aquatic systems into a component that also includes landforms for wetland systems while also having fewer classes (Table 2.2.5.1).

Input Data

The required input GIS data include: (1) NWI feature class with valid NWI codes, (2) a detailed stream linear reference network data with flow direction, (3) lake basin data (preferably delineated to the ordinary high-water mark), and

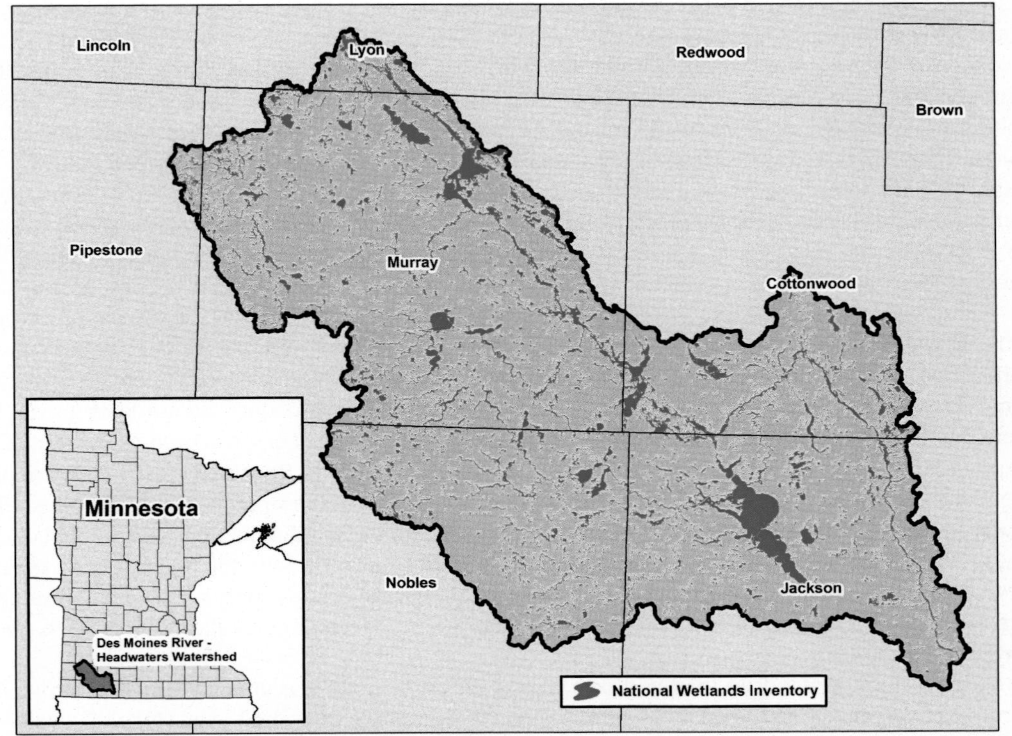

FIG. 2.2.5.1 The study site is located in southwestern Minnesota at the headwaters of the Des Moines River. Wetlands from the National Wetlands Inventory are shown in blue.

TABLE 2.2.5.1 Modified LLWW Classification System

Landscape Position	Landform/Waterbody	Water Flow Path
Lentic (LE) Lotic River (LR) Lotic Stream (LS) Terrene (TE)	Island (IL) Fringe (FR) Floodplain (FP) Basin (BA) Flat (FL) Slope (SL)	Inflow (IN) Outflow (OU) Throughflow (TH) Bidirectional nontidal (BI) Isolated (IS)/vertical (VE)
	Lake (LK) River (RV) Pond (PD)	

(4) a high-resolution digital elevation model (DEM) (preferably derived from LiDAR data). All these datasets were provided by the Minnesota Department of Natural Resources (MDNR) and all are made publicly available through the Minnesota Geospatial Commons (https://gisdata.mn.gov/).

Derived Data

Two additional layers were developed from the detailed stream data. The first layer was a *flooding stream layer* that was used later to define the extent of the lotic landscape component. The flooding stream layer was created by deleting all the drainage ditches, connectors, and small streams from the stream layer. Small headwater streams and artificial channels were generally found not to have developed floodplains. A visual review of the flooding stream layer was conducted and any stream without a visible floodplain was deleted. The second layer was a *complete stream layer* for the water flow path analysis. All streams were used and visually reviewed for accuracy. The visual review of the stream layer digitized any missing stream segment (e.g., where a stream did not completely connect with a lake) or any missed stream that would affect the water flow path analysis.

A landscape position mask was created in order to identify the landscape position component of the HGM class. The mask consists of identifying the lentic and lotic landscape positions and the remaining areas were considered terrene. The lentic and lotic landscape positions were created from the ordinary high-water mark delineation of lakes from the MDNR's *Public Water Inventory (PWI) basin layer* and the flooding stream layer as described below.

To identify the lentic landscape position, a *lentic layer* was derived from the PWI. The PWI basin layer contains both wetlands and open water bodies delineated to the approximate ordinary high-water mark. A spatial selection was done between the PWI layer and all lacustrine wetlands within the NWI. A reverse selection was performed to identify any feature within the PWI that was not associated with a lacustrine wetland within the NWI. These features were reviewed for accuracy and deleted. The remaining PWI basins were then used to create a *lentic landscape position feature class.*

To identify the lotic landscape position, a *riparian area layer* was derived from the flooding stream layer, as described above, and the DEM. A slope grid was first calculated on the DEM layer, which was used to create a fractional slope grid (rise over run). The fractional slope grid was used in conjunction with the flooding stream layer to run a cost-distance analysis where the flooding stream layer is the source feature and the fractional slope is used as the cost. The results from this analysis provide an approximation of the height above the nearest stream. A threshold classification approach was used to select a height above the nearest stream that corresponds to the extent of the riparian area. This threshold was selected based on visual inspection of the DEM, aerial photos, and FEMA floodplain delineation, where available.

A *landscape position classification layer* was made by combining the lentic and lotic layers. Lentic features were given priority over lotic features when they overlapped. Any areas within the project boundary not covered by either lotic or lentic features were assigned to a terrene class. A final inspection of the lotic feature class was performed to remove any small polygons surrounding the lentic features (Fig. 2.2.5.2).

In order to assign slope wetlands to the Landform type, a *2% slope layer* was created using the DEM. A percent slope grid was calculated from the DEM and all areas greater than 2% slope were identified. Any cell with a value of 2% slope or less was reclassified with a value of 1. Any cell with a value of greater than 2% was reclassified to a value of 2.

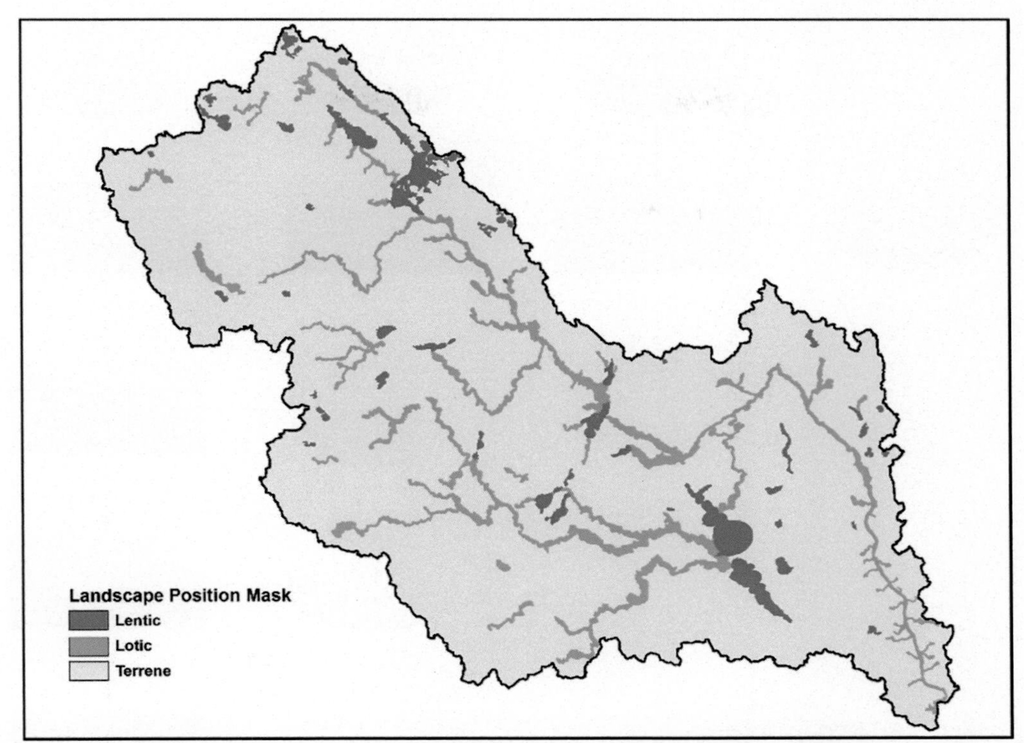

FIG. 2.2.5.2 A landscape position classification layer was created to assign landscape classification to the wetland features within the study area. Lentic areas associated with lake systems are shown in blue. Lotic areas associated with streams and rivers are shown in green. The remaining areas are classified as terrene.

Classification Process

The model inputs were: (1) the NWI layer, (2) a landscape position classification mask, (3) a 2% slope layer, (4) a complete stream layer, and (5) a lentic (lake) layer. The NWI layer included fields for the following attributes: landscape, landform, water flow path, secondary water flow path, and the HGM classification.

The model is composed of six primary steps: (1) assignment of landscape position, (2) assignment of landform classes within the lentic landscape position, (3) assignment of landform classes within the lotic landscape position, (4) assignment of landform classes within the terrene landscape position, (5) assignment of water flow path, and (6) quality control review and assignment of secondary water flow path (Fig. 2.2.5.3).

The landscape position mask was used to assign the lentic, lotic, and terrene landscape positions to all the NWI wetlands. A majority rule was used for wetlands intersecting multiple landscape positions within the mask. The lotic position was further subdivided by using the NWI riverine class (15 ft wide and larger) and assigning those wetlands a lotic river position with any remaining lotic positions being assigned lotic stream. Defining the landscape position restricts the potential landform classes that can be assigned to any given wetland. For example, lentic and lotic landscape positions cannot have slope landforms.

Wetlands within the lentic position were classified into fringe, basin, or flat landform types based on NWI water regime. Wetlands with a semipermanently flooded (F) water regime were assigned the fringe landform type. Wetlands with a seasonally flooded (C) water regime were assigned a basin landform type and any remaining lentic wetlands were assigned the flat landform type.

Wetlands within the lotic position were classified into fringe or floodplain landform types or reclassified into the terrene landscape position. Wetlands with an unconsolidated shore class or semipermanently flooded (F) water regime were assigned the fringe landform type. Wetlands with a temporarily flooded (A) or seasonally flooded (C) water regime were classified as floodplains. All wetlands within the lotic position with a saturated (B) water regime were reclassified to the terrene landscape position. The saturated water regime is characteristic of bog wetlands, which are mostly associated with the terrene landscape position.

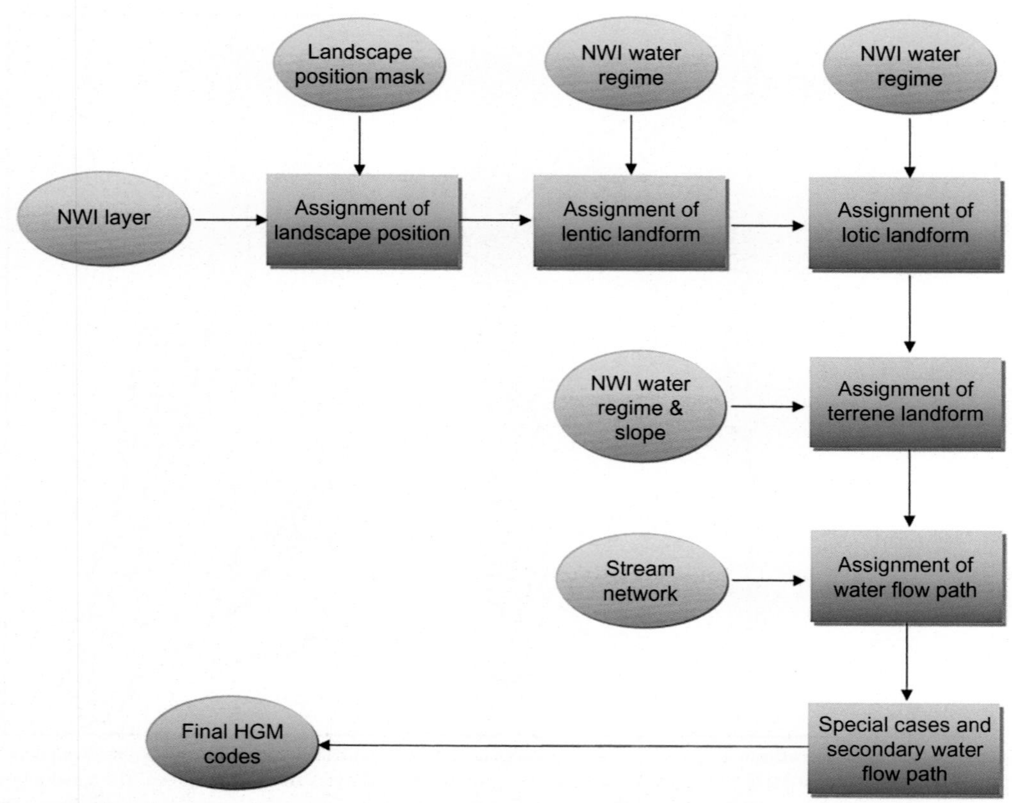

FIG. 2.2.5.3 The process flow diagram for the automated classification procedure shows the main inputs and key processing steps for adding the HGM classification.

Wetlands within the terrene position were classified into fringe, basin, flat, or slope landform types based on the water regime or the 2% slope layer. Wetlands with a semipermanently flooded (F) water regime were assigned the fringe landform type. Wetlands with a seasonally flooded (C) water regime were classified as basins and wetlands with a temporarily flooded (A) water regime were typed as flats. Wetlands with a slope greater than 2% were assigned the slope landform type (these wetlands should also generally have a saturated [B] water regime).

The water flow path was assigned by determining the location of the wetland in relation to the complete stream layer. If a stream ended at the wetland, it was assigned an inflow water flow path. If a stream started at the wetland or if the wetland was adjacent to a riverine wetland, it was classified as outflow. If the stream flowed through the wetland, it was assigned a throughflow water flow path. All wetlands without stream connection were classified with the vertical flow water flow path. All riverine wetlands were assigned a throughflow water flow path.

Finally, the classification process included a step to address some situations that were not adequately defined by the above steps. Any wetland that was surrounded by open water or aquatic bed were assigned the island landform type. In the case of the Minnesota version of the NWI, peatlands were identified through photo interpretation and assigned a special modifier in the Cowardin classification system. Any wetlands with this special modifier in the Cowardin classification system were also assigned the peatland landform type. Any lacustrine open water wetland was classified as a lake landform type and any palustrine open water wetland was assigned the pond landform type. Any riverine open water wetland was assigned the river landform type. Any slope wetland was assigned an outflow water flow path. Any fringe landform type with either a lentic or terrene landscape position was classified with a bidirectional water flow path and any wetland with a peatland modifier was assigned a terrene landscape position and vertical flow water flow path.

RESULTS

All 15,884 NWI wetland polygons within the project area were classified independently by both the automated and manual photo-interpretation methods (Tables 2.2.5.2 and 2.2.5.3). For the landscape position component of the HGM classification, there were 13,340 cases where the landscape position between both methods was in agreement (84%). For the landform

TABLE 2.2.5.2 Comparison of Landscape Position Class Between the Automated and Manual Classification Methods

	Automated Landscape Position Class				
Photo-Interpreted Landscape Position Class	**Lentic**	**Lotic River**	**Lotic Stream**	**Terrene**	**Total**
Lentic	421	9	18	194	642
Lotic River	45	3366	11	114	3536
Lotic Stream	36	42	275	609	962
Terrene	205	753	508	9278	10,744
Total	707	4170	812	10,195	15,884

TABLE 2.2.5.3 Comparison of Landform/Waterbody Class Between the Automated and manual Classification Methods

	Automated Landform/Waterbody Class									
Photo-Interpreted Landform Class	**Basin**	**Flat**	**Floodplain**	**Fringe**	**Island**	**Lake**	**Pond**	**River**	**Slope**	**Total**
Basin	2799	151	636	36	1		1		6	3630
Flat		4688	718							5406
Floodplain	33	76	3185	30	24		1			3349
Fringe		156		111	4				1	272
Island	2				11					13
Lake						117	3			120
Pond							1779			1779
River								28		28
Slope		447							840	1287
Total	2834	5518	4539	177	40	117	1784	28	847	15,884

component of the HGM classification, there were 13,558 cases where the landform class between both methods was in agreement (85%).

In general, the level of agreement between the two classification approaches is good with respect to landscape position and landform; however, there are some notable disagreements. For example, there was some disagreement on classification of lentic systems. The manual classification method only agreed with about 60% of the lentic classes assigned by the automated method and the automated method only agreed with about 66% of the lentic classes assigned by the manual method. The lotic stream classification agreement fared worse with only 34% and 29% agreement, respectively. Most commonly, these disagreements were due to differing estimates of the extent of influence of the lentic and lotic stream landscape positions and where exactly to draw the boundary between these systems and the adjacent terrene position. For the landform component of the classification, the point of disagreement between the two methods was generally confined to the fringe, island, and slope landforms. For example, while the manual method identified nearly all the wetlands with a slope landform

TABLE 2.2.5.4 Comparison of Water Flow Path Class Between the Automated and Manual Classification Methods Using the Primary Water Flow Path From the Automated Method

	Automated Water Flow Path Class					
Manual Water Flow Path Class	Bidirectional	Inflow	Outflow	Throughflow	Vertical Flow	Total
Bidirectional	52		5	32	305	394
Inflow	6	2	3	38	95	144
Outflow	28	25	960	1129	7021	9163
Throughflow	14	4	70	2299	1994	4381
Vertical flow	1	1	8	34	1758	1802
Total	101	32	1046	3532	11,173	15,884

identified by the automated method, the manual photo interpreters also assigned 447 other wetlands to the slope landform that were identified as flat landforms by the automated method. Also notable is the fact that the relative disagreement rates are higher for landform classes that are less frequently occurring, such as the island landform.

The agreement between the two methods diverged more significantly when examining the results of the water flow path component of the HGM classification (Table 2.2.5.4). The overall agreement between the manual photo interpretation of the water flow path and the primary water flow path from the automated method was only 32%. However, the automated classification used here potentially assigned a primary and a secondary flow path, where applicable. For example, if a lake had a fringe wetland, the primary water flow path would be bidirectional between the lake and the fringe. However, if the lake was also connected by a throughflow stream, then the fringe wetland would also receive a secondary flow path of throughflow. If we account for a possible match between the manually assigned water flow path and either the primary or secondary flow path from the automated method, the overall agreement did improve somewhat to 43%.

DISCUSSION

The results from the automated method compare favorably to the results from the manual methods for landscape position (84%) and landform (85%). The agreement between our automated classification results and the manual classification exceeds the classification accuracy reported by Dvorett et al. (2012). They reported 60% classification accuracy using an automated GIS method to apply HGM classes to NWI polygons in Oklahoma. Their accuracy assessment was limited to a four-class system including depressional, impounded depression, riverine, and lacustrine. Our classification scheme has four landscape positions and nine landform classes. The improved results from our effort were likely aided by having an accurate and recently updated NWI dataset based on high-resolution aerial imagery and LiDAR data.

Using automated spatial analysis to perform HGM classification of wetlands is not well documented. Efforts to automate the HGM classification process have been conducted by the Montana Natural Heritage Program (Newlon, 2013). A partially automated process is described by Tiner et al. (2013). However, there are many published efforts to use automated spatial analysis for related efforts, such as riparian area mapping (Baker et al., 2006; Johansen et al., 2011; Fernández et al., 2012) and terrain-based, automated landform classification (Drăguţ and Blaschke, 2006; MacMillan et al., 2003). The procedure described here borrows and builds on the technical basis provided by these previous efforts.

The strength of the automated method is that it is a fast, inexpensive, and repeatable process. However, the accuracy of this method is adversely impacted by inaccuracies and scale issues with input data. For example, an isolated wetland basin classified as "terrene flat outflow" by the photo-interpreted method was classified as "terrene flat vertical flow" by the automated method because the watercourse was not quite digitized to the wetland boundary (Fig. 2.2.5.4). This same issue was also noted by Lang et al. (2015). Using a manual photo-interpretation approach, an experienced photo interpreter can account for the inaccuracies in one or more of the input datasets. However, another potential advantage of the automated method is that it can be relatively easily rerun if improvements are made in one of the input datasets, thus potentially facilitating a faster and cheaper update process for the HGM classification.

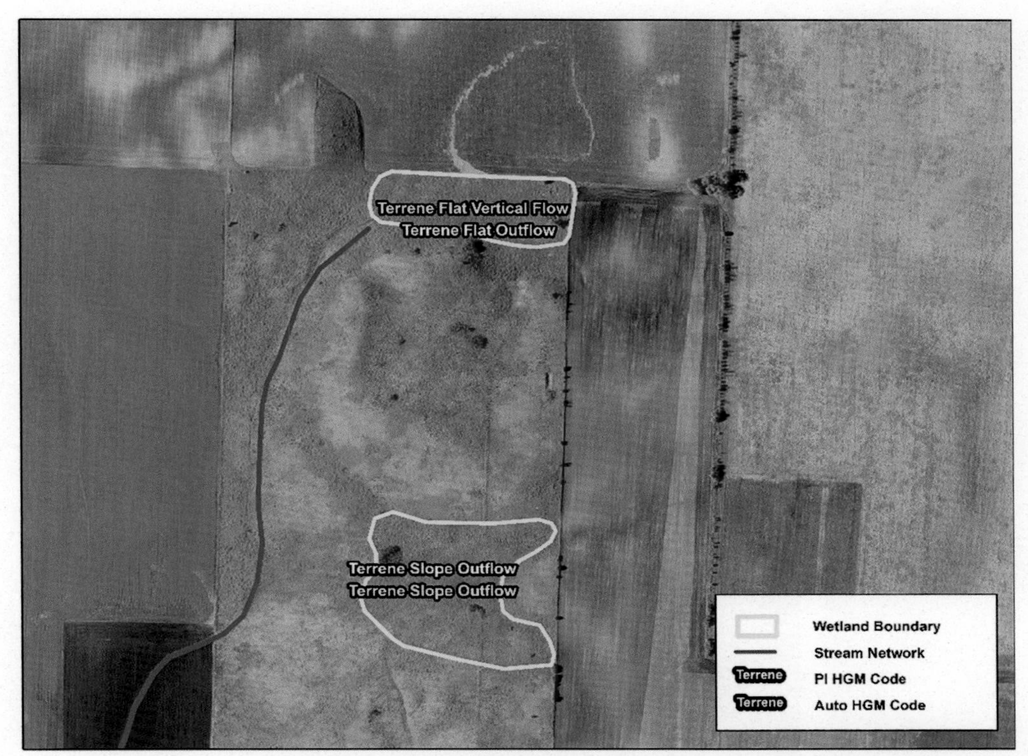

FIG. 2.2.5.4 This example illustrates a case where the stream layer *(blue line)* resulted in an error in the automated water flow path classification *(yellow text)* compared to the photo-interpreted water flow path classification *(white text)*. Because the stream feature doesn't intersect the wetland polygon, the automated classification algorithm classified this feature as vertical flow instead of outflow whereas the photo interpreter deduced that the stream feature was close enough to assume an outflow connection.

If the input data have adequate resolution and are accurate, there are cases where the automated model may perform better than a photo-interpreted classification. For example, we used a LiDAR-derived DEM to produce a 2% slope layer. All wetlands within that 2% layer were classified as slope wetlands using a majority rule. In some cases, the photo-interpreted wetland was classified as a slope wetland when it was clearly not within the 2% slope layer (Fig. 2.2.5.5). However, it is notable in this same example that the stream layer used in the automated classification was not detailed enough to assign the correct water flow path for the automated method.

One enhancement we incorporated into the automated classification method was to also assign a secondary water flow path based on the relationship between the watercourse layer and wetland complexes. The NWI delineates individual wetland polygons for each type within the Cowardin classification system, potentially resulting in multiple wetland polygons within a single wetland complex. As a result, wetland polygons within the complex that do not intersect the stream feature may not be assigned the correct water flow path. Therefore, we added an additional attribute for a secondary water flow path based on the relationship between the watercourse layer and the wetland complex using a GIS operation to aggregate individual NWI polygons into larger wetland complexes. For example, a headwater wetland may be classified as vertical flow in the automated HGM procedure because it doesn't intersect the watercourse layer whereas a manual photo-interpretation approach classified the same wetland polygon as outflow. However, the secondary water flow path for this wetland polygon was outflow due to its adjacency relationship to a pond that does intersect the watercourse layer (Fig. 2.2.5.6). In addition, it is notable that the photo-interpretation method appears to have incorrectly assigned a water flow path of outflow to the downstream wetland while the automated method classified the water flow path as throughflow.

Accuracy Assessment Issues

Conducting a true accuracy assessment for any data requires an agreed-upon reference dataset with a high level of accuracy. In this case, it isn't clear which of the two methods presented has superior accuracy. As such, we can only present our results as a comparison of the level of agreement between the two methods.

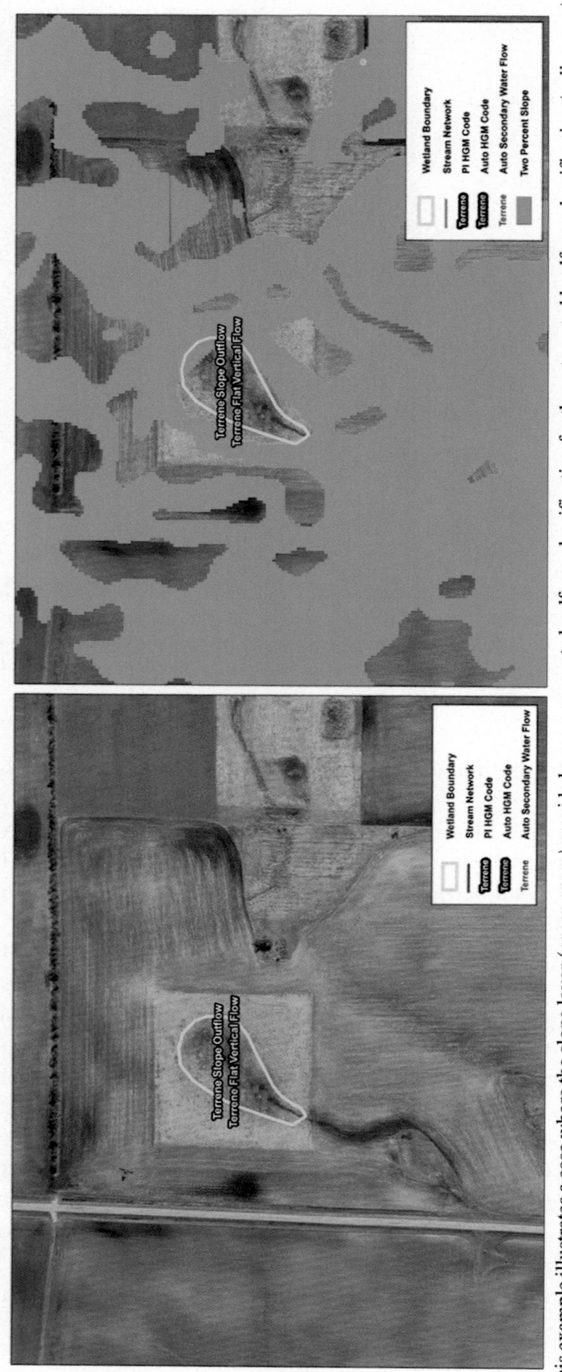

FIG. 2.2.5.5 This example illustrates a case where the slope layer *(green areas)* provided a more accurate landform classification for the automated landform classification *(yellow text)* when compared to the photo-interpreted landform classification *(white text)*.

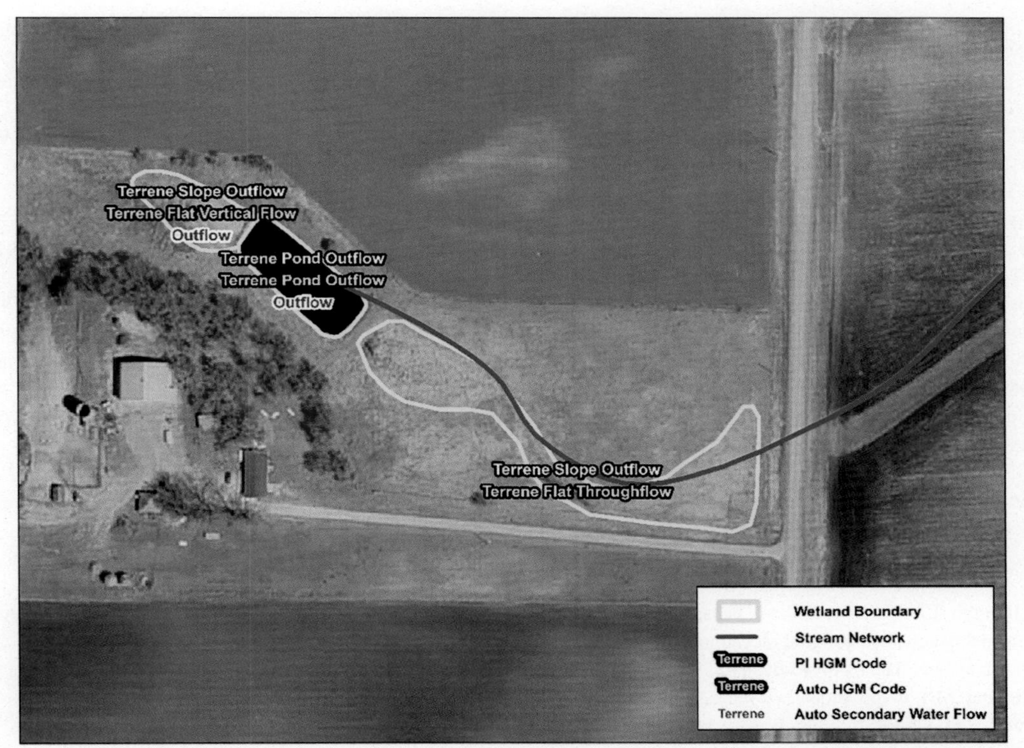

FIG. 2.2.5.6 The secondary water flow path *(blue text)* from the automated classification method occasionally provides a more accurate classification based on the relationship of the larger wetland complex to the stream layer. Note the secondary water flow path (outflow) of the far left wetland that does not intersect with the stream layer but is adjacent to a wetland that does intersect with the full stream layer.

An additional complication in conducting an accuracy assessment for HGM data is the difficulty associated with trying to create an independent set of "ground-truth" data. Unlike systems that classify wetlands solely on intrinsic information like the plant community, the HGM approach and the derived LLWW system require assessment of factors that are external to the wetland (Brinson, 1993). Assessing some of these external factors by site visit is generally not possible. For example, a seasonally flooded forested wetland cannot be accurately classified without understanding its connection to a larger floodplain complex. While there may be some field indicators of flooding, the extent of the floodplain is difficult to discern without additional geospatial data, such as detailed topography data, aerial imagery, and possibly hydrologic watershed modeling studies. This effectively means that there is no such thing as "ground-truth" data for HGM classification. HGM classification requires geospatial contextual data. However, it would be rare for a study to have two completely independent geospatial datasets for classification and validation. Therefore, it is unlikely that the classification data and the validation data are truly independent. We can and have applied two different procedures by two different analysts, but the source data were the same.

CONCLUSION

GIS analysis is well suited to assigning HGM classes due to the spatial relationships of the HGM attributes. There is a demonstrated scientifically sound basis for automated classification of this type; however, the scale and accuracy of the input data will limit the quality and accuracy of the resulting wetland classification. These data will be a potentially important part of future efforts in Minnesota to assess wetland functions at a landscape scale. Efforts are currently underway to develop predictive relationships for wetland function, particularly hydrologic function, using the updated and enhanced NWI data.

REFERENCES

Baker, C., Lawrence, R., Montagne, C., Patten, D., 2006. Mapping wetlands and riparian areas using Landsat ETM+ imagery and decision-tree-based models. Wetlands 26 (2), 465–474.

Bourdaghs, M., 2012. Development of a Rapid Floristic Quality Assessment. Minnesota Pollution Control Agency, St. Paul, MN. 56 pp. https://www.pca.state.mn.us/sites/default/files/wq-bwm2-02a.pdf. Accessed 8 March 2016.

Brinson, M.M., 1993. A Hydrogeomorphic Classification for Wetlands. U.S. Army Corps of Engineers, Waterways Experiment Station, Vicksburg, MS Technical Report WRP-DE-4.

BWSR, 2010. Comprehensive General Guidance for Minnesota Routine Assessment Method (mnRAM) Evaluating Wetland Function, Version 3.4 (Beta). Minnesota Board of Water and Soil Resources, St. Paul, MN. 77 pp. http://bwsr.state.mn.us/wetlands/mnram/MnRAM%20_Comprehensive_Guidance_3.4beta_2010.pdf. Accessed 8 January 2016.

Cole, C.A., Brooks, R.P., Wardrop, D.H., 1997. Wetland hydrology as a function of hydrogeomorphic (HGM) subclass. Wetlands 17 (4), 456–467.

Cowardin, L.M., Carter, V., Golet, F.C., LaRoe, E.T., 1979. Classification of Wetlands and Deepwater Habitats of the United States. US Department of the Interior, US Fish and Wildlife Service, 31 December 1979.

Drăguţ, L., Blaschke, T., 2006. Automated classification of landform elements using object-based image analysis. Geomorphology 81 (3), 330–344.

Dvorett, D., Bidwell, J.R., Davis, C.A., DuBois, C., 2010. An HGM Approach for Assessing Wetland Functions in Central Oklahoma: Hydrogeomorphic Classification and Functional Attributes (Doctoral dissertation). Oklahoma State University. 90 pp. https://www.ok.gov/wetlands/documents/2011.2.11%20HGM%20Classification%20Final%20Report.pdf. Accessed 25 August 2016.

Dvorett, D., Bidwell, J., Davis, C., DuBois, C., 2012. Developing a hydrogeomorphic wetland inventory: reclassifying National Wetlands Inventory polygons in geographic information systems. Wetlands 32 (1), 83–93.

EPA and USACE, 1990. Memorandum of Agreement Between the Environmental Protection Agency and the Department of the Army Concerning the Determination of Mitigation Under the Clean Water Act Section 404(b)(1)Guidelines, 6 February 1990. https://www.epa.gov/cwa-404/memorandum-agreement. Accessed 20 August 2016.

Fennessy, M.S., Jacobs, A.D., Kentula, M.E., 2007. An evaluation of rapid methods for assessing the ecological condition of wetlands. Wetlands 27 (3), 543–560.

Fernández, D., Barquín, J., Álvarez-Cabria, M., Peñas, F.J., 2012. Quantifying the performance of automated GIS-based geomorphological approaches for riparian zone delineation using digital elevation models. Hydrol. Earth Syst. Sci. 16 (10), 3851–3862.

Genet, J., 2015. Status and Trends of Wetlands in Minnesota: Depressional Wetland Quality Assessment (2007–2012). Minnesota Pollution Control Agency, St. Paul, MN. 61 pp. https://www.pca.state.mn.us/sites/default/files/wq-bwm1-08.pdf. Accessed 8 March 2016.

Gernes, M., Norris, D.J., 2006. A comprehensive wetland assessment, monitoring, and mapping strategy for Minnesota. Minnesota Pollution Control Agency and Minnesota Department of Natural Resources. 47 pp. https://www.pca.state.mn.us/sites/default/files/wq-bwm6-03.pdf. Accessed 8 January 2016.

Hruby, T., Granger, T., Brunner, K., Cooke, S., Dublanica, K., Gersib, R., Reinelt, L., Richter, K., Sheldon, D., Teachout, E., Wald, A., Weinmann, F., 1999. Methods for Assessing Wetland Functions Volume I: Riverine and Depressional Wetlands in the Lowlands of Western Washington. WA State Department Ecology Publication #99-115, 476 pp. https://fortress.wa.gov/ecy/publications/summarypages/99115.html. Accessed 25 August 2016.

Johansen, K., Tiede, D., Blaschke, T., Arroyo, L.A., Phinn, S., 2011. Automatic geographic object based mapping of streambed and riparian zone extent from LiDAR data in a temperate rural urban environment, Australia. Remote Sens. 3 (6), 1139–1156.

Kloiber, S.M., Macleod, R.D., Smith, A.J., Knight, J.F., Huberty, B.J., 2015. A semi-automated, multi-source data fusion update of a wetland inventory for east-central Minnesota, USA. Wetlands 35 (2), 335–348.

Lang, M.W., Purkis, S., Klemas, V.V., Tiner, R.W., 2015. Promising developments and future challenges for remote sensing of wetlands. In: Tiner, R.W., Lang, M.W., Klemas, V.V. (Eds.), Remote Sensing of Wetlands: Applications and Advances. CRC Press, Boca Raton, FL.

MacMillan, R.A., Martin, T.C., Earle, T.J., McNabb, D.H., 2003. Automated analysis and classification of landforms using high-resolution digital elevation data: applications and issues. Can. J. Remote Sensing 29 (5), 592–606.

Newlon, K., 2013. Final Report on Enhancing Wetland Classification for the FGDC Wetland Mapping Standard in Montana. https://www.fgdc.gov/grants/2012CAP/InterimFinalReports/144-12-5-MT-FinalReport.pdf. Accessed 28 June 2016.

NRCS. n.d. Rapid Watershed Assessment Des Moines Headwaters (MN) HUC: 07100001. http://www.nrcs.usda.gov/Internet/FSE_DOCUMENTS/nrcs142p2_022280.pdf Accessed 2 June 2016.

Omernik, J.M., 1987. Ecoregions of the conterminous United States. Ann. Assoc. Am. Geogr. 77 (1), 118–125.

Rokus, D., 2015. Technical Procedures for Updating the National Wetland Inventory for Southern Minnesota. St. Mary's University of Minnesota, Winona, MN. 84 pp. http://files.dnr.state.mn.us/eco/wetlands/nwi_smn_technical_documentation.pdf. Accessed 28 June 2016.

Tiner, R.W., 2002. Tiner, R. (Ed.), Enhancing wetlands inventory data for watershed-based wetland characterizations and preliminary assessment of wetland functions. Watershed-Based Wetland Planning and Evaluation: A Collection of Papers from the Wetland Millennium Event (August 6–12, 2000; Quebec City, Quebec, Canada). Association of State Wetland Managers, Inc., Berne, NY, pp. 17–39. https://www.fws.gov/northeast/ecologicalservices/pdf/wetlands/WatershedbasedWetlandPlanningandEvaluation_WetlandsMillineumConference.pdf.

Tiner, R.W., 2003. Keys to Waterbody Type and Hydrogeomorphic-Type Wetland Descriptors for U.S. Waters and Wetlands. U.S. Fish and Wildlife Service, Northeast Region, Hadley, MA. https://www.fws.gov/northeast/ecologicalservices/pdf/wetlands/dichotomouskeys0903.pdf.

Tiner, R.W., 2005. Assessing cumulative loss of wetland functions in the Nanticoke River watershed using enhanced National Wetlands Inventory data. Wetlands 25 (2), 405–419. https://www.fws.gov/wetlands/Documents//Assessing-Cumulative-Loss-of-Wetland-Functions-in-the-Nanticoke-River-Watershed-Using-Enhanced-NWI-Data.pdf.

Tiner, R.W., 2011. Keys to Waterbody Type and Hydrogeomorphic-Type Wetland Descriptors for U.S. Waters and Wetlands: Version 2.0. U.S. Fish and Wildlife Service, Northeast Region, Hadley, MA. https://www.fws.gov/northeast/ecologicalservices/pdf/wetlands/DichotomousKeys_090611wcover.pdf.

Tiner, R.W., 2014. Keys to Waterbody Type and Hydrogeomorphic-Type Wetland Descriptors for U.S. Waters and Wetlands: Version 3.0. U.S. Fish and Wildlife Service, Northeast Region, Hadley, MA. https://www.fws.gov/wetlands/Documents/Dichotomous-Keys-and-Mapping-Codes-for-Wetland-Landscape-Position-Landform-Water-Flow-Path-and-Waterbody-Type-Version-3.pdf.

Tiner, R.W., Herman, J., Roghair, L., 2013. Connecticut Wetlands: Characterization and Landscape-Level Functional Assessment. Prepared for the Connecticut Department of Environmental Protection, Hartford, CT. U.S. Fish and Wildlife Service, Northeast Region, Hadley, MA. http://www.ct.gov/deep/lib/deep/water_inland/wetlands/ct_wetland_characterization-functional_assessment_final-report_nov_2013.pdf.

Whigham, D.F., Deller Jacobs, A., Weller, D.E., Jordan, T.E., Kentula, M.E., Jensen, S.F., Stevens Jr., D.L., 2007. Combining HGM and EMAP procedures to assess wetlands at the watershed scale—status of flats and non-tidal riverine wetlands in the Nanticoke River watershed, Delaware and Maryland (USA). Wetlands 27 (3), 462–478.

Zedler, J.B., Kercher, S., 2005. Wetland resources: status, trends, ecosystem services, and restorability. Annu. Rev. Environ. Resour. 30, 39–74.

FURTHER READING

Tiner, R.W., 2015. Classification of wetland types for mapping and large-scale inventories. In: Tiner, R.W., Lang, M.W., Klemas, V.V. (Eds.), Remote Sensing of Wetlands: Applications and Advances. CRC Press, Boca Raton, FL, pp. 19–41 (Chapter 2).

Chapter 2.2.6

Developing a Functional Classification for the Wetlands of Colorado's Southern Rockies

Jeremy Sueltenfuss* and Joanna Lemly†

**Colorado State University, Fort Collins, CO, United States, †Colorado Natural Heritage Program, Fort Collins, CO, United States*

Chapter Outline

INTRODUCTION

Wetlands throughout the Mountain West provide essential ecosystem services, despite the small fraction of the landscape they comprise. In Colorado, wetlands store carbon, maintain streamflow, improve water quality, and provide habitat for a high percentage of the state's native plant and animal species (Culver and Lemly, 2013). High-elevation wetlands range from peat-accumulating fens where groundwater sustains saturated conditions throughout the year to wet meadows and shrub thickets adjacent and hydrologically connected to rivers and streams. Headwater wetlands, in particular, buffer excess streamflow following snowmelt and absorb high flows to attenuate downstream floods. However, as is true for many wetlands around the world, Colorado's wetlands have experienced a long history of degradation and conversion.

In Colorado's mountain valleys, anthropogenic alteration of wetlands began in the 1800s with the expanding fur trade. For millennia before European exploration and settlement, willow-dominated floodplains throughout the mountains were maintained by the work of beavers (*Castor canadensis*). The construction and maintenance of beaver dams over long periods of time profoundly shaped the geomorphology and vegetation of mountain valleys. Beaver dams spread water across floodplains, retaining snowmelt long into the summer months. Sediment deposited behind the dams built up floodplain elevations far above what would have been possible from flooding alone (Westbrook et al., 2011). The eradication of beavers through trapping in the early to mid-1800s converted many lower-energy, multithreaded river systems once impounded by beaver dams into higher energy, single-threaded channels lacking dams. As a result, many Colorado rivers became hydrologically decoupled from their historic floodplain. The new channels became incised, with banks dominated by drier herbaceous vegetation, prone to erosion. In the absence of beavers, disconnected floodplains support less wetland habitat and no longer provide the same hydrologic functions of floodplains with hydrologic connectivity to the river (Westbrook et al., 2006; Pollock et al., 2014).

Following beaver removal by trapping, early European settlers in the late 1800s began to clear woody vegetation and irrigate many mountain valleys to grow hay and forage for livestock. Water diverted from mountain streams and rivers was applied to floodplains to augment naturally occurring groundwater. While removing beavers narrowed the active floodplains, irrigation pulled water back up onto the old floodplains, but with different timing and duration and very different vegetation as a result. In some areas of Colorado, irrigation was applied to former or existing wetlands while in other areas, irrigation created new wetlands on former dry land (Sueltenfuss et al., 2013). The changing needs of water users continue to

Wetland and Stream Rapid Assessments. https://doi.org/10.1016/B978-0-12-805091-0.00022-0

impact Colorado wetlands to the present day. After a century or more of irrigated agriculture in the mountain valleys, many high-elevation water rights are being purchased by urban water districts to supplement municipal supplies. Thanks to a misguided rule in Colorado's water law, the transfer of a water right requires more from a rancher than simply not diverting water. Instead, the rancher must prove the water is no longer in use by fully draining previously irrigated land. While at lower elevations the land quickly dries out once irrigation water is no longer applied, many previously irrigated fields in high mountain valleys were originally naturally occurring floodplain wetlands with a naturally high groundwater table. These existing wetlands are therefore required to be ditched and drained to satisfy water right transfers. Through the removal of beavers, conversion to irrigated agriculture, and now the transfer of water rights downstream, the wetland resources of Colorado's mountain valleys have been highly altered and no longer provide the same suite of services they once did.

Colorado's mountains contain the headwaters of rivers used far beyond the state borders, and management decisions made here have a disproportionate effect downstream. To analyze the services provided by Colorado's wetlands, the Colorado Natural Heritage Program (CNHP) at Colorado State University has worked to characterize and evaluate existing wetland resources at the watershed scale. As a starting point, original wetland maps created in the 1970s and 1980s by the National Wetlands Inventory (NWI) program of the U.S. Fish and Wildlife Service (USFWS) were digitized and provided online, along with an ancillary dataset.[1] For specific priority watersheds, newly updated digital maps were created from high-quality aerial imagery following the Federal Geographic Data Committee's national wetland mapping standards (FGDC, 2009). All wetlands mapped in Colorado were attributed according to the most recent classification used by NWI, which is based on Cowardin et al. (1979). Although it is the national wetland classification standard for mapping nationwide, the Cowardin classification does not provide information on wetland functions. To supplement the Cowardin attribution, CNHP identified the Landscape, Landform, Water Flow Path, and Water Body (LLWW) classification (Tiner, 2003, 2014) as the most appropriate classification for landscape-level functional assessments. Through the generous support of an Environmental Protection Agency (EPA) Region 8 Wetland Program Development Grant as well as support from the Colorado Department of Transportation, the LLWW classification was modified and adapted for use in the Southern Rockies of Colorado. As a demonstration of its utility, the Southern Rockies LLWW (SR-LLWW) was applied to one high-elevation watershed in central Colorado, the Arkansas Headwaters Subbasin. Data from the Southern Rockies LLWW in this watershed will help agencies, nonprofit watershed groups, land trusts, and restoration practitioners to prioritize conservation and restoration decisions based on wetland functions.

METHODS

Developing an LLWW Classification for the Southern Rockies

The LLWW classification was originally developed to add functional attributes associated with the widely used hydrogeomorphic (HGM) classification (Brinson, 1993) to NWI data (Tiner, 2003). Development of an LLWW classification specific to the Southern Rockies started with this same premise, by defining the major HGM classes in Colorado's Southern Rocky Mountains, then crafting an LLWW-based classification scheme to tease them apart (Table 2.2.6.1). Wetland types vary across the country, necessitating a certain level of regionalization of the LLWW framework. Although the most recent iteration of the LLWW classification (Tiner, 2014) has become more inclusive to western landscapes, it is also more complex. CNHP decided to develop an LLWW classification specific to the Southern Rockies landscape in an effort to simplify and specify the classification for Colorado's needs.

Because of prominent differences in wetland hydrology, ecology, and geography in western landscapes compared to the eastern United States, the regional LLWW classification for the Southern Rockies differs from the national LLWW in a few ways:

- There are no marine or tidal systems within the Southern Rockies landscape, so their removal allows for a more streamlined classification.
- Instead of separating "lotic stream" and "lotic river" into two landscape positions, the SR-LLWW uses only one general "lotic" landscape position and includes the Strahler order for each stream segment, which can be obtained from the NHP+ dataset. Tiner (2014) recommends using the location where a mapped river changes from a single line to a double line/polygon on a 1:24,000 topo map to demarcate the split between lotic stream and lotic river. Although there are real

1. To view the Colorado Wetland Inventory online mapping tool, please see: www.cnhp.colostate.edu/cwic/location/viewSpatialData.asp.

TABLE 2.2.6.1 HGM, NWI, and SR-LLWW Codes for Different Wetland Types in Colorado

HGM	Wetland Description	NWI Class	Landscape	Landform	Water Flow Path
Riverine	Floodplain wetland	R3US/PEM/PSS/PFO/PUS/Pf	LO	FP/BA/FR/IS	TH
Slope	Fen, wet meadows, alpine seep	PEM/PSS/PFO/Pf	TE	SL	TH/OU/IN/VF
Depression	Vegetated depression	PEM/PFO/PSS/Pf	TE	BA	TH/OU/IN/VF
	Pond not in river system (natural or manmade)	PAB/PUB/PUS	TE	BA	TH/OU/IN/VF
	Pond in river system (oxbow, meander scar)	PAB/PUB	LO	BA	TH
Lacustrine Fringe	Natural and manmade lake fringe	L2US/L2EM/PSS/PFO/PUB/PAB	LE	FR/BA/IS	BI/TH
Irrigated Ag	Irrigated meadows, shrublands	PEM/PSS/Pf	LO	BA/SL	IN

ecological and functional differences between streams and rivers, identifying the correct line where a stream becomes a river is challenging in the field and almost impossible from imagery. Using the Strahler order provides a consistent approach to classifying stream reaches and adds more detail than the two classes of lotic stream and lotic river.

- To ensure clarity, the SR-LLWW specifies which landforms and flow paths can occur with each landscape position. This involved a simplification of the keys described by Tiner (2014) as well as the creation of keys specific to each landscape position. This was done in an effort to decrease the number of potential code combinations and eliminate code combinations that do not exist in the Southern Rockies landscape. The hydrology of mountain watersheds is snowmelt-driven, with significant groundwater discharge leading to the formation of wetlands at the base of high-elevation slopes or along the margins of valleys. Because many of these wetlands can be adjacent to streams, one might initially classify them within the lotic landscape position (riverine HGM class). However, they are maintained by groundwater discharge and not overbank flooding, so should instead be classified within the terrene landscape position (slope HGM class). Compared to earlier iterations of the LLWW guidance documents, Tiner (2014) recognizes this wetland type as important in western landscapes. Our use of this code is not different than Tiner (2014), but it is highlighted as a primary wetland type in high-elevation valleys.
- The dependence on snowmelt also creates strong seasonal fluctuations in the water table of wetlands and the flow of channels. Many ephemeral or intermittent channels flow in the spring due to snowmelt, but are dry most of the year. Where the channel gradient is low, wetlands can form within the banks of these otherwise dry channels but do not have an obvious code in the national LLWW key. The SR-LLWW key specifies wetlands within channels as lotic basins.

Finally, the role of irrigation on wetlands across the west is dramatically different than in eastern regions of the United States. Because of the regional aridity, agriculture largely depends on the diversion of streams or rivers for the delivery of irrigation water to otherwise dry land. The application of irrigation water often results in the incidental creation of wetland ecosystems where dry land previously existed. This novel type of wetland does not lend itself to easy placement within existing classifications. Because water is delivered from a flowing canal to a sloping section of irrigated land, the SR-LLWW classifies irrigation-maintained wetlands as lotic slope wetlands. In natural settings, the classifiers of lotic and slope would not co-occur, as slope wetlands are defined as being recharged by groundwater.

Coding for Wetlands and Waterbodies

Similar to the national LLWW classification, the Southern Rockies LLWW is meant to supplement NWI data. Newly mapped wetlands and waterbodies are first attributed using the Cowardin code. The next step in applying the SR-LLWW classification is to pull out waterbodies based on their Cowardin code and attribute them with an SR-LLWW waterbody type (river/lake/pond), water flow path, and, if necessary, a modifier. Lakes and ponds are further attributed with a landscape position.

TABLE 2.2.6.2 Southern Rockies LLWW Classification for Wetlands

Wetland = Landscape + Stream Order (LO only) + Landform + Flow Path + Modifier				
Landscape Position	**Stream Order (Lotic Only)**	**Landform**	**Flow Path**	**Modifier**
Lentic		Fringe (FR) Island (IS)	Bidirectional (BI) Through flow (TH)	Headwaters (hw) Beaver (b) Excavated (x) Impounded (h)
Lotic	Strahler order	Fringe (FR) Island (IS) Basin (BA) Floodplain (FP) Slope (SL)[a]	Though flow (TH) Inflow (IN)[a]	Headwaters (hw) Beaver (b) Excavated (x) Impounded (h) Farmed (f) Drained (d) Intermittent (int) Confined/unconfined valley (cv/uv) Intentionally irrigated (ii) Unintentionally irrigated (ui) Pond fringe (pd)
Terrene (TE)		Slope (SL) Basin (BA)	Vertical flow (VF) Inflow (IN) Outflow (OU) Through flow (TH)	Headwaters (hw) Beaver (b) Excavated (x) Impounded (h) Farmed (f) Drained (d) Pond fringe (pd) Peat accumulating (pt)

[a]*Lotic Slope and Lotic Inflow are only used for wetlands created from irrigation.*

Wetland polygons are then attributed with a landscape position, landform, and water flow path, and often a modifier (Table 2.2.6.2). Lotic wetlands are further described with their Strahler stream order. To make realistic code combinations obvious, the Southern Rockies LLWW classification defines the specific landforms, flow paths, and modifiers that are available for each landscape position. Though progress has been made to automate some of these classifiers, most codes are manually attributed. The exceptions are modifiers for valley confinement (cv/uv) and headwaters (hw). Unconfined valleys are automatically assigned using the valley confinement algorithm (VCA) developed by the U.S. Geological Survey (USGS) to identify unconfined valleys (Nagel et al., 2014). All lotic wetlands coincident with polygons created by the VCA are attributed with the unconfined valley modifier, and the rest are attributed as a confined valley. The headwater modifier is used on wetlands and waterbodies above an a priori determined elevation, above which wetlands are thought to provide disproportionate water quality and quantity benefits to the watershed.

Developing Functional Crosswalks

The primary intent of including the LLWW classification in the mapping of Colorado wetlands is to characterize the functional profile of the state's wetland resources (Tiner, 2003). This is predicated on having a functional crosswalk between LLWW codes and particular functions of interest. Using the wetland ecology literature and reports on the use of LLWWs in other parts of the country (Richtman et al., 2012; Vance et al., 2006), a crosswalk was developed between SR-LLWW codes and specific functions (Table 2.2.6.3). The functions characterized in this analysis included streamflow maintenance, surface water retention, nutrient cycling, carbon sequestration, temperature maintenance, sediment retention, shoreline stabilization, fish habitat, waterfowl habitat, shorebird habitat, amphibian habitat, and native plant maintenance. The longer-term goal of this effort is to solicit input from regional experts to refine the crosswalk between specific codes and the level of provisioning for these services. While this has not occurred to date, the preliminary crosswalk can be used to estimate the level of provisioning (high or moderate) of each function from individual wetland polygons.

TABLE 2.2.6.3 Crosswalk Between SR-LLWW and NWI Codes and the Level of Provisioning of Streamflow Maintenance

Streamflow Maintenance											
Functional Level	Wetland Codes										
		LLWW						NWI			
	Waterbody	Landscape Position	Stream Order	Landform	Waterflow Path	LLWW Modifier	CON	System	Class	Water Regime	NWI Modifier
High		LE		FR	TH						
		LO		FP			*And*		SS	*Not* A	*Not* x, d, h
									FO		
		TE		SL	OU	hw	*And*		EM	*Not* A	*Not* x, d, h
				BA	TH				SS		
									FO		
								#	#	#	b
		LE			BI	hw	*And*		EM		
					TH				SS		
									FO		
		LO		BA	OU	hw				*Not* A	*Not* x, d, h
					TH						
	PD				TH						
	LK				OU						
Moderate		LE			BI		*And*		EM		
					TH				SS		
									FO		
		LO		BA	OU					*Not* A	*Not* x, d, h
					TH						

Continued

TABLE 2.2.6.3 Crosswalk Between SR-LLWW and NWI Codes and the Level of Provisioning of Streamflow Maintenance—cont'd

Functional Level	Streamflow Maintenance										
	Wetland Codes										
		LLWW						NWI			
	Waterbody	Landscape Position	Stream Order	Landform	Waterflow Path	LLWW Modifier	CON	System	Class	Water Regime	NWI Modifier
		LO		FR			*And*	#	EM		
				IS					SS		
									FO		
		LO		FP			*And*	#	SS	A	*Not* x, d, h
									FO		
		LO		FP			*And*	#	EM	#	*Not* x, d, h
		TE		SL	OU		*And*	#	EM	*Not* A	*Not* x, d, h
				BA	TH				SS		
									FO		

CASE STUDY: APPLYING THE SOUTHERN ROCKIES LLWW TO THE ARKANSAS HEADWATERS

The Arkansas Headwaters served as a case study for testing and optimizing the Southern Rockies LLWW. The study area encompassed the majority of the Arkansas Headwaters Subbasin (HUC 11020001), located in the Southern Rocky Mountains of Colorado west of Colorado Springs (Fig. 2.2.6.1) and extended into neighboring watersheds based on the outlines of topographic quadrangles. The study area is characterized by a high-elevation valley bordered by steep mountains. Three of the tallest mountains in Colorado are located within the watershed, including Mount Elbert (14,439 ft), Mount Massive (14,428 ft), and Mount Harvard (14,421 ft). The watershed is the headwaters for the Arkansas River, which flows east out the mountains onto Colorado's High Plains and eventually discharges into the Mississippi River. Within the study area, Arkansas River water is largely derived from the melting of winter snowpack from the Sawatch and Mosquito ranges, and flows peak in the late spring and early summer; however, transbasin diversions augment local streamflow with water from the neighboring Colorado River basin. The economy of the study area is dominated by recreation and tourism. The river draws fisherman, boaters, and recreationalists of many types to the valley during the summer months. Hay fields on the valley floor also support local cattle operations. In the very upper reaches of the subbasin, mining historically played a major role in shaping the economy and the landscape.

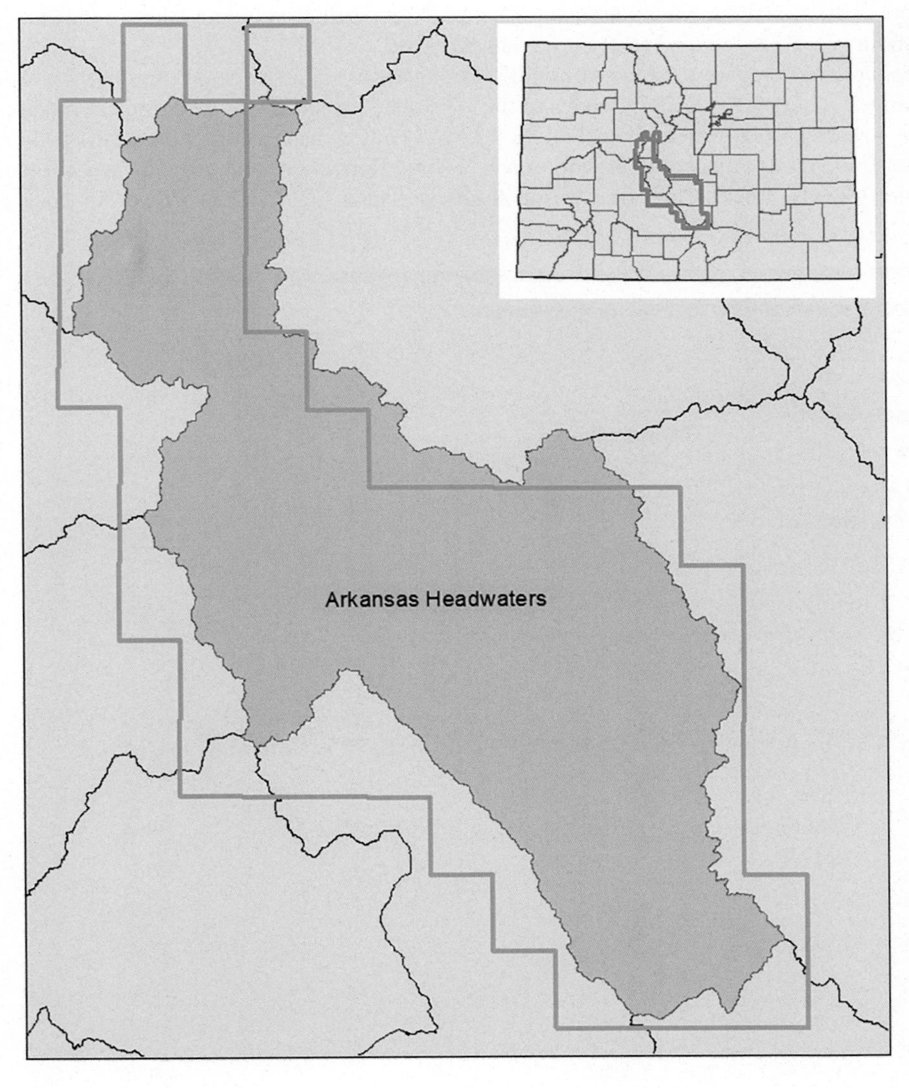

FIG. 2.2.6.1 Study area (outlined in *blue*) encompassing most of the Arkansas Headwaters in Colorado.

Methods

Wetlands and waterbodies within the study area were mapped in ESRI ArcGIS 10.2 at a scale of 1:4500 based on photo interpretation of 2011 National Agriculture Imagery Program (NAIP) color infrared (CIR) imagery. Photo-interpretation methods followed FGDC standards for wetland mapping (FGDC, 2009). Polygons were also created for nonwetland riparian features, following the USFWS riparian mapping classification (USFWS, 2009). To assist in wetland identification, many supplementary datasets were used, including National Agriculture Imagery Program images from 2005, 2009, and 2013, USGS topographic maps, soil survey data, hydrography data, and land-use data (Lemly et al., 2016). After wetlands were coded using the most recent coding rules of the NWI classification system, they were then coded according to the Southern Rockies LLWW as described in the previous section.

Results

Mapped aquatic resources in the Arkansas Headwaters were dominated by actual wetlands, with waterbodies (lake, river, and streams) and nonwetland riparian areas making up only a smaller fraction of the total mapped area (Fig. 2.2.6.2). Of the mapped wetlands, herbaceous and shrub-scrub wetlands were much more prevalent in this landscape than forested wetlands or ponds. Similar to other mountain valleys in Colorado, the primary wetland types in the study area were extensive snowmelt and groundwater-fed wetlands in the alpine and subalpine zones, expansive hay pastures in the valley bottoms, and willow-dominated stream corridors. Although there were a number of naturally occurring mountain lakes in the landscape, the largest lakes were human-built reservoirs created to store transbasin water.

These NWI data are a valuable resource to compare to national trends, although in isolation they do not describe the location or ecological processes performed. Adding SR-LLWW codes to the NWI data provides a more complete picture of the wetlands and waterbodies in the Arkansas Headwaters. Based on SR-LLWW coding, herbaceous wetlands (PEM in the Cowardin code) were split between floodplains adjacent to the larger rivers (lotic floodplain) and groundwater-driven slope wetlands (terrene slope) (Table 2.2.6.4). Results are similar for shrub wetlands (PSS in the Cowardin code).

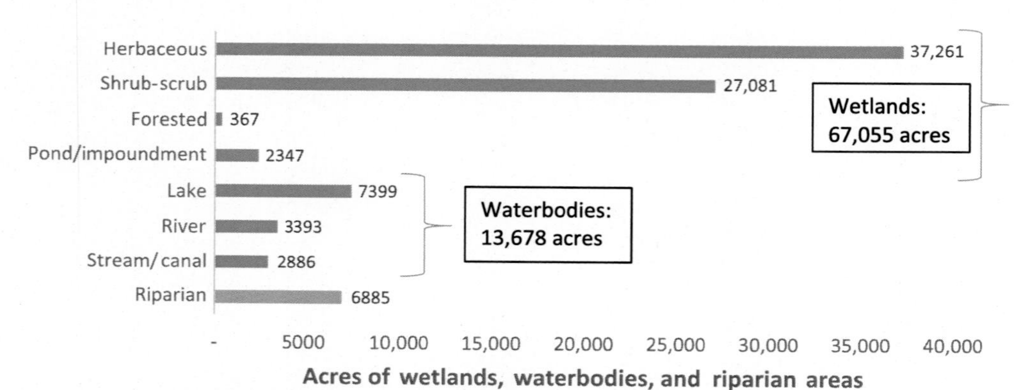

FIG. 2.2.6.2 NWI wetland *(green)*, waterbody *(blue)*, and riparian *(brown)* acres mapped in the study area.

TABLE 2.2.6.4 Mapped Wetland Acres in the Arkansas Headwaters, With Polygons Split by Both the Cowardin and SR-LLWW Codes

	PEM	PSS	PFO	(P/R/L)US	Total
LEFR	219	143		331	693
LOBA	1307	1084	68	29	2489
LOFL	16,517	10,654	190	17	27,378
LOFR				231	231
TESL	16,195	14,199	82	4	30,480
Total	34,237	26,080	340	612	61,270

TABLE 2.2.6.5 Level of Provisioning (Acres) of all Wetlands and Waterbodies in the Project Area for Each of 12 Functions

	Level of Provisioning (Percent of Total Acres)		
Functions	**High**	**Moderate**	**Total**
Surface water retention	27%	49%	76%
Shoreline stabilization	18%	38%	56%
Amphibian habitat	15%	37%	52%
Carbon sequestration	29%	23%	51%
Nutrient cycling	44%	7%	51%
Streamflow maintenance	34%	14%	47%
Native plant maintenance	32%	14%	46%
Fish habitat	27%	15%	42%
Sediment retention	30%	7%	37%
Waterfowl habitat	13%	23%	36%
Temperature maintenance	21%	4%	25%
Shorebird habitat	2%	18%	20%

Approximately half the shrub wetlands are within floodplains with the other half located in slope wetlands sustained by groundwater discharge (Terrene Slope).

The refined description of wetland types by combining the NWI codes with the SR-LLWW codes also allows for the approximation of wetland functions across the landscape. Although this can be subjective and based on best professional judgment, it is a more informed approach to functional approximation than simply using the NWI codes in isolation, and describes the hydrologic connectivity between and among wetlands. Surface water retention, shoreline stabilization, amphibian habitat, carbon sequestration, and nutrient cycling were the highest functions provided by wetlands and waterbodies in the Arkansas River Headwaters (Table 2.2.6.5). Waterfowl habitat, temperature maintenance, and shorebird habitat were the lowest functions provided.

Amphibian habitat, nutrient cycling, and carbon sequestration are all expected to be high in high-elevation watersheds. Terrene slope wetlands in the subalpine are hydrologically sustained by groundwater discharge continuously throughout the year, providing ideal habitat for various amphibian species, particularly the Boreal Toad (*Anaxyrus boreas boreas*), a species of concern within the Southern Rockies. These saturated conditions also lead to both the accumulation of peat, which sequesters carbon, and a significant cycling of nutrients. The high acreage of surface water retention as well as shoreline stabilization is likely an artifact of the increased waterbody acreage from the creation of reservoirs. These functions would not be expected to be as high in high-elevation watersheds without reservoir creation, although a historical analysis would be needed to confirm this. Although waterfowl and shorebird habitat were low, high-elevation watersheds are not well known for providing habitat for these avian assemblages. Many of these species prefer to occupy open water and shoreline habitat within the plains.

One function that may have been higher in the past is temperature maintenance. Streams and rivers lined with woody vegetation maintain cooler temperatures because of shading, which can be of high importance to native fish species (Sweeney et al., 2004). Because much of the woody vegetation has been removed adjacent to streams and rivers for agricultural practices, the provisioning of temperature maintenance is low, and rivers and streams may be warmer than in the past (Fig. 2.2.6.3). As stream temperatures are important for trout fisheries, streamside wetlands may therefore be an important restoration priority in this mountain watershed.

When using these data for management decision-making, it is important to note that some functions are expected to be higher than others in different landscapes. The goal should not be to have every acre in high levels of provisioning in every category, as some categories are mutually exclusive to others. High-elevation watersheds, for example, often do not have significant amounts of shoreline, leading to low levels of shorebird habitat. Viewing this as something to fix and "restore"

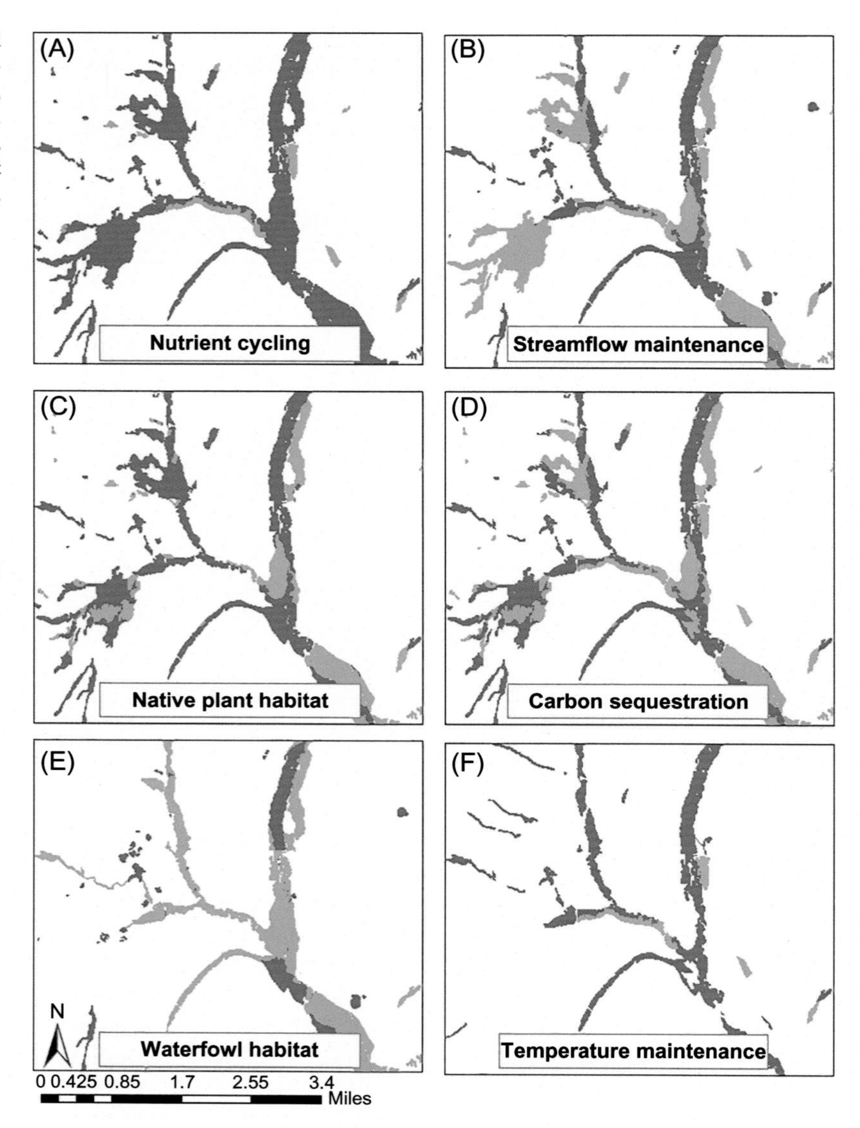

FIG. 2.2.6.3 Mapped wetlands and their associated level of provisioning for six different ecosystem functions. *Dark green* denotes high levels of provisioning, *light green* denotes moderate levels of provisions. Ecosystem functions include nutrient cycling (A), streamflow maintenance (B), native plant habitat (C), carbon sequestration (D), waterfowl habitat (E), and temperature maintenance (F).

would be erroneous and potentially ecologically disastrous. In an effort to increase shoreline, agencies might clear vegetation from streamsides and lakes, lowering the provisions of shoreline stabilization and sediment retention, two functions of high interest in many watersheds. However, the information can highlight areas in the watershed where the level of services provided are lower than other areas, for instance, stream reaches that lack riparian vegetation and, therefore, the services of temperature maintenance and shoreline stabilization.

COLORADO'S WATERSHED PLANNING TOOLBOX (CURRENT USE OF SR-LLWW)

The creation of the Southern Rockies LLWW classification was part of a larger project to create a watershed planning toolbox for use in Colorado. With assistance from an EPA Region 8 Wetland Program Development Grant and support from the Colorado Department of Transportation, CNHP intends to provide wetland restoration professionals, agencies, local governments, and interested citizens with an online watershed planning toolbox. As of the publication of this text, the toolbox is under development but will be a free online web-mapping portal that includes (1) existing wetland mapping data, (2) ecosystem services provided by existing wetlands, (3) sites suitable for wetland restoration, and (4) the potential ecosystem services provided by restored wetlands. While the toolbox is designed to grow into a statewide watershed planning tool, the Arkansas River Headwaters was the first iteration to serve as a pilot test of the concept. Once complete, organizations will be able to access and view the data online. Potential uses could include the identification of existing wetlands needing wetland protection, or wetlands in a degraded state for restoration. The Colorado Department of Transportation is interested in this effort to assist in their wetland mitigation requirements. Roads throughout the state often go through wetlands, requiring mitigation for the wetland impacts. Finding sites suitable for restoration to accomplish their mitigation requirements can be expensive and difficult without an easily accessible planning tool. The intention of this work is to provide enough data for organizations to identify conservation and restoration opportunities, and to prioritize those opportunities based on the functions needing to be improved across the landscape.

REFERENCES

Brinson, M.M., 1993. A Hydrogeomorphic Classification for Wetlands. U.S. Army Corps of Engineers, Waterways Experiment Station, Vicksburg, MS. Technical Report WRP-DE-4.

Cowardin, L.M., Carter, V., Golet, F.C., LaRoe, E.T., 1979. Classification of Wetlands and Deepwater Habitats of the United States. U.S. Fish and Wildlife Service, Washington, DC. FWS/OBS-79/31.

Culver, D.R., Lemly, J.M., 2013. Field Guide to Colorado's Wetland Plants. Identification, Ecology, and Conservation. Colorado Natural Heritage Program, Colorado State University, Fort Collins, CO.

Federal Geographic Data Committee (FGDC), 2009. Wetlands Mapping Standard. Report # FGDC-STD-015 2009, Federal Geographic Data Committee Wetlands Subcommittee. Available online: http://www.fgdc.gov/standards/projects/FGDC-standards-projects/wetlands-mapping/2009-08%20FGDC%20Wetlands%20Mapping%20Standard_final.pdf.

Lemly, J., Long, L., Smith, G., Sueltenfuss, J., 2016. National Wetland Inventory Mapping of the Arkansas Headwaters Subbasin. Report prepared for U.S. Environmental Protection Agency, Colorado Natural Heritage Program, Colorado State University, Fort Collins, CO.

Nagel, D.E., Buffington, J.M., Parkes, S.L., Wenger, S., Goode, J.R., 2014. A Landscape Scale Valley Confinement Algorithm: Delineating Unconfined Valley Bottoms for Geomorphic, Aquatic, and Riparian Applications. Gen. Tech. Rep. RMRS-GTR-321, USDA Rocky Mountain Research Station, Fort Collins, CO.

Pollock, M.M., Beechie, T.J., Wheaton, J.M., Jordan, C.E., Bouwes, N., Weber, N., Volk, C., 2014. Using beaver dams to restore incised stream ecosystems. Bioscience 64, 279–290.

Richtman, C., Anderson, J., Robertson, A., Rokus, D., 2012. Description of Existing Wetland Resources in the St. Croix River Headwaters Watershed. Report prepared for the U.S. Army Corps of Engineers, GeoSpatial Services, Saint Mary's University, Winona, MN.

Sueltenfuss, J.P., Cooper, D.J., Knight, R.L., Waskom, R.M., 2013. The creation and maintenance of wetland ecosystems from irrigation canal and reservoir seepage in a semi-arid landscape. Wetlands 33 (5), 799–810.

Sweeney, B.W., Bott, T.L., Jackson, J.K., Kaplan, L.A., Newbold, J.D., Standley, L.J., Horwitz, R.J., 2004. Riparian deforestation, stream narrowing, and loss of stream ecosystem services. Proc. Natl. Acad. Sci. U. S. A. 101 (39), 14132–14137.

Tiner, R.W., 2003. Keys to Waterbody Type and Hydrogeomorphic-Type Wetland Descriptors for U.S. Waters and Wetlands. U.S. Fish and Wildlife Service, Northeast Region, Hadley, MA.

Tiner, R.W., 2014. Dichotomous Keys and Mapping Codes for Wetland Landscape Position, Landform, Water Flow Path, and Waterbody Type Descriptors: Version 3.0. U.S. Fish and Wildlife Service, National Wetlands Inventory Program, Northeast Region, Hadley, MA.

USFWS, 2009. A System for Mapping Riparian Areas in the Western United States. U.S.Department of the Interior. Fish and Wildlife Services, Arlington, VA.

Vance, L., Kudray, G.M., Cooper, S.V., 2006. Crosswalking National Wetland Inventory Attributes to Hydrogeomorphic Functions and Vegetation Communities: A Pilot Study in the Gallatin Valley, Montana. Report to the Montana Department of Environmental Quality and the United States Environmental Protection Agency, Montana Natural Heritage Program, Helena, Montana 37 pp. plus appendices.

Westbrook, C.J., Cooper, D.J., Baker, B.W., 2006. Beaver dams and overbank floods influence groundwater-surface water interactions of a Rocky Mountain riparian area. Water Resour. Res. 42, 1–12.

Westbrook, C.J., Cooper, D.J., Baker, B.W., 2011. Beaver assisted river valley formation. River Res. Appl. 27, 247–256.

FURTHER READING

Jordan, T.E., Whigham, D.F., Hofmockel, K.H., Pittek, M.A., 2003. Nutrient and sediment removal by a restored wetland receiving agricultural runoff. J. Environ. Qual. 32, 1534–1547.

Newlon, K.R., Burns, M.D., 2010. Wetlands of the Gallatin Valley: Change and Ecological Condition. Report Prepared for the Montana Department of Environmental Quality and. U.S. Environmental Protection Agency. Montana Natural Heritage Program, Helena, MT.

Chapter 2.2.7

Wetland Mapping Provides Opportunity to Compare Landscape-Level Functional Assessments to Site-Level Wetland Condition Assessments in Delaware, United States

Mark A. Biddle

Delaware Department of Natural Resources and Environmental Control, Division of Watershed Stewardship, Wetlands Monitoring and Assessment Program, Dover, DE, United States

Chapter Outline

INTRODUCTION

The Delaware Wetland Monitoring and Assessment Program (WMAP) within the Department of Natural Resources and Environmental Control (DNREC) is charged with maintaining data and information on the health of wetlands statewide. Mapping Delaware's wetlands continues to be a major component of this program. Periodic updates to wetland spatial data are necessary to determine trends in wetland acreage, function, and overall health across watersheds statewide. In recent years, the WMAP has focused on updating wetland status and change reporting, economic valuation of wetland function, and both rapid and comprehensive site-level wetland condition assessments. Conducting periodic wetland mapping and regular site-level assessments provide invaluable data to assist in predicting wetland functions at the landscape level for the entire state.

In 2009, the State of Delaware, DNREC, and the National Wetlands Inventory (NWI) of the U.S. Fish and Wildlife Service (USFWS) partnered resources to update and enhance wetland mapping for the state. This new mapping effort created a single spatial data resource that federal agencies, the state, local communities, and the general public could use for planning, and essentially removed confusion by replacing previous separate mapping initiatives by the state and NWI. Wetland maps for this project were based on interpretations of the most recent leaf-off aerial imagery (2007), enabling better-informed resource management, status, and changes reporting, and landscape-level functional assessment of wetlands.

Wetland and Stream Rapid Assessments. https://doi.org/10.1016/B978-0-12-805091-0.00028-1

Need for Landscape-Level Assessment

Wetlands provide many functions that differ by wetland type and landscape setting. It is difficult to put an economic, societal, and environmental value on wetlands, and even more difficult to put a dollar figure on individual functions provided by different wetland types. Recognizing the increasing amount of scientific data and information on wetland functions, condition, and management practices, it remains a challenge to communicate the benefits of wetland functions to nonscientists. Most glaring is the ability to accurately transfer economic value of wetland functions to decision-makers at state and local levels to insure wise use of natural wetlands to maintain the ecological integrity of the state's wetlands and to protect the health and safety of citizens. Accurate prediction of wetland function is the basis for communicating and valuing wetland function.

The small size of Delaware facilitates production of statewide wetland maps in a single effort as well as periodic updates. Functional assessment would require a tremendous effort to evaluate each wetland individually for performed functions at ground level even if access to private property was not difficult. A landscape-level assessment allows for broad interpretation in a cost-effective manner that can be easily transferred to decision-makers considering economic land-use choices (e.g., the cost of acquiring and protecting wetlands in a watershed versus the cost of a new stormwater management facility), and to natural resource managers responsible for maintaining healthy and productive ecosystems.

An accurate landscape-level functional assessment tool predicting the vital environmental services performed by wetlands across the state advances wetland science and informs communities. The 2007 Statewide Wetland Mapping Project (SWMP) provided an opportunity to map wetlands in two ways: by major ecological type and by abiotic properties. This project included expanding the functionality of the state's wetland database to include hydrogeomorphic (HGM) descriptors to describe landscape position, landform, water flow path, and water body type (LLWW descriptors; Tiner, 2003a). Combining elements of the Cowardin classification system (Cowardin et al., 1979) with LLWW descriptors created the NWIPlus dataset that would be used to produce a statewide landscape-level assessment to predict the potential for individual wetlands to perform 11 wetland functions (Tiner, 2003b).

Funding Source

Project funding came from two sources. The NWI provided funding to map wetlands in Sussex County as well as the services of Ralph Tiner, Regional Wetland Coordinator. An Environmental Protection Agency Region 3 Wetland Program Development Grant provided funds to map wetlands in Kent and New Castle Counties and produce a report on the current status and recent trends in the state's wetlands (*Delaware Wetlands: Status and Changes from 1992 to 2007*; Tiner et al., 2011). The mapping was contracted through Virginia Tech's Conservation Management Institute (CMI).

METHODS

The SWMP update was designed to reflect the status of Delaware wetlands in 2007 (Fig. 2.2.7.1). Image analysts from CMI mapped the spatial extent and classified wetlands using a heads-up image analysis procedure (McGuckin, 2011). Although there are inherent limitations to using this technique to map wetlands, it is an efficient and cost-effective process for producing an approximation of the true wetland extent across the state. A full discussion of the advantages and disadvantages of this method can be found in Dahl et al. (2009).

Source Data/Technical Requirements

CMI updated the 1992 SWMP to meet or exceed Federal Geographic Data Committee wetland mapping standards (Wetlands Subcommittee, 2009). Delaware provided CMI with color infrared (CIR) 2007 orthophotos (4-band, 0.25 m). ESRI ArcGIS 9.3 was used to manually create all spatial and classification changes using standard photogrammetric techniques. All wetlands were identified to a minimum mapping unit of 0.5 acre with smaller, highly recognizable polygons (e.g., ponds) mapped down to approximately 0.10 acre. Image analysts typically worked at a scale from approximately 1:15,000 to 1:10,000, with delineations completed at 1:6000 and, occasionally, larger as necessary and as time permitted.

Several ancillary datasets were used to aid in interpretation, including the Soil Survey Geographic Database (SSURGO), 2002 orthoimagery from Delaware, the National Hydrography Dataset (NHD), the National Elevation Dataset (NED), previous NWI data, and the U.S. Geological Survey topographic data (digital raster graphics (DRGs)). After completing the delineation and attribution of the wetland polygons, datasets were inspected through a CMI in-house quality control process for spatial, classification, and topological errors. The wetlands data were then run through the NWI's

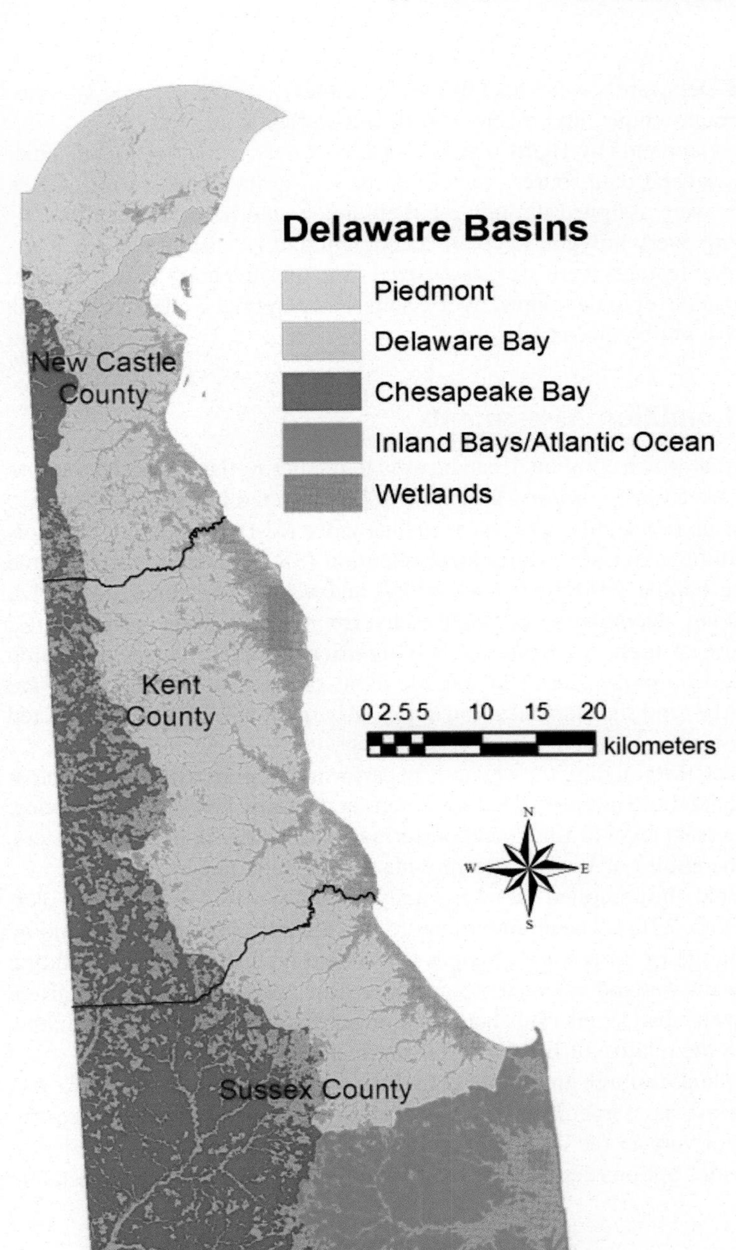

FIG. 2.2.7.1 Delaware Basins and Counties with wetland coverage.

verification tool within ArcGIS to further check for the accuracy of wetland codes. NWI personnel also reviewed the data for delineation and classification accuracy. DNREC provided review of and comment on the datasets prior to release of the final inventory products.

Wetland Classification

Wetlands can be classified in various ways, including by vegetation type, hydrology, landscape position, and other physical properties (Tiner, 2017). The federal standard for wetland classification—the Cowardin et al. (1979) classification system—groups wetlands into five ecological systems: marine, estuarine, riverine, palustrine, and lacustrine. Other characteristics are then added to these five systems such as vegetation type, water regime, and salinity to describe the diversity of wetlands within a geographic area. The majority of Delaware's wetlands fall into the estuarine and palustrine systems. Special Delaware modifiers were applied to identify rare, unique, and locally significant wetlands.

After preparing the basic inventory, CMI classified wetlands by HGM features—LLWW descriptors following Tiner (2003a). These classifications were interpreted using several data sources including the Cowardin classification, NHD, aerial imagery, and topographic maps. Initially, codes were assigned through a largely automated process developed by CMI and the USFWS followed by manually classifying every polygon to create a coherent and consistent dataset. Classifications were reviewed by DNREC and USFWS. Other features were also classified during this inventory (e.g., potential wetlands based on undeveloped hydric soils, landuse/landcover of developed hydric soils, and potential wetland restoration sites) but they are not the focus of this chapter and will not be discussed.

Developing Functional Correlations and Condition Assessments

The LLWW and Cowardin codes contain information that, when combined, can be used to predict wetland functions across large geographic areas (Tiner, 2003b). For the Delaware SWMP, wetland functions to be predicted included: (1) surface water detention (SWD), (2) coastal storm surge detention (CSS), (3) streamflow maintenance (SM), (4) nutrient transformation (NT), (5) carbon sequestration (CAR), (6) sediment and other particulate retention (SR), (7) bank and shoreline stabilization (BSS), (8) fish and aquatic invertebrate habitat (FAIH), (9) waterfowl and waterbird habitat (WBIRD), (10) other wildlife habitat (OWH), and (11) unique, uncommon, or highly diverse wetland plant communities (UWPC). Each wetland polygon was assigned a score of high, moderate, or "no significant value" for each function (Tiner, 2003b). Additional descriptors can be applied to provide clarity in specific areas of interest. For the Delaware project, "stream shading" potential was rated under FAIH, and significant "wood duck" (*Aix sponsa*) habitat was predicted under WBIRD.

The landscape-level functional assessment focused on the potential for wetlands to perform functions at the landscape or watershed scale. It is important to recognize that all wetlands do not provide all functions at the same level of performance. Site-level field assessment is needed to determine the actual level of functional performance as land use, natural disasters, and anthropogenic impacts can reduce or eliminate the ability of wetlands to provide natural functions.

Once mapping of Delaware's wetlands was completed, the compiled data were analyzed and reported in *Delaware Wetlands: Status and Changes from 1992 to 2007* (Tiner et al., 2011). The ability to report on wetlands over a 15-year window revealed many changes across the landscape, including gains, losses and changes to wetland type (Fig. 2.2.7.2). Adding functional classification to the project enabled Delaware wetland scientists to consider how changes in wetlands affect wetland function across the state and at the county or watershed levels. This landscape-level functional analysis is valuable for helping maintain and improve wildlife habitat and the quality of life for Delaware residents and visitors.

The WMAP has been assessing the condition of wetlands through ground-level investigations in Delaware for more than 15 years using a four-tiered assessment approach: intensive assessment, comprehensive field assessment, rapid assessment, and landscape assessment. The four tiers of assessment vary in the detail of data collected and the resources needed to perform an assessment. The multitiered approach provides options depending on the specific goals and resources available

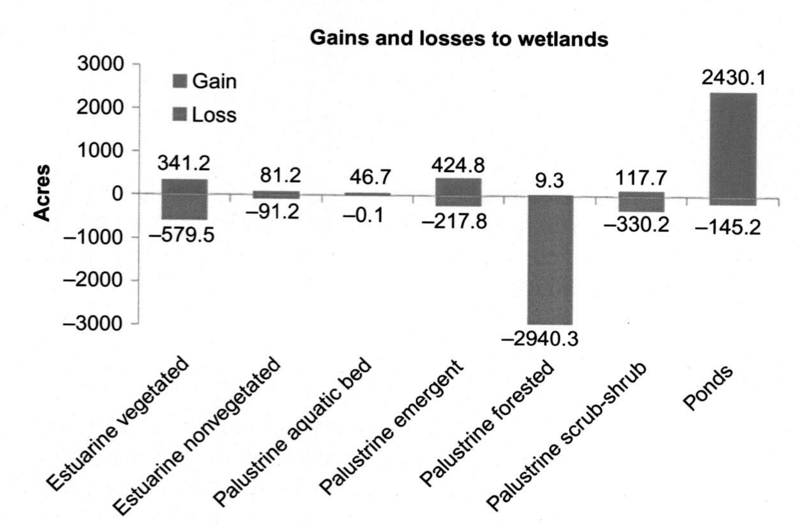

FIG. 2.2.7.2 Gross acreage gains and losses of Delaware wetland types, 1992–2007.

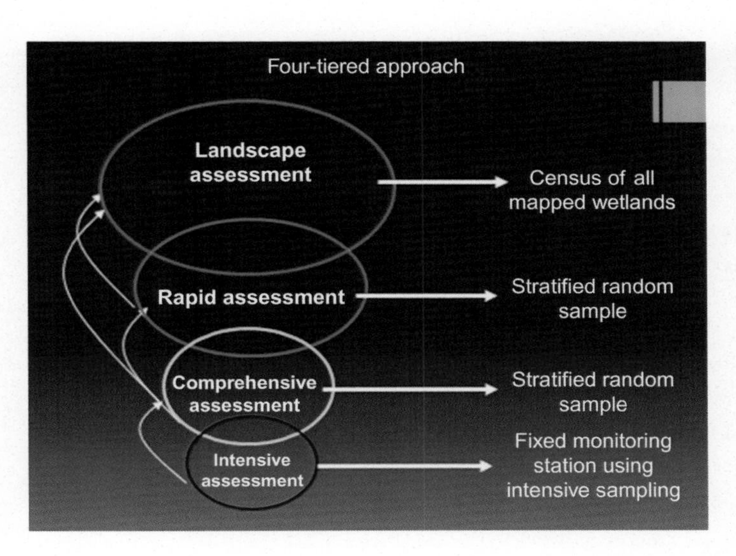

FIG. 2.2.7.3 Four-tiered approach for wetland assessment used in Delaware.

for a project (Fig. 2.2.7.3). The WMAP has produced reference wetland condition and functional data based on the Level 2 Delaware Rapid Assessment Procedure (DERAP) (Jacobs, 2010) and the Level 3 Delaware Comprehensive Assessment Procedure (DECAP) (Jacobs et al., 2009) (http://www.dnrec.delaware.gov/admin/delawarewetlands/pages/Portal.aspx). Additionally, the WMAP employs a tidal wetland condition and functional assessment using the Mid-Atlantic Tidal Rapid Assessment Method (MidTRAM) (Jacobs et al., 2008).

The creation of the DECAP and DERAP evolved from previously created methods. These include the Connecticut and New Hampshire methods (Amman et al., 1986; Amman and Stone, 1991), which use observable indexes of conditions and functions; the Wetland Evaluation Technique (Adamus et al., 1991); HGM classification and functional assessment work (Brinson, 1993; Brinson et al., 1995; Smith et al., 1995); and other resources. These assessment tools generally evaluate wetlands by Brinson's HGM type (Brinson, 1993) based on their existing condition in relation to stressors and level of disturbance (Fig. 2.2.7.4).

The DECAP and DERAP both evaluate the condition of individual wetlands by comparing them to the conditions found in an established set of reference wetlands (minimally disturbed). Measurements within the DECAP use intensive research-derived, multimetric indices to determine level of function and overall condition by wetland type. The variables use both degree of disturbance and ecologic integrity measurements. For example, *hydrology* for a forested flat wetland is affected by disturbance of fill or drainage whereas *plant community* is determined using floristic quality in vegetation strata

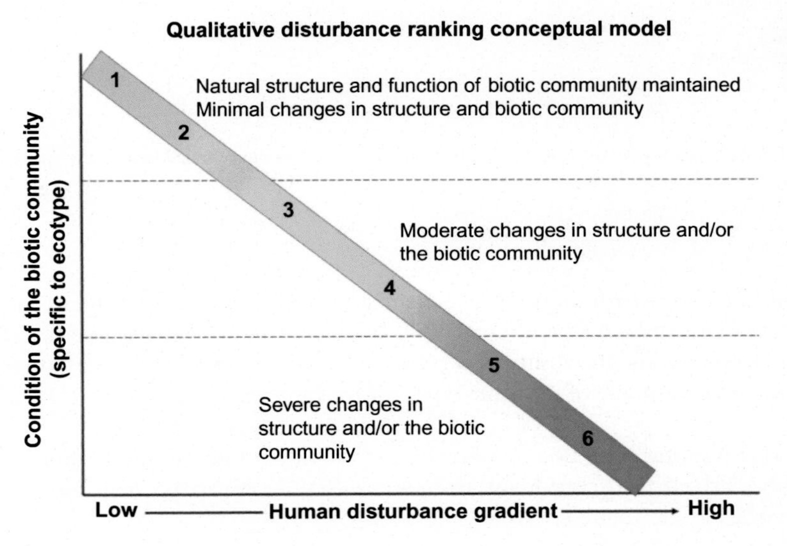

FIG. 2.2.7.4 Wetland qualitative disturbance rating used in Delaware wetland assessments.

Flats functions

Hydrology = $(0.25*V_{FILL}) + (0.75*V_{DRAIN})$

Biogeochemistry = $((V_{MICRO}+V_{CWD}+((V_{TBA}+V_{TREEDEN})/2)/3)*$Hydrology FCI

Plant Community = $(V_{TREESPP}+V_{HERB}+V_{RUBUS}+V_{SHRUBSPP})/4$

Habitat = $(V_{DISTURB}+((V_{TBA}+V_{TREEDEN})/2)+V_{SHRUBDEN}+V_{CWD})/4$

Buffer = $((2*V_{BUFFUSE200}+V_{BUFFBA}+V_{BUFFRD200})/4)*V_{BUFFIMP200}$

Index of Wetland Condition (IWC) = $50/7*(V_{DISTURB}+V_{HERB}+V_{RUBUS}+V_{TBA}+V_{TREEDEN}+V_{TREESPP}+V_{SHRUBDEN}) + 40/3*(V_{MICRO}+V_{FILL}+V_{DRAIN}) + 10/1*(V_{BUFFUSE200})$

Riverine functions

Hydrology = $\sqrt{}\ (((V_{STREAMIN} + (2*V_{FLOODPLAIN}))/3)*V_{CHANNEL\_OUT}*V_{HYDRO\_ALT\_OUT})$

Biogeochemistry = $((V_{TBA}+V_{MICRO})/2)*$(Hydrology FCI)

Plant Community = $(.8*((V_{INVASIVE}+V_{FQAI}+V_{TREESPP})/3) + 0.2(V_{RUBUS})$

Habitat = $(V_{TBA}+V_{MICRO}+V_{SHRUBDEN}+V_{DISTURB})/4$

Buffer = $[0.5(V_{DISTANCE\_TO\_ROADS}) + (0.5(V_{BUFFUSE200}+V_{BUFFERBA})/2)]$

Index of Wetland Condition (IWC) = $(40/2*(V_{FLOODPLAIN}+V_{CHANNEL\_OUT}))+(50/4*(V_{INVASIVE}+V_{RUBUS}+V_{TBA}+V_{TREESPP})) + (10/1*(V_{BUFFUSE200}))$

Depression functions

Hydrology = $[V_{HYDROALT}+((V_{DIST\ ROADS}+V_{LAND\%NATVEG})/2)]/2$

Biogeochemistry = $[((V_{TREEDEN}+V_{TBA}+V_{CWD})/3)+$Hydrology FCI$]/2$

Plant Community = $(V_{SHRUBSPP}+V_{TREESPP}+V_{NATIVE})/3$

Habitat = $[V_{SAPDEN}+((V_{TREEDEN}+V_{TBA})/2)+V_{SHRUBDEN}+V_{CWD}]/4$

Landscape = $[2*V_{LAND\%NATVEG}+V_{BUFFERBA}+V_{DIST\ ROADS}]/4$

Index of Wetland Condition (IWC) = $(10/1*V_{HYDROALT})+(50/6*\sum(V_{TBA}, V_{SHRUBDEN}, V_{SHRUBSPP}, V_{NATIVE}, V_{TREESPP}, V_{CWD})+(40*(V_{LAND\%NATVEG}+V_{BUFFERBA}))$

Variable definitions. All measurements taken within the Assessment Area (AA) unless indicated otherwise.

$V_{BUFFERBA}$ – tree basal area in 200 m buffer
$V_{BUFFIMP200}$ – impervious surface in 200 m buffer
$V_{BUFFRD200}$ – percent roads in 200 m buffer
$V_{BUFFUSE200}$ – high impact land use in 200 m buffer
$V_{CHANNEL\_OUT}$ – degree of channelization outside the AA
V_{CWD} – volume of course woody debris and density of snags
$V_{DISTANCE\_TO\_ROADS}$ – distance to nearest road
$V_{DISTURB}$ – timing and intensity of disturbance to natural vegetation
V_{DRAIN} – percent of AA impacted by ditching/draining
V_{FILL} – percent of AA covered by fill
$V_{FLOODPLAIN}$ – alterations to the floodplain
V_{FQAI} – floristic quality index (ecological conservatism and richness)
V_{HERB} – identification of all understory species <1 m in height in 4 subplots
$V_{HYDROALT}$ – alterations to hydrology of depression
$V_{HYDRO\_ALT\_OUT}$ – hydrology alteration outside the AA such as dams, culverts
$V_{INVASIVE}$ – invasive species abundance
$V_{LAND\%NATVEG}$ – percent of land in natural condition/vegetation
V_{MICRO} – amount of hummocks and tree-tip ups or large downed wood
V_{NATIVE} – percent of understory species that are native in the entire site
V_{RUBUS} – presence of Rubus spp. (4 in Delaware) indicate disturbed conditions
V_{SAPDEN} – density of saplings in forested zone of a depression
$V_{SHRUBDEN}$ – number of shrubs per hectare >.5 m (>1 m for depressions)
$V_{SHRUBSPP}$ – presence of indicator shrub species
$V_{STREAMIN}$ – stream incision ratio and alterations
V_{TBA} – tree basal area
$V_{TREEDEN}$ – trees with >15 cm diameter DBH per hectare
$V_{TREESPP}$ – presence of indicator tree species in canopy

FIG. 2.2.7.5 Delaware wetlands—DECAP functional condition definitions. Methods and individual variable protocols can be found here: (http://www.dnrec.delaware.gov/admin/delawarewetlands/pages/Portal.aspx).

(Fig. 2.2.7.5). The variables of the five functional categories (hydrology, biogeochemistry, plant community, habitat, and buffer) are combined to determine the overall Index of Wetland Condition (IWC). In essence, the IWC is a combined statistical measurement of the components of a wetland type and the ability to perform certain functions, which determines the condition of the wetland for comparison to a reference wetland of the same type, or how far removed a wetland is from peak natural condition and function.

Given the availability of this site-level data in measuring condition and function, it would be interesting to compare these findings with those from the independently derived landscape-level assessment. Such comparison may provide perspectives that could help improve both approaches in the future.

Alignment of Categorical Ratings to Numeric Scores for Comparison

Comparison of categorical functional ratings with numeric wetland condition assessments is challenging because they are looking at wetlands from different perspectives. The former is attempting to identify functional performance regardless of condition (pristine to highly altered) and wetland type. The latter is focused on stressors and assumes that the higher performance of functions that a given wetland type offers, the closer the subject wetland is to the "reference condition" (relatively pristine condition) for that type. It does not compare performance between different types.

The DECAP uses intensive research-derived, multimetric indices from the HGM approach and biologic assessments as the basis for assessing wetland functional condition. "Wetland functional condition" is the capacity of a wetland, at a specific snapshot in time, to perform the functions similar to the reference type for the given HGM wetland type. Scores from DECAP are derived from data collection on specific metrics that create a combination of weighted variables. Weighted variables are based on the contributions of three categories of stressors (hydrology, habitat, landscape) and the site assessment results in an IWC score relative to reference standard by wetland type.

The IWC scores the following wetland functions by type–riverine, flat, depression (Fig. 2.2.7.5): (1) wildlife habitat (amount of disturbance and a measure of trees, shrubs, herbs, microtopography, and course woody debris), (2) plant community (a measure of floristic quality, invasive species, and tree/shrub/herb components), (3) hydrology (a measure of fill, draining, alterations, quality of the floodplain, and in-stream condition), (4) biogeochemical cycling (a measure of tree basal area, microtopography, course woody debris, tree density, and the hydrology index), and (5) buffer integrity (a measure of surrounding land use, buffer width, distance to roads, percent of impervious surface, and native vegetation).

The DERAP final scores are calculated using multiple wetland functions supported by DECAP functional scores to create stressor weights based on multiple regression. Final overall scores in DERAP estimate the condition of a wetland based on stressor occurrence and intensity, which are validated and calibrated to functional scoring in DECAP. Although comparison of IWC scores across wetland types offers cursory evaluation, each wetland type has different variable measurements and to date no detailed analysis of comparison has been completed to determine statistical validity. In watershed evaluation, wetland types are kept separate for reporting out on condition.

In order to compare DERAP and DECAP scoring to the categorical landscape-level predictions, numerical scores were needed for landscape-level predicted functions for each function individually (a 1–10 scale). For this comparison, all functions categorized as "high" in NWIPlus were given a value of 10 while those categorized as "moderate" were given a value of 5. This allowed for summation of all predicted functions (functional sum) for comparison to site-level assessment final scores determined through metrics collected in the field during DERAP and DECAP assessments. Giving a numerical expression to the categorical ranks of the landscape-level assessment created a "common comparison" value versus site-level assessment value of "functional condition." Acknowledged is the fact that this is a simplified method of giving numeric value to landscape-level categorical ranks, and no weighting or other adjustments were made for this initial analysis but could be beneficial in future analysis.

RESULTS

Wetland Characterization

Ecological Features

The mapping identified 320,076 acres of wetlands across the state of Delaware. This total included 62,291 acres of undeveloped hydric soil map units that are naturally vegetated but did not exhibit a distinctive wet signature on aerial imagery (referred to as "H-wetlands" or "potential wetlands"). These are likely to support seasonally saturated wetlands, which are among the most difficult to identify by aerial image interpretation.

Examining wetland distribution by county, 47% were located in Sussex County, 38% in Kent County, and 15% in New Castle County. For a number of years, DNREC has managed natural resources at a watershed or drainage basin level. Describing the amount of wetlands in each major drainage basin, 42% were within the Delaware Bay Basin, 42% within the Chesapeake Bay Basin, 14% within the Inland Bays Basin, and 2% within the Piedmont Basin.

Delaware's wetlands are dominated by palustrine forests that account for 64% of the total wetland acreage. The next most abundant types are estuarine emergent wetlands covering 23% of all wetlands. The remaining wetlands are represented by lacustrine, riverine, and marine types, in descending order of acreage. More than 85% of Delaware's palustrine wetlands (freshwater wetlands) are forested while 7% are scrub-shrub types and 4% are emergent types. Salt marshes are the dominant emergent type of estuarine wetlands, comprising 96% of this type. There are 667 acres of marine wetlands along the state's ocean shoreline, with ponds (palustrine unconsolidated bottoms) and farmed wetlands comprising the state's remaining wetland acreage.

Abiotic Features

Nearly 75% of the palustrine wetlands in Delaware are characterized as terrene isolated (surrounded by upland) or terrene outflow (source of streams, and adjacent or upgradient of other wetlands) types. Lotic wetlands (i.e., those along rivers and streams and subject to out-of-bank flow) account for about 19% of the state's wetlands. Many of this type were found along small streams, usually less than 25 feet wide. Lentic wetlands (i.e., those associated with ponds, lakes, and impoundments) account for about 2% of the state's wetlands and 3% of palustrine wetlands.

Flats are the most common wetland landform in Delaware, comprising more than half (54%) of the wetland area. This is not surprising because more than 75% of Delaware is located on the Atlantic Coastal Plain, with its nearly level topography creating a large expanse for flats and interfluves. Being a coastal state, fringe wetlands comprised more than 20% of the wetlands, including a majority (81%) of the estuarine wetlands. Delaware also contains about 20% of basin wetlands (depressions and ponds), including some recognized as critical habitat for rare plant and amphibian species—they are called Coastal Plain Ponds or Delmarva Bays. Due to the minimal topographic differences across most of the state, floodplain wetlands associated with major rivers are rather limited, accounting for less than 2% of the state's wetland landforms.

Delaware wetlands mainly fall into three categories defining water flow: bidirectional flow (two-way), unidirectional (throughflow/outflow/inflow), or isolated (no surface water inflow or outflow and supplied mainly through groundwater rise or direct precipitation). The most common water flow paths are outflow wetlands that represent about 50% of the wetland area and bidirectional-tidal wetlands accounting for about 25%. Throughflow wetlands along rivers and streams comprised about 14% of wetlands. Isolated wetlands (i.e., geographically surrounded by upland with no apparent surface water connection) represent about 10%. These wetlands aren't necessarily hydrologically isolated when considering groundwater interactions. Less than 1% of Delaware's wetlands are bidirectional nontidal associated with lakes and reservoirs.

Statewide Landscape-Level Wetland Functional Assessment

Considering the 11 functions predicted, there are a few that have widespread implications either at a high or moderate level across multiple wetland types that are in relatively good condition. For instance, it is generally known that estuarine and palustrine lotic wetlands are important for storing surface waters from storms. The latter is dependent on a wide and unchannelized floodplain that adds holding capacity. Watersheds with a majority of historic wetlands and waterbodies intact have reduced flood flows from precipitation events due to the occurrence of these wetlands.

Nearly two-thirds or more of the state's wetlands were predicted to have the potential to perform the following functions at high or moderate levels: surface water detention, nutrient transformation, carbon sequestration, bank and shoreline stabilization, and provision of habitat for other wildlife. Other functions predicted to be provided by more than 40% of the state's wetlands were streamflow maintenance, sediment retention, habitat for waterfowl and waterbirds, and habitat for fish and shellfish. About one-third of the wetlands were deemed important for coastal storm surge detention, which aligns well with the extent of tidal wetlands—approximately 25% of Delaware's wetlands are estuarine, with freshwater tidal wetlands accounting for much of the remaining storage. About one-fifth of the wetlands have the potential to serve well as vital habitat for unique, uncommon, or highly diverse plant communities.

In the past, reporting on wetland losses, gains, and changes offered little insight into the impact on wetland functional services. The expanded abiotic classification (LLWW descriptors) used in the 2007 mapping assists scientists in determining the implications of wetland loss, gain, and change. Generally, changes to Delaware's wetlands from 1992 to 2007 increased some wetland types (ponds by 2285 acres), and decreased vegetated wetland types by nearly 3900 acres. The addition of ponds may have increased a few functions, namely surface water detention, provision of fish and invertebrate habitat, and provision of habitat for waterbirds and waterfowl, but these functions are no substitute for the multitude of functions provided by natural vegetated wetlands. Functions that experienced significant declines due to losses of wetlands from 1992 to 2007 were streamflow maintenance, provision of other wildlife habitat, nutrient transformation, carbon sequestration, sediment retention, and bank and shoreline stabilization (Table 2.2.7.1).

Current Use of Landscape-Level Functional Assessment

The data collected from the mapping project on wetland change and functional assessment provide valuable information that was not previously available to resource managers and decision-makers. The information led to many meetings to review regulatory permitting, better tracking of wetland impacts, and an increased use of assessing functions as part of

TABLE 2.2.7.1 Acreage of 2007 Wetlands (Including Ponds) Predicted to Perform Each Wetland Function at High or Moderate Levels, Percent of the State's 2007 Wetlands That They Represent, and the Wetland Acreage of Gain (+) or Loss (−) from 1992 to 2007

Wetland Function	2007 Acreage	% of DE's Wetlands Likely Performing at Moderate to High Levels in 2007	Acreage Change 1992–2007
1. Surface water detention (This function is limited to freshwater wetlands; the role of coastal wetlands in water storage is handled by the coastal storm surge detention function.)	171,045	66.5	−414
2. Coastal storm surge detention (This function includes tidal wetlands plus contiguous nontidal wetlands subject to flooding during storm surges.)	83,523	32.5	−180
3. Streamflow maintenance (These wetlands are sources of streams or along first order perennial streams or above.)	134,620	52.4	−5888
4. Nutrient transformation	246,847	96.0	−3422
5. Carbon sequestration	249,012	96.9	−3162
6. Sediment and other particulates retention	156,756	61.0	−2141
7. Bank and shoreline stabilization	182,105	70.7	−1383
8. Fish and aquatic invertebrate habitat	78,230	30.5	+123
Stream shading	36,935	14.4	−31
9. Waterfowl and waterbird habitat	80,920	31.5	+104
Wood duck	25,691	10.0	+147
10. Other wildlife habitat	248,090	96.5	−3119
11. Unique, uncommon, or highly diverse wetland plant communities (The following types are included in this category: estuarine aquatic beds, regularly flooded salt marsh (low marsh), slightly brackish tidal marshes, tidal freshwater flats (e.g., wild rice beds), marshes and shrub swamps, Atlantic white cedar swamps, bald cypress swamps, and lotic fringe wetlands.)	54,963	21.4	−372

Farmed wetlands and areas mapped as H-wetlands were not included in this assessment or in the state total for determining the percentages.
Source: Tiner, R.W., Biddle, M.A., Jacobs, A.D., Rogerson, A.B., McGuckin, K.G., 2011. Delaware Wetlands: Status and Changes From 1992 to 2007. Cooperative National Wetlands Inventory Publication. U.S. Fish and Wildlife Service/Delaware Department of Natural Resources and Environmental Control, Northeast Region, Hadley, MA/Dover, DE, 35 pp.

a permitting review process. Currently in Delaware, nontidal wetlands are regulated by the federal government (U.S. Army Corps of Engineers) while tidal wetlands are chiefly regulated by the state. Improvements to the regulatory process identified with the assessment of these data continue to be addressed, and wetland function is being considered increasingly in permitting and required compensation (mitigation).

Additionally, the data and information gleaned from the 2007 mapping assessment have opened new doors to discussion with state and local land use decision-makers. Many were not aware of wetland value and the functions that wetlands provide "free of charge" every day, but also were unaware of wetland condition and the continued impacts from human interaction. Tying wetland functional value to economic decisions on land-use projects is valuable to decision-makers and the public.

Another benefit of this mapping assessment is the ability to connect wetland function to the general public and communicate "value" to individuals. Social media continues to be an effective method for the Delaware WMAP in educating the public on wetland importance, how citizens can become involved, and how to use the information from this project and from watershed wetland condition assessments when managing their own lands or with advocacy.

Comparison With Field Evaluation of Wetland Condition

The DERAP is stressor-based to estimate wetland condition as a representative of function while the DECAP uses weighting of biological or physical components to report out variable scores that can be combined to report on predicted wetland functional condition. The WMAP used reference data to evaluate the relationship between landscape-level functional predictions and site-level condition assessment information using several approaches. Using data reconfiguration and analysis, a comparison was made of landscape-level functional estimates to site-level assessment condition data, DERAP value-added data, comprehensive IWC scores, and comprehensive functional scores (both from using DECAP) as a whole or by HGM type. Four techniques were used for this evaluation (Rogerson et al., 2015); they are discussed below.

The following comparisons are generally looking for corroboration of landscape functional prediction to site-level functional condition assessment, and it is understood that to a degree this comparison is not on equal terms. This investigation attempts to reveal how each wetland assessment method might assist and inform the other, and at a minimum, improve and refine the data collected within each assessment method.

Field Visits and Accuracy Verification

Sites were chosen in the Inland Bays watershed, and five field visits were completed in the Georgetown area. Initially, investigators noted the wetland characteristics and general condition and focused efforts on a few of the 11 functions in both riverine and flat landscape positions. This first evaluation set out to verify where the two assessment types (landscape-level and site-level) showed initial agreement. For more broad functions, such as surface water detention, carbon sequestration, and nutrient transformation, it was found that most wetland types that are in relatively good condition perform these functions at high or moderate levels; this was predicted by the landscape-level classification.

For other functions, such as fish and aquatic invertebrate habitat, unique plant communities, and even streamflow maintenance, inconsistencies were found between the landscape-level assessments and WMAP site-level assessments. The differences were due to highly specified anthropogenic impacts across parts of Delaware's landscape. For instance, the extent of ditching and drainage is not always evident on aerial imagery when performing photo-interpretive wetland mapping. In some cases, the ditches are small enough to cause impact to wetland hydrology but not large enough to pick up on aerial imagery, especially if under coniferous tree cover. This can lead to an incorrect wetland classification that affects the prediction of functions during landscape-level analysis. Perhaps referencing finely detailed NHD data during the image interpretation phase and/or performing the photo interpretation at a larger scale may help reduce this limitation, recognizing that these steps will likely add significant costs to future mapping efforts.

Confirming LLWW Predictions

While documenting the accuracy of LLWW classification, field biologists identified wetland classifications errors during the 2013 field season for tidal and nontidal wetlands in the Leipsic River and Little Creek watersheds. This was not due to the mapping process, but small anthropogenic impacts or other natural features that are not discernable during mapping. Also, some areas mapped as seasonally saturated wetlands (B water regime) were found to not possess hydric soil properties onsite (a concern with how soils are mapped and the dependence on soil data to aid wetland detection). For instance, in the Leipsic River watershed, 18 of 99 sites scheduled for assessment were dropped due to an incorrect LLWW classification, or the site was not a wetland when verified onsite. Predicted LLWW type sometimes labeled a wetland as terrene even though it was adjacent to an NHD feature and therefore should be considered a lotic wetland. The NHD feature onsite resulted in a former stream that had been partially straightened, but it still had a discernable floodplain and evidence of overbank flooding (Fig. 2.2.7.6). Further consideration should be given to the criteria that all ditched headwater wetlands are terrene flats, and then determine the best method for identifying a floodplain with out-of-bank flow into the adjacent wetland. Presumably, this would lead to an increase in functional prediction for these areas.

Mapping classifications for depressions appears to underestimate where depressions (basins) are identified within interfluves. Criteria to establish a specific depth to be considered a depression may help to better identify overlooked depression wetlands. By mapping convention, the NWI polygons were used as the units for enhanced LLWW classification. Consequently, depressions within large palustrine-forested polygons were not delineated for enhanced classification. Future mapping should also incorporate the reverse using LLWW criteria/features to help label/confirm NWI polygons. Automated modeling during the mapping process can be adjusted to account for more localized features. Highly detailed elevation data (from LiDAR) that was acquired recently can assist with finding shallow depressions across flat landscapes and should be useful for future inventories. The above examples illustrate the onsite nuances that can be difficult to recognize in a landscape-level effort on the flat coastal plain.

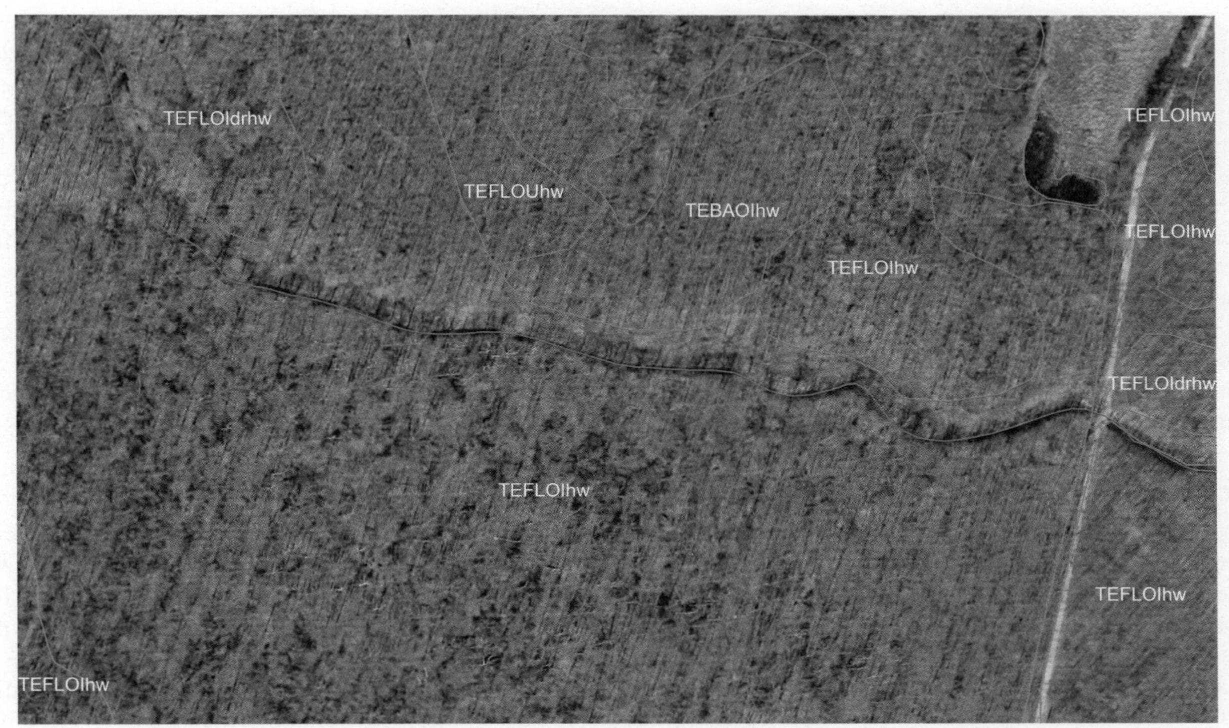

FIG. 2.2.7.6 An example of a mapped terrene wetland that has a straightened stream. Site investigation found a stream with out-of-bank flow that floods the adjacent wetland, which could be considered a lotic wetland. Further consideration is needed to clarify in future mapping.

Pilot Comparison Using Inland Bays Basin Data

Preliminary conclusions were drawn about the relationship between site-level and landscape-level data during the initial evaluation, identifying challenges for expanding the comparison to a larger dataset. The second comparison involved a pilot study involving headwater flat and riverine wetlands in the Inland Bays watershed that compared wetland functions predicted from landscape-level assessment with functional scores from DECAP field assessments. Site-level functional assessment scores for 58 headwater flat and floodplain riverine wetland sites were evaluated to see how NWIPlus rankings compared on a categorical basis. Based on best professional judgment (BPJ) and wetland ecology, a match was established for each of the 11 functions to 4 functional categories that WMAP's level 3/DECAP assessment produces: hydrology, biogeochemistry, habitat, and plant community (Fig. 2.2.7.7).

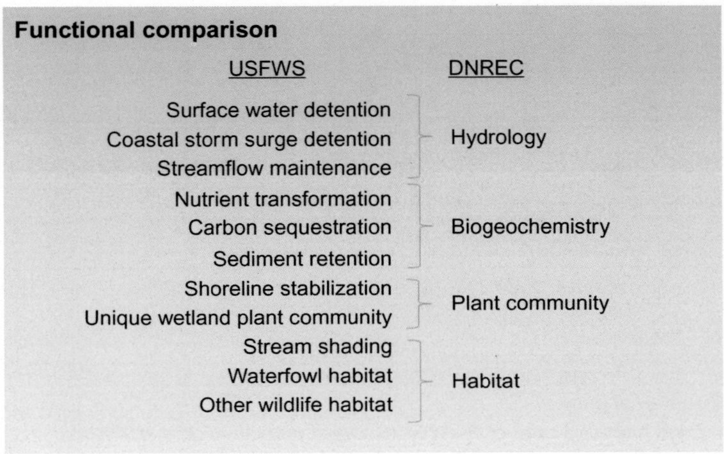

Functional comparison

USFWS	DNREC
Surface water detention Coastal storm surge detention Streamflow maintenance	Hydrology
Nutrient transformation Carbon sequestration Sediment retention	Biogeochemistry
Shoreline stabilization Unique wetland plant community	Plant community
Stream shading Waterfowl habitat Other wildlife habitat	Habitat

FIG. 2.2.7.7 Wetland ecology and best professional judgment used to compare site-level wetland assessment data collected by DNREC and USFWS landscape-level wetland functional assessment categories.

Sites were separated by HGM classification (headwater flats and riverine) to report on relationships. For headwater flats ($n=37$), hydrology and biogeochemistry NWIPlus landscape-level functional estimates and DECAP site-level functional scores were on even terms whereas NWIPlus underestimated plant community compared to DECAP functional scores. It is important to note that in the NWIPlus functional correlation report, unique plant community is admittedly a very conservative prediction. NWIPlus rankings estimated greater habitat function potential than the DECAP functional scores indicated. For riverine sites ($n=21$), habitat estimates were about even but, as expected, NWIPlus estimated lower unique plant community potential than site-level scores. Functional estimates for hydrology and biogeochemistry in riverine wetlands were greater than site-level functional scores indicated. This was attributed to site-specific riparian wetland condition factors that were not picked up on aerial imagery, such as a channelized stream, impediments in the floodplain, and streams that are disconnected from the natural floodplain.

This comparison highlighted the differences in HGM type and the relationship between NWIPlus functional predictions grouped to match the DECAP functional weighted scoring by matched grouping (Fig. 2.2.7.7). It also highlighted challenges in comparing information that is categorical (e.g., "high" and "moderate") to numerical, continuous scores (e.g., 0–100). Reconfiguring data during future investigations may produce more meaningful comparisons.

Comparison With IWC Scores

The third evaluation used DECAP/level 3 data and scores from 158 freshwater wetland sites from several watersheds. Specifically, focus went to the IWC score, which is derived from five functional variable score formulas that are calculated and weighted by HGM type following DECAP scoring protocols (Jacobs et al., 2009). IWC incorporates an overall functional assessment aspect derived from stressors and wetland composition related to condition. In order to provide a more meaningful comparison with the NWIPlus functional estimates, categorical rankings were again converted into scores. For example functions ranked as "moderate" were awarded 5 points and those rated as "high" were given 10 points. Wetlands with "stream shading" potential were awarded points based on HGM and the importance of that function (5 for flats, 10 for riverine wetlands). Wetlands with "wood duck" habitat potential were awarded 5 points and wetlands with "high" potential for waterbird habitat were awarded 10 points. These points were then summed for each wetland polygon and dubbed "functional sum."

A scatter plot of the IWC scores against the functional sum for all 158 sites revealed a weak relationship ($R^2=0.0092$) between IWC and functional sums, meaning wetland condition did not align with functional predictions. Many wetlands with high IWC scores had a low functional sum, meaning that NWIPlus underestimated the "functional potential" compared to field assessments of wetland condition (Fig. 2.2.7.8). It could also mean that some high-quality wetlands do not perform some functions at high levels. For example, a high-quality flatwoods is not going to perform flood storage at a high level. In an attempt to identify whether HGM type was a factor in the misalignment of functional rankings, flats

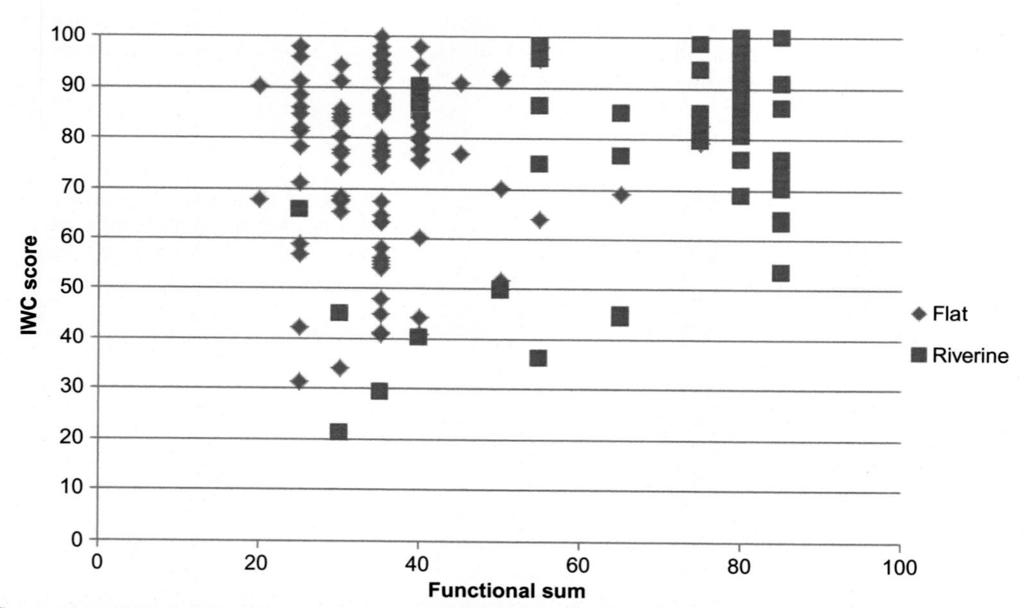

FIG. 2.2.7.8 Scatter plot of Index of Wetland Condition scores against functional sums of NWIPlus functional predictions ($R^2=0.0092$).

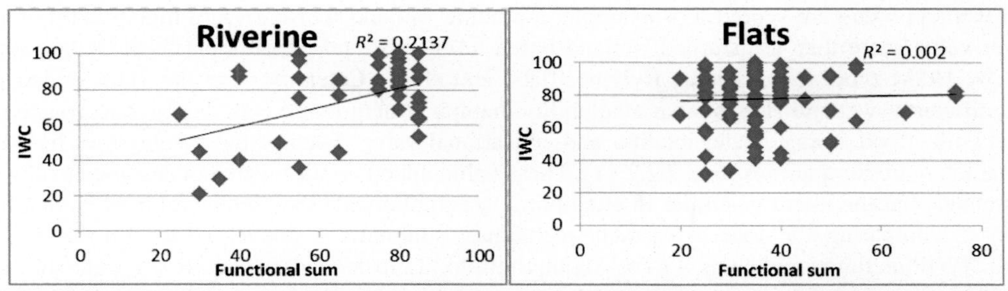

FIG. 2.2.7.9 NWIPlus functional predictions as functional sums against Index of Wetland Condition scores by HGM type (riverine and headwater flat).

and riverine wetlands were separated and repeated for this exercise. The results revealed that there is a difference based on HGM type; functional sums for riverine wetlands were more aligned with the IWC scores ($R^2=0.213$) but flats still showed a weak relationship ($R^2=0.002$) (Fig. 2.2.7.9). Although there were portions of the sample population in high and low condition that were well correlated with functional sums, there were also many sites that had high IWC scores that had low functional sums. Further investigation is needed to determine if features of classification could explain differences. Those that fell between high and low were scattered, meaning there was no bias either way. Creating a numerical functional sum tends to better correlate functional estimates and IWC scores, but not enough to be a viable comparison. Clearly, there are wetlands in good condition that perform some functions poorly, and disturbed wetlands that can produce a high level of function. Good condition doesn't necessarily translate to high function for all wetland functions.

Evaluating by Function

The fourth comparison looked for relationships between NWIPlus functional estimates (as functional sums) to condition assessment-fed function scores function by function (based on the above function matching) based on HGM type. This fourth comparison looked to confirm the third comparison with batched NWIPlus categories, and further validates the difficulty in comparing functional estimates with condition-based scoring. Once again, a general regression of functional sum to each function score indicated a stronger relationship within the floodplain riverine class and a weaker relationship within the headwater flats. For example, Fig. 2.2.7.10 shows a positive relationship, although not robust, for the riverine class. Within the flat category this is not exhibited. Often, there is a zero score for the functional sum, meaning NWIPlus did not rank them as having "moderate" or "high" potential for plant community components, which in this case is comprised of shoreline stabilization and unique wetland plant community. These results could be attributed to the selections made while matching the 11 functions with the WMAP's functional categories but best professional judgment was employed (Fig. 2.2.7.7). All attempts were made to find any method of comparison between functional predictions and condition-based assessments.

Value-Added Metrics

In 2014, Delaware incorporated additional metrics into the DERAP. These value-added metrics were created to capture "functional value" for nontidal wetlands remotely using GIS and in the field during the rapid assessments. Wetland values are based on the opportunity of the wetland to provide a function and the local significance of that function. Both wetland

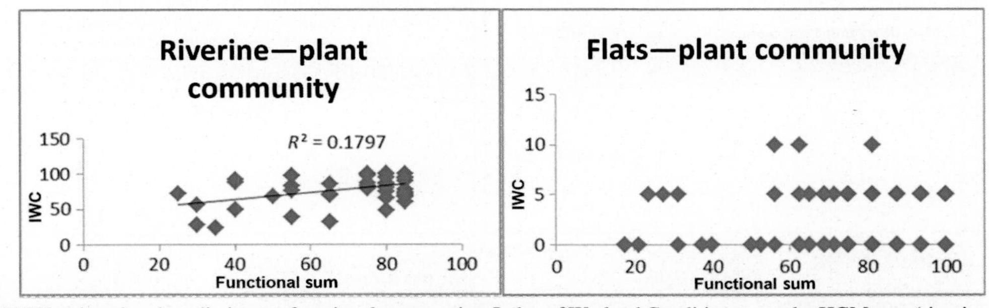

FIG. 2.2.7.10 NWIPlus functional predictions as functional sums against Index of Wetland Condition scores by HGM type (riverine and headwater flat) and by function (potential for high or moderate plant community).

condition and wetland value are scored. For example, a pristine wetland will receive a higher condition score but may receive a lower value score than a disturbed wetland based on the societal benefits provided by that system. Wetlands are classified by HGM type—depressions, riverine, flats, and slope. Categories for the value-added metrics include uniqueness/significance, wetland size, habitat availability, habitat structure and complexity, occurrence in the Delaware Ecological Network, flood storage/water quality, and educational value. Each of these categories has stratified scoring for specific characteristics or qualities (Fig. 2.2.7.11). These value-added metrics were not designed to align with NWIPlus functional estimates, but the intent is similar in attempting to communicate the human value of wetland functions.

In addition to comparing the four assessment techniques summarized above, DERAP value-added scores were compared with NWIPlus functional sums for the Appoquinimink, Leipsic, and Smyrna River watersheds. Over the three watersheds, 254 sites were chosen and analyzed, combined and separated by riverine, flat, and depression HGM classes. The combined sites data showed a weak relationship between the functional sums and value-added scores ($R^2=0.141$). However, this relationship was stronger than the relationship between the functional sums and the IWC scores ($R^2=0.0092$; see Fig. 2.2.7.8 for previous analysis), suggesting a slight improvement (Fig. 2.2.7.12).

DELAWARE WETLAND VALUE ASSESSMENT FORM Version 1.1

Site # ________ Site Name ________ Date ________

Observers ________ Lat / Long ________ AA moved from original location? yes / no

HGM ________ Cowardin ________ LLWW ________ Wetland AA size and shape ________

VALUE-ADDED METRICS — Points

1. UNIQUENESS/SIGNIFICANCE

- ☐ 20 pts Wetland is ecologically significant in DE
- ☐ 5 pts Wetland is rare in the given landscape
- ☐ 5 pts Wetland has been restored, established, or enhanced

Specify: ________

2. WETLAND SIZE ______ha

- ☐ 10 pts ≥ 300 ha
- ☐ 8 pts ≥ 150 to < 300 ha
- ☐ 6 pts ≥ 50 to < 150 ha
- ☐ 4 pts ≥ 15 to < 50 ha
- ☐ 2 pt ≥ 5 to < 15 ha
- ☐ 0 pts < 5 ha

3. HABITAT AVAILABILITY ______ha/6.16ha = ______%

- ☐ 10 pts 100% of buffer unfragmented and natural
- ☐ 8 pts ≥80 to <100% of buffer unfragmented and natural
- ☐ 6 pts ≥60 to <80% of buffer unfragmented and natural
- ☐ 4 pts ≥30 to <60% of buffer unfragmented and natural
- ☐ 2 pts ≥5 to <30% of buffer is unfragmented and natural
- ☐ 0 pts <5% of buffer is unfragmented and natural

4. DELAWARE ECOLOGICAL NETWORK

Select one of the following:

- ☐ 8 pts AA and buffer entirely within core area
- ☐ 6 pts AA entirely within core area, buffer partially within
- ☐ 4 pts AA partially within core area
- ☐ 0 pts None of AA within core area

Select all that apply:

- ☐ 2 pts AA partially in polygon with Final Score ≥0.50
- ☐ 4 pts AA partially in polygon that contains a BCD element occurrence

______ DEN Final Score value ______ # of BCD EO

VALUE-ADDED METRICS — Points

5. HABITAT STRUCTURE AND COMPLEXITY

2 pts for each structure present in AA

- ☐ Snags (≥15cm DBH, ≥45°) # ____
- ☐ ≥ 3 Large downed wood (≥15cm DBH, <45°) # ____
- ☐ Coarse woody debris (7.5-15cm DBH, <45°)
- ☐ Microtopographic relief (≥10% of AA)
- ☐ Surface water suitable for amphibians/macroinvertebrates
- ☐ Surface water suitable for fish
- ☐ Tree canopy gap est: ____% of AA

1 pt for each stratum present in AA

Plant Layers (≥10% of AA)

- ☐ Submerged aquatic vegetation
- ☐ Herb
- ☐ Shrub/Sapling
- ☐ Tree
- ☐ Vine

6. FLOOD STORAGE/WATER QUALITY

2 pts for each present

- ☐ AA is adjacent to surface waters
- ☐ Water pools on ≥ 50% of AA
- ☐ AA is 75% vegetated and has evidence of storm flow (wrack, sedimentation)

Complete with GIS (Cowardin and LLWW classifications):

- ☐ AA has water regime C or wetter
- ☐ AA rated 'Moderate' or 'High' for sediment retention
- ☐ AA rated 'High' for surface water detention

7. EDUCATIONAL VALUE

1 pt for each present

- ☐ AA is viewable from a public road
- ☐ AA is on public property with public access

For public property only:

- ☐ Parking available for ≥ 2 vehicles
- ☐ Trail system relatively close to AA
- ☐ Elevated boardwalk/trail through the AA

Y / N / NA Will proposed activity increase public access and/or opportunity for education?

COMMENTS:

FINAL SCORE:

Sum of values: ☐

Value Category: ☐ Rich ≥45 ☐ Moderate <45 ≥30 ☐ Limited <30

Refer to protocol for variable scoring and detailed descriptions

entered: ________

checked: ________

FIG. 2.2.7.11 DERAP field data sheet for value-added metrics.

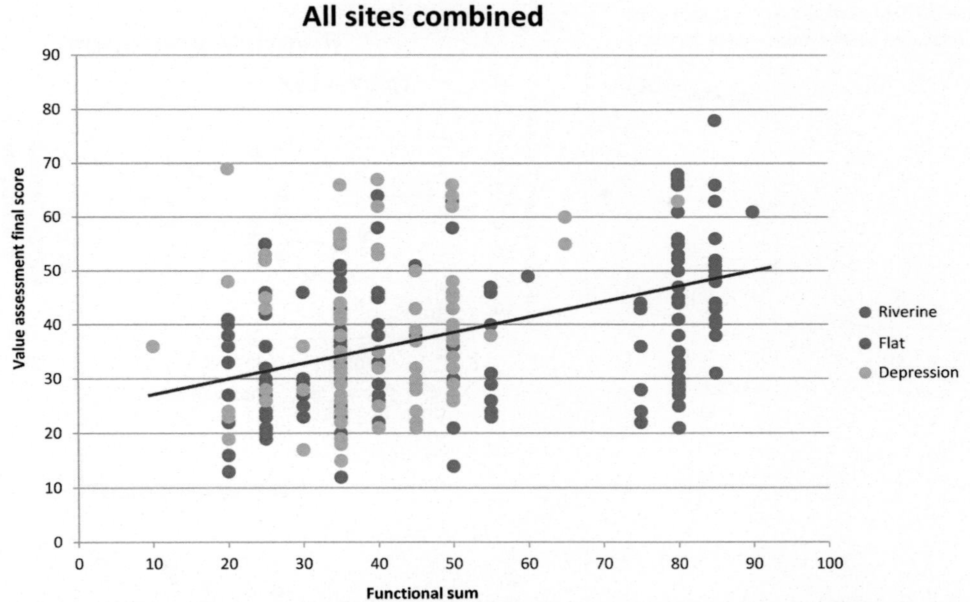

FIG. 2.2.7.12 NWIPlus functional predictions as functional sums against value-added assessment scores by HGM type (riverine, depression, and headwater flat).

Similar to the previous analysis when analyzing each of the wetland types separately for all three watersheds, riverine wetlands showed a weak relationship between the functional sums and the value-added final scores, but it was the strongest relationship of the three types ($R^2=0.1035$). This was the same pattern seen in the previous analysis where riverine wetlands had a stronger relationship between the functional sums and IWCs than the flats ($R^2=0.2137$). Flat wetlands showed the weakest relationship between the functional sums and value-added assessment scores ($R^2=0.0456$). Again, this was the same pattern seen in the previous analysis; however, the relationship here was slightly stronger than between the functional sums and IWCs ($R^2=0.002$). Depression wetlands showed a very weak relationship between the functional sums and value-added assessment scores, only slightly higher than that of the flat wetlands ($R^2=0.0697$) (Fig. 2.2.7.13). Depression wetlands were not evaluated in the previous analyses.

In adding colorization to combined site graphs by HGM type, it was evident that riverine wetlands score higher as functional sums than flat and depression wetlands (Fig. 2.2.7.12). This may be due to riverine wetlands functioning at a higher level, or it could be that the functional values placed in NWIPlus need adjustment to better recognize highly functioning flats and depressions. Further investigation will be employed.

Lessons Learned

Wetland mapping techniques are vital to producing consistent and accurate data. Unfortunately, supporting data used to assist identifying where wetlands exist can be limiting, depending on the age and accuracy of the data. These exercises identified some inherent challenges in comparing the landscape-level assessments designed to predict wetland functions and site-level assessments designed for addressing wetland condition to predict function (i.e., that presume high condition yields high function for the wetland type but does not address the level of performance for individual functions separately). Overall, landscape-level functional estimates for wetland evaluation are useful and site-level visits are valuable to identify alterations or features that are not picked up in wetland mapping and landscape-level analysis. The NWIPlus seems to reflect site-level assessments better for the riverine subclass than for headwater flats, and this should be an area of focus to determine why. In the current form, comparing individual functions to wetland condition does not give meaningful results. There are admitted limitations to the landscape-level analysis at this point to identify site level details, but it can still be useful in identifying potential wetlands of high functional value and helpful in quantifying value in current wetland acreage in a broader context. Site level assessments of wetland condition need to incorporate more direct measure of individual functional indicators to be able to report on the level of performance of different functions for comparison between different wetland types.

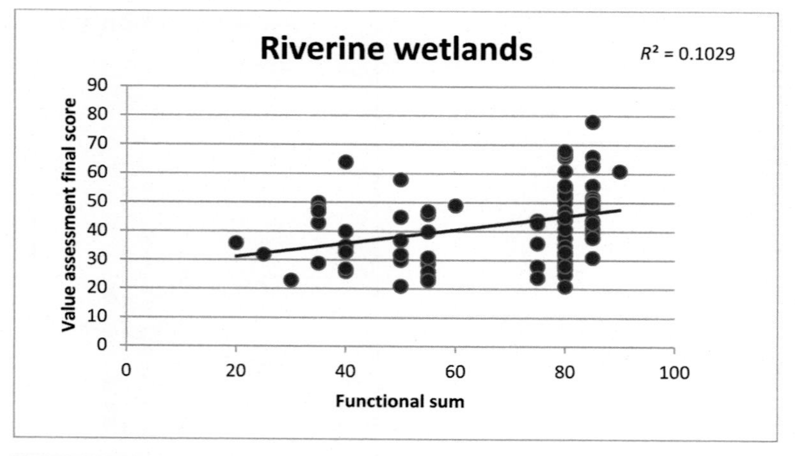

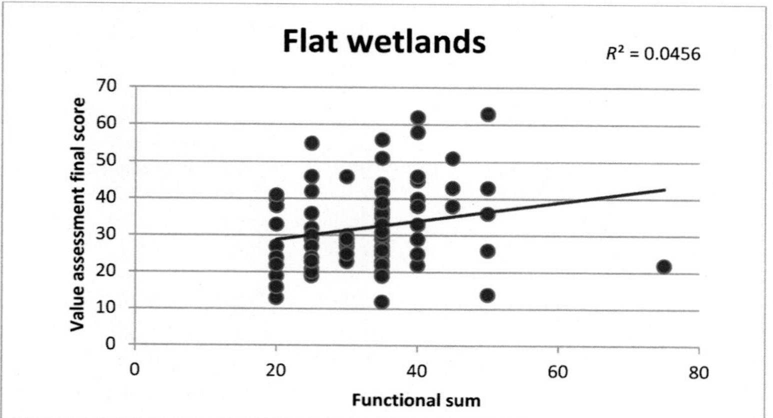

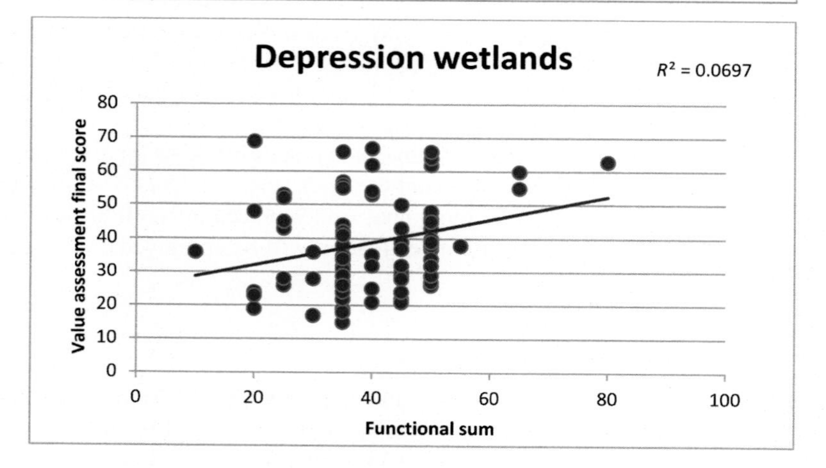

FIG. 2.2.7.13 Each HGM wetland type graphed separately using functional sum against value-added scores.

Suggested Next Steps

1. For the next Delaware Statewide Wetland Mapping Project (planned for 2017–2018), investigate improvements to mapping techniques used within the FGDC framework, and also review other new techniques that use spatial data and imagery analysis to determine wetland type and extent.
2. During wetland mapping, better identify HGM types and incorporate spatial data (i.e., digital elevation data) that reveal depressions and smaller anthropogenic factors (e.g., ditches). Also identify HGM types that are typically in unaltered condition, leading to more predictable functional estimates.
3. Consider more refined functional estimates versus only high and moderate categories. Also consider numerical scoring to allow better comparison.

4. Determine if new metrics or adjustments to existing metrics in DECAP and DERAP could improve functional value assessments. Collect more direct onsite evidence of function, such as water marks or water-carried debris for flood storage, during field investigations.
5. Review the USFWS/NWI wetland correlations document to identify differences between functional estimates from NWIPlus and findings from site-specific evaluations.
6. Investigate whether other wetlands assessment methods, such as Index of Biological Integrity (IBI), Floristic Quality Index (FQI), etc., can offer value to improve functional estimates during landscape-level or site-level functional assessment.
7. Evaluate the regional HGM classification for the Mid-Atlantic for use in predicting functional value rather than for simply predicting wetland condition and assuming wetland function based on condition (Brooks et al., 2011).

Future Use of Landscape-Level Assessment in Delaware

There are many wetland assessment methods being employed, whether for condition, function, or both. Some are very specific and measure specific wetland qualities such as habitat while others are more holistic. And many are designed for regional, state, or local application. Having a universal predictor of wetland function has its benefits, and the WMAP investigations validated that there is less certainty with remotely sensed analyses (and these largely offer generalized assessment), which is inherent to landscape mapping. Including more mapping variables tied to condition and more detailed separation of wetlands into units based on LLWW properties may produce more site-level type accuracy to landscape-level analysis. Possibly applying NWIPlus with more specificity to geomorphic setting may improve the accuracy of landscape-level functional estimation, especially when including site level assessment data. The conclusion is that, with a wealth of site-level condition data, the WMAP should identify methods to use that data to improve predictions of function at the landscape level, which may involve adding new metrics to site-level assessments that provide evidence of specific functions instead of mainly condition stressors. It must also be recognized that landscape-level assessment is a first-cut, preliminary evaluation and is not intended to replace the need for or to be as accurate at assessing wetlands in the field.

Moving forward, the WMAP would like to invest more time to improve the landscape-level assessment/mapping of the 11 functions based on findings during site-level assessments. The WMAP believes that adjustments to mapping methods and collecting new/additional data during site-level assessments would benefit landscape-level functional assessments and quite possibly the site-level assessments. These two different levels of assessments have the ability to assist each other, but this will require more comparison over more wetlands sites. The WMAP is committed to investing more time and effort into taking landscape-level assessment to the next level by bringing the knowledge gained from site-level assessments into the functional correlations used to predict levels of performance for individual functions. A highly accurate functional mapping protocol at the landscape-level would benefit not only Delaware, but many other states updating their wetland mapping to NWIPlus, by saving time and money to create a quality product with functional predictions while reducing the number of intensive site-level wetland assessments.

ACKNOWLEDGMENTS

Alison Rogerson and Erin Dorset of the Delaware Wetland Monitoring and Assessment Program assisted in data analysis for this chapter.

REFERENCES

Adamus, P.R., Stockwell, L.T., Clairain Jr., E.J., Morrow, M.E., Rozas, L.P., Smith, R.D., 1991. Wetland Evaluation Technique (WET). Volume I. Literature Review and Evaluation Rationale (Technical Report WRP-DE-2). U.S. Corps of Engineers Waterways Experimental Station, Vicksburg, MS. 297 pp.

Amman, A.P., Stone, A.L., 1991. Method of the Comparative Evaluation of Non-Tidal Wetlands in New Hampshire (NHDES-WRD-1991-3). New Hampshire Department of Environmental Services, Concord, NH. 97 pp.

Amman, A.P., Franzen, R.W., Johnson, J.L., 1986. Method for the Evaluation of Inland Wetlands in Connecticut: A Watershed Approach. Connecticut Department of Environmental Protection, Harford, CT. (Bulletin 9). 161 pp.

Brinson, M.M., 1993. A Hydrogeomorphic Classification for Wetlands (Wetland Research Program Technical Report WRP-DE-4). U.S. Army Corps of Engineer Waterways Experimental Station, Vicksburg, MS. 103 pp.

Brinson, M.M., Hauser, F.R., Lee, L.C., Nutter, W.C., Smith, R.D., Whigham, D.F., 1995. Guidebook for Application of Hydrogeomorphic Assessment to Riverine Wetlands (Wetlands Research Program Technical Report WRP-DE-11). U.S. Army Corps of Engineers Waterways Experimental Station, Vicksburg, Mississippi, USA. 220 pp.

Brooks, R.P., Brinson, M.M., Havens, K.J., Hershner, C.S., Rheinhardt, R.D., Wardrop, D.H., Whigham, D.F., Jacobs, A.D., Rubbo, J.M., 2011. Proposed hydromorphic classification for wetlands of the Mid-Atlantic region, USA. Wetlands 31, 207–219.

Cowardin, L.M., Carter, V., Golet, F.C., LaRoe, E.T., 1979. Classification of Wetlands and Deepwater Habitats of the United States. U.S. Fish and Wildlife Service, Washington, DC (Report No. FWS/OBS/-79/31).

Dahl, T.E., Dick, J., Swords, J., Wilen, B.O., 2009. Data Collection Requirements and Procedures for Mapping Wetland, Deepwater and Related Habitats of the United States. Division of Habitat and Resource Conservation, National Standards and Support Team, Madison, WI. 85 pp.

Jacobs, A.D., 2010. Delaware Rapid Assessment Procedure Version 6.0. Delaware Department of Natural Resources and Environmental Control, Dover, DE. 36 pp.

Jacobs, A.D., McLaughlin, E.N., O'Brien, D.L., 2008. Mid-Atlantic Tidal Rapid Assessment Method Version 3.0. Delaware Department of Natural Resources and Environmental Control, Dover, DE. 50 pp.

Jacobs, A.D., Whigham, D.F., Fillis, D., Rehm, E., Howard, A., 2009. Delaware Comprehensive Assessment Procedure Version 5.2. Delaware Department of Natural Resources and Environmental Control, Dover, DE. 72 pp.

McGuckin, K.G., 2011. Methods Used to Create Datasets for the Delaware State Wetlands Update. Virginia Tech Conservation Management Institute/ Prepared for the U.S. Fish and Wildlife Service/Delaware Department of Natural Resources and Environmental Control, Blacksburg, VA/Northeast Region, Hadley, MA/Dover, DE. 5 pp.

Rogerson, A.B., Biddle, M.A., Haywood, B.L., 2015. Protecting Ecosystem Services through Integration of Wetlands Assessment, Information Delivery, and Strategy Improvements in Delaware. Delaware Department of Natural Resources and Environmental Control, Dover, DE. (Final Grant Report to the Environmental Protection Agency, Region 3, Wetland Program Development Grants), 21 pp.

Smith, R.D., Ammann, A., Bartoldus, C., Brinson, M.M., 1995. An Approach for Assessing Wetland Functions Using Hydrogeomorphic Classification, Reference Wetlands, and Functional Indices. Technical Report WRP-DE-9, U.S. Army Engineer Waterways Experiment Station, Vicksburg, MS. 90 pp.

Tiner, R.W., 2003a. Dichotomous Keys and Mapping Codes for Wetland Landscape Position, Landform, Water Flow Path, and Waterbody Type Descriptors. U.S. Fish and Wildlife Service, National Wetlands Inventory Program, Northeast Region, Hadley, MA. 44 pp.

Tiner, R.W., 2003b. Correlating Enhanced National Wetlands Inventory Data With Wetland Functions for Watershed Assessments: A Rationale for Northeastern U.S. Wetlands. U.S. Fish and Wildlife Service, National Wetlands Inventory Program, Region 5, Hadley, MA. 26 pp.

Tiner, R.W., 2017. Wetland Indicators: A Guide to Wetland Formation, Identification, Delineation, Classification, and Mapping (Revised and Expanded), second ed. CRC Press, Boca Raton, FL. 606 pp.

Tiner, R.W., Biddle, M.A., Jacobs, A.D., Rogerson, A.B., McGuckin, K.G., 2011. Delaware Wetlands: Status and Changes From 1992 to 2007. Cooperative National Wetlands Inventory Publication. U.S. Fish and Wildlife Service/Delaware Department of Natural Resources and Environmental Control, Northeast Region, Hadley, MA/Dover, DE. 35 pp.

Wetlands Subcommittee, 2009. Wetland Mapping Standard (FGDC-STD-015-2009). Federal Geographic Data Committee, Reston, VA. https://www.fgdc.gov/standards/projects/FGDC-standards-projects/wetlands-mapping.

FURTHER READING

Tiner, R.W., 1985. Wetlands of Delaware. U.S. Fish and Wildlife Service, National Wetlands Inventory/Delaware Department of Natural Resources and Environmental Control, Wetlands Section/Cooperative Publication, Newton Corner, MA/Dover, DE. 77 pp.

Tiner, R.W., 2001. Delaware's Wetlands: Status and Recent Trends. U.S. Fish and Wildlife Service/Prepared for the Delaware Department of Natural Resources and Environmental Control, Watershed Assessment Section, Division of Water Resources/Cooperative National Wetland Inventory Publication, Northeast Region, Hadley, MA/Dover, DE. 19 pp.

Chapter 2.2.8

Virginia Wetland Condition Assessment Tool (WetCAT): A Model for Management

Kirk J. Havens*, Carl Hershner*, Tamia Rudnicky*, David Stanhope*, Dan Schatt*, Kory Angstadt*, Michelle Henicheck†, Dave Davis† and Donna M. Bilkovic*

**College of William and Mary, Virginia Institute of Marine Science, Gloucester Point, VA, United States, †Virginia Department of Environmental Quality, Richmond, VA, United States*

Chapter Outline

INTRODUCTION

The Clean Water Act requires that all U.S. waters be periodically assessed, and wetlands are included in this definition of waters. As described elsewhere in this book, a number of wetland assessment methods have been developed over the last two decades (Brinson, 1993; Karr and Chu, 1999; Lopez and Fennessy, 2002; Brooks et al., 2004; Wardrop et al., 2007; Rooney et al., 2012, and this book) requiring various levels of detail in data collection.

One approach assumes that beneficial wetland services do not all operate as a linked set. Instead, individual services (e.g., habitat ecosystem service or water quality ecosystem service) are controlled by specific sets of wetland characteristics, and therefore there may be no single optimal state. For convenience we refer to this as the service capacity impairment (SCI) model. Under the SCI model there are typically assessment metrics for each ecosystem service of interest. The difference between these two approaches may not seem significant at first, but it can have important implications for the structure of the assessment method and the kind of information the method can provide.

SCI models can differ from other assessment models in several ways. Perhaps the most basic is the description/definition of optimal conditions. Under the SCI model, each wetland ecosystem service can have a set of physical, biological, or chemical conditions that improves the wetland's capacity to perform. For example, conditions that optimize habitat ecosystem services may not be identical to those that are important for water quality ecosystem services. Identification of the optimal set of conditions for each ecosystem service is typically a conceptual rather than empirical effort. The model is defined based on best professional judgment or existing knowledge as a starting point. The utility of the model depends on the accuracy of these assumptions, and so calibration and validation are important steps.

SCI models generate several assessments for each wetland. The assessments are specific to ecosystem service (Hemond and Benoit, 1988; Preston and Bedford, 1988). Integrating ecosystem service assessments to provide an overarching characterization for a wetland or population of wetlands can be accomplished, but requires an explicit protocol that is well understood. Combining individual ecosystem service assessments inherently involves relative value. This is a management policy decision that cannot be ecologically based, and so should be very clear if undertaken.

The Virginia Institute of Marine Science, in collaboration with the Virginia Department of Environmental Quality, developed an SCI model—the Wetland Condition Assessment Tool (WetCAT)—that involves a three-part process. This provides for comprehensive wetland condition assessment analysis scalable from the individual wetland to the entire state. In WetCAT, the Level 2 and Level 3 sampling are intended to calibrate and validate the model that is applied to a Level 1 landscape assessment (Wardrop et al., 2013). The condition assessment of all mapped wetlands is conducted using remotely sensed landscape metrics. Surrounding landscape metrics are considered a reliable proxy for wetland condition; Rooney and others (2012)

Wetland and Stream Rapid Assessments. https://doi.org/10.1016/B978-0-12-805091-0.00002-5

suggest that detailed land cover data improves the accuracy of wetland assessments. In this model the three levels of data collection are not designed to operate independently and the goal is to characterize the capacity of the wetland to provide water quality and habitat ecosystem services using remotely sensed data. The underlying models are based on existing research and specify the combination of landscape-level parameters that are most likely to be predictive of these capacities (Larson et al., 1980; Klopatek, 1988; Lee and Gosselink, 1988; Guadagnin and Maltchik, 2007; Rooney et al., 2012; Herlihy et al., in press). The model application produces a score relative to the unstressed capacity of each wetland for each ecosystem service. The scores are then refined and calibrated by site visits to randomly selected wetlands. The relationship between stressors, landscape metrics, and ecosystem services is validated by intensive study of ecological service endpoints.

We present an example of how this approach can be used to assess wetland condition in decision-making from individual wetlands to all units of landscape such as watersheds, local government boundaries, physiographic provinces, or various hydrologic unit codes. We use the results of the methodology to describe the condition, from a habitat and water quality ecosystem service perspective, for a subset of Coastal Plain and Piedmont wetlands of Virginia. This information is useful for status and trends reporting under the Clean Water Act Section 305(b) and can also be used in permitting programs to assess cumulative impacts to wetlands within watersheds and establish mitigation/compensation ratios.

METHODS

In order to develop the assessment model for habitat and water quality ecosystem services, two separate geographic information systems (GIS) analyses were conducted. First, the watershed around each wetland was generated using the U.S. Geological Survey (USGS) National Elevation Dataset (NED). The NWI of the U.S. Fish and Wildlife Service was used as the wetlands frame. A census was conducted of surrounding landscape metrics of all nontidal NWI wetlands ($n = 167,004$), which represented approximately 1,640,284 nontidal wetland acres. The NWI and the NED were imported into ESRI ArcGIS software version 9.3.1. Isolated sinks in the NED were assumed to be anomalies and filled. The new NED was used to generate a "flow direction" GRID raster; the flow direction GRID assigns numeric values to individual cells in the GRID raster based on the flow direction in that cell. Each NWI wetland was converted into a GRID format, and the hydrologic tools available in ArcGIS were used with the flow direction GRID to generate a watershed GRID around the wetland. The USGS TIGER/Line roads data and the National Land Cover Database (NLCD) created by the MultiResolution Land Characteristics Consortium (MRLC) were combined with the drainage watersheds created above and the NWI wetlands data. All raster data were converted to vector data and analyses were run in Workstation ArcInfo.

In the second part of the analysis, each wetland was buffered 200m and combined with the NCLD land cover to determine relationships (Rooney et al., 2012). The NCLD land cover classifications were combined into four types for our analysis: natural, developed, pasture, and row crops (Table 2.2.8.1). Habitat and water quality ecosystem service determination was based upon the union of the drainage watershed and the 200m buffer (Fig. 2.2.8.1).

Model calibration was conducted on randomly selected wetlands utilizing a suite of anthropogenic stressors. Stressors are considered a good indicator of wetland condition (Crosbie and Chow-Fraser, 1999; Adamus et al., 2001; Otte, 2001; Houlahan and Findlay, 2004; Schlesinger et al., 2008) and were selected after a review of extant

TABLE 2.2.8.1 Combined Landcover Types

Land Cover Type	
Wetland Forest Water Grassland Unconsolidated shoreline	Natural
Pasture	Pasture
Cropland	Cropland
Bare rock/sand, Transition Residential Urban Industrial	Developed

FIG. 2.2.8.1 Example of wetland water quality stress condition with impaired waters layer and permit layer.

TABLE 2.2.8.2 Onsite Stressor List From Literature Review

Stressor	Reference
Sediment deposits Eroding banks	Detenbeck et al. (1996), Ewing (1996), Wardrop and Brooks (1998), and Pontier et al. (2004)
Active construction	Harris (1988) and Panno et al. (1999)
Potential point source discharge	Comeleo et al. (1996) and Chague-Goff and Rosen (2001)
Potential nonpoint source discharge	Hemond and Benoit (1988) and Comeleo et al. (1996)
Active agriculture/plowing	Gerakis and Kalburtji (1998), Butet and Leroux (2001), Zedler (2003), and Kalisinska et al. (2004)
Unfenced livestock	Kassenga (1997), Gerakis and Kalburtji (1998), Xu et al. (2002), and Hongo and Masikini (2003)
Active timber harvesting	Childers and Gosselink (1990), Conner (1994), Johnston (1994), and Aust et al. (1997)
Active clear cutting	Walbridge and Lockaby (1994) and McLaughlin et al. (2000)
Drain/ditch	Vickers et al. (1985), Conner (1994), and Richardson (1994)
Filling/grading	Bedford (1999) and Ehrehfeld (2000)
Dredging/excavation	Zedler (2003)
Stormwater inputs/culverts/input ditches	O'Brien (1988), Reinelt et al. (1998), Thurston (1999), and Groffman et al. (2003)
4-Lane paved road, 2-lane paved road, 1-lane paved road, gravel road, dirt road, railroad, other roadways (parking lots)	Forman and Alexander (1998), Trombulak and Frissel (2000), Houlahan and Findlay (2003, 2004), Kalisinska et al. (2004), Pontier et al. (2004), Roe et al. (2006), Hamer and McDonnell (2008), and Rooney et al. (2012)
Utility easement maintenance	Woo (1979), Magnusson and Stewart (1987), Nickerson et al. (1989)
Herbicide application	Marrs et al. (1992), Brown et al. (2004), and Edginton et al. (2004)
Dike/weir/dam	Thibodeau (1985) and Kingsford (2000)
Beaver dam	Burns and McDonnell (1998) and Gurnell (1998)
Mowing	De Szalay et al. (1996) and Rothenbücher (2005)
Brush cutting	Anderson et al. (1977), Oliver (1980), Chadwick et al. (1986), and Hanowski et al. (1999)
Excessive herbivory	Kauffman and Krueger (1984)
Invasive species present	Pimentel et al. (2000), Sakai et al. (2001), and Zedler and Kercher (2004)

literature and their suitability for management alteration (Table 2.2.8.2). NWI wetlands were stratified by physiographic province (Coastal Plain, Piedmont, Valley and Ridge, Blue Ridge, and Appalachian Plateau), wetland type, forested (FO), scrub/shrub (SS), emergent (EM), and 12-digit hydrologic unit code (HUC). Sites were selected for sampling by Generalized Random Tessellation Stratified (GRTS) sampling design (Stevens and Olsen, 2004) ($n=2126$). Additional sites were sampled in 2010, 2011, 2013, and 2014 for periodic calibration of land cover to stressor relationships ($n=351$).

Randomly selected wetlands were assessed for stressors from the wetland center point to within a 30m radius circle and between 30 and 100m radius circle and were tabulated using a programmed hand-held computer downloaded to a Microsoft Access database upon returning from the field (Fig. 2.2.8.2). Validation of the relationship between stressors, surrounding

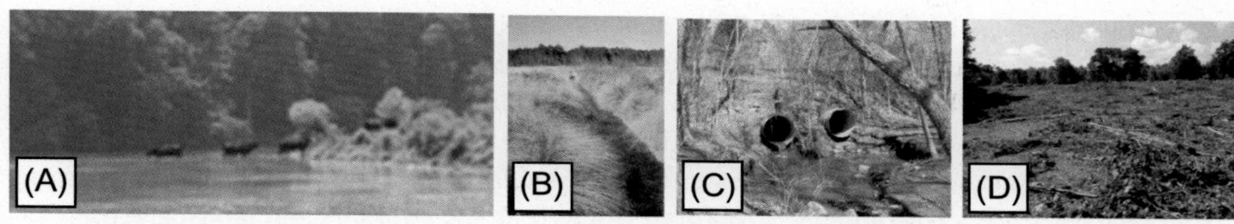

FIG. 2.2.8.2 Stressors impacting wetlands: (A) unfenced livestock access, (B) ditching, (C) culverts, and (D) brush cutting.

landscape metrics, and ecological service endpoints is a necessary step in any condition assessment model. Validation of stressor and surrounding landscape metric effects on habitat and water quality ecological services was conducted by randomly sampling 87 sites for intensive study of habitat and water quality ecosystem service.

Habitat ecosystem service was determined by assessing avian, amphibian, and plant community structure metrics. Avian community structure was determined by three rounds of stratified point count surveys. Surveys were conducted from 0.5 and 4.5 h after sunrise between late May and mid-July. Data collected at each point included site, date, start time, species of birds detected, distance from point center (within 50 and >50 m) of each detection, time period of detection (0–3, 3–5, 5–7, 7–10, and 10–15 min), and detection method (visual, aural, both) (Buskirk and McDonald, 1995; Hamel et al., 1996). Additional data collection included multiyear deployment of automatic recording devices during the summer breeding season for birds and in early spring for amphibians. Vocalizations of birds and amphibians were recorded for three consecutive days for 15 min at 6 a.m. and 15 min at 9 p.m., respectively. Amphibian community structure was also sampled by early and late spring 1 m sweeps of a D-ring dip net and the number per dip net sweep was used as a measure of relative abundance. Visual encounter surveys at night as well as day and night frog call surveys were also conducted. The avian and amphibian communities were determined by amphibian richness, priority wetland and neotropical migrant bird species abundance (Pashley et al., 2000), and the priority in flight index (Pashley et al., 2000; Mehlman et al., 2004).

Water quality ecosystem service was determined by potential disruption of water residence time in wetlands. In general, anaerobic conditions exist in hydric wetland soils, and this condition is necessary for denitrification processes to take place within these soils (Groffman et al., 2002). Much of the water quality improvement (at least with respect to nitrates) occurs in shallow groundwater within the rooting zone of the wetland vegetation and, generally, water quality improvement will be greatest when the water resides within this zone for longer periods of time (Mayo and Bigambo, 2005). Thus, assessing the capacity of a wetland to improve water quality depends on water table elevation and the residence time of the water in this zone. In addition, impervious surfaces can result in a large amount of water discharge, resulting in creek or stream incision, and alter wetland vegetation composition (Groffman et al., 2003). In our model we assumed that anthropogenic stressors that modify existing wetland hydrology are critical to the wetlands water quality service condition and those stressors that are associated with landscape metrics were targeted for analysis. We assessed water quality ecosystem service by analyzing total dissolved nitrogen (TDN), total dissolved phosphate (TDP), and total suspended solids (TSS) in water samples from wetlands where possible. Water samples were obtained from wetland systems monthly and after rain events. Samples were transported to the laboratory and analyzed using a SKALAR SANplus Continuous Flow Analyzer.

Additional habitat and water quality ecosystem service validation sampling included stream incision ratios and plant community composition. Stream incision ratios in headwater or riverine wetlands were calculated by sampling 1 m intervals along a 50 m stream segment and measuring bank full height and bank height (Rosgen, 2001). Wetland plant community composition was sampled by the Bitterlich plotless method (Basal Area Factor = 2) at three areas and by measuring tree species diameter at breast height (DBH). All woody stems >1 m tall and <5 cm DBH in three 1.9 m radius plots were recorded for each Bitterlich sample point. Wetland indicator status (IS) (Lichvar, 2012) was recorded for all species. To measure a potential shift in wetland plant community to "drier" plant communities over time due to stream incision, the incision ratio (stream bank height/bank full height) was compared with a tree and sapling wetland plant community wetland indicator status index (PCWIS). The PCWIS index was calculated as

$$\frac{\sum_{\#\text{sapling species} \times \text{indicator status}/\#\text{Saplings}}}{\sum_{\#\text{tree species} \times \text{indicator status}/\#\text{Trees}}}$$

where the obligate wetland species IS (Indicator Status) equals 1, facultative wetland species IS equals 2, facultative species IS equals 3, facultative upland species IS equals 4, and upland species IS equals 5. A PCWIS index below 1.0 indicates a shift toward more wet conditions while a PCWIS above 1.0 indicates a shift toward more dry conditions. The PCWIS index was compared with landscape metrics to determine associations.

Stressors were compared to the remotely sensed surrounding landscape metrics to determine the strength and direction of association using Pearson Product-Moment Correlation, where both the stressors and landscape metrics were considered ratio variables. Data were checked for normality using the Ryan-Joiner test and for nonnormal distributions, the Fisher's z' transformation or the Spearman Rank Order correlation was used (Minitab®, 2010). Avian and amphibian community sound signature similarities were examined with nonparametric multidimensional scaling (nMDS) and analysis of similarities (ANOSIM) in PRIMER 6.0 (Clarke and Warwick, 2001). MDS ordinates sites based on similarities in sound signature makeup, using rank order of distances to map out relationships. Sites with high similarity are placed close together on the MDS map. A Euclidean distance coefficient was used to calculate the similarity matrix. Factors were overlaid on the MDS plot to visualize community groupings in relation to land use and stress level. Subsequently, ANOSIM was used to test relationships among land use and stress level.

Classification and Regression Tree (CART) (JMP®, 2009) analysis was used to look for patterns between selected landscape metrics and ecosystem service endpoints and to weight landscape metrics. CART partitions were used to develop formula scoring thresholds and to cross-validate stressors with ecosystem endpoints.

RESULTS

Landscape metrics that showed a moderate to strong association ($r > 0.25$, $P < 0.05$) with stressors were selected for comparison to the ecosystem service endpoints of habitat and water quality. Some stressors showed weak or no association with surrounding landscape metrics and, because the model uses remotely sensed landscape metrics to determine condition score, those stressors were removed from consideration. Those stressors that showed a moderate to strong association with surrounding landscape metrics are listed in Table 2.2.8.3. Two stressors (sediment deposits and eroding banks) that did show a moderate to strong association with the surrounding landscape were removed after quality assurance checks revealed difficulty in distinguishing between anthropogenic activity and natural episodic events in the field.

Landscape metrics that showed moderate to strong association with stressors and with habitat and water quality ecosystem service endpoints are listed in Table 2.2.8.4. Stressor groups showed similar frequency across time within physiographic province (Fig. 2.2.8.3). Dominant stressors were vegetation alteration (brush cutting and mowing) and roads with a higher frequency of ditching and draining and filling in the Coastal Plain and higher frequency of unfenced livestock access in the Piedmont and Valley and Ridge, Blue Ridge, and Appalachian Plateau (Fig. 2.2.8.3).

CART partitions were used to develop wetland condition scoring thresholds (Table 2.2.8.5) and the scoring formulas. Dichotomous thresholds within individual landscape metrics were scored as 0.1 (most stressed) or 1.0 (least stressed) while multiple thresholds were standardized from 0.1 to 1.0 and scored linearly. Landscape metrics in the scoring formula were weighted by frequency of association with ecosystem service endpoints. For example the landscape metric "proximity to other wetlands" was associated with the two ecosystem service endpoints of amphibian richness and neotropical migrant bird species abundance and was weighted by a factor of two.

Overall wetland habitat and water quality condition stress levels were placed into four categories (slightly stressed, somewhat stressed, somewhat severely stressed, and severely stressed) determined by breaks in scoring distributions (Table 2.2.8.6). The use of distinct condition categories allowed the expression of wetland condition in terms of stress level, a concept useful for managers and the public (Van Sickle and Paulsen, 2008).

A subset of 128,422 NWI coastal plain and piedmont wetlands (920,084 acres) in Virginia was analyzed. Wetland condition scores for both habitat and water quality showed shifts over a 10-year period from 2001 to 2011. Overall wetland habitat considered severely and somewhat severely stressed increased from 24.2% in 2001 to 27.3% in 2011 with a concurrent decrease in wetlands considered somewhat stressed and slightly stressed from 75.8% in 2001 to 72.7% in 2011. Wetlands with water quality condition considered severely and somewhat severely stressed increased from 42.5% in 2001 to 43.5% in 2011 with a concurrent decrease in wetlands considered somewhat stressed and slightly stressed from 57.4% in 2001 to 56.6% in 2011 (Table 2.2.8.7). Condition scores averaged by 12-digit HUC are shown for three years (2001, 2006, and 2011) in Fig. 2.2.8.4.

TABLE 2.2.8.3 Stressors That Show a Moderate to Strong Association, $r > 0.25$ ($P < 0.05$) With Landscape Metrics

Landscape Metrics Stressors	Natural	Developed	Pasture	Row Crops	Road Density	Natural Drainage	Developed Drainage	Pasture Drainage	Row Crops Drainage	Road Density Drainage
Dam/dike/weir										+0.28 (0.01)
Timber harvest (1–5 year)	+0.37 (0.05)					+0.39 (0.05)				
Livestock access			+0.28 (0.01)	+0.66 (0.01)					+0.65 (0.01)	
Point source discharge	−0.62 (0.01)	+0.73 (0.01)				−0.51 (0.06)	+0.61 (0.01)			+0.48 (0.05)
Active agriculture/ plowing	−0.25 (0.06)		+0.79 (0.01)	+0.28 (0.01)		−0.46 (0.05)		+0.57 (0.01)		
Drain/ditch	−0.32 (0.01)	+0.34 (0.01)			+0.35 (0.01)		+0.25 (0.01)			+0.58 (0.01)
Filling/grading	−0.46 (0.05)	+0.79 (0.01)			+0.64 (0.01)	−0.37 (0.01)	+0.55 (0.01)			+0.40 (0.01)
Stormwater inputs, culvert and ditch inputs	−0.28 (0.01)	+0.30 (0.01)			+0.58 (0.01)					+0.55 (0.03)
Roads	−0.63 (0.01)	+0.64 (0.01)			+0.70 (0.01)	−0.53 (0.01)	+0.53 (0.01)			
Mowing	−0.67 (0.01)	+0.61 (0.01)	+0.37 (0.05)		+0.66 (0.01)	−0.62 (0.01)	+0.60 (0.01)	+0.64 (0.01)		+0.57 (0.01)
Brush cutting	−0.47 (0.01)		+0.38 (0.05)	+0.33 (0.01)		−0.60 (0.01)	+0.31 (0.01)	+0.72 (0.01)		+0.45 (0.01)
Invasive plant species		+0.35 (0.01)			+0.27 (0.01)					

Minus sign (−) equals a negative association and plus sign (+) equals a positive association.

TABLE 2.2.8.4 Ecosystem Service Endpoints Associated With Landscape Metrics

	Landscape Metrics Within 200 m									
Ecosystem Service Endpoints	Natural	Developed	Pasture	Row Crops	Road Density	Natural Drainage	Developed Drainage	Row Crops Drainage	Size	Proximity to Other Wetlands
Habitat										
Amphibian richness		X			X					X
Priority wetland bird species abundance			X			X	X		X	
Priority in flight index		X		X	X					
Neotropical migrant bird species abundance		X						X		X
Plant community wetland indicator status index		X								
Water quality										
Total dissolved nitrogen	X									
Total dissolved phosphate										
Total suspended sediment									X	
Plant community indicator status index		X								

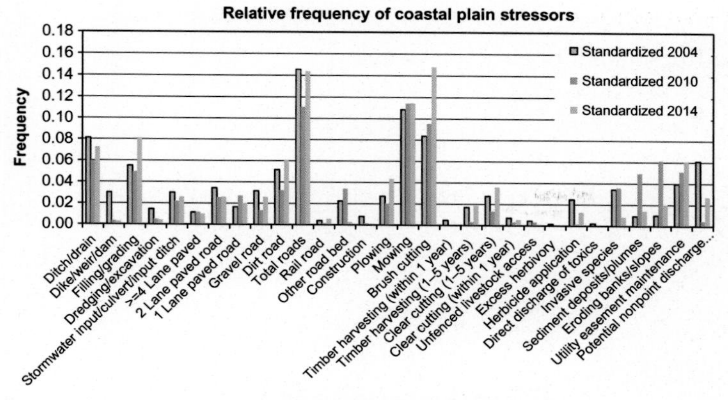

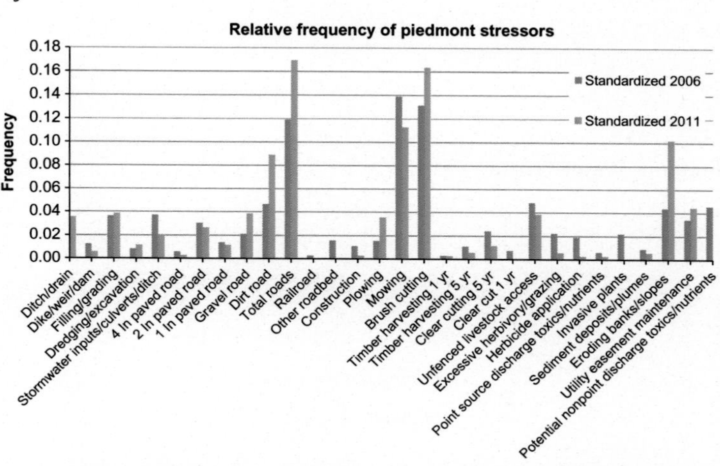

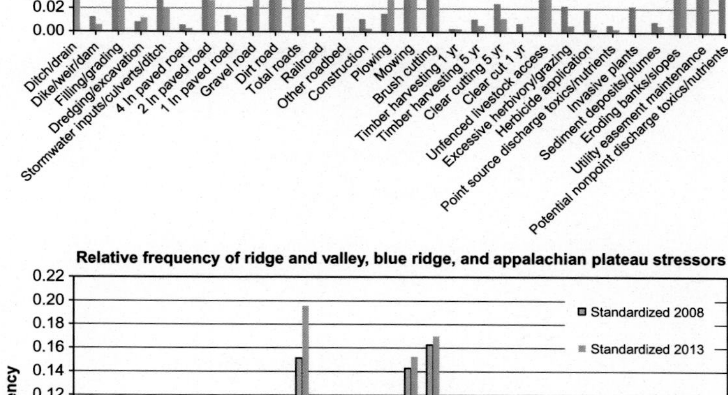

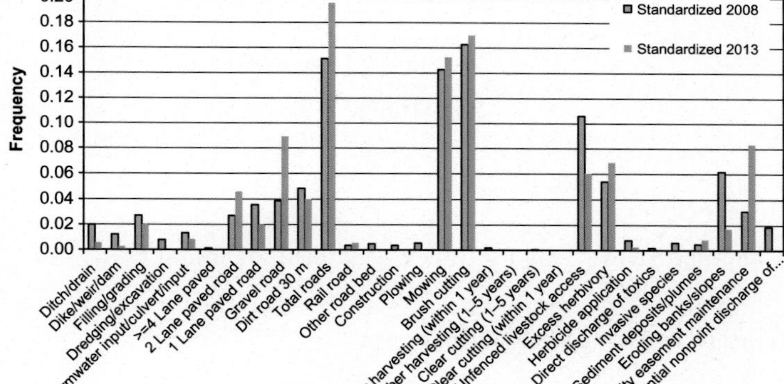

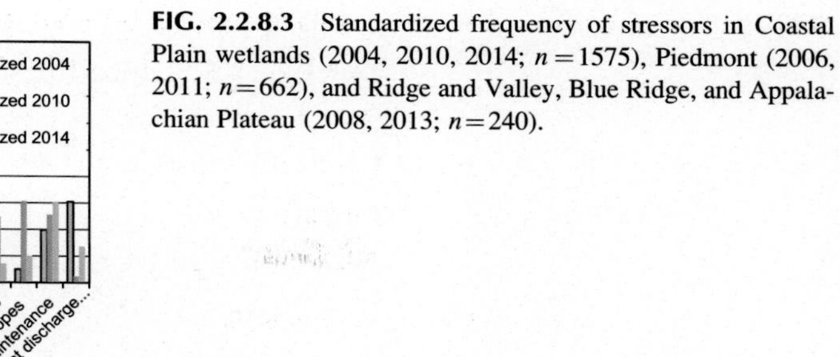

FIG. 2.2.8.3 Standardized frequency of stressors in Coastal Plain wetlands (2004, 2010, 2014; $n=1575$), Piedmont (2006, 2011; $n=662$), and Ridge and Valley, Blue Ridge, and Appalachian Plateau (2008, 2013; $n=240$).

TABLE 2.2.8.5 Scoring Thresholds for Landscape Metrics
Habitat
Within a 200 m buffer
Percent developed land (>1%, >9%, >21%)
Percent wetlands (<12%, <5%)
Percent pastureland (>2%)
Percent row crops (>5%)
Road density (>19, >31)
Size of the wetland (<5.4 acres)
Within a 200 m buffer in the contributing drainage

Continued

TABLE 2.2.8.5 Scoring Thresholds for Landscape Metrics—cont'd

Percent developed land (>1.4%)
Percent row crops (>3%)
Percent natural land (<72%)
Water quality
Within a 200 m buffer
Percent natural land (<71%)
Percent developed land (>1%)
Size of the wetland (<2.4 acres)

TABLE 2.2.8.6 Scoring Thresholds for Wetland Habitat and Water Quality Stress

Score	Stress Level
Habitat	
≥0.90	Slightly stressed
0.90< and ≥0.60	Somewhat stressed
0.60< and ≥0.30	Somewhat severely stressed
<0.30	Severely stressed
Water quality	
1.0	Slightly stressed
0.7	Somewhat stressed
0.4	Somewhat severely stressed
0.1	Severely stressed

TABLE 2.2.8.7 Wetland Habitat and Water Quality Condition Stress Level for the Coastal Plain/Piedmont Wetlands of Virginia for 2001, 2006, and 2011

	Water Quality 2001		Water Quality 2006		Water Quality 2011		
Condition Score	*n*	%	*n*	%	*n*	%	**Stress Level**
0.1	18,113	14.1	18,615	14.5	18,751	14.6	Severely stressed
0.4	36,478	28.4	36,750	28.6	37,055	28.9	Somewhat severely stressed
0.7	51,654	40.2	51,174	39.8	51,063	39.8	Somewhat stressed
1.0	22,147	17.2	21,883	17.0	21,553	16.8	Slightly stressed
Total	128,422	100	128,422	100.0	128,422	100	
	Habitat 2001		**Habitat 2006**		**Habitat 2011**		
Condition Score	*n*	%	*n*	%	*n*	%	**Stress Level**
0.1<0.30	5433	4.2	6684	5.2	6936	5.4	Severely stressed
0.30<0.60	25,679	20.0	28,169	21.9	28,104	21.9	Somewhat severely stressed
0.60<0.90	60,946	47.5	59,677	46.5	59,498	46.3	Somewhat stressed
0.90–1.00	36,364	28.3	33,892	26.4	33,884	26.4	Slightly stressed
Total	128,422	100	128,422	100.0	128,422	100	

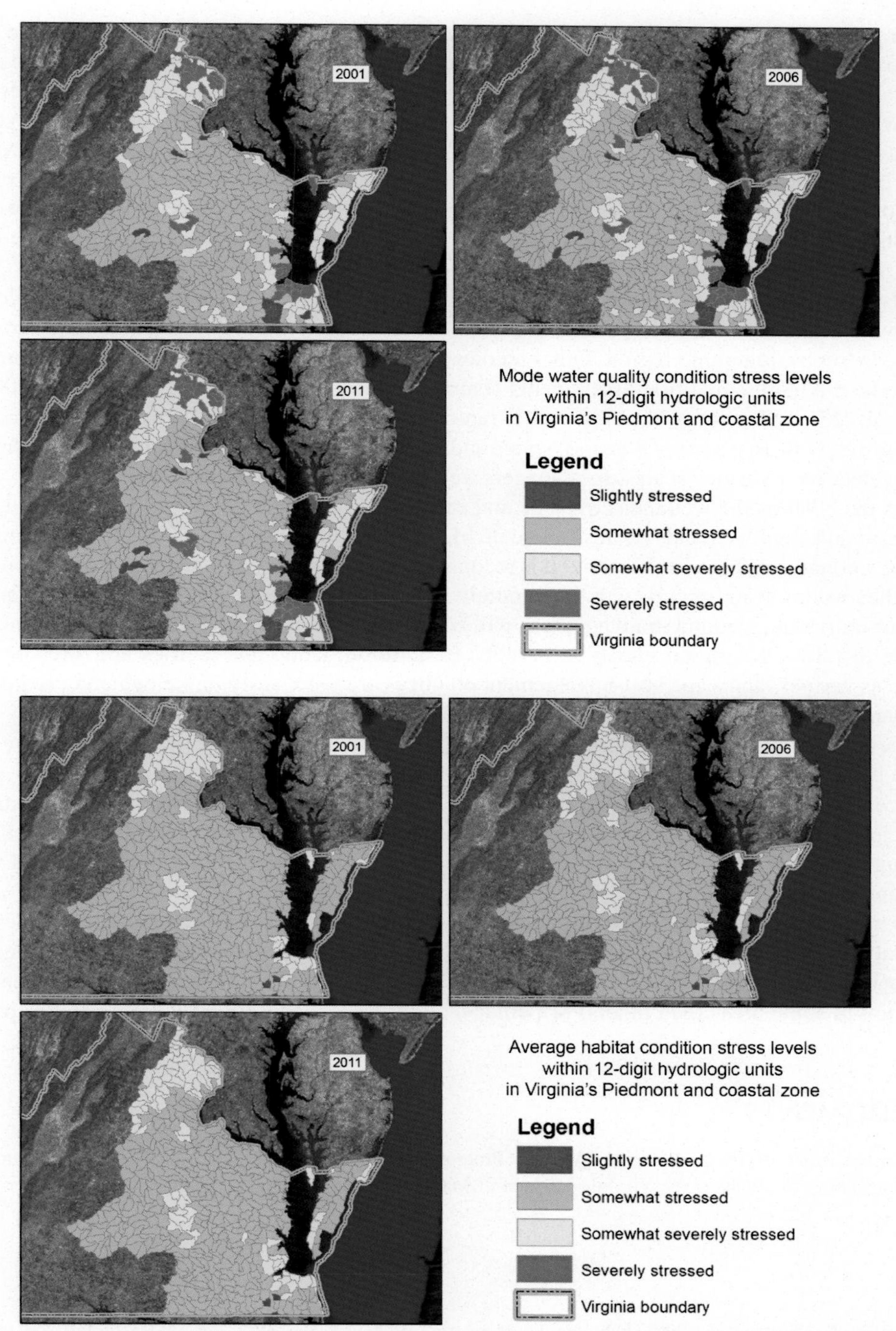

FIG. 2.2.8.4 Wetland condition scores by 12-digit HUC for habitat ecosystem service and water quality ecosystem service in the Coastal Plain/Piedmont wetlands of Virginia for the years 2001, 2006, 2011.

DISCUSSION

WetCAT uses a multiple services, multitiered, service capacity impairment approach to derive a wetland condition score that can be used to rate a wetland's capacity to perform habitat and water quality ecosystem services. This was developed specifically to inform management of nontidal wetland condition. Because the model is designed to work with remotely sensed information rather than direct field observations, it predicts the occurrence of relevant stressors in the areas surrounding a wetland. As such it focuses on factors that are or can be managed—namely the landscape stressors that affect a wetland's natural ecosystem service capacity. This assessment does not compare a wetland to some pristine optimum. Instead it seeks to assess the degree to which the wetland is functioning below its full natural capacity to provide ecosystem services, specifically for habitat and water quality services. The assessment does not seek to rate wetlands on the basis of the absolute rate at which those ecosystem services are provided. It assumes that wetlands are naturally very variable in the performance of services and the maximum potential level of performance also varies widely across seemingly similar wetlands. From this perspective, what matters most in the management of wetlands is avoidance or minimization of anthropogenic impacts that degrade a wetland's capacity to perform an ecosystem service, thus the scoring protocol rates wetlands based on the probability that conditions in the surrounding landscape will create stressors that reduce ecosystem service capacity.

We used landscape metrics from remotely sensed data coupled with ground calibration and ecosystem service endpoint validation to provide census-level condition assessment information for wetlands that can be scaled from individual wetlands to watersheds to physiographic region. Other studies have shown the utility of using GIS or remotely sensed land cover data to produce reliable condition assessments (Fairbairn and Dinsmore, 2001; King et al., 2005; Brazner et al., 2007; Rooney et al., 2012; Herlihy et al., in press). If remotely sensed data on land cover and landscape metrics can be obtained at regular intervals (e.g., every 5 years), status and trend analysis on wetland capacity to perform ecosystem services can be provided, over various scales, to managers, regulators, and other stakeholders.

WetCAT uses the NWI as the wetlands coverage and census frame. While the NWI has variable detection rates, particularly for isolated and small wetlands (Baldwin and de Maynadier, 2009; Leonard et al., 2012), it remains the only comprehensive digital wetland coverage for Virginia. It is recognized that the utility of the wetland condition assessment model is dependent on the resolution and accuracy of the remotely sensed data and the ability to acquire similar data in the future. The delineation of individual wetland drainage areas will be improved as more light detection and ranging (LIDAR) data become available to refine the digital elevation models. In addition, landscape metrics and stressor relationships may change over time as various landscape best-management practices are voluntarily implemented or required. Periodic recalibration of the landscape-stressor relationship is necessary to detect such a change if it occurs.

The time series coastal plain/piedmont assessment suggests that current management programs are probably far from achieving the no net loss of function goals set by both federal and state governments. For both water quality and habitat ecosystem services, the current management practices that focus on keeping disturbances out of wetland boundaries is probably insufficient to preserve functional capacity. The stressors affecting wetland ecosystem services are largely created in the surrounding landscape. This implies that a new focus on preserving or restoring buffers around wetlands may be a means for attaining the goal.

WetCAT provides the potential for targeting management efforts based on assessment results. These include the mitigation/compensation potential suggested by the buffer condition mentioned above or landscape setting (Bedford, 1996). Another issue is linking wetland condition and impaired water designations for stratified management efforts (more stringent protection in some areas than others) or perhaps, stratified compensation requirements for permitted impacts.

ACKNOWLEDGMENTS

This work was supported, in part, by the EPA Region 3 Wetland Program Development grants and by the Virginia Department of Environmental Quality. This is Contribution No. 3649 of the Virginia Institute of Marine Science, William & Mary.

REFERENCES

Adamus, P.R., Danielson, T.J., Gonyaw, A., 2001. Indicators for Monitoring Biological Integrity of Inland, Freshwater Wetlands: A Survey of North American Technical Literature (1990-2000) (EPA843-R-01). U.S. Environmental Protection Agency, Office of Water, Washington, DC. http://www.epa.gov/owow/wetlands/bawwg/monindicators.pdf.

Anderson, S.H., Mann, K., Shugart Jr., H.H., 1977. The effect of transmission-line corridors on bird populations. Am. Midl. Nat. 97, 216–221.

Aust, M.W., Schoenholtz, S.H., Zaebst, T.W., Szabo, B.A., 1997. Recovery status of a tupelo-cypress wetland seven years after disturbance: silvicultural implications. For. Ecol. Manag. 90, 161–169.

Baldwin, R.F., de Maynadier, P.G., 2009. Assessing threats to pool-breeding amphibian habitat in an urbanizing landscape. Biol. Conserv. 142, 1628–1638.

Bedford, B.L., 1996. The need to define hydrologic equivalence at the landscape scale for freshwater wetland mitigation. Ecol. Appl. 6, 57–68.

Bedford, B.L., 1999. Cumulative effects on wetland landscapes: links to wetland restoration in the United States and Southern Canada. Wetlands 19 (4), 775–788.

Brazner, J.C., Danz, N.P., Trebitz, A.S., Niemi, G.J., Regal, R.R., Hollenhorst, T., Host, G.E., Reavie, E.D., Brown, T.N., Hanowski, J.M., Johnston, C.A., Johnson, L.B., Howe, R.W., Ciborowski, J.J.H., 2007. Responsiveness of Great Lakes wetland indicators to human disturbances at multiple spatial scales: a multi-assemblage assessment. J. Great Lakes Res. 33, 42–66.

Brinson, M.M., 1993. A Hydrogeomorphic Classificiation for Wetlands. U.S. Army Corps of Engineers, Waterways Experiment Station, Vicksburg, MS (Technical Report WRP-DE-4).

Brooks, R.P., Wardrop, D.H., Bishop, J.A., 2004. Assessing wetland condition on a watershed basis in the Mid-Atlantic region using synoptic land cover maps. Environ. Monit. Assess. 94, 9–22.

Brown, C.D., Dubus, I.G., Fogg, P., Spirlet, M., Gustin, C., 2004. Exposure to sulfosulfuron in agricultural drainage ditches: field monitoring and scenario-based modelling. Pest Manag. Sci. 60, 765–776.

Burns, D.A., McDonnell, J.J., 1998. Effects of a beaver pond on runoff processes: comparison of two headwater catchments. J. Hydrol. 205 (3–4), 248–264.

Buskirk, W.H., McDonald, J.L., 1995. Comparison of point count sampling regimes for monitoring forest birds. In: Ralph, C.J., Sauer, J.R., Droege, S. (Eds.), Monitoring Bird Populations by Point Counts. United States Forest Service General, (Technical Report PSW-GTR-149), pp. 25–34.

Butet, A., Leroux, A.B.A., 2001. Effects of agricultural development on vole dynamics and conservation of Montagu's harrier in western French wetlands. Biol. Conserv. 100, 289–295.

Chadwick, N.L., Progulske, D.R., Finn, J.T., 1986. Effects of fuelwood cutting on birds in southern New England. J. Wildl. Manag. 50, 398–405.

Chague-Goff, C., Rosen, M.R., 2001. Using sediment chemistry to determine the impact of treated wastewater discharge on a natural wetland in New Zealand. Environ. Geol. 40, 1411–1423.

Childers, D.L., Gosselink, J.G., 1990. Assessment of cumulative impacts to water quality in a forested wetland landscape. J. Environ. Qual. 19, 455–464.

Clarke, K.R., Warwick, R.M., 2001. Change in Marine Communities: An Approach to Statistical Analysis and Interpretation, second ed. PRIMER-E, Plymouth, UK.

Comeleo, R.L., Paul, J.F., August, P.V., Copeland, J., Baker, C., Hale, S.S., Latimer, R.W., 1996. Relationships between watershed stressors and sediment contamination in Chesapeake Bay estuaries. Landsc. Ecol. 11 (5), 307–319.

Conner, W.H., 1994. Effect of forest management practices on southern forested wetland productivity. Wetlands 14, 27–40.

Crosbie, B., Chow-Fraser, P., 1999. Percentage land use in the watershed determines the water and sediment quality of 22 marshes in the Great Lakes basin. Can. J. Fish. Aquat. Sci. 56, 1781–1791.

De Szalay, F.A., Batzer, D.P., Resh, V.H., 1996. The use of mesocosm and macrocosm experiments to examine the effects of mowing in emergent vegetation on wetland invertebrates. Environ. Entomol. 25, 303–309.

Detenbeck, N.E., Taylor, D.B., Lima, A., Hagley, C., 1996. Temporal and spatial variability in water quality of wetlands in the Minneapolis/St. Paul, MN metropolitan area: implications for monitoring strategies and designs. Environ. Monit. Assess. 40, 11–40.

Edginton, A.N., Sheridan, P.M., Stephenson, G.R., Thompson, D.G., Boermans, H.J., 2004. Comparative effects of pH and Visiont herbicide on two life stages of four anuran amphibian species. Environ. Toxicol. Chem. 23, 815–822.

Ehrehfeld, J.G., 2000. Evaluating wetlands within an urban context. Ecol. Eng. 15 (3–4), 253–265.

Ewing, K., 1996. Tolerance of four wetland plant species to flooding and sediment deposition. Environ. Exp. Bot. 36, 131–146.

Fairbairn, S.E., Dinsmore, J.J., 2001. Local and landscape-level influences on wetland bird communities of the prairie pothole region of Iowa, USA. Wetlands 21, 41–47.

Forman, R.T.T., Alexander, L.E., 1998. Roads and their major ecological effects. Annu. Rev. Ecol. Syst. 29, 207–231.

Gerakis, A., Kalburtji, K., 1998. Agricultural activities affecting the functions and values of Ramsar wetland sites of Greece. Agric. Ecosyst. Environ. 70, 119–128.

Groffman, P.M., Boulware, N.J., Zipperer, W.C., Pouyat, R.V., Band, L.E., Colsimo, M.F., 2002. Soil nitrogen cycle processes in urban riparian zones. Environ. Sci. Technol. 36 (21), 4547–4552.

Groffman, P.M., Bain, D.J., Band, L.E., Belt, K.T., Brush, G.S., Grove, J.M., Pouyat, R.V., Yesilonis, I.C., Zipperer, W.C., 2003. Down by the riverside: urban riparian ecology. Front. Ecol. Environ. 1 (6), 315–321.

Guadagnin, D.L., Maltchik, L., 2007. Habitat and landscape factors associated with neotropical waterbird occurrence and richness in wetland fragments. Biodivers. Conserv. 16, 1231–1244.

Gurnell, A.M., 1998. The hydrogeomorphological effects of beaver dam-building activity. Prog. Phys. Geogr. 22 (2), 167–189.

Hamel, P., Smith, B.W.P., Twedt, D.J., Woehr, J.R., Morris, E., Hamilton, R.B., Cooper, R.J., 1996. A Land Manager's Guide to Point Counts of Birds in the Southeast (Gen. Tech. Rep. SO-120). U.S. Dept. of Agriculture, Forest Service, Southern Forest Experiment Station, New Orleans, LA. 39 pp.

Hamer, A.J., McDonnell, M.J., 2008. Amphibian ecology and conservation in the urbanizing world: a review. Biol. Conserv. 141 (10), 2432–2449.

Hanowski, J.A.M., Christian, D.P., Nelson, N.C., 1999. Response of breeding birds to shearing and burning in wetland brush ecosystems. Wetlands 19 (3), 584–593.

Harris, L.D., 1988. The nature of cumulative impacts on biotic diversity of wetland vertebrates. Environ. Manag. 12, 675–693.

Hemond, H.F., Benoit, J., 1988. Cumulative impacts on water quality functions of wetlands. Environ. Manag. 12, 639–653.

Herlihy A.T., Sifneos J.C., Lomnicky G.A., Nahlik A.M., Kentula M.E., Magee T.K., Weber M.H. and Trebitz A.S. The response of wetland quality indicators to human disturbance across the United States. Environ. Monit. Assess. in press.

Hongo, H., Masikini, M., 2003. Impact of immigrant pastoral herds to fringing wetlands of Lake Victoria in Magu district Mwanza region, Tanzania. Phys. Chem. Earth 28, 1001–1007.

Houlahan, J.E., Findlay, C.S., 2003. The effects of adjacent land use on wetland amphibian species richness and community composition. Can. J. Fish. Aquat. Sci. 60, 1078–1094.

Houlahan, J.E., Findlay, C.S., 2004. Estimating the 'critical' distance at which adjacent land-use degrades wetland water and sediment quality. Landsc. Ecol. 19, 677–690.

JMP®, 2009. JMP®, Version 8. SAS Institute Inc., Cary, NC.

Johnston, C.A., 1994. Cumulative impacts to wetlands. Wetlands 14, 49–55.

Kalisinska, E., Salicki, W., Myslek, P., Kavetska, K.M., Jackowski, A., 2004. Using the Mallard to biomonitor heavy metal contamination of wetlands in north-western Poland. Sci. Total Environ. 320, 145–161.

Karr, J.R., Chu, E.W., 1999. Restoring Life in Running Waters: Better Biological Monitoring. Island Press, Washington, DC.

Kassenga, G.R., 1997. A descriptive assessment of the wetlands of the Lake Victoria basin in Tanzania. Resour. Conserv. Recycl. 20, 127–141.

Kauffman, J.B., Krueger, W.C., 1984. Livestock impacts on riparian ecosystems and streamside management implications: a review. J. Range Manag. 37, 430–438.

King, R.S., Baker, M.E., Whigham, D.F., Weller, D.E., Jordan, T.E., Kazyak, P.F., Hurd, M.K., 2005. Spatial considerations for linking watershed land cover to ecological indicators in streams. Ecol. Appl. 15 (1), 137–153.

Kingsford, R.T., 2000. Ecological impacts of dams, water diversions and river management on floodplain wetlands in Australia. Austral Ecol. 25 (2), 109–127.

Klopatek, J.M., 1988. Some thoughts on using a landscape framework to address cumulative impacts on wetland food chain support. Environ. Manag. 12, 703–711.

Larson, J.S., Mueller, A.J., MacConnell, W.P., 1980. A model of natural and man-induced changes in open freshwater wetlands on the Massachusetts coastal plain. J. Appl. Ecol. 17, 667–673.

Lee, L.C., Gosselink, J.G., 1988. Cumulative impacts on wetlands: linking scientific assessments and regulatory alternatives. Environ. Manag. 12, 591–602.

Leonard, P.B., Baldwin, R.F., Homyack, J.A., Wigley, T.B., 2012. Remote detection of small wetlands in the Atlantic coastal plain of North America: local relief models, ground validation, and high-throughput computing. For. Ecol. Manag. 284, 107–115.

Lichvar, R.W., 2012. The National Plant List (No. ERDC-CRREL-TR-12-11). Engineer Research and Development Center, Cold Regions Research and Engineering Lab, Hanover, NH.

Lopez, R.D., Fennessy, M.S., 2002. Testing the floristic quality assessment index as an indicator of wetland condition. Ecol. Appl. 12, 487–497.

Magnusson, B., Stewart, J.M., 1987. Effects of disturbances along hydroelectrical transmission corridors through peatlands in northern Manitoba, Canada. Arct. Alp. Res. 19 (4), 470–478.

Marrs, R.H., Frost, A.J., Plant, R.A., Lunnis, P., 1992. The effects of herbicide drift on semi-natural vegetation: the use of buffer zones to minimize risks. Asp. Appl. Biol. 29, 57–64.

Mayo, A.W., Bigambo, T., 2005. Nitrogen transformation in horizontal subsurface flow constructed wetlands I: model development. Phys. Chem. Earth 30 (11–16), 658–667.

McLaughlin, J.W., Gale, M.R., Jurgensen, M.F., Trettin, C.C., 2000. Soil organic matter and nitrogen cycling in response to harvesting, mechanical site preparation, and fertilization in a wetland with a mineral substrate. For. Ecol. Manag. 129, 7–23.

Mehlman, D.W., Rosenberg, K.V., Wells, J.V., Robertson, B., 2004. A comparison of North American avian conservation priority ranking systems. Biol. Conserv. 120, 383–390.

Minitab®, 2010. Minitab 16 Statistical Software (Computer Software). Minitab, Inc., State College, PA.www.minitab.com.

Nickerson, N.H., Dobberteen, R.A., Jarman, N.M., 1989. Effects of power-line construction on wetland vegetation in Massachusetts, USA. Environ. Manag. 13 (4), 477–483.

O'Brien, A.L., 1988. Evaluating the cumulative effects of alteration on New England wetlands. Environ. Manag. 12, 627–636.

Oliver, C.D., 1980. Forest development in North America following major disturbances. For. Ecol. Manag. 3, 153–168.

Otte, M.L., 2001. What is stress to a wetland plant? Environ. Exp. Bot. 46, 195–202.

Panno, S.V., Nuzzo, V.A., Cartwright, K., Hensel, B.R., Krapac, I.G., 1999. Impact of urban development on the chemical composition of ground water in a fen-wetland complex. Wetlands 19, 236–245.

Pashley, D.N., Beardmore, C.J., Fitzgerald, J.A., Ford, R.P., Hunter, W.C., Morrison, M.S., Rosenberg, K.V., 2000. Partners in Flight: Conservation of the Land Birds of the United States. American Bird Conservancy, The Plains, VA.

Pimentel, D., Lach, L.R., Zuniga, R., Morrison, D., 2000. Environmental and economic costs of nonindigenous species in the United States. Bioscience 50, 53–65.

Pontier, H., Williams, J.B., May, E., 2004. Progressive changes in water and sediment quality in a wetland system for control of highway runoff. Sci. Total Environ. 319, 215–224.

Preston, E.M., Bedford, B.L., 1988. Evaluating cumulative effects on wetland functions: a conceptual overview and generic framework. Environ. Manag. 12, 565–583.

Reinelt, L., Horner, R., Azous, A., 1998. Impacts of urbanization on palustrine (depressional freshwater) wetlands—research and management in the Puget Sound region. Urban Ecosyst. 2, 219–236.

Richardson, C.J., 1994. Ecological functions and human values in wetlands: a framework for assessing forestry impacts. Wetlands 14 (1), 1–9.

Roe, J.H., Gibson, J., Kingsbury, B., 2006. Beyond the wetland border: estimating the impact of roads for two species of water snakes. Biol. Conserv. 130, 161–168.

Rooney, R.C., Bayley, S.E., Creed, I.F., Wilson, M.J., 2012. The accuracy of land cover-based wetland assessments is influenced by landscape extent. Landsc. Ecol. 27, 1321–1335.

Rosgen, D.L., 2001. In: A stream channel stability assessment methodology. A practical method of computing streambank erosion rate.7th Federal Interagency Sedimentation Conference, March 25–29, Reno, NV.

Rothenbücher, J., 2005. The Impact of Mowing and Flooding on the Diversity of Arthropods in Floodplain Grassland Habitats of the Lower Oder Valley National Park, Germany (Thesis/Dissertation). University of Göttingen, Germany.

Sakai, A.K., Allendorf, F.W., Holt, J.S., Lodge, D.M., Molofsky, J., With, K.A., Baughman, S., Cabin, R.J., Cohen, J.E., Ellstrand, N.C., McCauley, D.E., O'Neil, P., Parker, I.M., Thompson, J.N., Weller, S.G., 2001. The population biology of invasive species. In: Fautin, D.G., Futuyman, D.J., Shaffer, H.B. (Eds.), Annual Review of Ecology and Systematics. In: vol. 32. Annual Reviews, Palo Alto, CA, pp. 305–332.

Schlesinger, M.D., Manley, P.N., Holyoak, M., 2008. Distinguishing stressors acting on land bird communities in an urbanizing environment. Ecology 89 (8), 2302–2314.

Stevens Jr., D.L., Olsen, A.R., 2004. Spatially-balanced sampling of natural resources in the presence of frame imperfections. J. Am. Stat. Assoc. 99, 262–278.

Thibodeau, F.R., 1985. Changes in a wetland plant association induced by impoundment and draining. Biol. Conserv. 33, 269–279.

Thurston, K., 1999. Lead and petroleum hydrocarbon changes in an urban wetland receiving stormwater runoff. Ecol. Eng. 12, 387–399.

Trombulak, S.C., Frissel, C.A., 2000. Review of ecological effects of roads on terrestrial and aquatic communities. Conserv. Biol. 14 (1), 18–30.

Van Sickle, J., Paulsen, S.G., 2008. Assessing the attributable risks, relative risks, and regional extents of aquatic stressors. J. N. Am. Benthol. Soc. 27 (4), 920–931.

Vickers, C.R., Harris, L.D., Swindel, B.F., 1985. Changes in herpetofauna resulting from ditching of cypress ponds in coastal plains flatwoods. For. Ecol. Manag. 11, 17–29.

Walbridge, M.R., Lockaby, B.G., 1994. Effects of forest management on biogeochemical functions in southern forested wetlands. Wetlands 14, 10–17.

Wardrop, D.H., Brooks, R.P., 1998. The occurrence and impact of sedimentation in central Pennsylvania wetlands. Environ. Monit. Assess. 51, 119–130.

Wardrop, D.H., Kentula, M.E., Stevens, D.L., Jensen, S.F., Brooks, R.P., 2007. Assessment of wetland condition: an example from the upper Juniata watershed in Pennsylvania, USA. Wetlands 27 (3), 416–431.

Wardrop, D.H., Kentula, M.E., Brooks, R.P., Fennessy, M.S., Chamberlain, S.J., Havens, K.J., Hershner, C., 2013. Monitoring and assessment of wetlands: concepts, case studies, and lessons learned. In: Mid-Atlantic Freshwater Wetlands: Advances in Wetlands Science, Management, Policy, and Practice. Springer, New York, pp. 381–419.

Woo, M., 1979. Effects of power line construction upon the carbonate water chemistry of part of a mid-latitude swamp. Catena 6, 219–233.

Xu, Y., Burger, J.A., Aust, W.M., Patterson, S.C., Miwa, M., Preston, D.P., 2002. Changes in surface water table depth and soil physical properties after harvest and establishment of loblolly pine (*Pinus taeda* L.) in Atlantic coastal plain wetlands of South Carolina. Soil Tillage Res. 63, 109–121.

Zedler, J.B., 2003. Wetlands at your service: reducing impacts of agriculture at the watershed scale. Front. Ecol. Environ. 1, 65–72.

Zedler, J.B., Kercher, S., 2004. Causes and consequences of invasive plants in wetlands: opportunities, opportunists, and outcomes. Crit. Rev. Plant Sci. 23, 431–452.

FURTHER READING

Groffman, P.M., Hanson, G.C., Kiviat, E., Stevens, G., 1996. Variation inmicrobial biomass and activity in four different wetland types. Soil Sci. Soc. Am. J. 60, 622–629.

Reed, P.B., 1988. National List of Plant Species That Occur in Wetlands (Report No. 88(26.3)). U.S. Fish and Wildlife Service, Washington, DC.

Chapter 2.2.9

The Use of Landscape-Level Assessment for Producing a Decision-Support Tool for Puget Sound Watersheds

Stephen Stanley
Washington Department of Ecology, Shorelands and Environmental Assistance Program, Lacey, WA, United States

Chapter Outline

INTRODUCTION

Both nationally and within the state of Washington, there is a growing understanding that a watershed or landscape approach is necessary to protect the overall health of aquatic ecosystems (NRC, 1996, 2001; Spence et al., 1996; Dale et al., 2000; Roni et al., 2002; Simenstad et al., 2006; Beechie et al., 2010). This understanding is based on more than 20 years of watershed research demonstrating that physical processes, such as the movement of water and sediment, operating throughout a watershed at multiple spatial and temporal scales control the biological and physical functions at the site scale (Naiman and Bilby, 1998; Beechie and Bolton, 1999; Hobbie, 2000; Benda et al., 2004; Simenstad et al., 2006; King County, 2007). It has been argued that without this type of information, local governments will continue the traditional patterns of land-use development driven primarily by infrastructure needs/limitations. Over time, those patterns of development may significantly degrade areas important for sustaining watershed processes to the point that site-scale actions to protect or restore functions in downstream wetland and stream ecosystems become ineffective.

Beginning in 2004, the Washington Department of Ecology (WDOE) developed landscape-level methods that assessed and assigned a relative value of the condition of those broad-scale landscape processes that drive the structure and function of wetlands and stream ecosystems. The methods were initially applied to watersheds of the Puget Sound region. Acceptance of this approach was bolstered in January 2004 by the substantially revised shoreline guidelines (173-26 WAC, Part III) that required, in part, the use of a watershed-scale assessment method in the development of shoreline master programs. The addition of a watershed assessment method also complemented the WDOE wetlands rating system (Hruby, 2014) and provided a toolbox to help local governments assess wetlands functions and processes over multiple spatial scales.

With the help of grants from the U. S Environmental Protection Agency (EPA), the department published a series of documents (see Stanley et al., 2005; Stanley and Grigsby, 2008; Stanley et al., 2016) to help guide local governments and NGOs in the interpretation and application of landscape-scale information over multiple scales to their planning and permitting actions within Puget Sound. The later documents included a coarse-scale decision-support tool (not a decision-making tool) providing information for regional, county, and watershed-based planning. This tool consists of a set of water and habitat assessments that compares areas within a watershed for identifying the most suitable areas for restoration, protection, and development. The areas prioritized for protection, restoration, conservation, and development were also linked to specific key actions (e.g., solution templates) necessary to protect and restore watershed processes. The overall results are known as the Puget Sound Watershed Characterization.

Wetland and Stream Rapid Assessments. https://doi.org/10.1016/B978-0-12-805091-0.00023-2

It was anticipated that the resulting land-use patterns from application of this tool would better protect the health of Puget Sound's terrestrial and aquatic resources while also helping to direct limited financial resources to the highest priority areas for restoration and protection.

Overall, the objectives of the guidance and decision-support tool were to:

- Develop, prioritize, and implement solutions to environmental problems such as wetland fill or stream stabilization, based on an understanding of processes at watershed (water assessment) or landscape (habitat assessment) scales.
- Replace planning based on jurisdictional or statutory boundaries (e.g., city and county boundaries) with coordinated regional planning.
- Provide a watershed-scale context to help guide site-scale reviews of land-use changes, including alterations to wetland that not only meet regulatory requirements but also more fully achieve their intended outcomes.
- Move toward integrated resource planning and management grounded in a landscape-scale understanding of how ecosystems work.

CONCEPTUAL FRAMEWORK

In developing a scientifically valid decision-support tool for an area as large as Puget Sound (15,000 $mile^2$ or >40,000 km^2) within a reasonable period of time, it was necessary to determine the most important questions that might be answered by the tool and its assessments. Resource planners indicated that they needed to know where to locate new development in a watershed so that it minimized impacts to key processes and they needed to know the best locations for actions that protect or restore watershed processes. This type of output did not require the measurement of rates and quantities of water, sediment, or nutrients but rather the comparison of areas between individual watersheds that controlled their delivery, movement, and loss. Therefore, collection of data and calibration of models to address finer-scale questions such as frequency of flooding, which might have required 20 years of stream flow data, were unnecessary.

The type of controls on the landscape that regulate the delivery, movement, and loss of water, sediment, nutrients, metals, and toxins could be identified by using existing research, data, and input by experts. For example, areas of water storage (Fig. 2.2.9.1) primarily control the movement of surface water across the landscape. Indicators for storage areas include depressional wetlands, floodplains, and lakes. With the aid of digital data for land cover, soils, topography, wetlands, lakes, and streams, a uniform database was developed across Puget Sound to show the spatial extent and distribution of these storage areas. As a result, small watersheds, or assessment units (AUs), across Puget Sound could be compared to each other for their relative presence of these storage features. This ability to compare across watersheds or AUs also applied to the other controls and their indicators.

The decision-support tool consists of two submodels that measure both the importance of and level of degradation to the processes for delivery, movement and loss of water, sediment, nutrients, metals, and toxins. These two indices were combined using the matrix presented in Fig. 2.2.9.2 so that the tool results could be displayed in maps that were useful to planners.

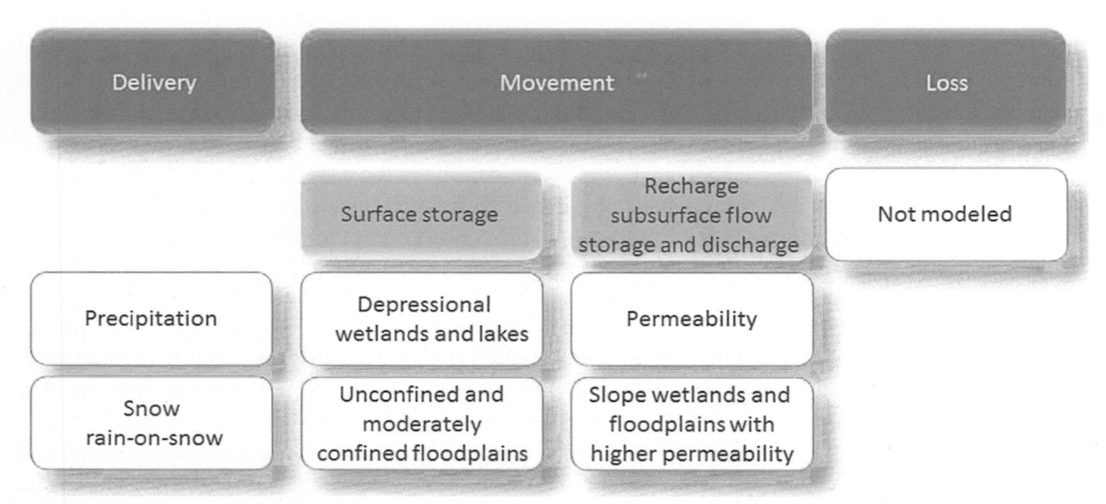

FIG. 2.2.9.1 Submodel for assessing the "importance" of the process for the delivery, movement, and loss of water. For each component (delivery, movement, and loss) the primary indicators were identified (e.g., for delivery: precipitation, snow, and rain on snow).

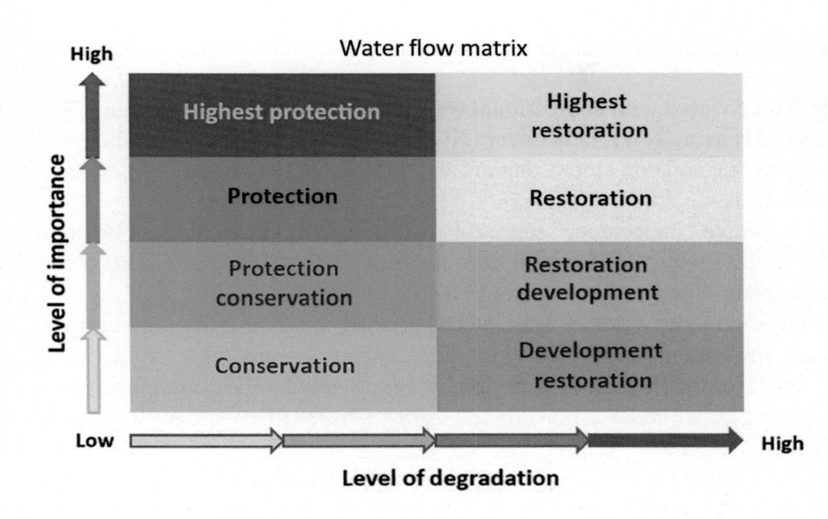

FIG. 2.2.9.2 Water flow matrix. Matrix for combining results of importance and degradation models in order to identify management actions of protection, restoration, or development.

The AUs used by the tool were aggregated from smaller watersheds mapped as part of the Salmon, Steelhead Habitat Inventory and Assessment Program (SSHIAP) and range from a few coastal units as small as 1 mile^2 to mountainous units more than 40 mile^2, with a median size of 3.4 mile^2 (8.8 km^2). The size of the AUs represented a tradeoff between increased resolution, analytical complexity, availability of other data at appropriate scale, and the intended uses of the results.

The decision-support tool evolved to a modeling effort using existing spatial data, research, and expert opinion. Because the models were not calibrated or validated, the Department of Ecology formed a watershed team in 2011 to assist local governments in interpreting and applying the results of the decision-support tool appropriately. The team consisted of experts in hydrology, water quality, geomorphology, watershed ecology, and fisheries and wildlife biology. This resulted in the application of the tool to a number of land-use plans, including the Gorst subarea plan, which is further discussed below. Both the tool and the guidance on its application have been modified according to outcomes from its applications in the field. Initially, measurement of success of the tool can be based on whether its application has actually changed land-use planning and zoning patterns. However, to properly test the model, a long-term monitoring program needs to be applied. This could consist of monitoring water flow and water quality processes in two relatively similar, relatively undeveloped watersheds, with one applying the tool to planning and the other following conventional planning avenues (e.g., application of a critical areas ordinance).

DESCRIPTION OF METHOD

The methods consist of two submodels, one for "importance" and one for "degradation," that are both applied to every AU across the Puget Sound region. As designed, they assess the relative status or condition within an individual watershed of the processes that control water quantity and water quality. In this chapter, the focus is on the models of water flow and, in particular, on processes related to storage of surface flows in depressional wetlands.

The fundamental assumption of the submodel that characterizes importance is that different parts of the landscape have intrinsic differences in their ability to support natural volumes, rates, and timing of delivery, storage, movement, and loss of water. Those areas that are most essential to maintaining natural flow regimes will presumably be those areas most critical to the support of aquatic biota that have evolved in concert with these natural conditions.

The submodel for importance evaluates each AU in its "unaltered" state—that is, based on its physical attributes of topography, soils, geology, and hydrology, and without any consideration of land-use changes or human modifications that may have occurred. It considers four fundamental groups of water flow processes: delivery, surface storage, movement (separated into recharge and discharge), and loss of water in each AU.[1] Fig. 2.2.9.1 presents both these fundamental groups of water flow processes and the indicators for each of these processes.

For delivery, the model considers the relative volume of precipitation and the timing of the delivery of that precipitation to lower-elevation areas as events potentially affected by rain occurring while ground surfaces are snow-covered. For the

1. In western Washington, the assumption is that all AUs have approximately the same rate of evapotranspiration in nondegraded conditions because they were all generally forested; thus evapotranspiration was not included as a variable in the Model 1 equation.

movement of water, the model considers the relative area of surface storage and the relative area contributing to subsurface flow, recharge, and discharge.

Wetland storage is based on the percentage of AUs covered with depressional wetlands (both upland and riverine). The percentage of possible wetlands is estimated for all AUs using NWI, land cover (NOAA), and the intersection of the topographic layer with the hydric soil layer. Areas with hydric soils on slopes that are less than 2% are considered to be areas where storage wetlands exist or have existed in the past.

Table 2.2.9.1 presents the steps in calculating a relative "importance" score for surface water storage that can be compared across AUs within a single watershed. AU "B" in Table 2.2.9.1 would have the highest score for the importance of wetland storage within a hypothetical watershed containing two AUs, "A" and "B."

The submodel for degradation evaluates the watershed in its "altered" state by considering the impact of human actions to the four water flow processes (delivery, storage, movement, and loss) across all landscape groups. This evaluation is based on the magnitude of human-affected land cover (for the Puget Sound region, this is assumed to be all nonforest land except those limited areas that are natural grassland and water bodies), constructed infrastructure (roads and rooftops), and measures of consumptive water extraction and use.

Degradation of wetland storage is modeled as the relative loss of surface storage of wetlands in an AU. The severity of wetland storage loss is characterized in terms of wetlands that are permanently degraded due to urbanization, and those temporarily degraded due to extensive ditching/tiling in agricultural and rural areas.

Table 2.2.9.2 presents the steps in calculating a relative score for degradation to water storage that can be compared across AUs within a single watershed. AU "B" in Table 2.2.9.2 would have the highest score for the degradation of wetland storage within a hypothetical watershed containing two AUs, "A" and "B."

By combining the two scores (importance and degradation) in a matrix (Fig. 2.2.9.2), the type of management action can be identified. AU "B" would have the highest score for both importance and degradation and thus wetlands located there would have a high priority for restoration actions. This approach has the ability, therefore, to predict the restoration

TABLE 2.2.9.1 Example of Importance Model Steps for Calculating Wetland Storage (WS) for Depressional Wetlands (DW) in Watershed With AUs "A" and "B"

Steps	Equation	Calculation	Score
Calculate percent storage in AU A	$WS = \frac{\text{Area DW A} \times 100}{\text{Total area in AU A}}$	$WS = \frac{10\text{ acres} \times 100}{100\text{ acres}}$	10
Calculate percent storage in AU B	$WS = \frac{\text{Area DW B} \times 100}{\text{Total area in AU B}}$	$WS = \frac{60\text{ acres} \times 100}{300\text{ acres}}$	20
Divide highest scoring AU in assessment watershed for percent storage: AU A	$WS^{\text{Normalized}} = \frac{\text{A WS Score}}{\text{Highest AU Score}}$	$WS^{\text{Normalized}} = \frac{10}{20}$	0.5
Divide highest scoring AU in assessment watershed for percent storage: AU B	$WS^{\text{Normalized}} = \frac{\text{B WS Score}}{\text{Highest AU Score}}$	$WS^{\text{Normalized}} = \frac{20}{20}$	1.00

TABLE 2.2.9.2 Example of Degradation Model Steps for Calculating Wetland Storage (WS) for Depressional Wetlands (DW) in Watershed With AUs "A" and "B"

Steps	Equation	Calculation	Score
Calculate degradation loss for urban wetlands in AU A	$WS = \frac{\text{Area DW A} \times 3}{\text{Total area in AU A}}$	$WS = \frac{10\text{ acres} \times 3}{100\text{ acres}}$	0.3
Calculate degradation loss for rural wetlands in AU B	$WS = \frac{\text{Area DW B} \times 2}{\text{Total area in AU B}}$	$WS = \frac{60\text{ acres} \times 2}{300\text{ acres}}$	0.4
Divide AU A score by highest scoring AU in assessment watershed	$WS^{\text{Normalized}} = \frac{\text{A WS Score}}{\text{Highest AU Score}}$	$WS^{\text{Normalized}} = \frac{0.3}{0.4}$	0.75
Divide AU B score by highest scoring AU in assessment watershed	$WS^{\text{Normalized}} = \frac{\text{B WS Score}}{\text{Highest AU Score}}$	$WS^{\text{Normalized}} = \frac{0.4}{0.4}$	1.00

potential of wetlands whereas other methods assessing wetland functions examine only the condition at the time of assessment. When local governments use wetland assessments to assign categories of protection required under critical areas ordinances, degraded wetlands (e.g., rural and agricultural wetlands) that have the potential to provide restoration of important watershed-wide storage processes and functions are typically assigned a lower level of protection (e.g., narrower buffers, more disturbance of wetland allowed).

Additional details on the scoring and normalization methods for the variables used in the importance and degradation models are provided on the Department of Ecology publications website: https://fortress.wa.gov/ecy/publications/parts/1106016part2.pdf

TIME SPENT IN DEVELOPING THE METHOD

From 2005 to 2010, the Department of Ecology developed the method and tested it by applying it to several counties as part of their shoreline master plan updates. This included Clark, Jefferson, Lewis, and Whatcom Counties. The models were then revised and updated based on the results of these applications, and internal review of results by ECOLOGY scientists plus two rounds of input from a technical review committee and external peer review of the methods (2005, 2009).

EXAMPLE OF METHOD APPLICATION

The 6570-acre Gorst Creek watershed (Fig. 2.2.9.3) is located on the west side of Puget Sound in the central portion of the Kitsap peninsula. The watershed contains 440 acres of wetlands, including 201 acres of slope wetlands and 239 acres of depressional wetlands (Fig. 2.2.9.4). These "potential" wetlands were identified using a combination of NWI, hydric soils, and land cover data (NOAA). In 2012, the Departments of Ecology and Fish and Wildlife prepared a watershed assessment of the Gorst watershed for the city of Bremerton in order to: (1) assist in developing a watershed-based management plan for the freshwater and terrestrial portions of the watershed, and (2) identify the best areas for protection, conservation,

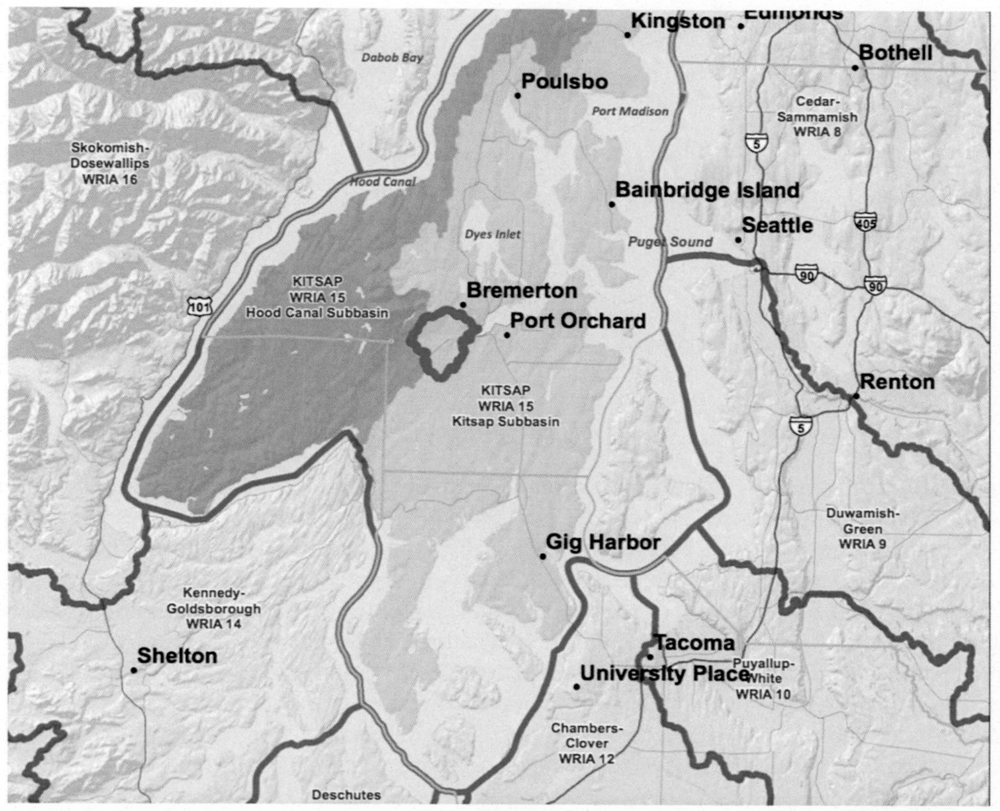

FIG. 2.2.9.3 The Gorst Creek watershed in the Puget Sound region of Washington (the outlined area adjacent to the City of Bremerton).

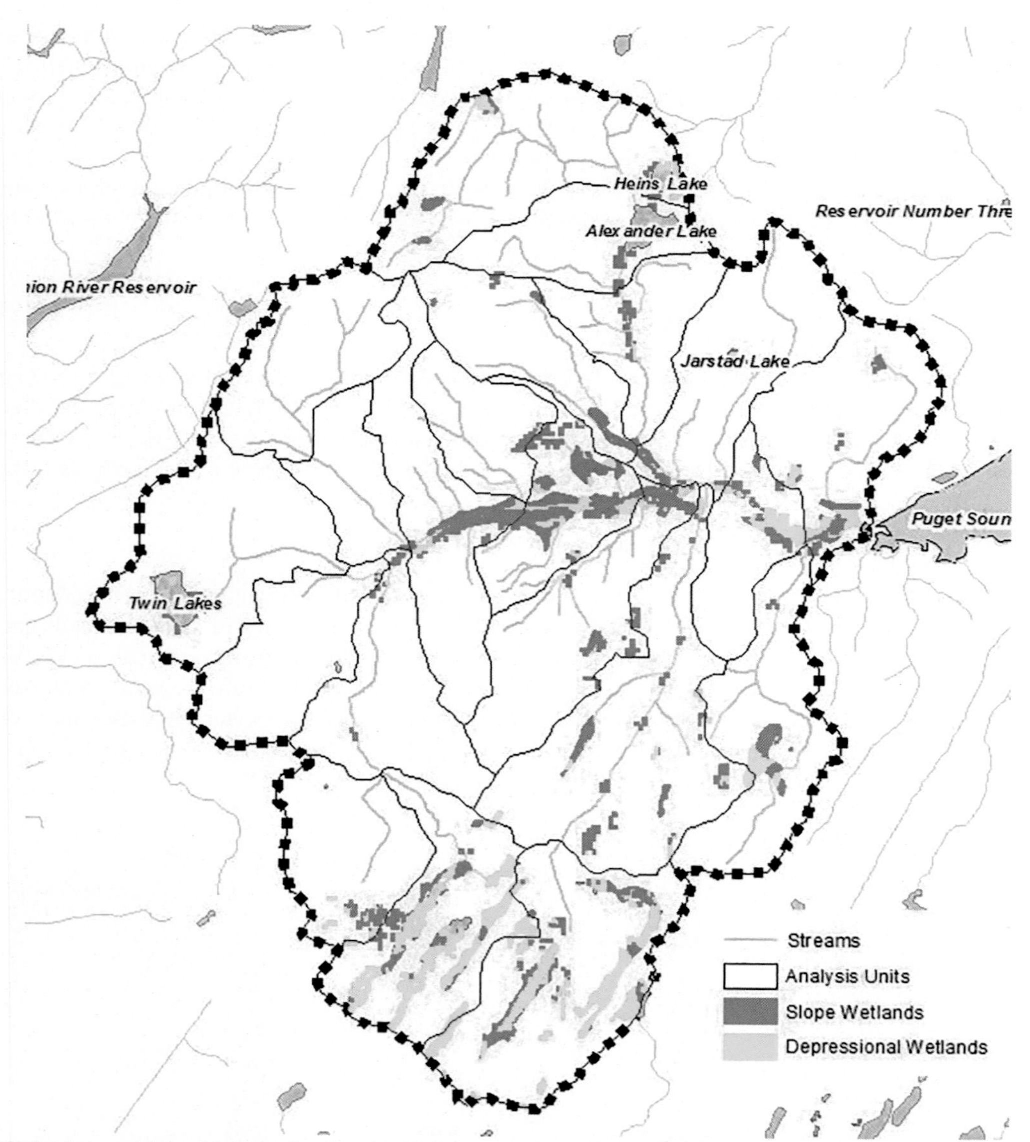

FIG. 2.2.9.4 Wetland cover (slope and depressional wetlands) for the Gorst watershed. Slope wetlands are used in the model as important areas for groundwater discharge. Depressional wetlands are important areas for surface water storage.

restoration, and development. Because a large portion of the relatively undeveloped Gorst watershed was within their urban growth boundary, the city of Bremerton and Kitsap County had the opportunity to site and design new development in a manner that would help protect, restore, and sustain the aquatic ecosystem of Gorst Creek.

The results for wetland storage, for example, helped identify a group of wetlands located in the southeast AU of the watershed as the highest ranked for the wetland storage function and for restoration (Figs. 2.2.9.5 and 2.2.9.6). In particular, this area within Kitsap County jurisdiction contained the area with the highest level of degradation and therefore was the top candidate for undertaking actions for restoring storage functions (Zone 2B in Fig. 2.2.9.7). Because the city limits for Bremerton do not extend into this AU, they are not required to implement these restoration actions along with the implementation of their watershed-based subarea plan within the city limits. This will require action by Kitsap County. Presently, the management plans for the Gorst watershed have not conducted assessments of wetland function (e.g., Washington State

Assessing Wetland Storage Function at a Landscape Scale
Gorst Watershed, Washington

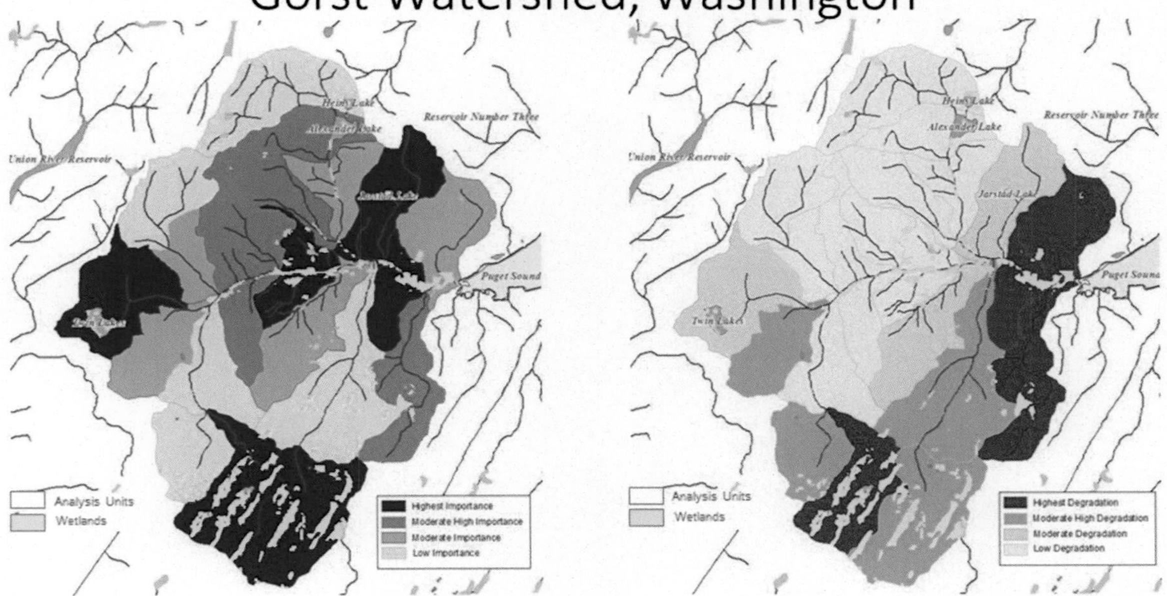

Depressional wetlands(e.g. yellow or lighter shaded linear features) with highest importance for storage function shown in darker shaded units.

Depressional wetlands with highest degradation for storage function shown in darker shaded units.

Greatest Restoration Potential = Highest ranking for restoration and degradation

FIG. 2.2.9.5 Wetlands rated by wetland storage function and degree of degradation. The left panel shows the individual wetlands (light-colored linear features) ranked for importance to the storage function. Wetlands within the darkest shaded assessment units have the highest importance for the storage function while wetlands within the lightest shaded units have the least importance for this function. The right panel indicates which assessment units have the highest level of degradation (darkest shading). The results suggest a high potential for restoring the storage function for wetlands located within the assessment areas with the highest ranking for both importance and degradation (see Fig. 2.2.9.6).

Wetland Rating System) in order to identify detailed restoration actions for restoring storage functions. Local jurisdictions typically do not require application of this rating system until development is proposed on properties containing wetlands.

Based on the assessment results for the individual water flow components (delivery, storage, recharge, and discharge) in the Gorst watershed, the 19 AUs displayed spatial patterns that suggested an overall distribution of regions broadly suited for restoration, protection, and development. Results from the individual components of the water flow model (delivery, storage, recharge, and discharge) helped identify the specific type of restoration and protection actions for those zones (Fig. 2.2.9.7).

TIME SPENT IN FIELD

The Gorst watershed characterization and similar ones require approximately one to two weeks to map the AUs and run the models and one to two days in the field to verify both the consistency of results and identify what type of protection and mitigation actions would be appropriate for each of the protection, restoration, and development zones. Field observations suggest whether inaccuracies in the spatial data are so great that another run of the model may be advisable. Overall, the modeling process, field work, and write-up typically take three to four weeks. Results can be delivered within days once all relevant spatial data layers have been acquired, placed on a common scale, and matched and programmed to access these databases automatically.

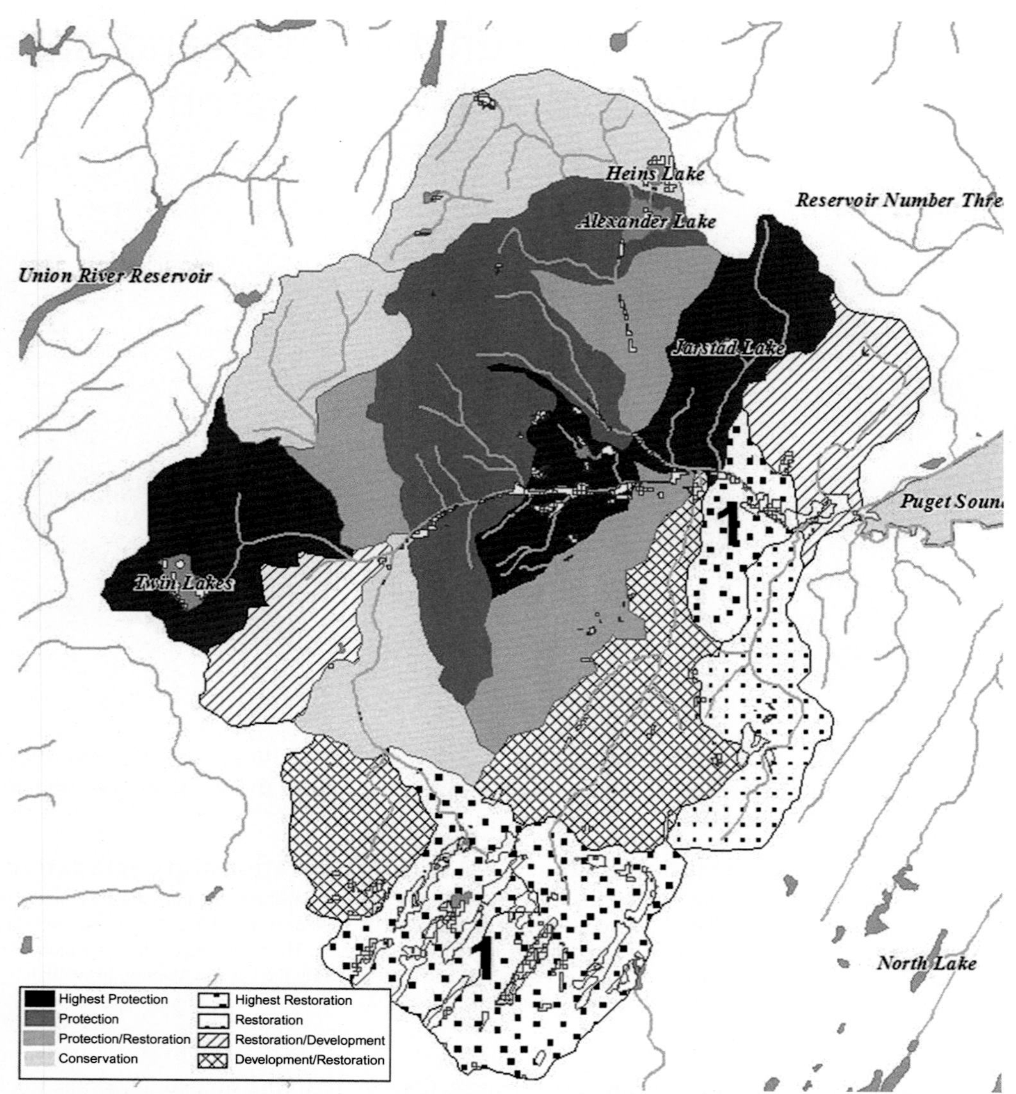

FIG. 2.2.9.6 Wetlands with the highest potential for restoring wetland storage function. By applying the water flow matrix in Fig. 2.2.9.2, the overall management actions for the Gorst watershed can be identified. Wetlands within areas marked with the number 1 had the highest potential for restoring this function.

HOW DATA WERE USED

Using the characterization data, the city developed the Gorst Creek Watershed Characterization and Framework Plan (City of Bremerton, 2013a,b) and subarea plan (Volume 3) for the Gorst watershed, and received a "Smart Communities" award (Judges Merit Category) from the governor's office in 2015.

The Gorst Creek Watershed Characterization and Framework Plan provides a common set of goals, policies, and best management practices intended for adoption and implementation by the city, which governs nearly two-thirds of the watershed in its city limits, and by Kitsap County, which governs unincorporated lands comprising the rest of the watershed.

WHAT WAS LEARNED

The use of watershed planning information by local governments in Puget Sound has revealed several key issues: (1) lack of coordinated planning across watersheds, (2) lack of guidelines and rules on how to implement watershed assessment results, and (3) lack of mid-scale assessments (hundreds of acres) to link site-scale response to watershed-scale actions.

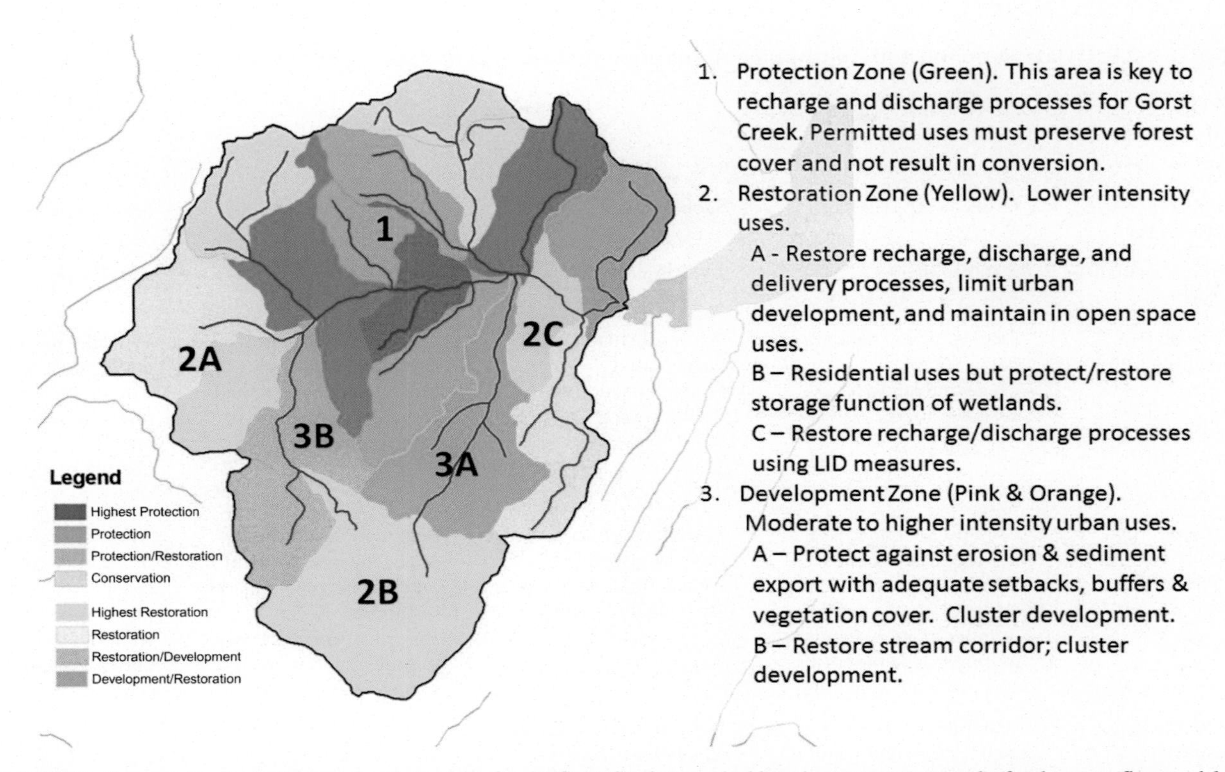

FIG. 2.2.9.7 Suggested watershed management strategies for the Gorst Creek watershed based on assessment results for the water flow model.

Most watersheds encompass several jurisdictions (e.g., cities and counties) that typically plan just within their political boundaries. However, many root causes of problems such as flooding may be located outside a jurisdiction's planning boundaries, making it difficult to develop and implement solutions. Further, there is no clear legal requirement for jurisdictions within an individual watershed to use watershed information in a coordinated manner. For example, the City of Bremerton created a subarea plan for its jurisdiction located within the lower portion of the Gorst watershed. The county, however, was not required to simultaneously develop and adopt a similar detailed management plan for the upper portion of the watershed, which is critical to the success of the city's plan. Multiple local governments must cooperate at the watershed scale in order for any watershed-based management plan to be fully successful. This is a serious challenge for natural resource managers.

Implementation of watershed assessment information is also hindered by the lack of a comprehensive set of guidelines and regulatory rules at the state level to assist locals, as is already the case with wetlands and streams (e.g., critical areas ordinances). A comprehensive watershed assessment approach across multiple spatial scales would help locals select the correct type of watershed analysis as well as understand how to interpret and apply results, including selection of the most effective regulatory and mitigation measures (Table 2.2.9.3).

Presently, results from assessments operating at the scale of several square miles cannot provide certainty that actions implemented at that scale will result in improvement of aquatic functions at the site scale. This underscores the need to better link the two scales with additional assessments operating at the mid-scale as well as to monitor the effect of mitigation installed at the broad scale upon restoring stream and wetland processes at the site scale. An appropriate monitoring effort could reveal the effect of mitigation across the watershed scale upon functions and processes at the site scale for wetlands and streams.

However, because the availability of the basic sets of spatial data to conduct assessments at multiple scales is not available throughout the United States and is absent in many parts of the world, it is difficult to conduct assessments that link broad, mid, and fine scales. If we are to be successful in protecting aquatic resources worldwide, landscape scientists need to help initiate programs to fund collection of watershed data over multiple scales through international consortiums of governmental entities or private organizations.

TABLE 2.2.9.3 Data Required and Appropriate Application of Results at Multiple Spatial Scales

Level of Information and Analysis	Course/General ⇨ Fine/Detailed			
Unit of organization	Basin, WRIA, Subbasin	Subbasin, valley segment	Reaches, water bodies	Segments/sites
Typical spatial scale	>100 mile^2	1–100 mile^2	100 acres–1 mile^2	<100 acres
Type of data acquisition	Existing spatial data from Puget Sound Characterization or similar source.	Existing spatial data from Puget Sound Characterization or similar source.	Use existing data or collect new data on biological, physical, and chemical conditions at these scales.	Typically requires collection of new data on biological, physical, and chemical conditions at these scales.
Type of application at each level	Land-use planning and zoning. This includes location, type, and intensity of development in order to avoid impacts to broad-scale processes and site-scale functions.	Refinements of coarse-scale assessment for application to land use planning and zoning.	Reach and watershed scale strategies for land and water protection and restoration. Reach specific actions or BMPs to protect and restore processes and functions.	Site and reach scale project designs for specific BMPs to protect and restore water bodies (functions and processes).
How the Puget Sound Characterization results could be applied	Water flow and water quality, fish, and wildlife habitat assessments are most applicable at this scale. Helps inform management decisions on the best locations for development and protection/restoration actions and strategies.	The water flow and water quality assessments provide information at the subbasin scale for informing management decisions implemented through land-use plans, policies, and regulations.	The Puget Sound Watershed Characterization does not provide results at these scales. However, the characterization's results should be used to confirm whether actions at these scales are appropriate. For example, installation of wood at a site or reach scale should not be undertaken if upper watershed delivery and storage processes are degraded.	

PROSPECTS FOR THE FUTURE

There is continuing interest in using watershed information and characterization to support and inform planning decisions. The EPA has provided funding to further refine the Puget Sound Characterization by developing mid-scale models (hundreds of acres), which includes a hydrologic condition and low flow indices. An interactive tool for using the characterization results can be accessed at the following site: https://fortress.wa.gov/ecy/coastalatlas/wc/landingpage.html

REFERENCES

Beechie, T., Bolton, S., 1999. An approach to restoring salmonid habitat-forming processes in Pacific Northwest watersheds. Fisheries 24, 6–15.

Beechie, T.J., Sear, D.A., Olden, J.D., Pess, G.R., Buffington, J.M., Moir, H., Roni, P., Pollock, M.M., 2010. Process-based principles for restoring river ecosystems. Bioscience 60, 209–222.

Benda, L., Poff, N.L., Miller, D., Dunne, T., Reeves, G., Pess, G., Pollock, M., 2004. The network dynamics hypothesis: how channel networks structure riverine habitats. Bioscience 54, 412–427.

City of Bremerton, 2013a. Gorst Creek Watershed Characterization & Framework Plan. vol. 1. Department of Community Development.http://www.ci.bremerton.wa.us/DocumentCenter/View/1467.

City of Bremerton, 2013b. Gorst Creek Sub-Area Plan. vol. 3. Department of Community Development.

Dale, V.H., Brown, S., Haeuber, R.A., Hobbs, N.T., Huntly, N., Naiman, R.J., Riebsame, W.E., Turner, M.G., Valone, T.J., 2000. Ecological principles and guidelines for managing the use of land. Ecol. Appl. 10, 639–670.

Hobbie, J.E. (Ed.), 2000. Estuarine Science: A Synthetic Approach to Research and Practice. Island Press, Washington, DC.

Hruby, T., 2014. Washington State Wetland Rating System for Western Washington: 2014 Update. Washington Department of Ecology, Olympia, WA. Publication 14-06-029.

King County, 2007. King County Shoreline Master Program, Appendix E: Technical Appendix. Shoreline Inventory and Characterization: Methodology and Results, http://www.metrokc.gov/)shorelines/shoreline-master-program-plan.aspx.

Naiman, R.J., Bilby, R.E. (Eds.), 1998. River Ecology and Management: Lessons from the Pacific Coastal Ecoregion. Springer-Verlag, New York, NY.

National Research Council (NRC), 1996. Committee on Protection and Management of Pacific Northwest Anadromous Salmonids. Upstream: Salmon and Society in the Pacific Northwest. National Academy Press, Washington, DC, p. 472.

National Research Council, 2001. Compensating for Wetland Losses Under the Clean Water Act. National Academy Press, Washington, DC.

Roni, P., Beechie, T.J., Bilby, R.E., Leonetti, F.E., Pollock, M.M., Pess, G.R., 2002. A review of stream restoration techniques and a hierarchical strategy for prioritizing restoration in Pacific Northwest watersheds. N. Am. J. Fish Manag. 22, 1–20.

Simenstad, C., Logsdon, M., Fresh, K., Shipman, H., Detheir, M., Newton, L., 2006. Conceptual Model for Assessing Restoration of Puget Sound Nearshore Ecosystems. Washington Sea Grant Program, University of Washington, Seattle, WA. Puget Sound Nearshore partnership report 2006-03, http://pugetsoundnearshore.org.

Spence, B.C., Lomnicky, G.A., Hughes, R.M., Novitzki, R.P., 1996. An Ecosystem Approach to Salmonid Conservation. ManTech Environmental Research Services Corporation, Corvallis, OR, Draft report TR-4501-96-6057.

Stanley, S., Brown, J., Grigsby, S., 2005. Protecting Aquatic Ecosystems: A Guide for Puget Sound Planners to Understand Watershed Processes. Washington State Department of Ecology, Olympia, WA. Publication 05-06-027.

Stanley, S., Grigsby, S., 2008. Watershed Assessment and Assessment of Clark County, Washington. Final. Washington State Department of Ecology Publication, Olympia, WA.

Stanley, S., Grigsby, S., Booth, D.B., Hartley, D., Horner, R., Hruby, T., Thomas, J., Bissonnette, P., Fuerstenberg, R., Lee, J., Olson, P., Wilhere, G., 2016. Puget Sound Characterization. Volume 1: The Water Resources Assessments (Water Flow and Water Quality). Washington State Department of Ecology, Olympia, WA. Publication 11-06-016.

Chapter 2.2.10

NovaWET— Basic Information for Assessing Wetland Functions in Nova Scotia, Canada

Ralph W. Tiner

Institute for Wetland & Environmental Education & Research, Leverett, MA, United States

Chapter Outline

INTRODUCTION

In 2007, the Environmental Goals and Sustainable Prosperity Act (EGSPA) mandated that the province develop a policy of no net loss by the end of 2009. Since then, the province has required applications for certain wetland alterations (Hilchey, 2015). With passage of the Environmental Act (2011), projects that involve 0.5–2.0 ha of wetland alteration or impact more than one wetland were required to complete an application that would include a wetland functional assessment among project details—project description, a wetland delineation for the site, and the credentials of the wetland specialist. The law specified that applicants use the Nova Scotia Wetland Evaluation Technique (NovaWET; http://www.gov.ns.ca/nse/wetland). Other approaches could be employed but they would have to be agreed upon by the NSE inspector beforehand. NovaWET evaluations are required for all wetland alteration approval applications where the proposed alter ation is 0.5 ha or larger in size and for projects that intend to alter multiple wetlands. The purpose of this chapter is to summarize NovaWET and reference a few examples of applications.

DEVELOPING THE METHOD

By 2009, after reviewing the environmental assessments of the impacts from projects requiring a wetland alteration or environmental assessment application, Nova Scotia Environment (NSE) was concerned that environmental consultants were not taking into account all the functions of wetlands when planning construction in and around wetlands, nor were they viewing impacts from the watershed perspective (Kathleen Johnson, personal communication, 2009). Peter Veneman and I had been teaching wetland delineation courses in the province through the Maritime College of Forest Technology and I was asked to develop an advanced wetland course that included functional assessment. I felt the best way to teach this would be to have a method for students to learn during the course and apply on their own afterward. This would also give the regulators a common approach for reviewing the assessments rather than having each consultant develop his or her own assessment method and reporting procedure. Because the province did not have a wetland functional assessment method or detailed studies characterizing the diversity of wetlands across the province, I reviewed some methods that were being used in the United States and pulled various aspects from them to develop a qualitative approach for teaching students both in the classroom and in the field during the course. The basic objectives were to: (1) provide information for wetland alterations or environmental assessment applications for proposed projects, (2) estimate the most significant functions of a wetland based on the literature and properties observed in the field, (3) evaluate the condition of the wetland buffer, (4) characterize the relationship between the project area's wetlands and neighboring wetlands and waters, and (5) provide a general assessment of the contributing watershed.

Wetland and Stream Rapid Assessments. https://doi.org/10.1016/B978-0-12-805091-0.00025-6

The first iteration of the wetland assessment method was called Nova Scotia Wetland Assessment Method (NovaWAM) but was later changed to NovaWET to avoid confusion with a provincial mapping project—Wet Areas Mapping (WAM). The method could be used by environmental consultants to provide NSE with basic information on project site wetlands, the surrounding landscape, and the contributing watershed to help evaluate the condition of the wetlands in a project area and the functions of the wetlands affected by the proposed wetland alteration.

NovaWET was designed as a two-step process involving offsite assessment using maps, aerial imagery, and other data plus onsite evaluation to confirm or modify the offsite interpretation and to record site-specific information on wetland characteristics and indicators of wetland functions for wetlands in the project area. The end result of the analysis is a characterization of the wetland condition and likely functions, the condition of the wetland buffer, the relationship between wetlands in a project area and neighboring wetlands and waterbodies, and a general assessment of the contributing watershed. The method was strictly qualitative because at the time of its development there was not any detailed information on the province's wetlands that was sufficient to develop a more quantitative method. Consequently, the method did not generate a numeric score for comparing the subject wetland with similar wetlands.

The offsite portion of this method begins with a landscape-level functional assessment adapted from one developed for the northeastern United States—the Watershed-based Preliminary Assessment of Wetland Functions (W-PAWF; Tiner, 2003, 2005, see Chapter 2.2.1). That method was based on wetland classification and available geospatial information (e.g., aerial imagery, maps, and geospatial data accessible via geographic information system, or GIS). It was intended to provide a preliminary assessment of wetland functions based on correlations between wetland characteristics and 10 functions: surface water detention, streamflow maintenance, nutrient transformation, sediment and particulate retention, carbon sequestration, coastal storm surge detention, shoreline stabilization, fish and shellfish habitat, waterfowl and waterbird habitat, and other wildlife habitat. The correlations were derived from the scientific literature and peer reviewed by professional wetland scientists. They were simplified for NovaWET based on the wetland classification used in mapping the province's wetlands. Eight types of wetlands were recognized: salt marsh, bog, fen, freshwater marsh, forested swamp, shrub swamp, coastal saline pond, and vernal pool. This landscape-level approach was expanded to include additional classification of wetland attributes plus a characterization of the wetland buffer and the contributing watershed, requiring an examination of existing aerial imagery or thematic maps.

This method was further expanded to include other desktop analyses and field observations adapted chiefly from Minnesota's rapid assessment method (MnRAM 3.2; Minnesota Board of Water and Soil Resources, 2008), but significantly broadened by contributions from other assessment methods developed by North Carolina (North Carolina Wetland Functional Assessment Team, 2008), Oregon (Adamus et al., 2009) and California (Collins et al., 2008) and from method reviews by Hanson and others (2008) and Fennessy and others (2004). This step addresses site-specific characteristics of the area of the proposed wetland alteration.

NovaWET was intended for use in evaluating vegetated wetlands and not for nonvegetated wetlands such as tidal flats, cobble-gravel streambanks, and similar areas devoid of vegetation. This information was intended to be included in a project's environmental assessment report and not to replace such a report as more detailed information on project design, alternatives analysis, and compensation/monitoring is required.

NovaWET (Version 3.0)

NovaWET was reviewed by NSE and others when NSE was considering it for inclusion in a provincial wetland conservation policy. It was modified and adopted as the province's standard wetland functional assessment technique in 2011 with passage of the Environmental Act. The main difference between the original NovaWET and later versions may be the ratings for various elements of the evaluation and the designation of significant functions. The ratings were reviewed and revised where necessary by NSE personnel working with wetland specialists across the province. The current version is NovaWET 3.0 (September 2011; https://www.novascotia.ca/nse/wetland/docs/NovaWET.3.0.pdf; forms available at http://www.novascotia.ca/nse/wetland/assessing.wetland.function.asp) and is briefly summarized below.

NovaWET 3.0 begins with a preliminary assessment of wetland functions based on wetland classification (Table 2.2.10.1; Appendix A in the methods document). This is done in the office before going to the field. The more detailed assessment on a function-by-function basis is derived from information interpreted from a desktop review of imagery and other sources coupled with field observations.

NovaWET is divided into 11 sections largely associated with key wetland functions (Table 2.2.10.2). Most sections include questions intended to help assessors determine the significance of specific functions associated with a particular wetland. Descriptive guidance is provided to aid in the interpretation of significant functions (SFs). Selection of the appropriate responses related to an SF rating (e.g., high, medium, or low) is largely based on the combined responses to the

TABLE 2.2.10.1 Predicted Wetland Functions Based on Wetland Type/Classification

Predicted Function	Performance Level	Wetland Types
Surface Water Detention	High	Salt marsh, lentic marsh, lentic swamp, lotic marsh, lotic swamp, terrene throughflow marsh, terrene throughflow swamp
	Moderate	Terrene outflow marsh, terrene outflow swamp, bog, fen (*Note*: Slope wetlands do not provide this function to any significant degree.)
Coastal Storm Surge Detention	High	Salt marsh, lotic tidal wetlands
	Moderate	Terrene basin and flat wetlands contiguous with salt marsh (these are bordering nontidal wetlands subject to infrequent or occasional tidal flooding during storms)
Streamflow Maintenance	High	Headwater wetlands that are not ditched
	Moderate	Ditched headwater wetlands, Lotic river marsh (nontidal), lotic river floodplain swamp (nontidal), lentic wetlands associated with a throughflow lake
Nutrient Transformation	High	Salt marsh, marsh, swamp, fen
	Moderate	Bog, any seasonally saturated or temporarily flooded wetland
Carbon Sequestration	High	Salt marsh, marsh, swamp, bog, fen (seasonally flooded or permanently saturated)
	Moderate	Any temporarily flooded or seasonally saturated swamp or fen
Sediment and Other Particulate Retention	High	Salt marsh, lotic marsh, lotic swamp, lentic marsh, lentic swamp (subject to seasonal flooding)
	Moderate	Lotic and lentic wetlands that are temporarily flooded
Shoreline Stabilization	High	Salt marsh, lotic wetlands, lentic wetlands
	Moderate	Terrene headwater marsh, terrene headwater swamp (*Note*: Wetlands that are islands in the waterbodies are excluded from this function.)
Fish and Shellfish Habitat	High	Salt marsh, lotic marsh, lentic marsh
	Moderate	Terrene pond marsh (pond >0.4 ha)
	Stream shading	Lotic swamp (shrub or treed, both along perennial streams)
Waterfowl and Waterbird Habitat	High	Salt marsh, lentic marsh, lotic river marsh
	Moderate	Terrene pond marsh (pond >0.4 ha)
	Cavity nesters	Lotic swamp (shrub and treed, both along perennial streams only)
Other Wildlife Habitat	High	Any vegetated wetland complex >8 ha, Wetlands 4–8 acres in size composed of two or more wetland types, Small isolated wetlands in dense cluster (>3 per ha) in a forest matrix (restrict to forest regions with woodland vernal pools)
	Moderate	Other vegetated wetlands

This is a preliminary assessment because it depends on remotely sensed data.
Adapted from Tiner, R.W., 2003. Correlating Enhanced National Wetlands Inventory Data With Wetland Functions for Watershed Assessments: A Rationale for Northeastern U.S. Wetlands. U.S. Fish and Wildlife Service, National Wetlands Inventory Program, Northeast Region, Hadley, MA. http://library.fws.gov/Wetlands/corelate_wetlandsNE.pdf; Tiner, R.W., 2005. Assessing cumulative loss of wetland functions in the Nanticoke River watershed using enhanced National Wetlands Inventory data. Wetlands 25(2), 405–419. http://library.fws.gov/Wetlands/TINER_WETLANDS25.pdf.

preceding questions for that section, but professional judgment and other reference material may also be considered. SF ratings highlighted in red on the field data sheet indicate critical wetland functions or watershed conditions that are highly degraded (Fig. 2.2.10.1). The critical wetland functions are often unique or rare or associated with high risk to the watershed if lost (e.g., flood control), so that minimizing or compensating for their loss may be difficult or impossible. Whenever a wetland is found to have red-highlighted SFs, the project's proponent is encouraged to contact NSE for specific advice on

TABLE 2.2.10.2 Evaluated Functions, Factors Considered, and Focus of Questions for Determining Significant Functions (SFs)

Feature of Interest (Section on Data Sheet)	Considerations and Focus of Significant Function (SF) Questions
Watershed Characteristics	Name/size of tertiary watershed
	Land cover percentage of forest, open natural land/old fields, pasture/hay, cropland, urban/commercial, roads, and other developed
	Percentage of wetland in watershed and relative percentage of each type (salt marsh, bog, fen, freshwater marsh, forested swamp, shrub swamp, coastal saline pond, and vernal pool)
	SF question 1: Watershed condition—H (highly modified; e.g., >20% impervious), M (modified; 5%–20%), and L (relatively unmodified; <5% impervious)
	SF question 2: Density of wetlands in watershed related to floodwater detention (H <10%, M 10%–20%, L >20%); when wetland densities exceed 20% flood storage of additional wetlands rapidly decrease
Site Description and Wetland Character	Wetland type, size, landscape position, landform, water flow path, and wetland origin
	Water regime
	Number of wetlands in project area by type
	Is wetland part of a complex
	Percentage of each wetland type in complex
	Bordering pond or lake, or within 100 m of such waterbodies
	Average standing water depth and percentage of wetland inundated
	Presence of inlet, outlet, or both
	Adjacent upland land use in the buffer zone (100 m of wetland boundary)
	Stressors in and directly adjacent to wetland (e.g., ditch, channelized watercourse, oil/chemical spill, sedimentation, discharge pipe, garbage, ATV trails, rip-rap, development, farms, etc.)
	Hydrology altered by ditching, tiles, dams, culverts, well pumping, and/or diversion of surface flow
	SF question 3: general condition/ecological integrity of the wetland (H, M, L)
Condition and Integrity of Adjacent Land	Average width of undeveloped land adjacent to wetland
	Water quality rating (related to buffer width; H=>15 m, M=8–15 m, L=<8 m)
	Wildlife habitat rating (H=>100 m buffer, M=15–100 m, L=<15 m)
	Describe cover, vegetation diversity and structure, and slope of adjacent land (within 15 m of wetland boundary)
	Adjacent land support of water quality functions
	Adjacent land support of wildlife habitat functions
	SF question 4: Rate overall condition/integrity of adjacent land (H, M, L)
Identification of Exceptional Features	If answer yes to any SF question=significant function
	SF question 5: Designated wetland of significance
	SF questions 6–7: Important for commercial/recreational fish or shellfish, rare/endangered/threatened/at-risk species
	SF question 8: Restored or preserved wetland under conservation or other agreement, or restored or created compensatory wetland for previous wetland alteration
	SF question 9: Calcareous fen, cedar or black ash swamp or large floodplain swamp

TABLE 2.2.10.2 Evaluated Functions, Factors Considered, and Focus of Questions for Determining Significant Functions (SFs)—cont'd

Feature of Interest (Section on Data Sheet)	Considerations and Focus of Significant Function (SF) Questions
	SF questions 10–12: Location—in public water supply system, in floodplain upstream or within a populated area, in federal/provincial/municipal area of interest (wildlife area, park, beach, Canadian heritage river, Ramsar wetland of international importance, historic site, etc.)
Hydrologic Condition and Integrity	Is wetland the source of stream or along headwater stream (perennial orders 1 and 2)?
	Geographically isolated (evidence of ponding?)
	Ability to maintain its characteristic hydrologic regime based on the degree of alteration of hydrologic regime
	Percentage wetland with ponding at different depths
	Signs of surface water retention (12 field indicators listed)
	Extent of observable/historical sediment delivery to the wetland from anthropogenic sources including agriculture and developed areas
	Soil disturbance
	Predominant upland soil type in surrounding area (sands, silts or loams, clays or shallow bedrock)
	Functional capacity of the wetland in retarding or altering flows based on the surface flow characteristics through the wetland
	Roughness coefficient of the potential surface water flow path in relation to wetland vegetation biomass, numeric density and plant morphology (not applicable if isolated)
	Characteristics of stormwater, wastewater, or concentrated agricultural runoff detention/water quality treatment prior to discharging into the wetland
	Water source (natural, mostly natural, partly altered/controlled, controlled)
	Tidal wetland hydrology (unrestricted, reduced, restricted)
	Coastal storm surge protection
	SF questions 13–15: wetland's hydrologic condition (natural, modified, significantly modified), wetland's importance for maintaining stream flow, and wetland's ability to detain surface water
Water Quality	Overflow or direct discharge of stormwater, wastewater, or concentrated agricultural runoff as primary source of water?
	Significant nutrients/sediment from adjacent land?
	Allows for retention of sediments?
	Has significant vegetative density to decrease water energy and allow settling of suspended materials?
	Acts as a filter based on wetland type and landscape position?
	SF questions 16–18: features to improve water quality, signs of excess nutrient loading/ contamination, does wetland contribute to maintaining downstream fish or water supply resources?
Groundwater Interactions	Mineral (=recharge) or organic soil (=discharge)
	Likely recharge or discharge site based on land use and runoff characteristics in upstream watershed
	Wetland size and hydrologic properties of upland soils within 200 m of the wetland
	Hydroperiod of wetland, wetland type, and landform
	Inlet/outlet configuration (i.e., no outlet = recharge; perennial outlet = discharge)
	Topographic relief surrounding wetland
	SF questions 19–20: likely recharge or discharge site?

Continued

TABLE 2.2.10.2 Evaluated Functions, Factors Considered, and Focus of Questions for Determining Significant Functions (SFs)—cont'd

Feature of Interest (Section on Data Sheet)	Considerations and Focus of Significant Function (SF) Questions
Shoreline Stabilization/Integrity	Fringing a waterbody; width of stream or river; exposed or sheltered location along pond, lake, estuary, or ocean
	Percentage rooted vegetation in shallow water
	Average wetland width along shore
	Prevalence of strong-stemmed emergent vegetation for marshes and fens along shorelines
	Shoreline erosion potential
	Vegetation condition upslope of waterline
	SF question 21: rate ability to stabilize shoreline (H, M, L, N/A)
Plant Community	Overall diversity compared to a healthy wetland of this type
	Regionally scarce or rare?
	Nonnative dominants and percentage cover
	Vegetation disturbance/type (e.g., ATV, grazing, mowing, filling, draining, etc.)
	Integrity (exceptional integrity, high, medium, and low integrity—based on disturbance and percentage cover of invasive species)
	SF questions 22–25: unique or rare plant community, diverse community, overall integrity (H, M, L), and presence of plant species of special status
Fish and Wildlife Habitat/ Integrity	Interspersion of types within complex for freshwater marshes and shallow open-water wetlands; total vegetation cover
	Degree of interspersion of vegetation types within other wetland complexes
	Presence of detritus (vegetative litter) in different stages of decomposition
	Degree of interspersion in other wetlands in the vicinity
	Fragmentation
	Noteworthy wildlife (including fish) observed
	Connection with waterbody that supports fish spawning and nursery habitat
	Wetland-landscape context—in large block of contiguous upland or wetland
	Habitat for amphibians, reptiles, waterfowl, waterbirds, mammals, fish, rare/endangered species?
	SF questions 26–28: associated with watercourse supporting fish or fish habitat, presence of rare, endangered, etc. fish and wildlife species, and overall habitat quality (H, M, L)
Community Use/Value	Visible from vantage point, aesthetics, commercial products, conservation/public ownership, public access, greenbelt, education, hiking, wildlife viewing, boating, hunting, fishing, plant gathering, berry picking, exploration, relaxation, other
	SF question 29: qualitative rank (H, M, L)

how to proceed because NSE is unlikely to approve alterations to wetlands that would affect these red-rated functions. An experienced professional with appropriate wetland expertise must be hired to complete all NovaWET evaluations; fieldwork must be performed between June 1 and September 30 unless approved by NSE. The importance of SFs for a particular wetland must be evaluated on a case-by-case basis as the watershed context is critical to this evaluation and the approval decision. NovaWET assessments provide NSE with basic information on wetlands located in the immediate vicinity of the proposed development area and the surrounding watershed as well as an assessment of the potential significance of wetland functions that may be lost if proposed development is approved. A completed NovaWET package must include: (1) digital and hard copy data sheets with required responses; (2) maps/aerial photographs of the site; (3) photos of the wetland(s) in question; (4) brief written description of wetland(s) (e.g., one short paragraph); and (5) brief conclusion/summary of results with a focus on significant functions.

APPENDIX C:	**Nova Scotia Wetland Evaluation Technique Field Data Sheet (September 2011)**									
	Project Name:	Evaluator:			GPS Coordinates:					
PID:	Site Address:									
	Sources and Dates of Mapping/Images:									
	Evaluation Date:	Site Visit Date:								
	Weather Conditions (past 48 hours):									
	Seasonal Weather Conditions:									
SECTION ONE: WATERSHED CHARACTERISTICS										
1	Watershed Name (tertiary):	Size: km^2								
2	% Watershed Land Cover	For:	Nat:	Past/Hay:	Crop:	Urb/Com:	Road:	Other Dev:		
3	% Watershed WL Cover and by Class	Total: %	SM:	BO:	FE:	FM:	FS:	SS:	CP:	VP:
SF1	***Watershed condition***	***H***	***M***	***L***						
SF2	***Proportion of WL area in watershed & opportunity for floodwater detention***	***H***	***M***	***L***						
SECTION TWO: WETLAND CHARACTERISTICS										
	Wetland Type:	WL size: hectares			Landform:			Landscape Position:		
	Water flow path:	Wetland Origin:								
1	Water Regime	PF	SF	TF	SS	PS	RfT	IfT	AF	
2	# WL's within 30m project area	Total#	SM:	BO:	FE:	FM:	FS:	SS:	CP:	VP:
3	Is WL part of complex	Yes	No							
4	% each wetland type in complex	SM:	BO:	FE:	FM:	FS:	SS:	CP:	VP:	
5	Is WL bordering or associated with a lake or pond?	bordering		within 100m		N/A		specify		
6	Standing water?	Yes	Avg Dep:		% Inundated:		No			
7	Inlet or Outlet (circle all that apply)?	Inlet	Outlet							
8	Adjacent Upland Land Use within 100m (%)	For:	Nat:	PasHay:	Crop:	UrbCm:	Road:	Other Dev:		
9	Are there stressors in WL or WL buffer area? Circle primary stressor(s).	DD__, CW__, WcS__, O/C__,EB__,DP__,F__,M__, ES__,NE__,DwP__,								
		M__,GC__,ATV__,DG__,EA__, R__,Rr__,U/CD__,F__,FA__, other (specify):								
10	Hydrology Altered (circle all that apply)?	Ditching	Dams	Tiles	Culvert	Well	Diversion	Other Specify:		
SF3	***Rate the general wetland condition/integrity***	**H**	**M**	**L**						
SECTION THREE: ADJACENT LAND CONDITION AND INTEGRITY										
1	Average width of adjacent naturalized buffer	___meters								
2	Widths for water quality	H >15	M 8-15	L <8						
3	Widths for wildlife habitat	H >100	M 15-100	L <15						
4	Adjacent area vegetation condition (list % in each category)	H	M	L						
5	Adjacent area diversity and structure (list % in each category)	H	M	L						
6	Adjacent Upland Slope (list % in each category)	Steep	Mod	Gentle						
7	Adjacent land supports water quality	*Yes*	*No*	Specify:						
8	Adjacent land supports wildlife habitat	*Yes*	*No*	Specify:						
SF4	***Rate the overall condition and integrity land adjacent to wetland***	***H***	***M***	***L***	**is buffer required to maintain red flag functions of wetland? If yes if no**					
SECTION FOUR: DOCUMENTED IMPORTANT FEATURES										
SF5	***Is the WL a WSS?***	***Yes***	***No***							
SF6	***Does the WL support commercial/recreational fish/shellfish?***	***Yes***	***No***							
SF7	***Species of concern (Fed/Prov)? Specify.***	***End***	***Thr***	***SpC***	***Red***	***Yellow***	***S1***	***S2***	***S3***	***N/A***
SF8	***Wetland has conservation/compensation agreements/activity?***	***Yes***	***No***	specify:						
SF9	***Wetland is calcerous fen, black ash or cedar swamp?***	***Yes***	***No***							
SF10	***Within Drinking Water Protected Area (designated watershed/wellfield)***	***Yes***	***No***	specify:						
SF11	***WL within a floodplain and upstream of or within of a populated area?***	***Yes***	***No***							
SF12	***Fed/Prov/Municipal area of interest?***	***Yes***	***No***	specify:						
SECTION FIVE: HYDROLOGIC CONDITION AND INTEGRITY										
1	Is WL source of stream or headwater(wc order 1 or 2)	Yes	No	Specify:						
2	Is WL geographically isolated?	Yes	No	Specify:						
3	WL ability to maintain characteristic hydrologic regime	High		Med		Low				
4	Water Storage Depth (list % in each class)	>30cm	15-30cm	up to 15cm		No ponding				
5	Signs of surface water retention observed?	SW__cm, WSL__, WCD__, WM__cm, SM__cm, SD__, AD__, ID__, PMT__, AI__, BT__, AR__, Other:								
6	Describe observable/historical anthropogenic sediment delivery	Low		Med		High				
7	Disturbance of WL soils	Low		Med		High				
8	Predominant soils adjacent to WL	Sand		Silt/loam		Clay/bedrock				
9	Capacity of WL to alter/retard flows	High		Med		Low				
10	Roughness coefficient for surface water flow path	High		Med		Low				
11	Stormwater/Wastewater/Agricultural runoff detention	High		Med		Low				
12	Water Source	Natural		Mostly natural		Partly altered		Controlled		
13	Hydrology of tidal wetlands	Unrestricted		Reduced		Restricted		***N/A***		
14	Coastal storm surge	Yes	No							
SF13	***WL hydrologic condition***	***Natural***	***Modified***		**Significantly Modified**					
SF14	***WL important for maintaining stream flow?***	***Yes***	***No***							
SF15	***WL ability to detain surface water***	***High***	***Med***	**Low**						

FIG. 2.2.10.1 The NovaWET data sheet.

(Continued)

	SECTION SIX: WATER QUALITY										
1	Stormwater/Wastewater/Agricultural runoff as water source?	High		Med		Low					
2	Nutrients/sediments from surrounding land	High		Med		Low					
3	Significant flood/stormwater attenuation	Yes	No								
4	Vegetation capacity to settle suspended sediments	High		Med		Low					
5	WL type /landscape position holds/filters runoff?	Yes	No								
SF16	***Wetland improves water quality?***	***Yes***	***No***								
SF17	***Evidence of excess nutrient loading/contamination?***	***Low***	***Med***	***High***							
SF18	***WL contributes to water quality in downstream resources***	***High***	***Med***	***Low***							
	SECTION SEVEN: GROUNDWATER INTERACTIONS										
1	Describe soils in wetland	Recharge		Discharge							
2	Land use / run off in subwatershed upstream	Recharge		Discharge							
3	Conditions of upland soils within 200m of wetland	Recharge		Discharge							
4	Hydroperiod of wetland	Recharge		Discharge							
5	Describe inlet/outlet configuration	Recharge		Discharge							
6	Characterize topographic relief surrounding wetland	Recharge		Discharge							
SF19	***WL serves as a recharge site***	***Yes***	***No***								
SF20	***WL serves as a discharge site***	***Yes***	***No***								
	SECTION EIGHT: SHORELINE STABILIZATION AND INTEGRITY										
1	Wetland fringing ocean/estuary/lake/pond/river/stream?	Yes	No	streamwidth >4m		streamwidth<4m		WB Exposed	WB Sheltered		
2	% cover of rooted vegetation in shallow water zone	H >50%	M 10-50	L <10%							
3	Avg veg WL width b/w shoreline/streambank & 2 m depth contour	H >10m	M 3-10	L <3m							
4	Prevalence of strong-stemmed emerg. veg (shoreline marshes and fens only)	High	Med	Low							
5	Describe shoreline erosion potential	High	Med	Low							
6	Shoreline/streambank veg condition upslope of water level	Low	Med	High	Artificial						
SF21	***WL ability to stabilize shoreline***	***H***	***M***	***L***	N/A						
	SECTION NINE: PLANT COMMUNITY										
1	Vegetation diversity	High	Med	Low							
1b	Dominant plant species and % cover in the WL	list:									
3	Dominant Non-native or Invasive species and % cover	Yes	No	specify: %							
4	Vegetation Disturbance	H	M	L	specify type(s) below						
5	Disturbance Types	H__,ATV__,G__,,M__,In__, D/D__, Im__, OAH__, li__, Sd__,E__,,other__,									
7	Vegetative Integrity of plant community	E	H	M	L						
SF22	***Is the plant community unique or rare regionally or provincially?***	***Yes***	***no***	***specify:***							
SF23	***Does the WL contain a diversity of plant communities***	***H***	***M***	***L***							
SF24	***Rate the overall integrity/quality of plant community?***	***H***	***M***	***L***							
SF25	***Are there any observed rare or endangered plant species? Specify.***	***End***	***Thr***	***SpC***	***Red***	***Yellow***	***S1***	***S2***	***S3***	***N/A***	
	SECTION TEN: FISH AND WILDLIFE HABITAT AND INTEGRITY										
1	Interspersion of open water and vegetation (open water types only)	H	M	L							
1b	% cover in vegetation verus open water	____%									
2	Interspersion that best fits entire wetland	H	M	L	N/A						
3	Wetland condition related to detritus	H	M	L	N/A						
4	Interspersion of other wetlands in vicinity	H	M	L							
6	Barriers/restriction between wetland and other habitat	L	M	H							
7	Noteworthy wildlife or evidence (birds, mammals, amphibians,etc)	Yes	No	list:							
8	Connected to permanent water (accessible to fish)?	Exceptional	High	Med	Low	N/A					
9	Fish species observed or evidence seen (list)	Yes	No	list:							
10	Wetland part of contiguous upland or wetland:	>50ha	25-50ha	10-25ha	<10ha						
11	WL provides habitat for:	Amphibians	Reptiles	Waterfowl	Waterbirds	Mammals	Fish	R/E species			
SF26	***Does wetland support fish/fish habitat?***	***Yes***	***No***	***specify:***							
SF27	***Rare or endangered fish/wildlife species found in the wetland?***	***End***	***Thr***	***SpC***	***Red***	***Yellow***	***S1***	***S2***	***S3***	***N/A***	
SF28	***Overall fish and wildlife habitat quality***	***H***	***M***	L							
	SECTION ELEVEN: COMMUNITY USE/VALUE										
1	Describe community use	VV__,CP__,CO__,PO__,PA__,AV__,GB__,E__,HI__, WV__, BO__,HU__, PG__, BP__,F__, E__, R__, Other:									
SF29	***Rate the wetland's community use/value***	***H***	***M***	L							

SF ratings highlighted in red indicate critical wetland functions or watershed conditions that are highly degraded. Whenever a wetland is found to have red-highlighted **SFs** the proponent is encouraged to contact NSE for advice about the approval because NSE is unlikely to approve alterations to wetlands that would affect these red-rated functions.

FIG. 2.2.10.1, CONT'D

EXAMPLES OF APPLICATIONS

Because NovaWET has been the recommended approach for characterizing wetlands and adjacent lands for projects that may significantly impact wetlands across the province, NovaWET results have been included in environmental impact assessment reports for projects such as mining and quarry operations (e.g., AMEC Environment and Infrastructure, 2014; Stantec, 2011, 2015), road construction (e.g., CBCL Limited, 2010), pipeline routes (e.g., Stantec, 2013), major facility construction (e.g., AMEC Environment and Infrastructure, 2013), and commercial development (e.g., McCallum Environmental Limited, 2015). While the method is mainly used in Nova Scotia, environmental consultants have also used it to evaluate wetlands elsewhere in the Maritimes (e.g., Kami Iron Ore Project in West Labrador; http://www.mae.gov.nl.ca/env_assessment/projects/Y2011/1611/Amendment_to_EIS_Volume_3_IRs_Chapter_1_February_2013.pdf).

NovaWET has allowed for an analysis of wetlands in the project areas and identification of differences in their functions. Besides the desktop functional assessment, the method required detailed descriptions and data on the wetland vegetation, soils, hydrology, and conditions of the wetland buffer plus observations of site conditions (e.g., stressors)—information that can be used to evaluate impacts and plan appropriate mitigation. Here are a few specific examples of conclusions reached from the functional assessments:

From the Goldsboro LNG Project Report (AMEC, 2013):

The functional assessment indicate that seven of the 13 wetlands perform red rated functions which elevate the relative importance of these wetlands in terms of the functions they provide to the surrounding watershed. All but one of the wetlands assessed with red rated functions occur along the unnamed stream that flows along the west end of the property and are associated with providing habitat for American Eel as well as maintaining stream flow. The one wetland (WL11) not associated with the unnamed stream may be a source of ground water recharge.

(Appendix E, p. 28)

From the Black Point Quarry Project Report (AMEC, 2014):

The functional assessment indicate that 12 of the 22 wetlands perform red rated significant functions which elevate the relative importance of these wetlands in terms of the functions they provide to the surrounding watershed. Six of the wetlands assessed with red rated significant functions (WL1, 3, 6, 8, 14 and 17) occur along or form the headwater of small watercourses throughout the site and as such are important in maintaining stream flow. Seven of the wetlands assessed as having red rated significant functions (WL 8, 9, 11, 16, 18, 19 and 20) may serve as groundwater recharge sites while one wetland (WL2) provides a red rated significant function of stabilizing the shoreline. Results of this study will be used to assess the potential impacts of the proposed Black Point Quarry on wetland habitat within the Project Study Area.

(Appendix F1, p. 13)

From the Brierly Brook Quarry Expansion Report (McCallum Environmental Ltd., 2015):

Four wetlands and four watercourses were identified in the Project Area. The wetlands are largely intact and undisturbed, yet the overall functional significance of each wetland is low, due to factors such as wetland size, wetland type, landscape position, and absence of priority species. However, given that overall wetland cover throughout the tertiary watershed is low (3%), each individual wetland provides floodwater detention to downstream areas. Four watercourses were also identified within the project area, none of which provide fish habitat.

(p. 49)

From the (CBCL Limited, 2010):

The wetlands within the study area are significantly altered as a result of being within close proximity to the Town of Antigonish. The larger North Site wetland complex is perhaps more valuable due to its large contiguous size compared to the smaller isolated wetlands found on the South Site. In summary, the main wetland functions of any significance for either site would be habitat, storm surge buffering, flood control and species diversity. The compensation project proposes to retain these functions, enhance the function of habitat by creating more wetland on site and adding the function of interpretation and wetland process education to the study area.

(pp. 34–35)

CURRENT WETLAND ASSESSMENT ACTIVITIES IN THE MARITIMES

While NovaWET has provided the NSE with more information on wetlands and the surrounding landscape for evaluating impacts from land development projects than it was receiving prior to its adoption, it is a wetland characterization and data collection method with a qualitative approach to functional assessment, thereby leaving some room for interpretation. In 2013 NSE began a search for an alternative wetland rapid assessment method (RAM) that could integrate information similar to NovaWET's into scores and ratings for functions of individual wetlands. Developing this type of RAM would require collecting data from hundreds of wetlands.

In 2014, the adjoining province of New Brunswick received a grant from Canada's National Wetland Conservation Fund to adapt a RAM used by Alberta's wetlands program (ABWRET) for all of Atlantic Canada. That RAM, in turn, had been based on the Wetland Ecosystem Services Protocol originally developed in the United States (WESP; see Chapter 4.3.2). The grant required that WESP's author (Paul Adamus) complete a version for Atlantic Canada (AC), termed WESP-AC, by collaborating with wetland programs not only in New Brunswick but also in Nova Scotia, Newfoundland-Labrador, and Prince Edward Island. With the assistance of wetland specialists in those provinces, over a three-year period WESP-AC has been calibrated to 121 New Brunswick wetlands, 156 Nova Scotia wetlands, 128 Newfoundland-Labrador wetlands, and 73 on Prince Edward Island. In each province, scores from tidal wetlands (salt marshes) were calibrated separately from those of nontidal wetlands. Now that field calibration of WESP-AC has been completed throughout the region, the supporting spreadsheet calculator automatically compares the function scores from any wetland assessed in the future to those from the large statistically chosen set of calibration wetlands from the same province. The final version of the WESP-AC spreadsheet calculator with accompanying manual and database will be available from the region's wetland programs in 2018. A comparison of the results obtained from NovaWET and the WESP-AC would be interesting.

ACKNOWLEDGMENT

Thanks to Paul Adamus for the update on wetland assessment in the Maritime provinces.

REFERENCES

Adamus, P., Morlan, J., Verble, K., 2009. Manual for the Oregon Rapid Wetland Assessment Protocol (ORWAP). Version 2.0. Oregon Department of State Lands, Salem, OR. http://www.oregon.gov/DSL/WETLAND/docs/orwap_manual_v2.pdf.

AMEC Environment & Infrastructure, 2013. 2013/2014 Wetland Field Survey, Delineation and Functional Assessment Report. Goldboro LNG Project—Natural Gas Liquefaction Plant and Marine Terminal, Dartmouth, NS. http://www.novascotia.ca/nse/ea/goldboro-lng/Appendix-E-WetlandDelineations-Part-1.pdf.

AMEC Environment & Infrastructure, 2014. 2010/2011/2014 Wetland Field Survey, Delineation and Functional Assessment Report. Black Point Quarry, Guysborough County, Dartmouth, NS. https://novascotia.ca/nse/ea/black-point-quarry/app_f_1_amec_wetland_baseline_survey_part_1.pdf.

CBCL Limited, 2010. NSTIR—Antigonish Wetland Compensation Project. Wetland Compensation Proposal & Baseline Report, Halifax, NS. http://www.hwy104antigonish.ca/wordpress/wp-content/uploads/2011/05/NSTIRAntigonishWetlandCompensationProposal.pdf.

Collins, J.N., Stein, E.D., Sutula, M., Clark, R., Fetscher, A.E., Grenier, L., Grosso, C., Wiskind, A., 2008. California Rapid Assessment Method (CRAM) for Wetlands. User's Manual. Version 5.0.2. San Francisco Estuary Institute/Southern California Coastal Water Research Project/California Coastal Commission/Moss Landing Marine Laboratories, Oakland, CA/Costa Mesa, CA/Santa Cruz/Moss Landing, CA (CRAM Field Books for Estuarine, Depressional and Riverine Wetlands). http://www.cramwetlands.org/.

Fennessy, M.S., Jacobs, A.D., Kentula, M.E., 2004. Review of Rapid Methods for Assessing Wetland Condition. U.S. Environmental Protection Agency, National Health and Environmental Effects Laboratory, Corvallis, OR (EPA/620/R-04/009). http://www.epa.gov/owow/wetlands/monitor/RapidMethodReview.pdf.

Hanson, A., Swanson, L., Ewing, D., Grabas, G., Meyer, S., Ross, L., Watmough, M., Kirkby, J., 2008. Wetland ecological functions assessment: an overview of approaches. Canadian Wildlife Service, Atlantic Region (Technical Report Series No. 497). http://www.wetkit.net/docs/WA_TechReport497_en.pdf.

Hilchey, K., 2015. In: How to avoid getting bogged down by the Nova Scotia wetland conservation policy.Presentation for the Environmental Services Association of Nova Scotia, October 28, 2015. http://www.esamaritimes.ca/uploads/5/5/7/9/55790621/presentation_oct_28_2015.pdf.

McCallum Environmental Ltd, 2015. Wetland Alteration Application. Bedford Common, Bedford, NS. https://www.halifax.ca/sites/default/files/documents/city-hall/boards-committees-commissions/170510rwab611add.pdf.

Minnesota Board of Water & Soil Resources, 2008. Comprehensive General Guidance for Minnesota Routine Assessment Method (MnRAM) Evaluating Wetland Functions. Versions 3.1 and 3.2. St. Paul, MN. http://www.bwsr.state.mn.us/wetlands/mnram/index.html.

North Carolina Wetland Functional Assessment Team, 2008. N.C. Wetland Assessment Method (NC WAM) User Manual. Version 1 April 30, 2008. N.C. Department of Environment and Natural Resources, Raleigh, NC.

Stantec, 2011. Final Report: Environmental Assessment Registration for Northumberland Rock Quarry Extension Project. Dartmouth, NS. https://novascotia.ca/nse/ea/northumberland.rock.quarry.extension/AlvaQuarry_EA.pdf.

Stantec, 2013. Environmental Assessment Focus Report for Alton Natural Gas Pipeline Project. Dartmouth, NS. https://www.novascotia.ca/nse/ea/alton.natural.gas.pipeline.project/Alton_Focus%20Report.pdf.

Stantec, 2015. Environmental Assessment Registration for the National Gypsum Mine Extension. Dartmouth, NS. https://www.novascotia.ca/nse/ea/national-gypsum-mine-extension-project/Registration-Document.pdf.

Tiner, R.W., 2003. Correlating Enhanced National Wetlands Inventory Data With Wetland Functions for Watershed Assessments: A Rationale for Northeastern U.S. Wetlands. U.S. Fish and Wildlife Service, National Wetlands Inventory Program, Northeast Region, Hadley, MA.http://library.fws.gov/Wetlands/corelate_wetlandsNE.pdf.

Tiner, R.W., 2005. Assessing cumulative loss of wetland functions in the Nanticoke River watershed using enhanced National Wetlands Inventory data. Wetlands 25 (2), 405–419. http://library.fws.gov/Wetlands/TINER_WETLANDS25.pdf.

FURTHER READING

Adamus, P., 2011. Manual for the Wetland Ecosystem Services Protocol for the United States (WESPUS). Beta Test Version 1.0 Draft. Adamus Resource Assessment, Inc., Corvallis, OR.

Chapter 2.2.11

Development and Preliminary Tests of Remotely Based Imagery, Digital Databases, and GIS Methods as Tools to Identify Wetlands and Selected Functions and Values in Ontario, Canada

Adam Hogg and Brian Potter

Ontario Ministry of Natural Resources and Forestry, Peterborough, ON, Canada

Chapter Outline

INTRODUCTION

First published in the early 1980s (Environment Canada and Ministry of Natural Resources, 1983), the Ontario Wetland Evaluation System (OWES) is a science-based system designed to define, identify, and map wetlands as well as assess their functions and values and rank them relative to one another. It supports Ontario's land use planning policy framework by helping determine which wetlands should be protected from development.

With more than 30 years of application, the OWES is recognized by practitioners for its comprehensiveness. Wetlands are evaluated against almost 50 scored criteria organized under four main components:

- Biological (e.g., biological productivity and diversity).
- Social (e.g., economic, recreational, and educational uses).
- Hydrological (e.g., flood attenuation, water quality improvement, and erosion control).
- Special Features (e.g., provision of habitat for rare and provincially important species).

There are two evaluation manuals, one for southern Ontario and one for northern Ontario. The differences between the two manuals are relatively minor and reflect variations in geomorphology, climate, human uses, and other factors. Evaluated wetlands are either single contiguous wetlands or wetland complexes, which are defined as two or more discrete wetlands that are functionally related to one another (Ontario Ministry of Natural Resources, 2013). Many evaluated wetlands are wetland complexes.

Wetland and Stream Rapid Assessments. https://doi.org/10.1016/B978-0-12-805091-0.00026-8

The evaluation system has significantly increased understanding of the values of Ontario wetlands and improved the province's wetland inventory. To date more than 600,000 ha of wetland have been evaluated, primarily in southern Ontario where land development pressures generally are highest.

Despite these benefits, evaluation challenges remain. The OWES, with its emphasis on field work, is considered to be labor-intensive and therefore costly. Limited resources for field work, lack of direct access to many wetlands on privately owned lands, and a high proportion of unevaluated wetlands throughout much of the province have prompted efforts to investigate a more rapid, remote-based evaluation system.

Developments in digital remote sensing and GIS technology, in the time since the OWES was created, provide the potential to design a more streamlined and less-expensive approach to wetland evaluation (Chisholm et al., 1997; Stow et al., 2006; Strobl et al., 2003).

One way to integrate remote mapping and evaluation would be to employ a tiered approach where evaluation is done remotely from a landscape perspective, supported in the field only when necessary. In the near term, four objectives to achieve this goal have been identified, three of which (one, two, and four) will be addressed in this chapter:

1. Develop remote sensing-based wetland mapping standards.
2. Automate selected OWES components amenable to remote and automated evaluation and test them statistically.
3. Investigate and select landscape-level spatial measurements applicable to the OWES.
4. Develop and test ground-truthing field protocols.

Geospatial Wetland Mapping and Evaluation

Historically, air photo interpretation guides have also provided remote wetland interpretation and mapping insights. For example, there are a few organizational reports in Ontario that are useful in this regard (Arnup et al., 1999; Jeglum and Boissonneau, 1977; Zsilinszky, 1966). While these documents are based on sound methods, guidance is focused on use of one out-of-date data source (i.e., analog black and white air photos).

Modern guidance has focused more on known and easily quantifiable aspects of mapping, such as appropriate scale, associated accuracies and precisions, minimum mapping units, and suggested useful geospatial data sources (e.g., orthophotography, soils mapping). There is extensive material regarding wetland mapping addressing standards for geospatial data collection, mapping scale and quality, vegetation classification, and information management. For example, the U.S. Fish and Wildlife Service offers extensive guidance (Dahl et al., 2015; FGDC, 2009). The OWES manuals also provide guidance on this topic from an Ontario perspective (OMNR, 2013).

Over the past 20 years, remote sensing and GIS have revolutionized the manner in which mapping and ecological assessments are done (see Tiner et al., 2015 for details). Furthermore, as land-use pressures mount in densely populated landscapes such as southern Ontario, there is a heightened need to protect wetlands. Results of wetland mapping and evaluation can trigger municipal hearings at which OWES wetland mapping and evaluation results are challenged and defended by scientists. Standardized, precise, and accurate wetland boundary delineation and assessment are paramount to the successful resolution of these debates. Fortunately, advancements in remote sensing and GIS technologies offer opportunities to do this work in a cost-effective manner.

Wetland mapping and evaluation methods presented in this chapter therefore focus on providing guidance on wetland boundary delineation, vegetative community attribution, and functional assessment using new technologies and data sources. Considering these tools and data as providing "models of the real world" (with error) is a key concept driving our approach, which works toward integrating both "bottom up" and "top down" methods into a rapid wetland mapping and evaluation approach.

METHODS

Bottom-Up Wetland Mapping Standards

Technical guidelines and standards for wetland identification and delineation were developed and field-tested in a number of locations in two study areas in southern Ontario (Fig. 2.2.11.1). They were developed by a working group composed of Ontario Ministry of Natural Resources and Forestry (OMNRF) wetland and mapping experts. Specifically, keys to the delineation of the outer boundaries of wetlands and identification of wetland types and vegetation communities were produced.

Boundary delineation guidelines provide direction on desktop mapping of external wetland boundaries using remotely sensed imagery, GIS data, and, where available, field observations. This boundary delineation process is "interface driven"

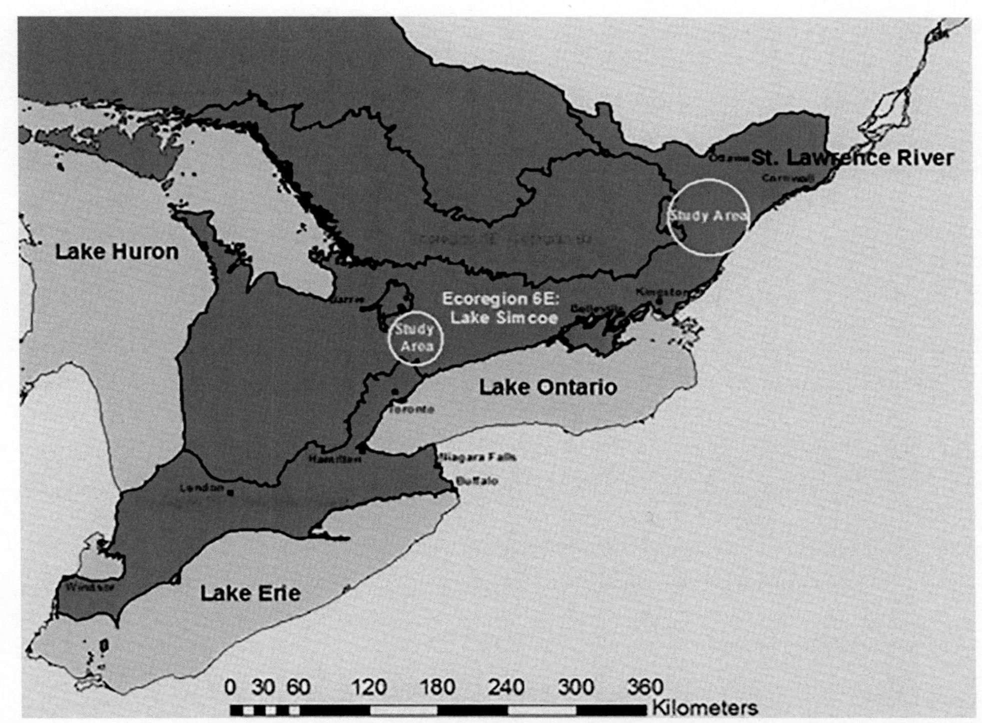

FIG. 2.2.11.1 Map of southern Ontario, Canada, showing two study areas (circled in white). Wetland identification and delineation guidelines and standards were field-tested at a number of locations within the two areas.

in that interpreters answer a series of questions designed to identify three wetland boundary interfaces commonly found in Ontario: anthropogenic, water, and upland forest. A dichotomous key (Fig. 2.2.11.2) is provided that requires the use of remote-based indicators of wetland hydrology, soils, and vegetation, as described in Ontario air photo interpretation literature (Arnup et al., 1999; Jeglum and Boissonneau, 1977; Zsilinszky, 1966) or by the expert working group. Indicators are categorized as either "clear" or "possible" indicators of wetland, 38 in total, examples of which are provided in Tables 2.2.11.1 and 2.2.11.2.

Boundary mapping is a crucial aspect of wetland function evaluation. Extensive amounts of time and effort have therefore been spent producing high-quality wetland boundary mapping based on air photo or digital orthophoto interpretation. This experience has led to the identification of key questions regarding standardization and efficiency. For example, wetland mapping may still require supporting expensive and time-consuming field work regardless of the remote sensing and/or GIS data sources used. The Wetland Boundary Delineation Key and associated indicators have been provided to standardize and expedite the processes of determining when field work will be required through use of a sequence of logical questions requiring yes/no responses.

The guide to wetland boundary delineation, including the interpretation key, was tested on a total of 44 potential wetland features located in one unique southern Ontario ecoregion in both sedimentary landscapes with deep soils and granite bedrock-dominated landscapes with thin soils. In summary, upfront delineation of wetland features containing similar image tones, textures, and patterns was performed first. Interpretation and delineation were completed in a digital stereo mode (i.e., digital equivalent to analog interpretation with a stereoscope) within the GIS environment using preleaf orthophotography, soils mapping, and a large-scale bare earth digital elevation model. This was followed up with field-based validation by interpreters and numerous expert field biologists with the OMNRF.

Candidate test sites were selected in the GIS environment using a stratified random approach, ensuring the sample population would address all aspects of wetland boundary delineation identified in the boundary delineation guide. The sample population was then refined based on local expert advice. Each site was composed of numerous vegetative and physiographic conditions (i.e., shallow aquatic, graminoid/emergent, shrub, and treed). Unique polygons were delineated for each homogeneous physiographic condition, the boundaries of which employed the logic in the interpretation key. Each polygon was then attributed with its respective interpretive indicators, as described in the key. After attribution, the combined indicators were assessed and assigned "clear" or "potential" status. Clear status was assigned when its respective polygon

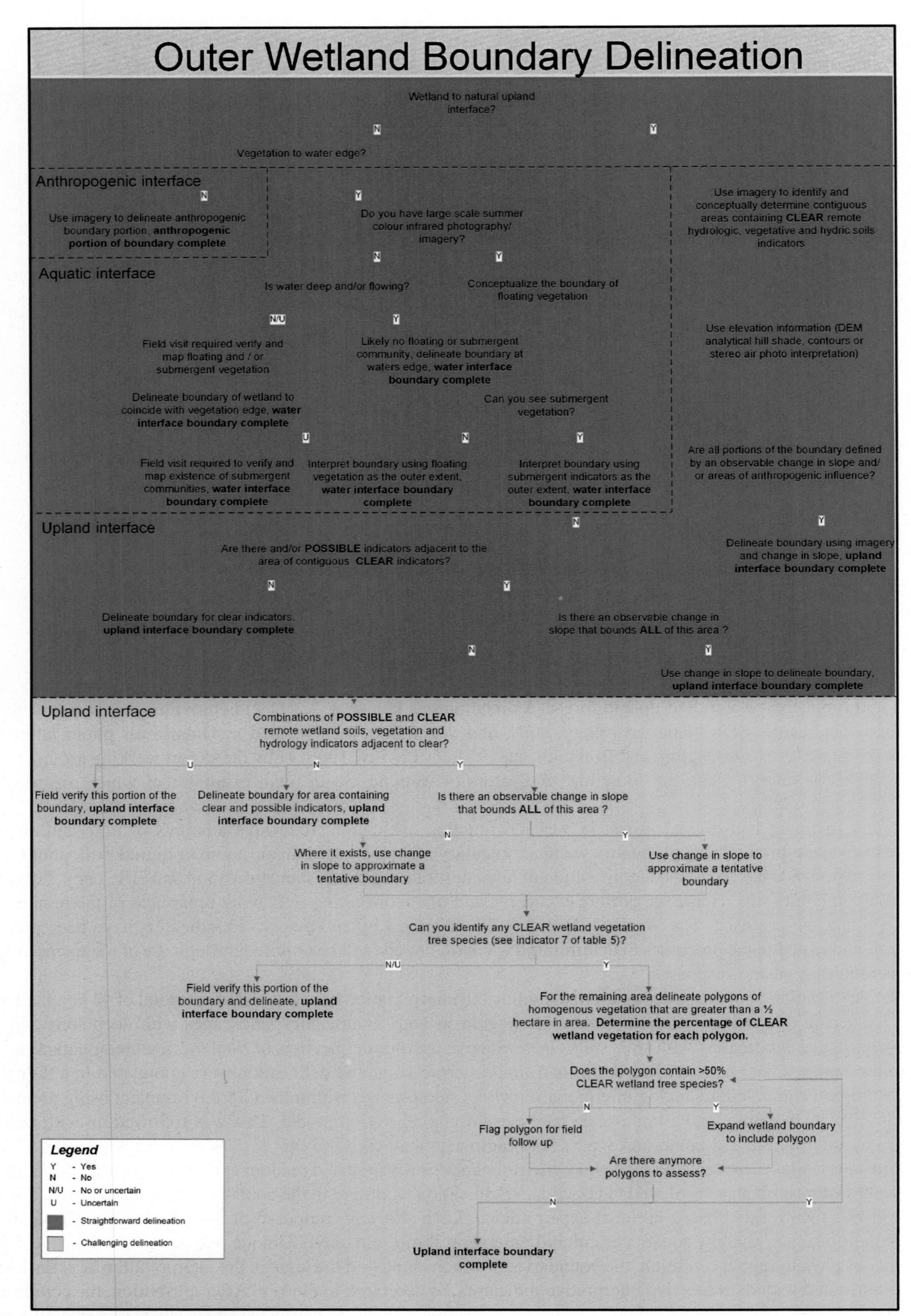

FIG. 2.2.11.2 Wetland Boundary Delineation Key.

TABLE 2.2.11.1 Example of Clear Wetland Indicators

Hydrologic	Flat or gently sloping low-lying area surrounded by distinguishable slopes, ridges, and/or plateaus in combination with numerous smaller drainage corridors oriented roughly parallel to the direction of water flow in either a net-like or delta pattern (Arnup et al., 1999)
Soil	Localized depressions in clay plains
Vegetation	Wetland trees including silver/Freeman's maple, black/green ash[a], tamarack, black spruce (mixed wood plains ecozone), pin oak, black gum

[a]*Wetland ash species tend to be in dense stands, almost a monoculture, for example, as one of two to three species. White ash is typically found in upland areas among many other species.*

TABLE 2.2.11.2 Example of Possible Wetland Indicators

Hydrologic	Floodplain areas up to the foot of a surrounding valley wall, for example, area adjacent to a watercourse (Jeglum and Boissonneau, 1977)
Soil	Level floodplain with coarse soil (sands) where water is observed and vegetation is moderately to highly productive
Vegetation	Irregular stand pattern (Jeglum and Boissonneau, 1977; Arnup et al., 1999; Zsilinszky, 1966)

contained at least one clear indicator from each of the three physiographic classes: hydrology, soil, and vegetation. In the event these criteria were not met, possible status was assigned. Finally, all features were visited in the field where wetland status and boundary delineations and indicators that had been interpreted remotely were checked for indicator and location accuracy.

Remotely interpreted and field-based data were next compared and summarized by wetland status (i.e., wetland/nonwetland), physiographic indicator, boundary interface, and assignment of clear or potential. The purpose of this summary was to document findings by providing a quantitative measure of the ability of remote interpretation processes to:

- Identify wetland indicators correctly.
- Separate those wetlands that can be mapped reliably from the desktop from those requiring field work.
- Measure success of delineation for the three different interface types (i.e., anthropogenic, open water, and natural wetland to upland).

Key to Wetland Types and Vegetation Communities

These guidelines are complementary to the Boundary Delineation Guide and Key, and provide direction on:

- Remote-based separation of wetland types and vegetation community types relevant to the OWES.
- Remotely identifiable vegetation species and/or community compositions.
- Determining where remote interpretation and mapping is possible, and recommending when field work is required.

A dichotomous key is provided (Fig. 2.2.11.3) along with a set of nodal descriptions as an aid to interpreting each decision point or box (node).

Remote-Based Identification of Bogs

Although very common in northern Ontario, bogs are rare in southern Ontario, and therefore were not included in the field testing of the draft "type and community" key. Remotely, bogs are distinguishable by unique multispectral responses of *Sphagnum* moss and lichen (Bubier et al., 1997). These communities, when dominant and large, are detectable by viewing a medium-scale Landsat satellite image near infrared, mid infrared and visible red composite. Stunted and sparse black spruce (Arnup et al., 1999; Jeglum and Boissonneau, 1977; Zsilinszky, 1966), the presence of a slightly raised peat dome or plateau (Jeglum and Boissonneau, 1977), and the absence of surface water flow either through direct observation of flow water (i.e., streams) or vegetative pattern (Arnup et al., 1999; Jeglum and Boissonneau, 1977) are all also indicators of this wetland type. The utility of these criteria was field tested at three sites near North Bay, Ontario.

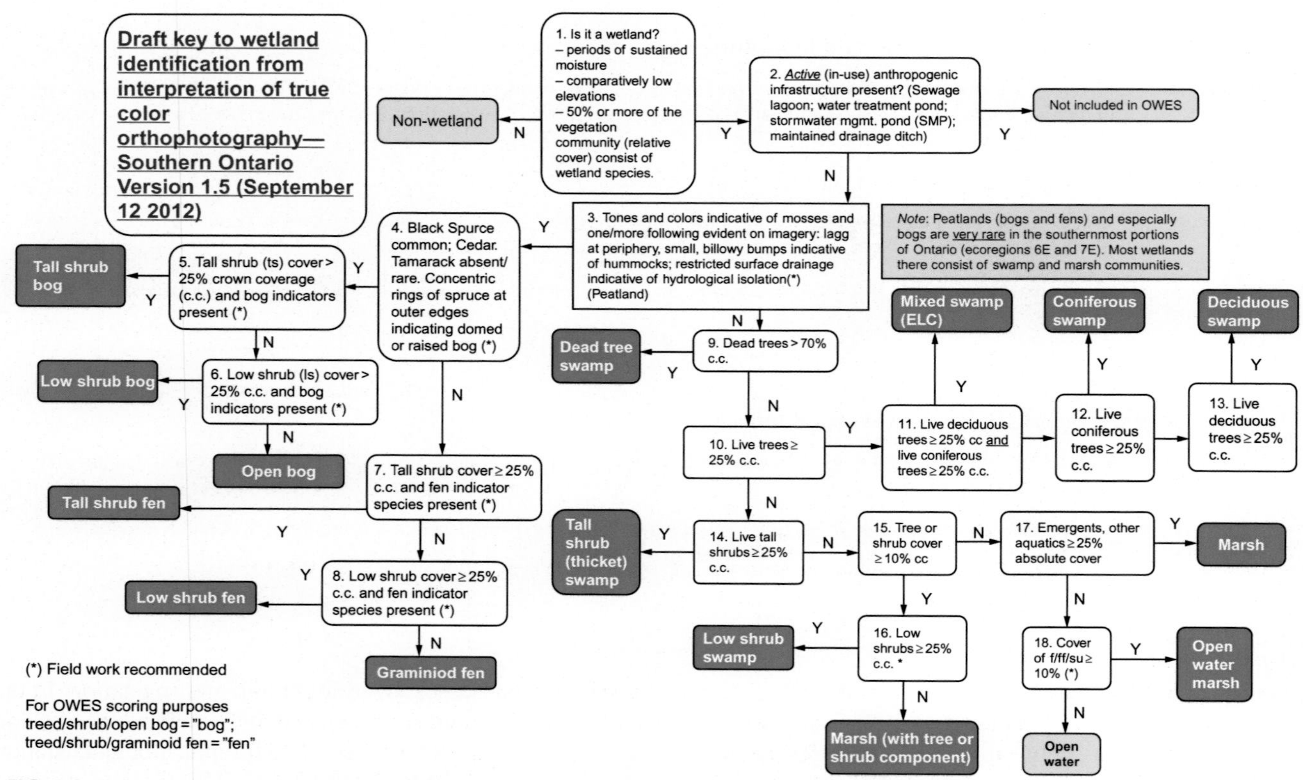

FIG. 2.2.11.3 Draft key to wetland identification from interpretation of true color orthophotography.

Automated Top-Down Wetland Evaluation Using Remote Sensing and GIS

A subset of OWES attributes was identified as potentially suitable for automated scoring based on experience and knowledge of the OWES, GIS, and remote sensing. Eight attributes were selected, including at least one attribute from each of the four OWES components as described in Table 2.2.11.3.

The objective of this work was to automate attribute scores using remote sensing and GIS, and statistically compare the results to field-derived scores using a paired *t*-test. It was hypothesized that scores automatically calculated with remote sensing and in GIS data/tools were not significantly different from those calculated using the original field-based OWES method. More than 50 wetland complexes from two study areas covering three unique ecoregions (St. Lawrence Lowlands, Frontenac Axis, and Lake Erie Lowlands), including more than 4000 wetland features, were included in the analysis.

RESULTS AND DISCUSSION

Bottom-Up Wetland Mapping Standards

The results of this work were analyzed and summarized to better understand two key aspects of the mapping process: (1) assignment of wetland/nonwetland using the interpretive key, and (2) the ability of clear or possible indicators, when considered individually, to identify the presence of wetland. Table 2.2.11.4 summarizes the field validation conducted for assessment of the remote interpretation boundary delineation key. The results illustrate an accurate mapping strategy erring on the side of caution. For example, 100% of the features mapped as having only clear remote wetland indicators and therefore remotely designated as wetland were validated as correct in the field (i.e., 22 out of 22 polygons). Conversely, 45% (10 of 22) of the sample features identified as having possible wetland indicators were confirmed as wetland in the field. This reflects the challenge of identifying some clear indicators remotely (tree species, moisture under canopy, etc.) and the need to consider possible indicators. This illustrates the fundamental limitations of mapping wetlands using spatially and temporally scaled-down models of the real word (i.e., two orthoimages acquired on varying dates). Fortunately for wetland conservation purposes, the erroneous "commission" of valid wetland features into a possible remote wetland

TABLE 2.2.11.3 Components and Attributes Assessed for Automated Remote Wetland Evaluation

OWES Component	Wetland Attribute	Data Source(s)	Method
Biological	*Proximity to other wetlands*: a measure of habitat connectivity—amount of wetland and water features at various "birds eye" distances	Land Information Ontario (LIO) wetlands, Ontario Hydrographic Network (OHN), single line streams and water bodies	Multiple buffers and selections
	Open water types: relative proportion and areal configuration of permanent open water adjacent to vegetation	LIO wetlands, high-resolution summer optical satellite imagery	Percentage automatically calculated, configuration qualitatively determined
	Interspersion: a measure of number of edges of internal vegetation community boundaries by counting number of boundary intersections with an artificial grid representative of wetland complex shape and size	LIO wetlands	GIS tool generates representative grid, overlays and counts intersections
Social	*Wood products*: total of treed area containing trees used for lumber, pulp, fencing, and firewood	LIO wetlands, provincial woodland layer(s)	Overlay and sum of area
Hydrology	*Flood attenuation*: assessment of a wetland's ability to attenuate flooding based on its area relative to its catchment and noncomplexed wetlands/water bodies within	LIO wetlands, OHN streams and water bodies, Provincial Digital Elevation Model	Automatic delineation of wetland complex outflow and catchment, overlay and calculation of fractional area
Special Features	*Rarity of wetland type*: rarity of type (i.e., marsh, fen, bog, swamp) within their broader respective landscape	LIO wetlands (no internal wetland type attribute, orthoimagery and stereo color infrared air photos)	Type was interpreted using imagery and wetland type interpretation key
	Winter cover for wildlife: identification treed coniferous area and cattail marsh	LIO wetlands, satellite imagery, orthoimagery	Vegetation communities automatically classified per satellite imagery and verified using orthophotos
	Identification of endangered or threatened species	LIO wetlands, Natural Heritage Information Center (NHIC) species occurrence observations	Overlay of species occurrences and wetlands

TABLE 2.2.11.4 Comparison of Remotely Interpreted and Field-Validated Wetland Features

		Field Validation			
		Wetland	Not-Wetland	Total	Omission Error (%)
Remote interpretation	Wetland	22	0	22	0
	Possible wetland	10	12	22	45
	Total	31	13	44	
	Commission error (%)	32	8		
Overall accuracy					75

Remote/field matches, error, and overall accuracy highlighted by dark gray squares with bolded numbers.

designation is preferable to a remote wetland designation validated as nonwetland in the field. Finally, the near-even split of remotely identified "possible" wetland polygon samples, 10 valid wetland and 12 nonvalid wetland, confirms the uncertainty of possible wetland indicators and the need for field follow up when designating wetland features using only these indicators.

The ability to remotely identify all documented hydrology, soils, and vegetation indicators was also considered. Soil indicators were the most difficult to identify and therefore most often obtained through consulting existing soils and/or surficial geology mapping. Vegetative and hydrologic indicators were observed on orthophotography and typically validated as correct in the field. Shallow aquatic, graminoid, emergent, and shrub wetland indicators and boundaries were all highly accurate. Anthropogenic and open water interfaces were also always correct. The natural wetland to upland, particularly in forested features, was least accurate. In particular, vegetative indicators based on tree species were most challenging to identify in these environments because of the history of anthropogenic disturbance and resultant nonecologically driven heterogeneous stand patterns. This issue is further compounded by the small stand sizes typically found in southern Ontario. This was expected as most vegetative interpretive indicators were drawn from guides that were calibrated in northern Ontario, where swamp wetland disturbances and stand sizes are much larger and more likely to be driven by natural changes.

A detailed review and analysis of the success of each individual clear and possible hydrology, soil, and vegetative remotely identified wetland indicator was also conducted; a summary is provided in Fig. 2.2.11.4. Results showed that samples containing at least one clear hydrologic, soil, and vegetation indicator had no nonwetland features erroneously mapped as wetland. Furthermore, when at least two clear indicators were present, misclassification error was an acceptable 5%. A total of the 44 samples (i.e., 70%) met this criteria. Error rates increased dramatically and far beyond acceptability (i.e., 80%) when only one clear indicator was identified. The results therefore indicate that a minimum of two clear indicators is required for reliable, remote-based mapping of wetland features.

Key to Wetland Types and Vegetation Communities

Testing to date indicates that the type-community key, used in conjunction with the remote-based wetland mapping standards, can be a useful tool for identification of marshes, swamps, and fens.

Application of the Outer Wetland Boundary Delineation Key sets the bounds on the land areas most likely to be wetland. Then, within these potential wetland areas, remotely sensed imagery is interpreted to identify wetland type and, where possible, OWES vegetation structure type (e.g., treed versus tall shrub) and individual species.

Limitations to the use of the key include resolution of digital imagery and acquisition date and year. The optimum remote mapping solution will have high resolution imagery (e.g., 40 cm resolution or better) and both spring and summer imagery, preferably in the same year. However, while swamp, marsh, and shallow aquatic communities can be reliably remotely identified, fen and bog separation still remains problematic. For example, field work was required to confirm the presence or absence of indicator species for wetland types such as bogs and fens. This was largely attributed to the ecological zones in which they reside (i.e., mixed wood plains and boreal shield). True bogs are rare in these ecozones and generally atypical in their size (i.e., small) and formation processes. As a result they are quite often on the bog side of the most frequent natural ecological transition from fen to bog and therefore exhibit vegetative conditions found in both wetland types. Conversely, bogs in the Hudson Bay Lowlands occur on a large plain with minimal surface water movement. As shown in Fig. 2.2.11.5 these features are much larger, have greater peat accumulation and thickness, and therefore exhibit more remotely distinguishable bog characteristics (e.g., expansive *Sphagnum* moss and/or lichen-dominated communities).

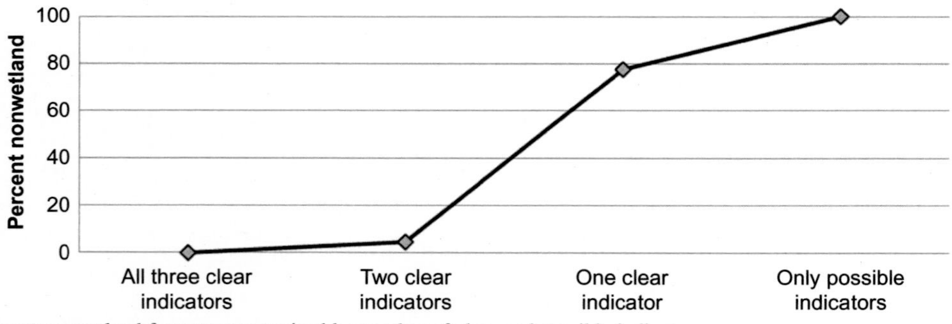

FIG. 2.2.11.4 Percent nonwetland features summarized by number of clear and possible indicators.

FIG. 2.2.11.5 Landsat image (30m resolution) of *Sphagnum* moss and lichen-rich bog domes drained by slowly flowing fen and eventually thicket swamp in the Hudson Bay Lowlands (left). The center image is a ground photo of a *Sphagnum* moss rich bog while the right image is that of a treed bog with lichen-rich hummocks.

FIG. 2.2.11.6 Open water marsh aerial image (left) with two accompanying ground photos from late summer showing portions of an open water marsh (center), and both emergent vegetation and dead trees (right).

Marshes

Marshes, especially those with an open water component, often are readily identifiable on both spring true and summer false color orthophotographic images, as can be seen in Fig. 2.2.11.6. In this example from eastern Ontario, the dark tones in Polygon 4 are indicative of open water. To the right of the open water, there is evidence of linear flow patterns (toward the beaver dam), and brown-colored tones indicative of senesced, graminoid, and emergent marsh vegetation. Light brown tones show last year's dry senesced vegetation while dark brown tones reflect the previous year's damp senesced vegetation. The color infrared imagery confirms the presence of shallow aquatic floating (white and cyan) and submerged (misty blue) communities and adds additional detail regarding dead tree, graminoid, and emergent vegetation. Linear cyan features in the upper right portion of the wetland can be observed both lying down on the wetland surface and standing/leaning. Red colors in this same area are indicative of the current year's dense graminoid vegetation (with portions of grass blades bent and facing the camera lens) while cyan tones indicate areas dominated by emergent vegetation (with blades vertical and therefore perpendicular to the camera lens). In the latter case, even during midsummer, only last year's senesced vegetation is visible from overhead. The two accompanying ground photos, from late summer, show portions of open water marsh (center and right) emergent vegetation and dead trees.

Swamps

For OWES purposes, swamps are defined as wooded wetlands with 25% or more of trees or tall shrubs. Exceptions to the 25% rule include low shrub swamps (50% or more of wetland area is comprised of low shrubs), and dead tree swamps where extensive stands of dead trees (more than 70% cover) occur. Trees are defined as woody vegetation greater than 6 meters in height while tall shrubs are considered to be woody vegetation 1–6m in height, including stunted and sapling tree species. The height of woody vegetation can be estimated using digital imagery and GIS applications, which facilitates differentiation of treed versus tall shrub communities. Fig. 2.2.11.7 includes a true color spring orthophoto, a summer color infrared image, and a field photo of a swamp in eastern Ontario. Application of the Wetland Boundary Delineation Key resulted in the identification of clear soil and vegetation indicator criteria while interpretation of color infrared imagery yielded identification of black spruce, a clear swamp habitat indicator in Southern Ontario, and a number of tree species that are possible indicators.

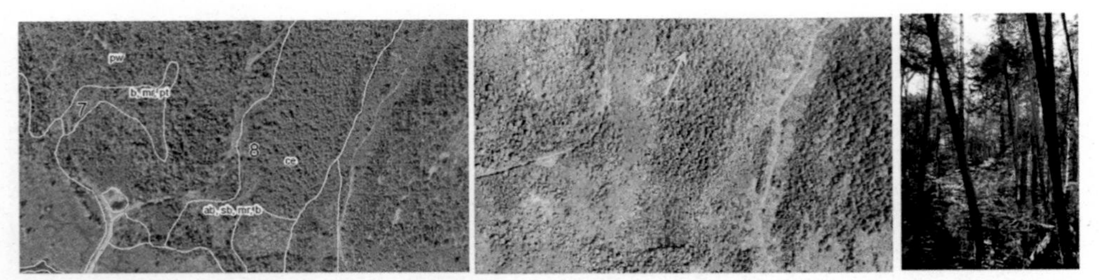

FIG. 2.2.11.7 Swamp wetland images. Polygon 8 in the true color, spring orthophoto (left) was identified as a wetland containing soil and vegetation clear indicator criteria in the Wetland Boundary Delineation Key. Interpretation of the same feature using summer color infrared imagery (center) allowed for identification of numerous possible vegetative wetland indicators Balsam fir (*Abies balsamea*), Eastern white cedar (*Thuja occidentalis*), red maple (*Acer rubrum*), paper birch (*Betula papyrifera*), and trembling aspen (*Populus tremuloides*). Black spruce (*Picea mariana*), a clear wetland indicator in southern Ontario, was also identified. Field photo on right shows a single large Balsam fir, a paper birch overstory with dense Eastern white cedar understory; the clearing was due to blow down (more susceptible in wetter soils).

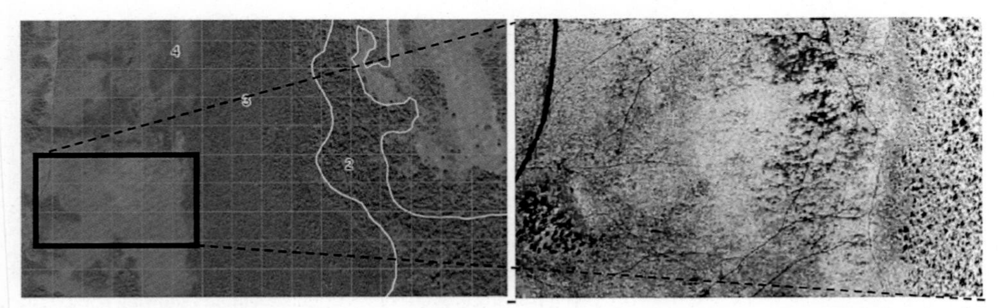

FIG. 2.2.11.8 Spring, true color orthophoto of a wetland identified remotely as a fen (left). The fine-textured and light-colored area in the lower left corner is indicative of a graminoid vegetation community. Linear water tracks lacking dendritic pattern were observed on false color composite orthoimage (right).

Fens

The OWES recognizes two major fen types: nutrient-rich fens that are usually groundwater-fed, and nutrient-poor fens characterized by reduced groundwater inputs, limited mineral inputs, and lower pH. Fens are usually covered by peat, partially decomposed plant material that accumulates under saturated soil conditions. Rich fens can also form directly on limestone fed by minerotropic groundwater discharge. Peatland is a general term that may refer to bogs or fens covered in peat (OMNR, 2013). Nutrient-poor fens often have a strong moss component, which typically displays as yellow tones in leaf-on digital imagery, for example, color infrared enhanced Forest Resource Inventory Imagery (eFRI). This is one of the most reliable attributes that can be used to identify this type of peatland. It is a very exclusive trait and can, for example, be used to distinguish between meadow marshes and fens; yellow tones often are evident in nutrient-poor fens but absent in marshes. If this imagery is not available, for example, where only spring (leaf-off) digital orthophotography is accessible, then interpretation of Landsat satellite imagery may assist in the identification of peatlands. For example, *Sphagnum* mosses appear as purple tones in Landsat 7 multispectral composite imagery (near infrared; mid infrared, and red). With this imagery, only relatively large and homogenous vegetation communities (i.e., greater than half a hectare) would be visible given the relatively low resolution of Landsat imagery (30 m).

In both true-color and false-color orthoimagery, nutrient-rich graminoid fens and marshes often exhibit similar *colors*, as grass communities may mask the presence of a moss ground cover layer. These two wetland types, as can also be seen in Fig. 2.2.11.8, may be distinguishable from one another based on differences in texture and surface water features. Fens tend to have a smoother appearance compared to marshes. Marshes often exhibit dendritic/linear dark tones indicating surface water flow and the presence of small stream channels, whereas fens generally lack these dendritic dark tones but may exhibit similar linear "water tracks" (pathways of depressed peat mat caused by repeated wildlife and/or human traffic) that do not generally align with the direction of flow but rather the shortest traversable distance across the wetland.

Bogs

A number of features in the North Bay area were remotely identified as bogs based on interpretation of Landsat and color infrared imagery. Three sites were field-checked to test the utility of four criteria: unique, multispectral responses of

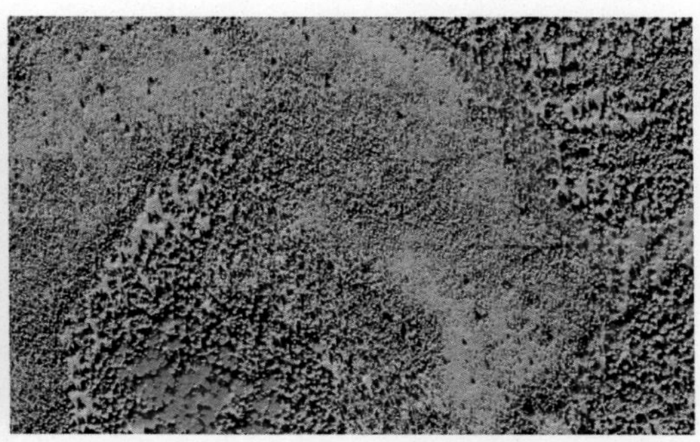

FIG. 2.2.11.9 Remote-based identification of bogs. Of the three sites that were visited, one was determined to be a bog—see digital imagery and ground photo. (Left) Well-developed hummocks in small bog. Black spruce was prevalent; Leatherleaf common, with some tamarack present, mostly at edge. (Right) The imagery at Site 1, in the above color infrared imagery, shows subtle orange tones (*Sphagnum* moss) where canopy is open, or where sparse, stunted trees with narrow canopies occur in low-lying areas with no observable surface water or evidence of flow.

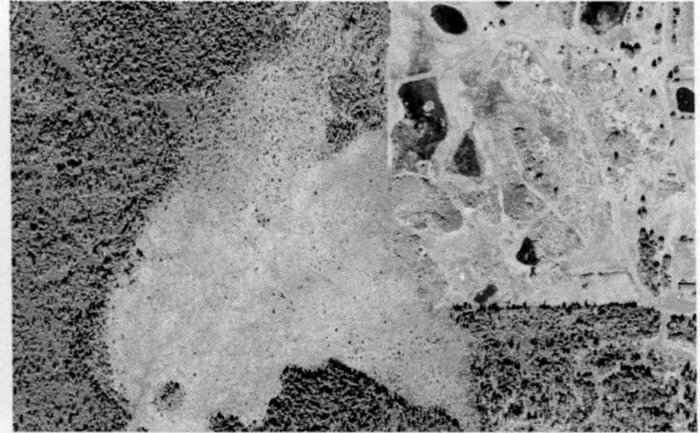

FIG. 2.2.11.10 Two sites were determined to be nutrient-poor fens. The images are from one of these two fens. (Left) Poor fen containing tamarack, ericaceous shrubs, and graminoids; moderate hummock development was also observed. (Right) Summer color infrared image with smooth texture, mottled pattern, and sparse scattered trees. Patches of subtle orange tones are indicative of *Sphagnum* moss.

Sphagnum mosses and lichens; presence of stunted and sparse Black Spruce; presence of a slightly raised peat dome or plateau; and absence of surface water flow.

One of the three sites was confirmed to be a bog. As seen in Fig. 2.2.11.9, subtle orange tones in color infrared imagery indicate the presence of *Sphagnum* moss. Also evident are an open canopy, or the presence of sparse, stunted trees, and lack of observable surface water flow and water tracks.

The other sites were confirmed to be nutrient-poor fens. Fig. 2.2.11.10 provides a color infrared image and a field photograph for one of these two sites. While subtle orange tones in the color infrared imagery again indicate the presence of *Sphagnum* moss, Tamarack, a species not commonly present in bogs, was found throughout the site. Also, the moderate hummock development is more typical of fens than bogs.

Automated Top-Down Wetland Evaluation Using Remote Sensing and GIS

Wetland function scores for more than 50 wetland complex samples, containing more than 4000 individually mapped wetland features, were derived using remote sensing/GIS tools and data and original field-based methods. Resultant scores were then statistically compared using a paired *t*-test with a null hypothesis that scores were not significantly different. Results are described in Table 2.2.11.5.

None of the GIS-derived biological component scores were significantly different from the field-assigned scores, although in all cases on average scores were found to be higher (Table 2.2.11.5). Similar results were obtained for the hydrology (flood attenuation) and special features components. Upon consultation with technical field experts, it was agreed that these desktop-generated scores were likely higher due to the GIS's ability to comprehensively consider all wetland, water, and topographic features at a landscape-scale rather than what was selected as practical at the time of each site-specific wetland evaluation.

The GIS-based scores for the social component (wood products) were significantly different from the original evaluation score and on average drastically lower than the average of the original field-based scores. Given that this was

TABLE 2.2.11.5 Results of Paired *t*-Test of Automated Remote and Manual Field-Derived OWES Attribute Scores ($P = 0.05$; $N = 50+$).

OWES Component	OWES Attribute	Statistically Significant Difference?	Average Difference From Field (%)	Automation Efficiency
Biological	Proximity—other wetlands	No	+15	Complete
	Open water types	No	+25	Partial
	Interspersion	No	+20	Complete
Social	Wood products	Yes	−50	Complete
Hydrology	Flood attenuation	No	+25	Complete
Special features	Rarity of wetland type	No	+5	Partial
	Winter cover	No	−5	Complete
	Endangered or threatened species	Yes	−45	Complete

conducted in GIS using a simple overlay of existing accurate and current wooded product mapping, forest loss since time of initial evaluation was assessed to be the primary reason for the discrepancy. Contrary to the other components of the OWES, much of the scoring for the social component is not as reliant on field work, so the lack of a remote method of scoring would not be a major setback in streamlining scoring. Although less drastic, the GIS winter cover attribute also scored lower on average than its field-based counterpart. This discrepancy was also attributed to forest loss.

Finally, the endangered or threatened special features attribute was significantly different and on average drastically lower than the field-based evaluation. This result illustrates the incompleteness of provincial rare and endangered species occurrence data and emphasizes the importance of site-specific and a comprehensive ground-based approach for evaluation of this attribute. The most recent version of the OWES manuals requires submission of a rare species sighting form for any previously unrecorded species scored in the wetland.

ADDITIONAL COLLABORATIVE WORK

Carleton University, in collaboration with the Ontario and Canadian governments, conducted research that focused on improving the efficiency of the Ontario Wetland Evaluation System using primarily a top-down remote sensing and GIS approach (Robertson, 2014). The application of high-resolution radar and multispectral optical imagery sources, (i.e., Radarsat 2 and Worldview 2 respectively), medium resolution multispectral optical imagery (i.e., Landsat 5), and existing provincial GIS layers was investigated. Material representing two different aspects of this research was published in two separate scientific journal articles. Both articles offer useful quantitative insights regarding the application of state-of-the-art remote sensing and GIS techniques for rapid remote wetland mapping and assessment. The first focused on evaluating spatial and seasonal variations in wetland characteristics in eastern Ontario (Robertson, 2014). Use of unprocessed c-band radar satellite imagery obtained by Radarsat 2 generally resulted in very poor wetland classification accuracies (40%–65%, overall). Spring (leaf-off) swamps, however, were distinguishable from other wetland types due to their more physically complex vegetation structure, and the manner in which this randomly scatters radio energy in all directions. Flooded emergent marshes and leaf-off swamps, at varying depths of flooding, were also easily detectable. In these environments the radar energy, which is emitted from the satellite and approaches its ground targets on an angle, often bounces off water and again off perpendicular vegetation structure, creating a distinguishable bright *corner reflection/double bounce* response. However, it was more difficult to distinguish between marsh and fen types, due to their structurally similar vegetation. The highest degree of overlap occurred between bog and fen given their similar vegetation structures, even though in some cases fens were dominated by grasses and bogs by low ericaceous shrubs.

The use of high-resolution, spring-season optical imagery, in conjunction with digital elevation data and polarimetric radar imagery, was assessed to determine if this would result in improved discrimination of wetland types, compared to the use of either optical/DEM or radar data alone (Robertson et al., 2015b).

WorldView-2 optical image variables were combined with Radarsat-2 image variables in an object-based image analysis (OBIA) classification of wetland type. Relative to the use of optical imagery, only, incorporation of radar polarimetric variables improved the classification accuracy for fens, bogs, and swamps. For example, the use of noncomposition filtered radar imagery plus high-resolution optical imagery improved the accuracy of swamp identification by 10%–25%. The study also determined that, compared to summer radar data, spring data generated more accurate results, and steep incidence angles were better than shallow incidence angles.

Robertson et al. (2015a) evaluated the benefits of using multidate Landsat 5 imagery and spectral mixture analysis (SMA) in wetland change detection. Individual land cover changes and anthropogenic impacts, and their timing, were detected using Landsat 5 TM and SMA fractions (vegetation, bare, moisture) in two-date, time-series assessments. Annual analyses allowed for the detection of sudden and dramatic changes in shallow aquatic vegetation. The abrupt nature of the changes suggested anthropogenic or beaver-related impacts. SMA did not distinguish brown fields versus senesced grasses, the latter being a common temporal response in marshes. The findings may be applicable to remote-based scoring of absence of human disturbance, one of the OWES attributes in the social component.

SUMMARY

GIS-based desktop interpretations of hydrology, soil, and vegetation characteristics can be used to identify wetland habitats and provide a first estimate of outer wetland boundaries. Based on testing to date, if at least one clear indicator from each of these three categories is present, then the misidentification error rate is zero (i.e., there were no nonwetland features erroneously mapped as wetland). Low levels of error (5%) can be achieved when at least two clear indicators are present. As a result, remote-based wetland identification and mapping can expedite wetland conservation efforts by reducing the amount of field work required. This is especially true in those areas of Ontario governed by land-use plans such as the Oak Ridges Moraine Conservation Plan; this stipulates that protection of some wetlands requires identification and mapping only, and not full evaluation.

Similarly, results for the wetland type-community key indicate that it can expedite the wetland conservation process by reducing the amount of field work required to complete a wetland evaluation. In many cases, the colors, tones, textures, sizes, and patterning evident in true-color and false-color orthoimagery can be accurately interpreted to distinguish among wetland types (e.g., marshes versus fens) and vegetation communities (e.g., treed versus tall shrub swamp). Limitations to the use of the key include the current resolution of digital imagery (e.g., not all wetland vegetation communities can be discerned remotely) and the need in some cases for field work to confirm the presence or absence of indicator species for wetland types such as bogs and fens. In addition, testing to date has focused on remote identification of discrete wetland types. Additional testing of the key to assess its ability to discriminate among mixed wetland types would be beneficial. For example, can fen-swamp boundaries be discerned?

Automated desktop scoring of selected OWES attributes also holds potential as a means of streamlining the wetland evaluation process. Most GIS-derived scores calculated in our analysis were higher than the ones based on the existing OWES methods while the results from Carleton University (which relied on different data sets) displayed a greater variability of results, that is, same scores, lower scores, higher scores, scoring not possible. In either case, GIS-based desktop approaches to scoring may serve as a useful screening tool to identify those wetlands most likely to be provincially significant and therefore eligible for protection.

DIRECTION FOR FUTURE WORK

Moving forward, the conservation of Ontario's wetlands will be strongly influenced by provincial government policy. The work described in this chapter has demonstrated useful strategies for improving wetland mapping and assessment to ensure that essential knowledge is available and used to make decisions.

Ontario has been active in wetland conservation since the early 1980s. Activity has largely been focused on completion of wetland mapping and assessment through the OWES by the public sector, NGOs, and the private sector. Throughout this period, the mapping assessment completed using the OWES was considered state of the art by the bioscience community. In the early 2000s, the improvement of remote sensing, spatial information management and analysis, and increased availability of spatial information initiated a paradigm shift in the scientific and planning/policy communities. Government and the broader scientific community are now beginning to adopt a holistic landscape management approach that requires timely consideration of multiple information sources covering large areas (e.g., the entire province). Ontario will be able to address this challenge but, from a wetland perspective, much mapping and assessment work is still required to meet the science and information needs that such an approach demands. For example, very little wetland assessment work has been conducted outside southern Ontario, which only represents 15% of the total landmass of the province. Furthermore, mapping and assessment work was conducted, as directed by the governments and scientific communities of the day, with a focus on small areas and reviewed, compiled, and standardized by local scale government offices. Mapping and assessment was therefore representative of the pressures affecting those offices but lacked consistency from a provincial perspective, making broad landscape-scale reporting and conservation difficult.

The work presented in this chapter is intended to provide a framework for future mapping with the intent of working toward a consistent, precise, and accurate wetland inventory for rapid assessment and planning purposes. The summary and testing of wetland interpretation guidelines revealed a systematic desktop mapping approach that promotes standardized boundary decision-making and judicious delegation of required field work. Such an approach could be incorporated into current government wetland information management and mapping methods (e.g., attributes reflecting number of clear and possible indicators identified and supported by field work). Furthermore, existing wetland mapping could be measured by this standard to ensure consistent broad landscape-scale reporting.

There are still many opportunities for exploration of new wetland mapping approaches. For example, computer automation reduces subjectivity and dramatically increases efficiency and consistency. While this work employed heads up digitizing in a GIS stereo environment to identify and attribute clear and possible indicators, this exercise could be automated using today's imagery, data, and image object-based software such as eCognition. Furthermore, these data and technologies are readily accessible to the public and private sectors, NGOs, and academia.

The ecological assessment of wetland function has proven to be more challenging. OWES uses more than 40 functional wetland attributes to assess ecological function. Scientific and technological efforts have shown a meaningful quantitative assessment of wetland function can still be done using a dramatically reduced set of functional wetland attributes, and that many of these attributes can be obtained using remote-sensing imagery and GIS analysis of spatial data. In fact, this work showed that there is no statistical difference between selected desktop-derived and field-based functional attributes covering three of the four major OWES components, and aspects of the fourth (social) component. Additionally, there are still many more functional attributes that could be automated in a similar manner. Despite these opportunities, one key issue still exists. The final functional score of a wetland "complex" when derived using the original field- and expert-driven approach is often statistically different from overall scores derived using desktop approaches. The challenge here resides in coming to terms with the differences between the group of wetland features included using the original complexing approach, which were sometimes derived using a narrow, nonecologically driven mapping study area, and those included in a broad landscape-scale approach using modern spatial hydrological and biological metrics. When addressing this, efforts could be directed toward redefining how wetlands are spatially and functionally grouped and assessed for conservation status using modern science and technology, rather than focusing on attempts to match overall wetland complex functional scores of two often very different spatial groupings of wetlands. Comparison of conservation status results yielded by old and new approaches could then be conducted at a landscape scale using functional units such as ecoregions and watersheds. Doing so would provide for a fair comparison of old and new overall wetland functional assessments and allow us to move closer to scientific acceptance of a desktop functional wetland assessment tool.

ACKNOWLEDGMENTS

We thank Joel Mostoway, Regina Varrin, Dr. Tom Whillans, and Rebecca Zeran for critical reviews of the manuscript.

REFERENCES

Arnup, R., Racey, G.D., Whaley, R.E., 1999. Training Manual for Photo Interpretation of Ecosites in Northwestern Ontario. Ontario Ministry of Natural Resources, Northwest Science and Technology, Thunder Bay, Ontario, 130 p.

Bubier, J.L., Barret, N.R., Crill, P.M., 1997. Spectral reflectance measurements of boreawetland and forest mosses. J. Geophys. Res. 102 (D24), 483–494.

Chisholm, S., Davies, J.C., Mulamoottil, G., Capatos, D., 1997. Predictive models for identifying potentially valuable wetlands. Can. Water Resour. J. 22 (3), 249–267.

Dahl, T.E., Dick, J., Swords, J., Wilen, B.O., 2015. Data Collection Requirements and Procedures for Mapping Wetland, Deepwater, and Related Habitats of the United States. Version 2: 92 p. Madison, WI. Retrieved from: http://www.fws.gov/wetlands.

Environment Canada and Ministry of Natural Resources, 1983. An Evaluation System for Wetlands of Ontario South of the Precambrian Shield, first ed. 144 p.

Federal Geographic Data Committee (FGDC), 2009. Wetland Mapping Standard (FGDC-STD-015-2009). https://www.fgdc.gov/standards/projects/wetlands-mapping/2009-08%20FGDC%20Wetlands%20Mapping%20Standard_final.pdf. 35 p.

Jeglum, J.K., Boissonneau, A.N., 1977. Air Photo Interpretation of Wetlands, Northern Clay Section, Ontario. Canadian Forest Service, Department of the Environment, Sault Ste. Marie, Ontario, 78 p.

Ontario Ministry of Natural Resources, 2013. Ontario Wetland Evaluation System, Southern Manual, third ed. (Version 3.2), 283 p.

Robertson, L.D., 2014. Evaluating Spatial and Seasonal Variability of Wetlands in Eastern Ontario Using Remote Sensing and GIS. (Ph.D. Thesis). Carleton University, Ontario, Canada.

Robertson, L.D., King, D.J., Davies, C., 2015a. Assessing land cover change and anthropogenic disturbance in wetlands using vegetation fractions derived from Landsat 5 TM imagery (1984-2010). Wetlands 35, 1077–1091.

Robertson, L.D., King, D.J., Davies, C., 2015b. Object-based image analysis of optical and radar variables for wetland evaluation. Int. J. Remote Sens. 36 (23), 5811–5841.

Stow, N., Pond, B., Jahncke, R., 2006. Evaluation of a Wetland Rapid Assessment Procedure for Northern Ontario. Ministry of Natural Resources, Peterborough, Ontario. (Internal report), 13 p.

Strobl, S., Beckerson, P., Hogg, A., 2003. Enhanced Wetland Mapping & Evaluation for Ontario's Forested Shield. Ministry of Natural Resources, Peterborough, Ontario. (Internal report), 58 p.

Tiner, R.W., Lang, M.W., Klemens, V.V. (Eds.), 2015. Remote Sensing of Wetlands: Applications and Advances. CRC Press, Boca Raton, FL. https://www.crcpress.com/Remote-Sensing-of-Wetlands-Applications-and-Advances/Tiner-Lang-Klemas/p/book/9781482237351.

Zsilinszky, V.G., 1966. Photographic Interpretation of Tree Species in Ontario, second ed. Ontario Department of Lands and Forests, Toronto, Ontario, 86 p.

Chapter 2.2.12

Maintaining the Portfolio of Wetland Functions on Landscapes: A Rapid Evaluation Tool for Estimating Wetland Functions and Values in Alberta, Canada

Irena F. Creed*[,†], David A. Aldred[†], Jacqueline N. Serran[†] and Francesco Accatino[†]

*University of Saskatchewan, Saskatoon, SK, Canada, [†]Department of Biology, Western University (Ontario), London, ON, Canada

Chapter Outline

INTRODUCTION

Problem Statement

People value wetlands variously for the functions they provide, including flood control, improvement of downstream water quality, biodiversity, and aesthetic and recreational benefits. Despite a policy in existence for many years to prevent net wetland area loss in Alberta, Canada, wetlands continue to be lost due to noncompliance with existing laws simultaneous with replacement mechanisms that have been judged to be inadequate. This policy failure has led to the development of a new policy and innovative approaches that are expected to better support protection of both wetland area and function. The new policy features simple and accessible tools to address planning and regulatory requirements. This case study relates the building and uses of a tool for simultaneously estimating the functions and values for more than 1 million wetlands in Alberta.

Background

Agricultural, industrial, and natural resource extraction activities have caused substantial loss and degradation of wetlands in the Canadian province of Alberta since the late 1800s. The loss is extensive, with claims that two-thirds of all wetlands in the southern and central settled areas of Alberta have been lost (Alberta Environment and Parks, 2013a).

Wetland and Stream Rapid Assessments. https://doi.org/10.1016/B978-0-12-805091-0.00027-X

In response to continuing wetland losses, the Government of Alberta (GOA) introduced a wetland policy in 1993 to manage wetlands in developing areas to "sustain the social, economic and environmental benefits that functioning wetlands provide, now and in the future" (Alberta Water Resources Commission, 1993). The policy's strategy was to achieve no net loss of total wetland area by conserving wetlands where possible or otherwise compensating for wetland degradation or removal. While Alberta was one of the first provinces to adopt a wetland policy, a mechanism to fully implement this policy was not developed until 1999 with the introduction of the Water Act, which created a legislative requirement to obtain a permit to conduct activities that negatively impact wetlands.

However, loss of wetland area has continued since 1999 (Clare and Creed, 2014), indicating that the wetland policy has been ineffective at meeting its no net loss of area goal. Although the policy emphasized avoidance as the preferred mechanism of wetland conservation when development occurs, Clare et al. (2011) found that the policy was often overlooked by developers or agricultural property owners, in part because the economic benefits of wetland removal were assumed to outweigh an inherent perceived low value of wetlands. Other key challenges were found to be: (1) lack of agreement on what constituted avoidance; (2) failure to identify and prioritize wetlands in advance of development; (3) a "technoarrogance" that assumes that construction of compensatory wetlands can fully replicate the quality of removed wetlands; and (4) inadequate enforcement of regulatory approval or compensation requirements (Clare et al., 2011).

With ongoing unpermitted wetland loss and emerging scientific literature regarding the importance of wetlands and their functions (e.g., Cohen et al., 2016), the GOA developed a new wetland policy in 2013 that shifts the focus of wetland protection from area to function. Instead of focusing on the no net loss of wetland area, the policy acknowledges that different wetlands function differently and that wetland areas are not all of equal value. Recognizing that land developers lack accessible information about the functions and values of wetlands that are being considered for development, the policy also requires the GOA to make assessments of individual wetland functions and values in advance for planning purposes and subsequent regulatory processes. The policy also strives for continuous improvement of assessment methods—as new science and technology emerge, this information will be incorporated into the policy.

As with the previous policy, the new policy emphasizes avoidance as the preferred action when developing. Where avoidance is not possible, the goal is to minimize impacts on cumulative wetland value over regional scales by enforcing replacement of wetland values within the same region. Replacement can take the form of restoring previously removed or degraded wetlands, constructing new wetlands, or contributing funds to help preserve, restore, or change wetland functions through educational outreach or research that advances wetland science (Alberta Environment and Parks, 2016). Replacement is based on the value of the removed wetland relative to those of other regional wetlands, with policy targets requiring a replacement of (1) a 1:1 area-for-area ratio for wetlands with low levels of functions and up to 8:1 area-for-area ratio for wetlands with high levels of functions, and (2) a total ratio of 3:1 for all wetlands in the settled areas of Alberta (also known as the "White Area"). The policy has been in effect for the White Area since June 2015.

Need for Landscape-level Assessment

The new wetland policy requires simultaneous development of rapid assessment tools to (1) provide estimates of wetland functions and values at broad regional scales for planning purposes, and (2) provide site-based assessments for regulatory approval. The GOA commissioned the Alberta Wetland Relative Value Evaluation Tool-Estimator (ABWRET-E) to deliver offsite estimates of relative value for all wetlands in the White Area of Alberta, and the ABWRET-Actual (ABWRET-A) to estimate levels of wetland functions at the site level based on integrating field observations with existing spatial data. These tools are built from the same logical foundation and are intended to complement each other (Fig. 2.2.12.1). The ABWRET-E is designed to facilitate avoidance at the beginning of the planning process by allowing planners and developers to understand the relative value of a particular wetland early in the development process. The ABWRET-A, on the other hand, is a regulatory tool that provides a standardized method for rapid on-the-ground assessment of wetland functions. Together, these tools enhance the potential for avoiding impacts to wetlands and, when wetlands must be removed, ensure that the relative levels of functions are known so that steps can be taken to make certain that value is replaced. We focus on ABWRET-E here; information on the ABWRET-A can be found in Government of Alberta (2015).

Funding Source

The development of the ABWRET tools by the authors was funded by Alberta Environment and Parks, Water Policy Branch.

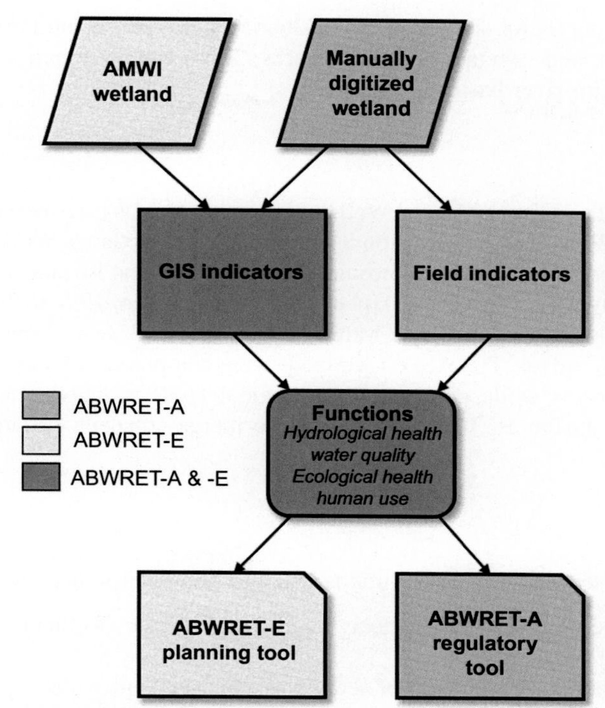

FIG.2.2.12.1 Relationship between ABWRET-E and ABWRET-A.

STUDY AREA

The province of Alberta has had one of the fastest growing economies in Canada since the 1990s. Resulting urban and agricultural development pressures have resided primarily in the southern and central areas of the province, set aside in 1948 for settlement and agriculture as the White Area. Comprising 39% of Alberta's 660,000 km^2, the White Area has more than 80% of the province's population living in its urban centers.

More than one million wetlands exist in the White Area; they are predominantly small shallow marsh wetlands in the Grassland and Parkland natural regions that correspond roughly to Alberta's portion of the Prairie Pothole Region. The Institute for Wetlands and Waterfowl Research has estimated that 80%–90% of wetlands adjacent to urban centers have been lost in the White Area, with losses continuing at a rate of approximately 0.5% annually (Badiou, 2013).

The ABWRET-E was developed and implemented for the White Area. The other major land designation in Alberta is the "Green Area," set aside for forest and resource production and wildlife protection. A version of the ABWRET-E for the Green Area has not been developed to date, although the GOA plans to extend the tool to this area. In the meantime, developers and regulators are expected to operate in accordance with wetland policy directives applied to the White Area.

METHODS

The ABWRET-E was designed to provide an efficient "desktop" mechanism for simultaneously estimating the functions of large numbers of wetlands at regional scales. The ABWRET-E differs from the ABWRET-A in that (1) it is a planning tool and not a regulatory tool, (2) it is completely automated and does not require a site visit, and (3) it can be continually improved as new data and models become available.

Relative Value Assessment Units

The relative value assessment approach required by Alberta's wetland policy uses spatial units (Relative Value Assessment Units or RVAUs) in which wetlands in similar landscapes are assessed against each other to provide relative estimates of their functions and values. RVAUs ensure that wetlands are assessed within common hydrological and ecological units in accordance with the spatial principles of water management frameworks (Creed et al., 2011). At the same time, the size of the RVAUs must be suitable to management purposes; the GOA recommended defining about 20 RVAUs for the province.

We defined RVAUs by classifying minor subwatersheds (Alberta Environment and Parks, 2014) to a provincial classification of climate, soil, landform, and vegetation (Alberta Parks, 2005) and then merging same-classification minor subwatersheds within the seven major river basins in Alberta.

Wetland Inventory

The basis of the ABWRET-E is the Alberta Merged Wetland Inventory (AMWI) (Alberta Environment and Parks, 2013b), which is classified to Canadian Wetland Classification System (National Wetlands Working Group, 1997) classes: marsh, swamp, bog, fen, and open water. The AMWI is a mosaic of multiple wetland inventories captured using different source data types (minimum size of polygons ranges from 0.04 to 1 ha) acquired from 1998 to 2009; it uses different technologies under a variety of initiatives. The AMWI wetlands were assigned to RVAUs by clipping to RVAU boundaries as these boundaries represent topographic ridges. The AMWI wetlands were composed of adjacent fragmented wetlands that were merged into single wetland objects while preserving the original areas and perimeters of each wetland class within the resulting objects as tabular attributes. The methods used to merge adjacent wetlands are available upon request to the first author.

Wetland Functions

The GOA's Wetlands Policy Intent identified four major function groups provided by wetlands to be assessed:

*Hydrologic health** (HH): the water storage and delay function provided by wetlands for impeding and desynchronizing the downslope movement of peak flows.
Water quality (WQ): the retention or removal of sediment or nutrients provided by wetlands for purifying receiving waters.
*Ecological health** (EH): the habitats for aquatic and terrestrial plants and animals provided by wetlands for enhancing biodiversity.
Human use (HU): the multiple human activities that are supported by wetlands, including recreation and education as well as the importance of wetlands to historical and current culture (this is not a wetland function per se, but its inclusion as a function was to ensure that wetlands that are highly used by people would receive a relatively high value).

*The GOA's use of the term "health" is not necessarily synonymous with wetland integrity or condition, as is the practice in the United States, but rather refers to the level of functions within these defined function groups.

Models derived from expert opinion and extensive literature review (Government of Alberta, 2015) were used to identify and combine indicators of wetland functions. Wetland HH was derived from water storage and stream flow support subfunctions. Wetland WQ was derived from water cooling, sediment retention, phosphorus retention, and nitrate removal subfunctions. Wetland EH was derived from organic nutrient export, fish habitat, invertebrate habitat, amphibian habitat, water bird nesting habitat, songbird, raptor and mammal habitat, and plant and pollinator habitat subfunctions. There were no subfunctions for HU. Subfunctions were combined into functions, and the functions were then aggregated into value scores (Fig. 2.2.12.2). The modeling steps are outlined in Fig. 2.2.12.3 and summarized in the subsequent sections. The GOA was prompted to take this function score approach as they realized that the relative functions and values of individual wetlands could not be reliably estimated based solely on the Canadian Wetland Classification System (National Wetlands Working Group, 1997).

Geographical Information System Database

The ABWRET-E was executed in a vector geographical information system (GIS) environment because (1) vector geometry allows finer delineation of wetland and other landscape features, meaning that less information is lost, and (2) vector geometry allows for a tabular data structure that provides easy cross-referencing in desktop or web-based applications. Vector methods were developed and tested to automatically extract data useful for estimating wetland values in ArcGIS (ESRI, 2011). These vector methods are available upon request to the first author.

Potential GIS data that could be used in the ABWRET-E were identified by searching provincial, national, and international databases. Where several different data layers for one indicator were identified, the layer that was of the highest quality was chosen, representing the "best in practice." The best in practice layer that provided complete spatial coverage of a given RVAU was used for that RVAU. From those best in practice layers, secondary GIS data layers were generated where required for indicator extraction using GIS reclassifications, feature selections or mergers, or digital terrain analyses.

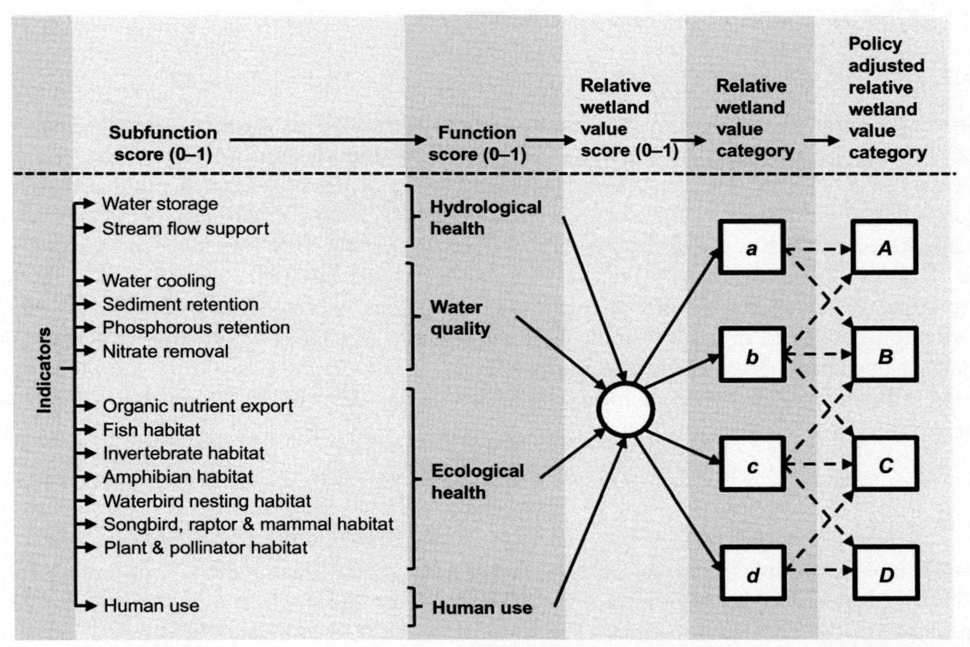

FIG 2.2.12.2 ABWRET-E and ABWRET-A combine indicators to wetland subfunction and function scores and bundle function scores to produce wetland scores.

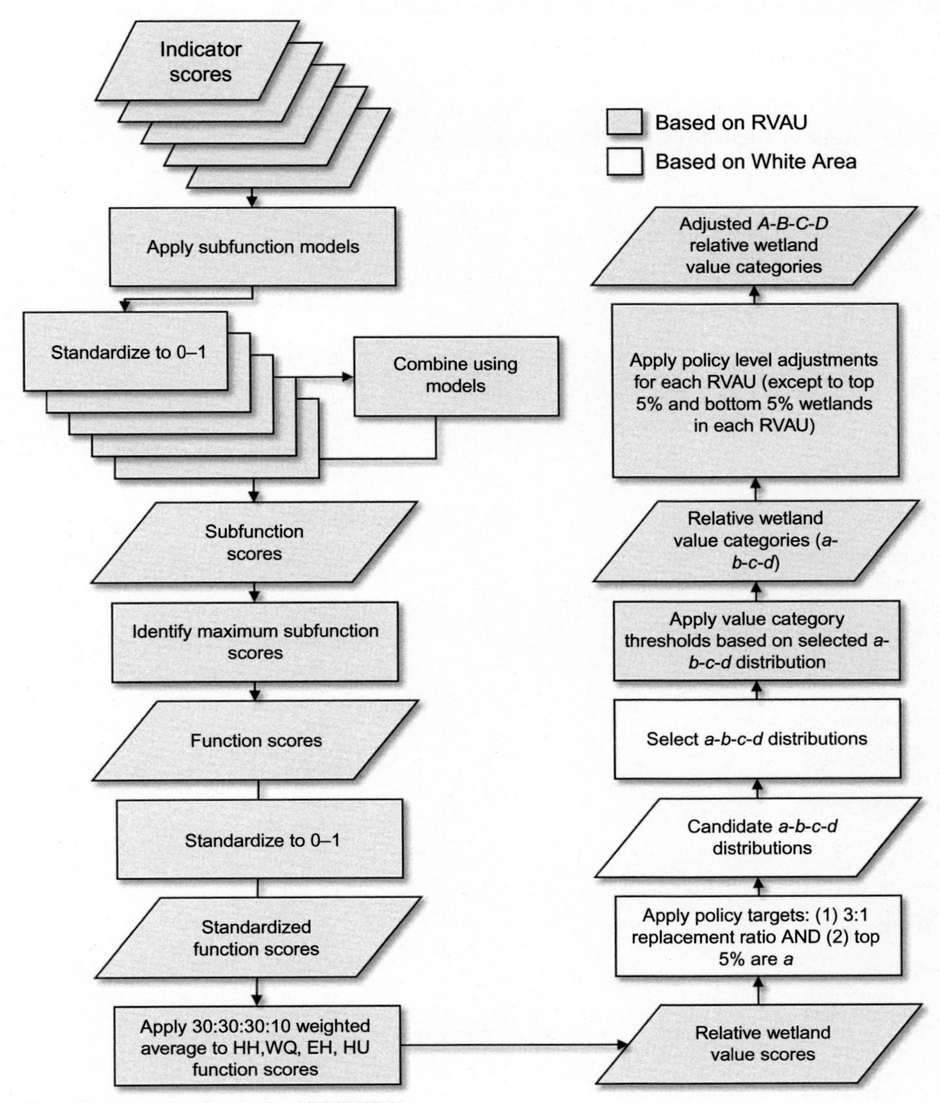

FIG. 2.2.12.3 Flowchart of steps used to apply the ABWRET-E.

A data confidence score of the best in practice data was determined to identify strengths and weaknesses in the ABWRET-E, and to set priorities for data acquisition to ensure continuous improvement of the tool. Spatial confidence scores from 1 (low) to 10 (high) were calculated by linear scaling of log-transformed minimum resolvable unit areas. The minimum resolvable unit area (the smallest area that can be detected on a map) was obtained from information published in metadata, or calculated as either the pixel resolution for a raster GIS data layer or the square of the minimum detectable size of a vector GIS data layer (i.e., the map scale divided by 1000 (Tobler, 1987)). Temporal confidence scores from 1 to 10 were assigned by placing the age of the data layers into a qualitative matrix of "temporal sensitivity" where the information contained in a given data layer is sensitive to changes over time (e.g., geology layers are not sensitive to changes over time, 30-year climate normal layers are moderately sensitive, and land-use maps are highly sensitive). An overall data confidence score for each data layer was determined by averaging the spatial and temporal confidence scores.

Wetland Indicators

Indicators are measurable variables that can be used to quantify specific parameters contributing to a wetland's subfunction. The GOA intended that the ABWRET tools (-E and -A) use the smallest number and simplest characterization of indicators possible to estimate subfunctions. The initial list of 200+ GIS- and field-based indicators identified by wetland experts (Wetland Assessment Working Group, 2011) was narrowed down to a final list of 73 indicators, each of which could be automatically extracted from GIS data layers (Tables 2.2.12.1–2.2.12.4). This final list of indicators captured wetland functions with a level of confidence acceptable to the GOA at the time of the ABWRET-E development.

TABLE 2.2.12.1 Indicators of Hydrologic Health Subfunctions

Indicator	Units	WS	SFS
Aquifer vulnerability index	Categorical	✓	
Channel connection	Yes/no	✓	✓
Class A or B (Alberta Water Act Codes of Practice Streams)	Yes/no	✓	✓
Elevation percentile in HUC8	Percentile	✓	✓
Floodways or riparian area	Yes/no	✓	✓
Hydrologic detention time (L/G index)	m	✓	
Percent fen	Percent	✓	✓
Percent open water	Percent	✓	✓
Percent wooded (forest and swamp)	Percent	✓	✓
Perimeter-area ratio	$m:m^2$	✓	✓
P-PET (1971–2000)	mm/year	✓	✓
Soil texture	Categorical	✓	
Springs or other groundwater discharge rea	Yes/no	✓	✓
Subzero days (1971–2000)	days/year	✓	✓
Water permanence probability	Percent	✓	✓
Wetland area	m^2	✓	
Wetland size percentile in HUC8	Percentile	✓	✓
Wetland vegetated area	m^2	✓	✓
Wind intensity—Summer	W/m^2	✓	✓

WS, water storage; *SFS*, stream flow support.

TABLE 2.2.12.2 Indicators of Water Quality Subfunctions

Indicator	Units	WC	SR	PR	NR
Aspect of wetland's 100m upslope buffer	Categorical				✓
Channel connection	Yes/no	✓	✓	✓	✓
Class A or B (Alberta Water Act Codes of Practice Streams)	Yes/no	✓	✓	✓	✓
Clumpiness index for vegetation and water	m^2		✓	✓	✓
Floodways or riparian area	Yes/no	✓	✓	✓	✓
Growing degree days (1971–2000)	days/year				✓
Hydrologic detention time (L/G index)	m	✓	✓	✓	✓
Organic soil content	Categorical			✓	✓
Percent fen	Percent	✓			
Percent marsh or swamp	Percent				✓
Percent open water	Percent	✓	✓	✓	
Percent wooded (forest and swamp)	Percent	✓			✓
Perimeter-area ratio	$m:m^2$				✓
Soil texture	Categorical			✓	✓
Springs or other groundwater discharge area	Yes/no	✓			
Slope in 500m buffer	Degrees		✓		
Subzero days (1971–2000)	days/year		✓	✓	✓
Water permanence probability	Percent	✓		✓	✓
Wetland vegetated area	m^2		✓	✓	

WC, water cooling; *SR*, sediment retention; *PR*, phosphorus retention; *NR*, nitrate removal.

TABLE 2.2.12.3 Indicators of Ecological Health Subfunctions

Indicator	Units	OE	FR	INV	AM	WB	SRM	PH+POL
Bog/fen/marsh/swamp uniqueness for 1 km buffer	Percent			✓			✓	✓
Channel connection	Yes/no	✓	✓					
Class A or B (Alberta Water Act Codes of Practice Streams)	Yes/no	✓	✓					
Clumpiness index for herbaceous and woody vegetation in the wetland	m^2			✓			✓	✓
Clumpiness index for vegetation and water	m^2	✓	✓	✓	✓	✓	✓	✓
Distance from wetland to nearest developed land or annual cropland	m			✓	✓	✓	✓	✓
Distance from wetland to nearest road	m		✓		✓		✓	✓
Fen, marsh, or swamp uniqueness in 1 km buffer	Percent				✓			
Fen or marsh uniqueness in 1 km buffer	Percent					✓		
Floodways or riparian area	Yes/no	✓	✓			✓	✓	✓

Continued

TABLE 2.2.12.3 Indicators of Ecological Health Subfunctions—cont'd

Indicator	Units	OE	FR	INV	AM	WB	SRM	PH+POL
Growing degree days (1971–2000)	days/year	✓						
Habitat connectivity to other wetlands within 1 km	Yes/no				✓	✓	✓	
Important bird area	Yes/no					✓		
Internal wetland type richness	Number (continuous)			✓			✓	✓
Key wildlife biodiversity zone	Yes/no						✓	
Nesting bird colony, piping plover water body, or trumpeter swan use area	Yes/no					✓		
Open water area	m^2		✓					
Organic soil content	Categorical	✓						
Percent bog, fen, or marsh	Percent	✓						
Percent fen, marsh, or swamp	Percent						✓	✓
Percent marsh	Percent			✓	✓	✓		
Percent natural cover within 1 km	Percent			✓			✓	✓
Percent open water	Percent		✓	✓	✓	✓	✓	✓
Percent perimeter adjoined by natural cover	Percent		✓	✓	✓	✓		✓
Percent undeveloped openlands within 1 km	Percent				✓	✓		
Rare plant species range	Yes/no							✓
Road density in 1 km buffer			✓				✓	✓
Sensitive amphibian range	Yes/no				✓			
Sensitive raptor nesting area	Yes/no						✓	
Shorebird staging wetland	Yes/no					✓		
Soil texture	Categorical	✓						
Subzero days (1971–2000)	days/year		✓					
Trumpeter swan area	Yes/no					✓		
Water permanence probability	Percent		✓					
Waterfowl breeding density	Number (continuous)					✓		
Waterfowl staging wetland	Yes/no					✓		
Wetland density (open water only) within 1 km	Percent					✓		
Wetland density within 1 km	Percent						✓	✓
Wetland density (no bogs) within 1 km	Percent			✓	✓			
Wetland type richness within 1 km	Number (continuous)			✓			✓	✓
Wetland vegetated area	m^2	✓			✓	✓	✓	✓
Wind intensity—Winter	W/m^2						✓	

OE, organic nutrient export; *FR*, fish habitat; *INV*, invertebrate habitat; *AM*, amphibian habitat; *WB*, waterbird nesting habitat; *SRM*, songbird, raptor and mammal habitat; *PH+POL*, plant and pollinator habitat.

TABLE 2.2.12.4 Indicators of Human Use Function

Indicator	Units
Access to trail network	Yes/no
Alberta culture listing of historical resources	Categorical
Distance from wetland to nearest road	m
Distance to nearest well-settled area	m
Ecological reserve or natural area	Yes/no
Fringe wetland	Yes/no
Lacustrine wetland	Yes/no
Ownership	Yes/no
Percent open water	Percent
Road density in 1 km buffer	
Water permanence probability	Percent
Wetland area	m^2

Indicator values may be categorical or continuous integer or floating point numbers depending on the data layer type used. For each wetland, indicator values were extracted using vector GIS methods and standardized to 0–1 indicator scores (where 1 indicates the best indicator performance within an RVAU) using classification or linear scaling according to subfunction models. Standardization of indicator values by linear scaling used the following formula:

$$y_i = \frac{x_i - min_i}{max_i - min_i} \tag{2.2.12.1}$$

where y is the standardized score for indicator i, x is the observed score, and *min* and *max* are the minimum and maximum scores from all wetlands within the RVAU. This linear scaling technique assigns a value of 0 to the minimum observation and a value of 1 to the maximum observation and assigns scores to all other observations based on their position between the minimum and maximum.

Wetland Subfunctions and Functions

Indicator scores were combined into subfunction scores from 0 to 1 (where 1 indicates the best subfunction performance within a RVAU) using a combination of scaling, weighted average, and "if-else" statements, according to the subfunction models (cf. Government of Alberta, 2015).

Subfunction scores were combined into function scores for each wetland by taking the highest of the subfunction scores associated with each function within an RVAU.

Function scores were then standardized to a 0–1 range within each RVAU using the same linear scaling method described for indicator scores.

Wetland Values

Relative wetland value scores describe the overall functioning of a wetland compared to all other wetlands in an RVAU. These overall scores were calculated by combining function scores into a single value between 0 and 1. Relative wetland value scores were calculated for each wetland by using a weighted average of the four standardized function scores, where weights were given as 0.3 for HH, WQ, and EH and 0.1 for HU. Thresholds in the frequency distribution of relative wetland value scores for the entire White Area were then used to define relative wetland value categories ranging from *a* to *d*, where *a* is the highest value category and *d* the lowest.

Policy Lever to Ensure Maintenance of Wetland Area (*a, b, c,* and *d* Scores)

While the GOA moved from an area- to a function-based wetland policy, the GOA still wanted to maintain the 3:1 replacement ratio target based on wetland area for the White Area (Alberta Water Resources Commission, 1993). Percentiles of relative wetland value categories were simulated to identify potential thresholds for a, b, c, and d value scores that would result in a 3:1 replacement ratio. To ensure that the highest functioning wetlands within the White Area are conserved, the top 5% of value scores were required to be assigned a relative wetland category of a. To ensure that options for compensation were maintained, the bottom 5% of value scores were required to be assigned the value category of d. These constraints are summarized in Eq. (2.2.12.2).

$$\begin{cases} a\times 8+b\times 4+c\times 2+d\times 1=3:1 \\ a+b+c+d=1 \\ 0.05\leq a<1, 0<b<1, 0<c<1, 0.05\leq d<1 \end{cases} \tag{2.2.12.2}$$

where a, b, c, and d represent the thresholds in the percentiles (expressed as decimal fractions) that were used to define value categories a, b, c, and d by wetland number only (not area).

All possible percentiles of b in intervals of 0.01 were solved for all percentiles of a in intervals of 0.01 from 0.05 to 0.29:

$$\begin{cases} 0<a\leq 0.167, -3\times a+0.5<b<-\frac{7}{3}\times a+\frac{2}{3} \\ \text{OR} \\ 0.167<a<0.29, 0<b<-\frac{7}{3}\times a+\frac{2}{3} \end{cases} \tag{2.2.12.3}$$

Progressing through each combination of a and b percentiles, we then calculated the answers to c and d using Eqs. (2.2.12.4) and (2.2.12.5):

$$c=-7\times a-3\times b+2 \tag{2.2.12.4}$$

$$d=6a+2b-1 \tag{2.2.12.5}$$

For example, if $a=0.10$, then b ranges from 0.21 to 0.43. If we let $b=0.30$, then $c=-7\times 0.10-3\times 0.30+2=0.40$ and $d=6\times 0.10+2\times 0.30-1=0.20$. All 438 possible combinations of a, b, c, and d percentiles meeting the requirements of the 3:1 replacement ratio and the minimum 5% a and d assignments were calculated.

The GOA selected an option where the wetland value categories approximated but did not meet the 3:1 ratio in the White Area. The selected option was where 10% of wetlands were classified as a, 20% as b, 30% as c, and 40% as d (i.e., a 2.6:1 replacement ratio). The selected 2.6:1 ratio provided by the 10-20-30-40 option offered a conservative estimate of value that would avoid ratios higher than 3:1 when the regulatory tool was applied. The GOA felt conservative estimates were appropriate for three reasons. First, ABWRET-E value scores included results from GIS-based indicators only; the GOA anticipated that field-based indicators in the ABWRET-A would raise maximum function scores by identifying features that cannot be detected in the GIS data layers alone. Second, the original 3:1 replacement ratio was based on the assumption that many replacement wetlands would fail, an assumption the GOA felt was no longer appropriate given increasing experience, research, and technology in replacing wetlands. Finally, the GOA used the 10-20-30-40 option during policy development when they were calculating the potential effects of their policy on the economy, and given the preceding qualifications, it made sense to stay with the 10-20-30-40 option.

Policy Lever to Ensure Protection of High Risk Areas (*A, B, C,* and *D* Scores)

Some RVAUs have experienced high rates of wetland loss where others have experienced lower rates; therefore, RVAUs should be treated differently when assigning wetland value scores. Alberta's new wetland policy estimates wetland loss—both number and area—using an area-frequency power function that was applied to each RVAU wetland inventory. Estimating wetland loss depends on the assumptions that (1) a negative linear relationship exists between the number and area of wetlands plotted on log-log scales on undeveloped landscapes, and (2) that there is a preferential loss of wetlands from small to large. The linear relationship between number and area of wetlands can be expected to break at a point equivalent to the largest wetlands that have been lost. We extracted this break in the linear relationship for each RVAU using piecewise linear regression. Wetland loss (both number and area) was estimated from the difference between (1) the linear relationship below the break extrapolated to small wetland areas from observed data above the break, and (2) the observed linear relationship below the break. See Serran and Creed (2016) for further details of the method used.

The relative wetland value score letters (*a-b-c-d*) were raised by one letter (not higher than *a*) in RVAUs with the highest loss of wetlands to increase wetland protection. Conversely, the relative wetland value score letters were lowered by one letter (not less than *d*) in RVAUs with the lowest loss of wetlands to reduce wetland protection. Adjusted letters were indicated by changing the letters to upper case (*A-B-C-D*). Wetlands in the top 5 and bottom 5% of value scores within each RVAU were exempted from this wetland loss adjustment of value categories.

RESULTS

Twenty-one RVAUs were defined for the province of Alberta. Eleven RVAUs had at least 10% White Area and were selected for development and application of the ABWRET-E. These 11 RVAUs ranged in area from 13,447 to 75,506 km^2, covering 95% of the White Area and featuring 1,326,194 wetlands covering a total of 5,619,934 ha. Wetland number per square kilometer ranged from 1.4 (RVAU #10) to 8.3 (RVAU #5) and wetland area in hectares per square kilometer ranged from 3.6 (RVAU #2) to 45.3 (RVAU #9). The lowest density of wetlands by area (but not by number) and the highest rates of wetland loss were found in the southern areas of the province. Wetland density by number is actually lowest in the north, but that is because of large peat wetlands. A single wetland in the north can be hundreds or thousands of square kilometers in size, yielding low density by number values (Table 2.2.12.5).

Wetland function scores were not distributed evenly throughout the White Area. Higher HH and EH function scores were distributed in the southern areas of the province, and higher WQ and HU function scores were distributed in the northern areas.

The distributions of relative wetland value scores (*a-b-c-d*) by number and area for each RVAU are shown in Table 2.2.12.6 and Fig. 2.2.12.4. The proportions of high functioning (i.e., *a* or *b*) wetlands were smaller by number in all RVAUs except two in the southern areas of the province, and smaller by area in all RVAUs. After adjustment of relative wetland value scores to increase value in RVAUs with the highest historic loss of wetlands (and to decrease value in RVAUs with the lowest historic loss), the proportions of high functioning (i.e., *A* or *B*) wetlands were found to be generally larger by both number and area in southern RVAUs where wetland loss was generally higher (Table 2.2.12.6; Fig. 2.2.12.5). The policy lever adjustment ensures that the costs of wetland removal will be higher in the southern areas of the province where historic wetland loss has been greatest. Distributions of wetland value categories by area were more highly skewed toward the bottom range of values than wetland value categories by number, both before and after policy lever adjustments,

TABLE 2.2.12.5 Current Wetland Number and Area and Estimated Percent Loss of Wetlands by Number and Area for Each RVAU

RVAU	Area (km^2)	Current Wetland Number	Current Wetland Area (ha)	Lost Wetland Number (%)	Lost Wetland Area (%)	Policy Lever Adjustment
1	26359.82	76,802	97662.32	82.50	40.11	+1
2	26567.35	57,009	94526.69	83.55	38.10	+1
3	43397.48	130,093	189115.86	84.89	40.42	+1
4	13447.24	57,254	137060.42	83.62	30.93	+1
5	32316.84	268,342	309229.13	70.08	28.26	0
6	25182.83	47,042	299035.24	81.35	23.59	0
7	23161.47	181,692	264544.74	43.54	10.18	−1
8	25020.79	134,053	481478.90	69.46	30.86	0
9	45204.45	183,071	2048024.09	63.07	15.85	−1
10	75506.06	108,719	1095703.03	75.38	15.08	0
11	24174.32	82,117	603553.29	85.06	30.22	+1

Loss estimates were used to develop policy lever adjustments to relative wetland value categories in each RVAU (−1 denotes lowering value categories by one letter, +1 denotes raising value categories by one letter, 0 denotes no adjustment).

TABLE 2.2.12.6 Distributions of Relative Wetland Value Categories for Each RVAU

RVAU	Policy Lever Adjustment	*a* Wetland Number (%)	*a* Wetland Area (%)	*A* Wetland Number (%)	*A* Wetland Area (%)	*b* Wetland Number (%)	*b* Wetland Area (%)	*B* Wetland Number (%)	*B* Wetland Area (%)
1	+1	16.99	14.08	60.36	43.67	43.38	29.60	18.60	9.14
2	+1	5.62	3.43	20.30	17.66	14.67	14.24	21.56	10.21
3	+1	2.33	2.76	41.56	36.16	39.23	33.40	13.51	8.53
4	+1	23.53	18.79	55.00	38.29	31.46	19.50	19.79	10.53
5	0	17.05	8.45	17.05	8.45	8.37	11.98	8.37	11.98
6	0	1.75	0.51	1.75	0.51	3.52	3.38	3.52	3.38
7	−1	21.55	8.85	5.00	4.40	12.74	13.94	16.55	4.45
8	0	3.42	0.77	3.42	0.77	2.95	2.49	2.95	2.49
9	−1	2.66	0.39	2.66	0.39	27.34	10.81	2.34	8.08
10	0	3.49	0.79	3.49	0.79	21.15	4.77	21.15	4.77
11	+1	1.05	0.40	37.87	10.42	36.82	10.02	31.04	5.96

RVAU	Policy Lever Adjustment	*c* Wetland Number (%)	*c* Wetland Area (%)	*C* Wetland Number (%)	*C* Wetland Area (%)	*d* Wetland Number (%)	*d* Wetland Area (%)	*D* Wetland Number (%)	*D* Wetland Area (%)
1	+1	18.60	9.14	16.03	40.10	21.04	47.18	5.00	7.08
2	+1	21.56	10.21	53.14	63.83	58.15	72.13	5.00	8.30
3	+1	13.51	8.53	39.92	44.63	44.92	55.31	5.00	10.68
4	+1	19.79	10.53	20.22	43.38	25.22	51.18	5.00	7.80
5	0	37.02	20.23	37.02	20.23	37.56	59.34	37.56	59.34
6	0	22.52	6.82	22.52	6.82	72.21	89.29	72.21	89.29
7	−1	44.76	16.75	12.74	13.94	20.94	60.46	65.70	77.21
8	0	19.32	5.24	19.32	5.24	74.31	91.51	74.31	91.51
9	−1	36.39	2.79	25.01	3.02	33.60	86.00	69.99	88.50
10	0	30.45	6.00	30.45	6.00	44.90	88.44	44.90	88.44
11	+1	31.04	5.96	26.09	78.82	31.09	83.61	5.00	4.79

Categories given in lower case letters describe the relative value of wetlands from *a* (high value) to *d* (low value). For each RVAU, policy lever adjustments were used to adjust relative wetland value category letters upward (+1), downward (−1) or no change (0); adjusted relative wetland value categories are given in upper case letters from *A* (high value) to *D* (low value).

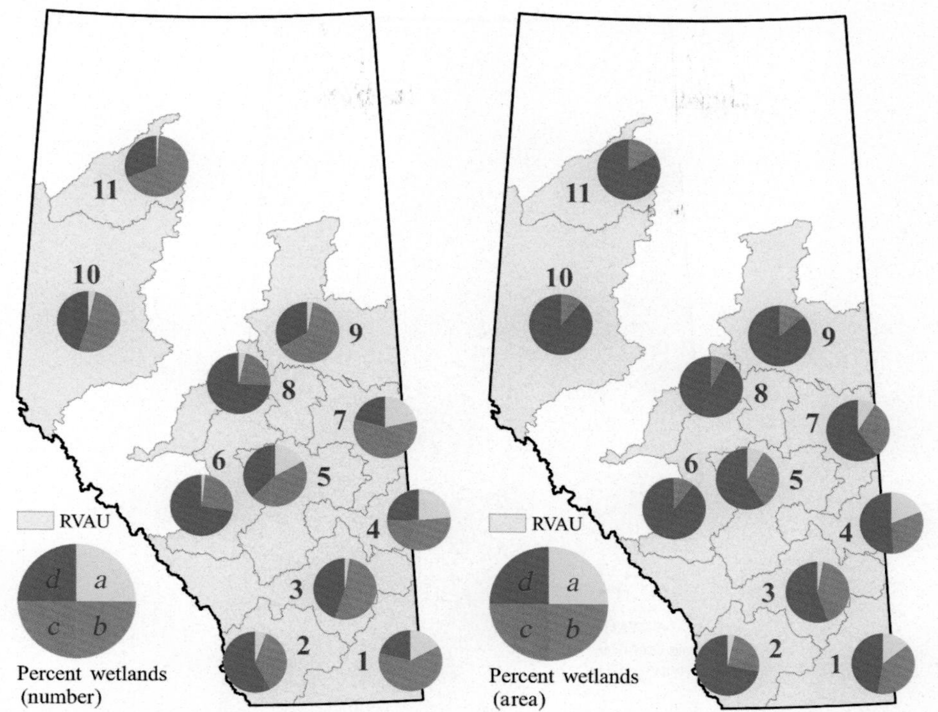

FIG. 2.2.12.4 Distribution of relative wetland value categories by number (left) and area (right) for each RVAU (see Table 2.2.12.6 for numerical summaries). Categories *a*, *b*, *c*, and *d* describe the relative value of wetlands from *a* (high value) to *d* (low value).

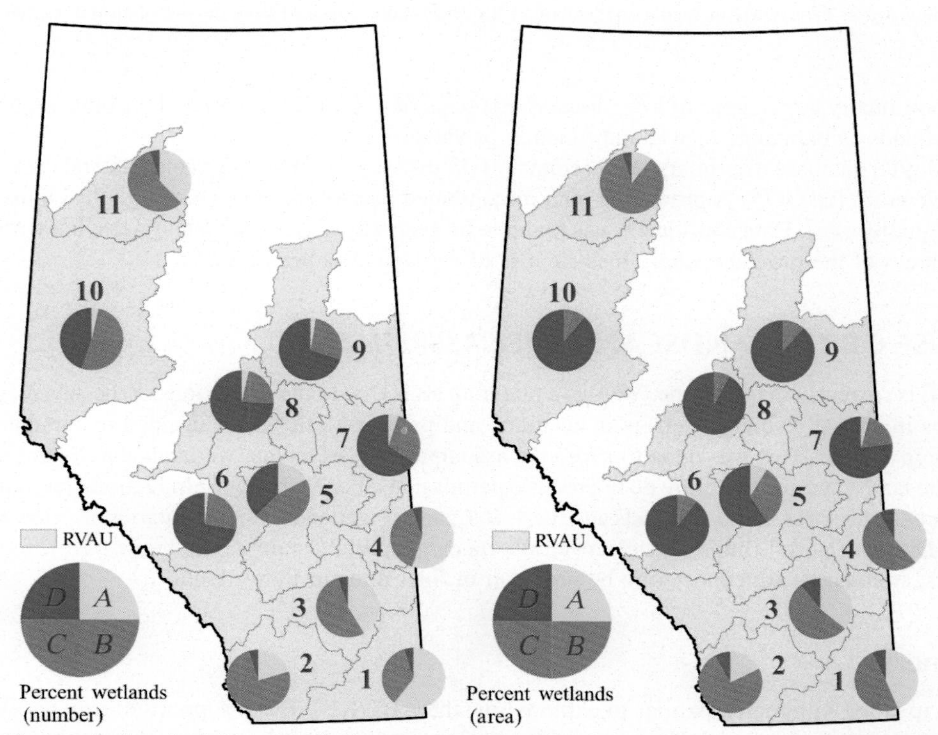

FIG. 2.2.12.5 Distribution of adjusted relative wetland value categories by number (left) and area (right) for each RVAU (see Table 2.2.12.6 for numerical summaries). Categories *A*, *B*, *C*, and *D* describe the relative value of wetlands from *A* (high value) to *D* (low value).

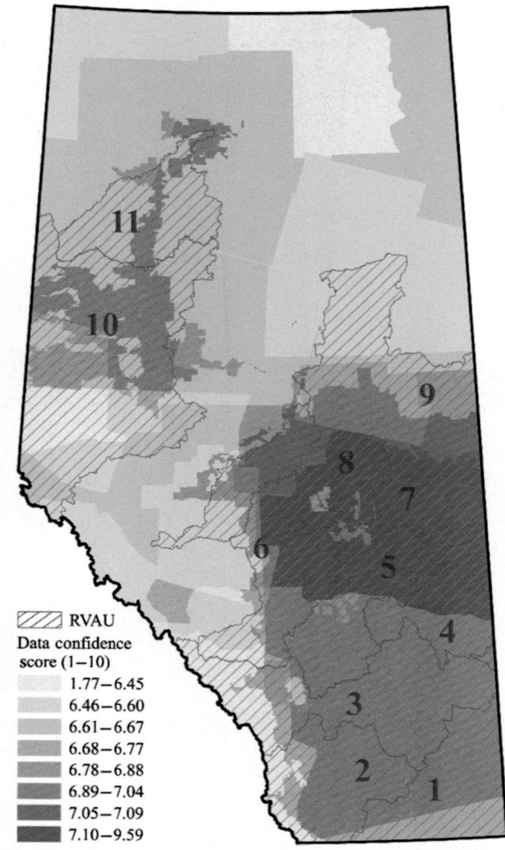

FIG. 2.2.12.6 Confidence scores for data layers used in relative wetland value assessment. Data confidence scores from 1 (low confidence) to 10 (high confidence) were calculated as the average of spatial and temporal confidence scores assigned from metrics of spatial resolution and data layer age, respectively.

resulting in proportionally larger areas of low functioning (i.e., *c* or *d*, C or D) wetlands. This indicates that there is no bias toward large wetlands as indicators of wetland function or value.

A complete digital database of primary and secondary GIS data layers, including metadata and data confidence assessments, was delivered to the GOA. A provincial map of combined data confidence (Fig. 2.2.12.6) illustrates where shortcomings in data quality exist. Data confidence was found to be generally high in the White Area, but with lower confidence in the southern areas of the province where historic loss of wetlands has been highest.

CURRENT USE OF THE LANDSCAPE-LEVEL ASSESSMENT

The ABWRET-E is currently in use by the GOA as a planning tool. Due to the limitations of the AMWI (i.e., classification biases, variations in spatial resolution, errors in accuracy and precision), the GOA decided to summarize wetland value scores by township section. This was done to avoid the assumption by users that wetlands that do not exist in the AMWI do not exist on the landscape and therefore do not require permissions. For each township section, the summary provides the total estimated area in hectares for all wetlands and for *A*, *B*, *C*, and *D* wetlands. The summaries are viewable online through the GOA's GeoDiscover portal (https://geodiscover.alberta.ca/geoportal/catalog/main/home.page). In the future, as data quality improves, summaries are expected to be provided in finer units or by wetland.

Use by Planners

The GOA plans to work with municipalities to explore how the ABWRET-E can support informed wetland planning and management decisions at the local scale. Relative wetland value categories can be used by planners to identify wetlands that should be prioritized for conservation under municipal legislation. In Alberta, the Municipal Government Act (2000) provides municipal governments control over bodies of water within their jurisdictions, allowing them to designate them as a conservation easement (where land owners retain ownership of the land but are required to protect the water body) or an environmental reserve (where municipalities take ownership of the land and can implement a buffer around the water body to prevent pollution and conserve the wetland's natural state). The ABWRET-E can inform planners of areas that should be targeted for protection (e.g., areas with the highest densities of high-value wetlands) or mitigation (e.g., areas with the lowest densities of high-value wetlands where restored wetlands have greater potential for high value). Further, in the event of an incident (e.g., an oil spill), the ABWRET-E results can assist in making a quick assessment to examine wetlands at risk and coordinate responses accordingly.

The ABWRET-E outputs can be scaled from individual wetland to regional scales, allowing for management at watershed scales. At the regional scale, results from the ABWRET-E can be used to feed into other policies and frameworks such as Alberta's land-use framework (Government of Alberta, 2008). Alberta's land-use framework strives to manage Alberta's land and natural resources sustainably, allowing for the achievement of Alberta's long-term economic, environmental, and social goals. The land-use framework has developed several strategies that involve wetlands to achieve "smart growth." These strategies include using cumulative effects management to manage impacts of development on water, developing policy instruments for conservation on private and public lands, and establishing data and knowledge systems so that sound decisions are made. The ABWRET-E can assist in developing these strategies as the ABWRET-E outputs (1) can be input into models to examine the cumulative effects of the loss of wetland functions at watershed scales, and (2) can help to prioritize areas for wetland protection or restoration to achieve watershed management targets.

Use by Regulatory Personnel

The ABWRET-E estimates of wetland value scores should not replace those of the ABWRET-A, which are required for regulatory purposes. However, the ABWRET-E can be used in the regulatory process to facilitate avoidance and to encourage, where appropriate, relocation of planned activity. To conduct work in and around a water body in Alberta, approval must be obtained under the Water Act (1999) and (if appropriate) the Public Lands Act (2000). The Water Act requires approval for any activity that may affect an aquatic environment. When officials are deciding whether to grant approval, the ABWRET-E can estimate the loss of wetland area and function for proposed activities, and can offer alternative sites if the loss in the proposed area is too high. The ABWRET-E can also identify priority areas for policy compliance inspections. Priority areas for inspection could include areas where the consequences of Water Act contraventions would have a severe impact on wetland value loss. Government officials can look at past and present development projects where there are a large number of high value wetlands to ensure that any wetland removal or degradation is permitted and the appropriate compensation is being conducted.

COMPARISONS WITH FIELD EVALUATIONS

It is crucial that rapid evaluation tools based on GIS data are representative of field data. Given the recent release of the ABWRET tools for implementation, only a small amount of the ABWRET-A field data is available for comparison. Once a larger repository of field data is collected, the GOA plans to assess the concordance between the ABWRET-E and ABWRET-A tools on a regular basis, (i.e., every 5 years). Results from this concordance exercise will help GOA continuously improve the ABWRET tools.

For this case study, we assessed the concordance between the ABWRET-E data and the limited ABWRET-A data that were available. The ABWRET-E methods differ from those of the ABWRET-A in several important ways. First, ABWRET-E identifies the spatial attributes of wetlands through GIS processing of AMWI polygons whereas ABWRET-A requires manual digitization of wetland boundaries from local aerial photography. Second, ABWRET-E is based on GIS data only whereas ABWRET-A uses combined GIS and field data. We compared ABWRET-E and ABWRET-A indicator values, subfunction, function, and value scores, and value categories in 48 manually digitized wetland polygons that intersected the ABWRET-E wetland polygons. The ABWRET-E and ABWRET-A tools were found to produce largely concordant results, demonstrated by strong and significant correlations as measured by Spearman rank correlation coefficients between (1) all pairs of indicator values, (2) all pairs of subfunction and function scores except one WQ subfunction score (water cooling) and three EH subfunction scores (organic nutrient export, invertebrate habitat, and amphibian habitat), and (3) EH function scores. These disagreements were reduced when function scores were combined into value scores; pairs of value scores were found to be concordant. However, this concordance did not result in strong agreements in *a-b-c-d* (37.5%) or adjusted *A-B-C-D* (51.1%) value categories.

Based on these concordance tests, the GOA introduced 10 modifications to the ABWRET-A in May 2016 to improve concordance with the ABWRET-E, including the removal or addition of indicators, modification of subfunction combining models, standardization of function scores, and the application of ABWRET-E thresholds for value score categorization. As data quality improvements are secured in the future from more accurate and recent data sources, concordance is expected to improve.

LESSONS LEARNED

Simplifying Indicators and Subfunction Models

The GOA recommended defining a set 20 indicators for the quick assessment of wetlands. The final list of indicators that was used in the ABWRET-E consisted of 73 indicators. We used a mathematical technique known as global sensitivity analysis to investigate how variations in outputs are affected by variations in input parameters to target the most influential indicators of HH, WQ, and EH subfunctions (Pianosi et al., 2016). Specifically, the conditional variance method of sensitivity analysis (Saltelli et al., 2008) calculates a sensitivity index for each indicator in a subfunction model, representing the variance obtained in subfunction scores using all different possible indicator values to identify the most influential indicators. Including the most influential indicators only would both reduce computational and data storage expense for the ABWRET-E and simplify the regulatory process for permit seekers who would be required to do field-based measurements of indicators for the ABWRET-A.

Not all indicators had the same degree of influence on the wetland subfunction scores. Of the original 73 indicators, four were exclusive to HU and were not used in the global sensitivity analysis as we focused on wetland function only (and not wetland use). Of the remaining 69 indicators, six were not measured in all RVAUs due to incomplete spatial coverage of the input data layers and therefore were excluded from the global sensitivity analysis. Of the remaining 63 indicators, 23 indicators (31% of all indicators) were identified that contributed at least 10% of the variance to at least one of the 13 subfunction models. The contributions of these 23 indicators totaled no less than 60% of the variance for 10 of the 13 subfunction models. The remaining 40 indicators may be considered as candidates for removal in simplified models, or the models could be reconfigured to inflate the relative contributions of the indicators according to their relative importance based on expert knowledge.

The global sensitivity analysis revealed that data type (i.e., continuous versus binary) and data order in the subfunction models had strong influences on whether an indicator was important. Indicators with binary (0 or 1) scores were usually more influential in the subfunction score results. Further, the higher an indicator appeared in the rule hierarchy of a subfunction model formula, the higher the influence it had. For example, a binary indicator of connectivity to surface water networks was given the first priority in the hierarchies of HH and WQ subfunction models, meaning that if the indicator value met a defined condition, then no other indicators contributed information to the subfunction score. We suggest that the results of the sensitivity analysis be discussed with the same experts that provided the knowledge for the construction of the combining models. This would create a process where the mathematical technique and the experts are used iteratively to avoid unintended consequences (e.g., some indicators that have a low importance and should have more).

Synergies and Tradeoffs

Unintended consequences can occur in the aggregation of functions into overall value scores and categories. Synergies may be produced where protection of one function has positive feedback for the protection of other functions; alternatively, tradeoffs may be produced where protection of one function comes at the cost of other functions.

Equal weights were given for HH, WQ, and EH scores (0.3) and 0.1 for HU scores in the ABWRET-E. This equal weighting of functions removes consideration of whether one or more functions may be a priority in a given RVAU. However, aggregation techniques could be chosen to prioritize some functions over others; for example, different weights could be assigned to characterize different degrees of importance. The choice of aggregation techniques requires careful consideration. For example, if a higher weight is given to HH function scores, will this lead to a general improvement in HH function on the landscape? If equal weight is given to all function scores, will the same improvement be found for all functions? Simulations made by Accatino et al. (2018) showed that the outcomes of aggregation techniques can lead to counterintuitive and unintended consequences. Some techniques based on weighted average led to synergies; for example, when a higher weight was assigned to HH function scores, improvement in WQ function was also detected as the two functions are highly correlated. However, other techniques led to tradeoffs; for example, threshold-based schemes based on a minimum, where the aggregated score is equal to the worst performing function, promoted monofunctional wetlands but multifunctional landscapes.

Static Versus Dynamic Assessments

Wetland function assessments and value assignments are currently made under the assumptions of static landscapes and human preferences. However, the implementation of the wetland policy will create feedback that must be considered. For example, indicators of wetland function are sensitive to changes in a surrounding landscape that evolve over time; indicators, functions, and values will periodically need to be recalculated and value distributions refreshed to capture potential shifts that may affect the value categories. Similarly, values are sensitive to changes in human preferences that can change over time; aggregation rules for combining functions will need to be revisited to capture potential shifts in human preferences (Ostrom, 2009). Agent-based models are recognized tools for simulating complex systems (Farmer and Foley, 2009), including landscape and human dynamics (Matthews et al., 2007). The potential consequences of implementation of Alberta's wetland policy should be simulated over time to capture this feedback. We plan to use agent-based models to explore the potential consequences of feedback between the landscape and human components that arise from the implementation of wetland protection and restoration strategies to meet the wetland policy objectives.

FUTURE USE OF THE LANDSCAPE-LEVEL ASSESSMENT

Refining the Wetland Inventory

The wetland inventory is the single most important data layer in the ABWRET-E, but confidence in wetland function estimates is negatively affected by its unreliability in many areas. The current AMWI contains numerous omission and commission errors as well as incorrect wetland boundaries, due in part to the contributions of different initiatives using different methods, specifications, and data sources to create the merged inventory. Future work will require improving the AMWI by developing methods to standardize and automate wetland inventory creation using fine spatial resolution data acquired at regular repeat intervals. The integration of accurate and high spatial resolution light detection and ranging (LiDAR) elevation data and color satellite or aerial photograph data obtainable at regular intervals will help to improve wetland identification, delineation, and classification results (e.g., Serran and Creed, 2016; Waz and Creed, 2017). Standard automated methods will provide replicable results that can be compared across regions and will allow for rapid reassessments in a changing landscape. Investment in the wetland inventory will also significantly improve confidence in wetland function estimates. All 73 indicators depend on the accurate identification of the presence or absence of wetlands, and 26 of the 73 indicators are obtained directly from attributes of the wetland inventory (e.g., wetland area, wetland type, wetland perimeter-to-area ratio).

Monitoring Extent and Recovery of Wetland Functions

Monitoring of existing wetland functions over time and the return of wetland functions postrestoration can determine whether the ABWRET tools effectively capture the processes that affect wetland function. The ABWRET-E, in combination with other emerging technologies, can assess the functions of existing and restored wetlands on a regular basis using indicator data that is regularly refreshed. For example, new and cost-effective remote sensing techniques using drones could be used to acquire accurate and repeatable indicator data on demand to monitor functions over time and space (Kaneko and Nohara, 2014). This information will assist the GOA in determining if its policy of preserving wetland function through wetland restoration is effective.

ACKNOWLEDGMENTS

The authors acknowledge Thorsten Hebben (Section Head, Surface Water Policy, Alberta Environment and Parks, 15POL800) for his commitment to developing tools for implementing wetland policy for the province of Alberta. The authors thank Susan Meilleur, Policy Project Management Specialist, and Matthew Wilson, lead on the technical working group for the wetland policy tools, for their important feedback during the development of the ABWRET-E and ABWRET-A. Finally, the authors acknowledge Dr. Paul Adamus for his leadership in developing wetland rapid evaluation tools, and providing the scientific basis that the ABWRET tools were based on.

REFERENCES

Accatino, F., Creed, I.F., Weber, M., 2018. Landscape consequences of aggregation rules for functional equivalence in compensatory mitigation programs. Conserv. Biol. in press.

Alberta Environment and Parks, 2013a. Alberta Wetland Policy, ISBN 978-1-4601-1287-8. Retrieved from: http://aep.alberta.ca/water/programs-and-services/wetlands/documents/AlbertaWetlandPolicy-Sep2013.pdf.

Alberta Environment and Parks, 2013b. Alberta Merged Wetland Inventory. Alberta Environment and Parks, Government of Alberta, Edmonton, Alberta. https://geodiscover.alberta.ca/geoportal/catalog/search/resource/details.page?uuid=%7BA73F5AE1-4677-4731-B3F6-700743A96C97%7D.

Alberta Environment and Parks, 2014. Hydrologic Unit Code Watersheds of Alberta. Alberta Environment and Parks, Government of Alberta, Edmonton, Alberta.https://geodiscover.alberta.ca/geoportal/catalog/search/resource/details.page?uuid=%7B657AA2A1-2914-422A-B914-5DC1E87CF1F0%7D.

Alberta Environment and Parks, 2016. Alberta Wetland Mitigation Directive. Government of Alberta, Edmonton, Alberta, ISBN 978-1-4601-3002-5. Retrieved from: http://aep.alberta.ca/water/programs-and-services/wetlands/documents/AlbertaWetlandMitigationDirective-Jul2016.pdf.

Alberta Parks, 2005. Natural Regions and Sub-Regions of Alberta. Alberta Parks, Government of Alberta, Edmonton, Alberta.http://www.albertaparks.ca/albertaparksca/management-land-use/current-parks-system.aspx.

Alberta Water Resources Commission, 1993. Wetland Management in the Settled Area of Alberta: An Interim Policy. Water Resources Commission, Edmonton, Alberta.

Badiou, P.H.J., 2013. In: Conserve first, restore later: a summary of wetland loss in the Canadian prairies and implications for water quality. Presentation at Prairie Flood Management and Mitigation Seminar, March 27, 2013, Yorkton, Saskatchewan. Retrieved from: https://www.yorkton.ca/news/2013/waterseminar/pdf/conserve_first_restore_later.pdf.

Clare, S., Creed, I.F., 2014. Tracking wetland loss to improve evidence-based wetland policy learning and decision making. Wetl. Ecol. Manag. 22, 235–245.

Clare, S., Krogman, N., Foote, L., Lemphers, N., 2011. Where is the avoidance in the implementation of wetland law and policy? Wetl. Ecol. Manag. 19, 165–182.

Cohen, M.J., Creed, I.F., Alexander, L., Basu, N., Calhoun, A., Craft, C., D'Amico, E., DeKeyser, E., Fowler, L., Golden, H.E., Jawitz, J.W., Kalla, P., Kirkman, L.K., Lane, C.R., Lang, M., Leibowitz, S.G., Lewis, D.B., Marton, J., McLaughlin, D.L., Mushet, D.M., Raanan-Kiperwas, H., Rains, M.C., Smith, L., Walls, S., 2016. Do geographically isolated wetlands influence landscape functions? Proc. Natl. Acad. Sci. 113, 1978–1986.

Creed, I.F., Sass, G.Z., Buttle, J.M., Jones, J.A., 2011. Hydrological principles for sustainable management of forest ecosystems. Hydrol. Process. 25, 2152–2160.

ESRI, 2011. ArcGIS Desktop: Release 10. Environmental Systems Research Institute, Redlands, CA.

Farmer, J.D., Foley, D., 2009. The economy needs agent-based modelling. Nature 460, 685–686.

Government of Alberta, 2008. Land-Use Framework. Government of Alberta, Edmonton, Alberta, ISBN 978-0-7785-7714-0. Retrieved from: https://landuse.alberta.ca/LandUse%20Documents/Land-use%20Framework%20-%202008-12.pdf.

Government of Alberta, 2015. Alberta Wetland Rapid Evaluation Tool-Actual (ABWRET-A) Manual. Water Policy Branch, Alberta Environment and Parks, Edmonton, Alberta.

Kaneko, K., Nohara, S., 2014. Review of effective vegetation mapping using the UAV (Unmanned Aerial Vehicle) method. J. Geogr. Inf. Syst. 6, 733.

Matthews, R.B., Gillbert, N.G., Roach, A., Polhill, J.G., Gotts, N.M., 2007. Agent-based land-use models: a review of applications. Landsc. Ecol. 22, 1447–1459.

National Wetlands Working Group, 1997. In: Warner, B.G., Rubec, C.D.A. (Eds.), The Canadian Wetland Classification System, second ed. Wetlands Research Centre, University of Waterloo, Waterloo, ON, p. 68.

Ostrom, E., 2009. A general framework for analyzing sustainability of social-ecological systems. Science 325, 419–421.

Pianosi, F., Beven, K., Freer, J., Hall, J.W., Rougier, J., Stephenson, D.B., Wagener, T., 2016. Sensitivity analysis of environmental models: a systematic review with practical workflow. Environ. Model. Softw. 79, 214–232.

Saltelli, A., Ratto, M., Andres, T., Campolongo, F., Cariboni, J., Gatelli, D., Saisana, M., Tarantola, S., 2008. Global Sensitivity Analysis: The Primer. John Wiley & Sons Ltd, Chichester, UK.

Serran, J.N., Creed, I.F., 2016. New mapping techniques to estimate the preferential loss of small wetlands on prairie landscapes. Hydrol. Process. 30, 396–409.

Tobler, W.R., 1987. In: Measuring spatial resolution.Proceedings, Land Resources Information Systems Conference, Beijing, pp. 12–16.

Waz, A., Creed, I.F., 2017. Automated techniques to identify lost and restorable wetlands in the Prairie Pothole Region. Wetlands 37, 1079–1091.

Wetland Assessment Working Group, 2011. Assessing Relative Wetland Function: A Discussion Paper. Government of Alberta, Edmonton, Alberta.

Section 3

Field-Level Rapid Assessment Methods: Overview and General Process for Developing or Regionalizing a RAM

Chapter 3.1

Process for Adapting or Developing a RAM

John Dorney* and Paul Adamus†
*Moffatt and Nichol, Raleigh, NC, United States, †Oregon State University, Corvallis, OR, United States

This section describes the overall process of adapting or developing a wetland or stream RAM.

Section chapters then elaborate on each of several major steps in that process. These chapters are authored by Adamus except as noted below. These are:

- Process for Adapting or Developing a RAM (by Dorney and Adamus, in this chapter)
- Developing Guidance for Delimiting the Assessment Areas or Reaches (Chapter 3.2)
- Selecting Indicators and Testing the Data Forms (Chapter 3.3)
- Creating Models for Rolling Up Indicator Data Into Scores (Chapter 3.4)
- Collecting Calibration Data (Chapter 3.5)
- Converting Scores to Ratings (Chapter 3.6)
- Converting to an Overall Site Score (Chapter 3.7)
- Analyzing the RAM's Repeatability and Sensitivity (Chapter 3.8)
- Analyzing the RAM's Accuracy (Chapter 3.9)
- General issues in the statistical testing of RAMs by Munoz, Savage, and Baker (Chapter 3.10)
- Training (Chapter 3.11)
- Applications: Using Field-Based Wetland and Stream RAMs in Decision-making (by Dorney and Adamus, Chapter 3.12)

Section 4 then provides case studies demonstrating the development, regionalization, and/or application of specific field-based wetland RAMs from North America. Section 5 provides case studies for non-North American methods.

The following narrative describes a process for developing and testing wetland and stream rapid assessment methods (RAM). The goal of this chapter is to provide a framework for authors of these methods to consider while designing their methods.

A few recent publications describe what has been learned in developing stream RAMs (Dorney et al., 2014; Somerville, 2010) with some having a strong emphasis on verification and calibration issues for streams (Nadeau et al., 2015). Several other publications describe some of the issues relevant to the development of wetland RAMs (CBD/Ramsar, 2006; De Groot et al., 2006; Fennessy et al., 2007; Sutula et al., 2006; Smith et al., 2013). A checklist in Table 3.1.1 summarizes much of this information in order to help RAM authors verify that they have considered pertinent factors before and during RAM development and testing.

At the most basic level, the process for developing or regionally adapting a RAM flows like this (adapted from Sutula et al., 2006):

1. Organize the RAM project.
2. Build a scientific foundation by assembling and reviewing existing relevant information.
3. Create an initial draft of the RAM, try it, and revise as necessary.
4. Optionally, visit and assess a number of reference sites (wetlands or stream reaches) to calibrate the RAM, then use those calibration data to adjust the RAM's raw scores or categories.
5. Optionally, test the adequacy of the RAM according to various criteria.
6. Publish and distribute the final version.

In this introductory chapter, we describe only the first step above (Organize the RAM project); the others are addressed in Chapters 3.2–3.9. Assuming that the RAM's purpose has been defined, the most important factors to consider are shown in Table 3.1.1 and summarized as follows:

Wetland and Stream Rapid Assessments. https://doi.org/10.1016/B978-0-12-805091-0.00003-7

TABLE 3.1.1 Factors to Consider in the Development of Wetland and Stream RAMs

Factor to Consider	Issues to Address
1. Type of method desired	Streams—assess flow duration (type)? Or condition? Or both? Wetlands—assess functions or condition ("health"), or both?
2. General use of the method	Will method be required or optional? In what situations? Inappropriate uses? Who are intended users and their typical skill levels?
3. Makeup of science team to develop method	Multiagency? Interdisciplinary? Agency staff only? Consultant only? Both? Are all relevant disciplines represented on the team?
4. Time to develop and test method	How much time is available to method authors? What will be the schedule?
5. Time available to conduct field evaluation	How much time is available to users?
6. Source of funding for method development and testing	Staff time donated by agency? By consultant? Expenses for travel? Materials? Consultants and peer reviewers?
7. Format of RAM results	Scores, categories, or other? Reference-based?
8. Selection of sites for beta-testing of the draft RAM	Schedule with subject matter (nonteam) experts after working draft has been completed but before final product is produced.
9. Selection of sites for RAM calibration	Select sites subjectively ("hand pick") or with statistical algorithm? Select sites by geography? Geomorphic type? Degree of prior disturbance? Other factors? Some combination? Has the sample population been mapped with regard to factors to be used in the selection?
10. Repeatability	How similar must the results be from different independent users who assess the same wetlands? How much can that be affected by season, climate, training, and other factors?
11. Verification of accuracy	Verification by opinions of experts? Or by detailed long-term measurements of ecosystem processes? Which variables and measurements should be used in the comparison with RAM outputs? Which statistical procedures and levels of confidence should be used to judge the adequacy of the RAM?
12. Long-term training	How much training will be required? Conducted by public sector? By private sector? Both? How to maintain quality?

- Amount of funding available and timing.
- Skills and knowledge possessed by typical users of the RAM.
- Skills and time availability of those who will be responsible for RAM development.
- Which attributes the RAM is intended to assess.
- Goals for scope, quantification, repeatability, sensitivity, and accuracy.

Together, these factors inform the critical decision about whether to create a new RAM or take one from another region, modify it where necessary, and calibrate it to the current region. For example, instructions for taking an existing RAM template and modifying and calibrating its data forms, indicators, and spreadsheet to assess services of wetlands in a new country, state, province, or other region are provided by Adamus (2016) based on experience transferring Oregon's method (ORWAP) to Alaska, Alberta, and the provinces of Atlantic Canada (see Chapter 4.3.2). Another example is California's CRAM (Collins et al., 2006), which is largely a modified and recalibrated adaptation of Ohio's ORAM (Mack, 2001), which in turn borrowed significantly from Washington's earliest method (1993) and, going back farther, from WET (Adamus et al., 1987). For streams, examples include SDAM (the Streamflow Duration Assessment Method for the Pacific Northwest, Chapter 4.1.1) and the Virginia (Chesapeake Bay) Method (Chapter 4.1.3). Both were modified partly from the NC Stream Identification Method (Chapter 4.1.2).

Funding: Without adequate funds, no credible RAM project can proceed. Cost will depend partly on the project's geographic and thematic scope (also called its domain); the extent of peer review, calibration, and testing desired; the number of staff or consultants involved; and whether a new RAM is being devised or only an existing RAM is being modified and calibrated to a new region. For modifying existing RAMs for a new region or wetland type, costs can be less than $100,000 US, including travel, equipment, salaries and benefits of the responsible scientists and data analysts. Adequately calibrating, testing, and documenting the RAM will add significantly to that amount. And if an entirely new method is to be developed, costs for the entire effort may be close to $1 million US. Funding limits are particularly acute in developing nations, so progress there may be slow or impossible without the support of international donors and/or volunteer assistance from local or foreign scientists. Developing, calibrating, and testing a credible new RAM can take up to 4 years whereas adapting and calibrating an existing one to a new region and/or wetland type normally takes 1–2 years. The time required depends partly on the RAM's purpose and geographic scope as well as the number of experts available to assist the RAM project.

Skills and knowledge possessed by typical users: Some RAMs specify the use of equipment that many potential users cannot afford or lack the skills to use. Other RAMs fall into the trap of being accurate only when used by persons having comprehensive and advanced skills in the identification of local flora and fauna or the interpretation of wetland hydrology and geomorphology. Those skills must be continually practiced, but many government and private consulting positions do not provide such regular opportunities. Moreover, each year fewer and fewer university programs provide extensive taxonomic and natural history training (Noss, 1996), so persons with those skills are often rare and in high demand from competing projects. Thus, no matter how elegant and scientifically robust a RAM may be, if it requires skills that few potential users possess and that cannot be taught sufficiently in a course lasting less than about a week, the RAM will be impractical to routinely use on large numbers of wetland or stream assessments.

Skills and time availability of RAM author(s): Early in the project planning stage, a decision must be made regarding which person or persons will bear primary responsibility for doing the work and/or administering the RAM project. There are advantages to having the work shared by multiple individuals ("team"), and advantages to having all or most of it done by a single expert. Both strategies have been used successfully. An obvious advantage of the team strategy is that the RAM may be more likely to include and integrate a wide range of scientific disciplines—assuming team members are chosen partly for that purpose. Disadvantages of the team strategy involve potentially higher costs and diluted responsibilities as well as scheduling difficulties. Compensating all team members for a sufficient number of hours can become very expensive, but involving each team member for too small a fraction of their time can muddy communications, dilute responsibilities, and stall progress. In addition, momentum can be lost as team members get pulled aside temporarily by competing projects, or even withdraw entirely. Quality RAMs cannot be developed only in a series of committee meetings. Highly focused efforts over unbroken periods of time involving extensive field visits to a variety of sites are essential. Broader acceptance by a wider variety of users often results from using a team approach.

In contrast, dedicating a single position to the project full-time is sometimes more likely to provide smooth and swift forward progress. This assumes an expert is available and willing to help, someone who can easily be shifted to the project or placed under contract. The expert should be grounded in a wide variety of scientific disciplines; be able to articulate relevant knowledge of those disciplines well when writing and speaking or (if desired) as computer code; and be able to demonstrate skill in matching scientific principles to the typical abilities of the intended RAM users without compromising sound science. If the single-expert strategy is used, it is advisable that drafts of the RAM developed or modified by the expert be reviewed independently at each stage, feedback given, and progress monitored. Regardless of which strategy is chosen, it also is critical that, beginning in the early stages of the RAM project, input is sought regularly from representatives of all agencies and groups that will use the resulting RAM. That involvement may impart a sense of ownership that ultimately can lead to the critically important effect of broader acceptance and use of the final version of the RAM. Moreover, it helps assure that, as the RAM evolves, it continues to meet its intended purpose.

If personnel from one or more agency will lead the development of the RAM, it still may be helpful to hire a consultant to assist the project technically as well as to keep track of decisions made by others involved in the RAM's development, and to assume responsibility for logistical planning of the field work. If a consultant is involved, either in support of agency staff or as the principal author, that role should be made clear to all participants at the project's outset.

Finally, a beta testing process often provides valuable input and helps address the inherent danger of "group think." Basically, a selected group of individuals with broad backgrounds (such as potential users) should be engaged to review the quasifinal product of the team or consultant to bring an educated, outside perspective to the RAM before it is introduced to a wider group of users. This is often a combination of an abbreviated office training exercise followed by field visits to representative sites. Subsequent feedback can be used to improve the speed and ease with which the RAM is applied while not compromising its technical integrity.

Which wetland or stream attributes the RAM will assess: Ultimately, the choice will depend on laws and policies—both regulatory and nonregulatory—of agencies or groups responsible for wetlands and streams as well as the skills, time, and equipment needed to implement that choice. The most common choices are functions, values (benefits), other attributes, or some combination. Any or all of these may be termed *condition*, although that term is sometimes used more narrowly to denote the degree to which a site has been impacted by humans. *Functions* are what wetlands and streams do naturally, which in the most general sense includes storing or conveying water, altering chemical and physical properties of waters they receive, and providing habitat. These functions are supported by a host of natural processes. If the intent is to design a RAM that will assess functions, then the particular functions it will address should be identified and defined. This will depend partly on the region and types of wetlands or streams being included, and partly on the availability of science-based variables that are reliable in estimating relative levels of the function. For example, in most regions there are many variables that are both rapid to use and can be applied with others to identify wetlands that are likely to be sinks or sources of sediment. In contrast, there are few if any variables that can reliably and rapidly indicate if a wetland discharges or recharges groundwater on a net annual basis. Similarly, the assessment of the biota of a stream can often be readily accomplished while the ability of a stream channel to process pollutants (say denitrification) can be difficult or impossible to estimate in a timely manner.

When society or a community explicitly recognizes benefits from the performance of stream or wetland functions, they are termed *values*. For example, storage of runoff (a function) may be interpreted as a value (flood control) when vulnerable infrastructure is located in a floodplain downriver from the wetland. If intrinsic characteristics of the particular wetland (e.g., soils, vegetation, morphology) make it highly functional for storing water, its juxtaposition with the described infrastructure (which provides an apparent opportunity to benefit from that function) results in what is termed the delivery of a *service*. It is important to distinguish between the function and value components of a wetland service. Otherwise, a wetland could be wrongly claimed to be supporting a water storage service merely because it is located upgradient from vulnerable infrastructure, despite its geomorphic characteristics suggesting it has little or no capacity to store water.

The levels and types of values that wetlands and streams provide, individually and collectively, are largely determined by the *opportunity* to provide a particular function and the local *significance* of that function (Adamus, 1983). For many hydrologic and water quality values, opportunity is determined by what is upslope of a wetland or stream (e.g., land use and buffers in the wetland and stream's contributing area) and significance is predicted partly by what is downslope (e.g., critical fish spawning areas, quality-limited water bodies). For example, if two wetlands are identical but one receives feed lot runoff from lands above it and also has a lake below that has experienced nuisance algal blooms as a result of overenrichment, then the nutrient processing function of that wetland may be considered to have greater value than that of the other wetland. That is because that function is supporting a service that is recognized as being especially beneficial in that context. Similarly, if a stream is in a developed area, its ability to manage flood flow properly in its floodplain may be more critical than the same water storage in an undeveloped area.

However, it is important to understand that even when wetlands and streams have been shown to be performing particular functions, differences exist among cultures and communities in how they regard the performance of that particular function, that is, whether it is indeed considered to be of value and thus provides a service (Gaucherand et al., 2015). For example, in many regions the capacity of wetlands to remove phosphorus is considered beneficial because the performance of that function protects downstream waters from problems associated with overenrichment. But in other regions, the concentrations of phosphorus in receiving waters are so low that they do not support the desired levels of productivity of valued resources such as fish. Also, the ability of ephemeral streams to carry water may be more critical in arid landscapes where such features can quickly become raging torrents of stormwater than in more humid landscapes where ephemeral features rarely (if ever) inundate the surrounding landscape.

In addition to assessing values associated explicitly with specific functions, many RAMs assign value scores based on all or some of the following:

- Nonconsumptive use (e.g., for recreation and scientific research).
- Consumptive sustainable use (e.g., harvesting of native hay or anadromous fish such as salmon).
- Cultural relevance (for instance the importance of sweetgrass or wild rice or the human and spiritual history of some desert springs for some native North American tribes).
- The level of regional or national endangerment of a site's component species.
- Relative scarcity locally and regionally of wetlands or streams of the same class.
- Special conservation designations previously assigned by agencies or groups of experts, for example, Ramsar Wetland, Wild and Scenic River.

Some RAMs automatically assign the highest possible rating to a site regardless of the levels of its other functions and values, if it has an endangered plant community or species. That happens where laws have been passed that recognize "endangerment" as a value that should trump all others.

Many RAMs assess other values. The most common ones are variously termed wetland or stream *integrity*, *quality*, *intactness*, or *health*. These terms are sometimes used interchangeably with *condition*. However, there exists no scientific consensus on how best to measure or represent these. For example, a site that is judged to be of poor integrity because it is overrun by invasive plants may nonetheless be of sustainably high quality and function for native fish, birds, and amphibians. Many other examples could be cited (e.g., Acreman et al., 2011; McLaughlin and Cohen, 2013). No single site, regardless of how intact or pristine it may be, can provide all functions at a high level (Hansson et al., 2005). That is because many functions operate naturally in opposing directions. For example, sites that retain sediment at a rate that is atypically high for their class may simultaneously export carbon at a rate that is atypically low for their class. Thus it is inappropriate to describe a site as being "highly functional" without also specifying the function or combination of functions or species to which one is referring and how they are being weighted.

Some disagreements exist among wetland and stream scientists (e.g., Hruby, 2014) over whether RAMs should attempt to represent integrity or "overall condition" of wetlands (e.g., USA-RAM) or focus on individual wetland functions (e.g., Smith et al., 2013) and services (Adamus, 2016). To the extent that assessments of functions, values, and services of natural systems reflect a human-focused "utilitarian" perspective whereas integrity of those systems reflects an ecologically focused "preservationist" perspective, this debate goes back to the earliest days of the conservation movement in the United States (Nash, 2014). The director of the national forest service at that time, Gifford Pinchot, argued that the nation's forests be managed mainly for timber production while preservationist John Muir argued that they be managed for their ecological integrity and spiritual value. Both men considered themselves to be conservationists (Fox, 1981).

However, to some extent these are false choices because often both can be accommodated. In the case of wetlands, these two types of RAMs are complementary and ideally should be used together in many situations while in other situations, the choice will depend on how one intends the results to be used. RAMs that focus on wetland integrity are often the preferred choice of monitoring programs, partly because many such RAMs are more sensitive to slight interannual changes in wetland flora and fauna. RAMs that focus on functions and/or services are often best for communicating wetland benefits to the general public, for prioritizing wetlands for development or conservation, and for deciding on an appropriate amount and type of compensatory mitigation to offset loss of functions case-by-case.

Many RAMs are reference-based in that they compare the condition of a particular wetland or stream channel to a relatively undisturbed system. However, it is assumed that the RAM's authors can make a robust and appropriate decision as to what is the relevant reference for a particular stream or wetland, after considering historical land use, physiographic region, local slopes, local soil types, etc. As long as the RAM's authors can provide clear definitions of types and boundaries for these types, then reference-based approaches are logical. Another approach would be to base the RAM on an estimate of the level of function. For example, for a stream method, the RAM's authors might consider the amount of water that is provided by baseflow versus stormflow or the numerical rating of the aquatic macrobenthos based on taxa, diversity, and sensitivity to pollutants while for wetland methods, the RAM's authors might consider the volume of water stored in a particular wetland. These approaches are not usually taken with RAMs because they are more time consuming and therefore not rapid.

Similarly, stream integrity is often affected by historical land use practices. The authors of RAMs need to consider such widespread past disturbances and how they may have affected present stream integrity and functioning. For example, streams in the humid Piedmont region of the southeastern United States have evolved over the past several centuries through a series of widespread sediment deposition and subsequent down-cutting such that many streams are not actively connected to their adjacent floodplain (Trimble, 1974). In addition, legacy sediments from past small head dams have resulted in extensive deposits of sediment in many small stream valleys in the mid-Atlantic region of the United States (Walter and Merritts, 2009).

Concepts such as quality or health are sometimes confused with *risk*. Risk to a site's functions is normally judged by considering potential pollution sources, width and other characteristics of adjoining buffers of natural vegetation, land use ordinances, local projections of population growth, and other factors that are mainly extrinsic to the site. The mere presence of high-risk conditions should never be automatically regarded as synonymous with poor site condition. Indeed, wetlands and streams that have excellent water quality and show no evidence of human impact can sometimes support levels of species richness, productivity, and element cycling that are atypically low for their wetland or stream geomorphic class. That is because it cannot be automatically assumed that humans are the main influencers of the performance levels of wetland and stream functions. Even within a single wetland or stream class, performance levels often are dictated primarily by topography, geology, climate, biological interactions, and other natural factors.

Other attributes (i.e., neither functions nor values) that are assessed by some RAMs include a site's exposure to stressors, its sustainability, and its sensitivity overall or to specific stressors.

In what format will RAM outputs be expressed? There are pros and cons to numerical versus categorical (high, medium, and low) formats for RAM outputs. Numerical outputs translate into ratios for compensatory mitigation more readily than do categorical systems. Categorical formats can be converted to numerical values when needed (e.g., high = 3, medium = 2, and low = 1). Users of numeric-based RAMs may misperceive a small difference in scores as being more functionally significant than it likely is. This concern can be partially addressed by calculating and presenting the statistical confidence intervals around output scores, based on results from prior repeatability testing (see Chapter 3.8).

Goals for RAM scope, quantification, repeatability, sensitivity, and accuracy. Most RAMs have been developed for a specific state, province, country, or other area defined by political boundaries. That is because programs and agencies that fund and administer RAM projects are typically organized in such a manner and cannot assume jurisdiction over resources beyond those boundaries. It is scientifically preferable that RAMs be organized by ecoregions (large areas with distinctive assemblages of species and similar climate, geology, and soils) regardless of political boundaries. If a RAM's geographic scope must be limited to a single state, province, or country, the RAM should at least account for differences in ecoregions within that area, and ideally, should be calibrated separately to each (see Chapter 3.5).

Early in the planning process, a decision must be made whether the RAM should address all wetland (or stream) types or only specific ones. The more inclusive the RAM, the more likely it is to be used often and widely once its development is complete. However, if all types are included, the RAM's ability to detect differences among sites of any given type, that is, its sensitivity, may be limited. In contrast, focusing on just one or a few wetland or stream types and/or ecoregions may increase the RAM's ability to detect (i.e., its sensitivity to) differences among sites within each type. If funding or other considerations allow RAM development and calibration for only one or a few types and/or ecoregions, those may be selected based on their risk from future development and likely distinctiveness with regard to water regime, flora, and fauna.

At project outset, it is important to decide on the desired degree of quantification. For instance, whether the outputs from the RAM will be nominal (e.g., good, fair, poor; or high, medium, low), ordinal (e.g., relative scores on a continuous scale of 0–1), or less often used by RAMs, actual measures or model predictions expressed on a continuous scale, for example, number of plant species per quadrant, projected kilograms of sediment eroded per square meter per year. Of course, nominal outputs can easily be converted to numerical (e.g., low = 1, moderate = 2, high = 3). Conversely, there are several ways to easily convert ordinal or continuous numerical outputs into categories. That can be done either within the RAM's structure or as separate administrative decisions. Those conversion procedures, along with factors to consider when deciding on an appropriate number of categories, are discussed in Chapter 3.6.

Repeatability, *sensitivity*, and *accuracy* together determine a RAM's *validity*. While planning the RAM project, expectations for the RAM should be discussed in terms of each of these, and consideration given to expressing in quantitative terms a goal for each. The degree of rigor expected of a RAM, as defined by these factors, will influence the cost and time requirements of the RAM project.

Repeatability (also called replicability, consistency) is the degree of correlation or agreement in RAM outputs among users who simultaneously but independently assess the same series of wetlands or stream reaches. Repeatability testing procedures are described further in Chapter 3.8.

Sensitivity is a RAM's ability to distinguish important differences across all sites, or within one site across multiple time periods, for example, before and after restoration, before and after impact. Sensitivity is described further in Chapter 3.8.

Accuracy is the "gold standard" of RAM validation. Accuracy is the degree to which the results from a RAM resemble reality, as usually defined by long-term, intensive monitoring. The process of comparing RAM outputs (scores, ratings) with some measure of reality is called *verification* or *validation*. Verification usually involves comparing the outputs with the opinions of experts whereas validation involves comparing the outputs with detailed long-term measurements of functions or condition. See Chapter 3.9 for further discussion.

REFERENCES

Acreman, M.C., Harding, R.J., Lloyd, C., McNamara, N.P., Mountford, J.O., Mould, D.J., Purse, B.V., Heard, M.S., Stratford, C.J., Dury, S.J., 2011. Trade-off in ecosystem services of the Somerset Levels and Moors wetlands. Hydrol. Sci. J. 56, 1543–1565.

Adamus, P.R., 1983. A Method for Wetland Functional Assessment. Vol. II. Methodology. Report No. FHWA-IP-82-24Federal Highway Administration, Washington, DC.

Adamus, P.R. 2016. Wetland Ecosystem Services Protocol (WESP) version 2.0. Available from: people.oregonstate.edu/~adamusp/WESP.

Adamus, P.R., Clairain, E.J., Smith, R.D., Young, R.E., 1987. Wetland Evaluation Technique (WET). Volume II. Methodology. US Army Corps of Engineers Waterways Experiment Station, Vicksburg, MS.

CBD/Ramsar, 2006. Guidelines for the Rapid Ecological Assessment of Biodiversity in Inland Water, Coastal and Marine Areas. Secretariat of the Convention on Biological Diversity, Montreal, Canada. CBD Technical Services No. 22 and the Secretariat of the Ramsar Convention, Gland, Switzerland, Ramsar Technical Report No. 1.

Collins, J.N., Stein, E.D., Sutula, M., Clark, R., Fetscher, A.E., Grenier, L., Grosso, C., Wiskind, A., 2006. California rapid assessment method (CRAM) for wetlands and riparian areas. Version 4 (3), 136.

De Groot, R.S., Stuip, M.A.M., Finlayson, C.M., Davidson, N., 2006. Valuing Wetlands: Guidance for Valuing the Benefits Derived From Wetland Ecosystem Servicers. Ramsar Technical Report No. 3/CBD Technical Services No. 27. Ramsar Convention Secretariat, Gland, Switzerland & Secretariat of the Convention on Biological Diversity, Montreal, Canada.

Dorney, J.R., Paugh, L., Smith, S., Lekson, D., Tugwell, T., Allen, B., Cusack, M., 2014. Development and testing of rapid wetland and stream functional assessment methods in North Carolina. Natl. Wetl. Newsl. 36 (4), 31–35.

Fennessy, M.S., Jacobs, A.D., Kentula, M.E., 2007. An evaluation of rapid methods for assessing the ecological condition of wetlands. Wetlands 27 (3), 543–560.

Fox, S., 1981. John Muir and His Legacy: The American Conservation Movement. Little Brown, Boston, MA.

Gaucherand, S., Schwoertzig, E., Clement, J.C., Johnson, B., Quétier, F., 2015. The cultural dimensions of freshwater wetland assessments: lessons learned from the application of US rapid assessment methods in France. Environ. Manag. 56 (1), 245–259.

Hansson, L., Bronmark, C., Nilsson, P.A., Abjornsson, K., 2005. Conflicting demands on wetland ecosystem services: nutrient retention, biodiversity or both? Freshw. Biol. 50 (4), 705–714.

Hruby, T., 2014. Washington State Wetland Rating System for Western Washington. Washington State Department of Ecology, Olympia, Washington.

Mack, J.J., 2001. Ohio Rapid Assessment Method for Wetlands v. 5.0, User's Manual and Scoring Forms. Ohio Environmental Protection Agency Technical Report WET/2001-1, Ohio Environmental Protection Agency. Division of Surface Water 401.

McLaughlin, D.L., Cohen, M.J., 2013. Realizing ecosystem services: wetland hydrologic function along a gradient of ecosystem condition. Ecol. Appl. 23 (7), 1619–1631.

Nadeau, T.-L., Leibowitz, S.G., Wigington Jr., P.J., Ebersole, J.L., Fritz, K.M., Coulombe, R.A., Comeleo, R.L., Blocksom, K.A., 2015. Validation of rapid assessment methods to determine streamflow duration classes in the Pacific Northwest, USA. Environ. Manag. 56 (1), 31–53.

Nash, R., 2014. Wilderness and the American Mind. Yale University Press, New Haven, CT.

Noss, R.F., 1996. The naturalists are dying off. Conserv. Biol. 10 (1), 1–3.

Smith, R.D., Noble, C.V., Berkowitz, J.F., 2013. Hydrogeomorphic (HGM) Approach to Assessing Wetland Functions: Guidelines for Developing Guidebooks (Version 2). ERDC/EL-TR-13-11, Environmental Lab, Engineer Research and Development Center, Vicksburg, MS.

Somerville, D.E., 2010. Stream Assessment and Mitigation Protocols: A Review of Commonalities and Differences. Prepared for the U.S. Environmental Protection Agency, Office of Wetlands, Oceans, and Watersheds (Contract No. GS-00F-0032M), Washington, DC. Document No. EPA 843-S-12-003.

Sutula, M.A., Stein, E.D., Collins, J.N., Fetscher, A.E., Clark, R., 2006. A practical guide for the development of a wetland assessment method: the California experience. JAWRA 42, 157–175.

Trimble, S.W., 1974. Non-Induced Soil Erosion on the Southern Piedmont. Soil Conservation Society of America, Ankeny, IA 70 pp.

Walter, R.C., Merritts, D.J., 2009. Natural streams and the legacy of water-powered mills. Science 319, 299–304.

FURTHER READING

Daniels, R.B., 1974. Soil erosion and degradation in the Southern Piedmont of the USA. In: Wolman, M.G., Fournier, F.G.A. (Eds.), Section 12 in: Land Transformation in Agriculture. John Wiley and Sons Ltd, Chichester and New York.

Hruby, T., 2001. Testing the basic assumption of the hydrogeomorphic approach to assessing wetland functions. Environ. Manag. 27 (5), 749–761.

Chapter 3.2

Developing Guidance for Delimiting the Assessment Areas or Stream Reaches

Paul Adamus
Oregon State University, Corvallis, OR, United States

When using a RAM, to what exact area or reach of stream do the scores or ratings apply? Answering that question precisely and transparently is essential to correctly interpret the results of a RAM application. For wetlands, the simplest situation is where an entire wetland—surrounded on all sides by dry upland or unvegetated deep water—was viewed. In that situation, the guidance for nearly all RAMs indicates that the *assessment area* (sometimes called the assessment unit or scoring boundary) is the *entire* wetland. However, examples of other situations where that may not be desirable or realistic include:

- The wetland is too large to view entirely during a single visit.
- Part of the wetland is unviewable due to safety considerations (e.g., toxic wastes, impenetrable vegetation, deep water, dangerous wildlife).
- Part of the wetland is unviewable because permission for access was not granted.
- Only part of the wetland—most commonly the part that has been or will be restored or impacted—is of interest, and a clear boundary can be drawn between the parts that have been restored or impacted and the parts that have not.
- Parts of the wetland have inarguably different functions than the rest, and a clear boundary can be drawn between the parts having those functions and the parts that do not.

For streams, distinct reaches usually need to be designated as well. Usually, observations of a change in condition are the key factor in determining the limits of an assessment reach. For instance, a change in flow regime (say from ephemeral to intermittent flow or intermittent to perennial flow) is a common reason to designate a different reach. Similarly, a change in condition such as the presence of a wooded buffer as opposed to a disturbed buffer (such as a parking lot) could be used to separately designate and assess a different assessment reach.

The guidance from different RAMs varies somewhat in how to deal with situations such as these. In no particular order, the options generally are:

Option 1. "Just do your best" to see as much of the wetland or stream as possible, and based on that, do a single assessment that purports to represent the whole. This also takes into account what you can tell from aerial imagery, existing spatial data, information from the landowner, and other sources. Some RAMs also suggest a minimum and/or maximum time limit for walking the site, such as 10 min or 1 h, or a minimum viewing percentage.

Option 2. Place a statistically random point (or series of points) in the wetland or along the stream and base your assessment only on what you can see from that point or series of points, or within a specified radius of those. The random sample may be further stratified by vegetation form, water regime, or other factors.

Option 3. Delimit the assessment area or stream reach boundary only according to the part you could access. If access was possible only to a fragmented patchwork of subunits within the site, delimit each as a separate assessment area and do multiple assessments, one for each accessible subunit.

Option 4. Delimit the assessment area or stream reach boundary-based subunits having different treatments (e.g., restoration, impact) or, rarely, dramatically different levels of functions, if that can be reliably perceived beforehand. Then delimit each subunit as a separate assessment area and do multiple assessments, one for each treatment or functionally distinct subunit.

Option 5. If consensus can be reached on spatial boundaries based on readily observable disturbances (such as land uses adjacent to a stream or soil rutting from past logging), these areas could be assessed separately. If the RAM user is having to imagine the "average" answer for several indicators, that may indicate the need for using levels of disturbance to define separate assessment areas of a reach.

Wetland and Stream Rapid Assessments. https://doi.org/10.1016/B978-0-12-805091-0.00029-3

The choice of options will depend somewhat on the assessment's purpose. For example, if the objective is only to determine if the cover of wetland vegetation throughout an entire region is, on average, more than 10% cover of invasive plants, then Option 2 is appropriate. However, if the objective is to know the levels of various functions of an individual wetland, as needed for many conservation and mitigation decisions, then Option 2 is not appropriate. Other challenges with Option 2 are described by Fennessy et al. (2007).

Even if it is possible to view an entire site and apply the RAM to it, it may still be desirable to apply the RAM to subunits within it, for reasons stated in Options 4 and 5, and then compare results to those for the entire site. RAM guidance should caution users to avoid making the subunits delimited by Options 3–5 so small that assessment accuracy may be compromised. A minimum size beyond which results become very distorted is not known but will depend on the wetland or stream type, configuration, and other factors.

Whatever option is being considered, the guidance prepared for the RAM might suggest that key stakeholders in the wetland or stream decision (e.g., regulatory agency personnel, consultants) first discuss the boundaries for the assessment area(s) and find agreement. Otherwise, resulting scores and ratings may be later challenged on those grounds, necessitating that the assessment be redone. Guidance for the RAM's use should specify that in all instances where the RAM is applied to areas that are subunits of an entire wetland, RAM users should draw the final boundaries chosen for each subunit (assessment area or stream reach) on a map or aerial image, and the assumptions supporting each of those boundaries should be documented. This also helps ensure that any future assessments of that site, especially any done by a different RAM user, are consistent with the original assessment. Tentative boundaries for assessment areas and reaches are typically first drawn on aerial images and then adjusted based on observations when visiting the wetland or stream.

Where wetlands border unvegetated open water such as lakes, rivers, and estuaries, the open water may sometimes be quite deep and technically not covered by laws that regulate wetlands. How much (if any) of the open water should be included in an assessment, and under what circumstances? If the open water is totally excluded, especially in small wetlands in depressions, estimates of several wetland habitat functions are likely to be erroneous. Guidance from various RAMs has dealt with this in one of two ways. One is for RAM authors to include questions on RAM data forms that ask users to evaluate the extent of open water adjacent to the assessment area, rather than drawing the assessment boundary to include it. This is analogous to asking users to evaluate the extent of buffers in nonjurisdictional uplands that adjoin a wetland assessment area. Another option is to direct RAM users to include open water in the assessment area if the open water covers less than a specified area whereas if larger than that, include only a specified portion, for example, shallower than a specified depth, or equal to the average width of the vegetated wetland.

Delimiting assessment areas consistently in the case of wetland *complexes* also poses a challenge. Complexes are areas comprised of many proximate wetlands separated by a usually lesser cumulative area of upland. The upland that separates the wetland members of the complex may be either natural topography or artificial berms. Vernal pools, for example, often occur as a complex of tiny wetlands. RAMs that provide explicit guidance recommend including all such pools or wetlands into a single assessment area if they are within a specified distance of one another, have similar water regimes and vegetation, have similar impacts (if any) of stressors, and if the area circumscribed by a line that connects the outer edge of the farthest members is predominantly wetland (i.e., some nonjurisdictional upland is included in the assessment area, but it does not predominate).

REFERENCE

Fennessy, M.S., Jacobs, A.D., Kentula, M.E., 2007. An evaluation of rapid methods for assessing the ecological condition of wetlands. Wetlands 27 (3), 543–560.

Chapter 3.3

Selecting Indicators, Creating and Testing the Data Forms

Paul Adamus
Oregon State University, Corvallis, OR, United States

Regardless of whether the RAM is intended to assess functions or some other attributes, the indicators (variables) that are intended to estimate that endpoint are the fundamental building blocks of the RAM. They are subsequently phrased as questions on the RAM's data forms, and are the basis for whatever scores and ratings the RAM generates. Thus, decisions about which indicators to use will ultimately influence the validity of the RAM.

The choice of indicators will, of course, depend on what the RAM is intended to assess. If interest centers around stream or wetland functions, a decision must be made regarding which functions the RAM should address. Table 3.3.1 shows and defines the functions most commonly attributed to wetlands and the services potentially associated with these. Table 3.3.2 shows additional attributes that wetlands support to varying degrees. Tables 3.3.3 and 3.3.4 present similar lists for streams. A RAM can be designed to provide a score and/or rating for each, as does the WESP template (Adamus, 2016), or just for a subset deemed most likely to occur within a region. Any of the habitat functions, such as waterbird habitat, could be broken down further by separating into different taxonomic or functional groups, for example, dabbling ducks versus diving ducks, shorebirds versus waterfowl, or breeding versus nonbreeding habitat. Those decisions will depend partly on whether rapid indicators are available that are capable of making such distinctions and are supported by science-based knowledge.

Alternatively, the functions may be defined more broadly, resulting in a fewer number of functions to be scored. For example, Washington's wetlands RAM (Hruby, 2004) yields scores and ratings just for hydrologic, water quality, and habitat functions. Providing scores to decision-makers from a long list of functions or other attributes can result in "information overload" and complicate decisions, but aggregating functions into broad groups ignores important distinctions in driving factors within those groups, potentially resulting in less accurate ratings. In North Carolina's stream and wetland RAMs (NC Stream Functional Assessment Team, 2015; NC Wetland Functional Assessment Team, 2016), separate scores are provided for what are termed subfunctions (for instance, surface versus subsurface water storage) in addition to the three main functions of hydrology, water quality, and habitat.

Scientific principles indicate that wetland characteristics that are good indicators of native plant habitat (for example) are not necessarily good indicators of waterbird habitat, and not all the characteristics that are good indicators of phosphorus retention (for example) are good indicators of nitrate removal. When scores or ratings are assigned only to broad groups or themes, it often is not clear *which* wetland functions were assumed to drive that output.

Ultimately, RAM project participants will need to discuss this issue and find a balance that best suits their purposes between "too many" score outputs (resulting in information overload) and "too few" (which clouds assumptions and reduces accuracy). One solution is to configure the RAM to compute scores for a longer list of functions and then combine those into summary function groups. That way, assumptions are more transparent. However, as described in Chapter 3.7, this requires deciding on a defensible rule or equation to use for "rolling up" the scores or ratings from different functions.

Strategies for selecting indicators can be characterized as top-down, bottom-up, or, preferably, some combination. A top-down strategy might involve first creating conceptual diagrams of major factors and processes that are known or believed to be causally linked to each endpoint the RAM is intended to assess, for example, to a specific wetland or stream function (Rosen et al., 1995; Smith et al., 2013). Then at least one indicator representing each causal connection is identified and included in the RAM. Often, however, it takes hours or days to prepare and find consensus on how to depict processes in a conceptual diagram, only to discover that there are no indicators to represent some of those linkages that can be assessed observationally during a single site visit.

In contrast, a bottom-up strategy involves first making a list of all features that can realistically be observed by typical users of the RAM during a single 1-day visit to a site, or which can be inferred indirectly from other observations.

Wetland and Stream Rapid Assessments. https://doi.org/10.1016/B978-0-12-805091-0.00030-X

TABLE 3.3.1 Wetland Functions Scored by WESP and the Services They Provide (See Chapter 4.3.2)

Function	Definition	Potential Services
Hydrologic and water quality maintenance functions:		
Water storage and delay	The effectiveness for storing runoff or delaying the downslope movement of surface water for long or short periods	Flood control, maintain ecological systems
Water cooling	The effectiveness for maintaining or reducing temperature of downslope waters	Support coldwater fish and other aquatic life
Sediment retention and stabilization	The effectiveness for intercepting and filtering suspended inorganic sediments, thus allowing their deposition, as well as reducing energy of waves and currents, maintaining natural erosion rates, and facilitating immobilization and/or detoxification of some contaminants	Maintain quality of receiving waters. Protect shoreline structures from erosion
Phosphorus retention	The effectiveness for retaining phosphorus for long periods (>1 growing season)	Maintain quality of receiving waters
Nitrate removal and retention	The effectiveness for retaining particulate nitrate and converting soluble nitrate and ammonium to nitrogen gas while generating little or no nitrous oxide (a potent greenhouse gas)	Maintain quality of receiving waters
Carbon sequestration	The effectiveness for retaining incoming particulate and dissolved carbon, and through photosynthesis converting carbon dioxide to organic matter and then retaining that on a net annual basis for long periods while emitting little or no methane	Maintain global climate
Organic nutrient export	The effectiveness for producing and subsequently exporting organic nutrients (mainly carbon), either particulate or dissolved	Support food chains in receiving waters
Habitat functions:		
Fish habitat	The capacity to support an abundance and diversity of native fish	Support recreational and ecological values
Aquatic invertebrate habitat	The capacity to support or contribute to an abundance or diversity of invertebrate animals that spend all or part of their life cycle underwater or in moist soil. Includes dragonflies, midges, clams, snails, water beetles, shrimp, aquatic worms, and others	Support fish and other aquatic life. Maintain regional biodiversity
Amphibian and reptile habitat	The capacity to support or contribute to an abundance or diversity of native frogs, toads, salamanders, and turtles	Maintain regional biodiversity
Waterbird habitat	The capacity to support or contribute to an abundance or diversity of waterbirds	Support hunting and ecological values. Maintain regional biodiversity
Songbird, raptor, and mammal habitat	The capacity to support or contribute to an abundance or diversity of native songbird, raptor, and mammal species and functional groups, especially those that are most dependent on wetlands or water	Maintain regional biodiversity
Wildfire barrier	The capacity to resist ignition by wildfire, thus limiting wildfire spread	Avoid damages to property and people from fire
Pollinator habitat	The capacity to support pollinating insects such as bees, wasps, butterflies, moths, flies, and beetles	Support productivity of commercial crops and other vegetation
Native plant habitat	The capacity to support or contribute to a diversity of native hydrophytic plant species, communities, and/or functional groups	Maintain regional biodiversity and food chains

This might include, for example, flood marks on trees as evidence of high water or a general characterization (as opposed to a detailed inventory) of the aquatic insect community in a stream. This also includes indicators whose status can be interpreted from aerial imagery, existing spatial data, and landowner interviews. However, the relationship of the potential indicators to the geochemical and biological processes that support wetland or stream functions, values, or integrity should constantly be kept in mind so the list is more than just an assortment of things that would be convenient to assess.

TABLE 3.3.2 Additional Attributes That Some Wetland RAMs Score or Rate

Public use and recognition	Capacity to support low-intensity outdoor recreation, education, research, or sustainable consumptive uses. Also, existence of prior official designations as some type of special protected area
Wetland sensitivity	A wetland's lack of intrinsic resistance and resilience to human and natural stressors
Cultural importance	The value of a particular wetland or wetland attribute to a local group's spiritual or cultural beliefs
Threatened and endangered species	Capacity to support species formally designated as threatened, endangered, or otherwise of conservation concern. This is sometimes subsumed in assessments of the value of various habitat functions listed in Table 3.3.1
Wetland ecological condition	Similarity of a wetland's structure, composition, and functions with that of a reference wetland of the same type and landscape setting, operating within the bounds of natural or historical disturbance regimes
Stressors	The degree to which a site is or has recently been altered by, or exposed to risk from, primarily human-related factors capable of reducing one or more of its functions
Wetland risk	Likelihood of a sensitive wetland being exposed to high levels of stressors. May also include projections of long-term risks from weak or absent protective ordinances, projections of population growth, climate change, and other factors that are mainly extrinsic to a site

TABLE 3.3.3 Functions Commonly Scored by Stream and Riparian RAMs and the Services They Provide

Function	Definition	Potential Services
Hydrologic and water quality maintenance functions:		
Water storage and delay	The effectiveness for storing runoff either in the stream channel itself or in the adjacent floodplain	Flood control, maintain ecological systems
Water cooling	The effectiveness of the adjacent riparian buffer for maintaining or reducing temperature of downslope waters	Support coldwater fish and other aquatic life
Sediment retention and stabilization	The effectiveness for intercepting and filtering suspended inorganic sediments, thus allowing their deposition, especially in the adjacent riparian area and floodplain, as well as reducing energy of waves and currents, maintaining natural erosion rates, and facilitating immobilization and/or detoxification of some contaminants	Maintain quality of receiving waters. Protect shoreline structures from erosion
Phosphorus retention	The effectiveness for retaining phosphorus for long periods (>1 growing season), especially in the adjacent riparian area and floodplain	Maintain quality of receiving waters
Nitrate removal and retention	The effectiveness for retaining particulate nitrate and converting soluble nitrate and ammonium to nitrogen gas while generating little or no nitrous oxide (a potent greenhouse gas), especially in the adjacent riparian area and floodplain	Maintain quality of receiving waters
Carbon sequestration	The effectiveness for retaining incoming particulate and dissolved carbon, and through photosynthesis converting carbon dioxide to organic matter and then retaining that on a net annual basis for long periods while emitting little or no methane, especially in the adjacent riparian area and floodplain	Maintain global climate
Organic matter input to the stream channel	The effectiveness for providing organic input to the stream channel (especially tree and shrub leaves, and woody debris such as sticks and logs) which then provide energy sources and habitat for aquatic organisms	Support food chains in receiving waters
Habitat functions:		
Fish habitat	The capacity to support an abundance and diversity of native fish.	Support recreational and ecological values.

Continued

TABLE 3.3.3 Functions Commonly Scored by Stream and Riparian RAMs and the Services They Provide—cont'd

Function	Definition	Potential Services
Aquatic invertebrate habitat[a]	The capacity to support or contribute to an abundance or diversity of invertebrate animals that spend all or part of their life cycle underwater or in moist soil. Includes dragonflies, midges, clams, snails, water beetles, shrimp, aquatic worms, and others	Support fish and other aquatic life. Maintain regional biodiversity
Amphibian and reptile habitat	The capacity to support or contribute to an abundance or diversity of native frogs, toads, salamanders, and turtles	Maintain regional biodiversity
Waterbird habitat	The capacity to support or contribute to an abundance or diversity of waterbirds	Support hunting and ecological values. Maintain regional biodiversity
Songbird, raptor, and mammal habitat	The capacity to support or contribute to an abundance or diversity of native songbird, raptor, and mammal species and functional groups, especially those that are most dependent on water and riparian habitats	Maintain regional biodiversity
Native plant habitat	The capacity to support or contribute to a diversity of native hydrophytic plant species, communities, and/or functional groups	Maintain regional biodiversity and food chains

[a]*Some RAMs define this function to include only crustacean habitat.*

TABLE 3.3.4 Additional Attributes That Some Stream and Riparian RAMs Score or Rate

Public use and recognition	Capacity to support low-intensity outdoor recreation, education, research, or sustainable consumptive uses. Also, existence of prior official designations as some type of special protected area
Cultural importance	The value to a local group's spiritual or cultural beliefs
Sub/surface transfer	Maintains exchange of water between surface and subsurface environments, often through the hyporheic zone. Provides base flow, recharges aquifers, exchanges nutrients/chemicals through hyporheic zone, moderates flow, and maintains soil moisture
Flow variation	Maintains daily, seasonal and interannual variation in flow. Influences channel dynamics, provides environmental cues for life history transitions, redistributes sediment, provides habitat variability (temporal), provides sorting of sediment and differential deposition
Stream or riparian sensitivity	A site's lack of intrinsic resistance and resilience to human and natural stressors
Threatened and endangered species	Capacity to support species formally designated as threatened, endangered, or otherwise of conservation concern. (This is sometimes subsumed in assessments of the value of various habitat functions listed in Table 3.3.1)
Ecological condition	Similarity of a site's structure, composition, and functions with that of a reference site of the same type and landscape setting, operating within the bounds of natural or historical disturbance regimes
Stressors	The degree to which a site is or has recently been altered by, or exposed to risk from, primarily human-related factors capable of reducing one or more of its functions
Risk	Likelihood that a sensitive wetland or stream is being exposed to high levels of stressors. May also include projections of long-term risks from weak or absent protective ordinances, projections of population growth, climate change, and other factors that are mainly extrinsic to a site

Whatever the strategy, if the RAM project involves more than just adapting an existing RAM from another region or template, then it is prudent to review other RAMs and consider using many of their indicators, especially if those RAMs are well-documented or validated to some degree. For instance, during the development of a wetland RAM in New Jersey, Hatfield et al. (2004) compared eight existing methods in the field. They concluded that this in-field comparison provided them with valuable insight into the strengths and weaknesses of the methods and helped frame the structure of their New Jersey RAM. Literature reviews of wetland RAMs have not been updated recently, but earlier compilations include those of

Adamus and Brandt (1990), Adamus (1992), Bartoldus (1999), Adamus et al. (2001), and Fennessy et al. (2007). Somerville (2010) prepared a similar compilation for stream RAMs.

It is equally or more important to review scientific literature specific to the region and wetland and stream types that are the focus of the RAM project. Keyword searches using Google Scholar and similar free search engines are a good starting point for identifying such information, but deeper searches using commercial bibliographic databases are recommended. Often overlooked by most search engines, graduate theses and dissertations from local universities can be very informative. Additionally, local scientists should be asked about other potentially useful reports and ideas for rapid indicators with promise for predicting functions, values, or integrity.

In the past few decades, an increasing number of spatial data layers have been created and made available to the public. Common themes are soils, geology, climate, land cover, wetlands (such as the National Wetlands Inventory in the United States), topography, and hydrography. For streams, the Geological Survey's StreamStat website (USGS 2017) allows one to make a rapid determination of watershed area and watershed characteristics such as percent wooded or percent impervious surface. As described extensively in Section 2, many of the features in these layers are directly relevant to predicting functions, values, and integrity of wetlands and streams, especially for sites that are inaccessible due to distance from roads or lack of landowner permission for visitation. Thus, an increasing number of RAMs are requiring users to complement (and only where necessary, substitute) their field observations with such data. That information then contributes to calculations of the site's scores or ratings. In situations where some potential users are anticipated to lack the skills or geographic information systems (GIS) software to manipulate the layers, the layers can at least be viewed with software that is freely available online (e.g., Q-GIS). Using the free Google Earth Pro, some layers also can be converted to a format (KML) that can be overlaid on aerial imagery in Google Earth. Oregon (Rempel et al., 2015) and Alaska (Homan and Adamus, 2016) have taken this one step further and created publicly accessible online portals where dozens of spatial data layers (both point data and polygons) can be overlaid interactively with aerial imagery and wetland maps, then used to answer up to about half the questions comprising the RAMs for these regions.

Use of existing spatial data in RAMs is appropriate only for layers that cover the entire geographic area that is the subject of the RAM; otherwise, the scores or ratings of sites in different parts of that area will not be comparable. Spatial data have many other limitations in the context of their use in a RAM. Some of those limitations were described in Section 2. Nonetheless, the position of many agencies that have sponsored RAMs has been that the costs of ignoring the rich potential of spatial data is greater than problems with using it selectively—provided RAM users and decision-makers pay heed to its limitations in specific applications.

Any indicator used in a RAM can either be estimated or measured. Some candidate indicators, such as distance from a wetland's edge to the nearest road or width of a wooded stream buffer, are easily measured from aerial images and therefore are commonly considered for use in RAMs. Other potential indicators, such as percent cover of invasive plants, percent clay content of soils, or an assessment of the aquatic life of a stream, are more difficult because they require substantially more time, skill, and/or equipment to measure reliably. Such indicators may either be excluded entirely or RAM users may be asked to do their best to estimate them visually during the one-time site visit, often using simplified data collection techniques. Measurement is typically perceived as being more precise and accurate than estimation and, for simple measures such as distance to the nearest road, it usually is. However, for more complex indicators, accuracy and precision are often influenced more by the numbers of sample points or plots and their spatial and temporal distribution. The statistical adequacy of these is seldom tested. Typically, RAM users are directed to simply place the points in "representative" locations within a site, or to locate them systematically or in a random or stratified random manner. For some indicators, if the number and distribution of points is inadequate, as it often is given the time limits imposed by the requirement for sampling to be rapid, then a faster and potentially more integrative visual estimate may be at least as accurate, especially in smaller sites.

The indicators selected to predict the level of a function need not be *determinants* or key drivers of that function. Indicators that are *correlates* of that function, if correlations are strong, are also appropriate. For example, the presence of certain types of thermal anomalies in wetlands or streams is an excellent correlate of groundwater exchange functions, but does not *determine* the degree of groundwater exchange. Similarly, evidence of overbank flooding (adjacent wrack lines of leaves and sticks) indicates the extent of these events but does not determine those frequencies, being clearly dependent on recent weather.

Once chosen for use in a RAM, each indicator is usually given a descriptive title. A space is provided for users to report results of its actual measurement, or its potential conditions are listed and presented in either narrative or numeric form with instructions to users to characterize the indicator by selecting one or more of several named conditions. If the number of categorical choices is few, both the variation in the indicator's status and the variation among RAM users assessing that indicator will have larger effects on the final score or ratings than if the number of choices is large. But if users must decide from among more than about six statements describing the status of an indicator, they may become frustrated at having to make such fine distinctions, especially if users must rely only on rapid visual estimation. Also, if the number of choices is an odd number, users may unconsciously tend toward choosing the middle choice.

When presenting each indicator on a data form, several items warrant attention. They include *reference for comparison*, *scale*, *thresholds*, and *temporal variation*. Suppose one question on a RAM data form asks about an indicator called "deciduous woody vegetation" (DWV). This could be expressed as:

- Percentage of the wetland containing DWV.
- Percentage of the wetland's vegetation that is DWV.
- Percentage of the wetland's woody vegetation that is DWV.

If the *reference for comparison* (the wetland, just its vegetation, or just its woody vegetation) is not stated, different users of the RAM will likely interpret this differently and the RAM's repeatability will be compromised. Thus, it is quite important to state an indicator's reference for comparison and caution users to pay careful attention to it in each question. Training is clearly critical in this regard as well.

Scale refers to the size of the wetland area or length of the reach within which an indicator should be estimated or measured. For example, if the function "amphibian habitat" was known to be influenced by the indicator "percentage cover of DWV," a determination must be made whether this is best measured:

- strictly within the site, or
- only along its perimeter (upland edge), or
- within a buffer of specified width surrounding the site, or
- in an entire landscape defined by lands within a specified radius of the site, or
- some weighted combination of the above.

The choice of which of these expressions to use should be informed by reviewing scientific literature. If no guidance is obvious, then experts may be asked for opinions based on, for example, published data on the home range or dispersal distance of amphibian species in the RAM's region. For indicators whose status is most relevant when assessed beyond the wetland or stream margin, it may be least demanding of RAM users to assess all such indicators at just one standard distance from the site. However, doing so would not reflect the different scales over which different indicators exert maximum effect or correlation with different functions (Rooney et al., 2012), so use of multiple scales of measurement is advisable.

For streams, watershed size is a commonly used indicator of stream function that can readily be determined from existing topographic maps or more recently in the United States via websites such as the Geological Survey's StreamStats (2017). Any such categories will need to be carefully chosen to have either ecological or regulatory relevance.

Thresholds refer to the breakpoints between different numerical choices for an indicator. For example, for an indicator called "percentage of wetland containing unvegetated open water" or "percentage of streambank that is unstable," the choices might be:

___ $<1\%$ or none
___ 1 to $<30\%$
___ 30 to $<70\%$
___ $>70\%$

Alternatively, the range of possibilities could include three or six choices instead of four, or thresholds in the mid-range could be set at 20% and 80% rather than 30% and 70%, or a graph could be provided that shows the indicator score that should be assigned for each possible status level of the indicator, as is done for many HGM methods (Smith et al., 2013). Again, a review of scientific literature may reveal the choice of thresholds that appear to correlate best with the function or functions being predicted (as it does here), and if no information is available, the range of possibilities might just be distributed evenly. Also, note that while one set of thresholds may relate well to a function's response in a 1 ha wetland, the same percentages of open water estimated in a 0.01 ha wetland or across an entire 1000 ha wetland may be nearly meaningless for estimating some functions. Similarly, "vegetated width" is one appropriate indicator of a headwater wetland's capacity to influence temperature in receiving waters, but much less so if the wetland is located along a very large river or estuary where its contribution is likely to be overshadowed by the much larger volume of water being exchanged in those systems. Finally, smaller stream systems may naturally have different thresholds than larger streams. For instance, logs in some smaller streams may provide less benefit than logs in larger streams. These situations should be thoroughly discussed by the team developing and testing the RAM. If an indicator is being calibrated numerically to a variety of streams across a region, the field work may also direct attention to these situations and suggest a need for different thresholds based on a particular stratifying factor, for example, stream size, elevation, ecoregion.

Temporal variation is another important consideration, especially in tidal wetlands and for streams in regions with drastic changes in precipitation and temperature. Depending on the indicator and the functions it is intended to predict, it may be advisable to have separate questions for wet season and dry season conditions, or daily low and high tide, monthly low and high tide, or drought and no-drought conditions (although the term "drought" will need to be defined). Because RAM users will visit a site for only one day, guidance should be provided on how conditions at other times might be inferred or determined, for example, by using landowner interviews, historical aerial imagery, local weather station data, or tide elevation tables.

Data forms and/or user manuals should also contain definitions of terms that are most likely to be interpreted differently by different users. A few examples are fen, bog, open water, aquatic vegetation, herbaceous vegetation, invasive plants, downed wood, streambank instability, and overbank flooding.

In deference to the time constraints of RAM users, it is generally best to use as few indicators as scientifically justifiable to estimate a particular wetland or stream attribute. At the same time, some intentional redundancy among indicators can be a good thing, provided the data for the indicators are being averaged rather than summed. Some redundancy can be beneficial because RAMs with too few variables and/or with too few choices of condition offered for each variable often generate results that are insensitive, meaning that nearly all wetlands or streams may be shown to have similar scores or ratings (Kusler, 1992, 2003). Use of indicators that correlate highly among themselves, perhaps representing the same ecological process from various angles, can improve a RAM's precision, especially when those indicators are ones that are based on estimation rather than measurement. For RAMs, achieving adequate repeatability (precision) of indicator estimates among users is often equally or more important that making accurate predictions of a wetland or stream function or other attribute. Thus, the usual modeling paradigm of "the best model is one that is most parsimonious in explaining a phenomenon" need not automatically apply to RAMs. Moreover, when estimation rather than measurement is the operation mode as it is in most RAMs, the difference in time required to assess (say) 20 indicators is usually only marginally less than the time required to assess 30, relative to the total time spent traveling to a site, walking around, delineating its boundary, and later processing the data.

Some RAMs, such as the HGM series (Smith et al., 2013) and Washington's Rating System (see Chapter 4.3.10), require users to first determine a wetland's class (e.g., depressional or slope, forested or emergent, or a more detailed classification such as the 16 wetland types defined by the NC Wetland Assessment Method, Chapter 4.3.1). Only then can users assess the indicators, and only the ones deemed applicable to that class. Other RAMs apply the same set of indicators to all wetlands regardless of class. If the goal is to have a RAM that is applicable to all wetland classes in a state, province, or region, then the first strategy potentially results in a large number of data forms (one per class) with considerable overlap in the indicators they use. It also assumes that typical RAM users are able to classify wetlands correctly using that classification, and it assumes that whichever wetland classification is used is the most functionally relevant entry point for the assessment, whereas climate, soils, or other factors may in some cases be more influential drivers of particular functions. However, implementing the alternative strategy (ignoring a wetland's class) can confuse and frustrate RAM users because the single data form will contain questions (indicators) that are clearly irrelevant in a particular context for estimating wetland functions or other attributes. For example, most bogs contain little or no persistent surface water so asking the percentage of the water occupied by submergent aquatic vegetation in such a wetland would confuse RAM users. Also, the indicators that are appropriate for some types of wetlands (e.g., tidal marshes) are vastly different from the indicators appropriate for other wetland types.

A possible solution to this dilemma (whether to create a data form for each wetland or stream class, or only one data form covering all classes) is to sequence the data form questions such that if a particular question (indicator) is answered a certain way, the user is instructed to skip a specified sequence of the questions that follow because they would be irrelevant to that described situation. As explained in Chapter 3.4, the model that then uses the question responses to predict a specified function should be scripted so that the skipped questions are automatically omitted from the score calculations for the function or other attribute, rather than being counted as zeros. In the example above, if a user indicated no submergent aquatic vegetation was present that response would unfairly reduce the bog's function score.

When interpreting RAM results, it's often helpful to keep in mind to which of the following four categories a particular indicator belongs:

1. *Onsite modifiable*. These indicators are features that may be either natural or human-associated and are relatively practical to manage. Examples are water depth, flood frequency and duration, amount of large woody debris, and presence of invasive species. More important than the simple presence of these are their rates of formation and resupply, but those factors often are more difficult to estimate and control.
2. *Onsite intrinsic*. These are natural features that occur within the site and are not easily changed or managed. Examples are soil type and groundwater inflow rates. They are poor candidates for manipulation when the goal is to enhance a particular wetland or stream function.

3. *Offsite modifiable*. These are human or natural features whose ability to be manipulated (to benefit a particular function) depends largely on property boundaries, water rights, local regulations, and cooperation among landowners. Examples are watershed land use, stream flow in tributaries, lake levels, and buffer zone conditions.
4. *Offsite intrinsic*. These are natural features such as a wetland or stream's topographic setting (catchment size, elevation), stream flow duration, and regional climate that in most cases cannot be manipulated. Still, they must be addressed in a RAM because of their sometimes-pivotal influence on wetland or stream functions.

The more that offsite and intrinsic indicators are used in a RAM, the less sensitive the RAM scores are likely to be in depicting functional changes in response to management actions such as invasive species control and operation of water-control structures. It is nonetheless important to include those offsite and intrinsic indicators because to exclude them would ignore scientific evidence of their influence on many functions.

Once the functions or other endpoints of the RAM are specified, their indicators selected, and data forms created containing at least one question for each indicator, the next step is to try the data forms on a diverse series of sites. At this stage, no models or rules have yet been determined for rolling up the indicator question responses into one or more scores. Therefore, the accuracy of the RAM is not being verified. The purposes of this trial are simply to determine if:

- The questions on the data form are interpreted correctly by persons other than their author.
- The choices listed for each question cover all conditions likely to occur in the region's wetlands or streams.
- The existing spatial data sources that are being required to answer some of the questions live up to expectations for coverage, spatial resolution, and accuracy.
- The order of the questions on the data form should be changed in ways that would allow more context-specific skipping of questions that are inapplicable in those contexts, thus reducing the time required to fill out the data form.

Following the completion of the edits to the data forms, there are at least two options for the next step of the RAM project:

Option 1. Develop or modify rules (models) for combining the indicator question responses into one or more scores. If time allows, organize 1-day workshops of subject experts to critique each scoring model; see Chapter 3.8 for details. Then visit and assess a series of sites in varying conditions to determine if the scores rank those sites in approximately the same order as several subject experts who also visit the sites and independently rank them based on their knowledge of biogeochemical and ecological principles, that is, a verification trial or "beta test;" see Chapter 3.8. Or, actually measure particular functions or integrity in a handful of sites using more intensive non-RAM procedures, that is, a validation trial. Use the results from that comparison with expert judgments (verification trial) or intensive measurements (validation trial) to adjust the RAM's indicators and models, then publish a final version *without calibrating the scores to a systematic sample of reference sites*.

Option 2. Delay the development or modification of rules (models) for combining the question responses into one or more scores. Instead, *collect calibration data* by systematically selecting a large number of reference sites (see Chapter 3.5), then visiting and collecting data from those sites. After the data forms have been filled out for all those sites, initiate work on the scoring models. When those models are done, use them to process the calibration data. Either finish the RAM project here and publish the final version, or (as in Option 1) verify the RAM by comparing a ranking of a subset of the sites based on their RAM scores with a ranking of those sites done by subject experts, or with a ranking based on applying more intensive non-RAM procedures to directly measure specific functions or other attributes of that subset of the sites.

Option 1 is normally preferred but Option 2 may be necessary if there are scheduling issues, for example, at the time project funding is awarded the end of the growing season is nearing so calibration data must be collected soon or progress may be delayed for another four months or more. Implementing the *verification trial* will depend on whether an expert can be recruited for each function or other attribute the RAM intends to assess, for example, nitrate removal, sediment dynamics, fish, amphibians, birds, plants, pollinators. Implementing the *validation trial* will depend on whether funds and time are sufficient to actually measure functions or the integrity of enough sites to allow for a statistically valid comparison with their RAM scores.

REFERENCES

Adamus, P.R., 1992. Data sources and evaluation methods for addressing wetland issues. In: Statewide Wetlands Strategies. World Wildlife Fund and Island Press, Washington, DC, pp. 171–224.

Adamus, P.R. 2016. Wetland Ecosystem Services Protocol (WESP) Version 2.0. Available from: people.oregonstate.edu/~adamusp/WESP.

Adamus, P.R., Brandt, K., 1990. Impacts on Quality of Inland Wetlands of the United States: A Survey of Indicators, Techniques, and Applications of Community Level Biomonitoring Data. EPA/600/3-90/073 (NTIS PB 113 837/AS), USEPA Environmental Research Lab, Corvallis, OR. http://www.epa.gov/owow/wetlands/wqual/introweb.html.

Adamus, P.R., Danielson, T.J., Gonyaw, A., 2001. Indicators for Monitoring Biological Integrity of Inland Freshwater Wetlands: A Survey of North American Technical Literature (1990–2000). Office of Water, U.S. Environmental Protection Agency, Washington, DC. EPA843-R-01, http://www.epa.gov/owow/wetlands/bawwg/monindicators.pdf.

Bartoldus, C.C., 1999. A Comprehensive Review of Wetland Assessment Procedures: A Guide for Wetland Practitioners. Environmental Concern, St. Michaels, MD, USA.

Fennessy, M.S., Jacobs, A.D., Kentula, M.E., 2007. An evaluation of rapid methods for assessing the ecological condition of wetlands. Wetlands 27 (3), 543–560.

Hatfield, C.A., Mokos, J.T., Hartman, J.M., 2004. Development of Wetland Quality and Function Assessment Tools and Demonstration. New Jersey Department of Environmental Protection, Trenton, NJ.

Homan, K., Adamus, P.R., 2016. WESPAK-SE (Wetlands) and NATAK-SE (Nearshore) Module. Southeast Alaska GIS Library, University of Alaska Southeast, Juneau, AK.http://seakgis.alaska.edu/flex/wetlands/.

Hruby, T., 2004. Washington State Wetland Rating System for Western Washington. Washington State Department of Ecology, Olympia, WA.

Kusler, J. (Ed.), 1992. State Perspectives on Wetland Classification (Categorization) for Regulatory Purposes. Association of State Wetland Managers, Windham, ME. ISBN: ASWM-SPWC-11.

Kusler, J., 2003. Reconciling Wetland Assessment Techniques. Institute for Wetland Science and Public Policy, Association of State Wetland Managers, Berne, NY.

North Carolina Stream Functional Assessment Team, 2015. N.C. Stream Assessment Method (NC SAM) User Manual. Version 2.1, Raleigh, NC. Available from: http://www.ncaep.org/resources/Documents/NCSAM/NC%20SAM%20User%20Manual%20v2.1.pdf. Accessed 6 April 2017.

North Carolina Wetland Functional Assessment Team, 2016. N.C. Stream Assessment Method (NC SAM) User Manual. Version 5.0, Raleigh, NC. Available from: http://www.ncaep.org/resources/Documents/NCWAM/NC%20WAM%20User%20Manual%20v5.pdf. Accessed 6 April 2017.

Rempel, M., Adamus, P., Kagan, J., 2015. Oregon Explorer—Oregon Rapid Wetland Assessment Protocol (ORWAP) Map Viewer: An Internet Tool for ORWAP Wetland Assessment Support and Data Archiving. Oregon State University Library and Institute for Natural Resources, Oregon State University, Corvallis, OR. Available from: http://tools.oregonexplorer.info/oe_map_viewer_2_0/Viewer.html?Viewer=orwap.

Rooney, R.C., Bayley, S.E., Creed, I.F., Wilson, M.J., 2012. The accuracy of land cover-based wetland assessments is influenced by landscape extent. Landsc. Ecol. 27 (9), 1321–1335.

Rosen, B.H., Adamus, P., Lal, H., 1995. A conceptual model for the assessment of depressional wetlands in the Prairie Pothole Region. Wetl. Ecol. Manag. 3 (4), 195–208.

Smith, R.D., Noble, C.V., Berkowitz, J.F., 2013. Hydrogeomorphic (HGM) Approach to Assessing Wetland Functions: Guidelines for Developing Guidebooks (Version 2). ERDC/EL-TR-13-11, Environmental Lab, Engineer Research and Development Center, Vicksburg, MS.

Somerville, D.E., 2010. Stream Assessment and Mitigation Protocols: A Review of Commonalities and Differences. Prepared for the U.S. Environmental Protection Agency, Office of Wetlands, Oceans, and Watersheds, Washington, DC. (Contract No. GS-00F-0032M). Document No. EPA 843-S-12-003.

FURTHER READING

Blocksom, K.A., 2003. A performance comparison of metric scoring protocols for a multi-metric index for Mid-Atlantic Highlands streams. Environ. Manag. 31 (5), 670–682.

Hruby, T., 2001. Testing the basic assumption of the hydrogeomorphic approach to assessing wetland functions. Environ. Manag. 27 (5), 749–761.

Trimble, S.W., 1974. Man-Induced Soil Erosion on the Southern Piedmont. Soil Conservation Society of America, Ankeny, IA.

Chapter 3.4

Creating Models for Rolling Up Indicator Data Into Scores

Paul Adamus
Oregon State University, Corvallis, OR, United States

For an assessed wetland or stream reach, some published RAMs include decision rules or equations that allow a user to roll up the collected data into a single overall score. Other RAMs do not generate a single overall score but provide scores for a series of individual wetland or stream functions, values, services, and/or other attributes. Still other RAMs are only descriptive. That is, they standardize the collection of data but then leave it to the RAM user to decide on an overall score or scores for those individual attributes.

Existing RAMs have used two general strategies for rolling up the data from their indicators:

Strategy A. A purely mathematical strategy. Data for each indicator is converted to a standardized score (usually in the range of 0–1) that represents its actual condition relative to a desired condition. The desired condition may be based on theory, scientific literature, and/or a reference data set. The RAM then combines the indicator scores using a mathematical equation, often a weighted average or sum. The weights are meant to reflect the relationship of each indicator to a specified function or other attribute. The weights are preassigned by the RAM project team or author and cannot be altered by users. For example:

$$\text{Function score} = [(4 \times \text{Indicator A}) + (2 \times \text{Indicator B}) + \text{Indicator C}]/7$$

An asset of this strategy is that virtually any potential combination of indicators can be represented and weighted, allowing the equation to be a faithful representation of scientific knowledge limited only by the skill and imagination of the equation writer. A drawback is the fact that some RAM users have difficulty understanding math as logic and so are unable to explain how the equation results in a particular score. Some RAMs attempt to alleviate this perceived lack of transparency by providing a detailed narrative description of what each equation is doing. It is important that RAM authors resist the urge to simply average or sum a series of indicators, if done only as a matter of convenience. For example, consider four indicators of fish habitat:

$$\text{WQ} = \text{water quality},\ S = \text{habitat structure},\ P = \text{aquatic productivity},\ A = \text{physical access}$$

Simply averaging or adding these indicators when scoring a site defies logic because without fish being able to access the site, any consideration of the other indicators is moot. In this case, access is known to be *controlling*. When that is the case, the equation might be configured like this:

$$\text{Fish Score} = A \times \text{Average}\,(\text{WQ}, S, P)$$

so that when access is lacking (0) the other indicators are multiplied by 0 and the Fish Score goes to 0. In other words, access (A) serves as a switching indicator.

Strategy B. Narrative logic or a Boolean strategy. The RAM authors specify one or more combinations of indicator conditions or narrative criteria that define a desired status of a function or other attribute. This may take the form of "if-then/true-false" statements. For example:

IF Indicator A $>$ 20% and
Indicator B = riverine and
Indicator C $<$ 35 ppm THEN
Function score = 1, ELSE:
Function score = 0.

An upside of this strategy is that the criteria used in the roll up are very transparent. A downside is that usually there are dozens of potential combinations that could result in a score (in this case) of 0 or 1. Listing all those combinations and

Wetland and Stream Rapid Assessments. https://doi.org/10.1016/B978-0-12-805091-0.00031-1

Boolean logic process — an example

Riverine swamp forest: hydrology surface storage and retention

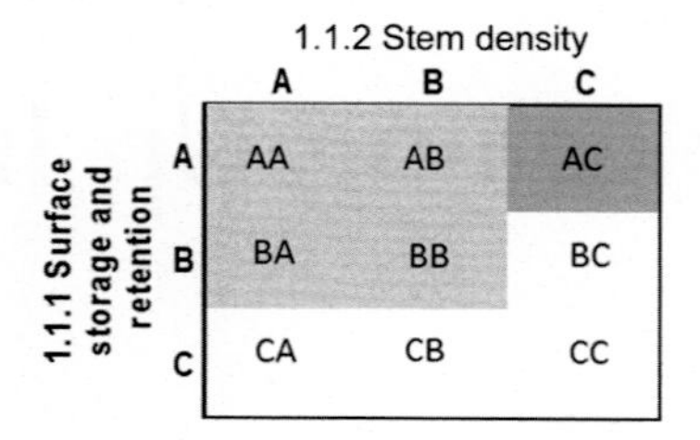

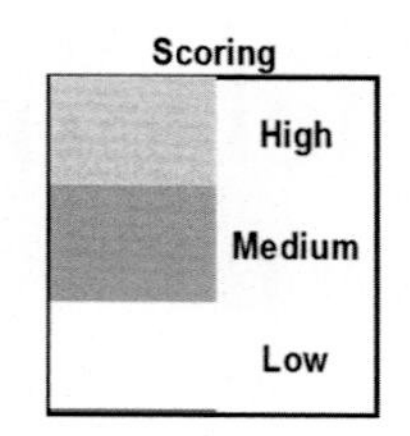

1.1.1 Surface storage and retention — assessment area condition metric

Consider both increase and decrease in hydrology. Compare to *reference wetland* if applicable (see User Manual). If not applicable, then rate based on evidence of disturbance to hydrology. Refer to the North Carolina Scope and Effect Guide for the zone of influence of ditches in hydric soils.

☐A Surface water storage capacity and duration not altered

☐B Surface water storage capacity or duration altered, but not substantially (typically, not sufficient to change vegetation).

☐C Surface water storage capacity or duration substantially altered (typically, alteration sufficient to result in vegetation change) (examples: diversion, man-made berms, intensive ditching, beaver dams, stream incision).

1.1.2 Stem density — assessment area condition metric

☐A Forested or contains dense shrubs

☐B Medium density shrubs, *canopy* may be present but disturbed

☐C Only herbaceous species or bare ground

FIG. 3.4.1 Example of Boolean logic process used by NC wetland assessment method (Chapter 4.3.1).

asking the user to scroll through and select the applicable one quickly becomes impractical. For instance, as a practical note, any combination beyond a 3-by-3 Boolean logic box (which has nine combinations) becomes very difficult to conceptualize (see Fig. 3.4.1 for an example). Usually 3-by-2 or 2-by-2 combinations are more practical.

Some of the more recent wetland RAMs use a hybrid of these strategies. They might include a series of equations like this:

IF WetlandClass = A, THEN Average(X, Z), ELSE:

IF WetlandClass = B, THEN Maximum (G,T,Z), ELSE:

$(3 \times X + T)/4$

Such equations can easily be scripted in a spreadsheet, allowing indicator data from dozens of sites to be processed almost instantaneously. The most challenging part is to first identify which indicators are the most likely drivers of particular functions in different wetland or stream types and situations and then assign them weights which are realistic in most situations. A useful approach is to conceptually break down a function into its component processes, write an equation to create a subindex representing each process, and finally, combine the subindex scores into a score for the function using a weighted average or other approach. Each relationship between an indicator and the function or other attribute it is estimating as well as relationships among individual indicators used in the estimation should be determined to be controlling, limiting, fully compensatory, partially compensatory, or cumulative. This determination should then be used as the basis for choosing the form of the scoring models, as demonstrated and explained in more detail by USFWS (1980) and Smith et al. (2013).

When considering relationships between a function and its indicators, the factors that contribute to the relationship can be categorized in three ways: (1) unknown influencers, (2) known influencers that are difficult to measure within a reasonable span of time, and (3) influencers that can be estimated visually during a single visit and/or from existing spatial data. Most RAMs provide an incomplete estimate of wetland or stream functions because they are only able to incorporate the last of these. Also, some of the indicators may be correlates of wetland functions rather than actual influencers. For example, changes in water levels are correlated with changes in nutrient cycling, but it is the difficult-to-measure changes in sediment oxygen and pH that induce the changes in nutrient cycling, not the water level changes themselves (which happen to correlate loosely with changes in oxygen and pH).

REFERENCES

Smith, R.D., Noble, C.V., Berkowitz, J.F., 2013. Hydrogeomorphic (HGM) Approach to Assessing Wetland Functions: Guidelines for Developing Guidebooks (Version 2). ERDC/EL-TR-13-11, Environmental Lab, Engineer Research and Development Center, Vicksburg, MS.

U.S. Fish and Wildlife Service, 1980. Habitat Evaluation Procedures (HEP) manual (102ESM). U.S. Fish and Wildlife Service. Department of the Interior, Washington, DC.

Chapter 3.5

Collecting Calibration Data

Paul Adamus

Oregon State University, Corvallis, OR, United States

Chapter Outline

Method calibration is a procedure that, once completed, provides a numerical context for interpreting the scores or ratings from any individual site. In the process of developing or adapting a RAM, several questions related to the RAM's calibration should be addressed:

- Should the RAM be calibrated?
- How many reference sites are needed to calibrate a RAM?
- By what criteria should reference sites be selected?
- How can sites meeting the selection criteria be found?
- Who should collect the reference data?
- How might the collected field data be analyzed?

SHOULD THE RAM BE CALIBRATED?

Under some circumstances, a RAM may contain data forms and scoring models but the scores that it produces for a particular assessed site are not transformed any further. Thus, the scores resulting from an assessment of that site can only be compared with scores from sites that had previously been assessed. It is unknown whether those are accurate representations of the range of conditions existing in a larger targeted population of sites. If time and funding do not allow a RAM to be calibrated to a valid sample of sites within the targeted region, the RAM is likely to be useful only for examining changes over time within a single site or a few sites. Thus, calibration of a new or adapted RAM to regional conditions is strongly advised (Nadeau et al., 2015).

HOW MANY REFERENCE SITES ARE NEEDED TO CALIBRATE A RAM?

The answer to this question will depend on the degree of variation among the region's wetlands or streams with regard to the target variable (i.e., more sites needed if a key variable varies greatly among sites), statistical characteristics of the RAM's models, and the degree of certainty (of capturing the full range of variation among wetlands) desired by the RAM's author or sponsor. The number of sites used for RAM calibration will often need to be a balance of statistical validity and budget. With regard to statistical validity, if some data for a variable (e.g., function score) have already been obtained from a number of sites, statistical simulation can be used to estimate the smallest number of sites that would need to be assessed in order for any of a RAM's measurements to "level off," that is, the point beyond which increasing the number of measured sites will not expand the captured range of variation much further.

To perform such a simulation,[1] it is necessary to randomly draw, from whatever data have been collected so far, a series of subsamples (each subsample representing one site) of progressively larger sizes. For each series, determine the

1. Because such a simulation is impossibly time consuming if done manually, an Excel macro that processes within minutes any data pasted into it has been created for this task and can be downloaded (for free) at people.oregonstate.edu/~adamusp/WESP.

Wetland and Stream Rapid Assessments. https://doi.org/10.1016/B978-0-12-805091-0.00032-3

TABLE 3.5.1 Leveling Off of a Variable's Range Derived From Multiple Sites (n = 150) Indicates the Minimum Number of Sites (Samples) Needed to Capture Variation in That Variable Among Wetland Sites

Number of Random Samples Drawn	Median of the Ranges From 100 Runs	Standard Deviation of the Ranges From 100 Runs
10	8.17	1.39
20	8.95	0.86
30	9.80	0.71
40	9.54	0.61
50	9.80	0.60
60	9.80	0.45
70	**10.00**	**0.46**
80	10.00	0.40
90	10.00	0.39
100	10.00	0.35
110	10.00	0.31
120	10.00	0.26
130	10.00	0.32
140	10.00	0.20
150	10.00	0.26

For this variable, that number was approximately 70.

maximum and minimum values measured for the variable of interest (e.g., function score), and then calculate the range among all sites in the series (maximum-minimum). For example, if data were collected from 150 sites, the range of values might first be determined among 10 of those sites drawn randomly. The draws should be repeated 100 times randomly (because each time a somewhat different 10 sites might result from the draw), each time calculating the range. Calculate the mean, median, and standard deviation across the 100 resulting ranges. Next, 20 sites could be drawn randomly from the 150, then the range of values among those 20 sites could be determined, repeated 100 times, and summary statistics calculated as before. Repeat this for subsamples consisting of 10, 20, 30, ..., n sites and visually examine the output statistics across all the draws to see at what point the key variable's range appears to level off, as shown in Table 3.5.1. As noted above, this will vary depending on the key variable being measured, and that is demonstrated by the results in Table 3.5.2. Also, note that it can be assumed that the sample range is within ± two standard deviations of the mean for 95% of samples taken at the corresponding sample size. Also understand that this provides only a rough estimate of the optimal sample size, and tends toward underestimation. That is because the true range of values for the entire population will remain unknown unless every wetland or stream reach in the population is assessed.

BY WHAT CRITERIA SHOULD REFERENCE SITES BE SELECTED?

Reference sites can serve several purposes: (1) places to try out early drafts of the RAM to help ensure its data forms are clear and relevant, (2) places that comprise all or part of the pool of sites used to calibrate the RAM, and (3) places where accuracy of the calibrated RAM is tested by comparison with long-term measurements of site condition or functions. Sites chosen for testing of the draft RAM as well as for its calibration and verification should span the full range of expected conditions, including climate, soils, wetland or stream type, and levels of disturbance in the study area.

With regard to this last factor, some RAMs use only human alterations to define reference condition while others include natural disturbances as well. Use of the disturbance gradient assumes that disturbances are a major driver of a site's functions, and that the relative levels of alteration (by various stressors, both historical and present) can be estimated accurately for the entire population of candidate reference sites. The scores from RAMs that use the human disturbance gradient as

TABLE 3.5.2 Minimum Number of Sites Required to Capture Score Variation Among New Brunswick (Canada) Wetlands Using WESP-AC (See Chapter 4.3.2)

Wetland Function Assessed by WESP-AC	Minimum Number of Sites to Level Off the Function Score (± 10)
Water storage and delay	70
Stream flow support	40
Water cooling	30
Sediment retention and stabilization	30
Phosphorus retention	30
Nitrate removal and retention	40
Carbon sequestration	70
Organic nutrient export	30
Anadromous fish habitat	40
Resident fish habitat	70
Aquatic invertebrate habitat	30
Amphibian and turtle habitat	70
Waterbird feeding habitat	70
Waterbird nesting habitat	50
Songbird, raptor, and mammal habitat	60
Pollinator habitat	50
Native plant habitat	50

This suggests that although 170 sites were assessed, adequate results could have been achieved for all assessed functions by visiting and assessing only 70 (approximately).

their reference would be expected to be more sensitive to showing slight changes from human impacts. When calibrating or verifying a RAM within a region, using only the reference sites that appear to be in the best condition is not appropriate because that will fail to capture the full range of conditions which future users of the RAM may encounter. Thus, selecting relatively disturbed sites as well as relatively unaltered ones is important to determine whether the RAM adequately can distinguish the effects of natural disturbances (such as saltwater intrusion from major storms) from man-induced disturbances (such as increased rates of runoff from impervious surfaces or invasive species).

However, wetland and stream functions are determined by far more than just disturbance level. Dozens of other attributes are important predictors, yet accounting for more than a few of those when handpicking reference sites becomes daunting. Thus, as an alternative to using disturbance gradients alone to define reference conditions, one could use existing spatial data available for an entire region to characterize the range of characteristics (including probable disturbance levels) of the region's wetlands or streams. This strategy attempts to account for both human and natural factors that could influence a site's functions.

This strategy was implemented for selecting calibration sites for RAMs in Alaska, Alberta, New Brunswick, and Nova Scotia (see Chapter 4.3.2). Sites were chosen in each of these places by first creating a geodatabase. That required intersecting wetland maps covering the entire area with all existing spatial data relevant to wetland functions: wetland class, ecoregion, soils, geology, climate, elevation, slope, aspect, hydrography, and others. Thus, every one of the thousands of wetlands in the geodatabase was characterized by dozens of attributes. The enormous geodatabase was then analyzed to determine how those data clustered statistically (a "cluster" is defined as a subset of sites that had similar attributes). At least one site from each of the resulting statistical clusters was chosen as a reference site and then visited and assessed using the RAM. After all sites had been visited, assessed, and scored, the range of raw scores was used as the basis for the calibration of all sites the RAM will assess in the future.

The Oregon Department of State Lands supported a variation of this strategy. Rudimentary models of wetland functions were constructed using only the attributes contained in the geodatabase (Paroulek, 2015). The geodatabase was then processed to yield a relative rating of high, moderate, or low for each major function of each of more than 500,000 wetlands. With the assistance of statisticians from the USEPA, a spatially distributed stratified random sample of 200 wetlands was drawn from the geodatabase, with attention to including sites that were at low and high ends of the spectrum for each function as suggested by the analyzed geodatabase. The sites were subsequently visited and scored using the field-based RAM, which assesses many indicators relevant to wetland functions but not in the geodatabase.

If lack of existing region-wide spatial data, time, and funds do not permit construction of a geodatabase beforehand, and knowledge of the types of wetlands and disturbance levels is limited, then a third option is to select the reference sites of varying conditions in a statistically random manner. This avoids assumptions about the accuracy of existing spatial data (or best professional judgment) to identify sites that collectively represent a range of functions or ecological integrity, and it may yield a credible sample if the population of wetlands or streams from which the sample is drawn is fairly small. However, if a purely random sample is used, in most instances the number of sites that are feasible to visit will be far too few to capture the true range of variation.

HOW CAN SITES MEETING THE SELECTION CRITERIA BE FOUND?

Regardless of how carefully the reference sites are selected, it will not be possible to visit many due to physical constraints (e.g., walking to them from the nearest road requires an hour or more) and/or lack of access permission from landowners. Thus, for the sake of convenient access, reference sites are often located on public lands or lands managed by conservation groups or land trusts that have arranged access permission. Although this might seem to introduce a bias toward sites unaltered by humans, even the sites on public conservation lands need to be examined, where possible, for lingering effects of past disturbances (for instance, relict ditching into and out of wetlands or the presence of historical dams). Similarly for streams, sites in protected areas are convenient but the condition of the entire watershed that drains to the stream is usually a critical factor in influencing stream condition. At the very least, the road proximity and ownership characteristics of the visited versus unvisited sites should be compared in a table that documents the RAM's use of reference sites. An additional advantage of publicly accessible sites is that they can be visited multiple times and used for training exercises.

WHO SHOULD COLLECT THE REFERENCE DATA?

To minimize variation in how the RAM's indicator questions are interpreted, it is preferred that the same person or team assesses all reference sites. If they did not author the RAM, they should undergo intensive training followed by an exam before they begin collecting the reference data.

HOW MIGHT THE COLLECTED FIELD DATA BE ANALYZED?

RAMs typically use the reference site with the highest score and the one with the lowest score to anchor the ends of the actual (as opposed to theoretical) scoring range. All scores from sites assessed in the future are compared to that range to see where the site falls within the range—or in some cases it may fall outside it. The simplest way to compare the raw score of an assessment to this range is to "normalize" the raw score by subtracting the score of the lowest-scoring site from it, then dividing by the scoring range (maximum-minimum). Iterative graphing procedures can also be used, as they are with many HGM methods (Smith et al., 2013); however, they are often quite time intensive.

Some methods normalize each site's final score or scores, some normalize only the scores of individual indicators, and some do both. Most often, scores of the indicators are normalized to a theoretically possible range while the scores of functions or other attributes are normalized to those found among the reference sites.

The status of the reference sites will inevitably change over time as a result of human and/or natural events, and this potentially weakens any future comparisons with the original reference data. Thus, if funding allows, a subset of the reference sites should be monitored with regard to changes in conditions that could affect their scores. For wetlands, connectivity to stream networks and groundwater as well as surface water depth and duration are strong influencers of many functions. For streams, gauge data as well as data collected to develop an understanding of the regularity of overbank flooding are important. Input and output of various chemical constituents (sediment, nutrients, metals, or bacteria) as well as water volume can suggest removal rates for these constituents. Wetlands, plants, aquatic insects, amphibians, birds, and small mammals can be measured by various techniques. Streams, aquatic macrobenthos, amphibians, and fish are commonly surveyed and algal communities may also reflect levels of functions. Monitoring for at least three years is advised to capture the natural year-to-year variation in rainfall or snowfall and temperature regimes.

After the RAM has been released to the public, users sometimes offer to contribute data from their assessments to the calibration database. This poses several problems. One, if new sites are continually being added, the normalized scores from prior assessments will no longer be valid. Each time a site is added, the normalization process must be repeated for all prior scores, and this also creates a "moving target." Second, if reference sites in the calibration database had been carefully selected to comprise a statistically valid sample of the population, adding new sites will disrupt that balance and render the calibration database less valid. For example, assessments contributed by RAM users tend to be disproportionately focused on urban or exurban areas because that is where development is mostly initiated. Third, the quality of assessments added to the calibration database by RAM users—even trained and experienced users—will remain unknown, thus potentially compromising the database's validity. A better use of these assessments may be to construct a regional database of evaluated sites that can be accessed via the Internet before conducting site evaluations in a particular study area. This database may also be useful for training purposes.

REFERENCES

Nadeau, T.-L., Leibowitz, S.G., Wigington Jr., P.J., Ebersole, J.L., Fritz, K.M., Coulombe, R.A., Comeleo, R.L., Blocksom, K.A., 2015. Validation of rapid assessment methods to determine streamflow duration classes in the Pacific Northwest, USA. Environ. Manag. 56 (1), 31–53.

Paroulek, M., 2015. Level 1 Landscape Assessment: Geospatial Analysis of Wetland Condition, Function, and Ecosystem Services in Oregon. Masters of Environmental Management Project Report, Portland State University, Portland, OR.

Smith, R.D., Noble, C.V., Berkowitz, J.F., 2013. Hydrogeomorphic (HGM) Approach to Assessing Wetland Functions: Guidelines for developing guidebooks (Version 2). ERDC/EL-TR-13-11, Environmental Lab, Engineer Research, and Development Center, Vicksburg, MS.

Chapter 3.6

Converting Scores to Ratings

Paul Adamus
Oregon State University, Corvallis, OR, United States

In regulatory applications, scores that represent the levels of a wetland's functions relative to those of other wetlands are sometimes combined mathematically with wetland area (acreage) to yield a number that some agencies have used as a credit or debit when deciding an appropriate amount of replacement. Also, scores intended to represent a stream or wetland's status are sometimes used to evaluate a restored site's progress in approaching a desired state. However, many citizens relate much better to simple words than to numerical values. Therefore, authors of some RAMs have, in addition to or in place of scores, represented the outputs of their RAM as ratings, using terms such as "low-moderate-high," "poor-fair-good-excellent," or "A-B-C-D." Technically, because RAM scores are nearly always relative rather than absolute, it would be more correct to use terms such as lower and higher rather than simply low and high, and some RAMs have done so.

How might scores be converted to ratings? Two procedures used most often for converting scores to ratings are *percentiles* and *break-point analysis*. With the percentile (or quantile) approach, some entity decides explicitly, for a particular wetland or stream function, which score ranges should be associated with each rating term or category. For example, that person or persons may decide that scores in the 90th percentile or above qualify for a rating of "A," those in the 40th through 89th get a rating of "B," etc. It is normally not the responsibility of RAM authors to select the particular thresholds, but rather to simply provide analyses of different options. They can do so by reporting, for example, that if the 90th percentile is chosen for defining "category A," then only three out of 60 sites (5%) might be in that category, but if the 80th percentile was chosen instead, 12 out of 60 sites (20%) would be in that category. Informed by such statistics and their graphical presentation, citizens or their representatives or agency administrators could then decide where to draw the lines. Inevitably, any percentiles that are chosen will have an element of subjectivity, so the process is often quite contentious. The approach could be reversed in a mathematical sense, wherein some entity decides how many or what percentage of the sites should belong to "category A," then iteratively adjusts the percentile criteria until that number or percentage is achieved among the calibration sites. Those percentiles would ultimately be applied to sites that were not part of the original RAM calibration, an approach that has been used in Alberta, Canada.

From a science perspective, the ideal separation points between categories might be based on well-researched ecological response thresholds, for example, stream invertebrate richness changes dramatically when impervious surface in the watershed exceeds, say, 7%. However, unbiased data to support the choice of such thresholds are often hard to come by.

In contrast to the percentile approach, the break-point approach is much more "hands off" because it implements statistical algorithms that choose the thresholds between categories. In some cases, that algorithm not only chooses the thresholds but indicates the number of categories there should be to achieve the best fit to the data (the data being the scores of the calibration wetland or stream). In other cases, the RAM author is asked to follow policies that specify there being three, four, or some other number of rating categories. Oregon's RAM uses three categories, Alberta's uses four, and North Carolina's stream and wetland RAMs both use the same three general categories. If the number of categories has been specified a priori in such a manner, then the Jenks optimization (Jenks, 1967) or other statistical procedures for breakpoint analysis (e.g., Hauer et al., 2013) can be implemented to iteratively define the best numeric values to separate the categories. "Best" is understood to mean groupings of scores that statistically are most alike within categories while being most dissimilar among categories. Software is available to do the Jenks optimization and provides statistics that quantify the homogeneity of the resulting categories. Additional ways to convert scores to ratings are described, for example, by Hughes et al. (1998), Hallett et al. (2012), and Hallett (2014).

Categorical ratings are typically intended to have legal implications, as is the case with decisions about whether to require a replacement ratio of 2-to-1 or 4-to-1 when a wetland is developed. Those decisions can have major economic implications so, understandably, the regulated public sometimes wishes to know how close a wetland or stream, now

Wetland and Stream Rapid Assessments. https://doi.org/10.1016/B978-0-12-805091-0.00033-5

categorized as Good (for example), is to being categorized as Excellent. One approach to address this concern is to specify the approximate statistical confidence intervals around the scores based on prior tests of the RAM's repeatability (see Chapter 3.8). For example, if 0.82 has been established as the threshold separating categories A and B wetlands with regard to sediment retention, and prior repeatability testing had determined that, on the average, the scores for the sediment retention function have a confidence interval of + or − 0.02, then if a particular site scores a 0.81, the actual category could be either A (0.81 + 0.02 = 0.83) or B (0.81 − 0.02 = 0.79) after taking into account the confidence interval. Separate administrative interpretation could be accorded to permittees whose site's score falls within the confidence interval, as is illustrated here.

REFERENCES

Hallett, C.S., 2014. Quantile-based grading improves the effectiveness of a multimetric index as a tool for communicating estuarine condition. Ecol. Indic. 39, 84–87.

Hallett, C.S., Valesini, F.J., Clarke, K.R., 2012. A method for selecting health index metrics in the absence of independent measures of ecological condition. Ecol. Indic. 19, 240–242.

Hauer, C., Unfer, G., Holzmann, H., Schmutz, S., Habersack, H., 2013. The impact of discharge change on physical instream habitats and its response to river morphology. Clim. Chang. 116 (3-4), 827–850.

Hughes, R.M., Kaufmann, P.R., Herlihy, A.T., Kincaid, T.M., Reynolds, L., Larsen, D.P., 1998. A process for developing and evaluating indices of fish assemblage integrity. Can. J. Fish. Aquat. Sci. 55 (7), 1618–1631.

Jenks, G.F., 1967. The data model concept in statistical mapping. Int. Yearb. Cartogr. 7, 186–190.

Chapter 3.7

Converting to an Overall Site Score

Paul Adamus
Oregon State University, Corvallis, OR, United States

RAMs that provide a score and/or rating for each of several functions are often necessary for a variety of purposes, such as supporting a goal of no net loss of stream or wetland function. However, sometimes administrators, politicians, and the public ask for a "bottom line," that is, is a particular site functional or not, *overall*? This requires that the scores for very disparate functions be combined in some manner, either by the RAM itself according to rules specified by its authors or through a separate process involving decision makers and/or the public. The same is true of RAMs that require that the condition of diverse types of potential stressors be combined into a single stressor index: there is no generally accepted, statistically sound, science-based procedure for doing so.

Almost without exception, RAMs combine the scores for different stressors or functions–sometimes as many as 20 for a single site—by *averaging or summing*. That is because these operators are viewed as the most practical and least complicated to explain. However, taking an average or sum assumes that (a) all functions will potentially have an equal mathematical effect on the overall site score, (b) scores of all functions follow a similar statistical distribution, and (c) all functions are of equal importance to a community and/or to other ecosystems.

The assumption that all functions potentially have an equal mathematical effect on the overall site score is seldom true. First, if a RAM is scoring seven functions that describe wildlife habitat but only one function that describes the capacity of a site to purify runoff, then a mathematical bias for wildlife habitat may exist implicitly, especially if "summing" is the chosen scoring rule. Second, some functions tend to be correlated inversely. For example, wetlands that score high for sediment retention tend to score low for organic nutrient export because a primary indicator that determines the score of both—an outlet connection to a stream—acts in opposite ways (wetlands that lack outlets are more likely to retain sediment but are less likely to export nutrients). If they are summed or averaged, their statistical tendency to "cancel out" may be considered undesirable. Similarly, if the scores of several functions tend to be positively correlated, those functions may implicitly exert undue mathematical influence if scores of all a site's functions are summed or averaged.

Also, the assumption that scores of all functions follow a similar statistical distribution is almost never true. Although the model for each function should be capable of generating a score at both ends of a scoring range (e.g., 0–10), in practice sites that have *every* condition that in theory is optimal for a given function may never exist in nature, and conversely, sites that have *no* conditions supporting a given function may not exist. Thus, for some functions the scores from even a large number of calibration sites may fail to completely span the theoretical 0–10 scale whereas for other functions the number of calibration sites may be sufficient to fill out the range. This mathematical skew can be mitigated, though only partially, by mathematically spreading out every function's range to fill the 0–10 scale by subtracting the score of the lowest scoring of the previously assessed calibration sites from a given site's score and then dividing by the scoring range (maximum-minimum) for that function as was determined from the data from all the calibration sites. However, the distribution may still be skewed if some function models unwittingly tend to produce higher scores than other function models only because criteria in the high-scoring function's model are more easily met (perhaps because they are easier to observe during a single site visit) than those of the consistently lower-scoring function model.

But clearly the most important reason why simple averaging or summing is problematic is that only rarely can all functions be considered equally important to a community or region. For example, in a watershed or region that is strongly dependent on commercial and sport fishing, river floodplains with high scores for fish habitat might be considered most important whereas in another watershed or region where many wells have been contaminated by excessive nitrate, the capacity of river floodplains to remove nitrate might be considered more important. Combining the scores of the two functions into one average ignores this important factor. One way to partly address this might be to adopt a policy stating that whichever function scores highest at a given site, that function or its rating will be used to represent the site generally. Another option is to calculate a weighted average wherein a weight is first assigned to each function that represents its

Wetland and Stream Rapid Assessments. https://doi.org/10.1016/B978-0-12-805091-0.00034-7

general importance in the region compared with the other functions. That determination is properly the responsibility of decision makers and the public rather than the RAM author or other scientists alone. A third option is to base a site's overall score or rating on some prespecified count of functions that were rated high, moderate, or low (or similar categories) for that site. Review of the distribution of function ratings among all calibration sites can serve as a foundation for deciding how many ratings of high (or combinations of highs and moderates) should be considered exceptional for a site and thus result in a site being categorized as highly functional overall. Again, such determinations should be the responsibility of decision makers and the public rather than the RAM author or other scientists alone. A fourth option (Christopher Hay-Jahans, University of Alaska Juneau, personal communication, January 20, 2017) is to specify that (say) 5 points be assigned to any of 15 functions scoring above a chosen percentile (e.g., 90th). Compare this to the number of functions that are at or above a smaller percentile (e.g., 66th) and retain whichever number is larger. Then assign (say) one-eighth of a point for each function larger than a still smaller percentile (e.g., 50th) and add that to the number from the preceding step. Finally, convert the resulting numbers from all sites to percentiles.

None of the above roll-up approaches completely eliminates statistical bias in the overall (rolled-up) score or rating, but some approaches may tend to offset implicit statistical bias more than others. In any case, what is apparent is that the least defensible approach for calculating an overall site score is to blindly average or sum the scores for functions or other attributes.

Another consideration is whether the same set of functions should be assessed for all sites when calculating a site's overall score. This issue arises when a significant proportion of a region's sites totally lack a particular function. For example, should the anadromous fish habitat function be scored (and the score set to 0) for wetlands with no connection to stream networks if most of a region's wetlands have that condition and thus cannot physically provide that function at any level? If sites with anadromous fish access have scores but those that don't lack scores, the overall scores of the sites from those two groups cannot be compared fairly because sites with different numbers of functions are being compared. But if all landlocked sites are assigned a score of 0 for that function, this dilutes and reduces the overall score of those sites systematically. A parallel situation occurs with some indicators in the function models. For example, when considering submerged wood that normally provides shelter to fish, should the absence of submerged wood in a small prairie stream reach—where wood characteristically is absent—be counted as a 0 or should wood not even be assessed, forcing the stream reach's score to be based on other indicators? Although developing different models for different stream or wetland types could alleviate this issue, budget and time limitations often restrict the capacity to do that.

For RAMs whose ratings for a site's *values* (benefits) are wisely kept independent of the scores or ratings for its functions, there is no generally accepted rule for how best to combine the function and value ratings as may be necessary if an overall site rating is required. However, consideration might be given to using something similar to the following decision matrix, applied to each function, where the rating in all-caps is the synthesis of each pairwise combination of function rating (row) and value (column) rating:

	Value Rating		
Function Rating	**Low**	**Moderate**	**High**
Low	LOW	LOW	MODERATE
Moderate	MODERATE	MODERATE	HIGH
High	MODERATE	HIGH	HIGH

Another approach in addressing the site-specific values of functions is to conduct a landscape-level analysis. For example, a GIS analysis of existing spatial data in Oregon (Paroulek, 2015) very roughly estimated the levels of functions of every wetland in all of Oregon's watersheds; this was used to identify which functions were most lacking in each watershed. Oregon's field-level RAM (ORWAP) then used that information such that if a particular wetland performs a given function at a high level according to ORWAP while existing in a watershed that is deficient in that function, a higher value score is assigned to that wetland.

In Alberta, wetland policies specify that one human value—the historical loss rate of wetlands in a given watershed—be used in conjunction with that province's wetland assessment tool to help calculate the amount of replacement required when a wetland is being altered. More replacement is required in regions with greater historical wetland losses (see Chapter 2.2.12). Few other RAMs require consideration of local wetland trends. That is largely due to a lack of fine-resolution trend data.

Some RAMs, such as Alberta's, intentionally do not score or rate the *values* associated with each function of a particular wetland. The provincial government acknowledges that for many functions, local governments and watershed groups understand best the context of a wetland and thus how to rank the needs and values in a community. Thus, provincial government encourages communities to apply their own value weights or rankings to the function scores that the RAM provides when determining the degree to which the level of protection accorded to some wetlands might be greater than what the provincial government requires.

Also, for purposes of site scoring, some RAMs consider the occurrence at a site of a threatened or endangered species to be a function, others treat it as a value (because the belief that rarity increases importance is a judgment placed by humans), and others do not include it as part of the RAM but rather as a separate "red flag" which in some cases is allowed to trump the ratings of all other functions and values. Those who develop a RAM should consider agency mandates when deciding whether to include species rarity as part of a RAM or other critical factor or to treat it as an outside-the-box consideration.

REFERENCE

Paroulek, M., 2015. Level 1 Landscape Assessment: Geospatial Analysis of Wetland Condition, Function, and Ecosystem Services in Oregon. Masters of Environmental Management Project Report, Portland State University, Portland, OR.

Chapter 3.8

Analyzing Repeatability and Sensitivity

Paul Adamus
Oregon State University, Corvallis, OR, United States

A large part of a RAM's credibility hinges on its *repeatability*. The goal for repeatability should be clearly stated and preferably quantified. At a minimum, the goal should be that, among its trained users, use of the RAM produces more consistency in the rankings of a series of wetlands than are obtained simply by the "best professional judgments" of several consultants or untrained users who also assess those sites independently of each other and without knowledge of the RAM results. If a RAM is intended to be used mostly by individuals working alone, then while testing the RAM, the testers should work without communicating or else the repeatability estimates are likely to be favorably biased.

Expecting that every indicator used in a RAM meet a repeatability goal would be a daunting requirement. More often, repeatability testing focuses on just the final score or rating achieving a prespecified goal. For example, a goal might be that, for at least 8 out of 10 functions being assessed, the variation in scores among the RAM testers visiting 10 test sites ("the noise") should be less than the variation among mean or modal scores of the test sites ("the signal", Elkum and Shoukri, 2008), or that scores of 10 independent users be within 5% of the group's mean or modal score for those sites. Table 1 presents guidance for doing repeatability tests.

Many factors—other than the RAM being tested—potentially contribute to lower repeatability among independent users. These include:

- Differences in user training and experience with the RAM (Herlihy et al., 2009).
- Not remembering key details from the manual or other supporting documents.
- Not consulting particular web sites or databases if required by the RAM.
- Delimiting their assessment area differently (see Chapter 3.2).
- Differences in the parts of the site that were walked or otherwise observed.
- Differences in the time of year (or phase of the tidal cycle) when the test sites were visited.
- Not noticing the spatial context of a question, for example, is it asking about the percent of the whole site or just the percent of the vegetated part of the site?
- Differences in visual interpretation.
- Differences in user skills, for example, at identifying plants, collecting aquatic insects, or texturing soils.
- Differences in prior knowledge of the particular site and region.
- Differences in willingness and ability to make informed judgments in the absence of actual measurements.
- Data entry errors.
- Fatigue.

If analysis of testing data shows that a RAM's outputs are not as repeatable as desired, then ideally the RAM should be modified and retested. Analysis and review of the data from the first round of testing should focus on identifying which particular indicators varied the most among testers at each site. However, the data for indicators (or their resultant scores when combined with other indicators) that have high among-tester variation at one site may have low among-tester variation at another site, depending on site characteristics. Indicators that perform best in discriminating differences among sites while having high consistency among users could be given more weight in subsequent revisions of the RAM. In some cases, high variation may be simply due to one tester or test site that was extremely different. Standard statistical tests can be used to objectively identify such "outliers" in the data set. If the "outlier" is a person who is much more experienced with the protocol and/or more knowledgeable about the test site, then removing that person's data may reduce the accuracy of the test results even while improving repeatability, and thus may not be advisable. Any decision of whether to reanalyze the test results with outliers removed has important implications.

Wetland and Stream Rapid Assessments. https://doi.org/10.1016/B978-0-12-805091-0.00035-9

TABLE 1 Suggested Guidance for Repeatability Testing of a Stream or Wetland RAM

1. Before testing the RAM, recruit as many testers as possible. Six is probably a minimum, but there are no statistical procedures available to specify exactly what this number should be.
2. All testers should be trained in the use of the RAM before they test it. The trainers should include the RAM's author or at least someone whose experience using it is extensive and diverse.
3. Testers should have all the skills required by the RAM, for example, the ability to identify nearly all common wetland plants in a region. This might be verified by a formal exam before field testing begins.
4. At the conclusion of the training, testers should take an exam to measure understanding of key terms, and retake it if necessary until an acceptable level of achievement is attained.
5. During the testing, all testers should apply the RAM to all the test sites. They should travel as a group to those sites, but once there, individuals should assess each site independently. Ideally, the same number of testers should assess each site.
6. The test sites should be relatively near each other (to minimize group travel time) yet be sites whose scores collectively are anticipated to span the RAM's full potential numeric range, for example, sites that potentially score in the low, mid, and high range for specified functions. That is necessary because the repeatability of a RAM may depend on the type of site to which it is applied.
7. Before field testing begins, the exact boundaries of the area to be assessed should be flagged (not just drawn on a map) and explained to all testers, unless there is a desire to include that frequently enormous variation in the repeatability estimates.
8. During the field testing time, no communication should be allowed among the testers or between the testers and any other person. However, testers may consult printed material such as the RAM guidance documents and field guides.
9. Before leaving each test site, each field tester should submit all the standard data forms, giving them to a single individual who will enter the data into a spreadsheet.

From Adamus, P., 2010. Strategies and Procedures for Testing Rapid Protocols Used to Assess Ecosystem Services. Report to the Willamette Partnership, Portland, OR. Available from: http://people.oregonstate.edu/~adamusp/WESP/.

If retesting is implemented, an iterative process should be used when modifying the protocol prior to the formal retesting. For example, this could involve trying out various rewordings of indicators, changing their weights or the ways they are combined, and then reviewing the results and modifying again as necessary (e.g., Blocksom, 2003). However, be aware that changes made to improve clarity of indicator questions to address the confusion of one set of users often will result in decreased clarity for another set of users. Also, changes should not be made solely to improve repeatability if they simultaneously are likely to have a major adverse effect on the scientific integrity of the resulting function score or rating. Examples of stream or wetland RAM repeatability studies that describe statistical procedures include Kaufmann et al. (1999), Whigham et al. (1999), Roper et al. (2002), Coles-Ritchie et al. (2004), Herlihy et al. (2009), and Adamus (2016).

A RAM's *sensitivity* also influences its credibility. A RAM that places 90% of wetlands or streams in a category of "Excellent" is of limited utility to decision makers who must prioritize individual sites, even if such a designation corresponds to the criteria that were used to define that status. Similarly, if a RAM's output scores can theoretically range from 0 to 10 but when the RAM is applied to a large number of diverse wetlands the scores are found to range only from 4 to 7, then that RAM may not be considered to be sufficiently sensitive. Depending on its intended purposes and uses, a goal for a RAM's sensitivity might be quite demanding (e.g., scores from 100 random sites should fit a statistically normal distribution) or fairly relaxed (e.g., scores from 100 sites only need to come close to filling completely the theoretical range of 0–10).

A RAM's actual scoring range for a particular function may be narrow if conditions of some of its indicators rarely or never occur together in the natural world. Output scores of many RAMs are unlikely to have the same statistical distribution for every function they score. That is, scores generated by models for some functions will skew high (e.g., more than half the time they will be above 8 on the 0–10 scale) whereas the scores generated by other models will skew low (e.g., more than half the time they may be 0). Because these are scoring models, not mechanistic equations, the high or low skew could be due to either (a) one function tending to be inherently less effective than another among the calibration sites generally, or (b) the relative conservativeness (or lack thereof) of the particular indicators and their criteria as used in a model for a particular function or other attribute.

Another component of evaluating a RAM's sensitivity involves identifying which component indicator or indicators tend, in general, to have the greatest influence on the final score for each function or other attribute. If a scoring model is relatively simple, a preliminary guess can be ventured by examining its structure. For example, in this equation:

$$F = 3 \times A + 2 \times B + C$$

indicator B mathematically could contribute 1/3 of the score 2/(3+2+1). If scoring equations are more complex or if missing values are allowed (e.g., some indicators are automatically dropped from an equation if certain other indicators are present), then one approach involves determining the statistical correlations of the scores of calibration sites versus the scores of their component indicators. The indicators with the lowest positive or negative Spearman rank correlation (rho value) may be assumed to have affected the function scores of *those* particular calibration sites the least.

In the strictest sense, applying correlation or regression analyses to ordinal data (scores), especially data that are not normally distributed, violates the assumptions of these statistical tests. However, the intent here is to just provide a preliminary evaluation of each indicator's relative influence. While it may be tempting to honor the principle of parsimony by deleting from subsequent versions of the equation those indicators that had the least overall influence as judged by their low rho, it is unwise to do so automatically. That is because a particular condition of an indicator may be absent in 99 of 100 sites assessed (thus resulting in a low correlation) but if present in the one other site, it could be ecologically pivotal at that site. Thus, the scientific as well as the statistical importance of each indicator should be taken into account.

If each indicator is comprised of categorical choices, an alternative approach to assess RAM sensitivity that does not involve examining correlations is to perform a Monte Carlo simulation, wherein all possible combinations of each of the function's component indicators are run and the resulting function scores noted. However, if there are more than a few indicators, each with several condition choices, this will demand considerable computer time. Moreover, not all potential combinations of conditions are likely to occur in nature, so the results may be unrealistic or will require considerable editing. Other analytical approaches are available for selecting an optimal number of indicators for a particular RAM model (e.g., Rooney and Bayley, 2010), but can require significant time and computer expertise as well as a large and statistically representative calibration data set. Ultimately, RAM authors need to consider whether reducing the length of their RAM's data forms by eliminating a few questions, thus reducing field time by perhaps only a few seconds or minutes, is really worth the effort.

REFERENCES

Adamus, P.R., 2016. Technical Supplement: Procedures Used to Refine, Re-Calibrate, and Test ORWAP Version 3.1. http://www.oregon.gov/dsl/WW/Documents/ORWAP_Technical_Supplement_v3_1.pdf.

Blocksom, K.A., 2003. A performance comparison of metric scoring protocols for a multi-metric index for Mid-Atlantic Highlands streams. Environ. Manag. 31 (5), 670–682. https://doi.org/10.1007/s00267-002-2949-3.

Coles-Ritchie, M.C., Henderson, R.C., Archer, E.K., Kennedy, C., Kershner, J.L., 2004. Repeatability of Riparian Vegetation Sampling Protocols: How Useful are These Techniques for Broad-Scale, Long-Term Monitoring? Gen. Tech. Rep. RMRS-GTR-138, U.S. Forest Service, Rocky Mountain Research Station, Fort Collins, CO.

Elkum, N., Shoukri, M.M., 2008. Signal-to-noise ratio (SNR) as a measure of reproducibility: design, estimation, and application. Health Serv. Outcome Res. 8, 119–133.

Herlihy, A.T., Sifneos, J., Bason, C., Jacobs, A., Kentula, M.E., Fennessy, M.S., 2009. An approach for evaluating the repeatability of rapid wetland assessment protocols: the effects of training and experience. Environ. Manag. 44, 369–377. https://doi.org/10.1007/s00267-009-9316-6.

Kaufmann, P.R., Levine, P., Robison, E.G., Seeliger, C., Peck, D.V., 1999. Quantifying Physical Habitat in Wadeable Streams. EPA/620/R-99/003, U.S. Environmental Protection Agency, Washington, DC, p. 102.

Rooney, R.C., Bayley, S.E., 2010. Quantifying a stress gradient: an objective approach to variable selection, standardization and weighting in ecosystem assessment. Ecol. Indic. 10 (6), 1174–1183. https://doi.org/10.1016/j.ecolind.2010.04.001.

Roper, B.B., Kershner, J.L., Archer, E.K., Henderson, R., Bouwes, N., 2002. An evaluation of physical habitat attributes used to monitor streams. J. Am. Water Resour. Assoc. 38, 1637–1646. https://doi.org/10.1111/j.1752-1688.2002.tb04370.x.

Whigham, D.F., Lee, L.C., Brinson, M.M., Rheinhardt, R.D., Rains, M.C., Mason, J.A., Kahn, H., Ruhlman, M.B., Nutter, W.L., 1999. Hydrogeomorphic (HGM) assessment—a test of user consistency. Wetlands 19, 560–569. https://doi.org/10.1007/BF03161693.

FURTHER READING

Adamus, P. 2010. Strategies and Procedures for Testing Rapid Protocols Used to Assess Ecosystem Services. Report to the Willamette Partnership, Portland, OR. Available from: http://people.oregonstate.edu/~adamusp/WESP/.

Chapter 3.9

Analyzing a RAM's Accuracy

Paul Adamus
Oregon State University, Corvallis, OR, United States

As explained earlier, a RAM's accuracy can be verified or validated. *Verification* involves comparing the outputs with the opinions of experts whereas *validation* involves comparing those outputs with detailed long-term measurements of functions or condition.

Verification can be conducted partly by organizing peer review of the RAM's indicators and models and their assumptions. If that is done, it is critical that the RAM's structure be explained to reviewers beforehand, preferably in a workshop setting, and that all reviewers accept the need for generating scores for wetland functions or condition using observational methods and visiting a site only once during a growing season. Regardless of their scientific qualifications, some academic researchers who might otherwise provide thorough peer review are unaccustomed and uncomfortable with the basic concept of RAMs. This needs to be considered when selecting peer reviewers.

As one example of a verification attempt, a series of workshops was organized in Alaska to peer review preliminary drafts of the WESPAK-SE tool (Adamus, 2015). Experts were asked to consider: (1) whether the indicators of each function and the choices of conditions provided for each indicator question were the best ones for that RAM, (2) the relationship between each function and each of its indicators, for example, whether nitrogen removal is more likely to increase or decrease in response to decreasing pH and under what circumstances that may involve other indicators (there were 28 such assumptions to consider and debate just for the nitrate removal function, and there were 22 other functions that similarly needed consideration and debate), and (3) whether the scores for the indicators were combined in a way that would likely result in a function score that best reflects relative levels of that function. Workshops were organized by general themes: Hydrology and Water Quality, Aquatic Habitat, Terrestrial Habitat. Local experts covering each theme were recruited to participate and the RAM's author led the discussion, allocating 3–5 min for the experts to discuss each indicator-function relationship and find consensus with regard to whether the RAM had characterized it properly. However, the uncompensated experts were unable to find time to review the RAM and its assumptions before or after the workshop, and the time allotted for each workshop (6 h) was far too little to accommodate meaningful discussion of many indicator-function relationships.[1]

Alternatively, subject experts could simply be driven to a series of diverse sites and asked to rank them, with or without discussion, based on their "gut feel" about the relative capacities of those sites to perform the function(s) the experts are most knowledgeable about. Using Spearman rank correlation procedures (available in nearly all statistical software), those numeric rankings can then be compared with rankings based on the function scores of the sites as generated by the draft version of the RAM. A correlation coefficient (rho) of greater than 0.6 and a P-value of less than 0.05 together indicate fairly good agreement between the RAM scores and rankings of the same sites by the experts. If categorical ratings (e.g., poor-fair-good-excellent) rather than scores are being compared for concordance with professional opinion, "confusion matrices" can be constructed—in this example, a 4-by-4 matrix with each cell containing a count of the number of sites where the experts and the RAM-assigned ratings agreed or disagreed (e.g., 10 sites that were rated poor by the experts were independently rated as good by the RAM). The degrees of concordance can be quantified and tested for statistical significance by the kappa index or others (Cohen, 1960; Fielding and Bell, 1997).

The state of Washington employed such an approach and used it to iteratively revise their RAM, in a sense calibrating it to expert opinion as applied to a series of sites they had chosen subjectively. However, in some states, provinces, and nations, it can be a challenge to find, recruit, and schedule appropriate experts on all functions covered by a RAM.

1. That RAM assesses 23 functions, values, and other attributes of the region's nontidal wetlands collectively using 118 indicators. Depending on the function, value, or other attribute being assessed, between 17 and 47 are used in any single equation.

Wetland and Stream Rapid Assessments. https://doi.org/10.1016/B978-0-12-805091-0.00036-0

And it should not be assumed that the opinions of experts would be a truer prediction of how a site is actually performing than are the outputs from a well-constructed model that consistently and systematically accounts for many factors.

The NC Stream Functional Assessment Team conducted a verification analysis of the NC Stream Assessment Method (Appendix B, North Carolina Stream Functional Assessment Team, 2015) to examine statistically whether the results of the method correctly produced the results intended by its authors.[2] The relationship of the rating for each of three stream attributes (hydrology, water quality, and habitat) to percent impervious surface in the watershed was examined statistically using Fischer's Exact Test and Spearman's ρ correlation coefficient. As expected, the ratings correlated positively with the percent impervious surface (lower water quality ratings with more impervious surface). In addition, the GENMOD (SAS) procedure showed an association between the overall rating and the three attributes (hydrology, water quality, and habitat) that are combined to yield the overall rating, suggesting that all three attributes are good predictors of the overall rating. Finally, the Bayesian information criterion and Akaike's information criterion were used to determine if any of the three general function ratings could be eliminated as predictors of the overall rating. The results suggested that the best model for the overall rating must contain all three general functions, as intended. Thus, NC SAM was considered to be validated and met its intended purpose. Long-term monitoring data are now being used to further validate the method.

Finally, the number and type of sites that should be visited to comprise a statistically meaningful verification is unknown and ultimately may have to be a compromise between statistical rigor and the available time and budget. To reduce time and cost, experts could simply be shown a photograph (aerial and ground) of a series of wetlands in a workshop setting and be asked to rank those with regard to the functions they know best. Of course, the validity of the results will depend partly on what is shown in each image.

A higher standard to attain involves actually *measuring* what the RAM purports to estimate. For example, if the RAM uses indicators of habitat structure to rank several wetlands according to their suitability as amphibian habitat, measures of accuracy might be obtained by conducting surveys of amphibians at different seasons in the same wetlands, analyzing the data to estimate amphibian richness, productivity, and survival in each wetland, and using the results to rank the sites. As before, validity is judged by applying the Spearman rank correlation procedure to the scores or the kappa statistic (Cohen, 1960) to the ratings.

Conducting a validation can be quite costly and time consuming for functions whose measurement requires laboratory analysis of samples, frequent visits to each site, or expensive equipment (e.g., many water quality parameters). Moreover, monitoring validation sites for at least three years is advisable to capture annual variations in weather. As was true for the calibration and verification processes, the number of sites used for RAM validation will often need to be a balance of statistical validity and budget. Also, it is not unusual for questions to be raised regarding whether the variables being measured are very complete representations of the intended function.

REFERENCES

Adamus, P.R. 2015. Wetland Ecosystem Services Protocol for Southeast Alaska (WESPAK-SE), Version 2.0. Report and Software for the Southeast Alaska Land Trust and US Fish and Wildlife Service, Juneau, AK. Available from: http://southeastalaskalandtrust.org/wetland-mitigation-sponsor/wespak-se/.

Cohen, J., 1960. A coefficient of agreement for nominal scales. Educ. Psychol. Meas. 20 (1), 37–46.

Fielding, A.H., Bell, J.F., 1997. A review of methods for the assessment of prediction errors in conservation presence/absence models. Environ. Conserv. 24 (01), 38–49.

North Carolina Stream Functional Assessment Team, 2015. N.C. Stream Assessment Method (NC SAM) User Manual. Version 2.1, Raleigh, NC. Available from: http://www.ncaep.org/resources/Documents/NCSAM/NC%20SAM%20User%20Manual%20v2.1.pdf. Accessed 6 April 2017.

2. See Chapter 4.3.1 of this book for a description of the NC Stream Assessment Method (NC SAM).

Chapter 3.10

General Issues in Statistical Analysis of RAMs

Breda Munoz*, **Rick Savage**[†] **and Virginia Baker**[‡]

**RTI International, Durham, NC, United States,* [†]*Carolina Wetland Association, Raleigh, NC, United States,* [‡]*Division of Water Resources, North Carolina Department of Water Quality, Raleigh, NC, United States*

Chapter Outline

INTRODUCTION

Decisions regarding management, regulation, and protection of water bodies often rely on results from RAMs. Those results may include indices, metrics, and/or scores. Because the results sometimes have implications for policies and regulations, their credibility can be strengthened by correctly applying appropriate statistical tests at several points during the development or modification of a RAM, and in some cases using those results to refine the RAM further.

This part of Section 3 describes statistical methods available for calibrating and then validating indexes, scores, and metrics. The described statistical methods are standard in the calibration and validation of models in many research areas. The goal of this chapter is to guide the identification and implementation of appropriate statistical methods as applied to RAM data for streams and wetlands as well as to aid in the interpretation of results. These methods are also appropriate for calibrating and validating indexes (e.g., Biotic Index) and components of longer-term assessments.

At the end of this chapter, an example demonstrates the calibration and validation process used by one categorical RAM (NC Wetland Assessment Method (NC WAM)—see Chapter 4.3.1). This could provide an analysis template for use in calibrating and validating similar RAMs. The example also includes programming code and annotated output.

Calibration

As described in Chapter 3.5, calibration refers to mathematical adjustments (e.g., scaling, standardization, or normalization) of the results from an initial version of a RAM so that subsequent versions encompass the range of scores or ratings expected within a particular region and/or wetland or stream type. The adjustments are typically made after first determining the minimum and maximum scores across a large number of sites. As described in Chapter 3.5, those sites may be selected using a statistical sampling procedure such as stratified sampling or cluster sampling, and/or may be hand-picked based on their having prior measurements or simply a belief that they collectively represent a wide range conditions of one or more environmental gradients of interest.

If the RAM's author intends the RAM's minimum and maximum scores (determined by applying the RAM to a large number of sites) to correspond to the minimum and maximum ranks of sites selected based either on more-intensive measurements or on expert opinion reflecting a desired range of a particular condition or function, then the RAM's range of scores may need to be adjusted mathematically to approximately match that range. The difference between the ranking determined from applying the RAM to all the calibration sites and the ranking based on the expected maximum and minimum is used to inform the need to adjust the internal scoring system of the RAM. This process continues until the

Wetland and Stream Rapid Assessments. https://doi.org/10.1016/B978-0-12-805091-0.00037-2

differences between them are zero or deemed negligible. For example, Wardrop et al. (2007) used a floristic quality index (one intensive measure of a wetland's condition) to calibrate a RAM for the Juniata watershed in south-central Pennsylvania.

Suppose that RAM authors prefer that output from a habitat assessment index be expressed as whole integers across a possible range of 0–3, where 3 denotes a site that the RAM authors consider to be in the best possible condition and 0 represents a site that the authors consider to be in the worst possible condition. Although the range of the potential scores was initially set to use whole integers between 0 and 3, if comparison of site rankings based on applying the RAM differs significantly from the ranking based on more intensive measures of habitat condition at those same sites, then it may be advisable to use fractional increments (e.g., the use of 0.5 or 0.25 increments) to represent a certain percentage loss of habitat condition. In contrast, if the RAM fails to discriminate well between low and high conditions, it may be necessary to try expanding the scale to a broader range such as 0–10.

Deviation measures, also known as goodness-of-fit measures, quantify the discrepancy between scores from the assessed calibration sites ("observed") and scores from the same sites based on expert opinion or more intensive measurements ("predicted"). Among the standard deviation measures for interval (numeric) variables are:

$$\text{Mean squared error (MSE)} = \frac{1}{n}\sum_{i=1}^{n}(x_i - y_i)^2$$

$$\text{Mean absolute percentage error (MAPE)} = \frac{1}{n}\sum_{i=1}^{n}\frac{|x_i - y_i|}{x_i} \times 100$$

$$\text{Mean absolute error (MAE)} = \frac{1}{n}\sum_{i=1}^{n}|x_i - y_i|$$

$$\text{Correlation coefficient} : R^2 = \frac{1}{n-1}\frac{\sum_{i=1}^{n}(x_i - \overline{x})(y_i - \overline{y})}{\sum_{i=1}^{n}(x_i - \overline{x})^2\sum_{i=1}^{n}(y_i - \overline{y})^2}$$

Theil's U_2 statistic (Theil, 1966):

$$U_2 = \frac{\sqrt{\frac{1}{n}\sum_{i=1}^{n}(x_i - y_i)^2}}{\sqrt{\frac{1}{n}\sum_{i=1}^{n}x_i^2}}$$

where

x_i is the predicted score for the ith reference site,
y_i is the observed score for the ith reference site,
$\overline{x}$ is the average predicted scores for the reference sites,
$\overline{y}$ is the average observed scores for the reference sites, and
n is the number of reference sites.

The MSE is the average of the squared deviations between the observed and predicted scores, and it is a standard measure of goodness of fit in modeling approaches. The closer the MSE to zero, the better the agreement between expert opinion or intensive measurements and the scores from the RAM. The MAPE and MAE are the average absolute percent errors by which the predicted score differs from the observed scores, and the average of absolute differences by which an observed value differs from the predicted value, respectively. Values close to 0 denote better predictions. The R^2 measures the linear association between the predicted and observed values. An R^2 closer to 1 denotes larger agreement between the RAM's predicted and observed output values, and a value closer to 0 denotes maximum entropy. The closer U_2 is to zero, the better the predictions and no calibration is needed.

Differences between the observed and predicted values should be examined to determine where the discrepancies are occurring. This examination will help to determine if the RAM is underestimating or overestimating a specific subset of sites, or if it is consistently predicting the lower, higher, or middle values. In addition, Spearman rank correlation can be applied to scores from a series of calibration sites to test whether the difference between the predicted and observed scores is statistically significant (e.g., Adamus, 1995).

In the case of nominal (e.g., land use type) and ordinal scales (e.g., low, medium, high), a two-way table between observed and predicted scores paired with interrater agreement analysis can provide insight into whether the scoring system needs calibration. Interrater agreement tests quantify the concordance of ratings of two different sources (Cohen, 1960). If concordance is weak, that indicates disagreement between the expert opinion (or intensive measurements) and the calculated RAM scores. All cases where disagreement occurs should be examined carefully to determine reasons and, where appropriate, the RAM should then be modified to improve concordance.

For nominal scales, ranks, or indices, Cohen's kappa is appropriate (Cohen, 1960) whereas for ordinal scales the *s* statistic based on the weighted kappa (Falotico and Quatto, 2010; Cohen, 1968; Marasini et al., 2016) is appropriate. The interpretation of Cohen's kappa follows a qualitative scale proposed by Landis and Koch (1977): <0.20 denotes poor agreement, 0.2–0.4 fair agreement, 0.4–0.6 moderate agreement, 0.6–0.8 good agreement, and >0.8 very good. A *P*-value for the s-statistic is based on the asymptotic distribution of the statistic (Fleiss, 1971). Critical values of the s-statistic are available through the application of Monte Carlo algorithms that implement the approach developed by Fleiss (1971) and later modified by Maraisini et al. (2014).

Also, statistical models can be used to assess the association between an outcome of interest and one or more parameters. Because a RAM's scores, indices, and/or rankings often result from combining several parameters, statistical modeling approaches can be used to assess the association between these and more-intensive measures. When using modeling approaches, keep in mind the assumptions required in the statistical analysis of any mathematical score or index. Selection of statistical measurements of association or correlation depends on the nature (e.g., continuous, binary, categorical: ordinal or nominal) of the RAM's score, index, and ratings. If the RAM's outputs are numerically continuous, then statistical methods such as regression models should be used to investigate the association between the outputs and field data that are surrogate measurements of the function or condition the RAM is assessing. In the case of ordinal scales (e.g., low, medium, high), multinomial logistic models are the appropriate approach to investigate the association between a RAM's values and the data.

Applying linear discriminant analysis (LDA) to a set of environmental parameters collected in the field can be used to obtain a classification rule for waterbody condition. Site classification can be compared with the classification provided by the RAM. LDA will identify the combination of environmental parameters that better explains specific site condition and function. If the parameters selected by LDA match those that define the different RAM functions, then this suggests that the RAM effectively measures the intended function.

Validation

To validate a RAM in whole or part, several authors (Stein et al., 2009, Wardrop et al., 2007; Sutula et al., 2006; Fennessy et al., 2007) recommend comparing RAM outputs with independent datasets from the same sites. Statistical tests described earlier can be used in the comparison. Desired characteristics of the independent datasets include long-term monitoring data, broad regional coverage, representation of a wide range of conditions across a gradient, and coverage of existing field conditions. The choice of an adequate sample size for the independent dataset is usually determined by cost and time constraints. The seminal work of Peduzzi et al. (1995, 1996) suggests that for each of the RAM's outputs (e.g., scores for different functions), the validation should be based on comparisons from at least 10 sites per model parameter. If this ideal number of sites is impractical for intensive sampling, or a small number of sites is only available, then variable reduction procedures such as factor analysis can be applied to reduce the number of model parameters. Also, variable selection procedures such as random forest (RF) (Breiman, 2001) can be applied to determine variable importance, and/or bootstrapping to reduce the effect of the small sample size. For example, for validating NC WAM, the same 23 sites were used to validate the overall rating as well as all three functions and nine subfunctions that had been derived from 22 metrics, and a RF method was used to reduce the number of model parameters (Dorney et al., 2015).

The considerable cost and time required to collect and analyze validation data is one of the biggest barriers to RAM validation. As a result, when validation is attempted at all, it is usually only for a few of several of the RAM's functions or estimates of condition, or is limited to one or a few wetland or stream types. For example, a portion of the California Rapid Assessment Method was validated using avian diversity, benthic macroinvertebrate indices, and plant community composition, but only for riverine and estuarine wetlands (Stein et al., 2009). Kentucky's Wetland Rapid Assessment was partially validated for forested riverine wetlands using vegetation, bird surveys, and landscape analysis (Polascik et al., 2015). The New England Rapid Assessment Method for coastal tidal wetlands was partially validated using intensive data that included vegetation, soils, and infauna (Wigand et al., 2010).

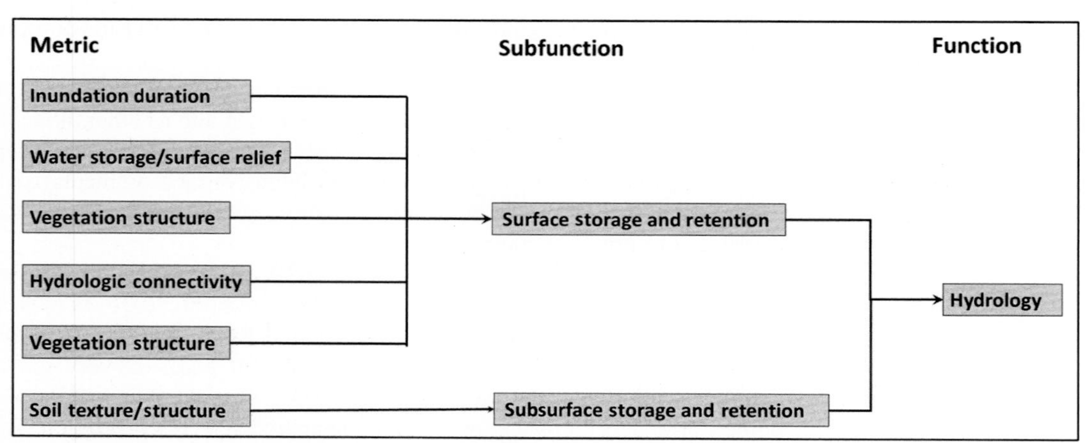

FIG. 3.10.1 NC WAM metric function diagram for hydrology for headwater forest wetlands.

The following case study, pertaining to headwater forest wetlands, describes how a portion of the NC WAM (Dorney et al., 2015 and Chapter 4.3.1) was validated using data from intensive long-term monitoring of headwater wetlands. NC WAM evaluates three wetland functions: hydrology, water quality, and habitat. Data for 22 metrics are summarized into three functions and nine subfunctions (Fig. 3.10.1 from Dorney et al., 2015). The validation data are from 23 sites located across the Piedmont and Coastal Plain of North Carolina where intensive monitoring occurred for up to 6 years (Baker and Savage, 2008). Data collected pertained to water quality, hydrology, soils, amphibians, macroinvertebrates, and plants. Baker and Savage developed several indices of biotic integrity for three biotic areas (amphibians, macroinvertebrates, and plants), and a land development index (LDI) was calculated for each site's watershed at 300 and 50 m buffers. LDI estimates the potential for impacts to wetlands from anthropogenic influences on land cover (Baker and Savage, 2008). Water quality and soil disturbance measures were calculated using 19 water quality parameters, and several parameters extracted from 10 soil samples were collected and analyzed, respectively (Baker and Savage, 2008).

Calibration Method for Ordinal Score

NC WAM functions, subfunctions, and overall rating are all ordinal measures (low, medium, high). In what follows, the process for calibrating an ordinal measure, index, or rank is illustrated with the NC WAM hydrology function.

Because the available data constituted a long-term monitoring dataset collected for validation purposes, no expert opinion data were available for these sites. To illustrate the calibration analysis, surrogate expert opinion parameters were constructed as modifications of available parameters to show discrepancies. Fig. 3.10.2 displays the R-code for performing the calibration analysis of the ordinal measure hydrology (variable name is Hydrology_NC). The expert-based hydrology variable was named obs_hydrology. Code statements following the symbol ">" denote R-commands, statements following the symbol "#" denote comments, and statements below the R-commands constitute output.

A perfect agreement and consequent no-need of calibration will occur when all observations fall in the diagonal of a two-way table. In this simulated exercise, output from the two-way table suggests NC WAM overestimated the hydrology function for 10 sites (nine in cell (medium, high), and one in cell (low, high) of the output). NC WAM also underestimated the hydrology for three sites, one in the cell (high, low) and two in cell (medium, low). The large discrepancies suggest the need to revise the internal scoring system of NC WAM for the hydrology function for this wetland type (headwater forest).

To quantitatively evaluate the agreement between NC WAM's hydrology score and observer hydrology, the R-function **wlin.conc** was fit using the two measures (predicted and observed). A dataset **hydro** was created, and the kappa s-statistic was produced. The nonsignificant result (P-value$=0.14>0.05$) suggests a high level of disagreement, which coincides with the observed discrepancies in the two-way table. This result suggests that the development team needs to calibrate the internal scoring of NC WAM to compensate for overestimation and underestimation for some sites for the hydrology function for this wetland type.

Calibration Method for Continuous Score

The Native Species Evenness Metric (NSEM) was selected to illustrate the example of calibration of a continuous score, index, or metric. Calibration of this metric is a key step in the calibration of the habitat function for headwater forests. The diversity of species in an area depends on both the number of species observed (species richness) and their total numbers, and evenness refers to the relative abundance of species. Evenness is high if all species have similar distribution

```
#load the raters library
>library(raters)

#prepare data for wlin.conc
>low1=ifelse(obs_hydrology=="Low" & Hydrology_NC=="Low",2,0)
>low1=ifelse(obs_hydrology=="Low" & Hydrology_NC!="Low",1,low1)
>low1=ifelse(obs_hydrology!="Low" & Hydrology_NC=="Low",1,low1)

>m1=ifelse(obs_hydrology=="Medium" & Hydrology_NC=="Medium",2,0)
>m1=ifelse(obs_hydrology=="Medium" & Hydrology_NC!="Medium",1,m1)
>m1=ifelse(obs_hydrology!="Medium" & Hydrology_NC=="Medium",1,m1)

>h1=ifelse(obs_hydrology=="High" & Hydrology_NC=="High",2,0)
>h1=ifelse(obs_hydrology=="High" & Hydrology_NC!="High",1,h1)
>h1=ifelse(obs_hydrology!="High" & Hydrology_NC=="High",1,h1)

# Create a dataset for the wlin.conc function
>hydro=data.frame(low1,m1,h1)

# Explore the two-way table of the two variables
>table(obs_hydrology,Hydrology_NC)
  Hydrology_NC
obs_Hydrology High Low Medium
 High        8  1  1
 Low         1  0  0
 Medium      9  2  1

# run the Inter-rater Kappa
>set.seed122345)
>kappa1=wlin.conc(hydro,test = "MC",alpha=0.05,B = 1000)
 S*          min         LCL         UCL         pvalue
 0.21739130 -1.25000000 -0.07608696 0.46195652 0.13800000
```

FIG. 3.10.2 Calibration of ordinal index.

```
# loading libraries
library(Metrics)
library(DescTools)
#calculating the statistics
mse1=mse(obs_NSEM,ncwam_NSEM)
mape1=mape(obs_NSEM,ncwam_NSEM
mae1=mae(obs_NSEM,ncwam_NSEM)
cor1=cor(obs_NSEM,ncwam_NSEM)
tu2=TheilU(obs_NSEM,ncwam_NSEM, type = 2, na.rm = TRUE)
# printing the results
data.frame(MSE=mse1,MAPE=mape1,R2=cor1,TheiU2=tu2)
MSE          MAPE       R2         TheiU2
1 0.2166596 0.08464848 0.9901232 0.07167272
```

FIG. 3.10.3 Example of calibration for continuous variable.

(i.e., similar population density) (Baker and Savage, 2008). The parameter NSEM was calculated as the ratio of the Simpson's Diversity Index and the maximum Simpson's Diversity Index (Baker and Savage, 2008). To create the expert opinion or observed NSEM, the calculated NSEM was altered by adding a random number between ±0.5 to each NSEM measurement.

The R-package **metrics** produces three of the deviation measures described previously (MSE, MAPE, MAE), R^2 is part of the default statistics in R, and the Theil's U_2 is implemented in the R-package **DescTools**. Fig. 3.10.3 displays the R-code needed to calculate these calibration statistics for validating NSEM. The low values of MSE, MAPE, and Theil's U_2 suggest that the predicted NC WAM values are very close to the observed (simulated) values and further calibration is not needed. The high R^2 value suggests almost perfect association between the observed and predicted values.

Validation of Ordinal Metrics

The validation process for an ordinal metric is illustrated with NC WAM's hydrology function. The first step requires a clear understanding of the function to be validated, followed by the identification of the parameters available in the validation dataset that capture or are surrogates for the different dimensions of the specific wetland function (hydrology).

The flow of water in and out of the wetland and the degree of soil saturation or inundation were chosen to represent the hydrology function (Baker and Savage, 2008). A diagram like Fig. 3.10.2 assists in identifying all dimensions of water hydrology and the parameters associated with each dimension. In this case, hydrology is determined by surface and sub-surface storage and retention subfunctions, which are in turn determined by metrics of inundation duration, water storage/surface relief, vegetation structure, hydrologic connectivity, and soil texture structure.

Given the large number of available parameters that measure different aspects of wetland hydrology functions, subfunctions, and metrics, a RF model was first fit to the data to identify the best predictors among all parameters and selected indices, using as outcomes scores for the hydrology function, subfunctions, and metrics. RF was first introduced by Breiman (2001) as a learning method for classification and regression. In RF, tree models are constructed using bootstrap samples of the original data. The RF calculates an estimate of variable importance, which is an estimate of the prediction capability of each predictor in each model in which the predictor was included. Variable importance is calculated as the change in prediction error between the model with and without the predictor while the other predictors are retained in the model (Liaw and Wiener, 2002).

Once the best predictors are determined, an evaluation of the multicollinearity between predictors is recommended prior to finalizing the validation models. Multicollinearity has an adverse effect on the regression coefficient estimates and their variances. In particular, regression coefficients may be too large and have the wrong signs. Also, the variance of the coefficients tends to be much larger for the explanatory variables involved in collinear relationships than for uncorrelated explanatory variables, affecting as a result the significance of the parameters. (Belsley et al., 1980; Mandel, 1982; Hocking, 2003). Variance inflation factor (VIF) is frequently used to report multicollinearity. It indicates how much larger the variance of the coefficient estimate will be for collinear data than for orthogonal data. A VIF <2.5 will indicate no serious multicollinearity issues. In the presence of multicollinearity (VIF >2.5), variable removal or variable reduction methods are recommended.

Fig. 3.10.4 displays the R-code necessary to run the RF model. The more relevant variables according to the RF are presented in Table 3.10.1. A multicollinearity analysis (VIF output) suggested no serious multicollinearity issues.

```
# loading libraries
library(randomForest)
rf1=randomForest(Hydrology_NC ~., data=ncwam, mtry=2, importance=TRUE)
print(rf1)
#assessing multicolinearity
fit <- lm(wd$S_Cu~V_Annual.Perennial.Metric+
   V_Fern.Cover.Metric+
   V_FQAI.Cover.Metric+
   V_FQAI.Species.Count.Metric+
   V_Invasive.Grass.Coverage.Metric+
   V_Moss.Coverage.Metric+
   M_Sensitive_Tolerant_Metric+
   V_Percent.Tolerant.Metric+
   V_Pole.Timber.Density.Metric+
   V_Invasive.Shrub.Coverage.Metric+
   M_Evenness.Metric+
   M_Percent_Crustacea_Metric+
   M_Percent_Predator_Metric+
   A_AQAI+
   V_Average.C.of.C.Metric, data=wd)
#Evaluate Collinearity
library(car)
vif(fit) # variance inflation factors
sqrt(vif(fit)) > 2.5 # if TRUE then problem
#Fit multinomial model
require(nnet)
hydro.m=multinom(y1~preds)
temp=round(hydro.m$fitted.values,1)
# Prepare dataset with predicted and observed
hydrores=data.frame(id=row.names(temp),temp)
y1res=data.frame(id=1:33,y1)
pred.res=merge(hydrores,y1res,by.x="id")
pred.res
#Two-way table
table(pred.res$hydro.pred,pred.res$y1)
```

FIG. 3.10.4 Example of validation for continuous variable.

TABLE 3.10.1 Variable Importance for Each Index/Predictor by NC WAM Function From Random Forest Analysis

	NC WAM Function			
Index/Predictor	**Overall**	**Hydrology**	**Habitat**	**Water Quality**
Vegetation index annual: perennial	11.96	14.45	10.40	4.71
Mean coefficient of conservation	13.10	13.59	11.01	
Fern percent cover	10.81	7.20	8.69	
FQAI cover metric	6.98	4.31	11.09	
FQAI species count metric	8.36	4.66	10.61	4.71
Percent cover exotic grass species		5.58		
Moss percent cover	4.19	3.54		
Percent sensitive (C>7)	3.90	5.65		
Percent tolerant (C<2)	4.84	3.53	11.58	
Pole Timber density		4.10		
Wetland shrub percent cover		4.92	6.43	
Macroinvertebrate index evenness	4.63	5.37		
Percent microcrustacea	5.52	6.54	7.96	
Percent crustacea				4.71
Percent decapoda				4.71
Percent predator		4.39	3.86	
Amphibian index species richness			8.99	4.71

Using the best noncollinear predictors identified by the RF and multicollinearity analysis, a log-linear model for ordinal data was fitted to the ordinal hydrology function variable using a jackknife drop-1 site approach. A jackknife drop-1 site approach systematically leaves out one observation from the dataset and fits the log-linear model to the remaining data. The probability of being classified as low, medium, and high was calculated for the site that was dropped from the fitting process. The agreement between the predicted hydrology level derived from the model and the NC WAM score was 90%. Separately, a discriminant analysis was used to classify the sites in low, medium, and high. The agreement between the classification provided by the NC hydrology score and the discriminant analysis was 85.61%. Therefore, the NC WAM hydrology function accurately predicts this aspect of the hydrology of the site for headwater forest wetlands.

A similar process could be used for other function scores as well as for the overall score in order to validate those components. For RAMs with continuous scores, the analyst will follow the same process but replace the log-linear model with a linear regression model. A discriminant analysis approach to assess agreement could be used as well in the case of a continuous variable, as long as the analyst categorizes the continuous variable in meaningful intervals.

ACKNOWLEDGMENTS

Support to write this section was provided by the RTI Fellows Program, RTI International.

REFERENCES

Adamus, P.R., 1995. Validating a habitat evaluation method for predicting avian richness. Wildl. Soc. Bull. 23, 743–749.

Baker, V., Savage, R., 2008. Development of a wetland monitoring program for headwater wetlands in North Carolina. NC Department of Environment and Natural Resources, NC Division of Water Quality, EPA Final Report CD 974260-01, Raleigh, NC.

Belsley, D.A., Kuh, E., Welsch, R.E., 1980. Regression Diagnostics: Identifying Influential Data and Sources of Collinearity. Wiley, New York.

Breiman, L., 2001. Random forests. Mach. Learn. 45, 5–32.

Cohen, J., 1960. A coefficient of agreement for nominal scales. Educ. Psychol. Meas. 20, 37–46.
Cohen, J., 1968. Weighted kappa: nominal scale agreement provision for scaled disagreement or partial credit. Psychol. Bull. 70, 213–220.
Dorney, J.R., Paugh, L., Smith, A.P., Allen, T., Cusack, M.T., Savage, R., Hughes, E.B., Munoz, B., 2015. The North Carolina Wetland Assessment Method (NC WAM): development of a rapid wetland assessment method and use for compensatory mitigation. Environ. Pract. 17, 145–155.
Falotico, R., Quatto, P., 2010. On avoiding paradoxes in assessing inter-rater agreement. Ital. J. Appl. Stat. 22, 151–160.
Fennessy, M.S., Jacobs, A.D., Kentula, M.E., 2007. A review of rapid methods for assessing the ecological condition of wetlands. Wetlands 27, 543–560.
Fleiss, J.L., 1971. Measuring nominal scale agreement among many raters. Psychol. Bull. 76, 378–382.
Hocking, R.R., 2003. Methods and Applications of Linear Models: Regression and the Analysis of Variance, second ed. Wiley Interscience, Hoboken, NJ.
Landis, J.R., Koch, G.G., 1977. The measurement of observer agreement for categorical data. Biometrics 33, 159–174.
Liaw, A., Wiener, M., 2002. Classification and regression by random forest. R News 2 (3), 18–21.
Mandel, J., 1982. Use of the singular decomposition in regression analysis. Am. Stat. 36 (1), 15–24.
Marasini, D., Quatto, P., Ripamonti, E., 2016. Assessing the inter-rater agreement for ordinal data through weighted indexes. Stat. Methods Med. Res. 25 (6), 2611–2633.
Peduzzi, P., Concato, J., Feinstein, A., Holford, T., 1995. The importance of events per independent variable in proportional hazards regression analysis: II. Accuracy and precision of regression estimates. J. Clin. Epidemiol. 48, 1503–1510.
Peduzzi, P., Concato, J., Kemper, E., Holford, T.R., Feinstein, A., 1996. A simulation study on the number of events per variable in logistic regression analysis. J. Clin. Epidemiol. 49, 1373–1379.
Polascik, J.R.A., 2015. Validating a Kentucky wetland rapid assessment method for forested riverine wetlands using vegetation, bird surveys, and landscape analysis. Online Theses and Dissertations. 305, http://encompass.eku.edu/etd/305.
Stein, E.D., Fetscher, A.E., Clark, R.P., Wiskind, A., Grenier, J.L., Sutula, M., Collins, J.N., Grosso, C., 2009. Validation of a wetland rapid assessment method: use of EPA's level 1-2-3 framework for method testing and refinement. Wetlands 29 (2), 648–665.
Sutula, M.A., Stein, E.D., Collins, J.N., Fetscher, A.E., Clark, R., 2006. A practical guide for the development of a wetland assessment method: the California experience. J. Am. Water Resour. Assoc. 42, 157–175.
Theil, H., 1966. Applied Economic Forecasting. North-Holland, Amsterdam.
Wardrop, D.H., Kentula, M.E., Stevens Jr., D.L., Jensen, S.F., Brooks, R.P., 2007. Assessment of wetland condition: an example from the Upper Juniata watershed in Pennsylvania, USA. Wetlands 27, 416–430.
Wigand, C., Carlisle, B., Smith, J., Carullo, M., Fillis, D., Charpentier, M., McKinney, R., Johnson, R., Heltshe, J., 2010. Development and validation of rapid assessment indices of condition for coastal tidal wetlands in southern New England, USA. Environ. Monit. Assess. 182, 31–46.

FURTHER READING

Brinson, M.M., Rheinhardt, R., 1996. The role of reference wetlands in functional assessment and mitigation. Ecol. Appl. 6 (1), 69–76.
Hruby, T., Cesaneck, W.R., Miller, K.E., 1995. Estimating relative wetland values for regional planning. Wetlands 15 (2), 93–107.
Hughes, R.M., Larsen, D.P., Omernik, J.M., 1986. Regional reference sites: a method for assessing stream potentials. Environ. Manag. 10, 629–635.
Karr, J.R., 1981. Assessment of biotic integrity using fish communities. Fisheries 6 (6), 21–27.
Kerans, B.L., Karr, J.R., 1994. A benthic index of biotic integrity (B-IBI) for rivers of the Tennessee Valley. Ecol. Appl. 4 (4), 768–785.
Sifneos, J.C., Herlihy, A.T., Jacobs, A.D., Kentula, M.E., 2010. Calibration of the Delaware rapid assessment protocol to a comprehensive measure of wetland condition. Wetlands 30, 1011–1022.

Chapter 3.11

Training as a Component of RAM Implementation

Paul Adamus
Oregon State University, Corvallis, OR, United States

Sponsors of nearly all RAMs require or at least offer training in their RAM. Training is especially important because RAMs are observationally based rather than measurement based, and training helps ensure consistency among users (Herlihy et al., 2009). Currently, we are aware of ongoing training programs for RAMs in North Carolina, Oregon, California, Alberta, and Atlantic Canada, though there undoubtedly are many more. Trainees (potential RAM users) typically are employed by other agencies, consulting firms, watershed partnerships, land trusts, professional organizations, colleges, and various nonprofit groups. Some educators who have been trained have subsequently used the RAM in field learning exercises with their students.

The most common duration of training appears to be 1–3 days with shorter (1 day) overview sessions, such as those for the RAMs in Alberta and North Carolina. These are often geared toward administrators who themselves are unlikely to apply the RAM but who need to understand its key components, strengths, and limitations sufficiently to ensure that their staffs or consultants are using it appropriately and that it is well suited to meeting laws and policies they are responsible for implementing. If drafts of a RAM are to be subjected to expert peer review, it is important that the reviewers first receive training in at least the basics of the RAM. If the RAM is revised significantly over the years, "refresher" classes for users previously trained may also be warranted.

From an accuracy perspective, the best trainer is usually the RAM's author or another person who was trained by the author and then has accumulated significant experience using the RAM. If there is a possibility that none of the most desirable trainers may be available over the long term to do the training, one option is to develop an e-learning module, as one group has done for Alberta's RAM. That consists of slides, animation, and other imagery narrated in learning segments by the author. After listening to the module, potential users must pass an exam and then join other trainees practicing the method in an intensive field session lasting at least one day. Also, several standard "learning sites" previously scored by the RAM author and accessible to the public can be publicized so that trainees may visit and assess them to see if the trainee's scores match those determined previously by the RAM author. In those instances, it is important that the learning site's boundaries be well marked so that everyone is viewing exactly the same features. Some RAM trainers distribute an exam at the end of each training to test trainee retention of key features, but to our knowledge no agency requires legal certification (training and passing an exam) before a user's site assessment is accepted by the agency. When users of Oregon's RAM submit a site assessment to state agencies, they must indicate whether they attended a training, when that occurred, and approximately how many assessments they have previously completed using that RAM. Agencies can use such information to suggest how carefully the assessment results should be scrutinized.

Costs of training users may be covered by the sponsoring agency or group. More commonly, each trainee is charged a fee such that the sum of fees from all those registering for the training is sufficient to cover the training costs. Trainings are typically instigated by the sponsoring agency, but in several instances the RAM author(s) themselves have instigated and organized a series of trainings. Trainings can also be organized or coordinated through professional societies or consulting firms, with occasional audits for accuracy by the entity that originated the RAM.

REFERENCE

Herlihy, A.T., Sifneos, J., Bason, C., Jacobs, A., Kentula, M.E., Fennessy, M.S., 2009. An approach for evaluating the repeatability of rapid wetland assessment protocols: the effects of training and experience. Environ. Manag. 44, 369–377.

Wetland and Stream Rapid Assessments. https://doi.org/10.1016/B978-0-12-805091-0.00038-4

Chapter 3.12

Applications: Using Field-Based Wetland and Stream RAMs in Decision-Making

John Dorney* and Paul Adamus[†]
*Moffatt and Nichol, Raleigh, NC, United States, [†]Oregon State University, Corvallis, OR, United States

Chapter Outline

As described in the introduction and Section 1.0, wetland and stream RAMs have been developed for various types of purposes. This chapter describes several ways that RAMs have been used to inform decisions about streams and wetlands. We also provide some brief examples of such uses in the hope that doing so will encourage wider use of RAMs for similar purposes as well as spur readers to imagine new situations where RAMs may assist wetland or stream management. In this manner, the hard work done to develop and test these RAMs will have tangible, practical benefits. This chapter does not attempt to describe all types of situations where RAMs have been or might be applied.

PLACING SITES INTO DECISION CATEGORIES

Once a RAM has been used to determine a score or rating for a wetland site or stream reach, that score or rating can be used by some agencies or other entities to assign the site to a category (e.g., A-B-C-D or I-II-III-IV) that reflects its relative level of functions, condition, and/or services. This is commonly termed *categorization*, not to be confused with *classification*, which recognizes different types of wetlands or streams without relating those types to specific required actions. Assignment of a category of "A" or "1," for instance, might require that the site be preserved and that all future impacts be avoided whereas a category of "D" or "4" might require that all proposed alterations be allowed. Intermediate categories might require specified amounts of compensation and/or actions to minimize damages from proposed alterations. Also, the number of categories may simply be two: "significant" or "other." For example, unrelated to any regulation or treaty, many individual wetlands throughout the world have been designated as "internationally important" by the Ramsar process (Ramsar Convention Secretariat, 2010). Even when scores and ratings generated by a RAM are not converted to categories, they can inform the public, governments, and conservation groups regarding which sites may be most worthy of protection through easements, purchase, or other mechanisms. For that purpose, scientists with the Nature Conservancy in western Colorado applied AREM, a method validated for use in estimating avian richness (Adamus, 1993, 1995), to help prioritize purchase or easement of wetlands in that region.

Categorization schemes based largely on RAM scores have been formally adopted by Juneau, Alaska (Adamus, 1987; City and Borough of Juneau, Alaska, 1997), Ohio (Mack, 2000; Ohio Environmental Protection Agency, 2015) and Washington (Hruby, 2004; Chapter 4.3.10), the province of Alberta (Chapters 2.2.12 and 4.3.2), and probably others. For example, Ohio requires the use of its RAM (Chapter 4.3.8) to separate wetlands into three tiers: Category 1 wetlands with minimal wetland function and/or integrity; Category 2 wetlands with moderate wetland function and/or integrity; and Category 3 wetlands with superior wetland function and/or integrity. Mack (2000) presented a table with ORAM scores and how they generally correlate to these categories.

Wetland and Stream Rapid Assessments. https://doi.org/10.1016/B978-0-12-805091-0.00040-2

On a more case-by-case basis, the NC Wetland Assessment Method (NC WAM) (Chapter 4.3.1) was used to assess degraded rice fields on a historic rice plantation near Wilmington, North Carolina, in order to determine which abandoned fields could be impacted and which could not be returned to active rice cultivation (NC Division of Water Quality, 2012a, b). The relict rice fields still qualified as jurisdictional wetlands. The eventual plan approved by the state's Division of Water Quality avoided impact to all wetlands with an overall rating of high. In addition, all approved impacts to wetlands were to wetlands identified as low overall quality, according to NC WAM, other than impacts directly associated with stabilizing the berm along the Cape Fear River and other minor dike repairs.

To translate RAM scores into decision categories, scores should first be determined for a large number of sites—preferably sites that have been selected to be statistically representative of the region's wetland or stream population—so that the actual range of score variation is known (Chapter 3.5). Once that range is known, agencies or other entities must define and then apply thresholds that dictate into which category a site will be assigned. The thresholds for the categories can be expressed numerically based on a site's score (e.g., any site with a score of >7 on a scale of 0–10 is assigned to the most protective category) or a percentile (e.g., any site whose score is lower than the 20th percentile of all calibration sites is assigned to the least protective category). Or if a RAM that assesses multiple functions and benefits has been used, the categorization might take into account a combination of a site's ratings (e.g., from a list of 15 assessed functions, any site that has a rating of high for any two functions *as well as* for the benefits of those functions is assigned to the most protective category). A decision matrix could be constructed that defines several alternative scenarios (rating combinations) that should result in assignment to a specified category (e.g., Dorney et al., 2014). Also, factors in addition to a site's RAM score or rating, such as local trends in loss of similar habitats (e.g., Serran et al., 2017; Chapter 2.2.12), are sometimes taken into account when assigning sites to categories.

In sum, categorization simplifies resource decision-making and makes it more consistent and predictable. However, when categorization systems are adopted, decisions about which thresholds to use to define the categories are properly the province of policymakers and the public, not exclusively the RAM authors. Adopting a categorization system is not essential to interpreting RAM results or making sound decisions about wetlands and streams. Some agencies, retaining an element of human judgment in individual decisions, have chosen not to adopt a system of categories and instead simply rely on staff interpretations of RAM scores or ratings (Kusler, 1992).

USE OF RAMs TO DETERMINE MITIGATION CREDITS AND DEBITS

In many parts of North America as well as other counties such as France (Chapter 5.2), governments adopt policies that use a site's category to prescribe whether to *avoid* impacts to the particular site entirely, the degree to which attempts should be made to *minimize* those impacts, and/or the extent of *replacement* (compensatory mitigation) that should be required as compensation or offset if the site is altered. Collectively, in North America these practices are termed "mitigation." Mitigation practices are generally categorized as follows (definitions paraphrased from US Environmental Protection Agency and US Army Corps of Engineers, 2008).

1. *Creation (establishment)* is the manipulation of the physical, chemical, or biological characteristics of an upland site to develop an aquatic resource that did not previously exist at the site. Establishment results in a gain in aquatic resource area and functions.
2. *Restoration* is the manipulation of the physical, chemical, or biological characteristics of a site with the goal of returning natural/historic functions to a former or degraded aquatic resource. For the purpose of tracking net gains in aquatic resource area, restoration is divided into two categories: reestablishment and rehabilitation.
   - **(a)** *Reestablishment* aims to *return* natural/historical functions to a former aquatic resource by rebuilding the former aquatic resource, resulting in a gain in aquatic resource area and functions.
   - **(b)** *Rehabilitation* aims to *repair* natural/historical functions to a degraded aquatic resource. This results in a gain in aquatic resource functions but does not result in a gain in aquatic resource area.
3. *Enhancement* is intended to heighten, intensify, or improve a specific aquatic resource function(s). Enhancement results in the gain of selected aquatic resource function(s) but may also lead to a decline in other aquatic resource function(s). Enhancement does not result in a gain in aquatic resource area.
4. *Preservation* is the removal of a threat to, or preventing the decline of, aquatic resources by an action in or near those aquatic resources. This includes activities commonly associated with the protection and maintenance of aquatic resources through the implementation of legal and physical mechanisms. Preservation does not result in a gain of aquatic resource area or functions.

RAMs are not intended to be the only tool for informing decisions about wetlands and streams. Particularly when the goal is to measure slight changes in condition or functions between consecutive years or between very similar sites, use of more labor-intensive procedures is warranted (e.g., Burton, 2008). However, information from RAMs may be a useful supplement to more intensively collected data. For instance, aquatic macrobenthic data and Bank Height Index measurements were used in addition to NC SAM scores (Chapter 4.2.3) to assess a floodplain and stream restoration project in Wilmington (Moffatt and Nichol, 2017). Results in this case were roughly comparable and supported the decision to restore the stream and its floodplain. RAM results tend to be simplifications of more complex processes, and as such they can often be used to assist with communicating more complicated data to a broader audience.

Agencies that have adopted categorization schemes commonly use a site's assigned category to determine the type and extent of mitigation required when a developer proposes to alter a particular wetland site or stream reach. For example, if a developer proposes to alter a Category C wetland, agencies might require only that an equal or greater area of another wetland be *enhanced*, whereas if a Category B wetland is proposed for alteration, agencies might require that either a much greater area of another wetland be enhanced, or that a former wetland of equal or slightly greater area be *restored*, or that an entirely new wetland of equal or greater area be *created* from upland.

"Mitigation ratios" are often applied that dictate the amount of each action. For example, a 2-to-1 ratio means every acre of a site that is to be altered must be compensated for by the creation, restoration, enhancement, or preservation of 2 acres of another site. The chosen ratio will vary according to government-prescribed rules that reflect the type of mitigation being proposed as well as the site's category as determined by a RAM. In addition to ratio requirements, agency policies may require that a credit site be of the same hydrogeomorphic and/or vegetation class as the debit site. When *preservation* of another site is the chosen strategy, the degree of threat that the credit site would otherwise be exposed to if not preserved may be taken into account in establishing the mitigation ratio.

An example of the use of a RAM to alter the amount and type of mitigation required is from the Army Corps of Engineers District in Charleston, South Carolina. The Charleston District's mitigation calculation method is widely used in the Southeastern United States (Corps, Charleston District 2010) to determine mitigation credit requirements for projects that impact qualifying wetlands in the District. The calculation method considers six "adverse impact" factors, one of which is the functional integrity of the wetland that is proposed to be impacted. Because the District has no existing RAM, a crosswalk was developed to translate the results of NC WAM (from the adjacent state) into the protocols for the existing Charleston guidance to determine the amount of compensatory mitigation needed to offset any particular impact (Dorney et al., 2014). This crosswalk resulted in four combinations of hydrology, water quality, and habitat as "fully functional," 13 combinations as "partially impaired," six as "impaired," and four as "very impaired." This crosswalk allowed the direct use of NC WAM evaluations during the review of impact sites and their required mitigation for a large development project in South Carolina (Dorney et al., 2014).

Much recent discussion has been focused on requiring that a credit site also have a function profile mostly similar to that of the debit site. For example, a credit site with a rating of high for water storage (based on application of a RAM) could normally not be allowed as compensation for alteration of a debit site with a rating of low for water storage despite being high for other functions. By implementing such a requirement, agencies can truly achieve a goal of no net loss of stream or wetland functions at a regional level over time. Unfortunately, RAMs that generate just a single score or rating for a site based on a definition of overall quality or health cannot provide such a level of resolution because they do not provide scores for individual functions and the services associated with each. Only the RAMs capable of providing explicit scores for multiple functions can satisfy that requirement.

The state of Oregon, in partnership with federal agencies in the region, is currently moving toward adopting such a requirement based on its function-based RAMs for wetlands and streams. Oregon will use specific ratios based on the type of mitigation practice as a starting point in credit and debit calculations: Restoration/creation/credit purchase = 1.0; Enhancement = 3.0; Preservation (case-by-case) = 10. However, agencies will consider the following factors that may change the minimum acreage required of the replacement (credit) site:

1. *Specific function and value replacement.* If the credit site does not match or exceed each of the function group ratings at the debit site, the acreage requirement will increase in proportion to the disparity of the matches. This adjustment will not be made if the RAM shows the credit site rates well for functions that agencies have judged to be a priority in a particular watershed.
2. *Function temporal loss.* If a significant time lag occurs between loss of the debit site and achievement of full function at the credit site, the acreage requirement imposed on the project applicant will increase. This factor accounts for (a) the time lag associated with replacement of the vegetation community and (b) the time lag for development of hydric soil structure and characteristics at the credit sites.

3. *Mitigation site protection and stewardship*. If the wetland or stream property at the credit site is not well protected legally over the long term, future onsite activities may harm its functions and thus compromise its worth as a replacement. To discourage this, a project applicant can obtain a reduction in the total mitigation requirements if they provide a level of stewardship beyond what is minimally required by the government.

Ratios will never be allowed to be less than 1-to-1, and the credit site normally must be within the same "service area" (typically the same river basin) as the debit site. In some other states, the required replacement acreage is decreased if the credit site is even closer to the impacted site.

When *rehabilitation or enhancement* projects are pursued, RAMs are typically applied to the site before and after project implementation to express numerically the actual or hypothesized changes in condition or function scores. The change in scores (postproject minus preproject) is commonly termed "lift" when those changes are considered beneficial, as they are intended to be. When scores generated by RAMs are used to help assess wetland creation or reestablishment projects, the baseline condition (i.e., site is currently an upland habitat) is typically assumed to have no functions and thus is assigned a score of 0. While this assumption is scientifically incorrect, the unavailability of RAMs appropriate for assessing those upland functions requires decision-makers to allow that assumption. Whichever mitigation strategy is used, the accounting can use either the scores or the ratings based on those scores. The latter is often preferred when the repeatability of the RAM's scores is known or anticipated to be insufficient (see Chapter 3.7).

When RAM scores for functions or condition are used in a mitigation accounting framework, they are sometimes multiplied by the acreage (wetlands) or linear feet (streams) of the debit site or its impacted portion. The resulting number must then be approximately matched by an eligible credit site after it has been created, restored, or enhanced. However, it cannot be assumed that simply adding more acres to a credit site will increase its low-scoring functions, and that is what is implied when function scores are multiplied by acres (Adamus, 2013a, b).

Use of *enhancement* at a credit site to account for loss of functions at a debit site can be problematic from an accounting perspective. Consider the following hypothetical example:

	Debit Site (Preimpact)	Debit Site (Postimpact)	Credit Site (Preimpact)	Credit Site (Postimpact)	Net Change in Function
Function A	3	0	7	10	−3 +3 = 0
Function B	4	0	8	9	−4 +1 = −3
Function C	6	0	6	3	−6 −3 = −9

In this case, we see that practices that enhanced Function A at the credit site unintentionally resulted in a decline in Function C at that site. In any case, the gains from that enhancement were insufficient to counterbalance the huge loss of function that occurred from filling the debit site. This does not mean that efforts to enhance one or more functions at a site are without value—only that enhancement projects, because they seldom are capable of creating much functional lift, may not be enough to fully offset losses and thus further an overall goal of no net loss of functions.

Use of an alternative accounting method could yield results more supportive of enhancement as a mitigation strategy in at least some cases. For example, to represent ecological lift that was anticipated to occur postenhancement in a hydrologically degraded wetland, Dorney et al. (2014) used categories determined by NC WAM (Chapter 4.3.1). They then converted the categories to numbers (low = 0.5, medium = 1, high = 2) and inserted them into an equation created to also account for the acreage of the enhanced site and North Carolina's prescribed ratio of 2:1 for all enhancement projects. The enhancement consisted of reconnecting flow from an existing adjacent ditch so it would again flow into an existing wetland whose natural water regime had been impaired by the historical diversion of flow from that ditch. Use of the alternative equation yielded 0.69 ha of credit from the 1.3 ha site (NC Division of Water Quality, 2009). Without use of this equation, the economic case for the enhancement of the site would have been so weak that no enhancement would have been done. A possible refinement of this approach would have been to calculate the uplift separately for each of the three main types of functions (hydrology, water quality, habitat) rather than using their combined score. Such an approach would highlight and partially take into account any conflicts in the ways those three types of functions responded to the enhancement.

USE OF RAMs IN WETLAND OR STREAM DESIGN

Whether to meet government mitigation rules or simply to provide greater benefits to people and ecosystems, attempts are sometimes made to engineer new wetlands or to enhance (increase) particular functions of existing wetlands. Doing so successfully requires an element of design. While RAMs alone cannot provide all components of the design, their indicator variables and the ways they are used in their models highlight some of the features that might be manipulated, and the direction in which they should be manipulated, to achieve greater performance of selected functions, as explained in a book by Marble (1991). Increasingly, RAM models are being encoded in spreadsheets that facilitate analysis of different alternative design scenarios by instantly showing how each function may change in response to altering any combination of site characteristics. This has led some scientists to worry that we may be creating "designer wetlands" that, in a quest to optimize functions, resemble natural systems only remotely. An assumption is that natural systems are inherently more self-sustaining, but given rapid changes in global climate and increased opportunities for accidental transport of species globally, that assumption remains generally unproven. Moreover, it is inconceivable that wetland creation, restoration, and enhancement projects with intentionally optimized designs will ever become so widespread that they will comprise a significant proportion of all wetlands in a watershed or region.

USE OF RAMs TO SUMMARIZE THE CONDITION FOR A PARTICULAR CLASS OF WETLAND OR STREAM

RAMs can also be used to provide a general characterization of a wetland or stream type of interest. For instance, NC WAM (Chapter 4.3.1) was used to rapidly assess the condition of geographically isolated wetlands, a highly vulnerable wetland type (RTI International et al., 2011). Of the 150 geographically isolated wetlands that were visited, 67% were rated as high overall quality, 30% as medium overall quality, and 3% as low overall quality.

USE OF RAMs TO CHARACTERIZE ALL WETLANDS AND STREAMS ACROSS WATERSHEDS OR REGIONS

Perhaps less often, a RAM is applied simultaneously to *all* sites across an entire watershed, community, or region. This helps identify and map sites whose protection may provide the most benefits to communities as well as sites that may gain the most from restoration or other management action, perhaps as part of a mitigation program. For example, in the Indian/Howard Creeks watershed in central and western North Carolina, a field-level RAM—the NC WAM (Chapter 4.3.1)—was used to estimate that 64% of 33 wetlands were of overall high quality, 12% of overall medium quality, and 24% of overall low quality. As a prelude to the assessment, a GIS analysis of several existing spatial data layers was used to predict locations of previously unmapped wetlands; this enhanced the existing wetland maps.

High-quality sites were considered good candidates for preservation while low-quality sites were better candidates for enhancement. The RAM also documented that excessive grazing by cattle was the stressor most likely to be associated with a rating of low quality. This allowed management practices to target wetlands most degraded by that stressor as part of a comprehensive watershed management plan (NC Ecosystem Enhancement Program, 2010; Dorney, 2015).

Similarly, in several watersheds within Juneau, Alaska, interpretation of LiDAR and color orthophoto imagery, paired with GIS analysis of other existing spatial data layers, was used to predict locations of unmapped wetlands, most of them forested. More than 300 of the wetlands were then visited and 15 of their functions were scored using WESPAK-SE, a method based on the WESP template (Chapter 4.3.2) that had been previously calibrated specifically to Southeast Alaska (Adamus, 2013a, b). The resulting information is being used locally to guide decisions regarding future development (Bosworth Botanical Consulting, 2016).

Also, simply knowing the locations and proportions of various wetland classes or stream types present in a watershed, community, or region can be useful for planning infrastructure development or prioritizing sites for restoration or protection. For example, the NC Stream Identification Method (Chapter 4.1.2) was used to approximate the origin points of intermittent and perennial channels in the Triassic ecoregion of North Carolina. That allowed Russell et al. (2015) to develop statistical models to predict intermittent and perennial streams. This statistical modeling approach is being combined with field data for each of the 27 Level IV ecoregions of the state (Griffith et al., 2002) to produce an enhanced stream map covering the entire state (Andrew Kiley, NC Division of Water Resources, personal communication, Nov. 14, 2017). This will be used as a more accurate reference for riparian buffer protection rules the state has adopted for particular river basins or local watersheds (for instance, NC Environmental Management Commission, 2007). These maps will also help public and private users locate projects where they will minimize impacts to streams.

If only a preliminary estimate is needed of the relative condition (quality), functions, or services of *all* the wetlands or streams in a watershed or region, landscape-level methods (Section 2.0) can provide that estimate, usually at lower cost per site than using the field-level RAMs that would require visiting every wetland or a statistical sample thereof. At watershed or regional scales, perhaps the most widely used wetlands RAM is NWI-Plus (Chapter 2.2.1 and various other chapters of Section 2.0). It has been applied in several regions to provide, for example, an initial estimate of the cumulative contributions of wetlands to important functions within entire watersheds. It cannot normally be used in regulatory programs where the functions of an individual site must be estimated with accuracy achievable only with a site visit.

USE OF RAMs TO GUIDE SCIENTIFIC RESEARCH

Many of the empirical relationships ("correlations") between indicator variables and functions that RAMs express in their scoring models are based not on widely proven causal relationships but rather on assumptions—albeit usually well-founded ones—and informed scientific opinion. For example, the WESP template (Adamus, 2016; Chapter 4.3.2) has documentation, including citation of peer-reviewed literature wherever possible, for more than 700 indicator-function relationships for wetlands. Most of these assumptions were initially articulated in Adamus and Stockwell (1983) and have been updated in many successive documents by the same authors as well as repeated by authors of most other wetland RAMs that followed (e.g., Hruby et al., 1993; Tiner, 2003). These assumptions can largely be viewed as hypotheses and thus are ripe for testing with standard research protocols. Doing so will contribute enormously over time to the improvement of RAM models and thus resource decisions. Rather than diverting attention and funding away from badly needed research, RAMs have added impetus to wetland research by focusing public attention on wetlands and their importance to society.

RAMs can also assist in diagnosing environmental conundrums. For instance, Henry and Dorney (2016) found that streams in the Triassic region of the Carolinas had less diverse and less numerous aquatic macrobenthic communities than those in an adjacent region, irrespective of the predominant land use in the watersheds of the streams. Results from an NC SAM (Chapter 4.2.3) evaluation showed that streams in the species-poor region had significantly less in-stream habitat, and that helped explain the discrepancies in the aquatic communities.

USE OF RAMs AS EDUCATIONAL TOOLS

Even when agencies or other groups decide to not require use of an existing RAM, the RAM may still be used extensively by consultants if it is the only one calibrated to the region or has other desired attributes. And even when not used by consultants or governments, a RAM can serve as a tool for teaching students and educating the general public about the functions of wetlands and streams. Many citizens do not appreciate the relevance to their lives of worthwhile but complex concepts such as floristic quality, landscape connectivity, or macrobenthic trophic diversity. However, they are often quick to relate to RAM scores or ratings for water purification, water storage, pollinator habitat, or overall quality. This increase in the broad knowledge of and appreciation for the functions and benefits of wetlands and streams can be the spark needed to support the creation or expansion of policies, projects, and programs that preserve and manage these resources.

REFERENCES

Adamus, P.R., 1987. Juneau Wetlands: An Analysis of Functions and Values. USEPA and Department of Community Development, Juneau, Alaska. http://oregonstate.edu/~adamusp/Puget/Juneau_Wetlands.pdf.

Adamus, P.R., 1993. User's manual: avian richness evaluation method (AREM) for lowland wetlands of the Colorado Plateau. EPA/600/R-93/240, U.S. Environmental Protection Agency, Environmental Research Laboratory, Corvallis, OR. http://www.epa.gov/owow/wetlands/wqual/arem_man/.

Adamus, P.R., 1995. Validating a habitat evaluation method for predicting avian richness. Wildl. Soc. Bull. 23, 743–749.

Adamus, P.R. 2013. Wetland functions: not only about size. Natl Wetl. Newsl., September–October 2013. pp. 18–19.

Adamus, P.R. 2013. Wetland ecosystem services protocol for Southeast Alaska (WESPAK-SE). Report and software for the Southeast Alaska Land Trust and US Fish & Wildlife Service, Juneau, AK. http://southeastalaskalandtrust.org/wetland-mitigation-sponsor/wespak-se/

Adamus, P.R. 2016. Wetland ecosystem services protocol (WESP) version 2.0. people.oregonstate.edu/~adamusp/WESP.

Adamus, P.R., Stockwell, L.T., 1983. A method for wetland functional assessment. Vol. I. Critical review and evaluation concepts. Report No. FHWA-IP-82-23, Federal Highway Administration, Washington, DC.

Bosworth Botanical Consulting. 2016. Juneau wetland management plan update. Report for Community Department, City and Borough of Juneau, Alaska. http://www.juneau.org/cddftp/WRB/documents/JWMPVolume1FinalApril2016.pdf

Burton, E.R., 2008. Classifying Wetlands and Assessing Their Functions: Using the NC Wetlands Assessment Method (NC WAM) to Analyze Wetland Mitigation Sites in the Coastal Plain Region (M.S. thesis). University of North Carolina, Wilmington, NC.

City and Borough of Juneau, Alaska, 1997. Revised City and Borough of Juneau Wetlands Management Plan. .

Dorney, J.R., 2015. In: Wetland and stream functional assessment in North Carolina—a new regulatory world?.North Carolina APWA Stormwater Conference, Wilmington, NChttp://northcarolina.apwa.net/Content/Chapters/northcarolina.apwa.net/File/Conference%20Presentations%2FJohn%20Dorney%20-%20NC%20WAM%20and%20NC%20SAM%20in%20NC%20-%20%20Sept%202015%20Revised.pdf. Accessed 27 October 2017.

Dorney, J.R., Paugh, L., Smith, S., Lekson, D., Tugwell, T., Allen, B., Cusack, M., 2014. Development and testing of rapid wetland and stream functional assessment methods in North Carolina. Natl Wetl. Newsl. 36 (4), 31–35.

Griffith, G.E., Omernick, J.M., Comstock, J.A., Schafale, M.P., McNab, W.H., Lenat, D.R., MacPherson, T.F., 2002. Ecoregions of North Carolina. US Environmental Protection Agency, Corvallis, OR (map scale 1:1,500,000).

Henry, A., Dorney, J.R., 2016. The effects of stream water quality and landscape characteristics on aquatic insects in the northern outer piedmont and triassic ecoregions in NC.Presented at EnviroMentors National Fair and Awards Ceremony, June 2, 2017 in the USDA Building, Washington, DC Poster Prepared for the NC Environmentor Program, NC State University, College of Natural Resources, Raleigh, NChttps://www.ncseglobal.org/environmentors. Accessed 31 October 2017.

Hruby, T., 2004. Washington State Wetland Rating System for Western Washington. Washington State Department of Ecology. Updated. 2014, Available from: https://fortress.wa.gov/ecy/publications/documents/1406029.pdf. Accessed 27 March 2018.

Hruby, T., McMillan, A., Toshach, S., 1993. Washington State Wetlands Rating Systems, second ed. Washington Department of Ecology, Lacey, WA. Pub #93-74.

Kusler, J. (Ed.), 1992. State perspectives on wetland classification (categorization) for regulatory purposes.Proceedings of a National Workshop. Association of State Wetland Managers, Berne, NY.

Mack, J.J. 2000. ORAM v. 5.0 quantitative score calibration: Last Revised: August 15, 2000. State of Ohio Wetland Ecology Unit, Environmental Protection Agency, Division of Surface Water. Columbus, Ohio. http://epa.ohio.gov/portals/35/401/oram50sc_s.pdf. Accessed 31 October 2017.

Marble, A.D., 1991. A Guide to Wetland Functional Design. Lewis Publishers, Boca Raton, FL.

Moffatt and Nichol. 2017. Clear run stream restoration and flooding management project—draft environmental assessment. Dated February 8, 2001. Page 10 of document. Submitted to the US Army Corps of Engineers and NC Division of Water Resources. http://edocs.deq.nc.gov/WaterResources/0/doc/489838/Page1.aspx. Accessed 30 October 2017.

NC Division of Water Quality, 2012a. Orton plantation—public notice response. Dated 7 February 2012, http://edocs.deq.nc.gov/WaterResources/0/doc/159218/Page1.aspx?searchid=092a7bd2-1738-4bf0-a0b3-3e69bf5279c0. Accessed 29 October 2017.

NC Division of Water Quality, 2012b. Orton plantation—401 water quality certification. Dated 21 August 2012, http://edocs.deq.nc.gov/WaterResources/0/doc/167933/Page1.aspx?searchid=4bdb211e-7d62-4ff3-a033-0d979759b6d2. Accessed 29 October 2017.

NC Division of Water Quality (NCDWQ). 2009. 401 Water Quality Certification Number 3771 as Modified. Issued to Mr. Ross Smith, PCS Phosphate, Inc. January 15, 2009. Condition 8. Department of Environment and Natural Resources, NC Division of Water Quality, Raleigh, North Carolina. http://edocs.deq.nc.gov/WaterResources/0/doc/58492/Page1.aspx?searchid=1b028841-4b61-4bb2-998b-bcb9a598d954. Accessed 30 October 2017.

NC Ecosystem Enhancement Program. 2010. Indian creek and howards creek local watershed plan. Watershed Assessment Report. April 2010. https://ncdenr.s3.amazonaws.com/s3fs-public/Mitigation%20Services/Watershed_Planning/Catawba_River_Basin/Indian_Howards_Creek/Indian-Howards%20PhaseII_Report_Final_20100412.pdf. Accessed 27 October 2017.

NC Environmental Management Commission. 2007. Neuse river basin protection and maintenance of existing riparian buffers. 15A NCAC 02B.0233. http://reports.oah.state.nc.us/ncac/title%2015a%20-%20environmental%20quality/Chapter%2002%20-%20environmental%20management/subChapter%20b/15a%20ncac%2002b%20.0233.pdf. Accessed 1 November 2017.

Ohio Environmental Protection Agency. 2015. Ohio EPA Wetland Program Plan 2011-2015. https://www.epa.gov/sites/production/files/2015-10/documents/oh_wpp.pdf. Accessed 31 October 2017.

Ramsar Convention Secretariat, 2010. Handbook 17–Designating Ramsar Sites, fourth ed. http://www.ramsar.org/sites/default/files/documents/pdf/lib/hbk4-17.pdf. Accessed 29 October 2017.

RTI International, NC Department of Environment and Natural Resources, South Carolina Department of Health and Environmental Control, University of South Carolina, 2011. Assessing geographically isolated wetlands in North and South Carolina—the southeast isolated wetlands assessment (SEIWA). Final Report, http://www.northinlet.sc.edu/training/media/2011/06142011isolatedwetlands/resources/seiwa_final_report.pdf. Accessed 30 October 2017.

Russell, P.P., Gale, S.M., Muñoz, B., Dorney, J.R., Rubino, M.J., 2015. A spatially explicit model for mapping headwater streams. J. Am. Water Resour. Assoc. 51 (1), 226–239.

Serran, J.N., Creed, I.F., Ameli, A.A., Aldred, D.A., 2017. Estimating rates of wetland loss using power-law functions. Wetlands. https://doi.org/10.1007/s13157-017-0960-y.

Tiner, R.W., 2003. Correlating enhanced National Wetlands Inventory data with functions for watershed assessments: A rationale for northeastern U.S. wetlands. U.S. Fish and Wildlife Service, National Wetlands Inventory Program, Hadley, MA. https://www.fws.gov/northeast/ecologicalservices/pdf/wetlands/CorrelatingEnhancedNWIDataWetlandFunctionsWatershedAssessments.pdf.

US Environmental Protection Agency and US Army Corps of Engineers, 2008. Compensatory mitigation for losses of aquatic resources; final rule. 10 April 2008, Fed. Regist. 73 (70), 19594–19705. https://www.epa.gov/sites/production/files/2015-03/documents/2008_04_10_wetlands_wetlands_mitigation_final_rule_4_10_08.pdf. Accessed 30 October 2017.

FURTHER READING

Adamus, P.R., Clairain Jr., E.J., Smith, D.R., Young, R.E., 1992. Wetland evaluation technique (WET). Vol. I. Literature Review and Evaluation Rationale. U.S. Army Corps of Engineers, Waterways Experiment Station, Vicksburg, Mississippi.

Dorney, J.R., Paugh, L., Smith, A.P., Allen, T., Cusack, M.T., Savage, R., Hughes, E.B., Muñoz, B., 2015. The North Carolina Wetland Assessment Method (NC WAM): development of a rapid wetland assessment method and use for compensatory mitigation. Environ. Pract. 17, 145–155.

Government of Alberta, 2015. Alberta Wetland Rapid Evaluation Tool-Actual (ABWRET-A) Manual. Water Policy Branch, Alberta Environment and Parks, Edmonton, Alberta.

NC Association of Environmental Professionals, 2017. NC WAM & SAM for Managers. http://www.ncaep.org/event-2482138. Accessed 29 October 2017.

NC Division of Water Resources. 2017a. Approval of Individual 401 Water Quality Certification, Additional Conditions—Conditions Modified. Clear Run Restoration. http://edocs.deq.nc.gov/WaterResources/0/doc/554509/Page1.aspx?searchid=c44180f1-8559-4ddd-9863-52b0c5b63dfd. Accessed 30 October 2017.

NC Division of Water Resources. 2017b. Surface Water Identification Training and Certification (SWITC) Course. https://deq.nc.gov/about/divisions/water-resources/water-resources-training/training/surface-water-identification-training. Accessed 29 October 2017.

Section 4

Case Studies—Rapid Field-Based Approaches

This chapter describes various RAMs developed across North America over the past several decades. It is not comprehensive but rather includes RAMs selected by the editors to represent the broad diversity of RAMs that have been developed. Chapter 5 is a companion chapter for non-North American methods. Table 4.1 lists all the RAMs included in this book and may help the reader understand the organization of this chapter as well as quickly compare these RAMs with regard to what they attempt to estimate. "Functions" are what natural systems do naturally, such as store water. "Benefits" account for the context in which a function is performed, for example, water storage is more beneficial to society if stored upslope from infrastructure that otherwise would be ravaged by floods. "Condition" mainly describes the degree to which a site is or was exposed to disturbances from humans as indicated largely by its biological or hydrological characteristics, for example, invasive plant cover or flashiness of the hydrograph.

TABLE 4.1 Field-Level Stream and Wetland RAMs Described in This Book

			Scores and/or Ratings Provided Separately for:		
Acronym of the RAM	**Where Calibrated**	**Chapter**	**Functions**	**Benefits (Values, Services)**	**Condition (Health)**
Streams—flow duration N. America					
Pacific Northwest	Pacific Northwest				
NC stream identification	North Carolina				
Qualitative indicators of perennial streams	Virginia				
Streams—condition N. America					
NC SAM	North Carolina				X
WSWVM	West Virginia				X
VUSM	Virginia				X
Streams—other regions					
(none)	Nepal				X
Wetlands—N. America					
NCWAM	North Carolina				X
WESP (ORWAP, ABWRET, WESP-AC, and WESPAKse)	Oregon, Alberta, Atlantic Canada, Alaska		X	X	X
CRAM	California				X
MIRAM	Michigan				X
UMAM	Florida				X
NEWFA	New England		X		
NMRAM	New Mexico				X
ORAM	Ohio				X
OWES	Ontario		X	X	
Wetland rating system	Washington		X		
UMARKA	Mid-Atlantic states				X
USA-RAM	United States				X
Wetlands—other regions					
(None)	France		X		
(None)	New Zealand				X
(None)	Jamaica				X
(None)	Costa Rica				X
WET-Health	South Africa				X

Section 4.1

Stream Identification and Flow Duration Methods

Chapter 4.1.1

North Carolina Division of Water Quality Methodology for Identification of Intermittent and Perennial Streams and Their Origins

John Dorney* and Periann Russell†

*Moffatt and Nichol, Raleigh, NC, United States, †Division of Mitigation Services, NC Department of Environmental Quality, Raleigh, NC, United States

Chapter Outline

INTRODUCTION

The methodology for identifying intermittent and perennial streams was developed as a regulatory tool to be used by the North Carolina Division of Water Quality (NCDWQ) beginning in 1999 to determine the origins and presence of ephemeral, intermittent, and perennial streams in the Neuse River Basin. This was required in order to implement the riparian buffer protection rules adopted by the state's Environmental Management Commission (15A NCAC 2B 0.0242), which mandated protection and permitting for unavoidable impacts to buffers along intermittent and perennial streams throughout the basin. Because the method is intended for regulatory use, it is designed to be rapid and reproducible while also reflecting the processes related to intermittent and perennial streams in North Carolina. The method has since been successfully used statewide for almost two decades.

A stream technical advisory committee (TAC) was established by the NCDWQ in 1998 to provide technical and scientific input related to the definitions of streams and waterbodies in the Neuse River Basin for use in applying the riparian buffer rules. The TAC created the narrative stream definitions described below and evaluated and approved a stream identification methodology that evaluates geomorphic, hydrologic, and biological stream features to determine the origins of intermittent streams.

The system of scoring stream features and developing minimum total scores for stream identification was established based on the results from more than 300 individual field trials conducted in the Piedmont and Coastal Plain portions of the Neuse River Basin. Field testing consistently indicated that a total score of 19 would distinguish ephemeral streams from intermittent streams. Scores less than 19 indicated ephemeral streams whereas scores of 19 or greater provided sufficient evidence that a stream flowed at least intermittently. A score of 30 or more is one method to use to determine the presence of

☆ The information in this chapter was mainly drawn from NC Division of Water Quality (2010) with updates.

Wetland and Stream Rapid Assessments. https://doi.org/10.1016/B978-0-12-805091-0.00014-1

a perennial stream. Alternate procedures for perennial stream identification are documented in Section 3—Guidance for the Determination of Perennial Streams. The method does not consider stream condition but rather flow classification (ephemeral, intermittent, and perennial). The method also focuses on streams rather than ditches because the underlying riparian buffer protection rules apply only to streams. However, the rules (and the method) address "modified natural streams," which are defined explicitly in the manual as "an onsite channelization or relocation of a stream and its flow as evidenced by topographic alterations in the immediate watershed" (15A NCAC 02B 0.0233(2)(h)).

The following narrative definitions are critical to the methodology and are taught in the associated 4-day training class (NCDWQ, 2010). *Ephemeral streams* are features that carry "only stormwater in direct response to precipitation with water flowing only during and shortly after large precipitation events. An ephemeral stream may or may not have a well-defined channel, the aquatic bed is always above the water table, and stormwater runoff is the primary source of water. An ephemeral stream typically lacks the biological, hydrological, and physical characteristics commonly associated with the continuous or intermittent conveyance of water." *Intermittent streams* are streams with "a well-defined channel that contains water for only part of the year, typically during winter and spring when the aquatic bed is below the water table. The flow may be heavily supplemented by stormwater runoff. An intermittent stream often lacks the biological and hydrological characteristics commonly associated with the continuous conveyance of water." *Perennial streams* are streams that have "a well-defined channel that contains water year round during a year of normal rainfall with the aquatic bed located below the water table for most of the year. Groundwater is the primary source of water for perennial streams, but the streams also are fed by stormwater runoff. A perennial stream exhibits the typical biological, hydrological, and physical characteristics commonly associated with the continuous conveyance of water."

Since the adoption of the first version of the stream identification manual in 1999, improvements and clarifications have been made based on scientific literature and investigation as well as on experience and recommendations from users. The current version (4.11) was adopted in 2010. Prior to the implementation of a revised manual, the manual is submitted for a 60-day public review period. All comments and suggestions collected over the review period are considered and incorporated when applicable. Following revisions, the final version is adopted with an effective date and made available for all users.

PURPOSE OF DEVELOPING THE METHOD

As described above, the method was developed in 1999 in response to the Neuse River riparian protection rules, which protected riparian buffers along intermittent and perennial streams throughout the basin as part of a comprehensive strategy to reduce nutrients in the downstream estuary. The rules were initially adopted in 1997 but were followed by many revisions with extensive stakeholder involvement when it became clear to agency staff and the general public that a more repeatable method of stream identification was essential in order to make the rules effective. The North Carolina Stream Identification Method was developed to address this regulatory need for a repeatable and consistent method to identify intermittent and perennial stream origins in the piedmont and coastal plain portions of the basin. The method is not designed to determine stream function or condition but rather to address the flow classification of the particular channel.

CONCEPTUAL FRAMEWORK FOR THE METHOD

A stream contains surface water in a channel resulting from one or more of five potential sources of water entering the stream from the adjacent landscape (Poff and Ward, 1989; Richter et al., 1996; Walker et al., 1995). Baseflow or normal low flow in a stream between rainfall events is provided by two of those sources: *groundwater discharge* into the channel and *unsaturated drainage from the soil moisture zone* above the water table to the groundwater zone. During and shortly after rainstorms, the increased flow in the channel known as stormflow is provided by *direct channel precipitation,* surface runoff as *overland flow*, and rapid unsaturated flow through the soil (*interflow*) directly to the stream or to the groundwater zone. Increased groundwater discharge also contributes to stormflow.

Streams exhibit both stormflow and baseflow characteristics as they flow from their origins to their destinations. However, the seasonal baseflow defines a stream as intermittent while continual baseflow during a year of normal rainfall defines a stream as perennial. The North Carolina stream definitions do not require water to be flowing, but only prescribe that water be present to meet the definition of intermittent or perennial flow for regulatory purposes. Also, within the regulatory framework, an intermittent or perennial stream origin is defined as a specific location along a stream channel. However, in many cases, streams originate as transition zones in which the location and length of the zone is subject to fluctuations in groundwater levels and precipitation. Typically, streams change from ephemeral to intermittent and intermittent to perennial along a gradient or continuum–sometimes with no single distinct or stable point demarcating these transitions. Regardless, a determined point is necessary for regulatory purposes because the protected riparian buffer begins at these points.

In North Carolina, some streams follow a pattern and transition from ephemeral to intermittent to perennial, but in many cases, they do not. The transition varies by landscape and general trends can be observed by geologic province, that is, mountains, piedmont, and coastal plain as well as Level IV ecoregion (Griffith et al., 2002). For instance, in the mountains, streams often begin as springs and do not have an ephemeral or intermittent reach. In various piedmont ecoregions (such as the Triassic basin), long intermittent reaches are common while in other ecoregions (such as the Northern Outer Piedmont) intermittent reaches are generally much shorter. In the coastal plain, streams often have their origins below large wetlands at the edge of interstream divides.

In general, stream systems can be characterized by interactions among hydrologic, geomorphic, and biological processes. Similar to the downstream continuum of ephemeral to intermittent to perennial stream flow, physical and biological characteristics often follow the same pattern in response to flow volume and classification. Variations in physical and biological characteristics along the length of a stream can help distinguish what source of water predominately contributes to flow.

As baseflow becomes more persistent in the downstream direction, stream discharge–both stormflow and baseflow–increases and stream characteristics related to geomorphic, hydrologic, and biological processes are more readily observed. Stream bedforms, such as gravel bars and pool-riffle sequences, are much more defined in perennial streams than in intermittent streams due to increased sediment supply as well as transport and depositional processes. Furthermore, aquatic organisms respond to the availability of habitat formed and maintained by geomorphic and hydrologic processes and vary depending on the persistence of water and streamflow.

Stream characteristics and commonly observable features resulting from geomorphic, hydrologic, and biological processes are used in this methodology to produce a numeric score. Attributes serve as indicators that can be observed, although they are not intended to independently determine stream flow classification. The indicators used in the method represent stream processes at varying scales. The total score of all indicators provides the means for stream determination. The score is then used to assign a stream type of ephemeral, intermittent, or perennial to the stream reach being evaluated.

DESCRIPTION OF THE METHOD

The method contains 26 metrics divided into one of three general categories–geomorphology, hydrology, and biology (Fig. 4.1.1.1). Metrics are scored on a four-tier scale: absent, weak, moderate, or strong (Table 4.1.1.1). Metrics either range from 0 to 3 points or 0–1.5 points, depending on the perceived importance of that metric to the determination of stream flow classification. Some metrics ("second order or greater channel" and "soil-based evidence of seasonal high water table") are given either 0 or 3 points. Points are totaled for the particular stream reach. Based on extensive field work, calibration, and subsequent field experience, scores of 19 or greater were determined to provide "sufficient evidence that at least an intermittent stream is present." A score of 30 or more "is one criterion that may be used to classify a stream reach as perennial. Another criterion for classifying a stream reach as perennial requires the presence of a later instar larvae of more than one benthic macroinvertebrate species that characteristically requires water for its entire life cycle. A list of these taxa is provided in the North Carolina manual and their proper identification is stressed in the 4-day training class. See the discussion under "Validation/calibration efforts undertaken with the method" in this chapter below.

Field equipment required for the proper use of the method includes a soil auger, a collection net and shallow white pan, GPS to determine the location in the field, camera, Munsell soil color charts to identify hydric soil characteristics, and copies of the most recent US Geological Survey (USGS) topographic map and county NRCS soil survey. The National Plant List for plants that occur in wetlands (US Army Corps of Engineers (USACE), 2012) and the Field Indicators of Hydric Soils of the United States (US Department of Agriculture (USDA), 2010) or their updated equivalents as well as the North Carolina manual are key references to include in the field evaluation. Field forms are also necessary. A recent upgrade to the method is a data dictionary that has been developed for Trimble GPS equipment to aid in the electronic capture of data in the field.

The North Carolina manual contains narrative descriptions and either figures or field photographs of each of the 26 metrics. Each description starts out with a copy of the metric language, followed by a description of the metric and then a description of what "absent, weak, moderate, or strong" means for that particular metric. Figures or photographs are included to illustrate the narrative. For instance, the language for Metric 1 "Continuity of Channel Bed and Bank" is copied below (NCDWQ, 2010, page 12):

NC DWQ Stream Identification Form Version 4.11

Date:	Project/Site:	Latitude:
Evaluator:	County:	Longitude:
Total Points: *Stream is at least intermittent if ≥ 19 or perennial if ≥ 30**	**Stream Determination (circle one)** **Ephemeral Intermittent Perennial**	**Other** *e.g. Quad Name:*

A. Geomorphology (Subtotal = _________)	Absent	Weak	Moderate	Strong
1[a] Continuity of channel bed and bank	0	1	2	3
2. Sinuosity of channel along thalweg	0	1	2	3
3. In-channel structure: ex. riffle-pool, step-pool, ripple-pool sequence	0	1	2	3
4. Particle size of stream substrate	0	1	2	3
5. Active/relict floodplain	0	1	2	3
6. Depositional bars or benches	0	1	2	3
7. Recent alluvial deposits	0	1	2	3
8. Headcuts	0	1	2	3
9. Grade control	0	0.5	1	1.5
10. Natural valley	0	0.5	1	1.5
11. Second or greater order channel	No = 0		Yes = 3	
[a] artificial ditches are not rated; see discussions in manual				
B. Hydrology (Subtotal = _________)				
12. Presence of Baseflow	0	1	2	3
13. Iron oxidizing bacteria	0	1	2	3
14. Leaf litter	1.5	1	0.5	0
15. Sediment on plants or debris	0	0.5	1	1.5
16. Organic debris lines or piles	0	0.5	1	1.5
17. Soil-based evidence of high water table?	No = 0		Yes = 3	
C. Biology (Subtotal = _________)				
18. Fibrous roots in streambed	3	2	1	0
19. Rooted upland plants in streambed	3	2	1	0
20. Macrobenthos (note diversity and abundance)	0	1	2	3
21. Aquatic Mollusks	0	1	2	3
22. Fish	0	0.5	1	1.5
23. Crayfish	0	0.5	1	1.5
24. Amphibians	0	0.5	1	1.5
25. Algae	0	0.5	1	1.5
26. Wetland plants in streambed	FACW = 0.75; OBL = 1.5 Other = 0			

*perennial streams may also be identified using other methods. See p. 35 of manual.

Notes:

Sketch:

FIG. 4.1.1.1 Stream identification form version 4.11.

TABLE 4.1.1.1 Guide to Scoring Categories

Category	Description
Absent	The character is not observed
Weak	The character is present but you have to search intensely (i.e., 10 or more minutes) to find and evaluate it
Moderate	The character is present and observable with brief (i.e., 1 or 2 min) searching and evaluation
Strong	The character is easily observable and quickly evaluated

"1. Continuity of Channel Bed and Bank.

Throughout the length of the reach, is the stream clearly defined by having a discernable bank and streambed?

The bed of a stream is the channel bottom and the physical confine of the "normal" baseflow or low water flow. Streambanks are vertical or sloped areas rising from the bed of the channel and are the lateral constraints (channel margins) of flow during all stages but flood stage. Flooding occurs when a stream overflows its banks and partly or completely fills its floodplain. As a general rule, the bed is that part of the channel at or near "normal" flow, and the banks are that part above the water line. However, because discharge varies, this differentiation is subject to local interpretation. Usually the bed is clear of terrestrial vegetation while the banks are subjected to water flow only during high stages, and therefore can support vegetation much of the time. This indicator will lessen and may diminish or become fragmented upstream as the stream becomes ephemeral.

Strong—*The stream has a well-developed channel with continuous bed and bank present throughout the length of the reach.*
Moderate—*The majority of the stream channel has a continuous bed and bank. However, there are obvious interruptions.*
Weak—*The majority of the stream channel has obvious interruptions in the continuity of bed and bank. However, there is still some representation of the bed and bank sequence.*
Absent—*The stream has a very poorly developed channel in which little or no bed and bank can be distinguished."*

An intensive 4-day class has been developed to teach the proper use of the method. The class has been taught (as of 2016) more than 75 times in the past decades, including more than 1500 students trained (NC DWR, 2016a). The class is a combination of office lecture and field practice over the 4 days with a strong emphasis on the proper identification of aquatic macroinvertebrates as well as proper identification and understanding of all 26 metrics. A written test and field test must be successfully completed in order for the trained individual to be listed on the North Carolina Division of Water Resources list of trained individuals (NCDWR, 2016b). Some local governments (for instance, Durham, 2016) require certification before submission of local site plans for review for development projects. This process as well as the intensive training process helps ensure proper use of the manual in the field. The NCDWR retains the ultimate regulatory authority for final decisions of stream flow classification subject to an internal appeal process and then ultimately a legal appeal process.

Validation/Calibration Efforts Undertaken With the Method

The method was not formally validated or calibrated before its first adoption in 1999. Rather, it received extensive testing across the state by the technical advisory team. Subsequently, validation has been done by Ftitz et al. (2013) and Lampo (2014), who reported that the method was accurate in terms of distinguishing between ephemeral and intermittent channels in South Carolina and southern Illinois but not between intermittent and perennial channels. However, these validation tests did not use the part of the methodology (Section 3 of Version 4.11 of the methodology) that states that the presence of specified long-lived aquatic species can be used to determine the presence of a perennial stream, regardless of the numerical score. Therefore, the state of North Carolina considers the method properly validated.

Time Spent in Developing/Testing the Method

The method was initially developed in 1988, followed by about 6 months of field testing on more than 300 sites before final adoption in 1999. Since then, the manual has had three major updates (September 2004, February 2005, and finally in September 2010 (the current Version 4.11)). It has been widely used across the state in all three major ecoregions (mountains, piedmont, and coastal plain) in natural, agricultural, and urban landscapes on thousands of stream channels for both regulatory and stream mapping purposes (Russell et al., 2014).

Sample Application of Method in the Field

In order to properly use the method, evaluators are provided the following general instructions:

- Do not evaluate the channel within 48 h of a runoff-producing rainfall event.
- Review map information on the stream and its watershed. In particular, the 1:24,000 USGS topographic map and the written version of the NRCS (or SCS) county soil survey should be consulted as well as high resolution geology maps or aerial photography.
- Attend the 4-day training class approved by the NCDWR and successfully pass the written and field tests.
- Walk to the upstream extent of the feature when feasible.

- Evaluate at least 100 ft (30 m) of the stream to determine average conditions.
- Divide the stream reach into sections with apparently uniform flow conditions, taking care to capture any change in bed profile or flow characteristics.
- Complete the North Carolina Stream Identification Form and include any written notes on the form and/or digital photographs as appropriate.
- Locate the origin points using flagging and GPS equipment.

Time Spent to Apply the Method in the Field

Not including the time spent on the steps described above, it will normally take at least 15 min, including sampling for aquatic macroinvertebrates.

How Were/Are the Data Being Used

The results of stream evaluations using this method are primarily used to determine the origins of intermittent and perennial streams for the purposes of various riparian protection rules. In addition, the method has been used to develop statistically valid models of stream origins by Level IV ecoregions (Russell et al., 2014) with models now developed for six of the 24 Level IV ecoregions in the state (Sarah Schwarzer, NCDWR, personal communication, April 10, 2016). Finally, the method has been used to distinguish between intermittent and perennial streams for stream mitigation requirements in the 404/401 permit program managed by the USACE and NCDWR. Finally, the NC Stream Assessment Method (Chapter 4.2.3 of this book) requires different evaluation procedures for intermittent and perennial streams and the classification method is explicitly described as the way to make that determination.

What Was Learned

The method has proven to be a robust, well-understood, and widely accepted method to determine stream flow classification across the state for a variety of regulatory purposes. Because a large cadre of professionals has been trained in the use of the method over the past decades, there is now a thorough understanding of the difference between stream types. In addition, the method has been successfully defended in administrative appeals over the past years, which has helped to further cement its validity to the public. Disagreements still arise from time to time concerning the flow classification of a particular channel, but the use of this method has narrowed those disagreements and focused them on more precise issues of concern. The method has proven to be useful for a wide variety of landscape settings (urban, suburban, agricultural, and forested) as well as a wide variety of physiographic regions. As such, the method is an integral part of the riparian buffer protection rules that aim to protect buffers along intermittent and perennial streams, mainly for their nutrient removal abilities.

Prospects for the Future

The method has been updated five times since it was first developed in 1999, mainly to reflect lessons learned during implementation of the method or new science that has been published. Each update is then reflected in the 4-day class authorized by the NCDWR; this sometimes requires additional training of previously trained individuals. A proposed draft of the revised method is sent out for public notice and comment before being finalized administratively. This updating process has recently (2015) been made more restrictive by the North Carolina general assembly's recent legislative prohibition on policy making. Therefore, additional updates to the manual will probably require a formal rule-making process undertaken by the state's Environmental Management Commission, which normally takes 2–3 years.

REFERENCES

City of Durham, NC, 2016. Department of Public Works (DPW) Pre-submittal Meeting Checklist. Available at: http://durhamnc.gov/DocumentCenter/View/1051. Accessed 11 February 2016.

Ftitz, K.M., Wenerick, W.R., Kostich, M.S., 2013. A validation study of a rapid field-based rating system for discriminating among flow permanence classes of headwater streams in South Carolina. Environ. Manag. 52, 1286–1298. https://doi.org/10.007/s00267-013-0158x.

Griffith, G.E., Omernick, J.M., Comstock, J.A., Schafale, M.P., McNab, W.H., Lenat, D.R., Mac Pherson, T.M., 2002. Ecoregions of North Carolina. U.S. Environmental Protection Agency, Corvallis, OR (map scale 1:1,500.000).

Lampo, M., 2014. A Validation Study of the North Carolina Rapid Field-Based Rating System for Discriminating Flow Permanence Classes of Headwater Streams in Agriculture Basins in Southern Illinois. MS Thesis, Department of Geography and Environmental Resources, Graduate School, Southern Illinois University Carbondale, Illinois.

NC Division of Water Quality, 2010. Methodology for Identification of Intermittent and Perennial Streams and Their Origins. Version 4.11, September 1, 2010. Raleigh, NC. Available at: http://portal.ncdenr.org/web/wq/swp/ws/401/waterresources/streamdeterminations. Accessed 26 January 2016.

NC Division of Water Resources, 2016a. 401 & Buffer Unit, Stream ID Classes. Available at: http://portal.ncdenr.org/web/wq/swp/ws/401/waterresources/streamdeterminations/streamclass. Accessed 11 February 2016.

NC Division of Water Resources, 2016b. Surface Water Identification and Training Certification Course Individuals Certified in Method 4.11. Available at: http://portal.ncdenr.org/c/document_library/get_file?uuid=0457891c-1124-4e9a-b6eb-8f3d68aba33f&groupId=38364. Accessed 11 February 2016.

Poff, N.L., Ward, J.V., 1989. Implications of streamflow variability and predictability for lotic community structure: a regional analysis of streamflow patterns. Can. J. Fish. Aquat. Sci. 46, 1805–1818.

Richter, B.D., Baumgartner, J.V., Powell, F., Braun, D.P., 1996. A method for assessing hydrologic alteration within ecosystems. Conserv. Biol. 10, 1163–1174.

Russell, P.P.S., Gale, M., Muñoz, B., Dorney, J.R., Rubino, M.J., 2014. A spatially explicit model for mapping headwater streams. J. Am. Water Resour. Assoc., 1–14. https://doi.org/10.1111/jawr.12250.

US Army Corps of Engineers, 2012. National Wetland Plan List. Available at: http://rsgisias.crrel.usace.army.mil/NWPL/. Accessed 24 May 2016.

US Department of Agriculture, Natural Resources Conservation Service, 2010. Field indicators of hydric soils of the United States: a guide for identifying and delineating hydric soils, version 7.0. L.M. Vasilas. In: Hurt, G.W., Nobles, C.V. (Eds.), USDA, NRCS, in Cooperation With the National Technical Committee for Hydric Soils. Washington, DC. Available at: http://soils.usda.gov/use/hydric/. Accessed 9 July 2010.

Walker, K.F., Sheldon, F., Puckridge, J.T., 1995. A perspective on dryland river ecosystems. Regul. Rivers Res. Manag. 11, 85–104.

Chapter 4.1.2

A Rapid Assessment Method for Classifying Flow Permanence of Stream Reaches in the Pacific Northwest, United States

Tracie-Lynn Nadeau
United States Environmental Protection Agency, Region 10 (Pacific Northwest), Portland, OR, United States

Chapter Outline

INTRODUCTION

Flow permanence—whether a stream is ephemeral, intermittent, or perennial—is an important characteristic informing ecological assessment and management of streams. Perennial streams typically flow year-round, receiving appreciable quantities of water from numerous sources but having consistent groundwater inputs throughout the year (Winter et al., 1998; Winter, 2007). Where groundwater aquifers are unable to supply sufficient quantities of water as baseflow, intermittent streams cease to flow during dry periods (Mosley and McKerchar, 1993; Rains and Mount, 2002; Rains et al., 2006). Ephemeral streams flow only in direct response to precipitation such as rainstorms, rain on snow events, or snowmelt. They do not receive appreciable quantities of water from any other source, and their channels are, at all times, above local water tables (Gordon et al., 2004; McDonough et al., 2011).

Enacted in 1972, the US Clean Water Act (CWA) has the broad mandate to "restore and maintain the chemical, physical, and biological integrity of the nation's waters." Longstanding regulations defined "waters of the United States," those waters that are regulated under the CWA, as traditional navigable waters, interstate waters, all other waters that could affect interstate or foreign commerce, impoundments of waters of the United States, tributaries, the territorial seas, and adjacent wetlands. U.S. Supreme Court rulings in 2001 (*Solid Waste Agency of Northern Cook County* v. *U.S. Army Corps of Engineers*, 531 US 159; *SWANCC*) and 2006 (*Rapanos v. United States*, 547 US 715; *Rapanos*) created uncertainty regarding federal CWA authority over certain waters (Downing et al., 2003; Nadeau and Leibowitz, 2003; Wood, 2004; Nadeau and Rains, 2007), and established new data and analytical requirements for determining CWA jurisdiction (Leibowitz et al., 2008; Nadeau et al., 2015). CWA Section 404, which requires a permit for the discharge of dredged or fill material into US waters, triggered these Supreme Court cases. But it is critically important that the definition of "waters of the United States" affects all CWA programs (Downing et al., 2003).

In response, the US Environmental Protection Agency (EPA) and the US Army Corps of Engineers (Corps), the coadministering agencies of CWA Section 404, issued guidance to EPA and Corps field staff implementing the *SWANCC* (USEPA/USACE, 2003) and the *Rapanos* (USEPA/USACE, 2008) decisions. Under the 2008 *Rapanos* guidance, the federal agencies continue to assert jurisdiction over nonnavigable tributaries of traditional navigable waters that are

Wetland and Stream Rapid Assessments. https://doi.org/10.1016/B978-0-12-805091-0.00041-4

"relatively permanent" (in general practice, streams considered perennial and intermittent). For the agencies to assert jurisdiction over nonnavigable tributaries that are not relatively permanent (in general practice, streams considered ephemeral), the 2008 guidance stated that such waters must have a significant nexus to a traditional navigable water. A "significant nexus" determination is an assessment of whether the stream significantly influences the chemical, physical, and biological integrity of downstream navigable waters. Implementation of the *Rapanos* decision has posed a challenge to EPA and Corps field staff and to the regulated community in determining whether a particular water is jurisdictional under the CWA (USEPA, 2009; Caruso, 2011), partly due to the lack of widely accepted rapid assessment methods to determine flow permanence classes of nonnavigable tributary streams.

The Streamflow Duration Assessment Method (SDAM; Nadeau, 2015) was developed to meet this need, guiding natural resource practitioners in evaluating the described indicators of flow permanence to help distinguish between ephemeral, intermittent, and perennial streams. While a primary driver behind method development is to provide technical guidance for identifying waters that may be subject to regulatory jurisdiction under the CWA, SDAM can also inform implementation of state and local mandates and ordinances (e.g., riparian buffer requirements), improve predictability for ecological assessment of streams to set appropriate water quality expectations, and support prioritization of restoration and protection efforts by providing a rapid flow classification tool.

DESCRIPTION OF THE METHOD

This method is used to distinguish between perennial, intermittent, and ephemeral streams in a single site visit. For the purposes of this method, the descriptor "stream" is attached to the channel—an area that contains flowing water that is confined by banks and a bed—and applies regardless of whether flow dries up seasonally or otherwise.

Reach Selection

SDAM assesses indicators of flow permanence. Recognizing that in many streams flow exists on a continuum, and flow characteristics vary along the length of a stream, choosing the reach on which to conduct an assessment can influence flow classification. Assessments are made for a representative reach rather than at one point of a stream. A representative reach for stream assessments is equivalent to 35–40 channel widths of the stream (Peck et al., 2006). For narrow streams, assessment reach length is a minimum of 30 m. Reach length is measured along the thalweg. If the subject reach is near a culvert or road crossing, the assessment reach should begin a minimum of 10 m from the culvert or road-crossing feature. If the reach is not uniform, two or more assessments are recommended to fully describe the changes along the reach.

Conducting Assessments

Assessments begin by first walking the length of the channel, as feasible, from the stream origin to the downstream confluence with a larger stream. Walking alongside, rather than in, the channel if possible for the initial review is preferable to avoid unnecessary disturbance to the stream and maximize the opportunity to observe single indicator organisms (i.e., fish and herpetological species). This initial site review allows the assessor to examine the overall form of the channel, surrounding landscape, and parent material, and variation of these attributes. Walking the channel enables the assessor to document watershed characteristics such as land use and sources of flow (e.g., stormwater pipes, springs, seeps, and upstream tributaries) that may not be well characterized in desktop sources. Once made, these stream channel observations inform identification of abrupt changes in flow, and indicate whether the stream segment is generally uniform or might best be assessed as two or more distinct reaches.

To apply this method, all indicators are first evaluated, and the field assessment form (Fig. 4.1.2.1) completed. The indicators are then considered sequentially, similar to using a dichotomous key (Fig. 4.1.2.2). The answers to each step of the key determine the relevant indicator for the next step. All pertinent observations that could influence the indicators are noted on the field form, including evidence of stream modifications or hydrologic alterations, and noting a clear and repeatable way of identifying reach boundaries. Observed hydrology (surface or hyporheic) is also recorded, using the described protocol (Nadeau, 2015). When possible, the method should be applied during the growing season for best results.

Streamflow Duration Field Assessment Form

Project # / Name	Assessor
Address	Date
Waterway Name Reach Boundaries	Coordinates at downstream end *(ddd.mm.ss)* Lat. N Long. W
Precipitation w/in 48 hours (cm) / Channel Width (m)	☐ Disturbed Site / Difficult Situation (Describe in "Notes")

Observed Hydrology
% of reach w/observed surface flow_______
% of reach w/any flow (surface or hyporheic) _______
# of pools observed_______

Observations

Observed Wetland Plants (and indicator status):

Observed Macroinvertebrates:

Taxon	Indicator Status	Ephemeroptera?	# of Individuals

Indicators

1. Are aquatic macroinvertebrates present?	☐ Yes	☐ No
2. Are 6 or more individuals of the Order Ephemeroptera present?	☐ Yes	☐ No
3. Are perennial indicator taxa present? *(refer to Table 1)*	☐ Yes	☐ No
4. Are FACW, OBL, or SAV plants present? *(Within ½ channel width)*	☐ Yes	☐ No
5. What is the slope? *(In percent, measured for the valley, not the stream)*	_____ %	

Conclusions

Single Indicators:
☐ Fish
☐ Amphibians

Finding:
☐ Ephemeral
☐ Intermittent
☐ Perennial

Notes: (explanation of any single indicator conclusions, description of disturbances or modifications that may interfere with indicators, etc.)

Difficult Situation:
☐ Prolonged Abnormal Rainfall / Snowpack
☐ Below Average
☐ Above Average
☐ Natural or Anthropogenic Disturbance
☐ Other: ________________

Describe situation. For disturbed streams, note extent, type, and history of disturbance.

Additional Notes: (sketch of site, description of photos, comments on hydrological observations, etc.) Attach additional sheets as necessary.

Ancillary Information:
☐ Riparian Corridor
☐ Erosion and Deposition
☐ Floodplain Connectivity

Observed Amphibians, Snake, and Fish:

Taxon	Life History Stage	Location Observed	Number of Individuals Observed

FIG. 4.1.2.1 Field assessment form for the Streamflow Duration Assessment Method for the Pacific Northwest.

Indicators

Stream reaches are categorized as perennial, intermittent, or ephemeral on the basis of five indicators, or stream attributes: four are biological—three aquatic macroinvertebrate indicators and the presence of hydrophytic plants—and the fifth is the abiotic channel slope. Indicator assessment is based on direct observation and should not include predictions of what could or should be present.

Macroinvertebrate Indicators (1–3)

Some aquatic macroinvertebrates are useful indicators of flow permanence because they require aquatic habitat, and in many instances flowing water, to complete specific life stages. The three macroinvertebrate indicators used here are

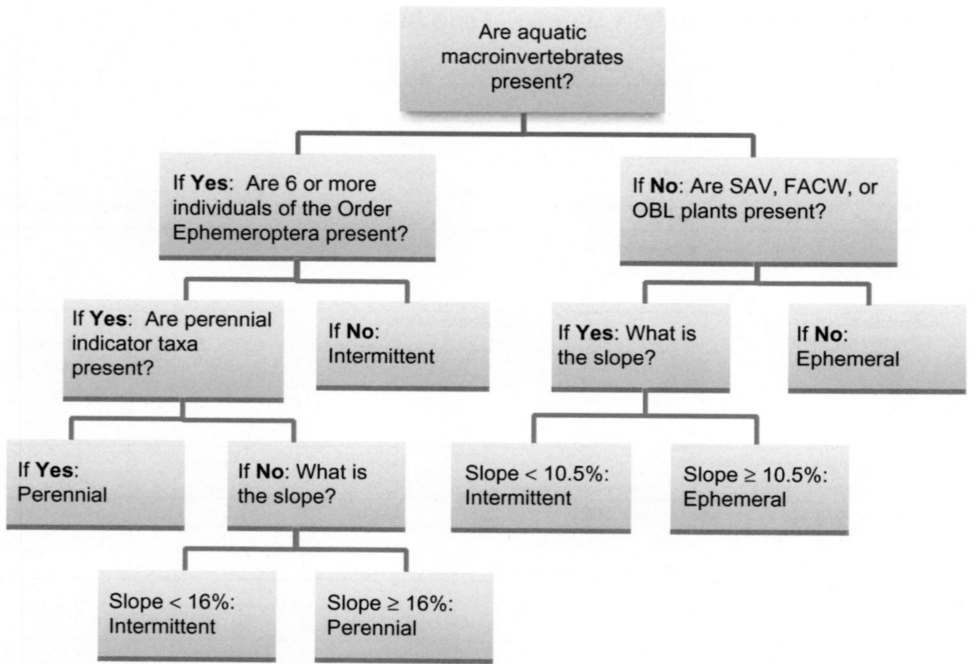

FIG. 4.1.2.2 Decision tree used in Streamflow Duration Assessment Method for the Pacific Northwest (Nadeau, 2015). Key: SAV, submerged aquatic vegetation; FACW, facultative wetland plants; OBL, obligate wetland plants.

assessed within the defined reach using a single search. The assessment for all three requires a minimum 15-min search time and at least six samples across the range of habitats present, using the described sampling methodology that includes dry channel assessment (Nadeau, 2015). The 15-min collection estimate does not reflect time spent on identifying individuals; rather, it is wholly focused on the searching and gathering effort. These indicators do not differentiate between live organisms and shells, casings, and exuvia (i.e., the shedded exoskeletons of aquatic arthropod larvae).

1. *Presence of aquatic macroinvertebrates*: Are there aquatic macroinvertebrates in the assessment reach?
   If at least one macroinvertebrate in an aquatic life stage (or macroinvertebrate shell, casing, or exuviae) is present, the answer is "yes." This indicator includes the range of macroinvertebrates typically associated with stream habitats including: Coleoptera (aquatic beetles), Diptera (true flies), Ephemeroptera (mayflies), Megaloptera (dobsonflies and alderflies), Mollusca (snails and clams), Odonata (dragonflies and damselflies), Plecoptera (stoneflies), Trichoptera (caddisflies), and Astacoidea (crayfish). If the only aquatic macroinvertebrate present is Culcidae (mosquito) larvae/pupae, which is found in ephemeral water, an exception is made and the answer is "no."
2. *Presence of six or more Ephemeroptera*: Are six or more larvae of the Order Ephemeroptera (mayflies) present in the assessment reach?
   If at least six Ephemeroptera are present, the answer is "yes." These may be represented by a single species or multiple species.
3. *Presence of perennial indicator taxa*: Are there perennial indicator aquatic macroinvertebrate taxa in the assessment reach?
   If at least one individual in an aquatic life stage (or macroinvertebrate shell, casing, or exuviae) of such taxa is present, the answer is "yes." Certain macroinvertebrate taxa are associated with the prolonged presence of water. Based on a literature review and synthesis (Mazzacano and Black, 2008; Blackburn and Mazzacano, 2012), several taxa and life stages of macroinvertebrates occurring in Pacific Northwest streams have been identified as "perennial indicators." These are further described in a compact companion field guide for identification of aquatic macroinvertebrates, developed specifically for use with this method (Mazzacano and Blackburn, 2012, https://www.epa.gov/sites/production/files/2015-10/documents/macroinvertebrate_field_guide.pdf). The field guide describes aquatic macroinvertebrates and their occurrence in perennial, intermittent, and ephemeral streams of the Pacific Northwest.

Additional Indicators (4 and 5)

4. *Wetland plants in or near streambed*: Within the assessment channel, and within one-half channel width of the stream on either bank, are there plants with a wetland indicator status of facultative wet (FACW) or obligate (OBL), or is there submerged aquatic vegetation (SAV) present?

   If hydrophytic plants are present, the answer is "yes." The presence of hydrophytic vegetation can be used as an indicator of the duration of soil saturation in or near stream channels. The recently revised US National Wetland Plant List (USACE, 2016) provides hydrophytic plant indicator status. This indicator is assessed based on the *single* most hydrophytic plant (SAV > OBL > FACW) found in, or within one-half channel width of, the assessed reach, even if that plant is not a dominant species. Abundance and prevalence throughout the reach is not a factor in documenting this indicator. Several aquatic plant species are protected by state and federal laws, and should not be collected or taken offsite for identification. If protected status is in question, photographs are recommended to facilitate further identification to be done offsite, if necessary.
5. *Slope*: What is the "straight line" slope, as measured with a clinometer, from the beginning of the reach to the end of the reach? Is it greater than or equal to 10.5%? To 16%?

   Channel slope is measured as percent slope between the lower and upper extent of the assessment reach, as described (Nadeau, 2015). If direct line of site from the bottom to top of the reach is not possible, the slope of the longest representative portion of the reach should be "line-of-site" evaluated.

Ancillary Information

The presence of these features is noted and briefly described, if applicable, as indicated on the assessment form. These, along with hydrological observations of the site, provide context and additional site information for the reviewer.

Riparian corridor: Is there a distinct change in vegetation between the surrounding uplands and the riparian zone, or corridor, along the stream channel? Intermittent and perennial streams often support riparian areas that contrast markedly with adjacent upland plant communities, and may indicate the presence of seasonal moisture.

Erosion and deposition: Does the channel show evidence of fluvial erosion in the form of undercut banks, scour marks, channel downcutting, or other features of channel incision? Are there depositional features such as bars or recent deposits of materials in the stream channel? Erosion and deposition are indicative of fluvial processes.

Floodplain connectivity: Is there an active floodplain at the bankfull elevation? A floodplain is a level area near a stream channel constructed by the stream and overflowed during moderate flow events if there is still connectivity. Floodplain areas can be absent or restricted to within the channel itself in incised streams. Floodplains are a depositional landform that is indicative of fluvial processes.

Drawing Conclusions

Results of the field evaluation, applied to the assessment decision tree (Fig. 4.1.2.2), are used to make a finding as to whether the assessed stream reach has perennial, intermittent, or ephemeral streamflow.

In addition, the method indicates that a stream is at least *intermittent* when either of two single Indicator criteria is met, for the presence of fish or for the presence of specific herpetological species. If present, single Indicators can be used as standalone, more rapid field evaluation alternatives to support a determination of "at least intermittent."

1. One or more fish are found in the assessment reach.

   Fish are an obvious indicator of flow presence and duration. The strongly seasonal precipitation pattern in the Pacific Northwest means intermittent streams may flow continuously for several months; thus, some native fish species have evolved to use intermittent streams for significant portions of their lifespan (e.g., Wigington et al., 2006). Nonnative fish, with the exception of mosquito fish (*Gambusia* spp.) that have been placed in streams as a vector control, are also included in assessing this indicator.
2. One or more individuals of an amphibian or snake life stage (adult, juvenile, larva, or eggs) identified as obligate or facultative wet are present in the assessment reach.

   Amphibians, by definition, are associated with aquatic habitats, and some require aquatic habitat for much or all of their lives. In the Pacific Northwest, there are likewise three snake species that require aquatic habitat for significant portions of their life cycle. This indicator focuses on herpetological species that require aquatic habitat using designations of life history stages for these species as facultative (FAC), facultative wet (FACW), or obligate (OBL), based on a review of the scientific literature and current understanding of their life history stages. The search protocol and indicator status for these salamanders, frogs, toads, and snakes is as described (Nadeau, 2015), and can be conducted concurrently with the macroinvertebrate search (Indicators 1–3) for greatest efficiency.

Several fish and amphibian species are protected by state and federal laws. Vertebrates must be identified at the assessment site and left at the site following identification. Photographs can be taken of vertebrates to allow further identification to be done offsite, if necessary.

Recorded hydrological observations provide independent corroboration of the predictor variables (indicators). The presence of a riparian corridor, erosional/depositional features, or floodplains ("ancillary information") also can indicate the presence of seasonal hydrology or fluvial processes.

METHOD DEVELOPMENT AND VALIDATION

Interim Method

SDAM was initially developed for Oregon based partly on the progress of the North Carolina Division of Water Quality's protocol (NCDWQ, 2010, see Chapter 4.1.1 of this book for a description of the method) and through best professional judgment (BPJ) informed by results of a single-season field test including >170 streams from both the humid and semiarid sides of the Cascade Range. The SDAM Interim Method (Topping et al., 2009) uses ordinal scoring of 21 geomorphic, hydrologic, and biologic stream attributes based on abundance and prominence to which an attribute is observable. Many of the 21 indicators are similar to those included in the NCDWQ method, modified for Pacific Northwest streams based on professional understanding of these systems; some new indicators were also developed based on expert opinion. Flow classification is based on the additive score of the assessed stream attributes compared to threshold values that separate perennial, intermittent, and ephemeral classes. Thresholds were determined using data from the 170 streams combined with BPJ. In addition, the Interim Method, like the NCDWQ method, classifies streams as at least intermittent (i.e., intermittent or perennial) based on the presence of single Indicator measures: fish or water-dependent life stages of specific herpetological and macroinvertebrate species. Species and life stages included as single Indicators were also modified, through expert opinion and current scientific understanding, for applicability to Pacific Northwest streams.

The Interim Method was made available and training was provided through field workshops held across the state. This allowed practitioners the opportunity to provide comment on their experiences using the method during a two-year field validation study of the Interim Method in Oregon. We then constructed a new method (revised) based on statistical analysis of the Oregon data, identifying the best indicators to discriminate flow permanence classes. In the second phase of the study, we conducted a one-year field study of the Interim and Revised Methods in Washington and Idaho to evaluate the applicability of the methods developed in Oregon to other areas of the Pacific Northwest. The study (Nadeau et al., 2015) is briefly described below.

Phase I, Oregon

A field validation study of the Interim Method was essential to meet our objective of developing a rapid flow permanence assessment method that is consistent, robust, and repeatable. The study included 178 streams ranging across the hydrologic settings of Oregon (Fig. 4.1.2.3), with an approximately equal distribution of streams from the humid west and semiarid east side of the Cascade Range, and in the perennial, intermittent, and ephemeral classes. Study design maximized representation of a diversity of hydrologic landscapes, based on a hydrologic classification framework that includes indices of annual climate, seasonality, aquifer permeability, terrain, and soil permeability (Wigington et al., 2013). Method evaluation compared results with actual flow permanence classes. Each study reach was assigned an independently determined, objectively defined flow permanence class based on a series of direct hydrological observations, and in 40% of the reaches we deployed electrical resistance (ER) data loggers.

The first phase of the study addressed several primary questions: (1) What is the accuracy of the Interim Method? (2) Is it equally applicable in different (wet/dry) seasons? (3) Is it equally applicable in different hydrologic landscapes across the state? (4) Are these 21 stream attributes the most predictive indicators of flow permanence? (5) Can results be improved by developing an alternative method (statistical analysis of data)?

The study included both wet and dry season sampling; in the Pacific Northwest, where the delivery of precipitation is generally greatest during the winter months, these correspond to wet winter/spring and dry summer seasons. Supplemental data were also collected at each site, including slope of reach, percentage streambed bedrock, and macroinvertebrate abundance.

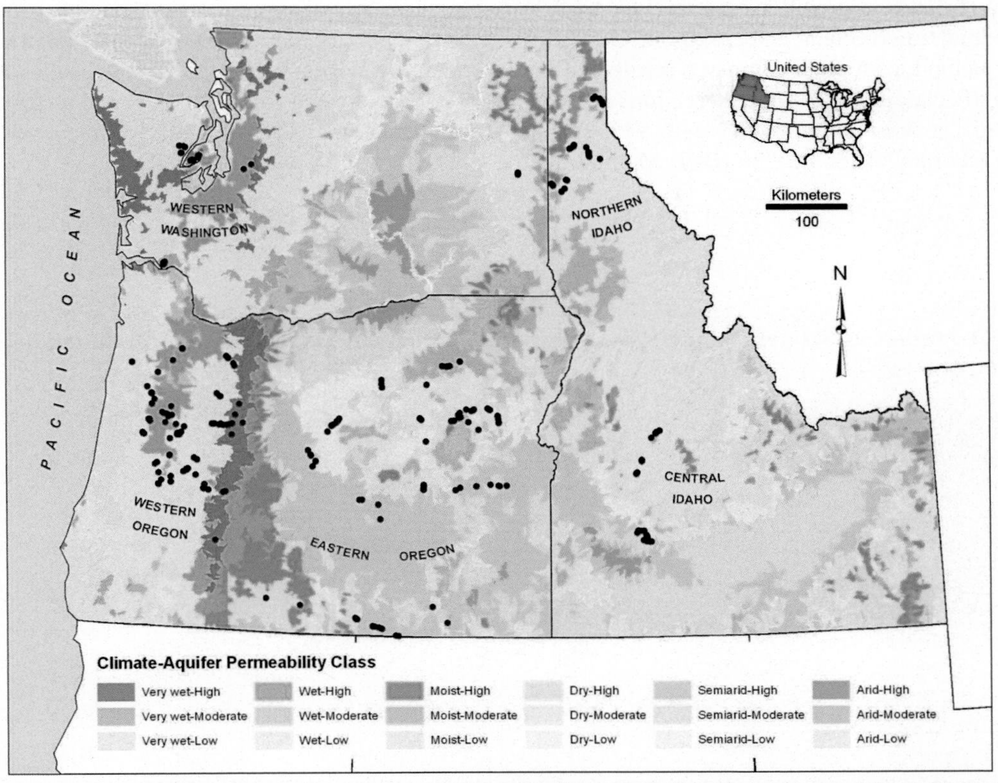

FIG. 4.1.2.3 Three-state study area. Black dots indicate study stream sites in the five study regions: Eastern Oregon (OR_e); Western Oregon (OR_w); Western Washington (WA_w); and Northern Idaho, including eastern Washington sites (ID_n), and Central Idaho (ID_c). Climate and aquifer permeability classes from Wigington et al. (2013) for Oregon and from draft versions of maps for Washington and Idaho. *(From Nadeau, T.-L., Leibowitz, S.G., Wigington Jr., P.J., Ebersole, J.L., Fritz, K.M., Coulombe, R.A., Comeleo, R.L., Blocksom, K.A., 2015. Validation of rapid assessment methods to determine streamflow duration classes in the Pacific Northwest, USA. Environ. Manag. 56 (1), 34–53.)*

The Interim Method agreed with the known flow permanence class for 62% of Oregon observations. The accuracy rate for distinguishing between ephemeral and "at least intermittent" (i.e., intermittent or perennial) streams was 81%. The high error rate of the Interim Method as applied in Oregon highlighted the need for an alternative method to more accurately determine flow permanence. Analyses of the Oregon data, using classification tree techniques (random forests, multivariate regression trees, cross-validation), found that a subset of the Interim Method and supplemental indicators appeared to have the strongest explanatory power in separating the perennial, intermittent, and ephemeral stream classes. This led to development of the Revised Method, comprised of five indicators—wetland plants in/near streambed, reach slope, and three aquatic macroinvertebrate indicators. Two of the five (wetland plants, presence of aquatic macroinvertebrates) are from the Interim Method, one is modified from an Interim Method single indicator (macroinvertebrate perennial indicator taxa), and two (slope, Ephemeroptera abundance) are new indicators resulting from the supplemental data collected. The Revised Method correctly classified 307 of the 356 Oregon observations, which is 86% correct compared with 62% accuracy of the Interim Method. Additionally, accuracy rates for distinguishing between ephemeral and "at least intermittent" classes (i.e., intermittent or perennial) rose from 81% to 95% with the Revised Method.

The Revised Method, significantly more accurate ($P < .0001$) than the Interim Method for predicting flow permanence classes and for "at least intermittent" accuracy, subsequently became the basis for the *Final Streamflow Duration Assessment Method for Oregon* (Nadeau, 2011), in which the five indicators are evaluated using a decision tree, similar to using a dichotomous key (Fig. 4.1.2.2). Additionally, the presence of certain vertebrate organisms that require the sustained presence of water for their growth and development are included as single indicators that a stream has at least intermittent flow.

Phase II, Idaho and Washington

We evaluated the regional applicability of the Interim and Revised Methods developed in Oregon through testing of 86 study reaches across a variety of hydrologic landscapes, both similar and dissimilar to those included in the Oregon phase of the study, in Idaho and Washington (Fig. 4.1.2.3). Again, there was an approximately equal distribution of streams from the perennial, intermittent, and ephemeral classes. As in the first phase, study streams were tested in both wet and dry seasons, and method evaluation compared results with independently determined flow permanence classes. The Revised Method correctly classified 84% of observations from the three-state study area (Table 4.1.2.1) and distinguished between ephemeral and "at least intermittent" with 94% accuracy, compared with 62% overall accuracy and 82% "at least intermittent" accuracy of the Interim (BPJ) Method.

TABLE 4.1.2.1 Accuracies of Interim and Revised Classification Methods as Determined From Confusion Matrices (Not Shown)

Accuracy (%)		Method	
		Interim	Revised
All		62.3	83.9
		(81.6)	(93.8)
Region	ID_c	63.3	80.0
		(78.3)	(88.3)
	ID_n	62.5	73.2
		(87.5)	(89.3)
	OR_e	59.7	91.5
		(81.3)	(97.2)
	OR_w	64.4	81.1
		(81.1)	(92.8)
	WA_w	62.5	83.9
		(82.1)	(96.4)
Climate class	Semiarid	66.7	91.7
		(73.8)	(94.0)
	Dry	53.9	86.8
		(78.9)	(96.1)
	Moist	67.2	91.4
		(98.3)	(100.0)
	Wet	60.1	77.4
		(79.8)	(89.9)
	Very Wet	66.7	84.3
		(84.3)	(96.1)
Season	Dry Summer	65.9	83.7
		(80.7)	(92.8)
	Wet Winter/Spring	58.7	84.1
		(82.6)	(94.7)

Data presented for all observations (Oregon, Washington, and Idaho) and by region, climate class, and season. Upper number is overall accuracy and lower parenthetical number is intermittent/perennial accuracy, both in percent.
OR_E=eastern Oregon; OR_W=western Oregon; WA_w=western Washington; ID_n=northern Idaho; ID_c=central Idaho.

RELEVANT CONCLUSIONS/WHAT WAS LEARNED

Based on results of our three-state study (Nadeau et al., 2015), the Revised Method is applicable across the Pacific Northwest and is the method described in this chapter (Nadeau, 2015). Developed through statistical analyses of field data, then validated using a spatially extensive, independent data set, it provides a faster, more simplified approach with significantly higher accuracy than the additive, weighted scale 21 indicator (Interim) method developed through best professional judgment. The decision-tree (Fig. 4.1.2.2) method is based on stream attributes—four biological and one physical—that are measurable, rather than subjective. Several assumptions were made in the inclusion and subsequent scoring and weighting of the 21 indicators included in the BPJ method, and our results indicate that many were not providing independent value as indicators of flow permanence.

Because of the diverse hydrology, climatic regimes, and distinct winter-wet and summer-dry seasons of the Pacific Northwest, we also explored method accuracy in different regions, climate classes, and seasons. Method performance does vary somewhat in different hydrological settings and at different times; for instance, it performs better during the spring for semiarid and very wet climate classes while classification is more accurate during the fall for wet climates (Table 4.1.2.1). However, overall accuracy for determining "at least intermittent" status is nearly 90% or greater in all categories. SDAM more accurately and consistently discriminates ephemeral streams from those that are "at least intermittent" than distinguishing the three separate flow classes, corroborating findings of other studies (Fritz et al., 2008, 2013) that used other methods for evaluating indicators of flow permanence.

Examining the accuracy of single indicators—organisms that require the sustained presence of water for their growth and development—at all study sites showed that while the absence of single indicator measures is not indicative of flow duration, their presence is strongly predictive. In all instances where fish were found at a study stream, the stream was intermittent or perennial, and 97% of the time that described herpetological organisms were found at a study stream, that stream was likewise "at least intermittent." While the classes of organisms that make up the single indicator measures are often not found in streams assessed as perennial or intermittent, when they are found they are a very accurate indication of perennial or intermittent status, and thus useful as rapid, standalone measures of determining "at least intermittent" status.

User accuracy—accuracy of the method when applied by a user in the field—was 92% for the described method in determining the ephemeral class of streams. There was also a high level of repeatability between duplicate assessments ($n=35$), but that may be due to, and point to the importance of, the level of field crew training in method application.

While the described method resulting from a statistically based modeling approach is significantly more accurate and rapid than the initial BPJ (Interim) method, identifying which indicators to test was based on BPJ, illustrating the importance of field observations and experience to inform management decisions and development of decision-support tools. The process from which this case study developed—from existing work and hypotheses (BPJ), incorporating local knowledge, and designing and executing a rigorous validation study with sufficient scope and statistical power—resulted in a meaningful tool. BPJ is valuable in establishing a testable hypothesis; this study (Nadeau et al., 2015) and others (Fritz et al., 2008, 2013; Johnson et al., 2009) highlight the importance of quantitative field testing of rapid assessment methods and associated indicators prior to routine implementation, and prior to application outside the region of development.

FIELD APPLICATION/CURRENT USE OF METHOD

A rapid assessment method, SDAM generally takes under 2 h on average for a trained team of two to apply at an assessment reach. If a particular stream requires more than one assessment reach, or there are multiple streams at a site, it would take accordingly longer. Time spent on plant and animal identification varies depending on assessor proficiency.

SDAM has been made available via joint public notice by Corps Portland, Seattle, and Walla Walla Districts and the EPA, and informs CWA jurisdictional determination, including support of CWA 404 enforcement cases in the three-state area where the method has been validated. It may also be used to inform state jurisdictional determinations, and has been made available via joint public notice by the Oregon Department of State Lands. Accurate classification of flow permanence is fundamental to water monitoring programs and watershed models, and to set appropriate expectations for the ecological status of intermittent and ephemeral streams, such as for water quality standards. Thus, SDAM is useful where knowledge of flow permanence improves ecological assessment, management, and decision-making.

PROSPECTS FOR THE FUTURE

Given that SDAM is an accurate, rapid assessment method for determining flow duration of streams and has proven useful in the Pacific Northwest, there is interest in adapting it to other regions of the country. We are currently testing the feasibility of adapting the method (model) for use outside the Pacific Northwest validation study area via a pilot study in the southwestern United States.

All relevant SDAM documents, including updates, are available here: https://www.epa.gov/measurements/streamflow-duration-assessment-method-pacific-northwest

ACKNOWLEDGMENTS

With thanks to coauthors of the study that resulted in the described method, and to colleagues across the Pacific Northwest and beyond who provided thoughtful input and shared their experiences during the years this method was under development. The author thanks the editors of this volume for the invitation to contribute, and appreciates constructive comments from Ken Fritz, Brian Topping, and Rose Kwok on an initial draft of this manuscript. The information in this document was funded entirely by the EPA. The content of this chapter represents the personal views of the author, and does not necessarily reflect official policy of the EPA or any other agency.

REFERENCES

Blackburn, M., Mazzacano, C., 2012. Using Aquatic Macroinvertebrates as Indicators of Streamflow Duration: Washington and Idaho Indicators. Report prepared by the Xerces Society for Invertebrate Conservation, 18 pp. http://www.xerces.org/wp-content/uploads/2009/03/Streamflow_duration_indicators_IDWA_2012_Final_06072012.pdf. Accessed 5 June 2017.

Caruso, B.S., 2011. Science and policy integration issues for stream and wetland jurisdictional determinations in a semi-arid region of the western U.S. Wetl. Ecol. Manag. 19 (4), 351–371.

Downing, D.M., Winer, C., Wood, L.D., 2003. Navigating through Clean Water Act jurisdiction: a legal review. Wetlands 23 (3), 475–493. https://doi.org/10.1672/0277-5212(2003)023[0475:ntcwaj]2.0.co;2.

Fritz, K.M., Johnson, B.R., Walters, D.M., 2008. Physical indicators of hydrologic permanence in forested headwater streams. J. N. Am. Benthol. Soc. 27 (3), 690–704. https://doi.org/10.1899/07-117.1.

Fritz, K.M., Wenerick, W.R., Kostich, M.S., 2013. A validation study of a rapid field-based rating system for discriminating among flow permanence classes of headwater streams in South Carolina. Environ. Manag. 52, 1286–1298. https://doi.org/10.1007/s00267-013-0158-x.

Gordon, N.D., McMahon, T.A., Finlayson, B.L., Gippel, C.J., Nathan, R.J., 2004. Stream Hydrology: An Introduction for Ecologists, second ed. John Wiley & Sons, LTD, Chichester, West Sussex.

Johnson, B.R., Fritz, K.M., Blocksom, K.A., Walters, D.M., 2009. Larval salamanders and channel geomorphology are indicators of hydrologic permanence in forested headwater streams. Ecol. Indic. 9 (1), 150–159.

Leibowitz, S.G., Wigington Jr., P.J., Rains, M.C., Downing, D.M., 2008. Non-navigable streams and adjacent wetlands: addressing science needs following the supreme Court's Rapanos decision. Front. Ecol. Environ. 6 (7), 364–371. https://doi.org/10.1890/070068.

Mazzacano, C., Black, S.H., 2008. Using Aquatic Macroinvertebrates as Indicators of Streamflow Duration. Report prepared by the Xerces Society for Invertebrate Conservation, 33 pp. http://www.xerces.org/wp-content/uploads/2009/03/xerces_macroinvertebrates_indicators_stream_duration.pdf. Accessed 5 June 2017.

Mazzacano, C., Blackburn, M., 2012. Macroinvertebrate Indicators of Streamflow Duration for the Pacific Northwest: Companion Field Guide. Xerces Society for Invertebrate Conservation. 22 pp. https://www.epa.gov/sites/production/files/2015-10/documents/macroinvertebrate_field_guide.pdf. Accessed 5 June 2017.

McDonough, O.T., Hosen, J.D., Palmer, M.A., 2011. Temporary streams: the hydrology, geography, and ecology of non-perennially flowing waters. In: Elliot, H.S.M.L. (Ed.), River Ecosystems: Dynamics, Management and Conservation. Nova Science Publishers, Hauppauge, pp. 259–289.

Mosley, M.P., McKerchar, A.I., 1993. Streamflow. In: Maidment, D. (Ed.), Handbook of Hydrology. McGraw-Hill, USA, pp. 8.1–8.39.

Nadeau, T.-L., 2011. Streamflow duration assessment method for Oregon (revised). EPA/910/R-11/002, U.S. Environmental Protection Agency, Region 10, Seattle, WA. https://www.epa.gov/sites/production/files/2016-01/documents/streamflow_duration_assessment_method_oregon_final_2011.pdf. Accessed 5 June 2017.

Nadeau, T.-L., 2015. Streamflow Duration Assessment Method for the Pacific Northwest. EPA/910/K-14/001, U.S. Environmental Protection Agency, Region 10, Seattle, WA. https://www.epa.gov/sites/production/files/2016-01/documents/streamflow_duration_assessment_method_pacific_northwest_2015.pdf. Accessed 5 June 2017.

Nadeau, T.-L., Leibowitz, S.G., 2003. Isolated wetlands: an introduction to the special issue. Wetlands 23 (3), 471–474.

Nadeau, T.-L., Rains, M.C., 2007. Hydrological connectivity between headwater streams and downstream waters: how science can inform policy. J. Am. Water Resour. Assoc. 43, 118–133.

Nadeau, T.-L., Leibowitz, S.G., Wigington Jr., P.J., Ebersole, J.L., Fritz, K.M., Coulombe, R.A., Comeleo, R.L., Blocksom, K.A., 2015. Validation of rapid assessment methods to determine streamflow duration classes in the Pacific Northwest, USA. Environ. Manag. 56 (1), 34–53.

NCDWQ, 2010. Methodology for Identification of Intermittent and Perennial Streams and Their Origins, Version 4.11. North Carolina Division of Water Quality, Raleigh, NC. http://portal.ncdenr.org/c/document_library/get_file?uuid=0ddc6ea1-d736-4b55-8e50-169a4476de96&groupId=38364. Accessed 5 June 2017.

Peck, D.V., Herlihy, A.T., Hill, B.H., Hughes, R.M., Kaufmann, P.R., Klemm, D.J., Lazorchak, J.M., McCormick, F.H., Peterson, S.A., Ringold, P.L., Magee, T., Cappaert, M.R., 2006. Environmental Monitoring and Assessment Program: Surface Waters Western Pilot Study—Field Operations Manual for Wadeable Streams. EPA/620/R-06/003, U.S. Environmental Protection Agency, Washington, DC.

Rains, M.C., Mount, J.F., 2002. Origin of shallow ground water in an alluvial aquifer as determined by isotopic and chemical procedures. Ground Water 40 (5), 552–563.

Rains, M.C., Fogg, G.E., Harter, T., Dahlgren, R.A., Williamson, R.J., 2006. The role of perched aquifers in hydrological connectivity and biogeochemical processes in vernal pool landscapes, Central Valley, California. Hydrol. Process. 20 (5), 1157–1175.

Topping, B.J.D., Nadeau, T.-L., Turaski, M.R., 2009. Interim Version, Oregon Streamflow Duration Assessment Method. U.S. Environmental Protection Agency and U.S. Army Corps of Engineers. https://www.epa.gov/sites/production/files/2016-01/documents/streamflow_duration_assessment_method_oregon_interim_2009.pdf. Accessed 5 June 2017.

U.S. Army Corps of Engineers, 2016. National Wetland Plant List, version 3.3. U.S. Army Corps of Engineers Engineer Research and Development Center Cold Regions Research and Engineering Laboratory, Hanover, NH. http://rsgisias.crrel.usace.army.mil/NWPL/. Accessed 5 June 2017.

U.S. Environmental Protection Agency, 2009. Congressionally Requested Reports on Comments Related to Effects of Jurisdictional Uncertainty on Clean Water Act Implementation. Report No. 09-N-0149, U.S. Environmental Protection Agency, Office of Inspector General. https://www.epa.gov/office-inspector-general/report-congressionally-requested-report-comments-related-effects. Accessed 5 June 2017.

U.S. Environmental Protection Agency/U.S. Army Corps of Engineers, 2003. Legal Memoranda Regarding Solid Waste Agency of Northern Cook County (SWANCC) v. United States. 68 Federal Register 1995–1998 (January 15, 2003), https://www.epa.gov/sites/production/files/2016-04/documents/swancc_guidance_jan_03.pdf. Accessed 5 June 2017.

U.S. Environmental Protection Agency/U.S. Army Corps of Engineers, 2008. Clean Water Act Jurisdiction Following the U.S. Supreme Court's Decision in Rapanos v. United States and Carabell v. United States. U.S. Environmental Protection Agency and U.S. Army Corps of Engineers. http://www.usace.army.mil/Portals/2/docs/civilworks/regulatory/cwa_guide/cwa_juris_2dec08.pdf. Accessed 5 June 2017.

Wigington Jr., P.J., Leibowitz, S.G., Comeleo, R.L., Ebersole, J.L., 2013. Oregon hydrologic landscapes: a classification framework. J. Am. Water Resour. Assoc. 49, 163–182. https://doi.org/10.1111/jawr.12009.

Wigington, P.J., Ebersole, J.L., Colvin, M.E., Leibowitz, S.G., Miller, B., Hanson, B., Lavigne, H.R., White, D., Baker, J.P., Church, M.R., Brooks, J.R., Cairns, M.A., Compton, J.E., 2006. Coho salmon dependence on intermittent streams. Front. Ecol. Environ. (10), 513–518.

Winter, T.C., 2007. The role of ground water in generating streamflow in headwater areas and in maintaining base flow. J. Am. Water Resour. Assoc. 43 (1), 15–25.

Winter, T.C., Harvey, J.W., Franke, O.L., Alley, W.M., 1998. Ground Water and Surface Water: A Single Resource. Circular 1139, U.S. Geological Survey, Denver, CO.

Wood, L.D., 2004. Don't be misled: CWA jurisdiction extends to all non-navigable tributaries of the traditional navigable waters and to their adjacent wetlands. Environ. Law Rev. 34, 10187–10193.

Chapter 4.1.3

Qualitative Indicators for Perennial Stream Determinations in Virginia

Douglas A. DeBerry[*,†] **and Travis W. Crayosky**[‡]

[*]Environmental Science and Policy, Integrated Science Center, College of William and Mary, Williamsburg, VA, United States, [†]*VHB, Inc., Williamsburg, VA, United States,* [‡]*Stantec, Williamsburg, VA, United States*

Chapter Outline

INTRODUCTION

The framework for perennial stream determinations described in this chapter was developed in the late 1990s and early 2000s at the request of the Virginia Department of Conservation and Recreation (DCR), the state agency at that time responsible for regulatory oversight of the Chesapeake Bay Preservation Act (Bay Act) in Virginia.[1] For the Virginia Bay Act program, nontidal streams held sway in ways that were also beginning to surface in the state and federal wetland protection programs at the turn of the century (e.g., Section 404 of the Clean Water Act, Virginia State Water Control Law), particularly with respect to intrinsic water quality value and the aquatic resource functions that streams perform. Stream function was (and continues to be) a tacit consideration in the Bay Act regulations; however, the primacy of nontidal stream channels in Bay Act jurisdiction has more to do with the mandatory upland buffers placed around perennial streams[2] and their adjacent wetlands, and the implications that these buffers have on private and public land-use decisions.[3]

As the Bay Act regulations were modified around this timeframe to admit "scientifically valid" means of determining perennial flow in streams, it became increasingly important for Virginia landowners and regulators to have defensible methods for distinguishing between perennial and intermittent or ephemeral streams on their properties. The authors had been developing field techniques for stream determinations for several years leading up to the late 1990s, and we were also aware of similar approaches being used by scientists at the Virginia Institute of Marine Science (VIMS) for various purposes. The DCR request initiated a collaboration between the authors and VIMS, and the product of that collaboration was the Qualitative Indicators for Perennial Steam Determinations described below.

---

1. The Bay Act program in Virginia has since been remanded to the Virginia Department of Environmental Quality.
2. For the purposes of this approach, a perennial stream is defined as a stream channel that maintains a quantifiable level of surface flow throughout a normal precipitation year (i.e., flows year-round, only drying up during drought years), and an intermittent stream is defined as a stream channel that lacks quantifiable surface flow at some point during a normal precipitation year (i.e., typically dries up during most years, and may only flow continuously during extremely wet precipitation years) (Gordon et al., 1992).
3. During this timeframe, reauthorizations of the Section 404 Nationwide Permits (NWP) administered by the US Army Corps of Engineers as well as the Virginia General Permit regulations also differentiated perennial streams from intermittent and ephemeral streams, with implications for general permit applicability to certain activities (i.e., permittable activities in perennial streams were more restrictive). Subsequent reauthorizations removed this distinction, but the inclusion of perennial streams at that time only strengthened the resolve of the regulatory community to develop defensible approaches for perennial stream determinations.

Wetland and Stream Rapid Assessments. https://doi.org/10.1016/B978-0-12-805091-0.00013-X

Our primary objective was to lay the groundwork for a field technique by first identifying qualitative (i.e., nonmetric) indicators that could be used to determine the upper limits of perennial stream flow in the absence of actual flow records. Unlike other approaches being used or developed at that time, the intent was not to incorporate a point system or a numerical index, but rather to develop a set of indicators that could be applied with best professional judgment to build corroborative evidence toward a defensible conclusion about flow regime. In practice, this approach is analogous to a wetland delineation wherein practitioners are enjoined to demonstrate the presence of indicators in the well-known "three-parameter approach" to wetland boundary identification (Environmental Laboratory, 1987; Wakeley, 2002). An additional objective was to develop a strategy for perennial stream determinations that could be applied across the various physiographic regions in Virginia, rather than just focusing on stream systems in the Coastal Plain where Bay Act jurisdiction is most prevalent.

The following sections provide an overview of the Qualitative Indicators approach as developed and eventually promulgated in 2002, followed by relevant stream research within the past 15 years that has advanced new ideas and a more nuanced understanding of some of these indicators. The evaluation has been designed to be performed by one reviewer in the field. The average review time will vary depending on the length of the assessment reach and site conditions. A relatively routine review of a hypothetical 100 m reach should take the average assessor approximately 30 min.

QUALITATIVE INDICATORS FOR PERENNIAL STREAM DETERMINATIONS (SEE FIG. 4.1.3.1)

Streamflow

The presence or absence of flowing water may be used as an indicator of flow regime dependent upon antecedent moisture conditions. The concept of flow regime is a relatively clear and historically well-defined basis for describing a stream along its profile (Hewlett, 1969; Dunne et al., 1975; Gordon et al., 1992). But how do we quantify flow characteristics given a point in time determination and/or limited available data? Long-term stream gauging data are typically scarce to nonexistent for headwater streams where the transition from intermittent to perennial flow occurs. At this landscape position, an important question is whether the observation of streamflow is reflective of perennial stream conditions.

Streamflow is dependent upon many basin-wide parameters, including antecedent basin wetness and diurnal fluctuations (Hewlett, 1969). How specific components of stormflow (channel precipitation, overland flow, and subsurface stormflow) interact with baseflow (groundwater) determine not only the episodic response, but also the frequency and duration of streamflow (Leopold et al., 1964). Drainage basin morphology, channel bed and bank soils, and geologic strata also affect flow conditions (Rosgen, 1994). For our purposes, the most critical component of streamflow is groundwater, which forms the basis for defining intermittent and perennial streams and, ultimately, determines effluent (gaining) or influent (losing) stream conditions. Streamflow and water quality characteristics (physical, chemical, and biological) are also influenced by urbanization and the overall land-use composition of the watershed (McMahon and Cuffney, 2000).

Without long-term data, we currently must rely on rules of thumb and best professional judgment to discern whether a stream is perennial or intermittent. One such example is the tacit understanding that observation of streamflow during drought conditions or the late summer months of normal climatic years suggests perennial flow. In another example, the rule of thumb that 1 square mile of drainage area produces an average annual baseflow of one cubic foot per second is commonly used in hydrologic assessments in the Coastal Plain. These generalizations, however, may be less applicable in headwater systems where seeps and other site-specific factors can influence streamflow (e.g., channel alteration or hydromodification from urbanization). Flow may be present but influenced by recent rainfall or conversely absent during "drier" climatic or drought conditions. Channel downcutting in response to increased flows may intercept the water table and significantly reduce the "typical" drainage area threshold for perennial stream determinations.

It is interesting to note that although streamflow is perhaps the least exact of the indicators due to the variability described above, recent research has suggested otherwise. For example, discriminant analysis by Fritz et al. (2008) across several US sites identified the presence of water as the *most* important factor in differentiating flow regime, and another study in South Carolina identified baseflow as the most accurate determinant of flow status (Fritz et al., 2013a).

Notwithstanding these results, the complex nature of streamflow is still an important consideration in flow regime evaluation. In practice, the presence of water alone does not provide conclusive evidence that the stream is perennial, nor does the absence of flowing water provide conclusive evidence that a stream is intermittent. Our position, therefore, is that the presence of streamflow should be used more as corroborative evidence in support of the perennial stream determination, or weighted according to time of year observations and antecedent moisture conditions.[4]

4. The Palmer Drought Severity Index (PDSI) is a useful, web-available resource for gauging regional antecedent moisture conditions. PSDI can be researched for specific regions at the National Weather Service Climate Prediction Center (http://www.cpc.ncep.noaa.gov/products/monitoring_and_data/drought.shtml).

Qualitative Indicators for Perennial Stream Determinations

Date: **Project ID:**

Evaluator(s):   **Reach ID:**

To the User: *Positive evidence of a majority –* ***4 or more out of 7*** *– of these indicators provides corroborative evidence that the assessment reach likely has a perennial flow regime. Check the corresponding box for each indicator where corroborative evidence of perennial flow is present based on professional judgement.*

1. **Streamflow** – *The presence or absence of flowing water may be used as an indicator of flow regime dependent upon antecedent moisture conditions.* Dry season flow, or flow observed during extended periods of low rainfall, should be considered corroborative evidence of a perennial flow regime. Dry channel conditions during the wet season, or during extended periods of sustained rainfall, should be considered corroborative evidence of a non-perennial flow regime. ☐
2. **Channel Geometry** – *Perennial streams typically exhibit a defined and consistent geometric shape, whereas intermittent streams typically lack defined and consistent internal geometry (indicator observed in cross-section and along longitudinal axis of channel).* ☐
3. **Streambed Soils** – *Perennial streams with fine-textured substrates tend to lack soil morphological characteristics associated with a fluctuating water table (redoximorphic features). Perennial streams with sandy substrates tend to show a relatively even distribution of organic material throughout the profile. Intermittent streams with fine-textured substrates may show redoximorphic features indicative of a fluctuating water table. Intermittent streams with sandy substrates may show organics unevenly distributed throughout the matrix, organic streaking evident in the surface horizon, and/or development of organic pans (i.e., accumulation of organic materials in bed layers).* ☐
4. **Instream Vegetation** – *Perennial streams tend to exhibit cross-sectional variability with respect to instream vegetation. The presence of stable depositional features (e.g., point bars, benches, islands, etc.) provides a variety of substrates for colonization of plant species that are frequently different from those found on the adjacent floodplain or along the top-of-bank. Intermittent streams do not tend to exhibit cross-sectional variability due to the lack of stable substrate for colonization in the more erosional setting of an intermittent channel.* ☐
5. **Macroinvertebrates** – *Perennial streams are typified by the presence of macroinvertebrate species having an aquatic life cycle of greater than one year.* For areas with regional lists of obligate perennial taxa, presence of these taxa should be used as corroborative evidence of perenniality irrespective of aquatic life cycle duration. Presence of intermittent taxa should be interpreted with caution and should only be used as corroborative evidence of intermittency if a dominance of these species is present (along with few to no perennial taxa). ☐
6. **Vertebrates** – *Perennial streams may support populations of fish and other vertebrates with an aquatic life cycle of greater than one year; however, care should be taken to consider the ability of certain vertebrate species to migrate upstream and persist in deep pools of intermittent streams.* ☐
7. **Offsite Resources** – *Among available offsite resources, USGS 7.5-minute topographic quadrangle maps, testimony from long-term residents and local interest groups (e.g., Isaac Walton League, etc.), local professionals (e.g. hydrologists, county agents, Natural Resources Conservation Service technicians, surveyors, foresters, field engineers, wetland ecologists), county soil survey maps, and aerial photography are particularly useful as corroborative evidence of flow regime.* Additional resources could include GIS-level data from localities and or agencies, as well as the National Hydrography Dataset if available. ☐

Final Determination (check one):

Perennial ☐   **Non-Perennial** ☐

Rationale (include relevant observations for all indicators):

Developed by: Douglas A. DeBerry and Travis W. Crayosky
Version 3.0 (March 2017)
Contact Info: dadeberry@wm.edu
travis.crayosky@stantec.com

FIG. 4.1.3.1 Field checklist for Qualitative Indicators.

Channel Geometry

Perennial streams typically exhibit a defined and consistent geometric shape whereas intermittent streams typically lack defined and consistent internal geometry (indicator observed in cross-section and along longitudinal axis of channel). Channel networks evolve to dispose potential and kinetic energy of flowing water in the most efficient manner possible (Hewlett, 1969; Leopold et al., 1964). Rosgen (1994) notes that, "...natural channels attempt to maintain a dynamic balance between sediment load and available energy such that the stream exhibits natural adjustments in sinuosity that result in maintaining a slope such that the stream neither aggrades nor degrades." The resultant stream pattern is based on the sine curve, which produces a symmetrical meander path along a stream's longitudinal profile.

Montgomery and Buffington (1993), Schumm (1977), and Rosgen (1994) have utilized these principles to establish stream classification methods that determine site-specific relationships, predict stream behavior, and/or provide the basis for comparison against other streams exhibiting similar characteristics. Relationships between radius of curvature, meander wavelength, sinuosity (Langbein and Leopold, 1966), and width/depth ratios at bankfull conditions as noted in Rosgen (1996) are widely accepted and applied in the discipline of fluvial geomorphology.

Based upon these relationships, width, depth, discharge, and sinuosity tend to increase as streams flow in the downstream direction. Slope and sediment load size, however, typically decrease in the downstream direction (Leopold et al., 1964; Rosgen, 1994). Perennial streams also tend to exhibit better-developed bed and bank conditions and more pronounced natural levees and floodplains due to the dynamic equilibrium established along the stream profile under a continual flow regime. These features are generally absent in intermittent streams, which are influenced by episodic, non-steady flow conditions. This distinction is reflected in the erosional-to-depositional geomorphic setting transition (Tabacchi et al., 1998), which roughly coincides with the intermittent-to-perennial continuum. Although direct channel and riparian buffer disturbances or changes in streamflow and sediment regimes can alter these characteristics at any point along this continuum, in general, streams that exhibit defined and consistent geometry are reflective of perennial flow conditions.

We have seen no significant changes in the recent scientific literature that would substantively change the context of this indicator. Some studies from Midwestern states (KY, IN, OH) have indicated that channel geometry was among the most important factors in predicting flow characteristics of streams (Svec et al., 2005; Johnson et al., 2009), and these observations only corroborate the use of channel geomorphic concepts for flow regime determinations.

Streambed Soils

Perennial streams with fine-textured substrates tend to lack soil morphological characteristics associated with a fluctuating water table (redoximorphic features). Perennial streams with sandy substrates tend to show a relatively even distribution of organic material throughout the profile. Intermittent streams with fine-textured substrates may show redoximorphic features indicative of a fluctuating water table. Intermittent streams with sandy substrates may show organics unevenly distributed throughout the matrix, organic streaking evident in the surface horizon, and/or development of organic pans (i.e., accumulation of organic materials in bed layers).

The presence of redoximorphic features such as iron oxide concentrations in a shallow soil profile typically indicates a fluctuating water table (USDA-NRCS, 1998). In waterlogged soil, oxygen is generally absent, having been overutilized by soil microbes beyond supply. The reintroduction of atmospheric oxygen to pore spaces creates a concentration gradient along which free oxygen is bound by reduced (ferrous) iron, creating the classic rust-colored zones in a soil matrix (Vepraskas, 1994). For this to occur, it follows that at some point the water table must be below the soil surface or, in the case of streams, below the stream bed, and that the stream ceases to flow during those times. By contrast, soils under constant inundation are likely to support gleyed or very low chroma matrices (chroma/1 or less in Munsell soil color notation, Kollmorgen Instruments Corporation, 1990).

An important caveat to chroma/redoximorphic feature indicators is that these soil attributes are most readily identifiable in sandy loam to clay soils, and are not easily observed in loamy sand or coarser soils due to a low surface-to-volume ratio and lack of stable macropore formation (Vepraskas, 1994). In sandy soils, iron coatings more readily leach from particle surfaces and move through the porous medium (usually downward), leaving no available iron for the redox reactions that create mottling. However, in sandy soils other indicators such as organic streaking (signifying translocation of fine particulate organic matter as the water table drops) and organic pans or bedding (leaf litter in layers below the surface sediments) would suggest that the stream ceases flow during drier periods. Conversely, an even distribution of organic coatings throughout the soil profile would indicate a uniformly wetted medium (Environmental Laboratory, 1987)—one that would likely develop under a perennial flow regime.

Because of the high variability in streambed texture and the overlap in substrate conditions from intermittent to perennial streams (Gordon et al., 1992), the applicability of texture analyses in perennial flow determinations is somewhat limited. This is due in large part to the notion that substrate texture is more dependent on basin morphometry and local geology/sedimentation rates than flow regime (Rosgen, 1996). For example, streams in the western regions of Virginia that formed over shallow bedrock typically have substrates composed of gravel or larger-sized rock materials. In these conditions, the soil indicators above cannot be used. However, in situations where gravel/cobble represents the dominant texture class, perennial streams tend to exhibit greater sorting of the finer particle sizes (sand, silt, clay) in the streambed, with removal from riffles and accumulation in pools (Leopold et al., 1964). By contrast, in intermittent streams with gravel/cobble-dominant substrates, finer-textured particles show minimal sorting.

One condition that we feel needs to be addressed at least theoretically if not empirically is the "oxyaquic" moisture regime. An oxyaquic regime is defined as a situation in which soils are saturated but not chemically reduced (Rabenhorst and Parikh, 2000). Therefore, under oxyaquic conditions, a soil could be saturated year-round but lack redoximorphic features due to the prevalence of dissolved oxygen in the water (Vaughan et al., 2009). In many cases, the hydrology in these systems is derived from oxygenated groundwater. In Virginia, this type of regime could theoretically occur in perennial streams with high dissolved oxygen content, either due to oxygenated groundwater or from incorporation of atmospheric oxygen at the water surface. Under these conditions, the abundance of dissolved oxygen would presumably preclude the development of redoximorphic features throughout the portions of the streambed where water is flowing; thus, soil samples from a perennial streambed would not be expected to show redoximorphic feature development (e.g., iron-oxide concentrations), which is consistent with the current interpretation of this indicator. In these circumstances, the other types of soil observations (organics, sorting) could be used to support a decision on the streambed soils indicator.

Instream Vegetation

Perennial streams tend to exhibit cross-sectional variability with respect to instream vegetation. The presence of stable depositional features (e.g., point bars, benches, islands, etc.) provides a variety of substrates for colonization of plant species that are frequently different from those found on the adjacent floodplain or along the top of bank. Intermittent streams do not tend to exhibit cross-sectional variability due to the lack of stable substrate for colonization in the more erosional setting of an intermittent channel.

There are two axes of organization for instream vegetation communities along the intermittent-perennial continuum in headwater streams: (1) longitudinal, and (2) cross-sectional (Hupp, 1983; Tabacchi et al., 1998). These axes are not mutually exclusive, but rather provide different lines of evidence for the same geomorphic and hydrologic phenomena that are occurring along this continuum (Hupp and Osterkamp, 1996; Bendix and Hupp, 2000). As noted above, although hydrology is variable, geomorphic changes are more predictable and can provide corroborative evidence of flow regime, suggesting that the instream vegetation patterns observed as a result of these changes may also be used to this end.

The longitudinal transition from extreme headwaters to lower stream reaches is characterized by a change in geomorphic setting from erosional (E) to transitional (T) to depositional (D) (Tabacchi et al., 1998). Intermittent streams are most frequently found in the E region at a steeper longitudinal gradient (>4%) than the T (1%–4%) and D (<1%) regions. As a result, intermittent streams tend to be structurally constrained by V-shaped valleys with minimal lateral adjustment, resulting in a net removal and export of substrate materials; therefore, intermittent channels lack the stable instream depositional features that characterize the downstream perennial reaches. The implication for intermittent riparian vegetation communities is a general lack of vegetation within the channel due to the unstable substrate conditions and destructive, higher-velocity flows, or a dominance by plants along the banks that are compositionally similar to those inhabiting the active floodplain (Sheath et al., 1986; Tabacchi et al., 1998). By contrast, perennial streams are usually found in the T and D zones where the lower gradient reduces the in-channel kinetic energy necessary to keep heavier materials in suspension. The resultant formation of instream geomorphic landforms such as bars, benches, and islands creates an increase in available niche space for a wider variety of species to colonize (Bendix and Hupp, 2000). Another factor is the unidirectional flow along the continuum, which results in a net transport of propagules (seeds and other reproductive parts) in the downstream direction. This is manifested in a net export of in-channel propagules from the headwaters (E zones) toward habitats lower in the watershed (T and D zones). Thus, there is a general pattern of increase in species richness and diversity along the transition from headwaters to downstream reaches attributed to changes in geomorphic setting, gradient, flow, and positive feedback from the inhabiting vegetation (Tabacchi et al., 1998; Tabacchi et al., 2000; Bendix and Hupp, 2000).

For a perennial stream determination, the above phenomena are most efficiently observed along the cross-sectional axis of the channel. This is due in part to the reality that field technicians will rarely have access to the entire longitudinal intermittent-perennial continuum described above, and by necessity will focus field efforts on target assessment reaches where property access has been granted. Fortunately, the phenomena described above are readily observable in the cross-sectional direction—lateral instream vegetation zonation is affected by landscape position (E-T-D) as a direct consequence of the presence/absence of stable depositional substrates for colonization, available propagules, destructive flows, and variable geomorphic landforms (Hupp, 1983).

Studies on Northern Virginia perennial streams in the Piedmont and Ridge and Valley physiographic provinces have documented consistent vegetation patterns corresponding to the following features: (1) depositional bars, inundated about 40% of the time, which lack persistent woody vegetation but are colonized by grasses and forbs adapted to live in the high-energy, frequently flooded environment of the lowest channel elevations; (2) active-channel shelves, inundated about 10%–25% of the time, which are colonized by a low shrub thicket community type with woody species tolerant of energetic flows such as willows (*Salix* spp.), alders (*Alnus* spp.), and shrubby dogwoods (*Cornus* spp.); and (3) the floodplain, which has a 1–3 year flood recurrence and a diverse flora characterized by species such as black walnut (*Juglans nigra*), hickories (*Carya* spp.), silver maple (*Acer saccharinum*), and hackberry (*Celtis occidentalis*) (Hupp, 1983; Osterkamp and Hupp, 1984; Hupp and Osterkamp, 1985). Although species found in Coastal Plain perennial streams would be expected to be different from those of the Piedmont and Ridge and Valley physiographic provinces, the general community transitions found on riparian landforms would be similar—floodplain species are expected to be less tolerant of destructive flooding than depositional bar or in-channel shelf species, providing a discernable zonation in the cross-sectional direction (Bendix and Hupp, 2000). This pattern is effectively absent in the more erosional environment of intermittent streams as described above. Thus, riparian vegetation within the intermittent reach is characterized by floodplain species that may extend channelward as far as the bank, with in-channel vegetation typically lacking due to the erosional setting and the lack of instream variability (Hupp and Osterkamp, 1996).

Given the importance of geomorphology to the vegetation patterns noted above, one could question whether the instream vegetation indicator is not simply duplicating the geomorphic landform considerations in Indicator 2 above. In our experience, observations of vegetation on instream depositional features give a direct indicator of substrate stability, an important factor in the assessment of flow regime. Although there are techniques that could be used to make assessments of bedform, bar, and shelf stability, instream vegetation provides an accessible and readily apparent analog for this type of consideration. We expect that as scientists and managers put these concepts into practice, regional species lists will be developed that will provide practitioners with additional information on which to assess this indicator. Along these lines, recent syntheses on vegetation community development in fluvial hydrosystems have produced plant strategy models (Bornette et al., 2008) and probabilistic species response curves (Merritt et al., 2010) that support the theoretical approaches outlined in this indictor.

Macroinvertebrates

Perennial streams are typified by the presence of macroinvertebrate species having an aquatic life cycle of greater than 1 year. At face value, this indicator is test-positive: a population of organisms with an aquatic life cycle of greater than 1 year (e.g., gilled aquatic mollusks, certain nonburrowing aquatic arthropods, etc.) should clearly demonstrate the presence of water throughout the year. However, this concept is limited in application as several aquatic macroinvertebrates may not even persist for a year (e.g., univoltine insects), much less retain an aquatic life cycle to that extent (Merritt and Cummins, 1996; Peckarsky et al., 1990; Pennak, 1989). Also, we recognize that a stream system that lacks a macrobenthic community with the above characteristics, or a stream that lacks a community altogether, could reflect poor water quality conditions rather than flow regime per se. Unfortunately, biological metrics in general are sensitive to water quality conditions, so we caution users to consider watershed conditions and stormwater inputs with using this indicator. Accordingly, the macroinvertebrate indicator should be used as corroborative evidence in combination with other observations.

In efforts to refine and broaden the scope of this indicator, colleagues in the region have begun to develop lists of macroinvertebrates at reasonably manageable taxonomic levels (e.g., Order or Family) that are found in perennial (or conversely, intermittent) streams. To this end, a few concepts deserve note: (1) among insect Orders, Neuroptera (spongilliflies), Lepidoptera (moths and butterflies), Plecoptera (stoneflies), and Odonata (dragonflies) frequent perennial streams while Hemiptera (water bugs), Coleoptera (beetles), Diptera (midges and flies), and select taxa within Ephemeroptera (mayflies) and Trichoptera (caddisflies) frequent temporary streams (Williams, 1996); (2) perennial streams typically support a more diverse assemblage of species than intermittent streams (Feminella, 1996; Williams, 1996; Vannote et al., 1980); (3) the hyporheic zone (zone of soil saturation beneath the stream bed) may serve as a refuge for macroinvertebrate organisms during dry periods (del Rosario and Resh, 2000; Miller and Golladay, 1996); (4) some characteristically "perennial" macroinvertebrates may possess several adaptations allowing successful colonization of intermittent streams [e.g., migration to pools or the hyporheic zone, diapause (physiological dormancy) in the larval or egg stages, temperature-linked development] (Williams, 1996; Delucchi, 1989); and (5) the relative abundance of certain macroinvertebrates may be habitat-specific rather than stream-specific, depending on the nature and complexity of habitat available (Grubaugh et al., 1996).

Since 2002, much attention has been given to the use of macroinvertebrate indicators for streamflow designation. Studies conducted in North Carolina (Williams, 2005), South Carolina (Fritz et al., 2013b), New England (Santos and Stevenson, 2011), the Pacific Northwest (Blackburn and Mazzacano, 2012; Nadeau et al., 2015), British Columbia (Price et al., 2003), and Europe (Smith and Wood, 2002; Wood et al., 2005; Bonada et al., 2007; Datry, 2012), among others, report variable results when using stream macroinvertebrates for flow regime determinations. A common theme is that perennial flow indicator organisms are regionally specific and/or species-specific; therefore, although it would be desirable to generalize across larger geographic territories or at higher taxonomic levels as outlined above, such generalizations may not be tenable with current scientific understanding. This is important because identification of stream macroinvertebrates to the Species or even Genus level requires advanced technical skill that many practitioners do not have. Thus, in rapid field assessments, the macroinvertebrate community is often documented at the Family level (or higher). For the Qualitative Indicators, a better approach may be to acknowledge functional group assemblages based on general characteristics of a community sample, making use of the river continuum concept (Vannote et al., 1980). To the extent that indicator organisms at higher taxonomic levels can be identified and validated empirically for perennial stream determinations in Virginia, the applicability of the macroinvertebrate indicator will only strengthen.

Vertebrates

Perennial streams may support populations of fish and other vertebrates with an aquatic life cycle greater than 1 year; however, care should be taken to consider the ability of certain vertebrate species to migrate upstream and persist in deep pools of intermittent streams. The presence of fish and other gilled vertebrates (e.g., some aquatic amphibians, Martof et al., 1980) is a reliable indicator of perennial flow in streams, based on the aquatic life history of most stream vertebrates (greater than 1 year). Most research on fish distribution and stream size has been presented in the context of stream order (Matthews, 1986; Jenkins and Burkhead, 1994), with a focus on higher-order streams with perennial flow. The few studies that have been conducted in first-order intermittent streams concluded that localized populations generally persist in downstream perennial reaches during dry periods and recolonize intermittent reaches annually (Maurakis et al., 1987; Matthews, 1998). In addition, small species and young of the year may be found in reaches upstream or downstream of permanent impoundments, presumably colonizing intermittent reaches from the permanent pool of the reservoir during higher flows. For these reasons, it may be important to ascertain the source of the population encountered at a given point in a stream. If we acknowledge that some species can migrate into intermittent waters on an annual cycle, it may be appropriate to weight fish data seasonally (i.e., the presence of certain fish during dry summer periods may be more conclusive evidence of perennial flow than during high spring flows). The concepts for this indicator have been corroborated by more recent research, with the same caveats as noted above (e.g., Johnson et al., 2009).

Offsite Resources

Among available offsite resources, US Geologic Survey (USGS) 7.5-min topographic quadrangle maps, testimony from long-term residents and local interest groups (e.g., Isaac Walton League, etc.), local professionals (e.g., hydrologists, county agents, Natural Resources Conservation Service technicians, surveyors, foresters, field engineers, wetland ecologists), county soil survey maps, and aerial photography are particularly useful as corroborative evidence of flow regime. A reliable account from a long-term resident is probably the best-case scenario for offsite review. However, because reliable local resident accounts are often difficult to acquire, references such as USGS maps represent at least a first approximation (USGS, 2000). Instances where site conditions may diverge from existing mapping are common in the Coastal Plain physiographic province where drainage anomalies can occur, particularly those caused by extensive groundwater discharge from water-bearing stratigraphic layers (i.e., gravel outwash plains or shell-marl deposits; Johnson et al., 1993). It is important to note that drainage area, although useful during project review, is a poor analog for flow regime. Because discharge is related to basin morphometry, geology and soils, local climate and antecedent moisture, and channel precipitation (Leopold et al., 1964; Crayosky et al., 1999), drainage area tends to vary widely among perennial streams and even between perennial and intermittent streams.

Given the advancements in technology and accessibility of web-available GIS data, we believe that this indicator could be updated with links to resources that could be used as corroborative evidence of channel flow status. Some recent research has focused on use of the National Hydrography Dataset (NHD) for this purpose (see Fritz et al., 2013a; Nadeau et al., 2015). Although the implication is that the NHD data has a tendency to overestimate the extent of perennial streams in general, the consensus is that such data are still useful as corroborative evidence of flow status, which is consistent with our recommended approach to offsite resources. In Virginia, many localities have updated their GIS systems to include hydrographic data, and some are based on field reconnaissance; the inclusion of these types of data sets in the Qualitative Indicators would strengthen the offsite resources indicator accordingly.

SUMMARY

Qualitative field indicators of perennial flow in streams are presented as follows: (1) stream flow, (2) channel geometry, (3) streambed soils, (4) instream vegetation, (5) macroinvertebrates, (6) vertebrates, and (7) offsite resources. These indicators have been developed and tested over several years by the authors and others through field application of scientific concepts related to perennial and intermittent flow, although there has been no formal validation/calibration process for the method. The selection of indicators has taken into account the need for efficient and effective data collection as well as manageable and easily interpreted results. In field application, positive evidence of a majority (four or more out of seven) provides corroborative evidence that the assessment reach likely has a perennial flow regime. The updated version of the method presented here is currently being reviewed by DCR and the Virginia Department of Environmental Quality (DEQ) for inclusion as an approved perennial stream determination approach within their respective regulatory programs. It is important to note that while these concepts unify several disciplines, practical and objective application of the indicators must be conducted by qualified professionals with experience in such studies.

REFERENCES

Bendix, J., Hupp, C.R., 2000. Hydrologic and geomorphic impacts on riparian plant communities. Hydrol. Process. 14, 2977–2990.

Blackburn, M., Mazzacano, C., 2012. Using Aquatic Macroinvertebrates as Indicators of Streamflow Duration: Washington and Idaho Indicators. Report Prepared by the Xerces Society for Invertebrate Conservation.

Bonada, N., Rieradevall, M., Prat, N., 2007. Macroinvertebrate community structure and biological traits related to flow permanence in a Mediterranean river network. Hydrobiologia 589, 91–106.

Bornette, G., Tabacchi, E., Hupp, C., Puijalon, S., Rostan, J.C., 2008. A model of plant strategies in fluvial hydrosystems. Freshw. Biol. 53, 1692–1705.

Crayosky, T.W., DeWalle, D.R., Seybert, T.A., Johnson, T.E., 1999. Channel precipitation dynamics in a forested Pennsylvania headwater catchment (USA). Hydrol. Process. 13, 1303–1314.

Datry, T., 2012. Benthic and hyporheic invertebrate assemblages along a flow intermittence gradient: effects of duration of dry events. Freshw. Biol. 57, 563–574.

del Rosario, R.B., Resh, V.H., 2000. Invertebrates in intermittent and perennial streams: is the hyporheic zone a refuge from drying? J. N. Am. Benthol. Soc. 19, 680–696.

Delucchi, C.M., 1989. Movement patterns of invertebrates in temporary and permanent streams. Oecologia 78, 199–207.

Dunne, T., Moore, T.R., Taylor, C.H., 1975. Recognition and prediction of runoff-producing zones in humid regions. Hydrol. Sci. Bull. 3, 305–327.

Environmental Laboratory, 1987. Corps of Engineers Wetlands Delineation Manual. Technical Report Y-87-1, U.S. Army Engineer Waterways Experiment Station, Vicksburg, MS.

Feminella, J.W., 1996. Comparison of benthic macroinvertebrate assemblages in small streams along a gradient of flow permanence. J. N. Am. Benthol. Soc. 15, 651–669.

Fritz, K.M., Johnson, B.R., Walters, D.M., 2008. Physical indicators of hydrologic permanence in forested headwater streams. J. N. Am. Benthol. Soc. 27, 690–704.

Fritz, K.M., Hagenbuch, E., D'Amico, E., Reif, M., Wigington Jr., P.J., Leibowitz, S.G., Comeleo, R.L., Ebersole, J.L., Nadeau, T.-L., 2013a. Comparing the extent and permanence of headwater streams from two field surveys to values from hydrographic databases and maps. J. Am. Water Resour. Assoc. 49, 867–882.

Fritz, K.M., Wenerick, W.R., Kostich, M.S., 2013b. A validation study of a rapid field-based rating system for discriminating among flow permanence classes of headwater streams in South Carolina. Environ. Manag. 52, 1286–1298.

Gordon, N.D., McMahon, T.A., Finlayson, B.L., 1992. Stream Hydrology: An Introduction for Ecologists. John Wiley and Sons, Chichester, West Sussex, England. 526 pp.

Grubaugh, J.W., Wallace, J.B., Houston, E.S., 1996. Longitudinal changes of macroinvertebrate communities along an Appalachian stream continuum. Can. J. Fish. Aquat. Sci. 53, 896–909.

Hewlett, J.D., 1969. Principles of Forest Hydrology. University of Geogia Press, Athens, Georgia. 183 pp.

Hupp, C.R., 1983. Vegetation pattern on channel features in the Passage Creek Gorge, Virginia. Castanea 48, 62–72.

Hupp, C.R., Osterkamp, W.R., 1985. Bottomland vegetation distribution along Passage Creek, Virginia, in relation to fluvial landforms. Ecology 66, 670–681.

Hupp, C.R., Osterkamp, W.R., 1996. Riparian vegetation and fluvial geomorphic processes. Geomorphology 14, 277–295.

Jenkins, R.E., Burkhead, N.M., 1994. Freshwater Fishes of Virginia. American Fisheries Society, Bathesda, MD, p. 48.

Johnson, G.H., Beach, T.A., Burkhart, P.A., Harris, M.S., Herman, J.D., Autrey, P.I., 1993. The Geology Along the Lower James Estuary, Virginia. Department of Geology, College of William and Mary, Williamsburg, VA. 41 pp.

Johnson, B.R., Fritz, K.M., Blocksom, K.A., Walters, D.M., 2009. Larval salamanders and channel geomorphology are indicators of hydrologic permanence in forested headwater streams. Ecol. Indic. 9, 150–159.

Kollmorgen Instruments Corporation, 1990. Munsell Soil Color Charts. Munsell Color, Baltimore, MD.

Langbein, W.B., Leopold, L.B., 1966. River Meanders—Theory of Minimum Variance. U.S. Geological Survey Professional Paper No. 422-H, 15 pp.

Leopold, L.B., Wolman, M.G., Miller, J.P., 1964. Fluvial Processes in Geomorphology. Dover Publications, Inc., New York, NY. 522 pp.

Martof, B.S., Palmer, W.M., Bailey, J.R., Harrison III, J.R., 1980. Amphibians and Reptiles of the Carolinas and Virginia. University of North Carolina Press, Chapel Hill, NC. 264 pp.

Matthews, W.J., 1986. Fish faunal 'breaks' and stream order in the eastern and central United States. Environ. Biol. Fishes 17, 81–92.

Matthews, W.J., 1998. Patterns in Freshwater Fish Ecology. Chapman and Hall, New York, NY, pp. 607–617.

Maurakis, E.G., Woolcott, W.S., Jenkins, R.E., 1987. Physiographic analyses of the longitudinal distribution of fishes in the Rappahannock River, Virginia. ASB Bull 34, 1–14.

McMahon, G., Cuffney, T.F., 2000. Quatifying intensity in drainage basins for assessing stream ecological conditions. J. Am. Water Resour. Assoc. 36(6).

Merritt, R.W., Cummins, K.W., 1996. An Introduction to the Aquatic Insects of North America. Kendall/Hunt, Dubuque, IA.

Merritt, D.M., Scott, M.L., Poff, N.L., Auble, G.T., Lytle, D.A., 2010. Theory, methods and tools for determining environmental flows for riparian vegetation: riparian vegetation-flow response guilds. Freshw. Biol. 55, 206–225.

Miller, A.M., Golladay, S.W., 1996. Effects of spates and drying on macroinvertebrate assemblages of an intermittent and perennial prairie stream. J. N. Am. Benthol. Soc. 15, 670–689.

Montgomery, D.R., Buffington, J.M., 1993. Channel classification, prediction of channel response, and assessment of channel condition. University of Washington, p. 107.

Nadeau, T.-L., Leibowitz, S.G., Wigington Jr., P.J., Ebersole, J.L., Fritz, K.M., Coulombe, R.A., Comeleo, R.L., Blocksom, K.A., 2015. Validation of rapid assessment methods to determine streamflow duration classes in the Pacific Northwest, USA. Environ. Manag. 56, 34–53.

Osterkamp, W.R., Hupp, C.R., 1984. Geomorphic and vegetative characteristics along three northern Virginia streams. Geol. Soc. Am. Bull. 95, 1093–1101.

Peckarsky, B.L., Fraissinet, P.R., Penton, M.A., Conklin Jr., D.J., 1990. Freshwater Macroinvertebrates of Northeastern North America. Cornell University Press, Ithaca, NY.

Pennak, R.W., 1989. Freshwater Invertebrates of the United States, Third Edition: Protozoa to Mollusca. John Wiley and Sons, New York, NY.

Price, K., Suski, A., McGarvie, J., Beasley, B., Richardson, J.S., 2003. Communities of aquatic insects of old-growth and clearcut coastal headwater streams of varying flow persistence. Can. J. For. Res. 33, 1416–1432.

Rabenhorst, M.C., Parikh, S., 2000. Propensity of soils to develop redoximorphic color changes. Soil Sci. Soc. Am. J. 64, 1904–1910.

Rosgen, D.L., 1994. A classification of natural rivers. Catena 22, 169–199.

Rosgen, D.L., 1996. Applied River Morphology. Wildland Hydrology, Pagosa Springs, CO.

Santos, A.N., Stevenson, R.D., 2011. Comparison of macroinvertebrate diversity and community structure among perennial and nonperennial headwater streams. Northeast. Nat. 18, 7–26.

Schumm, S.A., 1977. The Fluvial System. Wiley and Sons, New York. 338 pp.

Sheath, R.G., Burkholder, J.M., Hambrook, J.A., Hogeland, A.M., Hoy, E., Kane, M.E., Morison, M.O., Steinman, A.D., Van Alstyne, K.L., 1986. Characteristics of softwater streams in Rhode Island. III. Distribution of macrophytic vegetation in a small drainage basin. Hydrobiologia 140, 183–191.

Smith, H., Wood, P.J., 2002. Flow permanence and macroinvertebrate community variability in limestone spring systems. Hydrobiologia 487, 45–58.

Svec, J.R., Kolka, R.K., Stringer, J.W., 2005. Defining perennial, intermittent, and ephemeral channels in eastern Kentucky: application to forestry best management practices. For. Ecol. Manag. 214, 170–182.

Tabacchi, E., Correll, D.L., Hauer, R., Pinay, G., Planty-Tabacchi, A.M., Wissmar, R.C., 1998. Development, maintenance and role of riparian vegetation in the river landscape. Freshw. Biol. 40, 497–516.

Tabacchi, E., Lambs, L., Guilloy, H., Planty-Tabacchi, A., Muller, E., Decamps, H., 2000. Impacts of riparian vegetation on hydrological processes. Hydrol. Process. 14, 2959–2976.

USDA-NRCS (U.S. Department of Agriculture, Natural Resources Conservation Service), 1998. In: Hurt, G.W., Whited, P.M., Pringle, R.F. (Eds.), Field Indicators of Hydric Soils in the United States. Version 4.0. USDA, NRCS, Ft. Worth, TX. 30 pp.

USGS (U.S. Geological Survey), 2000. The National Hydrography Dataset-Concepts and Contents. U.S. Geological Survey, Reston, VA.

Vannote, R.L., Minshall, G.W., Cummins, K.W., Sedell, J.R., Cushing, C.E., 1980. The river continuum concept. Can. J. Fish. Aquat. Sci. 37, 130–137.

Vaughan, K.L., Rabenhorst, M.C., Needelman, B.A., 2009. Saturation and temperature effects on the development of reducing conditions in soils. Soil Sci. Soc. Am. J. 73, 663–667.

Vepraskas, M.J., 1994. Redoximorphic Features for Identifying Aquic Moisture Regimes. North Carolina Agricultural Research Service, Raleigh, NC (Technical Bulletin 301). 33pp.

Wakeley, J.S., 2002. Developing a 'Regionalized' Version of the Corps of Engineers Wetlands Delineation Manual: Issues and Recommendations. ERDC/EL TR-02-20, U.S. Army Engineer Research and Development Center, Vicksburg, MS.

Williams, D.D., 1996. Environmental constraints in temporary fresh waters and their consequences for the insect fauna. J. N. Am. Benthol. Soc. 15, 634–650.

Williams, N.B., 2005. Relationship Between Flow Regime and Aquatic Macroinvertebrate Abundance in Headwater Streams in the Piedmont Region of North Carolina (MS Thesis). North Carolina State University, Raleigh, NC.
Wood, P.J., Gunn, J., Smith, H., Abas-Kutty, A., 2005. Flow permanence and macroinvertebrate community diversity within groundwater dominated headwater streams and springs. Hydrobiologia 545, 55–64.

FURTHER READING

Hupp, C.R., 1982. Stream-grade variation and riparian-forest ecology along Passage Creek, Virginia. Bull. Torrey Bot. Club 109, 488–499.
Hupp, C.R., 2000. Hydrology, geomorphology, and vegetation of coastal plain rivers in the southeastern United States. Hydrol. Process. 14, 2991–3010.
Mazzacano, C., Black, S.H., 2008. Using Aquatic Macroinvertebrates as Indicators of Streamflow Duration. Report Prepared by the Xerces Society for Invertebrate Conservation.
McCafferty, W.P., 1983. Aquatic Entomology. Jones and Bartlett, Portola Valley, CA.

Section 4.2

Stream Condition Methods

Chapter 4.2.1

The West Virginia Stream and Wetland Valuation Metric (WVSWVM) Crediting Procedures and Assessments in Developing a Stream and Wetland Mitigation Banking Site

Dane Cunningham, Walter Veselka and Ryan Ward
AllStar Ecology LLC, Fairmont, WV, United States

Chapter Outline

INTRODUCTION

The West Virginia Stream and Wetland Valuation Metric (WVSWVM) is the regulatory mechanism that serves to establish a value, in terms of credits, for aquatic impacts to streams and wetlands as well as the mitigation or restoration credits to offset these impacts. Prior to the creation of the WVSWVM in 2010, there was no tool in West Virginia that could be used to make both consistent comparisons between water bodies being impacted, and (if mitigation was required) to measure the success or quality of the restoration or enhancement activity.

The WVSWVM was developed and its implementation overseen by the Interagency Review Team (IRT), which consists of state and federal members from the West Virginia Division of Natural Resources (WVDNR), West Virginia Department of Environmental Protection (WVDEP), US Fish and Wildlife Service (USFWS), US Army Corps of Engineers (USACE), and the Natural Resources Conservation Service (USACE, 2010a,b). It is meant to assess and determine the value of a waterbody based on the "physical, chemical, and biological" conditions of a stream or wetland. The WVSWVM is specific to the state of West Virginia and is used to determine the value of existing aquatic features onsite as part of the compensatory mitigation/permitting for development projects. It is also used to help determine appropriate mitigation for aquatic features that will be impacted for both onsite restoration and mitigation banking projects in the state.

Rather than being a completely new method for evaluating and rating water body conditions, this method incorporates numerous peer-reviewed and scientifically accepted existing assessments, including the West Virginia Stream Condition Index (WVSCI) (Gerritson et al., 2000), the Rapid Bioassessment Protocols (RBP) (Barbour et al., 1999), the

Wetland and Stream Rapid Assessments. https://doi.org/10.1016/B978-0-12-805091-0.00012-8

Hydrogeomorphic (HGM) Functional Stream Assessment (USACE, 2010a,b), and basic water chemistry testing. The main change with the WVSWVM is that these earlier methods only quantified a portion of stream function and were not integrated together and combined with all the assessment methods to generate an overall index score. The WVSWVM is a scoring matrix that "normalizes" the range of assessment values and arbitrarily places them into scoring bins for comparison. Without the ability to assign a quality rating to a stream based on physical, chemical, and biological data, it is difficult to quantify the baseline conditions of a stream proposed to be impacted as well as the ecological lift of restoration activities necessary to offset associated impacts. This is important in West Virginia where a stream may physically score well due to intact riparian floodplain and hydrological cycles, yet be severely impacted chemically by acid mine drainage or other biological limiting factors (e.g., fecal coliform, TDS, conductivity). The other main improvement with the WVSWVM is that the method addresses wetlands as well as streams.

The WVSWVM was released and implemented in West Virginia as a compensatory mitigation tool in February 2010. The assessment tool is a Microsoft Excel spreadsheet that combines the results of physical, chemical, and biological assessment methodologies into the normalized scoring bins to generate an ecological currency (i.e., credit/debit). Credits or debits are based on the results of four different parameters for streams extracted from other methods: Rapid Bioassessment Procedure–RBP scores (Barbour et al., 1999), HGM (USACE, 2010a,b), WVSCI (Gerritson et al., 2000), and water chemistry (pH, dissolved oxygen, and specific conductivity). The HGM is only utilized for high-gradient intermittent/ephemeral streams with a channel slope of 4% or greater. The results of these assessments are entered into the spreadsheet to calculate either a baseline index score for restoration activities and associated credits, or an index score for a stream proposed to be impacted and associated debits. By incorporating existing and proven assessment methodologies, the IRT endeavored to create a tool that was scientifically and legally defensible and validated (see Figs. 4.2.1.1 and 4.2.1.2).

The current process for incorporating wetlands into the WVSWVM is based on the Cowardin classification (Cowardin et al., 1979). There are no indices used to measure the wetland function, nor are there any considerations given to the geomorphic setting of the wetland.

DETAILED DESCRIPTION OF WVSWVM

Streams

The data required to use the WVSWVM is divided into three different aspects of stream health and quality: physical, chemical, and biological. In addition, there are "impact" factors that consider the time frame or temporal loss between resource loss, and the creation of a new aquatic resource through mitigation. For example, there is a debit multiplier for impacts if the required mitigation is not already completed (i.e., in lieu fees versus existing mitigation bank). This biological, chemical, and physical data combined with impact factors (temporal loss between construction impact and replacement of the resource, temporal loss between the new resource being constructed and before it reaches maturity, and ensuring long-term protection via deed restrictions or conservation easements) and mitigation considerations/incentives (complexity and extent of restoration as well as riparian buffer zone width) comprise all the information necessary to use the WVSWVM.

Wetlands

With all the complexities associated with determining stream debits and credits (as described below) in terms of biological, chemical, and physical value, the wetland portion of the WVSWVM is exceedingly simplistic. Currently, wetlands are evaluated solely based on acreage and the associated Cowardin classification (Cowardin et al., 1979) of emergent, scrub-shrub, or forested. There is no criterion that incorporates any of the recognized wetland functions, including but not limited to water attenuation, provision of carbon sequestration, sediment and pollutant filtration, etc. The current WVSWVM bases wetland compensatory mitigation on acreage ratios (1:1 for open water, 1:2 for emergent, and 1:3 for scrub- shrub/forested wetlands). This means if one acre of forested wetland is removed or impacted, three acres of forested wetland must be created or the payment for three acres of wetlands must be made to the in lieu fee mitigation fund. The position of the wetland in the landscape, whether it is a floodplain system or remnant of a cattail slurry pond, are scored the same and values are only based on vertical strata of the vegetative communities within the wetlands.

Field	Value
USACE FILE NO./Project Name:	USACE File # xxx-xxxx-xxxx-xxx/Example Onsite Mitigation Project
IMPACT COORDINATES: (in Decimal Degrees)	Lat. 39.322758 / Lon. -80.826452
WEATHER:	75-degrees, fair skies
DATE:	February 11, 2015
IMPACT STREAM/SITE ID AND SITE DESCRIPTION: (watershed size (acreage), unaltered or impairments)	Stream 6, 12-acre watershed, impairments due to relic logging road crossings
MITIGATION STREAM CLASS./SITE ID AND SITE DESCRIPTION: (watershed size (acreage), unaltered or impairments)	Stream 12, 24.48-acre watershed, impairments due to cattle trampling
Comments:	6/30/2016
STREAM IMPACT LENGTH:	212
FORM OF MITIGATION:	RESTORATION (Levels I-III)
MIT COORDINATES: (in Decimal Degrees)	Lat. 39.378546 / Lon. -80.445328
PRECIPITATION PAST 48 HRS:	0
Mitigation Length:	315

	Points Scale	Range	Column No. 1- Impact Existing Condition (Debit)	Column No. 2- Mitigation Existing Condition - Baseline (Credit)	Column No. 3- Mitigation Projected at Five Years Post Completion (Credit)	Column No. 4- Mitigation Projected at Ten Years Post Completion (Credit)	Column No. 5- Mitigation Projected at Maturity (Credit)
Stream Classification:			Intermittent	Intermittent	Intermittent	Intermittent	Intermittent
Percent Stream Channel Slope			6.5	4.8	4.8	4.8	4.8
HGM Score (attach data forms):							
Hydrology			0.29	0.44	0.96	0.96	0.96
Biogeochemical Cycling			0.53	0.49	0.81	0.81	0.9
Habitat			0.44	0.38	0.57	0.64	0.8
Average			0.42	0.436666667	0.78	0.80333333	0.8866667
PART I - Physical, Chemical and Biological Indicators (Site Score)							
PHYSICAL INDICATOR (Applies to all streams classifications)							
USEPA RBP (High Gradient Data Sheet)		0-1					
1. Epifaunal Substrate/Available Cover	0-20		8	5	13	13	13
2. Embeddedness	0-20		8	6	14	14	14
3. Velocity/ Depth Regime	0-20		10	10	13	13	13
4. Sediment Deposition	0-20		9	6	16	16	16
5. Channel Flow Status	0-20		16	11	17	17	17
6. Channel Alteration	0-20		16	13	20	20	20
7. Frequency of Riffles (or bends)	0-20		11	6	17	17	17
8. Bank Stability (LB & RB)	0-20		10	14	18	18	18
9. Vegetative Protection (LB & RB)	0-20		10	14	16	18	18
10. Riparian Vegetative Zone Width (LB & RB)	0-20		12	16	18	18	20
Total RBP Score			Marginal 110	Marginal 101	Suboptimal 162	Suboptimal 164	Optimal 166
Sub-Total			0.55	0.505	0.81	0.82	0.83
CHEMICAL INDICATOR (Applies to Intermittent and Perennial Streams)							
WVDEP Water Quality Indicators (General)		0-1					
Specific Conductivity	0-90		601 (600-749 - 40 points)	328 (300-399 - 70 points)	328 (300-399 - 70 points)	328 (300-399 - 70 points)	328 (300-399 - 70 points)
pH	0-80 / 5-90		7.65 (6.0-8.0 = 80 points)	6.91 (6.0-8.0 = 80 points)	6.91 (6.0-8.0 = 80 points)	6.91 (6.0-8.0 = 80 points)	6.91 (6.0-8.0 = 80 points)
DO	10-30		9.96 (>5.0 = 30 points)	10.04 (>5.0 = 30 points)	10.04 (>5.0 = 30 points)	10.04 (>5.0 = 30 points)	10.04 (>5.0 = 30 points)
Sub-Total			0.75	0.9	0.9	0.9	0.9
BIOLOGICAL INDICATOR (Applies to Intermittent and Perennial Streams)							
WV Stream Condition Index (WVSCI)	0-100	0-1	52.38 (Fair)	50.23 (Fair)	50.23 (Fair)	50.23 (Fair)	50.23 (Fair)
Sub-Total			0.4238	0.4023	0.4023	0.4023	0.4023
PART II - Index and Unit Score							
Index			0.497	0.51955	0.74205	0.755383333	0.798716667
Linear Feet			212	315	315	315	315
Unit Score			105.4276	163.65825	233.74575	237.94575	251.59575

FIG. 4.2.1.1 Streams part I and II.

PART III - Impact Factors
(See instruction page to insert default values for MITIGATION BANKING and ILF)

Temporal Loss-Construction

Note: Reflects duration of aquatic functional loss between the time of an impact (debit) and completion of compensatory mitigation (credit).

Years	1
Sub-Total	0

Temporal Loss-Maturity

Note: Period between completion of compensatory mitigation measures and the time required for maturity, as it relates to function (i.e. maturity of tree stratum to provide organic matter and detritus within riparian stream or wetland buffer corridor).

% Add. Mitigation	Temporal Loss-Maturity (Years)
30%	30
Sub-Total	0.19892

Long-term Protection

% Add. Mitigation and Monitoring Period	Long-Term Protection (Years)
0 + 5/10 Year Monitoring	101
Sub-Total	0

PART IV - Index to Unit Score Conversion

Final Index Score (Debit)	Linear Feet	Unit Score (Debit)	ILF Costs (Offsetting Debit Units)
0.69622	212	147.59864	$118,078.91

PART V- Comparison of Unit Scores and Projected Balance

Final Unit Score (Debit) [No Net Loss Value]	147.59864	Mitigation Existing Condition - Baseline (Credit)	163.65825	Mitigation Projected at Five Years Post Completion (Credit)	233.74575	Mitigation Projected at Ten Years Post Completion (Credit)	237.94575	Mitigation Projected At Maturity (Credit)	251.59575
FINAL PROJECTED NET BALANCE					70.0875		74.2875		87.9375

Part VI - Mitigation Considerations (Incentives)

Extent of Stream Restoration

*Note1: Reference the Instructional handout to determine the correct Restoration Levels (below) for your project
*Note2: Place an "X" in the appropriate category (only select one).

☐ Restoration Level 1	
☐ Restoration Level 2	
☑ Restoration Level 3	

Compensatory Mitigation Plan incorporates HUC 12-based watershed approach? (Yes or No) *Note: HUC 12-based watershed approach required to obtain Stream Restoration incentive	Yes

Site	Impact Unit Yield (Debit)	Mitigation Unit Yield (Credit)
Onsite Mit. Ex.: Stream 6 (Imp.), Stream 12 (Mit.)	147.59864	[illegible]

Extended Upland Buffer Zone

*Note[1]: Reference Instructional handout for the definitions of the Buffer Zone Mitigation Extents and Types (below)
*Note[2]: Enter the buffer width for each channel side (Left Bank and Right Bank)
*Note[3]: Select the appropriate mitigation type

Buffer Width	Left Bank	
50	0-50	Preservation and Re-vegetation
	51-150	Preservation and Re-vegetation
Buffer Width	**Right Bank**	
50	0-50	Preservation and Re-vegetation
	51-150	Preservation and Re-vegetation
Average Buffer Width/Side	50	

	Straight Preservation Ratio
Final Mitigation Unit Yield	
162.684375	

FIG. 4.2.1.2 Streams part III–VI.

Clean Water Act Parameters

Physical

Stream assessment methodologies utilized in the WVSWVM that evaluate the physical and functional components of stream ecology consist of the HGM and RBP methods. The HGM approach is a "method for developing functional indices and the protocols used to supply these indices to the assessment of ecosystem functions at a site-specific scale" and is utilized only on ephemeral/intermittent streams with a 4% or greater slope (USACE, 2010a,b). This approach was developed by USACE and the "Operational Draft Regional Guidebook for the Functional Assessment of High-Gradient Ephemeral and Intermittent Streams in Western West Virginia and Eastern Kentucky" was released to the public in July 2010. In addition to the guidebook and supplemental data form, a Microsoft Excel spreadsheet is also required to input data and calculate functional capacity index (FCI) scores for hydrology, biogeochemical cycling, and habitat (see Fig. 4.2.1.1, HGM Score).

The FCI scores that are calculated after assessing a 100-ft stream reach and inputting data into the spreadsheet are relative to reference sites; scores range from 0 to 1 with a 1 representing reference or ideal conditions. The 12 parameters assessed in the field include canopy cover, channel substrate embeddedness, channel substrate size, channel bank erosion, large woody debris, riparian/buffer zone tree diameter, riparian/buffer zone snag density, riparian/buffer zone sapling/shrub density, riparian/buffer zone species richness, riparian/buffer zone soil detritus, herbaceous cover, and watershed land use.

Parameters that affect the hydrology FCI score include substrate embeddedness, substrate size, bank erosion, large woody debris, and watershed land use. Parameters that affect the biogeochemical cycling FCI score include channel substrate embeddedness, large woody debris, tree diameter at breast height, riparian/buffer zone sapling/shrub density, percent cover of herbaceous vegetation, riparian/buffer zone soil detritus, and watershed land use. The habitat FCI score is affected by all parameters except bank erosion. The three different FCI scores are then input into the WVSWVM, with the average of the three FCI scores affecting 50% of the overall index score.

The WVSWVM only utilizes the habitat assessment and physiochemical characterization field data sheet from the RBPs (see Fig. 4.2.1.1, USEPA RBP). The RBPs were originally developed in 1989 by the United States Environmental Protection Agency (USEPA) and were reissued in 1999 for public use (Barbour et al., 1999). This data sheet is broken up based on slope with a high-gradient version and low-gradient version and focuses on assessing 10 physical habitat aspects of stream ecology over a 100 m reach: epifaunal substrate/available cover, embeddedness (pool substrate characterization-low gradient version), velocity/depth regime (pool variability-low gradient version), sediment deposition, channel flow status, channel alteration, frequency of riffles or bends (channel sinuosity-low gradient version), bank stability, vegetative protection, and riparian vegetative zone width. Each individual parameter of the RBP data sheet is weighted equally in the WVSWVM with the RBP affecting 16.7% of the overall index score. However, the RBP affects 33% (double) of the overall index score if utilizing the WVSWVM for a low-gradient stream (<4%) where the HGM approach is not applicable.

Chemical

At each stream reach, specific conductivity, pH, and dissolved oxygen are sampled. These water chemistry values are important to biological function. While samples represent only a snapshot of current conditions, if a stream is chemically impaired (e.g., high acidity), it is often very difficult to gain ecological lift from the restoration and enhancement of physical features. Conversely, when a stream that is chemically impaired is impacted, despite its physical form, it would be valued significantly less in terms of ecological debits.

These indicators were historically utilized by the WVDEP and comprise 16.7% of the overall index score for high-gradient streams (≥4%) and 33% (double) of the overall index score for low-gradient streams (<4%).

Biological

Biological assessment methodologies are utilized in the WVSWVM on perennial and intermittent streams and consist of sampling, sorting, and identification methods for aquatic insects using methods developed by the WVDEP's Watershed Assessment branch. The method developed, known as WVSCI, generates an overall score that ranges between 0 and 100. This score is input into the WVSWVM and comprises 16.7% of the overall index score for high-gradient streams (≥4%) and 33.3% of the overall index score for low-gradient streams (<4%). However, if the stream is ephemeral and WVSCI data cannot be obtained, an index score will be generated based only upon physical and chemical data.

TABLE 4.2.1.1 Overall Scoring Process for Stream Impacts for the West Virginia Stream and Wetland Valuation Metric (WVSWVM)

Stream Type	HGM	RBP	Biological	Chemical
Low gradient (<4% slope)	NA	33.3%	33.3%	33.3%
High gradient (≥4% slope)	50%	16.7%	16.7%	16.7%

Summary of Scoring Approach

The RBP, HGM, biological, and chemical scores are weighted in the following manner to yield the overall index score for the stream reach proposed to be impacted or restored (see Table 4.2.1.1).

Impact Factors

WVSWVM takes into account three additional factors to help ensure that any aquatic function impacted is fully offset with mitigation efforts (see Fig. 4.2.1.2).

Temporal Loss Construction

To take into account any delay between impacts and construction of mitigation, WVSWVM uses the "temporal loss construction" factor. This factor represents "the number of years reflecting the duration of aquatic functional loss between the time of impact (debit) and completion of compensatory mitigation (credit)" (USACE, 2010a,b). This parameter incentivizes the completion of mitigation in a timely manner so that the temporal loss of stream function is offset; otherwise more credits are required to offset temporal loss debits.

Temporal Loss Maturity

This factor takes into account "the number of years representing the period between completion of compensatory mitigation measures and the time required for maturity, as it relates to function" (USACE, 2010a,b). A mitigation scenario that requires fully revegetating the riparian buffer will typically have a longer temporal loss maturity timeframe in comparison to mitigation in a forested setting where only supplemental plantings are necessary. This parameter incentivizes the use of restoration techniques that result in faster maturity, including utilizing larger planting stock and preserving existing trees.

Long-Term Protection

This factor takes into account "the number of years representing the period of protection for the proposed mitigation site" (USACE, 2010a,b). Protection measures consist of recording a conservation easement or deed restriction along the mitigation stream and associated riparian buffer in order to ensure ecological success. This factor incentivizes a long-term protection period (in perpetuity when feasible); otherwise more credits are required to offset debits.

WVSWVM Applications for Determining Stream Impacts (Debits)

To calculate the amount of stream debits that will be incurred as part of an impact to a stream, the user must obtain the following information and enter it into the WVSWVM:

- RBP, HGM (where applicable), WVSCI, and water chemistry data along a representative reach of the impacted stream.
- Length of impact (linear feet).
- Stream classification (perennial, intermittent, or ephemeral).
- Percent stream channel slope.
- Time between impact and mitigation construction (temporal loss construction factor).
- Time between completion of compensatory mitigation measures and time required for maturity (temporal loss maturity factor).
- Period of protection proposed for the mitigation site (long-term protection factor).

The results of the aforementioned variables calculate stream debits, which must be offset. An impacted resource with a higher ecological value (i.e., higher index score) will result in more debits when compared to a resource with a lower ecological value (e.g., native trout stream versus acid mine drainage impacted stream). Therefore, by assigning individual index scores to each stream, impacts can be quantified and offset to achieve "no net loss" of aquatic function.

WVSWVM Applications for Stream Mitigation (Credits)

To calculate the amount of stream credits that will be generated as part of stream restoration, enhancement, and/or preservation activities, the user must obtain the following information and enter it into the WVSWVM:

- RBP, HGM (where applicable), WVSCI, and water chemistry data along a reach that is representative of the baseline conditions of the stream proposed for restoration, enhancement, and/or preservation.
- Proposed RBP, HGM (where applicable), WVSCI, and water chemistry scores at maturity that reflect the proposed restoration, enhancement, and/or preservation activity.
- Length of mitigation (linear feet).
- Stream classification.
- Percent stream channel slope.
- Extent of stream restoration (Restoration Levels 1–3).
- Extended upland buffer zone width.
- Buffer zone mitigation type (none, preservation, preservation and supplemental planting, preservation and revegetation)

The results of the aforementioned variables calculate stream credits that are used to offset debits. A stream with a lower baseline index (e.g., channelized stream in an agricultural setting) score will have greater restoration potential and potentially generate more credits. Therefore, by accurately calculating a baseline index score and proposed index score at maturity, ecological lift through stream restoration, enhancement, and/or preservation is quantified using the same unit of measurement as stream debits, thus ensuring accurate accounting to achieve no net loss.

WVSWVM User Training and Calibration

User training and calibration is a critical step in the application of WVSWVM in order to ensure that credit/debit numbers calculated accurately represent the ecological lift gained (credit) or lost (debit). As with any ecological assessment methodology, attendance at trainings is essential for user calibration. Given the multifaceted nature of WVSWVM, numerous assessment methodology trainings are necessary to proficiently use the tool, including HGM training, RBP training, and benthic macroinvertbrate sampling, sorting, and identification training.

In September 2015, the USACE provided a workshop on the application of the HGM approach while also providing a short summary of WVSWVM applications. This 2-day training included 1 day of lecture focusing on the development of the method as well as background on the individual assessment components. The second day was spent in the field applying the method to numerous stream assessment reaches and discussing how to properly measure/evaluate individual components of each reach.

Formal trainings on the use of the RBP are not regularly provided by the USEPA. However, state agencies may periodically provide training on the methodology, particularly in order to help local watershed groups as part of stream assessment programs.

The WVDEP offers training on benthic macroinvertebrate sampling annually as part of their Watershed Assessment branch standard operating procedures training. Trainings such as this are beneficial when applying for a scientific collection permit, which is required by the WVDNR in order to sample benthic macroinvertebrates. Therefore, the steps necessary to generate a WVSCI score (sampling, sorting, and identifying) typically require a high level of training and experience.

The time spent applying WVSWVM to a stream assessment reach can vary greatly depending on numerous factors, including user experience, terrain, weather, vegetative cover, etc. However, once trained and calibrated, a typical practitioner will likely be able to complete the required field data collection along one stream assessment reach in 1–2 h. This data then needs to be entered into the spreadsheet and presented in a reviewable format, which could take an additional 1–2 h. This estimate does not take into account the sorting and identification of benthic macroinvertebrates in order to generate a WVSCI score. The time required for this process is variable and highly dependent upon the experience level of the sorter and identifier. In summary, with the exclusion of WVSCI score generation, a trained and competent practitioner could complete the field and office work necessary for WVSWVM to determine a credit or debit number along one stream assessment reach in 3–4 h.

WVSWVM Development, Validation, and Future Prospects

Streams

Stream assessment components for WVSWVM crediting were developed by federal and state regulatory agencies. Before the use of these methodologies, a peer review and vetting process has already taken place. Therefore, the state saved significant money and time by utilizing protocols familiar to environmental specialists to implement WVSWVM.

Since the release of WVSWVM Version 1.0 in February 2010, two additional versions of the metric have been released by the WV IRT via joint public notice:

Version 2.0

WVSWVM v2.0 was released on February 1, 2011, and incorporates several revisions, including mitigation site location data, removal of no net loss default, projected scores at 10 years column, extent of stream restoration incentive, and buffer zone width incentive. Additionally, the HGM approach was also incorporated into WVSWVM v2.0.

Version 2.1

WVSWVM v2.1 has been available as a working draft since September 1, 2012, and was released via joint public notice on August 19, 2013. Several revisions were incorporated into this version including header revisions to incorporate stream class and percent slope, recalibrated RBP values for ephemeral streams, added descriptions of restoration incentive levels, added 12-digit HUC watershed approach, and added extended upland buffer incentive for wetlands. Additionally, WVSWVM v2.1 is capable of assessing a stream as "sole preservation" with no mitigation activities required to generate credit; however, a stream must exhibit a baseline index score of 0.8 or greater to generate credits through sole preservation.

In addition to the release of updated WVSWVM versions, the USACE has also announced the intended release of the "Operation Draft Regional Guidebook for the Functional Assessment of High-Gradient Headwater Streams and Low-Gradient Perennial Streams in Appalachia" via a public notice issued on August 11, 2015 (USACE, 2015). The incorporation of a low-gradient HGM into the WVSWVM would provide additional quantitative data with regard to the physical component of the metric on perennial streams. Since the initial draft of this chapter, the Operational Draft Regional Guidebook for the Functional Assessment of High-Gradient Headwater Streams and Low-Gradient Perennial Streams in Appalachia [ERDC/EL TR-17-1] was released via joint Public Notice on September 8, 2017. An updated WVSWVM v2.1 was also released which incorporates the low-gradient HGM (USACE, 2017).

Wetlands

Currently, there is no criterion that incorporates any of the recognized wetland functions; however, the West Virginia Wetland Rapid Assessment Procedure (WVWRAP) is in development and expected to be incorporated into the WVSWM. The WVWRAP is "designed to quantify wetland conditions in terms of functional capacity and biological integrity" (ASWM, 2015) and will provide the quantitative data with regard to wetland function and value that is currently missing from the WVSWVM.

CREDIT GENERATION BY A MITIGATION BANK: THE BEAR KNOB MITIGATION BANK AND USE OF THE WVSWVM

The Bear Knob property is located in Upshur County, West Virginia, and is a mixture of agricultural pasture, mixed-mesophytic hardwood forest, and historic surface coal mine with a pronounced ridgetop along the eastern side. The site was acquired by AllStar Ecology, LLC, in March 2014 to develop as a mitigation bank for unavoidable stream and wetland impacts within the West Fork (HUC# 05020002) watershed. The following case study outlines the procedures and objectives for developing a mitigation bank while utilizing the WVSWVM to calculate credit generation.

Site Selection

The Bear Knob property consists primarily of open bottomland pasture, which has significantly degraded the quality of the streams and surrounding wetlands. In essence, the stream was straightened and allowed to downcut, effectively drying out riparian wetlands to maximize pasture for hay production. This is typical of many farms in the Appalachian region of the United States, where steep topography historically limited agricultural opportunities. The resulting pastured or cleared land not only makes stream reconstruction more feasible but also allows for woody vegetation planting. This presents a financial incentive in terms of mitigation banking in West Virginia because revegetation generates additional credit in the WVSWVM in comparison to supplemental planting or no planting at all in forested areas.

In addition to land-use practices such as grazing, historic agricultural practices such as the channelization of bottomland streams and the draining of wetlands through tile drains also occurred. Historically channelized streams are not only beneficial for stream restoration from a constructability standpoint, but also are beneficial in generating more stream credits. Credits are generated through ecological lift occurring over the proposed temporal loss maturity time period. The amount of lift is dependent on the level of work proposed, ranging from supplemental riparian buffer planting to full-scale Rosgen Priority Level 1 Restoration (Rosgen, 1997) with revegetation planting in the riparian buffer. The latter would allow for a greater increase in RBP and HGM (where applicable) scores because the proposed work will affect all geomorphic aspects of the stream, in comparison to just enhancing the riparian buffer and increasing vegetative protection.

A large benefit in terms of financial incentive for practicing full-scale restoration on a historically channelized stream is in the increased length that can be added through restoration of sinuosity. Restoring a channelized stream with a sinuosity value of 1.0 to a Rosgen C Stream Type (Rosgen, 1996) with a sinuosity value of 1.4 increases total stream length by 40%. Therefore, by selecting a mitigation site with a historically channelized stream, higher ecological lift can result in increased credits along the new length created through restoration of sinuosity.

Existing resources of high ecological value must also be accounted for when selecting a mitigation site. As stated before, past land use and its effects on the current condition of the land play an integral role in mitigation site selection. Mitigation sites with an abundance of high quality streams and wetlands are generally not preferred in West Virginia due to the low potential ecological lift and construction-related issues in environmentally sensitive areas. The Bear Knob site consists primarily of degraded streams and wetlands in the more heavily grazed bottomland areas with relatively stable resources along the forested hillsides. This allows for optimal ecological lift in combination with construction that will have little to no adverse environmental impacts.

Design and Implementation

Baseline assessments were conducted along all streams proposed for restoration, enhancement, and/or preservation at the Bear Knob site. These assessments were then summarized and tabulated by reach (see Table 4.2.1.2).

This is the first step in developing restoration objectives and goals because it gives project personnel an accurate assessment of the current physical, chemical, and biological conditions of a stream. Baseline assessments consist of utilizing all aspects of the WVSWVM where applicable. Entire streams are usually represented by one baseline reach that is representative of the overall conditions of the entire stream; however, multiple reaches may be utilized for one stream if significant changes in riparian buffer and vegetative protection occur. Therefore, if baseline assessments are broken up based on changes in riparian buffer, more credit can be generated due to increased ecological lift along nonforested streams with a riparian buffer dominated by herbaceous vegetation and saplings/shrubs.

Once baseline assessments were conducted, the results of the individual assessments were input into the WVSWVM and a baseline index score was calculated. Specific restoration goals were based on mimicking conditions identified at reference sites in the local watershed. Streams lacking a forested riparian buffer with poor to marginal in-channel conditions were selected to receive restoration, with the level of restoration depending primarily on slope and stream classification. Projected ecological lift for restoration streams was based on high-quality reference streams that were also used as natural channel design references. All RBP and HGM (where applicable) parameters were increased on streams receiving restoration, considering that the channel will be reconstructed through restoration techniques. Streams that had a forested riparian buffer but exhibited marginal in-channel conditions consisting primarily of unstable streambanks were selected

TABLE 4.2.1.2 Summary of Baseline WVSWVM Parameters by Reach

	Baseline Parameters								Reference Reach Used
	HGM FCI Scores								
Reach	Hydrology	Biogeochemical Cycling	Habitat	RBP	Conductivity	pH	DO	WVSCI	
Reach 12-01	0.44	0.49	0.38	101 (marginal)	328	6.31	10.04	50.23	B

to receive enhancement measures focusing on bank stabilization techniques. Streams that had a forested riparian buffer and exhibited stable in-channel conditions with the exception of vegetative protection, slight bank instability, and riparian vegetative zone width were selected to receive enhancement measures focusing on strategic supplemental plantings. Projected ecological lift for enhancement streams was based on the geomorphic parameter being enhanced. For example, a stream lacking sufficient vegetative protection and bank stability would only receive ecological lift through the increase of two RBP parameters (bank stability and vegetative protection). Reference streams were not used in projecting ecological lift for streams receiving enhancement and the parameters being increased were resultant of the proposed enhancement activities.

With baseline assessments conducted and credit generation proposed, the next step in developing a mitigation project is permit approval and construction. For the Bear Knob site and most stream and wetland restoration projects, obtaining a federal permit from the USACE is the critical step in turning years of planning and design into reality. Most stream and wetland restoration projects are authorized under Nationwide Permit 27; however, various other permits as well as coordination with state and federal agencies may also be required, including, but not limited to, a stream activity permit (WVDNR), a county building permit, NPDES permitting (WVDEP), USFWS Section 7 consultation, West Virginia State Historic Preservation Office (WVSHPO) Section 106 consultation, and floodplain coordination (County Floodplain Manager).

It is advisable to provide every component of supporting data that was utilized in the formation of the mitigation site plan, regardless of how large the plan becomes. The majority of the complex data involved with stream and wetland restoration projects results from natural channel design and WVSWVM crediting. Presenting this data in an organized manner is critical in order to ensure timely review of the project. For example, making the WVSWVM crediting portion of the plan match the construction drawings with regard to stream stationing helps the reviewer to cross reference proposed crediting and proposed restoration construction activities. Moreover, providing additional quantitative support for portions of the plan that may seem cautionary to reviewers could help to avoid delays through back and forth communication and revisions and ensure timely review.

Success Criteria and Monitoring

After the stream and wetland restoration areas reach final grade, but prior to planting, a postgrading survey will be conducted. The stream lengths and dimensions calculated during the postgrading survey are used as a baseline in future monitoring and credit calculations. For example, if a design proposes that a stream will be 1250 linear feet after restoration activities, but the postgrading survey reveals that the stream is only 1240 linear feet, future credit calculations (based on the credit release schedule) are revised to include the postgrading survey length. Additionally, future monitoring will also be based off the postgrading survey. Submittal of the postgrading survey and associated visual/photo monitoring is typically required to be submitted to the IRT within 60 days postconstruction. Essentially, the postgrading survey acts as a benchmark and all future monitoring cross sections and longitudinal profiles will be compared to the postgrading survey in determining whether the stream meets success criteria.

Yearly monitoring is the next step in demonstrating whether success criteria were met, and if not, proposing adaptive management measures. Ten years of monitoring is the new standard for mitigation projects in West Virginia, with varying degrees of intensity throughout the monitoring period. Year one monitoring begins during the first growing season and is typically one of the more intensive monitoring years, consisting of visual/photo monitoring, vegetation monitoring, geomorphic assessments, hydrology monitoring, wetland delineation, invasive species monitoring, and WVSWVM assessments. WVSWVM assessments, including all the associated components, are utilized during the monitoring period to demonstrate that the project is succeeding and progressing toward the proposed 5- and 10-year WVSWVM scores. At the Bear Knob site, WVSWVM assessments were proposed for years 1–5 and again at year 10 with the anticipation that increasing scores during the first 5 years will negate the need for these intensive assessments during years 6–9. WVSWVM assessments at year 10 should, at a minimum, meet the 10-year projected HGM (where applicable), RBP, water chemistry, and WVSCI scores.

REFERENCES

Association of State Wetland Managers, 2015. West Virginia State Wetland Program Summary. http://www.aswm.org/pdf_lib/state_summaries/west_virginia_state_wetland_program_summary_083115.pdf. Accessed 17 February 2017.

Barbour, M.T., Gerritsen, J., Snyder, B.D., Stribling, J.B., 1999. Rapid Bioassessment Protocols for Usein Streams and Wadeable Rivers: Periphyton, Benthic Macroinvertebrates and Fish, second ed. EPA 841-B-99-002. U.S. Environmental Protection Agency; Office of Water, Washington, DC.

Cowardin, L.M., Carter, V., Golet, F.C., LaRoe, E.T., 1979. Classification of Wetlands and Deepwater Habitats of the United States. U.S. Department of the Interior, Fish and Wildlife Service/Northern Prairie Wildlife Research Center, Washington, DC/Jamestown, ND.

Gerritson, J., Burton, J., Barbour, M.T., 2000. A Stream Condition Index for West Virginia Wadeable Streams. Tetra Tech, Inc., Owings Mills, MD.

Rosgen, D.L., 1996. Applied River Morphology. Wildland Hydrology, Pagosa Springs, CO.

Rosgen, D.L., 1997. In: A geomorphological approach to restoration of incised rivers. Proceedings of the Conference on Management of Landscapes Disturbed by Channel Incision. Wildland Hydrology, Pagosa Springs, CO.

U.S. Army Corps of Engineers, 2010a. Guidance on the West Virginia Interagency Review Team Initiatives Administered in Accordance With the 2008 Final Rule on Compensatory Mitigation for Losses of Aquatic Resources Within the U.S. Army Corps of Engineers, Huntington and Pittsburgh Districts. Public Notice No. LRH-2009-WV IRT Initiatives, U.S. Army Corps of Engineers, Huntington, WV.

U.S. Army Corps of Engineers, 2010b. Operational Draft Regional Guidebook for the Functional Assessment of High-Gradient Ephemeral and Intermittent Headwater Streams in Western West Virginia and Eastern Kentucky, ERDC/EL TR-10-11. U.S. Army Engineer Research and Development Center, Vicksburg, MS (FCI Calculator).

U.S. Army Corps of Engineers, 2015. Public Workshop for the Hydrogeomorphic Functional Assessment of High-Gradient Headwater Streams (Ephemeral and Intermittent) and Low-Gradient Perennial Streams in Appalachia. Public Notice No. LRH-2015-HGM/SWVM, U.S. Army Corps of Engineers, Huntington, WV.

U.S. Army Corps of Engineers, 2017. Implementation of the Hydrogeomorphic Approach for Application Within West Virginia, United States (U.S.) Army Corps of Engineers (Corps), Huntington and Pittsburgh Districts. U.S. Army Corps of Engineers, Huntington, WV. Public Notice No. LRH 2017-HGM.

FURTHER READING

U.S. Army Corps of Engineers, 2011. West Virginia Stream and Wetland Valuation Metric v2.0. U.S. Army Corps of Engineers, Huntington, WV.

U.S. Army Corps of Engineers, 2013. West Virginia Stream and Wetland Valuation Metric v2.1. U.S. Army Corps of Engineers, Huntington, WV.

Chapter 4.2.2

Virginia Unified Stream Methodology Case Study

Bettina Rayfield*[,a] and Jeanne Richardson[†,b]

**Virginia Department of Environmental Quality, Richmond, VA, United States, † U.S. Army Corps of Engineers, Norfolk District, Lynchburg, VA, United States*

Chapter Outline

INTRODUCTION

Prior to the development of the Unified Stream Methodology (USM), two methods were being used in Virginia, one developed by the USACE, the Stream Attribute Assessment Methodology (SAAM), and one developed by the VDEQ, the Stream Impact and Compensation Assessment Manual (SICAM). Prior to January 2007, the regulated community was required to conduct both assessment methodologies and comply with the most restrictive. Based on feedback from the regulated community and the USACE and VDEQ's desire to provide certainty to those parties seeking permits for stream impacts as well as developers of stream mitigation banks and in-lieu fee fund sites, USACE and VDEQ staff initiated the development of the USM.

The USM was developed to rapidly assess what the stream compensation requirements would be for permitted stream impacts and the amount of "credits" obtainable through implementation of various stream compensation practices. The USM is used to assess the condition of the stream to be impacted in order to determine the appropriate compensation requirement (CR), and the existing condition of the stream on which compensation is proposed to determine the compensation credits (CC) to be obtained from applicable compensation activities.

CONCEPTUAL FRAMEWORK FOR THE METHOD

The USM was designed for use on wadeable, intermittent or perennial, nontidal streams throughout Virginia. Ephemeral streams lack the consistent flow that typically leads to the channel-forming events or habitat characteristics observed in intermittent and perennial streams. Ephemeral stream function is heavily influenced by the condition of the riparian buffer (leaf input, shade, and structural support provided by the riparian buffer). Therefore, ephemeral streams can be assessed using the USM by evaluating the riparian buffer parameter only. The USM was not intended to apply to concrete, gabion, or riprap lined channels.

When assessing the stream impact site to determine the CR, the USM requires identification of a stream assessment reach (SAR), the assessment of the stream, the calculation of the Reach Condition Index (RCI), the determination of a stream impact factor (IF), and the identification of the length of SAR to be impacted.

---

a. The article expresses the views of the author and does not necessarily represent official policy of the Virginia DEQ.

b. The article expresses the views of the author and does not necessarily represent official policy of the US Army Corps of Engineers.

Wetland and Stream Rapid Assessments. https://doi.org/10.1016/B978-0-12-805091-0.00004-9

The length of the SAR is determined by significant changes in one or more of the four condition parameters listed below. The scoring of each parameter is based on the percentage of the reach with that condition allowing for variations in how the reaches are determined. The following four parameters are assessed for each reach:

(1) *Channel condition parameter*: This is a visual evaluation to determine the current condition of the channel cross-section, as it relates to the stream channel evolutionary process and to make a correlation to the current state of stream stability. Due to the influence of the stream channel condition to water quality and stream habitat functions, this parameter is weighted twice that of the other three parameters. This parameter ranges from optimal channel condition to severe channel condition. The optimal channel condition includes those channels that have not experienced erosion and those that have developed stable point bars and bankfull benches, typically at a lower level, by developing through the channel evolutionary process. Severe channel condition includes those channels that are deeply incised with vertical and/or lateral instability and are likely to continue to incise or widen.

(2) *Riparian buffer parameter*: This is a qualitative evaluation of the cover types and the percentage of the riparian buffer those cover types comprise. The 100 ft riparian buffer is assessed on both sides of the stream and then averaged to determine the condition index for this parameter. For ephemeral streams, the riparian buffer parameter is divided by two to obtain the RCI.

(3) *In-stream habitat parameter*: This is a visual assessment of the suitability of physical elements within the SAR to support aquatic organisms. The in-stream habitat parameter provides examples of physical elements for high- and low-gradient streams and requires the assessor to consider the appropriate habitat features for the type of stream being assessed. Examples of habitat features include a varied mix of substrate sizes, low embeddedness, a varied combination of water velocities and depths, presence of leaf packs and woody debris, and shade protection provided by overhanging vegetation.

(4) *Channel alteration parameter*: This is a visual observation of the direct impacts to the SAR from anthropogenic sources and an evaluation of whether the entire SAR has been impacted or only a portion has been impacted.

This information is recorded on the Stream Assessment Form (Form 1) (Fig. 4.2.2.1). With the Channel Condition Parameter counting twice, all the parameter scores are averaged by five, resulting in the RCI.

The RCI is then multiplied by the IF and the length of the reach to determine the CR of that reach. The IF is determined by an evaluation of the varying levels of impairment that occur to a stream as a result of permitted impacts. The USM identifies numerous types of impacts and categorizes them into four impact classifications: severe (complete elimination of stream through filling, impoundment, or hardening), significant (hardening of stream banks only), moderate (bridge piers in the channel), and negligible (no permanent impact). These IFs range from impacts that would result in a complete or near-complete loss of stream channel to activities that would result in negligible impacts to the stream channel.

In order to determine the CR, the following equation is used:

$$\mathrm{CR} = \mathrm{L_I} \times \mathrm{RCI} \times \mathrm{IF}$$

where CR = compensation credits required; L_I = length of impact (in linear feet); RCI = Reach Condition Index (Form 1); and IF = impact factor.

When a project consists of more than one reach, the CRs are added together to determine the total compensation requirement (TCR).

The USM also is used to assess a potential compensation site, whether it is a permittee-responsible site, a mitigation bank site, or an in lieu fee site, in order to determine the number of CCs to be obtained from the proposed compensation plan. Methods typically employed include preserving, enhancing, stabilizing, or restoring stream channels and their associated buffers. Credit is given for the type of activity conducted with the most credit for stream restoration (1 CC per foot of restoration) and less credit for stream enhancement (from 0.3 to 0.09 CC per foot). Stream restoration includes Priority 1, 2, and 3 (Rosgen, 1997) restoration practices. This process should be based on a reference condition/reach for the stream valley type and should include restoring the appropriate geomorphic dimension (cross-section), pattern (sinuosity), and profile (channel slope). Stream enhancement credits are applied when the full stream restoration is not required and includes the following activities: in-stream structures, habitat structures, bankfull bench creation, laying back the banks, bioremediation techniques (stabilizing the banks with biological material such as coir logs or brush mats), and planting the stream banks.

Credit for riparian buffer activities is based on the area of buffer being included in the project site and the activities conducted within that area. The calculation is designed to give 100% of the credit for the area defined by stream length and a 100 ft wide riparian buffer. Less credit is given for buffer activities in areas outside the 100 ft buffer due to the diminished effect on stream stability and water quality. Buffer activities include buffer reestablishment (invasive species removal

Stream Assessment Form (Form 1)

Unified Stream Methodology for use in Virginia

For use in wadeable channels classified as intermittent or perennial

Project #	Project Name	Locality	Cowardin Class.	HUC	Date	SAR #	Impact/SAR length	Impact Factor

Name(s) of Evaluator(s)	Stream Name and Information

1. Channel Condition: Assess the cross-section of the stream and prevailing condition (erosion, aggradation)

	Conditional Category					
	Optimal	**Suboptimal**	**Marginal**	**Poor**	**Severe**	
Channel Condition	Very little incision or active erosion; 80-100% stable banks. Vegetative surface protection or natural rock, prominent (80-100%). AND/OR Stable point bars/bankfull benches are present. Access to their original floodplain or fully developed wide bankfull benches. Mid-channel bars, and transverse bars few. Transient sediment deposition covers less than 10% of bottom.	Slightly incised, few areas of active erosion or unprotected banks. Majority of banks are stable (60-80%). Vegetative protection or natural rock prominent (60-80%) AND/OR Depositional features contribute to stability. The bankfull and low flow channels are well defined. Stream likely has access to bankfull benches, or newly developed floodplains along portions of the reach. Transient sediment covers 10-40% of the stream bottom.	Often incised, but less than Severe or Poor. Banks more stable than Severe or Poor due to lower bank slopes. Erosion may be present on 40-60% of both banks. Vegetative protection on 40-60% of banks. Streambanks may bevertical or undercut. AND/OR 40-60% of stream is covered by sediment. Sediment may be temporary/transient, contribute instability. Deposition that contribute to stability, may be forming/present. AND/OR V-shaped channels have vegetative protection on > 40% of the banks and depositional features which contribute to stability.	Overwidened/incised. Vertically/laterally unstable. Likely to widen further. Majority of both banks are near vertical. Erosion present on 60-80% of banks. Vegetative protection present on 20-40% of banks, and is insufficient to prevent erosion. AND/OR 60-80% of the stream is covered by sediment. Sediment is temporary/transient in nature, and contributing to instability. AND/OR V-shaped channels have vegetative protection is present on > 40% of the banks and stable sediment deposition is absent.	Deeply incised (or excavated), vertical/lateral instability. Severe incision, flow contained within the banks. Streambed below average rooting depth, majority of banks vertical/undercut. Vegetative protection present on less than 20% of banks, is not preventing erosion. Obvious bank sloughing present. Erosion/raw banks on 80-100%. AND/OR Aggrading channel. Greater than 80% of stream bed is covered by deposition, contributing to instability. Multiple thread channels and/or subterranean flow.	CI
Score	3	2.4	2	1.6	1	
NOTES>>						

2. RIPARIAN BUFFERS: Assess both bank's 100 foot riparian areas along the entire SAR. (rough measurements of length & width may be acceptable)

	Conditional Category							NOTES>>
	Optimal	**Suboptimal**		**Marginal**		**Poor**		
Riparian Buffers	Tree stratum (dbh > 3 inches) present, with > 60% tree canopy cover and a non-maintained understory. Wetlands located within the riparian areas.	**High Suboptimal:** Riparian areas with tree stratum (dbh > 3 inches) present, with 30% to 60% tree canopy cover and containing both herbaceous and shrub layers or a non-maintained understory.	**Low Suboptimal:** Riparian areas with tree stratum (dbh > 3 inches) present, with > 30% tree canopy cover and a maintained understory. Recent cutover (dense vegetation).	**High Marginal:** Non-maintained, dense herbaceous vegetation with either a shrub layer or a tree layer (dbh > 3 inches) present, with <30% tree canopy cover.	**Low Marginal:** Non-maintained, dense herbaceous vegetation, riparian areas lacking shrub and tree stratum, hay production, ponds, open water. If present, tree stratum (dbh >3 inches) present, with <30% tree canopy cover with maintained understory.	**High Poor:** Lawns, mowed, and maintained areas, nurseries; no-till cropland; actively grazed pasture, sparsely vegetated non-maintained area, recently seeded and stabilized, or other comparable condition.	**Low Poor:** Impervious surfaces, mine spoil lands, denuded surfaces, row crops, active feed lots, trails, or other comparable conditions.	
		High	Low	High	Low	High	Low	
Condition Scores	1.5	1.2	1.1	0.85	0.75	0.6	0.5	

1. Delineate riparian areas along each stream bank into Condition Categories and Condition Scores using the descriptors.
2. Determine square footage for each by measuring or estimating length and width. Calculators are provided for you below.
3. Enter the % Riparian Area and Score for each riparian category in the blocks below.

Ensure the sums of % Riparian Blocks equal 100

Right Bank	% Riparian Area>							0%		
	Score >									
									CI= (Sum % RA * Scores*0.01)/2	
Left Bank	% Riparian Area>							0%	Rt Bank CI > 0.00	CI
	Score >								Lt Bank CI > 0.00	0.00

3. INSTREAM HABITAT: Varied substrate sizes, water velocity and depths; woody and leafy debris; stable substrate; low embededness; shade; undercut banks; root mats; SAV; riffle poole complexes, stable features.

	Conditional Category				NOTES>>
	Optimal	**Suboptimal**	**Marginal**	**Poor**	
Instream Habitat/ Available Cover	Habitat elements are typically present in greater than 50% of the reach.	Stable habitat elements are typically present in 30-50% of the reach and are adequate for maintenance of populations.	Stable habitat elements are typically present in 10-30% of the reach and are adequate for maintenance of populations.	Habitat elements listed above are lacking or are unstable. Habitat elements are typically present in less than 10% of the reach.	CI
Score	1.5	1.2	0.9	0.5	

FIG. 4.2.2.1 Stream Assessment Form (Form 1).

(Continued)

Stream Impact Assessment Form Page 2

Project #	Applicant	Locality	Cowardin Class.	HUC	Date	Data Point	SAR length	Impact Factor
							500	1

4. CHANNEL ALTERATION: Stream crossings, riprap, concrete, gabions, or concrete blocks, straightening of channel, channelization, embankments, spoil piles, constrictions, livestock

NOTES>>

	Conditional Category					
	Negligible	Minor		Moderate		Severe
Channel Alteration	Channelization, dredging, alteration, or hardening absent. Stream has an unaltered pattern or has naturalized.	Less than 20% of the stream reach is disrupted by any of the channel alterations listed in the parameter guidelines.	20-40% of the stream reach is disrupted by any of the channel alterations listed in the parameter guidelines.	40 - 60% of reach is disrupted by any of the channel alterations listed in the parameter guidelines. If stream has been channelized, normal stable stream meander pattern has not recovered.	60 - 80% of reach is disrupted by any of the channel alterations listed in the parameter guidelines. If stream has been channelized, normal stable stream meander pattern has not recovered.	Greater than 80% of reach is disrupted by any of the channel alterations listed in the parameter guidelines AND/OR 80% of banks shored with gabion, riprap, or cement.
SCORE	1.5	1.3	1.1	0.9	0.7	0.5

REACH CONDITION INDEX and STREAM CONDITION UNITS FOR THIS REACH

NOTE: The CIs and RCI should be rounded to 2 decimal places. The CR should be rounded to a whole number.

THE REACH CONDITION INDEX (RCI) >>	0.00
RCI= (Sum of all CI's)/5	
COMPENSATION REQUIREMENT (CR) >>	0
CR = RCI X LF X IF	

INSERT PHOTOS:

DESCRIBE PROPOSED IMPACT:

FIG. 4.2.2.1, cont'd

and heavy planting) at 0.4 inner 100 ft/0.2 outer 100 ft, heavy planting (planting at 400 stems or more per acre) at 0.38 inner 100 ft/0.19 outer 100 ft, light buffer planting (supplemental planting or planting at less than 400 stems per acre) at 0.29 inner 100 ft/0.15 outer 100 ft, and preservation (no activity within the buffer other than preservation in perpetuity). Preservation credit for the inner 100 ft buffer is based on the condition of the stream channel using Form 1 of the USM and includes the credit for preservation of the stream channel. Preservation credit for the inner 100 ft ranges from 0.14 for high quality streams (RCI of 1.25–1.5) to 0.07 for low quality streams (RCI of 1–1.24).

The USM also identifies adjustment factors (A_F) that are used to account for exceptional or site-specific circumstances associated with compensation sites and result in increased credits being derived from the site. Each A_F activity is credited within a prescribed range in order to account for the variations in activities and conditions. There are three A_F activities:

Rare, Threatened, and Endangered Species or Communities: Increased CC is warranted for sites that show a significant improvement in restoring, enhancing, or preserving communities or individuals of federally or state-listed rare or threatened and endangered species. The range of credit provided for this activity is a 0.1–0.3 multiplier.

Livestock Exclusion: Increased CC is warranted for sites that exclude livestock either by installing fencing or removing cattle or other livestock from the site entirely. The range of credit provided for this activity is a 0.1–0.3 multiplier.

Watershed Preservation: Increased CC is warranted if the compensation site incorporates legal mechanisms that preserve the entire watershed or subwatershed of the compensation site. The range of credit provided for this activity is a 0.1–0.3 multiplier.

The credit to be obtained from each methodology employed at the site and the A_F credits are recorded on Form 3 (Fig. 4.2.2.2). In order to determine the CC to be obtained from the compensation reach, the following equation is used:

$$\text{Total CC} = \text{Sum}(\text{Restoration Credit} + \text{Enhancement Credit} + \text{Riparian Buffer Credit} + \text{Adjustment Factor Credit})$$

Just like the assessment portion of the USM, if more than one compensation reach is proposed, the credits obtained are added together to determine the CC for the whole site.

DESCRIPTION OF THE METHOD

During the USM development period, site visits were made with the development team, which included representatives of the USACE, VDEQ, and the EPA, to calibrate the method to meet the overarching goals and to achieve currently acceptable compensation ratios and agency practices. Reliance on previously tested and enacted methods allowed the team to "spot check" the application of particular parameters. Where new conditions or parameters were developed, field calibration was conducted, including the application of the new condition parameter within the methodology as a whole.

VALIDATION/CALIBRATION EFFORTS UNDERTAKEN WITH THE METHOD

The USM team undertook numerous validation/calibration efforts before the methodology was implemented in order to ensure consistent application of the method. Validation/calibration efforts occurred during USM development and for several months after the USM was implemented. Due to the limited preimplementation timeframe, validation relied heavily on the previously approved methods. The USM was applied to existing projects that had been previously assessed with SICAM or SAAM and site visits were conducted to sites already known to the development team.

An essential aspect of the verification process included comparing the USM with previously approved methods and agency policies in terms of stream restoration methods and the "scale" of credits. The team strove to keep the CRs and CCs within the range of what was already accepted by the regulated community in Virginia.

USM implementation included a public comment period (30 days). Comments were received and assessed. Minor changes were made based on the comments and revised forms were released. A "frequently asked questions" document was created and made available to the public to clarify some questions but no significant changes were made based on comments. No changes to the manual have been made since the implementation of the USM. A supplement form was developed for the assessment of ephemeral streams that eliminates all other condition parameters except the riparian buffer. Because the USM is only an assessment and crediting tool, changes in policy as it applies to site selection or stream restoration success criteria are made in other policy documents, therefore not affecting the USM manual.

\

Compensation Crediting Form (Form 3)

Unified Stream Methodology for use in Virginia

Project #	Project Name	Locality	Cowardin Class.	HUC	Date	Reach #	Reach Length
							0

Name(s) of Evaluator(s)	Steam Name and Information

Project Credits

Restoration: Includes Priority 1, 2, and 3 restoration activities. Does not include buffer width.

List Reaches that will receive full Restoration:	Total length of Full Restoration		Credit per foot	Project Credits
	Credits = Stream Length X 1.0		1	

Enhancement With Instream Structures: Addressing Streambank Stability, Grade Control (Vanes, Weirs, Step-Pools), Constructed Riffles

Discuss Length Affected by Instream Structures (justify length):	Length Affected by Instream Structures		Credit per foot	Project Credits
	Credits = Stream Length X 0.3		0.3	0

Enhancement: Addressing Streambank Stability, Entrenchment Ratios, Access to Floodplain

Mitigation Categories

	Mechanical Bank Work			Biological Bank Work	
	Credit Per Length	Pick One Per Length		May Be Cumulative Per Length	
Activities	**Habitat Structures**	**Create Bankfull Bench**	**Lay Back Banks**	**Bio-Remediation Techniques**	**Stream Bank Plantings**
Credit per foot per bank	0.1	0.15	0.1	0.1	0.09
Right Bank Length					0
Credit>					
Left Bank Length					0
Credit >					

CREDITS			Project Credits
Rt Bank >	0.00	Credit	
Lt Bank >	0.00	*SUM of banks*	0

Σ(Length X Credit) for all areas (banks done separately)

Riparian Areas: Assess the proposed 100 foot buffer on both banks based on the activity proposed. Enter the percentage of area and the credit below. (Widths of buffer above 100' will be determined below)

Activities	Buffer Re-establishment (removal of invasives)	Buffer Planting - Heavy	Buffer Planting - Light	Preservation *High Quality, Restoration, Enhancement*	Preservation *Low Quality*	Buffer area not within preservation width
Credit for 0'-100'	0.4	0.38	0.29	0.14	0.07	0
Credit for beyond 100'	0.2	0.19	0.15	0.07		0

Calculation of "Goal" riparian buffer for each side (SAR length times 100') >>>> 0 square feet

WITHIN FIRST 100' - Mitigation Categories

One vegetative community maintained	Subtract 0.03
Two vegetative communities maintained	Subtract 0.06

Ensure the sums of % Riparian Blocks equal 100

Right Bank	Area #	1	2				
	Sq, Footage						
	% Area						
	Credit>						
Left Bank	Area #						
	Sq, Footage						
	% Area						
	Credit>						

CREDITS			Project Credits
Rt Bank >	0.00	Credit	
Lt Bank >	0.00	0.00	0

Σ(% Area X Credit) for all areas (banks done separately)
AVE of credit for banks X length of project

Outside First 100' - Mitigation Categories

One vegetative community maintained	Subtract 0.03
Two vegetative communities maintained	Subtract 0.06

Right Bank	Area #						
	Sq, Footage						
	% Area						
	Credit>						
Left Bank	Area #						
	Sq, Footage						
	% Area						
	Credit >						

CREDITS			Project Credits
Rt Bank >	0.00	Credit	
Lt Bank >	0.00	0.00	0

Σ(% Area X Credit) for all areas (banks done separately)
AVE of credit for banks X length of project

Adjustment Factors: These factors are applied as a multiplier to length of a reach for which they apply

Adjustment Factor Categories

Activity	Rare, Threatened, or Endangered Species or Communities	Livestock Exclusion	Watershed Preservation
Credit	0.1 - 0.3	0.1 - 0.3	0.1 - 0.3
Stream Length Affected			
Credit>			

Credits are cumulative and can apply to more than one reach. Each reach can have more than one Adjustment Factors

Record AF length /credit beneath the AF activity. Provide a narrative explanation of the applicable site conditions that warrant an adjustment and justify the AF credit chosen.

Credits >	0

ΣLength X Credit) for all areas

Total Compensation Credit Provided by Project	0

FIG. 4.2.2.2 Compensation Crediting Form (Form 3).

TIME SPENT IN DEVELOPING/TESTING THE METHOD

The USM was developed over a 4-month period. Development relied heavily on previously tested and approved methods, allowing an expedited development timeframe. Field testing was conducted during this time to refine the descriptions of stream and buffer conditions and to refine the scoring numbers.

SAMPLE APPLICATION OF METHOD IN THE FIELD

The USM was designed to be rapid and repeatable. The USM Stream Impact Site Assessment includes field forms that accompany the manual; this contains brief descriptions of each parameter's conditions with fillable fields for the scores. Field comments can also be recorded on these forms. The forms are meant to be completed electronically when the assessor is back in the office, such that the forms automatically calculate the results. However, formulas for all calculations are provided in the manual. The stream assessor would conduct an overview of the entire project and determine the number of reaches to be assessed based on the homogeneity of the system. The USM works equally well when a system is separated into several reaches or lumped into just a few. Measurements of reach length and riparian buffer width are necessary but all other assessments can be done visually.

The USM compensation crediting assessment forms include brief descriptions of each proposed practice and are used to evaluate the stream design plan based on the field conditions present and the stream and riparian buffer restoration methods employed. The stream length is calculated by the restored stream length and the buffer crediting is calculated on the plan view area of the buffer (versus along the ground surface, including slopes). This allows the reviewing agencies to verify the compensation crediting during plan review. Stream and riparian buffer restoration methods are verified in the field by the reviewing agencies.

TIME SPENT TO APPLY THE METHOD IN THE FIELD

The USM was designed to only take approximately 10 minutes per reach to apply in the field. As stated above, some assessors may split the system into numerous reaches, thereby extending the amount of time it takes to complete the USM for the entire site.

HOW WAS/IS THE DATA BEING USED

The data was designed and is currently being used to determine the CR for unavoidable impacts to streams that would occur as a result of development and to determine the CC to be obtained from a proposed stream compensation site. The USM calculations are used to determine compensation requirements for permitted impacts. The USM calculations are also used to credit permittee-specific, mitigation bank and in-lieu fee fund sites (Fig. 4.2.2.2).

WHAT WAS LEARNED

Factors considered in assessment and crediting methodologies can drive site selection. For example, cattle exclusion and watershed protection result in additional credits and those sites are often sought after for their higher restoration potential and available credits. Requiring more compensation for high-quality stream systems has resulted in additional avoidance and minimization of those reaches on permitted sites.

All the adjustment factors were left open such that their application would develop over time. The application of the threatened and endangered species adjustment factor has proven particularly difficult to apply and sometimes difficult to obtain. Stream credit can be given for terrestrial species and determining how much stream length would apply is not always clear. Providing credit for species is only given when the species is present, but this is rare for aquatic threatened and endangered species, especially in severely degraded streams. Reliance on the assistance of state and federal wildlife agencies is critical in determining credits for this adjustment factor.

The USM has proven through more than 11 years of application that has met the goals established in its development to be repeatable, rapid, and consistent. It has generally been accepted as useful in Virginia's regulated community. The regulatory agencies have appreciated its flexibility and ability to adjust credits based on the success of individual aspects of a compensation plan. In the 5 years following the implementation of the USM (2007–2012), 27 mitigation banks with stream credits were approved as compared to only 17 mitigation banks with stream credits approved in the previous 8 years (1999–2006), supporting the need of the regulated community for a predictable method.

Professional judgment has always been important and will remain important in its implementation and the USM is able to continue to calculate compensation requirements and compensation credits through the ever-evolving study of stream restoration. The USM is an assessment and crediting methodology and does not replace best professional judgment of the appropriateness of site selection, the stream and buffer restoration methods used, or the application of the USM.

PROSPECTS FOR THE FUTURE

There is and has been consideration for revising the USM from a condition-based assessment to a more function-based assessment. The challenge will be to translate a functional assessment methodology into credits while retaining a high level of predictability, repeatability, and consistency. Any future changes will need to consider the existing mitigation banks and in lieu fee fund sites that are credited with the USM, ensuring a smooth transition and equitable value for various restoration activities.

REFERENCES

US Army Corps of Engineers, http://www.nao.usace.army.mil/Missions/Regulatory/UnifiedStreamMethodology.aspx.

Virginia Department of Environmental Quality, http://www.deq.virginia.gov/Programs/Water/WetlandsStreams/Mitigation.aspx.

Schumm, S., Harvey, D., Watson, C., 1984. Incised Channels: Morphology Dynamics and Control. Water Resources Publications, Fort Collins, CO.

Stream Impact Manual for the Northern Virginia Stream Bank, Version 1.3, 2006. Prepared by Wetland Studies and Solutions, Inc.

Rosgen, D., 1997. In: A geomorphological approach to restoration of incised rivers.Proceedings of the Conference on Management of Landscapes Disturbed by Channel Incision. 11 pp http://www.wildlandhydrology.com/html/references_.html.

Virginia Department of Environmental Quality, 2006. Stream Impact and Assessment Methodology.

North Carolina Stream Assessment Methodology, https://ribits.usace.army.mil/ribits_apex/f?p=107:27:11274013788068::NO:RP:P27_BUTTON_KEY:20.

U.S. Army Corps of Engineers, 2004–2006. Norfolk District-Stream Attribute Assessment Methodology.

References to USM by Other States

Pennsylvania. http://www.elibrary.dep.state.pa.us/dsweb/Get/Document-99535/310-2137-003.pdf.

Alaska. http://www.alaskawatershedcoalition.org/wp-content/uploads/2015/01/SAMF-Symposium-Flyer.pdf.

Texas. http://www.swf.usace.army.mil/Missions/Regulatory/Permitting/ApplicationSubmittalForms.aspx.

Chapter 4.2.3

North Carolina: The North Carolina Stream Assessment Method (NC SAM)

John Dorney*, **LeiLani Paugh**[†] **and Sandy Smith**[‡]

**Moffatt and Nichol, Raleigh, NC, United States, †North Carolina Department of Transportation, Raleigh, NC, United States, ‡Axiom Environmental, Raleigh, NC, United States*

Chapter Outline

INTRODUCTION

NC SAM uses indicators of stream condition relative to a stream in reference condition (if appropriate) as a surrogate for stream function.[1] In effect, observed stream condition is used to infer stream function. These indicators are general measures (metrics) of the level of condition for the stream. NC SAM defines a reference stream as a typical, representative, or common example of that particular stream type without, or removed in time from, substantial human disturbance following the approach suggested by Sutula et al. (2006). For the purposes of NC SAM, the term "reference stream" includes a range of biotic and abiotic characteristics within each recognized stream type and is synonymous with "relatively undisturbed."

CONCEPTUAL FRAMEWORK

As noted above, NC SAM assesses the level of stream condition as an alternative to direct assessment of stream function. The method of determining the condition of a specific stream is to answer a series of questions (metrics) concerning the observed level of condition for the stream as well as stressors. A list of metrics specific to each general stream type was generated by the SFAT. Metrics corresponding to stream types with a reference standard are designed to assess the departure of a stream from the reference standard. All metrics for each of the 29 stream types were field tested and revised at multiple test sites representing various levels of disturbance. Following initial field testing, state and federal agency personnel participated in beta-testing exercises focused on the applicability of metrics for all stream types. Beta testing included lectures and field exercises followed by an opportunity for comments. A formal public notice and comment period was conducted before formal adoption of the final method.

Stream categories were determined using a combination of four NC SAM zones (geographic location), two valley shapes, and four watershed sizes, as shown in Table 4.2.3.1. In addition, a tidal marsh stream category was developed for those streams that are subject to lunar or wind tides and are bounded by tidal marsh. For this category, watershed size was not included. Finally, intermittent streams are treated in two different manners by the NC SAM method; the choice of which process to use is up to the regulatory agency given its varied regulatory interest. In the "US ACE/All Streams" approach, intermittent streams utilize the same evaluation procedure as perennial streams and therefore receive lower ratings when water is not present. In the "NCDWR Intermittent" approach, intermittent steams are considered to be a

1. The discussion of NC SAM was largely taken from Dorney et al. (2014) with updates to make it more current (as of 2016).

Wetland and Stream Rapid Assessments. https://doi.org/10.1016/B978-0-12-805091-0.00042-6

TABLE 4.2.3.1 NC SAM Stream Categories Based on NC SAM Zones (Geographic Location), Valley Shape, and Watershed Size

			Watershed Size (square miles)			
			<0.1	0.1–<0.5	0.5–<5.0	≥5.0
NC SAM Zone	Geomorphic Valley Shape	Designation	Size 1	Size 2	Size 3	Size 4
Outer Coastal Plain	Broad	Oa	Oa1	Oa2	Oa3	Oa4
Inner Coastal Plain	Broad	Ia	Ia1	Ia2	Ia3	Ia4
	Narrow	Ib	Ib1	Ib2	Ib3	Ib4
Piedmont	Broad	Pa	Pa1	Pa2	Pa3	Pa4
	Narrow	Pb	Pb1	Pb2	Pb3	Pb4
Mountain	Broad	Ma	Ma1	Ma2	Ma3	Ma4
	Narrow	Mb	Mb1	Mb2	Mb3	Mb4

different stream category when flow is not present and have a distinctive set of Boolean logic. Therefore, the lack of water does not affect the rating of the stream because that is considered to be a normal seasonal condition for intermittent steams.

The user's manual contains detailed explanations of all 25 metrics used in NC SAM on the field assessment form (Table 4.2.3.2). Each of these metrics provides a series of check boxes for evaluation of one or more multiple-choice questions intended to evaluate that metric and its deviation from reference or the effect of a variety of stressors (see Fig. 4.2.3.1 below for an example).

Each metric is evaluated within one or more specific areas, which are carefully defined in the manual (Section 4.1). The four areas are (1) assessment reach, (2) streamside area, (3) assessment area, and (4) watershed. The assessor is instructed to determine each of these areas for a particular site before conducting the NC SAM evaluation. The "assessment reach" is the reach of the stream that is subject to the NC SAM evaluation. Assessment reaches change based on a change in stream or streamside conditions as well as changes in stream category (for instance, watershed size). Other specific examples of reasons to change the assessment reach are outlined in the manual. The "streamside area" is the land contiguous to the assessment reach extending perpendicular to the elevation contour lines to a distance of 300 ft or the natural topographic high point (whichever is closest). The distance of 300 ft was chosen because that is a distance that is normally visible through the wooded vegetation most commonly encountered along North Carolina streams and is most intimately involved in various riparian buffer-related effects. The "assessment area" of a stream is the assessment reach and its associated streamside area. Finally, "watershed" is the area that contributes surface drainage to a specific point on a channel—normally, the lowest point of the assessment reach. The 4-day training class spends considerable time stressing the complex situations that can develop in the field for these four terms.

An important complexity comes into play when locating the streamside area for a particular project that happens to cross at the confluence of streams of different watershed sizes (Section 5.3 of the manual). The manual presents several examples of how to determine the extent of the streamside area in these situations, which can get quite complex. The use of very precise LiDAR mapping along with site observation of local topography greatly aids in this determination.

Metric 12 (aquatic life) is a critical metric that is used in multiple places throughout the method and for all stream types except tidal marsh stream (Table 4.2.3.3). The aquatic life categories listed in Metric 12 (see Fig. 4.2.3.1) were carefully vetted to account for the most commonly encountered aquatic species (mostly macrobenthos) in North Carolina. They are presented alphabetically and by common name in the method for ease of use. The 4-day training class (office and field) puts considerable emphasis on the correct identification of these organisms. The method then sorts these organisms mainly in terms of their sensitivity to pollution based on the guidance provided by the NC Division of Water Resources biological staff (NC Department of Environmental Quality, 2016).

A critical tool needed for data analysis to generate stream ratings in NC SAM is the rating calculator. The rating calculator utilizes a Boolean logic chain of reasoning to convert metric evaluation results into ratings. The Boolean logic process was developed by the SFAT following site visits and then extensive discussions regarding possible interactions between and among various metrics and subfunctions. These results were evaluated at numerous field sites by the SFAT.

TABLE 4.2.3.2 List of Field Metrics Used in NC SAM

1.	Channel Water
2.	Evidence of Flow Restriction
3.	Feature Pattern
4.	Feature Longitudinal Profile
5.	Signs of Active Instability
6.	Streamside Area Interaction
7.	Water Quality Stressors
8.	Recent Weather
9.	Large or Dangerous Steam (to assess)
10.	Natural In-Stream Habitat Types
11.	Bedform and Substrate
12.	Aquatic Life
13.	Streamside Area Ground Surface Condition
14.	Streamside Area Water Storage
15.	Wetland Presence
16.	Baseflow Contributors
17.	Baseflow Detractors
18.	Shading
19.	Buffer Width
20.	Buffer Structure
21.	Buffer Stressors

12. Aquatic Life – assessment reach metric (skip for Tidal Marsh Streams)

12a. ☐Yes ☐No Was an in-stream aquatic life assessment performed as described in the User Manual?
If No, select one of the following reasons and skip to Metric 13. ☐No Water ☐Other: ______________________

12b. ☐Yes ☐No Are aquatic organisms present in the assessment reach (look in riffles, pools, then snags)? If Yes, check all that apply. If No, skip to Metric 13.

1	>1	Numbers over columns refer to "individuals" for Size 1 and 2 streams and "taxa" for Size 3 and 4 streams.
☐	☐	Adult frogs
☐	☐	Aquatic reptiles
☐	☐	Aquatic macrophytes and aquatic mosses (include liverworts, lichens, and algal mats)
☐	☐	Beetles (including water pennies)
☐	☐	Caddisfly larvae (Trichoptera [T])
☐	☐	Asian clam (*Corbicula*)
☐	☐	Crustacean (isopod/amphipod/crayfish/shrimp)
☐	☐	Damselfly and dragonfly larvae
☐	☐	Dipterans (true flies)
☐	☐	Mayfly larvae (Ephemeroptera [E])
☐	☐	Megaloptera (alderfly, fishfly, dobsonfly larvae)
☐	☐	Midges/mosquito larvae
☐	☐	Mosquito fish (*Gambusia*) or mud minnows (*Umbra pygmaea)*
☐	☐	Mussels/Clams (not *Corbicula*)
☐	☐	Other fish
☐	☐	Salamanders/tadpoles
☐	☐	Snails
☐	☐	Stonefly larvae (Plecoptera [P])
☐	☐	Tipulid larvae (Cranefly)
☐	☐	Worms/leeches

FIG. 4.2.3.1 Metric 12 for aquatic life from the NC SAM manual.

TABLE 4.2.3.3 The Number of Times that the Aquatic Life Metric (Metric 12) is Used in NC SAM by Stream Size and Major Function

Watershed Size	Hydrology	Water Quality	Habitat
Size 1 Streams	0	1	0
Size 2 "USACE All Streams"	1	2	1
Size 2 "NCDWQ Intermittent Streams"	0	1	0
Size 3 Streams	1	2	1
Size 4 Streams	1	2	1
Tidal Marsh Streams	0	0	0

The Boolean logic was written into the rating calculator's computer program, which then generates ratings for stream metrics, subfunctions, functions, and the overall stream condition. The rating calculator is an Excel macro with 71 unique Boolean logic combinations across all stream types. However, each stream category has a unique Boolean combination. The Boolean process proceeds by using selected metric descriptors to sequentially generate ratings for subfunctions, functions, and then overall stream condition. Each level of function subsumes the next, effectively serving as building blocks for the levels that follow. For instance, of the four levels of functional assessment, the metric level has the narrowest purview. By themselves, metrics pertain to very specific aspects of the steam. Collectively, metrics are organized into subfunctions. The combination of the descriptors of all metrics within a particular subfunction produces a subfunction rating that offers a broader account of stream condition. Ratings generated for all subfunctions corresponding to a particular stream function (such as the hydrology function) are combined to produce a function rating for hydrology, water quality, or habitat. Ultimately, individual stream condition ratings are combined to produce an overall stream rating. This overall stream rating is the most comprehensive of the four levels of function–an aggregate of all functional levels considered in NC SAM. The assessor completes the form within the rating calculator by selecting proper boxes and option buttons. The program generates functional ratings from the completed form.

The use of NC SAM results in an overall rating for each assessed stream rating for each of the three specific functions and up to eight subfunctions (seven for tidal streams) as well as documentation of field conditions that contribute to the ratings. The product resulting from completion of NC SAM includes, but is not limited to, a completed field assessment form (with assessor notes), a completed stream rating sheet, a site map, site photographs, and additional notes, if appropriate.

VERIFICATION/VALIDATION/CALIBRATION EFFORTS

NC SAM was verified[2] before training began. Appendix B of the NC SAM manual (NC SFAT, 2015) provides an in-depth discussion of that validation as summarized below. For NC SAM, samples of 55 stream sites visited by the SFAT were used to analyze two main issues—(1) to determine if all three major functions (hydrology, water quality, and habitat) were needed to develop the overall score, and (2) whether the rating for the water quality function was inversely correlated with percent impervious surface as expected. Based on a statistical analysis of these 55 study sites and their NC SAM ratings, the analysis showed that all three major functions are needed to develop the overall rating and that percent impervious surface was inversely correlated with the water quality rating as expected because as impervious surface increases, water quality is expected to decrease. Therefore, the SFAT considered NC SAM as being verified because the method evaluates sites as the team expects.

To date (mid-2016), NC SAM has neither been formally validated[3] nor calibrated.[4] An initial attempt to conduct this validation was conducted by Gale (2014) for her MS thesis. She compared long-term hydrological and biological

2. Verification is a comparison of the results of a RAM to the expected results from a suite of study sites. In other words, does the RAM produce results that a designated group of people expects based on their experience and/or judgment?
3. Validation is the use of actual measurements (rather than simply group opinions) to compare with the results of a RAM across a broad range of sites.
4. Calibration is the mathematical conversion ("normalization") of the score or scores from a specific site to a number (e.g., percentile) that represents the site's ranking relative to a larger set of sites within a specified area.

(macrobenthos) data from 98 NC SAM evaluations from 70 sites across North Carolina. She concluded that, "the combination of predominantly high ratings and low rate of detection of stream impairment suggests that NC SAM was unable to identify water quality or habitat issues that were significant enough to cause use impairment" (as defined by regulatory criteria under the Clean Water Act). However, she also concluded that, "NC SAM functional ratings did show similar responses to watershed conditions such as land use, drainage area, and slope, as other measures (bioclassification, habitat assessment, water chemistry)." In general, she concluded that, "in many cases, NC SAM was found to be a promising tool for screening-level or similar studies (e.g., watershed characterizations), and in many cases will provide a rapid, cost-effective assessment." Finally, she concluded that, "accuracy of NC SAM as compared to in-stream conditions could be improved by greater weighting by the model of some of the information and observations already provided during assessments. It is believed that accuracy could also be improved by consideration of watershed conditions rather than simply local conditions" (especially impervious coverage in the watershed). These results indicate that NC SAM is an effective rapid assessment tool but that it could benefit from additional verification and calibration. However, this will require additional funding, which is not available at this time.

TIME SPENT TO DEVELOP THE METHOD

NC SAM was developed as part of a collaborative effort by representatives of the US Army Corps of Engineers (USACE), US Department of Transportation Federal Highway Administration (USFHWA), the Environmental Protection Agency (EPA), US Fish and Wildlife Service (USFWS), NC Division of Coastal Management (NCDCM), NC Department of Transportation (NCDOT), NC Division of Water Quality (NCDWQ), NC Ecosystem Enhancement Program (NCEEP), NC Natural Heritage Program (NCNHP), and NC Wildlife Resources Commission (NCWRC). Between 2003 and 2011 (not counting the 3-year hiatus in the middle), the SFAT met 27 times in the office and spent 33 days in the field examining streams across North Carolina, visiting a total of 280 stream sites for about 700 person-days spent developing and testing the method before its completion in April 2011. All decisions were made by consensus of WFAT members. Since then, the NC SAM user's manual (NC SFAT, 2015) has been refined several times based on field and teaching experience of the instructor team and SFAT members and a four-day training course has been developed and conducted 14 times (as of early 2016) in order to train students in the proper use of the method.

SAMPLE APPLICATION OF METHOD IN FIELD

The initial step in the field application of NC SAM is to identify the stream type or types found at the site. The next step is to decide if there is more than one assessment area on the site based on overall site condition. In order to make these determinations, the assessor must walk throughout the assessment area, actively observing conditions of the stream and its adjacent streamside area. After selection of the most appropriate stream type and the assessment area, the field assessment form is completed by checking off the most appropriate response for each of the 25 metrics and recording notes on stream condition. Examination of geographic information systems (GIS) sources such as aerial photographs via cell phone or maps printed before site access is needed in order to address some metrics. A particularly important tool to use for the determination of watershed area and general watershed characteristics is the online StreamStats tool developed by the US Geological Survey (USGS, 2015). This tool allows the user to locate a study site on a USGS mapped stream and then readily determine the watershed area as well as watershed characteristics such as land use/land cover or percent impervious surfaces that are used in NC SAM. As noted above, the stream channel must be depicted on the USGS 1:24,000 topographic map but StreamStats is an essential tool for NC SAM in rapidly and accurately determining watershed area. Alternatively, watershed area can be determined from analysis of the watershed draining to a particular location by drawing lines perpendicular to the contour lines. Appendix F of the NC SAM user's manual provides written instructions based on a report from the New Hampshire NRCS (NC SFAT, 2015), which has proved to be particularly useful for this purpose when StreamStats is not applicable.

TIME SPENT TO APPLY METHOD IN THE FIELD

Completion of a stream functional assessment is typically a six-step process: (1) becoming familiar with regional features through offsite research [mostly map analysis and use of the StreamStats online tool (USGS, 2015) as described above]; (2) conducting an onsite investigation to determine separate stream types; (3) determining the boundaries of one or more assessment areas within the proposed project or study area; (4) conducting a rapid onsite evaluation of each assessment area; (5) conducting an in-office map/GIS evaluation if needed; and (6) using the rating calculator to generate assessment ratings. The NC SAM method is designed to take about 15 min in the field on an average site after delineation and training of the evaluator. In addition, some office GIS work may be needed, depending on the site. Finally, the data have to be entered into the NC SAM rating calculator, which generates the final ratings. Therefore, the total time to collect the field data, conduct any necessary GIS analysis, and develop a computer-generated rating is probably about 30 min for the average project for a trained evaluator.

USE OF NC SAM

On April 21, 2015, the Wilmington District of the USACE published a public notice concerning the implementation of NC SAM in North Carolina (USACE, Wilmington District, 2015). This public notice stated that the USACE will begin to utilize NC SAM for internal review of permit applications, including decisions regarding the amount and type of compensatory mitigation, avoidance and minimization of impacts, or other decisions pertaining to aquatic resource quality and functions. The public notice states that the results of NC SAM can be considered along with other factors to adjust the typical 2:1 mitigation ratio to account for high- or low-quality aquatic resources. The public notice also states that NC SAM will be useful to meet the goals outlined in the joint mitigation rule (USACE and USEPA, 2008) and may be used as a screening tool to determine if potential mitigation sites are viable. However, NC SAM will not be used to determine mitigation success because the method was not designed for that purpose. The USACE will decide to require NC SAM on a case-by-case basis in the review of 404 Permit applications.

PROJECT REVIEW, PERMITTING, AND MITIGATION

The NCDOT has used NC SAM during the planning process for a wide variety of transportation projects. Impact sites were evaluated using NC SAM across multiple alignments on large transportation projects to characterize the quality and quantity of the impact. This allowed for consideration of alternatives with more length of stream impact but overall lower functional loss in the selection of the least environmentally damaging practicable alternative as part of the 404/401 Permit process. NC SAM has also been used in permitting decisions to help select the most appropriate mitigation from among several options. Comparison of impact site type and functions to mitigation site type and functions provided data for development of alternative mitigation packages that include adjusted ratios, out-of-kind mitigation, and mitigation opportunities in adjacent watersheds.

Compensatory Mitigation for a Large Private Mitigation Project

NC SAM has been used for a proposed large private compensatory mitigation bank in the mountains of North Carolina, as required by USACE for its 404 Permit review (Turnpike Mitigation Bank in Buncombe and Haywood Counties) (Moffatt & Nichol, 2015). A total of 17 NC SAM evaluations were completed on the site—15 mitigation reaches, one reference reach, and one reach on adjacent (undisturbed) property. In general, the segments proposed for mitigation were rated overall as medium (medium for hydrology, high for water quality, and medium for habitat), mainly due to heavy impact from sedimentation that had filled the pools with sediment and degraded habitat. The streamside area vegetation was largely intact. In comparison, the reference and offsite reaches were rated as overall high. This information is being considered by the regulatory agencies as they determine whether this site is a valid compensatory stream mitigation bank and in their determination of the appropriate mitigation ratios and credits for the site.

FUTURE PROSPECTS

NC SAM has been modified several times since 2010 as a result of experience gained from using the method in the field in addition to feedback from students during the classes that have been and continue to be taught. Any changes in the user's manual have to be approved by the remaining members of the SFAT (essentially the NCDOT and USACE). Changes to date have been fairly minor and mainly to improve the clarity of the method. Any major changes to the Boolean logic or additions/deletions of field metrics would require reestablishment of the SFAT in order to thoroughly discuss the results of validation and options to amend the method to address any shortcoming followed by public notice and comment. To date, this has not been necessary. However, this process would likely be an extensive undertaking by the SFAT agencies, and it is not clear whether the funding for this effort would be available.

REFERENCES

Dorney, J.R., Paugh, L., Smith, S., Lekson, D., Tugwell, T., Allen, B., Cusack, M., 2014. Development and testing of rapid wetland and stream functional assessment methods in North Carolina. Natl. Wetl. Newsl. 38 (4), 31–35.

Gale, S., 2014. Linkages Between Watershed Characteristics, Hydrologic Responses, and Instream Conditions (Masters of Science Thesis). Natural Resources, North Carolina State University, Raleigh, NC (Dr. Ryan Emanuel, Advisor).

Moffatt & Nichol, 2015. Summary of NC SAM (NC Stream Assessment Method) Evaluation for proposed Turnpike Mitigation Bank in Haywood and Buncombe Counties, NC. October 12, 2015. Available at: http://edocs.deq.nc.gov/WaterResources/DocView.aspx?dbid=0&id=328434&page=34&cr=1. Accessed 15 April 2015.

NC Department of Environmental Quality, Division of Water Resources, 2016. Standard Operating Procedures for the Collection and Analysis of Benthic Macroinvertebrates. NC Department of Environmental Quality, Division of Water Resources, Raleigh, NC. Available at: https://ncdenr.s3.amazonaws.com/s3fs-public/Water%20Quality/Environmental%20Sciences/BAU/NCDWRMacroinvertebrate-SOP-February%202016_final.pdf. Accessed 21 June 2016.

NC Stream Functional Assessment Team, 2015. N.C. Stream Assessment Method (NC SAM) User Manual. Version 2.1, August 2015. Available at: https://ribits.usace.army.mil/ribits_apex/f?p=107:150:7617504708947::NO::P150_DOCUMENT_ID:5210 through "Wilmington District" and "Assessment Tools" (Accessed 06 January 2016).

Sutula, M.A., Stein, E.D., Collins, J.N., Fetscher, A.E., Clark, R., 2006. A practical guide for the development of a wetland assessment method: the California experience. J. Am. Water Resour. Assoc. 42 (1), 157–175.

United States Geological Survey, 2015. StreamStats Program. Available at: http://water.usgs.gov/osw/streamstats/. Accessed 14 April 2015.

US Army Corps of Engineers and US Environmental Protection Agency, 2008. Compensatory Mitigation for Losses of Aquatic Resources; Final Rule. April 10, 2008. Federal Register 73(70):19594-19705. 33 CFR Parts 235 and 332; 40 CFR Part 230. US Army Corps of Engineers and US Environmental Protection Agency, Washington, DC. Available at: http://www.epa.gov/sites/production/files/2015-03/documents/2008_04_10_wetlands_wetlands_mitigation_final_rule_4_10_08.pdf. Accessed 31 December 2015.

US Army Corps of Engineers, Wilmington District, 2015. Wilmington District Implementation of the North Carolina Stream Assessment Method and North Carolina Wetland Assessment Method. Available at: http://www.saw.usace.army.mil/Missions/RegulatoryPermitProgram/PublicNotices/tabid/10057/Article/585625/implementation-of-nc-sam-and-nc-wam.aspx. Accessed 6 January 2016.

Section 4.3

Wetland Assessment Methods

Chapter 4.3.1

North Carolina: The North Carolina Wetland Assessment Method (NC WAM)

John Dorney*, **LeiLani Paugh**[†] **and Sandy Smith**[‡]

**Moffatt and Nichol, Raleigh, NC, United States,* [†]*North Carolina Department of Transportation, Raleigh, NC, United States,* [‡]*Axiom Environmental, Raleigh, NC, Unites States*

Chapter Outline

INTRODUCTION

North Carolina Wetland Assessment Method (NC WAM) uses indicators of wetland condition relative to a reference wetland (if appropriate) as a surrogate for wetland function. In effect, observed wetland condition is used to infer wetland function. These indicators are general measures (metrics) of the level of condition for the wetland. A condition metric examines inherent wetland characteristics that affect the wetland's ability to perform a given function. NC WAM defines a reference wetland as a typical, representative, or common example of that particular wetland type without, or removed in time from, substantial human disturbance, as suggested by Sutula et al. (2006). For the purposes of NC WAM, the term "reference wetland" includes a range of biotic and abiotic characteristics within each recognized wetland type and is synonymous with "relatively undisturbed." Some wetland types do not have a reference for a variety of reasons and the condition of those wetland types is generally evaluated based on stressors.

CONCEPTUAL FRAMEWORK

As noted above, NC WAM assesses the level of wetland condition as an alternative to direct assessment of wetland function. The method of determining the condition of a specific wetland is to answer a series of questions (metrics) concerning (a) the observed level of condition for the wetland and (b) in a few relatively rare instances related to the water quality function, the opportunity for modification of wetland functions due to land use disturbances in the watershed that then drain to the wetland. A list of metrics specific to each general wetland type was generated by the Wetland Functional Assessment Team (WFAT). Metrics corresponding to wetland types with a reference standard are designed to assess the departure of a wetland from the reference standard. All metrics for each of the 16 wetland types were field tested and revised at multiple test sites representing various levels of disturbance. Following initial field testing, state and federal agency personnel participated in beta-testing exercises focused on the applicability of metrics for all general wetland types. Beta testing included exercises and a provision for comments. A formal public notice and comment period was conducted before formal adoption of the final method.

☆ The discussion of NC WAM was mostly taken from Dorney et al. (2015), with updates to make it more current (as of 2016).

Wetland and Stream Rapid Assessments. https://doi.org/10.1016/B978-0-12-805091-0.00015-3

TABLE 4.3.1.1 List of Field Metrics Used in NC WAM

1.	Ground Surface Condition/Vegetation Condition—assessment area condition metric
2.	Surface and Subsurface Storage Capacity and Duration—assessment area condition metric
3.	Water Storage/Surface Relief—assessment area/wetland type condition metric
4.	Soil Texture/Structure—assessment area condition metric
5.	Discharge into Wetland—assessment area opportunity metric
6.	Land Use—opportunity metric
7.	Wetland Acting as Vegetated Buffer—assessment area/wetland complex condition metric
8.	Wetland Width at the Assessment Area—wetland type/wetland complex condition metric
9.	Inundation Duration—assessment area condition metric
10.	Indicators of Deposition—assessment area condition metric
11.	Wetland Size—wetland type/wetland complex condition metric
12.	Wetland Intactness—wetland type condition metric
13.	Connectivity to Other Natural Areas—landscape condition metric
14.	Edge Effect—wetland type condition metric
15.	Vegetative Composition—assessment area condition metric
16.	Vegetative Diversity—assessment area condition
17.	Vegetative Structure—assessment area/wetland type condition metric
18.	Snags—wetland type condition metric
19.	Diameter Class Distribution—wetland type condition metric
20.	Large Woody Debris—wetland type condition metric
21.	Vegetation/Open Water Dispersion—wetland type/open water condition metric
22.	Hydrologic Connectivity—assessment area condition metric

The user manual contains detailed explanations of all 22 metrics used in NC WAM on the field assessment form (Table 4.3.1.1). Each of these metrics provides a series of check boxes for evaluation of one or more multiple-choice questions intended to evaluate that metric and its deviation from reference or the effect of a variety of stressors.

A critical tool needed for data analysis to generate wetland functional ratings in NC WAM is the rating calculator. This utilizes a Boolean logic chain of reasoning to convert metric evaluation results into ratings. The Boolean logic process was developed by the WFAT following site visits and then extensive discussions regarding possible interactions between and among various metrics and subfunctions. These results were evaluated at numerous field sites by the WFAT. The Boolean logic was written into the rating calculator's computer program, which then generates ratings for wetland metrics, subfunctions, functions, and the overall wetland. The rating calculator is an Excel macro with 71 unique Boolean logic combinations across the 16 general wetland types. However, each wetland type has a unique Boolean combination. The Boolean process proceeds by using selected metric descriptors to sequentially generate ratings for subfunctions, functions, and then overall wetland. Each level of function subsumes the next, effectively serving as building blocks for the levels that follow (see Dorney et al., 2015 for examples). For instance, of the four levels of functional assessment, the metric level has the narrowest purview. By themselves, metrics pertain to very specific aspects of the wetland. Collectively, metrics are organized into subfunctions. The combination of the descriptors of all metrics within a particular subfunction produces a subfunction rating that offers a broader account of wetland function. Ratings generated for all subfunctions corresponding to a particular wetland function (such as the hydrology function) are combined to produce a function rating for hydrology, water quality, or habitat. Ultimately, individual wetland function ratings are combined to produce an overall wetland rating. This overall wetland rating is the most comprehensive of the four levels of function—an aggregate of all functional levels considered in NC WAM. The assessor completes the form within the rating calculator by selecting proper boxes and option buttons. The program generates functional ratings from the completed form.

The use of NC WAM results in an overall rating for each assessed wetland and ratings for each of the three specific functions and up to 10 subfunctions as well as documentation of field conditions that contribute to the ratings. The product resulting from completion of NC WAM includes, but is not limited to, a completed field assessment form (with assessor notes), a completed wetland rating sheet, a site map, site photographs, and additional notes if appropriate.

VALIDATION/CALIBRATION EFFORTS

NC WAM has been validated for one of the 16 General Wetland Types—Headwater Forest (Dorney et al., 2015) as well as Chapter 3.10 of this book. NC WAM was validated using independent measures of wetland condition calculated from intensive wetland monitoring data. Headwater forest wetlands were selected for this validation exercise based on availability of appropriate intensive monitoring data. Additional validation of NC WAM will occur in the future when intensive monitoring data are available for other wetland types. For Headwater Forest, a total of 33 sites have been studied intensively with long-term monitoring data for up to 6 years across the Piedmont and Coastal Plain of North Carolina (Baker and Savage, 2008). The monitoring consisted of groundwater well levels; surface water chemistry; soil descriptions and soil chemistry; amphibian and aquatic macroinvertebrate diversity, presence, and abundance; vegetation analysis; and quality of the surrounding buffer. To this end, the association was examined between NC WAM overall rating and NC WAM functions (water quality, hydrology, and habitat) and independent measures of wetland conditions. A random forest classification was first performed to identify the best predictors among all the indices for each of the three main NC WAM functions and the NC WAM overall rating. Using the best predictors identified by the random forest analysis, discriminant analysis was then used to classify the wetlands into low, medium, and high categories using amphibian, macroinvertebrates, and vegetation metrics as well as the abiotic variables (LDI index and soil variables). The agreement was calculated between the classification provided by the intensive data and NC WAM overall and main functions. Using all measures, NC WAM correctly classified the overall wetland condition 89.43% of the time; the habitat condition was correctly predicted 95.45% of the time, followed by water quality and hydrology (91.67% and 85.61% of the time, respectively). In general, Dorney et al. (2015) concluded that the NC WAM overall rating as well as the ratings for the three main functions are significantly related to and predicted by the intensive monitoring data, thereby calibrating NC WAM for the Headwater Forest wetland type. A more detailed validation for this wetland type and NC WAM is described in Chapter 3.10 of this book. This more in-depth analysis confirmed the overall conclusion from Dorney et al. (2015) but identified some issues with particular subfunctions (notably pathogen change) that may warrant review and revision by the WFAT. Finally, similar long-term monitoring data are available in North Carolina for one additional wetland type—Basin Wetland. Because these two wetland types with long-term monitoring data may represent other wetland types, the WFAT will have to decide if it is appropriate to make changes in the method for other wetland types based on the calibration results from these two types. Validation of other types will require additional funding for data collection.

TIME SPENT TO DEVELOP THE METHOD

NC WAM was developed as part of a collaborative effort by representatives of several federal groups, including the Army Corps of Engineers (USACE), Department of Transportation Federal Highway Administration (USFHWA), Environmental Protection Agency (USEPA), and the Fish and Wildlife Service (USFWS) as well as several groups in North Carolina: the Division of Coastal Management (NCDCM), Department of Transportation (NCDOT), Division of Water Quality (NCDWQ), Ecosystem Enhancement Program (NCEEP), Natural Heritage Program (NCNHP), and Wildlife Resources Commission (NCWRC). Between 2003 and 2007, the WFAT met 27 times in the office and spent 33 days in the field examining wetlands across the state of North Carolina, visiting a total of 280 wetland sites for about 700 person-days spent developing and testing the method before its completion in April 2008. All decisions were made by consensus of WFAT members. Since then, the NC WAM user manual (NC WFAT, 2010) has been refined several times and a 4-day training course has been developed and conducted 14 times (as of early 2016) in order to train students in the proper use of the method.

SAMPLE APPLICATION OF METHOD IN FIELD

The initial step in the field application of NC WAM is to identify the wetland types found at the site. To this end, NC WAM uses a dichotomous key. If the assessor has evidence that a wetland can reasonably fit into more than one wetland type, the assessor may rate the wetland as each potential wetland type with a final decision up to the regulator. If there is evidence suggesting that the wetland is a type other than the keyed type, the assessor should document this evidence and then classify the wetland accordingly. The next step is to decide if there is more than one assessment area on the site based on overall site condition. In order to make these determinations, the assessor must walk throughout the assessment area, actively observing conditions of the wetland. After selection of the most appropriate wetland type and the assessment area, the field assessment form is completed in the field by checking off the most appropriate response for each of the 22 metrics and recording notes on wetland condition. Examination of GIS sources such as aerial photographs via cell phone or maps printed before site access is needed in order to address some metrics.

TIME SPENT TO APPLY METHOD IN THE FIELD

Completion of a wetland functional assessment is typically a six-step process: (a) becoming familiar with regional features through offsite research (mostly map analysis); (b) conducting an onsite investigation to determine separate general wetland types; (c) determining the boundaries of one or more assessment areas within the proposed project or study area; (d) conducting a rapid onsite evaluation of each assessment area; (e) conducting an in-office map/GIS evaluation if needed; and (f) using the rating calculator to generate assessment ratings. The NC WAM method is designed to take about 15 min in the field on an average site after delineation and training of the evaluator. In addition, some office GIS work may be needed, depending on the site. Finally, the data have to be entered into the NC WAM rating calculator, which generates the final ratings. Therefore, the total time to collect the field data, conduct any necessary GIS analysis, and develop a computer-generated rating is probably about 30 min for the average project for a trained evaluator.

USE OF NC WAM

On April 21, 2015, the Wilmington District of the USACE published a public notice concerning the implementation of NC WAM in North Carolina (USACE, Wilmington District, 2015). This public notice stated that the USACE will begin to utilize NC WAM for internal review of permit applications, including decisions regarding the amount and type of compensatory mitigation, avoidance and minimization of impacts, or other decisions pertaining to aquatic resource quality and functions. The public notice states that the results of NC WAM can be considered along with other factors to adjust the typical 2:1 mitigation ratio to account for high or low quality aquatic resources. The public notice also states that NC WAM will be useful to meet the goals outlined in the joint mitigation rule (USACE and USEPA, 2008) and may be used as a screening tool to determine if potential mitigation sites are viable. However, NC WAM will not be used to determine mitigation success since the method was not designed for that purpose. The USACE will decide to require NC WAM on a case-by-case basis in the review of 404 Permit applications.

PROJECT REVIEW, PERMITTING, AND MITIGATION

Permitting for a Private Project

NC WAM was used to determine the existing level of wetland condition for a large, private project near Wilmington, where a historic rice plantation required state and federal permitting in order to refurbish and renovate dikes, water control, and conveyance structures for four remaining impoundments as well as prepare areas within those impoundments to grow rice. The original plantation was established in 1725 when rice fields were diked and water control structures installed to prevent intrusion of salt water from the Cape Fear River and adjacent salt marsh as well as management of water levels in the fields. Dikes on many of the fields long ago collapsed or otherwise succumbed to storms or lack of maintenance, but four fields survived into the 21st century with their dikes and water control structures largely intact. Rice production on a commercial scale ceased before the mid-20th century, but the new owner wanted to restore the rice fields and water conveyance system. This work required issuance of 404 Permits by the USACE and a Coastal Area Management Permit by the NCDCM.

The NC WAM work consisted of evaluating the condition of the wetlands onsite after subdividing the site into assessment areas with uniform conditions. A total of 17 NC WAM forms were completed and the site's wetland condition mapped. This information was then relayed to the permitting agencies for their decision-making in terms of allowable impacts and required compensatory mitigation. The final permit avoided impact to all wetlands rated as high for overall quality while allowing 4.83 acres of permanent wetland impact (mostly for dike repairs), 334 acres of temporary impact (to convert these areas back into rice plantings) in return for 2.57 acres of mitigation with the state in lieu fee program for the permanent wetland impact, and 187.5 acres of preservation of high quality, riverine swamp wetlands as compensatory mitigation for the temporary impacts (NCDWR, 2015), thus demonstrating the value of NC WAM for this project.

Permitting for Public Projects

The NCDOT has used NC WAM during the planning process for a wide variety of transportation projects. Impact sites were evaluated using NC WAM across multiple alignments on large transportation projects to characterize the quality and quantity of the impact. This allowed for consideration of alternatives with higher impact acres but overall lower functional loss in the selection of the least environmentally damaging practicable alternative as part of the 404/401 permit process. NC WAM has also been used in permitting decisions to help select the most appropriate mitigation from among several options. Comparison of impact site type and functions to mitigation site type and functions provided data for development of alternative mitigation packages that included adjusted ratios, out-of-kind mitigation, and mitigation opportunities in adjacent watersheds.

Compensatory Mitigation for a Large Private Project

NC WAM has been used to determine the amount of compensatory mitigation needed for a large Section 404 project in South Carolina (USACE, Charleston District, 2011). The USACE Charleston District uses a detailed calculation method (USACE, Charleston District, 2010) to determine mitigation credit requirements for Section 404 projects in the District. The calculation method considers six "adverse impact" factors, one of which is the functional integrity of the wetland that is proposed to be impacted. Since the District has no existing rapid assessment method, a crosswalk was developed to translate the results of NC WAM into the protocols for existing District guidance (Table 4.3.1.2). In this way, wetland function was addressed in the avoidance, minimization, and compensatory mitigation review for this project. Most wetlands were of overall medium or high quality. The permit for this project has since been issued by the USACE (Matt Cusack, Atkins North America, personal communication, Dec. 16, 2015).

FUNCTIONAL UPLIFT

Use of NC WAM to calculate the level of functional uplift from wetland enhancement was described in Dorney et al. (2015). Functional uplift is defined as the determination of the level of increase of wetland function from activities conducted on existing, non fully functional wetlands. Wetland enhancement is defined as "the manipulation of the physical, chemical, or biological characteristics of an aquatic resource to heighten, intensify, or improve a specific aquatic resource function(s)" without a gain in wetland area (USACE and USEPA, 2008). This rule requires regulatory documentation of uplift from compensatory mitigation and no net loss of acreage and function. The method to document that uplift is left to the individual USACE Districts. In order to determine functional uplift from enhancement, an NC WAM evaluation can be conducted on the wetland in its present state and then an NC WAM evaluation is completed on the site based on the projected outcome after the proposed enhancement activity or based on a relevant reference wetland for comparison. Long-term monitoring will probably then be required to document this potential condition. Since determination of mitigation ratios is inherently a numerical process, the NC WAM results needed to be converted to a numerical ranking system.

The NCDWQ 401 Certification rules [15A NCAC 2H.0506 (h) (6)] require 1:1 restoration or creation in order to achieve no net loss of wetlands, although the rules also allow the director to waive that portion of the rule if the "public good would be better served by other types of mitigation." In addition, the joint mitigation rule of the USACE and USEPA (2008, p. 19594) defines no net loss for wetland acreage *and* function (emphasis added). This use of NC WAM provides

TABLE 4.3.1.2 Crosswalk of NC WAM Scores by Function and Charleston, South Carolina, District Mitigation Guidelines

NC WAM Functions (Hydrology, Water Quality, and Habitat)	Charleston District Guidelines—Existing Condition
At least two of three functions rated *high*.	Fully functional
At least two of three functions rated *low*.	Partially impaired
If one function is rated *high*, then two are rated *low*. Otherwise, at least two of the three functions are rated *medium*	Impaired
No function rated *high*.	Very impaired

a consistent mechanism to calculate the degree of functional uplift from wetland enhancement that could then be used to calculate functional replacement for unavoidable impacts to address these regulatory requirements. Therefore, this method provides a means to calculate functional uplift from wetland enhancement and thereby achieve no net loss of function that otherwise would receive little to no mitigation credit. The analysis described below utilizes the overall wetland rating (as approved by the regulatory agencies) but could easily be calculated using any or all of the three functional ratings.

Bonnerton Hardwood Flat Mitigation Site

As a condition of the 401 Water Quality Certification issued by the NCDWQ to PCS Phosphate for an open-pit phosphate mine expansion on Jan. 15, 2009 (NCDWQ, 2009), a wetland mitigation effort will be conducted on the site by restoring natural flow to the relict stream channel, adjacent Headwater Forest, and adjacent Hardwood Flat through filling of an existing ditch and redirecting it to flow via its natural pattern into or adjacent to these wetlands. An NC WAM evaluation was completed for the Hardwood Flat before and after mitigation. The existing level of function for these wetlands was rated as medium, mainly as a result of the hydraulic alteration. The future level of function for this wetland was projected to be high once the ditch was filled and flow redirected into the relict stream channel. Using the above equation, the site yielded 0.69 ha of restoration equivalents from the 1.3 ha Hardwood Flat wetland. This amount of mitigation credit was explicitly acknowledged in the 401 Water Quality Certification issued by the NCDWQ (Condition number 8: Porter Creek Enhancement; NCDWQ Quality, 2009) and incorporated by reference by the USACE in its 404 Permit for the project. In addition, the possibility of functional uplift from the headwater forest wetland as well as stream restoration credit is provided in the 401 Water Quality Certification if PCS Phosphate provides additional monitoring that documents that uplift.

FUTURE PROSPECTS

NC WAM has been modified several times since 2010 as a result of experience gained from using the method in the field in addition to feedback from students during the classes that have been taught. Any changes in the user manual have to be approved by the remaining members of the WFAT (essentially the NCDOT and USACE). Changes to date have been fairly minor and mainly to improve the clarity of the method. Any major changes to the Boolean logic or additions/deletions of field metrics would require reestablishment of the WFAT in order to thoroughly discuss the results of validation and options to amend the method to address any shortcoming followed by public notice and comment. To date, this has not been necessary. However, this process would likely be an extensive undertaking by the WFAT agencies, and it is not clear whether the funding for this effort would be available.

REFERENCES

Baker, V., Savage, R., 2008. Development of a Wetland Monitoring Program for Headwater Wetlands in North Carolina. NC Department of Environment and Natural Resources, NC Division of Water Quality, Raleigh, NC. EPA Final Report CD 974260-01.

Dorney, J.R., Paugh, L., Smith, A.P.(.S.)., Allen, T.(.B.)., Cusack, M.T., Savage, R., Hughes, E.B., Muñoz, B., 2015. The North Carolina Wetland Assessment Method (NC WAM): development of a rapid wetland assessment method and use for compensatory mitigation. Environ. Pract. 17, 145–155.

N.C. Division of Water Quality (NCDWQ), 2009. 401 Water Quality Certification Number 3771 as Modified. Issued to Mr. Ross Smith, PCS Phosphate, Inc., January 15, 2009. Department of Environment and Natural Resources, NC Division of Water Quality, Raleigh, NC.

NC Division of Water Resources, 2015. LaserFiche File. Available from: http://edocs.deq.nc.gov/WaterResources/DocView.aspx?id=159218&searchid=c843f4ea-7919-416d-a17c-561ae41dcd44&dbid=0. Accessed 20 January 2016.

NC Wetland Functional Assessment Team. 2010. N.C. Wetland Assessment Method (NC WAM) User Manual. Version 4.1, October 2010. Available from: https://ribits.usace.army.mil/ribits_apex/f?p=107:150:7617504708947::NO::P150_DOCUMENT_ID:5210 through "Wilmington District" and "Assessment Tools" (accessed 06.01.16).

Sutula, M.A., Stein, E.D., Collins, J.N., Fetscher, A.E., Clark, R., 2006. A practical guide for the development of a wetland assessment method: the California experience. J. Am. Water Resour. Assoc. 42 (1), 157–175.

U.S. Army Corps of Engineers, Charleston District, 2010. Compensatory Mitigation Guidelines (Working Draft, Subject to Change): Guidelines for Preparing a Compensatory Mitigation Plan. Available from: http://www.sac.usace.army.mil/Portals/43/docs/regulatory/Guidelines_for_Preparing_a_Compensatory_Mitigation_Planf.pdf (last revised Oct. 7, 2010).

U.S. Army Corps of Engineers, Charleston District, 2011. Joint Public Notice: P/N #2009-122-SIR-Proposed William States Lee Nuclear Station. Available from: http://www.sac.usace.army.mil/Portals/43/docs/regulatory/SAC-2009-122-SIR_Cherokee[1].pdf.

U.S. Army Corps of Engineers, Wilmington District, 2015. Wilmington District Implementation of the North Carolina Stream Assessment Method and North Carolina Wetland Assessment Method. Available from: http://www.saw.usace.army.mil/Missions/RegulatoryPermitProgram/PublicNotices/tabid/10057/Article/585625/implementation-of-nc-sam-and-nc-wam.aspx. Accessed 6 January 2016.

US Army Corps of Engineers and US Environmental Protection Agency, 2008. Compensatory mitigation for losses of aquatic resources; Final rule. Fed. Regist. 73 (70), 19594–19705. 33 CFR Parts 235 and 332; 40 CFR Part 230. Washington, DC. April 10, 2008. Available from, http://www.epa.gov/sites/production/files/2015-03/documents/2008_04_10_wetlands_wetlands_mitigation_final_rule_4_10_08.pdf. Accessed 31 December 2015.

FURTHER READING

Dorney, J.R., Paugh, L., Smith, S., Lekson, D., Tugwell, T., Allen, B., Cusack, M., 2014. Development and testing of rapid wetland and stream functional assessment methods in North Carolina. Natl. Wetl. Newsl. 38 (4), 31–35.

Chapter 4.3.2

WESP (Wetland Ecosystem Services Protocol): A Suite of Regionalized RAMs

Paul Adamus
Oregon State University, Corvallis, OR, United States

Chapter Outline

CONCEPTUAL FRAMEWORK

The Wetland Ecosystem Services Protocol (WESP, Adamus 2016) template is a spreadsheet containing 106 questions that together take 1–3 h per wetland to complete. Each is used to assign a score to one or more of 20 wetland attributes, including functions and their associated values. Formulas in the spreadsheet do the calculations automatically and each formula (model) is described in the accompanying manual. A rationale is provided for each indicator-function relationship. Users are not required to first classify the wetland being assessed; characteristics normally used for that purpose are instead incorporated into function models. The manual prescribes steps for calibrating the template to any region. Once that has been done, raw scores from the spreadsheet are converted automatically to normalized scores and ratings that reflect the assessed wetland's rank, function by function, relative to other wetlands in the region.

HOW ASSESSMENTS ARE DONE WITH WESP METHODS

Method users locate their wetland in an aerial image (e.g., Google Earth) and first answer a series of questions on an "office" data form. Those questions require making distance and area measurements from the aerial image, plus obtaining data from provided tables and specified web sites. One-stop Internet portals were created to centralize and thus expedite acquisition of the office data for Oregon (ORWAP) and Southeast Alaska (WESPAK-SE). Users are directed to change their responses to the office-level questions if later onsite observations or different delineations of wetland boundaries dictate a need for such.

For the versions of WESP in Alberta (ABWRET-A), users must request answers to all office questions from Alberta Environment and Parks. For the target wetland, that agency uses geographic information systems (GIS) to query spatial data layers, some of which are not available to the public.

For all the regionalized WESPs, trained users subsequently visit the target wetland and complete two data forms, one to evaluate potential wetland stressors and the other to characterize conditions of several dozen variables relevant to predicting the relative levels of the wetland's functions and associated values. No sampling or measurement is required; users are instructed to base their answers on observations at the time of visit, extrapolation from physical indicators, and communications with the owner or manager of the wetland. Users normally assess an entire wetland but may also assess just the portion of a wetland expected to be altered or the only portion that is accessible. If a wetland contains multiple wetland classes or has well-defined subareas that differ with regard to past or ongoing disturbances or expected level of some functions, separate assessments may be done for each class or subarea.

As soon as users enter all office and field data into a spreadsheet, scores and ratings calculate automatically. Table 4.3.2.1 is an example of output from WESP-AC, and Table 4.3.2.2 defines the assessed wetland functions and other attributes.

Wetland and Stream Rapid Assessments. https://doi.org/10.1016/B978-0-12-805091-0.00011-6

TABLE 4.3.2.1 Example of Output From the WESP for Nontidal Wetlands of Atlantic Canada (WESP-AC)

Results for This Assessment Area (AA)	Function Score (Normalized)	Function Rating	Benefits Score (Normalized)	Benefits Rating	Function Score (Raw)	Benefits Score (Raw)
Wetland functions or other attributes						
Surface water storage (WS)	8.58	Higher	2.86	Moderate	7.70	2.50
Stream flow support (SFS)	0.76	Lower	0.00	Lower	0.41	0.00
Water cooling (WC)	0.50	Lower	0.00	Lower	0.33	0.00
Sediment retention and stabilization (SR)	4.77	Moderate	1.13	Lower	5.29	0.67
Phosphorus retention (PR)	2.65	Moderate	0.00	Lower	4.56	0.28
Nitrate removal and retention (NR)	4.28	Moderate	10.00	Higher	6.66	10.00
Carbon sequestration (CS)	2.21	Lower			5.85	
Organic nutrient export (OE)	6.41	Moderate			5.27	
Anadromous fish habitat (FA)	0.00	Lower	0.00	Lower	0.00	0.00
Resident fish habitat (FR)	5.88	Moderate	4.45	Higher	4.08	3.16
Aquatic invertebrate habitat (INV)	3.88	Moderate	3.94	Moderate	4.92	3.43
Amphibian and turtle habitat (AM)	6.10	Higher	2.55	Lower	6.32	2.31
Waterbird feeding habitat (WBF)	5.34	Moderate	2.73	Moderate	4.45	1.50
Waterbird nesting habitat (WBN)	3.61	Moderate	0.00	Lower	3.24	0.00
Songbird, raptor, and mammal habitat (SBM)	3.06	Moderate	0.00	Lower	2.50	0.00
Pollinator habitat (POL)	6.43	Higher	0.00	Lower	5.23	0.00
Native plant habitat (PH)	2.92	Moderate	4.35	Moderate	6.17	2.57
Public use and recognition (PU)			1.37	Lower		1.31
Wetland sensitivity (Sens)			5.79	Moderate		4.02
Wetland ecological condition (EC)			8.16	Higher		8.13
Wetland stressors (STR) (higher score means more)			3.15	Lower		2.71
Summary ratings for grouped functions						
HYDROLOGIC Group (WS)	8.58	Higher	2.86	Moderate	7.70	2.50
WATER QUALITY SUPPORT Group (max+avg./2 of SR, PR, NR, CS	2.65	0.00	7.87	Higher	6.13	6.83
AQUATIC SUPPORT Group (rnax+avg./2 of SFS, INV, OE, WC)	3.46	0.00	2.25	Lower	4.00	2.29
AQUATIC HABITAT Group (max+avg./2 of FA, FR, AM, WBF, WBN)	5.29	0.00	2.98	Moderate	4.97	2.28

TABLE 4.3.2.1 Example of Output From the WESP for Nontidal Wetlands of Atlantic Canada (WESP-AC)—cont'd

Results for This Assessment Area (AA)	Function Score (Normalized)	Function Rating	Benefits Score (Normalized)	Benefits Rating	Function Score (Raw)	Benefits Score (Raw)
TRANSITION HABITAT Group (rnax+avg./2 of SBM, PH, POL)	3.15	0.00	3.48	Moderate	5.40	1.72
WETLAND RISK (average of Sensitivity and Stressors)			4.36	Moderate		3.37

Note: A score of 0 does not mean the function or benefit is absent from the wetland. It means only that this wetland has a capacity that is equal to or less than the lowest scoring one for that function or benefit from among the 98 NB calibration wetlands that were assessed previously.

TABLE 4.3.2.2 Example of Function Definitions Used by WESPAK-SE

Function or Attribute	Definition	Values Related to Context
Water storage and delay	The effectiveness for storing runoff or delaying the downslope movement of surface water for long or short periods	Flood control, maintain ecological systems
Stream flow Support	The effectiveness for extending the period of surface water presence in streams with only seasonal or ephemeral flow	Support fish and other aquatic life
Water cooling	The effectiveness for maintaining or reducing temperature of downslope waters	Support coldwater fish and other aquatic life
Water warming	The effectiveness for increasing the temperature of downslope waters	Maintain late-season ice-free conditions
Sediment retention and stabilization	The effectiveness for intercepting and filtering suspended inorganic sediments, thus allowing their deposition, as well as reducing energy of waves and currents, resisting excessive erosion, and stabilizing underlying sediments or soil	Maintain quality of receiving waters. Protect shoreline structures from erosion
Phosphorus retention	The effectiveness for retaining phosphorus for long periods (>1 growing season)	Maintain quality of receiving waters
Nitrate removal and retention	The effectiveness for retaining particulate nitrate and converting soluble nitrate and ammonium to nitrogen gas while generating little or no nitrous oxide (a potent greenhouse gas)	Maintain quality of receiving waters
Carbon sequestration	The effectiveness for retaining both incoming particulate and dissolved carbon and converting carbon dioxide gas to organic matter (particulate or dissolved), and then retaining that organic matter on a net annual basis for long periods while emitting little or no methane (a potent "greenhouse gas")	Reduce risk of global climate warming
Organic nutrient export	The effectiveness for producing and subsequently exporting organic nutrients (mainly carbon), either particulate or dissolved	Support food chains in receiving waters. Facilitate transfer of iron to marine waters
Anadromous fish habitat	The capacity to support rearing or spawning habitat of fish species that migrate from marine waters into freshwater streams to spawn	Support commercial, subsistence, sport, and ecological values. Infuse uplands with marine nutrients
Resident fish habitat	The capacity to support an abundance and diversity of native fish (both resident and visiting species) that are not anadromous	Support commercial, subsistence, sport, and ecological values
Invertebrate habitat	The capacity to support or contribute to an abundance or diversity of invertebrate animals that spend all or part of their life cycle underwater or in moist soil. Includes dragonflies, midges, clams, snails, water beetles, shrimp, aquatic worms, and others	Support salmon and other aquatic life. Maintain regional biodiversity

Continued

TABLE 4.3.2.2 Example of Function Definitions Used by WESPAK-SE—cont'd

Function or Attribute	Definition	Values Related to Context
Amphibian habitat	The capacity to support or contribute to an abundance or diversity of native frogs, toads, and salamanders	Maintain regional biodiversity
Waterbird feeding habitat	The capacity to support or contribute to an abundance or diversity of waterbirds that migrate or winter but do not breed in the region	Support subsistence, sport, and ecological values. Maintain regional biodiversity
Waterbird nesting habitat	The capacity to support or contribute to an abundance or diversity of waterbirds that nest in the region	Maintain regional biodiversity
Songbird, raptor, and mammal habitat	The capacity to support or contribute to an abundance or diversity of native songbird, raptor, and mammal species and functional groups, especially those that are most dependent on wetlands or water	Maintain regional biodiversity
Pollinator habitat	The capacity to support pollinating insects such as bees, wasps, flies, butterflies, moths, and beetles	Maintain forest productivity and food chains
Native plant habitat	The capacity to support or contribute to a diversity of native, hydrophytic, vascular plant species, communities, and/or functional groups	Maintain regional biodiversity and food chains
Public use and recognition	Prior designation of the wetland by a natural resource or environmental protection agency as some type of special protected area. Also, the potential and actual use of a wetland for low-intensity outdoor recreation, education, or research	Commercial and social benefits of recreation. Protection of prior public investments
Wetland ecological condition	The integrity or health of a wetland, as defined operationally by its vegetation composition and richness of native species. More broadly, the similarity of a wetland's structure, composition, and function with that of reference wetlands of the same type and landscape setting, operating within the bounds of natural or historical disturbance regimes	(This is a value, not a function, and does not necessarily correlate with levels of functions)
Wetland sensitivity	A wetland's lack of intrinsic resistance and resilience to human and natural stressors (higher score = more sensitive)	(This is an attribute, not a function or value)
Stress potential	The degree to which a wetland is, or has recently been altered by or exposed to, risk from factors capable of reducing one or more of its functions and which are primarily human-related	(This is an attribute, not a function or value)

The equations comprising the models used to generate these scores are detailed and most are documented with literature citations in each of the regional WESPs.

BACKGROUND

Most of the correlations between variables (indicators) and the functions and values that WESP methods score were identified and described in the author's seminal document (Adamus, 1983), sponsored initially by the Federal Highway Administration and later the US Army Corps of Engineers (Wetland Evaluation Technique, Adamus et al., 1987; Adamus et al., 1992). That has served as the source of variable-function correlate information used by nearly every RAM published in North America since 1983. The author initially incorporated those function-variable correlations into hydrogeomorphic (HGM) methods addressing riverine, slope/flat, vernal pool, and tidal wetland HGM classes in Oregon (Adamus and Field, 2001; Adamus, 2006; Adamus et al., 2006) with support from the Environmental Protection Agency (EPA) and Oregon Department of State Lands (ODSL). However, with each HGM method requiring several years to develop and calibrate, Oregon officials grew concerned that it would take an unacceptably long time for the state to create and test HGM methods for every important wetland class within the state, as continued use of the HGM development framework would have required. Therefore, ODSL asked the author to develop and field-calibrate a somewhat different framework that would be applicable to all wetland classes and regions of the state, and obtained funds from the EPA to support that. The resulting Oregon Rapid Wetland Assessment Protocol (ORWAP, Adamus et al., 2016) consisted of a spreadsheet calculator, manual, and website. ORWAP has since been refined and updated (Adamus et al., 2016) and its use is required for most wetland permits in Oregon.

The WESP RAMs differ from another prolific series of function-focused RAMs—the HGM series (Smith et al., 2013)—in several fundamental ways (Table 4.3.2.3).

TABLE 4.3.2.3 Comparison of HGM and WESP Approaches for Wetland Function Assessment

Attribute	HGM	WESP	Explanation (See Below)
Uses simple mathematical logic in an Excel spreadsheet to calculate scores on an ordinal scale	Yes	Yes	1
Must first assign a wetland to a class	Yes	No	2
Can compare a wetland only to others in its class	Yes	No	3
Indicators and/or models must be field-calibrated by sponsor before version can be used by others	Yes	Yes	4
Every model variable is calibrated to least-altered site(s)	Yes	No	5
Function scores are automatically normalized	No	Yes	6
Scores are also converted to ratings	No	Yes	7
Number of functions that are scored (varies by regional version)	4–12	14–18	8
Provides scores for attributes other than wetland functions	No	Yes	9
Average time (per wetland) to collect all data needed for calculations	1–2 days	2–4 h	10
Main intended purpose	Assess impacts	Prioritize among wetlands	11

1. For both methods, most function scores are calculated using averages, weighted averages, maximums, or multiplications of groups of variables that comprise each model. In WESP methods, many calculations are conditioned (with logical if-then statements) on wetland class or the presence/absence of a specified condition of another variable.
2. WESP methods use a default set of variables applicable to all wetland classes and then make distinctions among wetland classes only where necessary to discern differences in function. The distinctions are embedded transparently in the models WESP uses. Also, where certain questions are not applicable to a given wetland class or situation, WESP data forms instruct users to skip them and those variables are ignored in subsequent score calculations. In contrast, the HGM approach requires a separate method for each HGM class, each with a partly different set of variables (questions) and models. In addition, HGM method accuracy is conditioned on users being able to assign a particular wetland to its class and this can often depend on site conditions.
3. In WESP methods, the scores of any wetland class can be compared with those of any other wetland class. HGM methods do not allow that comparison.
4. "Calibration" means visiting and assessing a range of wetlands and using their resulting score range to adjust model outputs to a common scale, usually 0–1 or 0–10.
5. In HGM methods, the condition of each variable that is assigned the most weight is usually the one that was found in wetlands believed to have been altered the least by humans, whereas in WESP methods the highest weight is assigned to the condition believed to support the highest rate or level of the named wetland function. HGM methods often provide graphs showing the hypothesized or measured relationship of each variable to the function it is believed to measure, whereas WESP methods do not provide graphs but rather express that relationship in multiple-choice questions (with different weights for different responses).
6. The function scores produced by most HGM methods are not subjected to further mathematical adjustment, whereas those produced by WESP methods are automatically adjusted (normalized) so that for each wetland function, the score from the highest-scoring calibration wetland is reset to 10 if it was greater than 10, that of the lowest-scoring calibration wetland is reset to 0 if it was less than 0, and intervening scores are adjusted proportionally. This is done independent of judgments about the degree to which the lowest and highest scoring wetlands have been altered by humans.
7. This is done because many decision-makers relate better to ratings than to numeric scores. WESP methods uniquely use Jenks Optimization to objectively convert each function score to a rating. They use a three-level categorization (e.g., higher, moderate, lower) but can accommodate a four-level or other categorization of scores.
8. WESP methods are generally regarded as addressing the most comprehensive series of important wetland functions. If one of the functions performed by some wetlands in a region is not performed by a specified wetland class, WESP assigns a score of 0 to wetlands of that class, whereas most HGM methods in that situation simply ignore the function. Neither WESP nor HGM methods roll up the scores of the functions they assess into a single number or rating.
9. HGM methods only score wetland functions. In contrast, most WESP methods additionally calculate a "value" or "benefit" score to each function. This is based partly on the degree to which a wetland is being given an opportunity to perform that function and partly on the significance of other valued features downslope and/or near the wetland that can benefit directly from the performed function, that is, a wetland's socioeconomic and ecological context. By addressing both functions and the contextual values of the functions, WESP methods are rightfully considered to be ecosystem service assessment tools, and are perhaps the only field-based tools that purport to assess many of the ecosystem services of wetlands. WESP methods also provide scores for wetland sensitivity, ecological condition, stressors, and public use.
10. WESP methods only require measurements from aerial imagery and other maps, with the balance of the variables based on field observations rather than the more time-consuming field measurements. Thus, despite WESP methods using many more variables than HGM methods, assessments typically require less time.
11. Because WESP methods mostly ask users to select from among a few categorical condition choices rather than make high-resolution measurements, WESP methods may be less likely than HGM methods to clearly portray temporal changes in functions.

In 2011, the author was asked by the Southeast Alaska Land Trust to develop and field-calibrate a WESP for that region. With funding from the US Fish and Wildlife Service and Alaska Department of Commerce, that resulted in the Wetland Ecosystem Services Protocol for Southeast Alaska (WESPAK-SE, Adamus, 2013). In 2012, the government of Alberta approached the author to regionalize WESP for the southern portion of that province, resulting in a method initially called WESPAB (Adamus, 2013) and later ABWRET-A (Government of Alberta, 2015). Shortly thereafter, the provincial government issued a formal wetland policy that specified the development of two wetland function assessment methods. One of those methods (see Chapter 2.2.12) was created for internal governmental use in planning and only uses GIS queries of existing spatial data. The other, termed ABWRET-A (Alberta Wetland Rapid Evaluation Technique Actual), uses nearly the same spatial data queries but integrates those data with data from onsite field observations. With matching funds from the North American Waterfowl Management Plan (NAWMP), the author prepared and helped field-calibrate one version of ABWRET-A for the southern part of the province (the "White Area"), partly by modifying WESPAB, and then prepared and helped field-calibrate a different version for northern Alberta wetlands (the "Green Area"). The government of Alberta is currently creating an ABWRET-D (modified from ABWRET-E, and similar in its desktop-only reliance on existing spatial data) to help estimate the required extent of wetland replacement where impacts to a wetland are judged to be permanent but minimal. Outputs from ABWRET-D have not been systematically compared with those from the same wetlands using ABWRET-A.

In 2014, the government of New Brunswick contacted the author to develop and field-calibrate WESP there, using a matching grant from Canada's National Wetland Conservation Fund (NWCF). Adjoining provinces of Atlantic Canada took note and requested to also be included in separate field calibrations for wetlands of Nova Scotia, Prince Edward Island, and Newfoundland-Labrador. All these RAMs (Adamus, 2018a, b, c, d, e) are available from the author's web site.

PROCEDURES AND RESOURCES FOR DEVELOPING THE REGIONALIZATIONS

Nearly the same procedure was followed to develop each WESP regionalization:

(1) Define boundaries of the region (reference domain). Do keyword searches of commercial bibliographic databases as well as Google Scholar to identify technical literature pertaining to wetland and landscape functions in that region. Obtain and read the literature.
(2) Identify and obtain spatial data layers potentially relevant to predicting the functions and values of the region's wetlands (Table 4.3.2.2). Whatever the imperfections of these spatial data layers, intersect the wetlands they have mapped with all other spatial data relevant to wetland functions, thereby creating a database.
(3) Based on available time, budget, and qualified personnel, estimate the number of wetlands (k) that are feasible to visit and assess in order to calibrate the regionalized WESP. Using statistical software, select calibration wetlands using a statistically stratified approach by conducting a k-means cluster analysis of the database after addressing issues with missing data and nominal variables. Select at least one wetland from each of the resulting statistical clusters. Base the selection on accessibility and a goal of encompassing the extremes of each database variable, for example, all wetland types in both headwater and lowland locations in all applicable subregions of the state or province. Note the coordinates (or create a kmz file) of the selected calibration sites and several backup sites to facilitate finding them while in the field using Google Earth. The k-means clustering algorithm, while not perfect and with accuracy depending largely on the temporal and geographic coverage and resolution adequacy of a region's spatial data, provides a means of selecting calibration wetlands that is more objective and systematic than simply hand-picking sites based on presumed differences in form, function, and disturbance level.
(4) Simultaneously, examine the existing WESP template (data forms and calculator) to determine what should be modified for this particular region. Especially consider the variables and their data sources on the "office" data form, the weights assigned to the various conditions of each indicator variable, and the structures of the models. Make those adjustments.
(5) Train a small number of wetland technicians to use the draft regionalized WESP and determine who will visit and assess each calibration wetland. Conduct the onsite assessments and enter the data into the regionally modified Excel calculator.
(6) For each visited calibration site, also answer the "office" questions using the database created in #2, Google Earth, and other sources referenced on the data form.
(7) Normalize all the scores from the collected data by identifying the sites with the highest and lowest scores for each function, then applying this formula to each set of function scores:

(score for site A–minimum of all sites)/(maximum of all sites–minimum of all sites)

(8) Prepare the regional manual, post that and the calculator online, and train future users.

The cost and time required to develop a regionalized WESP will, of course, depend on:

- Size of the region and accessibility of its wetlands.
- Extent of peer review and repeatability testing desired.
- Availability and cost of one or more persons skilled in wetland assessment, GIS, and simple programming using Excel software.

The range of costs for WESP regionalizations described in this chapter is about $50,000–$100,000 US dollars, not counting the time contributed by government employees, which has varied greatly among regions. It does not include costs for peer review, repeatability testing (Chapter 3.8), or training users (Chapter 3.11). This cost range assumes 100–200 calibration wetlands are visited and assessed. While the statistical sufficiency of that number for capturing the full range of variation in a region's wetland functions has not been verified, it would seem to be the minimum necessary to achieve a goal of addressing variation in function among all wetland types in all landscape settings and all disturbance levels within a large region.

WESP APPLICATIONS AND PROSPECTS FOR THE FUTURE

Because the WESP RAMs are relatively new, accounts of use by consultants are mainly anecdotal and no data have been compiled systematically on frequency of use in any region. Two communities (Creswell, Oregon, and Juneau, Alaska) and one watershed (Upper Deschutes, Oregon) are known to have used the applicable WESP RAMs to assess a large proportion of wetlands within their jurisdictions. Regulatory agencies responsible for wetlands in Southeast Alaska and the Atlantic Maritimes have not yet decided when or whether to require permit applicants to use the WESPs developed in those regions, but are training potential users. Nonetheless, even without official endorsements from higher levels of government, many entities are likely to use a WESP because it is the only well-documented RAM available that has been customized and calibrated for their region.

In Oregon, the 220 calibration sites that Oregon used to normalize the scores originally were almost entirely on public land and had not been selected in a statistically rigorous manner due to budget and time constraints in 2008–2009. However, in 2013 ODSL obtained an EPA grant to select 200 new calibration wetlands using a more defensible stratified random sampling scheme, and subsequently visited and assessed those and used their score range to establish new normalizing formulas for the revised ORWAP. Sponsors of the WESP methods in several regions are currently developing formulas for linking normalized scores to mitigation credits and debits (see Chapter 3.12). Applying such formulas routinely in regulatory programs would significantly advance our ability to ensure no net loss of wetland function, not just no net loss of wetland area.

The extent to which agencies in other states, provinces, and regions will choose to use the WESP template to develop their wetland RAMs remains unknown. That may depend on whether administrators at various levels of government in a particular area have a legal mandate for requiring that assessments of wetland functions—if required at all—be standardized (Arnold, 2012, 2014). It will also depend on gaining information about the relative cost and efficiency of using the WESP template to create their wetland RAM. A challenge to regionalizating WESP elsewhere is the paucity of experts familiar with the processes behind the full range of wetland functions while also being competent in constructing fairly complex but transparent models that reflect those processes within the limits of Excel software coding.

LESSONS LEARNED

As expected, the distribution of function scores varies from region to region and is directly a product of the particular combination of variables and model structures that WESP uses in each region. In a dry region with many hydrologically isolated wetlands, for example, scores for the fish habitat function will skew toward the low end because many of those wetlands are inaccessible to fish and naturally dry up completely during most years.

Repeatability (consistency) of WESP scores obtained independently from the same wetland by multiple trained users has been formally tested in Oregon, Alberta, and Alaska, with generally excellent results. That is, the variability of scores determined among users was in most cases found to be less than the variability of those scores among the wetlands assessed, and the statistical confidence bands around scores in the 0–10 range were about ±0.5. See Chapter 3.8 for additional description of testing procedures.

As is true of many wetland assessment methods, to date none of the output scores from any of the regionalized WESPs have been compared with actual measurements of wetland functions. This is due to the enormous cost and time requirements of doing so in a statistically valid number of wetlands. This has not been a large concern because the main objective of WESP is to increase the standardization and consistency of function assessments, not necessarily to provide quantitative assessments of wetland functions that are accurate in every case.

In the absence of opportunities for meaningful validation of function scores, peer review is useful for improving a method and contributing to its credibility. ORWAP and WESPAK-SE were subjected to peer reviews by function experts, the latter using the format of several half-day workshops on the particular functions that are scored. Suggestions were incorporated into subsequent revisions (see Chapter 3.9 for further discussion). Feedback from users (mainly consultants) is also essential, especially to enhance the clarity and user-friendliness of the data forms. Most such feedback comes during trainings, but Oregon also sent questionnaires to persons known to have used that WESP on many wetlands. Suggestions were incorporated into subsequent revisions of ORWAP.

ACKNOWLEDGMENTS

While hundreds of people have provided useful feedback on the regionalized WESPs, the following people have also been instrumental in supporting it administratively and/or as major participants in field calibration and testing within their region:

Oregon (ORWAP): Janet Morlan, Kathy Verble, Melody Rudenko, Yvonne Vallette, Dana Field

Southeast Alaska (WESP-AK): Neil Stichert, Diane Mayer, Allison Gillum

Alberta (WESPAB, ABWRET-A): Gillian Kerr, Matthew Wilson, Marsha Trites-Russell, Thorsten Hebben, Michael Barr

Atlantic Canada (WESP-AC): John Brazner (NS), Krista Hilchey (NS), Christie Ward (NB), Jonathan Sharpe (Newfoundland), Garry Gregory (Prince Edward Island)

REFERENCES

Adamus, P.R., 1983. A Method for Wetland Functional Assessment. Vol. II. Methodology. Report No. FHWA-IP-82-24, Federal Highway Administration, Washington, DC

Adamus, P.R. 2006. Hydrogeomorphic (HGM) Assessment Guidebook for Tidal Wetlands of the Oregon Coast. Manual, data forms, and spreadsheet calculator for the Oregon Department of State Lands, USEPA, and Coos Watershed Association.

Adamus, P.R. 2013. Wetland Ecosystem Services Protocol for Southern Alberta (WESPAB). Report and software for Alberta Environment and Water, Government of Alberta, Edmonton, AB.

Adamus, P.R. 2016. Manual for the Wetland Ecosystem Services Protocol (WESP). Version 1.3. people.oregonstate.edu/~adamusp

Adamus, P.R., 2018a. Wetland Ecosystem Services Protocol for Atlantic Canada (WESP-AC). Non-Tidal Wetlands, Manual, data forms, spreadsheet calculator, and supporting files for New Brunswick, New Brunswick Dept. of Environment and Local Government, Fredericton, NB.

Adamus, P.R., 2018b. Wetland Ecosystem Services Protocol for Atlantic Canada (WESP-AC). Non-tidal Wetlands, Manual, data forms, spreadsheet calculator and supporting files for Nova Scotia, New Brunswick Dept. of Environment and Local Government, Fredericton, NB.

Adamus, P.R., 2018c. Wetland Ecosystem Services Protocol for Atlantic Canada (WESP-AC). Non-Tidal Wetlands. Manual, data forms, spreadsheet calculator, and supporting files for Prince Edward Island, New Brunswick Dept. of Environment and Local Government, Fredericton, NB.

Adamus, P.R., 2018d. Wetland Ecosystem Services Protocol for Atlantic Canada (WESP-AC). Non-Tidal Wetlands, Manual, data forms, spreadsheet calculator, and supporting files for Newfoundland-Labrador, New Brunswick Dept. of Environment and Local Government, Fredericton, NB.

Adamus, P.R., 2018e. Wetland Ecosystem Services Protocol for Atlantic Canada (WESP-AC). Tidal Wetlands. Manual, data forms, spreadsheet calculator, and supporting files. New Brunswick Dept. of Environment and Local Government, Fredericton, NB.

Adamus, P.R., Field, D., 2001. Guidebook for Hydrogeomorphic (HGM)-Based Assessment of Oregon Wetland and Riparian Sites. I. Willamette Valley Ecoregion, Riverine Impounding and Slope/Flat Subclasses. Manual, data forms, and spreadsheet calculator for Oregon Department of State Lands, Salem, Oregon.

Adamus, P.R., Clairain, E.J., Smith, R.D., Young, R.E., 1987. Wetland Evaluation Technique (WET). Volume II. Methodology. US Army Corps of Engineers Waterways Experiment Station, Vicksburg, MS.

Adamus, P.R., Clairain Jr., E.J., Smith, D.R., Young, R.E., 1992. Wetland Evaluation Technique (WET). Volume I. Literature Review and Evaluation Rationale. U.S. Army Corps of Engineers, Waterways Experiment Station, Vicksburg, MS.

Adamus, P.R., Pakenham-Walsh, M., McCarten, N., 2006. Hydrogeomorphic (HGM) Method for Assessing Functions and Values of Vernal Pool Wetlands of the Agate Desert, Medford, Oregon. Manual, data forms, and spreadsheet calculator for Oregon Dept. of State Lands, Salem, OR.

Adamus, P., Morlan, J., Verble, K., Buckley, A., 2016. Oregon Rapid Wetland Assessment Protocol (ORWAP, revised): Version 3.1 calculator spreadsheet, manual, databases, and data forms. Oregon Dept. of State Lands, Salem, OR.

Arnold, G., 2012. Assessing Wetland Assessment: Understanding State Bureaucratic Use and Adoption of Rapid Wetland Assessment Tools. PhD dissertation, Indiana University.

Arnold, G., 2014. Policy learning and science policy innovation adoption by street-level bureaucrats. J. Public Pol. 34 (03), 389–414.

Government of Alberta, 2015. Alberta Wetland Rapid Evaluation Tool-Actual (ABWRET-A) Manual. Water Policy Branch, Alberta Environment and Parks, Edmonton, AB.
Smith, R.D., Noble, C.V., Berkowitz, J.F., 2013. Hydrogeomorphic (HGM) Approach to Assessing Wetland Functions: Guidelines for Developing Guidebooks (Version 2). No. ERDC/EL-TR-13-11. Engineer Research & Development Center, Environmental Lab, Vicksburg, MS

INTERNET DOWNLOAD LOCATIONS

WESP template

http://people.oregonstate.edu/~adamusp/WESP/.

Oregon

ORWAP calculator, manual, etc.: http://www.oregon.gov/dsl/WW/Pages/ORWAP.aspx
Supporting internet portal: http://oregonexplorer.info/content/oregon-rapid-wetland-assessment-protocol-orwap-map-viewer

Southeast Alaska

WESPAK-SE calculator, manual, etc.: http://southeastalaskalandtrust.org/wetland-mitigation-sponsor/wespak-se/
Supporting internet portal: http://seakgis.alaska.edu/flex/wetlands/

Alberta

http://aep.alberta.ca/water/programs-and-services/wetlands/documents/RapidEvaluationTool-Jun01-2015.pdf.

Atlantic Canada

http://people.oregonstate.edu/~adamusp/Atlantic Canada Wetland Assessment Tools.

Chapter 4.3.3

California Rapid Assessment Method for Wetlands and Riparian Areas (CRAM)

Josh Collins* and Eric D. Stein†

**San Francisco Estuary Institute and Aquatic Science Center, Richmond, CA, United States, †Southern California Coastal Water Research Project, Costa Mesa, CA, United States*

Chapter Outline

INTRODUCTION

The California Rapid Assessment Method (CRAM) for wetlands and riparian areas is an integral part of the Wetland and Riparian Area Monitoring Plan (WRAMP) produced by the Wetland Monitoring Workgroup of the California Water Quality Monitoring Council. WRAMP has two main objectives: to enable local, state, and federal agencies in California to consistently assess (1) the distribution, abundance, diversity, and condition of wetlands in the watershed context, and (2) the performance of public policies, programs, and projects intended to restore and protect California wetlands (Fig. 4.3.3.1).

WRAMP is based on the three-level framework for wetland assessment embodied in the US Environmental Protection Agency's (USEPA) Core Elements of an Effective State and Tribal Wetlands Program (USEPA, 2008), where Level 1 consists of map-based inventories, Level 2 consists of field-based rapid assessments of wetland overall health, and Level 3 consists of intensive field-based measures of particular health aspects. CRAM is the state's primary Level 2 tool.

CRAM assumes that the condition of a wetland is a manifestation of many processes that together control the kinds and levels of wetland functions, such that the overall functional capacity of a wetland can be assessed based on its overall condition.

CRAM meets a broadly expressed need in California for a standard, scientifically sound, and affordable way to assess the overall condition or functional capacity of wetlands in a watershed context. The need is amplified by the watershed approach to compensatory mitigation required under the federal Clean Water Act (CWA, Section 404 (USACE, 2008), and the state's intent to take a complimentary watershed approach under CWA Section 401 (SWRCB, 2016). The state has also recognized that CRAM can help meet its reporting requirements under CWA Sections 303(d) and 305 (b) while also helping to evaluate the governor's Wetland Conservation Policy (CRA, 1993).

The state conducted a peer review of CRAM as part of its adoption process (SWRCB, 2011), and the USACE also conducted a review of CRAM relative to its use in mitigation planning and evaluation. A key finding of the state's review was that CRAM should be subject to ongoing revision to assure its continued efficacy. The Level 2 Committee of the Wetland Monitoring Workgroup serves this objective while also overseeing a statewide CRAM training program and online database. A key recommendation from the USACE was to quantify the relationship between the age and condition of wetland restoration projects based on CRAM that could be used to forecast future project conditions relative to ambient, reference, or target conditions.

Wetland and Stream Rapid Assessments. https://doi.org/10.1016/B978-0-12-805091-0.00044-X

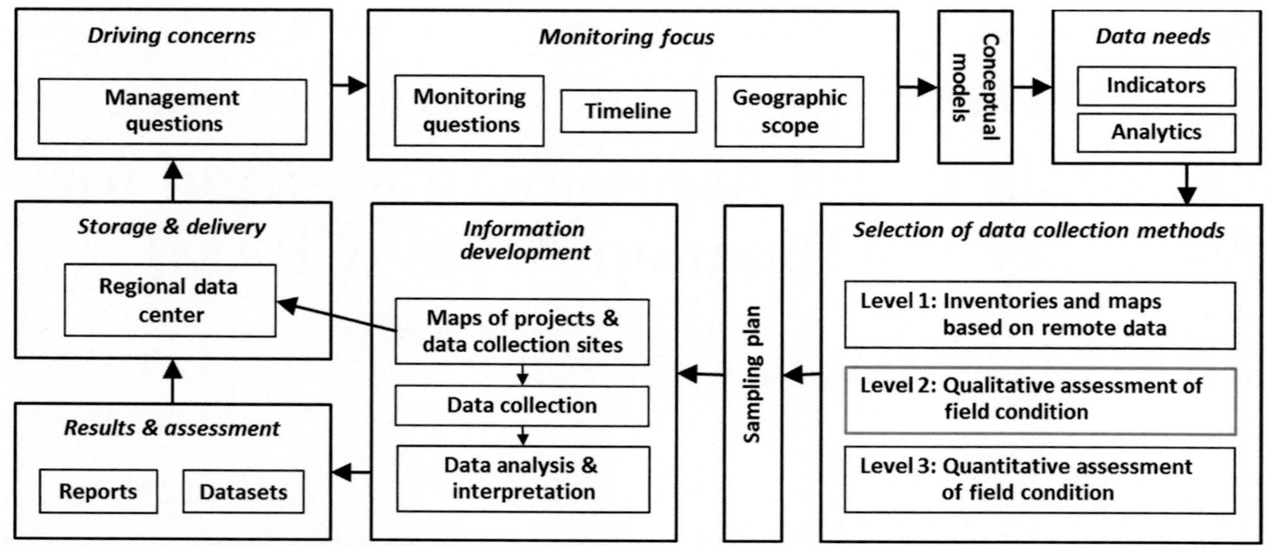

FIG. 4.3.3.1 Diagram of the WRAMP framework. CRAM is the main Level 2 tool in the box outlined in *red*.

CONCEPTUAL FRAMEWORK FOR THE METHOD

Development of CRAM has incorporated concepts and methods from other wetland assessment programs in California and elsewhere, including the Washington State Wetland Rating System (WADOE, 1993), New Mexico Rapid Assessment Method (MRAM, Burglund, 1999), and Ohio Rapid Assessment Method (ORAM, Mack, 2001). CRAM also draws on concepts from stream bioassessment and wildlife assessment procedures of the California Department of Fish and Wildlife, the different wetland compliance assessment methods of the San Francisco Bay Regional Water Quality Control Board and the Los Angeles Regional Water Quality Control Board, the "releve method" of the California Native Plant Society, and various hydrogeomorphic method (HGM) guidebooks that are being used in California.

CRAM is intended for application to streams and all classes of wetlands throughout California. Although centered on coastal watersheds through much of the initial development process, it has now spread inland to the Central Valley, Inland Empire, and Tahoe regions. CRAM development to date has involved scientists and managers from most regions of the state to account for the variability in wetland type, form, and function that occurs with physiographic setting, latitude, altitude, and distance inland from the coast. Future refinements of CRAM will be used to adjust CRAM metrics as needed to remove any systematic bias against any particular kind of wetlands or their settings.

CRAM was developed according to a set of underlying conceptual models and assumptions about the meaning and utility of rapid assessment, the best framework for managing wetlands, the driving forces that account for their condition, and the spatial relationships among the driving forces. These models and assumptions are explicitly stated in this section to help guide the interpretation of CRAM scores.

The management framework for CRAM is the pressure-state-response model (PSR) of adaptive management. The PSR model states that human operations such as agriculture, urbanization, recreation, and the commercial harvesting of natural resources can be sources of stress or pressure affecting the condition or state of natural resources. The human responses to these changes include any organized behavior that aims to reduce, prevent, or mitigate undesirable stresses or state changes. The approach used by CRAM is to focus on condition or state. A separate stressor checklist is then used to note which, if any, stressors appear to be exerting pressure affecting condition. It is assumed that managers with knowledge of pressures and states will exact more effective responses.

CRAM is based on six major assumptions about how the condition of a wetland is determined by interactions among internal and external hydrologic, biologic (biotic), and physical (abiotic) processes. First, CRAM assumes that the condition of a wetland is mainly determined by the quantities and qualities of water and sediment (both mineral and organic) that are either processed onsite or that are exchanged between the site and its immediate surroundings. Second, the supplies of water and sediment are ultimately controlled by climate, geology, and land use. Third, geology and climate govern natural disturbance whereas land use accounts for anthropogenic stress. Fourth, biota (especially vegetation) tends to mediate the effects of climate, geology, and land use on the quantity and quality of water and sediment (Fig. 4.3.3.2). For example, vegetation can stabilize stream banks and hillsides, entrap sediment, filter pollutants, provide shade that

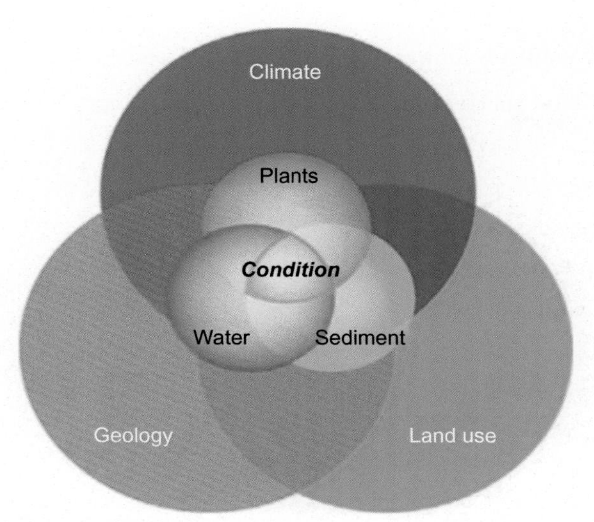

FIG. 4.3.3.2 Spatial hierarchy of factors that control wetland.

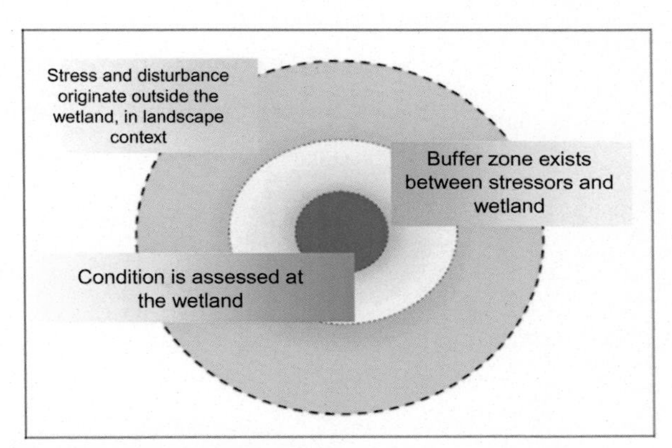

FIG. 4.3.3.3 Spatial hierarchy of stressors, buffers, and wetland condition.

lowers temperatures, reduce winds, etc. Fifth, stress usually originates outside the wetland in the surrounding landscape or encompassing watershed. Sixth, buffers around the wetland can intercept and otherwise mediate stress (Fig. 4.3.3.3).

Three major assumptions govern how wetlands are scored using CRAM. First, it is assumed that the societal value of any wetland function (i.e., its ecological services) matters more than whatever intrinsic value it might have in the absence of people. This assumption does not preclude the fact that the support of biological diversity is a service to society. Second, it is assumed that the value depends more on the diversity of services than the level of any one service. Third, it is assumed that the diversity of services increases with structural complexity and size. CRAM therefore favors large, structurally complex examples of each type of wetland.

California has a great diversity of wetlands owing to the state's coastal, montane, desert, and other biomes arrayed across very broad climatic gradients (Griffith et al., 2016). CRAM recognizes six major wetland classes and 12 subclasses (Table 4.3.3.1), each of which has its own unique set of assessment metrics for the standard four attributes.

The assessment area (AA) is the portion of a wetland that is assessed using CRAM. An AA might include a small wetland in its entirety; however, in most cases, the wetland will be larger than the AA.

The delineation of an AA follows strict guidelines. The intent is for AAs to encompass most of the internal workings of the wetland that account for its homeostasis, such that assessed changes in condition are likely to be due to external drivers such as management or climate change rather than internal short-term, stochastic variability. Therefore, to the degree possible, the delineation of an AA is based on prescribed hydrogeomorphic considerations. If these considerations are not applicable, or if the resulting AA is more than about 25% larger than the preferred size, delineation is guided by size alone. The preferred and minimum AA sizes vary by wetland class.

TABLE 4.3.3.1 Classification of Wetlands for the Purpose of CRAM Assessments

Wetland Classes	CRAM Subclasses	Wetland Classes	CRAM Subclasses
Riverine	Confined riverine	Estuarine	Perennial saline estuarine
	Nonconfined riverine		Perennial nonsaline estuarine
	Episodic riverine		Bar-built estuarine
Depressional	Individual vernal pools	Lacustrine	No subclasses
	Vernal pool systems	Slope	Seeps and springs
	Depressional		Forested slope
Playas	No subclasses		Wet meadows

TABLE 4.3.3.2 CRAM Attributes, Their Metrics, and Submetrics

Attributes		Metrics and Submetrics
Buffer and landscape context		Aquatic area abundance or steam corridor continuity
		Stream corridor continuity (bar-built estuaries only)
		Aquatic area in adjacent landscape (bar-built estuaries only)
		Marine connectivity (bar-built estuaries only)
		Buffer
		Percent of AA with buffer
		Average buffer width
		Buffer condition
Hydrology		Water source
		Hydroperiod or channel stability
		Hydrologic connectivity
Structure	Physical	Structural patch richness
		Topographic complexity
	Biotic	Plant community
		Number of plant layers present or endemic species richness (vernal pools only)
		Number of codominant species
		Percent invasion
		Horizontal interspersion
		Vertical biotic structure

CRAM is organized into four universal attributes that represent the three-dimensional structure of the wetlands, its hydrology, and its near- and far-field setting. Each attribute is represented by a set of metrics (Table 4.3.3.2), which can differ between wetland classes or their subclasses.

DESCRIPTION OF THE METHOD

Each application of CRAM follows a fixed procedure (Table 4.3.3.3). However, the sampling plan depends on the purpose of the assessment. Ambient surveys of baseline conditions are ideally based on probabilistic sampling designs (USEPA,

TABLE 4.3.3.3 Steps for Using CRAM.

Step 1	Assemble background information about the management of the wetland.
Step 2	Classify the wetland using the CRAM typology.
Step 3	Verify the appropriate season and other timing aspects of the field assessment.
Step 4	Estimate the boundary of the AA in the office (subject to field verification).
Step 5	Conduct the office assessment of stressors and onsite conditions of the AA.
Step 6	Conduct the field assessment of stressors and onsite conditions of the AA.
Step 7	Complete CRAM assessment scores and QA/QC procedures.
Step 8	Upload CRAM results into statewide information data management system (eCRAM).

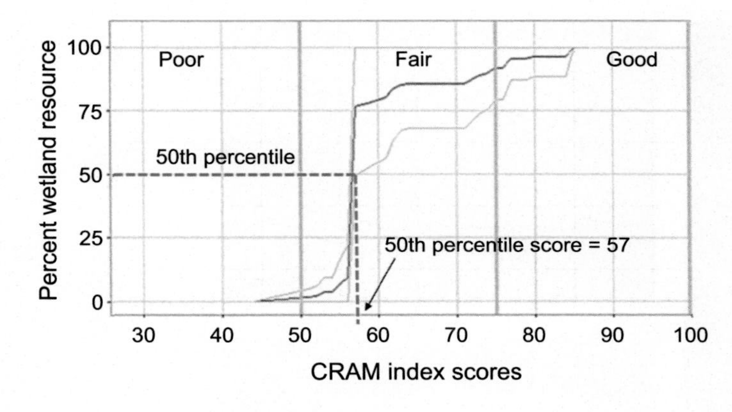

FIG. 4.3.3.4 CDF of CRAM index scores from a probabilistic survey of tidal marsh condition in one region of the California coast. The condition categories (poor, fair, good) are the tertiles of the full range of possible scores from 25 to 100. *Red lines* are the 95% confidence interval. Fifty percent of the wetland area has a CRAM index score ≤ 57.

2002) using the California Aquatic Resource Inventory (CARI is the preferred Level 1 tool) as the sample draw (SFEI, 2017). In such surveys, there is generally one AA per wetland or reach of stream. The results are summarized as a cumulative distribution function (CDF) that quantifies the proportion of the total area of the surveyed wetland class above or below any attribute or index score (Fig. 4.3.3.4).

The CDF of CRAM scores can be used to set performance criteria for projects at the watershed or regional scale. For example, unless projects score above the 50th percentile score for a watershed, they degrade its condition profile, as represented by the CDF. To improve the profile, projects should score above the 50th percentile. Higher scores for larger projects will improve the profile more because they represent more of the wetland resource.

Given that wetland projects evolve as habitat over time, their contributions to the condition profile will also change. This can be assessed using habitat development curves (HDCs) based on CRAM. HDCs are produced by plotting wetland condition against wetland age and reference condition (e.g., Kentula et al., 1992; Fong et al., 2017). When the HDC is based on CRAM, it quantifies the rate of habitat development as the increase in CRAM scores over time (Fig. 4.3.3.5). The HDC can therefore be used to estimate a project's future score and thus its future contribution to the profile based on its current score.

The CRAM scores are generally accepted by the user community as useful indices of the capacity of wetlands to provide high levels of their intrinsic functions (Fig. 4.3.3.6). This acceptance is due to the validation studies, the breadth of the community, and the clear relationship between the scores and visible field conditions (Fig. 4.3.3.6). As a consequence, CRAM metrics are beginning to be used as design guidelines to help ensure that projects score on or above the HDC, and thus have a greater likelihood of improving the wetland condition profile (Fig. 4.3.3.7).

Multiple AAs are necessary to assess large wetlands, such as large restoration projects. In these applications, randomly distributed AAs are sequentially assessed until their index scores, when plotted in order of occurrence, are asymptotic. The project can be assessed as the range in index or attribute scores, or as the average or median scores.

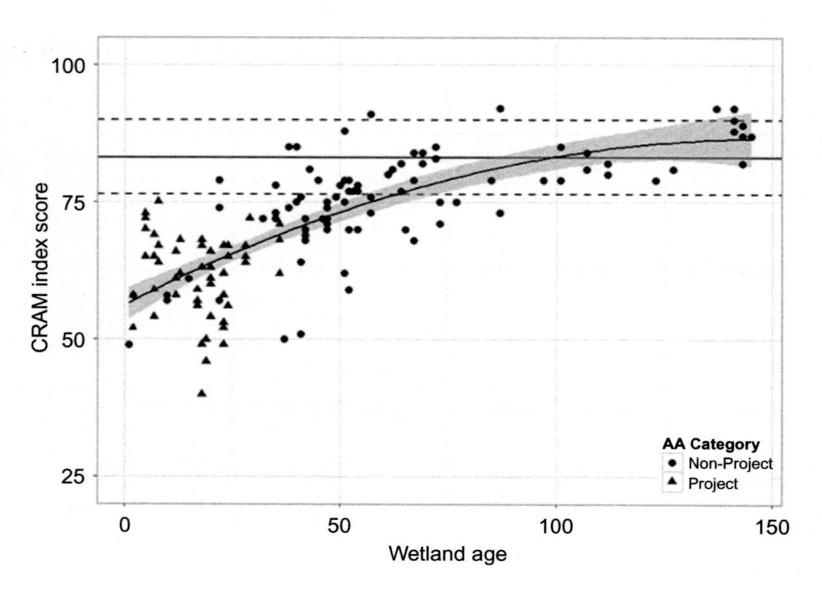

FIG. 4.3.3.5 Habitat development curve for California estuarine wetlands (SFEI, 2014). Solid and dashed horizontal lines represent the median and range of index scores for reference conditions. Shaded area is the 95% confidence interval for the curve. The variability in scores mainly represents differences in the design and operation of restoration and mitigation projects, all of which are <50 years old.

FIG. 4.3.3.6 Examples of CRAM assessment areas in estuarine wetlands of obviously different condition. (A) Good condition (score = 79). (B) Medium condition (score = 61). (C) Poor condition (score = 40).

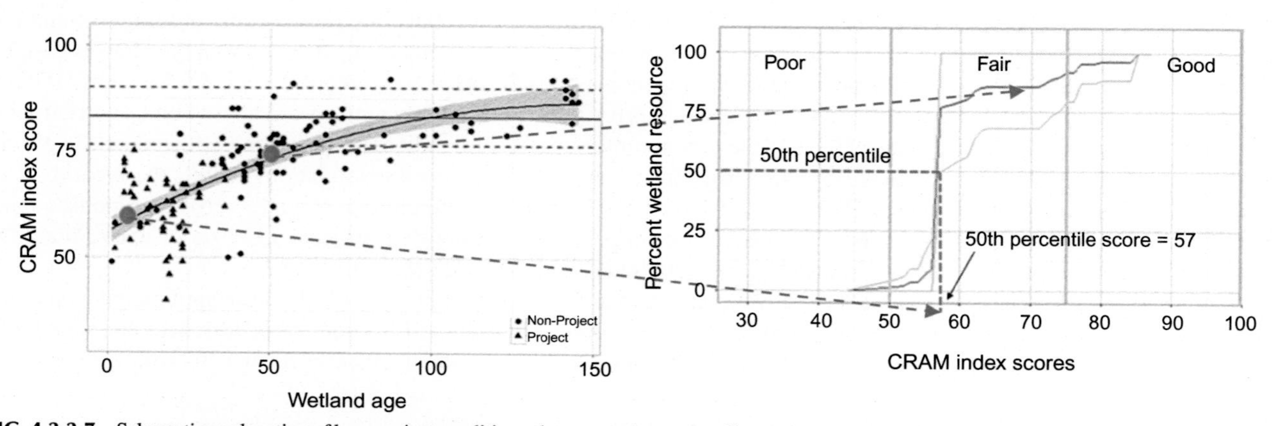

FIG. 4.3.3.7 Schematic explanation of how project condition relates to project age and hence its contribution to the condition profile. In this hypothetical case, a project scores 15 points higher after 40 years of maturation, and moves from the low to the high range of fair condition.

CALIBRATION AND VALIDATION

Each CRAM module is calibrated to a gradient of wetland stress. In most cases, the gradient is statewide. Calibration can involve adjustments in the metrics, their narrative alternative states, and their field indicators. Once calibrated, a module is added to the training curriculum and deemed ready for use.

Validation involves regressing CRAM scores on selected Level 3 data from the independent stress gradient (Stein et al., 2009). The results are compared to expected relationships between the CRAM scores and the wetland functions represented by the Level 3 data. Adjusting CRAM to increase its correlation with any particular function tends to decrease its correlation with other functions. This is because CRAM is designed to assess the overall capacity of a wetland to support all its intrinsic functions, some of which are uncorrelated or inversely correlated to each other. The criterion for validation is therefore a general agreement between the expected and observed regressions.

TIME SPENT IN DEVELOPING AND TESTING THE METHOD

The concerted effort to develop CRAM began in 2005 with the formation of a statewide steering committee and regional technical teams. During the ensuing 12 years, the effort to develop and update CRAM has intensified, with CRAM becoming an integral component of the Wetland and Riparian Area Monitoring Plan under the direction of the multiagency Wetlands Monitoring Workgroup of the legislated California Water Quality Monitoring Council.

HOW THE METHOD IS BEING USED

Applications of CRAM are growing in diversity and number (Table 4.3.3.4). As of 2016, >1000 practitioners have completed >5000 assessments throughout the state, and >300 of these assessments support state or federal permits.[1] The integration of CRAM into procedures issued by the South Pacific Division of the USACE for mitigation crediting and evaluation under USCWA Section 404 is especially noteworthy (USACE, 2015), as is the comparable intended use of CRAM by the state's USCWA Section 401 program.

TIME SPENT TO APPLY THE METHOD IN THE FIELD

Each CRAM assessment consists of the office and field work needed to establish one AA, complete the CRAM fieldbook for the AA, and upload the AA map and associated CRAM results to the online database. For ambient surveys, office and field time vary with AA remoteness. In general, each assessment requires one working day, although field time tends to be less than 2 h. For most project assessments, multiple assessments are completed in 1 day by a two- or three-person team.

TABLE 4.3.3.4 One Example of Each of Four Major Categories of CRAM Applications

Application	Example
Ambient assessment	Surface Water Ambient Monitoring Program, State Water Resources Control Board. http://www.waterboards.ca.gov/water_issues/programs/swamp/docs/cabw2016/1_ode_expand_univ.pdf
Mitigation evaluation	CA High Speed Rail EIR/EIS https://www.hsr.ca.gov/Programs/Environmental_Planning/final_fresno_bakersfield.html
Restoration evaluation	Montezuma Wetlands LLC, Monitoring and Reporting Program http://www.swrcb.ca.gov/rwqcb2/board_info/agendas/2012/October/MWRP/TO.pdf
Education and research	Morris, L. 2014. Evaluating the ability of the CRAM to capture seasonal variability within the Kachituli Oxbow Wetland Restoration Site. MS Thesis, Sacramento State University. http://www.csus.edu/envs/documents/theses/fall%202014/826.california%20rapid%20assessment%20method-the%20kachituli%20oxbow%20wetland%20restoration%20site.pdf

Numerous additional examples exist and others are forthcoming.

1. Use statistics provided by the CRAM website, http://www.cramwetlands.org/.

WHAT WAS LEARNED

The primary lesson from the ongoing effort to develop a wetland rapid assessment method applicable across multiple public policies and programs is that it's possible. Despite the unsurpassed diversity of wetland ecosystems in California, the large number of agencies at all levels of government responsible for wetland protection, and the healthy skepticism within the wetland science community about the efficacy of rapid assessment in general, CRAM has proven to provide meaningful and useful information about the condition of wetlands throughout the state.

The success of CRAM is due to many factors, not least of which is the capability of dozens of scientists who have contributed to its development. An essential aspect was the concerted top-down and bottom-up approaches. Early establishment of a multiagency statewide steering committee with regional or biome representation was foundational to all subsequent development efforts. Other aspects of CRAM development that were especially key include:

- Developmental intellectual framework grounded in science and based on clear assumptions.
- A comprehensive user's manual subject to ongoing revision as a living document.
- Strategic focus on serving statewide regulatory and ambient assessment programs.
- Repeated peer review as required to achieve and maintain credibility.
- Development of an online database for visualizing and sharing data.
- Development of a rigorous training program.

PROSPECTS FOR THE FUTURE

The greatest need at this time is funding to maintain the online CRAM software and database, and to support occasional revisions to the method as recommended by its user community. The challenge is to fund a method used by everyone and belonging to no single client agency or user group. A business plan that focuses on data sharing agreements and user fees has been produced on behalf of the Water Quality Monitoring Council for implementation by multiple client agencies.

An effort was made during the initial development of CRAM to separate assessments of condition from assessments of stress. The intent was twofold. First, the separate assessments could be used to explore the hypothesis that the amount of correlations between stress and wetland condition would decrease as buffer condition increased (see Fig. 4.3.3.2 above). Second, for cases with low buffer scores, the stressors could be ranked in importance based on their contribution to the correlations between stress and condition. Independent assessments of stress and conditions have not yet been achieved, however. For example, the plant community metric of the biotic structure attribute includes a submetric about the relative abundance of nonnative plant species, although biological invasion is usually considered a significant stressor. Until indicators of stress and condition are more completely separated, some autocorrelation between them can be expected. This autocorrelation diminishes the utility of CRAM for evaluating buffer effectiveness. There is an emerging consensus within the CRAM user community to convert the stressor checklist into a separate stress index, such that CRAM provides independent assessments of buffer condition, stress, and wetland condition.

REFERENCES

Burglund, J., 1999. Montana Wetland Assessment Method. Montana Department of Transportation and Morrison-Maierle, Inc., Helena, MT.

CRA, 1993. California Wetlands Conservation Policy. California Resources Agency, Sacramento, CA. http://resources.ca.gov/wetlands/policies/governor.html.

Fong, L.S., Stein, E.D., Ambrose, R.F., 2017. Development of restoration performance curves for streams in Southern California using an integrative condition index. Wetlands 37, 289–299.

Griffith, G.E., Omernik, J.M., Smith, D.W., Cook, T.D., Tallyn, E., Moseley, K., Johnson, C.B., 2016. U.S. Geological Survey Open-File Report 2016–1021, With Map. https://pubs.er.usgs.gov/publication/ofr20161021.

Kentula, M.E., Brooks, R.P., Gwin, S.E., Holland, C.C., Sherman, A.D., Sifneos, J.C., 1992. Wetlands: An Approach to Improving Decision Making in Wetland Restoration and Creation. Wetlands Research Program, U.S. Environmental Protection Agency, Environmental Research Laboratory, Corvallis, OR.

Mack, J.J., 2001. Ohio Rapid Assessment Method for Wetlands, Manual for Using Version 5.0. Ohio EPA Technical Bulletin Wetland/2001-1-1. Ohio Environmental Protection Agency, Division of Surface Water.

SFEI, 2014. Developmental Trajectory for California Tidal Marsh Restoration and Mitigation Projects. San Francisco Estuary Institute, Richmond, CA. http://www.sfei.org/documents/developmental-trajectory-california-tidal-marsh-restorationand-mitigation-projects.

SFEI, 2017. California Aquatic Resource Inventory (CARI) Standard Operating Procedures (SOP). http://www.sfei.org/cari#sthash.ycYh8UKG.dpbs.

Stein, E.D., Fetscher, A.E., Clark, R.P., Wiskind, A., Grenier, J.L., Sutula, M., Collins, J.N., Grosso, C., 2009. Validation of a wetland rapid assessment method: use of EPA's level 1-2-3 framework for method testing and refinement. Wetlands 29 (2), 648–665.

SWRCB, 2011. Staff Analysis Scientific Peer Review of California Rapid Assessment Method (CRAM). California State Water Resources Control Board, Sacramento, CA. http://www.waterboards.ca.gov/water_issues/programs/peer_review/docs/cram/cram_staff_report2.pdf.

SWRCB, 2016. Procedures for Discharges of Dredged or Fill Materials to Waters of the State [Proposed for Inclusion in the Water Quality Control Plans for Inland Surface Waters and Enclosed Bays and Estuaries and Ocean Waters of California]. June 17, 2016 Final Draft v1, State Water Resources Control Board. http://www.waterboards.ca.gov/water_issues/programs/cwa401/docs/dredge_fill/fnl_drft_prcdrs_20161706.pdf.

USACE, 2008. Compensatory Mitigation for Losses of Aquatic Resources. Federal Register/Vol. 73, No. 70/Thursday, April 10, 2008. Rules and Regulations, Department of Defense, Department of the Army, Corps of. Engineers.https://www.gpo.gov/fdsys/pkg/FR-2008-04-10/pdf/E8-6918.pdf.

USACE, 2015. Regional Compensatory Mitigation and Monitoring Guidelines for South Pacific Division. U.S. Department of Defense. U.S. Army Corps of Engineers, South Pacific Division. http://www.spd.usace.army.mil/Portals/13/docs/regulatory/mitigation/MitMon.pdf.

USEPA, 2002. Guidance on Choosing a Sampling Design for Environmental Data Collection for Use in Developing a Quality Assurance Project Plan. EPA QA/G-5S, U.S. Environmental Protection Agency, Washington, DC. https://www.epa.gov/sites/production/files/2015-06/documents/g5s-final.pdf.

USEPA, 2008. Core Elements of an Effective State and Tribal Wetlands Program. U.S. Environmental Protection Agency, Washington, DC. https://www.epa.gov/sites/production/files/2015-10/documents/2009_03_10_wetlands_initiative_cef_full.pdf.

WADOE, 1993. Washington State Wetlands Rating System. Technical Report 93-74, Washington State Department of Ecology, Seattle, WA.

FURTHER READING

CWMW, 2013. California Rapid Assessment Method (CRAM) for Wetlands, Version 6.1. California Wetlands Monitoring Workgroup. http://www.cramwetlands.org/sites/default/files/2013-04-22_CRAM_manual_6.1%20all.pdf.

CWMW, 2016. CRAM Classification Flow Chart. California Wetland Monitoring Workgroup. http://www.cramwetlands.org/sites/default/files/UPDATED%20CRAM%20Flowchart_July2016.pdf.

Sutula, M.A., Stein, E.D., Collins, J.N., Fetscher, A.E., Clark, R., 2006. A practical guide for the development of a wetland assessment method: the California experience. J. Am. Water Resour. Assoc. 42 (1), 157–175.

Chapter 4.3.4

Michigan Rapid Assessment Method for Wetlands (MiRAM)

Todd Losee*, Keto Gyekis[†], Susan Jones[†] and Anne Garwood[†]
Niswander Environmental, Brighton, MI, United States, [†]Michigan Department of Environmental Quality, Lansing, MI, United States

Chapter Outline

INTRODUCTION

The Michigan Department of Environmental Quality (MDEQ), formerly the Michigan Department of Natural Resources and Environment, developed the Michigan Rapid Assessment Method for Wetlands (MiRAM) for routine use in the evaluation of wetland sites across a broad spatial, geological, and ecological range in Michigan. MiRAM was developed under a federal grant from the US Environmental Protection Agency (USEPA). MiRAM evaluates the extent to which a given site provides the public benefits, functions, and values identified in Michigan's wetland protection laws (Part 303, Wetlands Protection, of the Natural Resources and Environmental Protection Act, PA 451 of 1994, as amended). MiRAM is designed to provide a numeric score that reflects the "functional value" of a wetland, which includes a wetland's ecological condition (integrity) and its potential to provide ecological and societal services (functions and values). In developing MiRAM, the MDEQ utilized the framework of several other rapid assessment methods (RAM) in use in the Great Lakes region. In particular, MiRAM follows the organization and many of the technical aspects of the *Ohio Rapid Assessment Method for Wetlands, v.5.0* (ORAM) (Mack, 2001).[1]

CONCEPTUAL FRAMEWORK FOR THE METHOD

MiRAM was developed to allow comparison of the diverse types of wetlands that exist across the broad spatial, geological, and ecological range in Michigan. Although MiRAM was developed primarily by the MDEQ, a panel of wetland experts representing local governments, state and federal agencies, university professors, and private consultants (MiRAM Development Committee) was established to provide input from a variety of perspectives. The development committee reviewed the use and effectiveness of other existing RAMs that had been developed for the Midwest, such as the ORAM, and the *Minnesota Routine Assessment Method, Evaluating Wetland Function, version 3.1* (MnRAM) (MBWSR, 2007). In addition, considerable time was spent reviewing the scoring boundary determination section of the *Washington State Wetland Rating System for Western Washington* (Hruby, 2004) to assist in developing and standardizing MiRAM scoring boundary protocol. After extensive evaluation of these existing RAMs, the development committee determined that the ORAM was the best overall framework to use as a guide for MiRAM development. ORAM's framework established

1. Note that ORAM is described in Chapter 4.3.8 of this book.

Wetland and Stream Rapid Assessments. https://doi.org/10.1016/B978-0-12-805091-0.00043-8

an easy to follow, metric-based system that focused on a wetland's ecological functions, societal values, and overall condition. In addition, MnRAM offered several relevant aspects, including the wetland vegetation community key and additional metrics for recreation and aesthetics.

One of the primary goals of MiRAM development was to ensure that all the metrics adhered to permit application review criteria outlined in the state of Michigan's wetland protection laws. This would allow the potential use of MiRAM during permit application reviews and would aid reviewers in identifying wetlands of high ecological value and/or with significant functional value. However, the intent of developing MiRAM was to provide a tool for assessing the functional value and condition of a wetland and not to replace the need for professional reviewers in the permit application review process. MiRAM is not intended to replace the need for documentation and quantification of criteria such as feasible and prudent alternatives that avoid and minimize the impacts to wetlands, as required in permit application reviews by state and federal statutes.

MiRAM has two main components, the rating form and the user's manual. The rating form contains a brief explanation of the method and its limitations. It also contains the background information section, the field datasheet, the narrative rating section, the quantitative rating section, and the rating summary. MiRAM user's manual compliments the rating form by providing detailed instructions and explanations, including instructions on how to use MiRAM, properly determine scoring boundaries, and document site conditions in the field and from a desktop as well as a section on literature cited. In addition, the user's manual provides example scenarios and detailed explanations for each metric.

MiRAM Boundary Determination Guidelines

One of the most significant challenges with using the ORAM (or any other existing RAM) in Michigan was defining what area to evaluate (i.e., wetland evaluation area). For example, without clear guidelines it was difficult for users to determining when to include wetland complexes as one scoring unit or at what point users cease evaluating a wetland that may literally continue for miles (e.g., river corridors in southern Michigan or the vast white cedar, *Thuja occidentalis*, swamps of Michigan's Upper Peninsula). The Washington State Wetland Rating System for Western Washington was consulted to aid in development of a relatively simple set of scoring boundary determination guidelines. These guidelines clearly identify when wetlands should be scored separately or together; to limit and standardize field time, a maximum review area size was established. The guidelines require a user to identify a proposed project site (see Fig. 4.3.4.1 for example site). Once the site is established, the scoring boundaries can be determined. The scoring unit guidelines establish a 100-ft rule that is used to determine if wetlands are separated by uplands (e.g., if the upland is <100 ft wide, then the wetlands are scored together) or have narrow wetland corridor connections. In addition, when wetlands are adjacent to open water bodies (river or lakes), a 100-ft (width) strip of the water body is normally scored as part of the wetland. To make sure review times are limited and can still be considered "rapid," a maximum review area of 50 acres was established. These scoring boundary guidelines reduce confusion among users and allow different users to assess the same standardized wetland area.

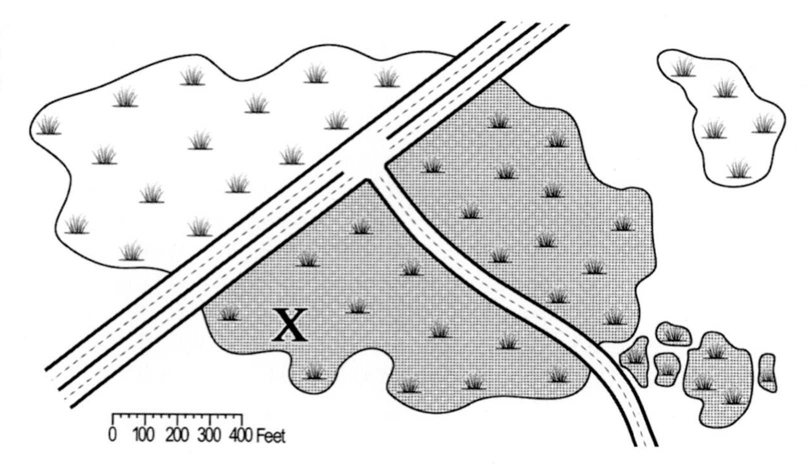

FIG. 4.3.4.1 Example MiRAM determination boundary. The "X" indicates the proposed project site. The *shaded gray* indicates the MiRAM scoring unit which includes wetlands on both sides of small road and a series of small wetlands.

DESCRIPTION OF THE METHOD

MiRAM includes three distinct steps for evaluating a wetland area; they are (1) evaluation of background information, (2) narrative rating, and (3) quantitative rating.

Steps to a MiRAM wetland evaluation:

(A) In-office review:
- **(1)** Obtain aerial photographs, National Wetland Inventory (NWI) maps, US Geological Survey topographic maps, Natural Resources Conservation Service (NRCS) soil survey maps and data, and other useful resource information.
- **(2)** Using all the available resource information and any onsite delineations, determine the approximate MiRAM boundary, following the guidelines in Chapter III of the MiRAM manual. See Fig. 4.3.4.2 showing an in-office review map utilizing NWI to determine the MiRAM boundary, 150-ft buffer, and 1000-ft surrounding land-use evaluation area.
- **(3)** Complete the background information form.

(B) On-site review:
- **(4)** Walk the entire wetland evaluation area and complete the field datasheet (modification of the MiRAM boundary may be necessary based on field observations).
- **(5)** Complete the narrative rating (if the wetland is identified as high functional value, the quantitative rating does not need to be completed).
- **(6)** Complete the quantitative rating. Carefully follow the instructions listed for each metric and submetric. Failure to properly consider all quantitative rating metrics and submetrics may result in an incorrect evaluation and score.
- **(7)** Complete the MiRAM summary.

Narrative Rating

The Narrative Rating helps users identify if they are in one of four wetland types identified as having *exceptional ecological value*. These include (1) wetlands within US Fish and Wildlife Service (USFWS) identified critical habitat, (2) wetlands containing state or federal threatened or endangered species, (3) wetlands defined as being a rare natural community type, and (4) Great Lakes coastal wetlands. Wetlands meeting any of these criteria are considered to have *exceptional ecological value* and, therefore, are automatically rated as having *high functional value* in MiRAM. If a wetland has *high functional value,* completing the quantitative rating is not necessary. If the wetland type is not one of the types listed in the narrative rating, users proceed to the quantitative rating.

Quantitative Rating

The quantitative rating is a series of seven metrics with 19 submetrics used to evaluate the functional value of the wetland (Table 4.3.4.1). It is designed to provide a numerical score that reflects the total functional value of a wetland, which includes a wetland's ecological condition (integrity) and its potential to provide ecological and societal services (functions and values).

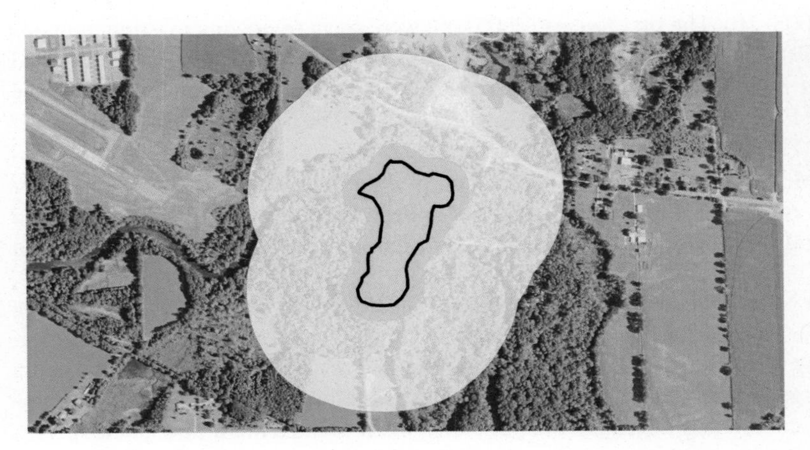

FIG. 4.3.4.2 Example MiRAM in-office evaluation map. The inner most polygon indicates the wetland boundary based on the NWI. The first outline surrounding the wetland is the 150-ft (width) buffer zone, the second outline indicates the 1000-ft (width) land use zone. All three areas are analyzed in the MiRAM quantitative rating.

TABLE 4.3.4.1 List of Metrics/Submetrics and the Assigned Maximum Value of Each

Metric	Title	Submetric	Submetric Maximum	Metric Maximum
1	Wetland size and distribution	*1a.* Wetland size	6	9
		1b. Wetland scarcity	3	
2	Buffers and surrounding land use	*2a.* Average buffer width around the wetland's perimeter	6	12
		2b. Intensity of surrounding land use	6	
3	Hydrology	*3a.* Sources of water	10	26
		3b. Connectivity	13	
		3c. Duration of inundation/saturation	4	
		3d. Alterations to natural hydrologic regime	8	
4	Habitat alteration and habitat structure development	*4a.* Substrate/soil disturbance	4	20
		4b. Habitat alteration	9	
		4c. Habitat structure development	7	
5	Special situations	*5a.* High ecological value	10	10
		5b. Forested wetland	5	
		5c. Low-quality wetland	−10	
6	Vegetation, interspersion, and microtopography	*6a.* Wetland vegetation communities	9	20
		6b. Open water component	3	
		6c. Coverage of highly invasive plant species	1	
		6d. Horizontal interspersion	5	
		6e. Habitat features	12	
7	Scenic, recreational, and cultural value	–	3	3
			Total	**100**

VALIDATION/CALIBRATION EFFORTS UNDERTAKEN WITH THE METHOD

MiRAM was developed over a 6-year period (2005–10). The first 2 years (2005–06) were spent reviewing existing methods and the last 4 years were spent developing MiRAM, conducting field testing, and calibration. For the purposes of field testing, the state was divided into two distinct regions, separated by the floristic tension zone (FTZ). Wetlands in the southern half of the Lower Peninsula, below the FTZ, were sampled during the first season. The Northern Lower and Upper Peninsula were sampled in the second season. The effort included testing a variety of wetland sites with a range of ecological and landscape conditions using MiRAM, ORAM, and a modified version of the Floristic Quality Assessment for Michigan (FQA) (Herman et al., 2001). The overall goal of the field testing was to determine whether MiRAM produced results similar to ORAM (which identifies several wetland functions and values) and similar results to the FQA, which is an indicator of a wetland's floristic condition.

In 2007, in collaboration with the Michigan Natural Features Inventory (MNFI) ecologists, a modified FQA sampling protocol was developed, producing a site specific Floristic Quality Index (FQI) that fit the time and budgetary constraints of the MiRAM project. Specifically, strict sampling timeframes were established that limited review to a maximum of 4 hours per site. Each vegetation community type (emergent, scrub/shrub, forested, mixed) was limited to a maximum 1 hour sampling period. In addition, sampling within any community ceased if the period between identifying a new plant species reached 10 min. The sampling was done following the random meandering method within each community type present within the wetlands. All species not easily identified in the field were collected for later laboratory identification. Significant time was required to ensure proper identification of all collected species.

MiRAM Versus FQI

During the 2-year sampling period, 83 sites were evaluated throughout Michigan using the MiRAM and the FQI utilizing the modified FQA method. A Spearman's rank-order correlation ($r = 0.855$, P-value $< .01$) showed a significant positive association between MiRAM score and FQI, indicating that higher MiRAM scores often are indicative of increased vegetation quality and condition.

MiRAM Versus ORAM

Because the ORAM has a long history of testing and development (see Chapter 4.3.8 of this book) and was chosen as the primary model for the initial MiRAM development, ORAM scores were produced for comparison to MiRAM at many field testing locations. During field testing, the MiRAM boundary determination guidelines were also adhered to when using ORAM, to hold scoring area as a constant. Results showed an apparent correlation, although site-specific variation was occasionally relatively high.

The differences between MiRAM and the ORAM scoring results were mostly attributed to the significant modifications that were made to metric 5, special situations. In particular, submetrics were added in MiRAM for forested wetlands, critical habitat, and urban/suburban wetlands; this also followed MNFI's natural community rating systems for rare and imperiled wetland communities. The ORAM tended to score wetlands within the uppermost scoring range slightly higher than associated MiRAM scores. This is likely due to differences in the relative "rarity" (and therefore functional value) of many types of wetland ecological communities and large wetlands in Michigan versus Ohio. For example, the ORAM awards points for all bogs while MiRAM only allows points for bogs located south of Michigan's FTZ. The ORAM also tended to score wetlands within the lower range slightly lower than MiRAM scores, reflecting the adjusted functional value points some urban/suburban wetlands receive during the MiRAM scoring process. Testing outside the prime late-spring through late-summer growing season tended to result in lower overall MiRAM and ORAM scores, which was attributed to the inability of users to properly identify vegetation diversity, structural complexity, and some habitat features. The modifications of MiRAM do not appear to have significantly changed the overall assumptions and function that have been tested and verified in ORAM.

TIME SPENT TO APPLY THE METHOD IN THE FIELD

Following the MiRAM scoring boundary determination guidelines required users to limit the evaluation area to 50 acres or less. The majority of wetlands sampled were <50 acres, which requires the evaluator to assess the entire wetland. The time required for the in-office review was less than 1 hour while the field evaluation took up to a maximum of 4 hours for the larger, more complex wetland systems. The time spent in the field varied significantly based on the complexity of the wetland system. For instance, a one-acre vernal pool dominated by open water with only a few vascular plant species can be assessed in 30 min or less while a similar-sized forested swamp with a diversity of plant species may take up to 2 hours due to the difficulty of accessing the entire area.

HOW WAS/IS THE DATA BEING USED

MiRAM is a useful tool for the comparison of wetland functional values. Because a scoring categorization has not been developed for MiRAM, it requires the scoring of multiple sites for comparison. Because of this, MiRAM is not being used in the regulatory program for single wetland impacts; however, some permit applicants have used MiRAM to assist in documenting alternatives. MiRAM has value in assessing alternative routes of linear projects (i.e., roads, pipelines, etc.) where multiple wetlands may be impacted and a MiRAM comparison can assist permit applicants in the process of selecting the least impacting routes. In these cases, MiRAM is simply one tool in the regulatory process and is not the sole basis for making regulatory decisions. In addition, in the regulatory process, MiRAM has been used to determine priority areas for wetland preservation or restoration by focusing these efforts on wetlands with higher functional value.

The *State of Michigan Wetland Monitoring and Assessment Strategy* (MDEQ, 2015) primarily utilizes MiRAM to quickly assess wetland functional value, regardless of ecological type, throughout the state. Currently MiRAM is one component of statewide wetland monitoring being implemented in an effort to understand the condition of wetlands across the state, by region, and/or by wetland type with a goal of improved wetland management on a statewide scale.

WHAT WAS LEARNED

MiRAM can be an efficient, standardized, cost-effective tool for evaluating wetland functional value. When scored properly, a wetland system's functional value can generally be compared to other wetlands within any local region of Michigan. Within the first decade of the initial release of MiRAM, it was utilized by ecological consulting firms when evaluating proposed wetland impacts among commercial development sites and alternative locations throughout Michigan. The largest documented use of MiRAM by the private sector to date was by a consulting firm that analyzed >75 wetlands along a proposed transportation corridor and several alternative corridors. In addition, MiRAM has been used by the private sector to compare wetland functional value among sites being proposed as potential wetland preservation areas to satisfy wetland mitigation requirements.

Since the release of MiRAM in 2010, the MDEQ (in conjunction with the Michigan Wetlands Association) has trained >150 professionals in the use of MiRAM. These training courses included classroom and field instruction to ensure proper understanding of the individual metrics and the use of MiRAM. Without proper training, significant variation is observed in MiRAM scores. In theory, an untrained MiRAM practitioner could download the user's manual, study it extensively, follow its guidelines, and become proficient with MiRAM. However, it has become apparent that the user's manual is often not utilized sufficiently, and therefore untrained practitioners of MiRAM are less likely to obtain accurate results. Based on the need for training and instruction, users are required to use the rating form, which includes brief explanatory directions for each scoring submetric as well as useful diagrams and graphics. The MDEQ drafted a "short" scoring form (i.e., a highly condensed two-page rating form) during the MiRAM development process, but the form was never finalized so that users must use the "long" rating form to encourage consistent use of the metrics. In addition to professional users, MiRAM has been included in the curriculum of several local university courses as an effective tool for students to better understand the functions and values of wetlands and the stressors that impact those functions and values.

The MiRAM user's manual is a robust document that explains in detail the scoring procedures of each submetric. The user's manual also describes in detail the fundamental process of determining the boundary limits of the MiRAM evaluation area. Following the MiRAM evaluation area scoring boundary protocol is especially important when a proposed project site is located within very large wetlands, complexes, and wetland riparian corridors or is divided by roads.

During MiRAM development and testing, it was discovered that some relatively rare and/or ecologically important wetland types or some wetland types with high individual functions often scored too low compared to their actual recognized functional value. For example, a coastal plain marsh (Kost and Penskar, 2000), ranked within Michigan as S2 (i.e., imperiled), might apparently not have high enough habitat complexity to score high in MiRAM; however, these areas have significant societal value because they contain important habitat for a multitude of rare species. Or, in a situation where a large wetland overrun by weedy or invasive species may yet provide substantial hydrologic cleansing and buffering functions for the local watershed. In both cases, the MiRAM score that reflects the wetland's overall functional value may be lower than the significant societal and ecological value of these wetlands. Users are cautioned that wetlands with significant single functions or values may have a low MiRAM score that does not reflect the overall value of the wetland. In these cases, additional studies or the use of alternative assessment methods may be required. In summary, MiRAM is not a substitute for trained wetland scientists, "best professional judgment," and local knowledge.

PROSPECTS FOR THE FUTURE

MiRAM has been incorporated into the state of Michigan's wetland monitoring program and is administered by the MDEQ on approximately 30 wetland sites per year. The MDEQ is compiling a statewide spatial geographic information systems-based database tracking recent MiRAM sites and scores.

The MDEQ has trained wetland regulatory staff on the use of MiRAM for occasional use in regulatory reviews and by the private sector in permit applications. MiRAM can be used to assist in review of potential alternative designs for large projects with significant proposed wetland impacts, or to document impacts associated with unauthorized activities. MiRAM is also used in evaluating proposed wetland preservation options to address required wetland mitigation.

In 2015, the MDEQ partnered with the Michigan Wetlands Association to host the first external training on MiRAM available to private consultants and wetland professionals, outside of MDEQ. This 2-day class was well attended with a variety of attendees from the private sector, state and federal agencies, conservation groups, and watershed management organizations. The MDEQ and MWA received positive feedback after this training, and several requests for additional training. Both organizations are interested in hosting similar trainings in future years.

SUPPORTING DOCUMENTS

MiRAM user's manual: http://www.michigan.gov/deq/0,4561,7-135-3313_3687-240071–,00.html
MiRAM rating form: http://www.michigan.gov/deq/0,4561,7-135-3313_3687-240071–,00.html

REFERENCES

Herman, K.D., Masters, L.A., Penskar, M.R., Reznicek, A.A., Wilhelm, G.S., Brodovich, W.W., Gardiner, K.P., 2001. Floristic Quality Assessment with Wetland Categories and Examples of Computer Applications for the State of Michigan—Revised, second ed. Michigan Department of Natural Resources, Wildlife, Natural Heritage Program, Lansing, MI. 19 pp. + Appendices.

Hruby, T., 2004. Washington State Wetland Rating System for Western Washington—Revised. Washington State Department of Ecology Publication No. 04-06-025, Washington State Department of Ecology, Olympia, WA.

Kost, M.A., Penskar, M.R., 2000. Natural Community Abstract for Coastal Plain Marsh. Michigan Natural Features Inventory, Lansing, MI. 6 pp.

Mack, J., 2001. Ohio Rapid Assessment Method for Wetlands v.5.0, User's Manual and Scoring Forms. Ohio EPA Technical Report WET/2001-1, Ohio Environmental Protection Agency, Division of Surface Water, 401/Wetland Ecology Unit, Columbus, OH.

Minnesota Board of Water and Soil Resources (MBWSR), 2007. Minnesota Routine Assessment Method, Evaluating Wetland Functions, Version 3.1. St. Paul, MN.

Michigan Department of Environmental Quality (MDEQ) Water Resources Division, State of Michigan Wetland Monitoring and Assessment Strategy, 2015, Lansing, MI, 33 pp.

Chapter 4.3.5

Florida Uniform Mitigation Assessment Method

Kelly C. Reiss*,† and Erica Hernandez†
*American Public University, Charles Town, WV, United States, †University of Florida, Gainesville, FL, United States

Chapter Outline

INTRODUCTION

The Florida Administrative Code (FAC) Chapter 62-345 Uniform Mitigation Assessment Method (UMAM), which went into effect on February 2, 2004, is used to identify the amount of compensatory wetland mitigation necessary to offset impacts to wetlands and surface waters; it also includes wetland mitigation banking credit awards and deductions. The development of UMAM was mandated under Section 373.414(18) of the Florida statutes (FS), establishing a method for the determination of the amount of wetland mitigation needed. The applicant, whether proposing impacts under an environmental resource permit or wetland resource permit or proposing a mitigation bank or regional mitigation permit application, is responsible for presenting supporting information for the UMAM. The agency reviewing the permit application will verify the supporting information and apply the UMAM, thus making a determination on the amount of mitigation needed to offset proposed impacts or the mitigation potential for a bank or regional mitigation area. Since UMAM became effective in 2004, it has been revised several times, most recently in June 2016.

UMAM is described as a functional assessment method with consideration of current condition (Ch 62-345.500(6), FAC), hydrologic connection (Ch 62-345.400(1)(d), FAC), uniqueness (Ch 62-345.400(1)(f), FAC), location (Ch 62-345.400(1); Ch 62-345.500(7), FAC), fish and wildlife utilization (Ch 62-345.400(1)(h), FAC), time lag (Ch 62-345.600(1), FAC), and mitigation risk (Ch 62-345.600(2), FAC). The method is based largely on reasonable scientific judgment (e.g., 373.414(18), FS; Ch 62-345.100(2), FAC; Ch 62-345.400(1), FAC; Ch 62-345.500(4), FAC), The name of the method includes the terms *uniform*, meaning consistent and reliable; mitigation, meaning only applicable in situations when mitigation is needed; *assessment*, meaning based on reasonable scientific judgment; and *method*, meaning a process (Bersok, 2011). In practice, UMAM asks for best professional judgment by the evaluator with a description of condition in three broad categories: location and landscape Support, water environment, and community structure.

CONCEPTUAL FRAMEWORK

Within the state of Florida, wetlands permitting adheres to a target goal of no net loss in functions for wetland or other surface water (10.1, Applicant's Handbook [AH] St. Johns River Water Management District [SJRWMD], 2013). The concept of no net loss refers to the maintenance and protection of the chemical, physical, and biological integrity of the nation's waters, and mitigation intends to replace losses from unavoidable impacts to the chemical, physical, and biological integrity (collectively ecological integrity) of wetland resources permitted under Section 404 of the U.S. Clean

Wetland and Stream Rapid Assessments. https://doi.org/10.1016/B978-0-12-805091-0.00045-1

Water Act. More specifically in Florida, 373.403(18) FS identifies functions performed by uplands, wetlands, and other surface waters as, but not limited to, "providing cover and refuge; breeding, nesting, denning, and nursery areas; corridors for wildlife movement; food chain support; and natural water storage, natural flow attenuation, and water quality improvement, which enhances fish, wildlife, and listed species utilization." UMAM does not measure function itself (Hull et al., 2011), but rather intends to calculate the value of functions. As the Florida Department of Environmental Protection (DEP, 2013a) notes, functions occur through time and are typically measured as rates, so that functional assessments require long-term data. In contrast, UMAM is intended as a RAM, which relies on measuring indicators of function. Indicators can be described as capturing only a moment in time with presence/absence or qualitative data collected in a single site visit.

Within Florida, UMAM is used when mitigation is necessary for permit issuance, when mitigation has been proposed, and when the anticipated mitigation is deemed suitable to offset impacts (Bersok, 2011). Suitable forms of mitigation under UMAM include wetland creation, restoration, enhancement, preservation, or upland preservation. The UMAM framework addresses impacts that require mitigation and provides for a statewide standardization of a RAM across plant and benthic community types with the intent of consistency across assessors (Bardi et al., 2007). The UMAM rule evaluates any type of legally defined wetland in Florida, including wetlands and surface waters with plant cover (e.g., salt marsh, cypress dome, floodplain swamp, seagrass beds, mangroves) and benthic communities (e.g., oyster bars, hardbottom communities). UMAM can also be used to evaluate uplands if they are proposed for mitigation preservation in support of protecting adjacent wetlands.

METHODS

UMAM has two parts that contribute to evaluating indicators of wetland function, loosely described in the rule as condition. The site characterization in Part I, qualitative description, provides a frame of reference and context for what is appropriate and expected to guide the evaluation in Part II, the quantification of assessment area. Part II enumerates condition in terms of indicators of function, and assigns a score relative to the target community type identified in Part I. Both parts rely on the assessor's familiarity with the wetland type being evaluated and best professional judgment. Other considerations in UMAM include application of a preservation adjustment factor, adjustment for time lag and risk, calculation of the delta (the degree of ecological change), and application of the UMAM formulas.

Completion of Part I includes gathering locational information on the assessment area to include Florida Land Use, Cover, and Form Classification System (1999) (FLUCC) codes; watershed location; special classification; geographic relationship to and hydrologic connection with wetlands, other surface waters, and uplands; significant nearby features; uniqueness; functions; previous mitigation; anticipated use by wildlife; anticipated use by listed species; and observed evidence of wildlife use. Each section is further described in the UMAM rule and some sections also refer the reader to outside references for further consideration. A specific form, identified within the rule and available through the Florida DEP, is provided (Fig. 4.3.5.1).

Within Part II, evaluation focuses on three categories, scored numerically on a scale from 0 to 10 (where 0 indicates not present and 10 indicates optimal; Hull et al., 2011). The first category (0.500(6)(a) location and landscape support) reviews the ecological context within which the system operates. The second category (0.500(6)(b) water environment) evaluates hydrologic alteration and water quality impairment, though this category is not scored for upland assessment areas. The third category (0.500(6)(c) community structure) is separated into (1) vegetation and/or (2) benthic Community (Fig. 4.3.5.2). Scores are based on visual observation of existing species present, assemblage, and wetland structure.

Sampling Scheme

UMAM is not a formalized method because the assessor can choose biological and physical attributes listed in the rule or provide their own. Indicators are weighted using best professional judgment and observed based on variable field effort from cursory to extensive as the assessor deems necessary, based on personal experience. Risk and preservation adjustment factors are applied in a subjective way (Florida DEP, 2013b). To remedy the opportunity for assessor bias, a standardized field protocol is suggested as guidance in an online training tool for UMAM (Bardi et al., 2007) but is not part of the actual rule. The guide recommends that the entire perimeter of the area receiving direct impacts or proposed for preservation should be inspected. Also, the wetland interior should be investigated with 30 m transects perpendicular to the hydrologic gradient with the total number of transects based on the size of the wetland. The rule itself offers no specific guidance and has no enforcement mechanism for sampling design or effort.

PART I – Qualitative Description
(See Section 62-345.400, F.A.C.)

Site/Project Name	Application Number	Assessment Area Name or Number

FLUCCs code	Further classification (optional)	Impact or Mitigation Site?	Assessment Area Size

Basin/Watershed Name/Number	Affected Waterbody (Class)	Special Classification (i.e.OFW, AP, other local/state/federal designation of importance)

Geographic relationship to and hydrologic connection with wetlands, other surface water, uplands

Assessment area description

Significant nearby features	Uniqueness (considering the relative rarity in relation to the regional landscape.)
Functions	Mitigation for previous permit/other historic use
Anticipated Wildlife Utilization Based on Literature Review (List of species that are representative of the assessment area and reasonably expected to be found)	Anticipated Utilization by Listed Species (List species, their legal classification (E, T, SSC), type of use, and intensity of use of the assessment area)

Observed Evidence of Wildlife Utilization (List species directly observed, or other signs such as tracks, droppings, casings, nests, etc.):

Additional relevant factors:

Assessment conducted by:	Assessment date(s):

Form 62-345.900(1), F.A.C. [effective date 02-04-2004]

FIG. 4.3.5.1 UMAM Part I qualitative description.

Parts of the qualitative depiction of the wetland can be evaluated remotely using scientific literature, technical reports, and geographic information systems (GIS) mapping resources. The descriptions in Part I contribute to understanding the context for what is expected for that type of wetland in that particular landscape as well as describe some of the characteristics relevant to wetland regulation. Describing the area of interest in Part I can be completed using online mapping tools or GIS to define geographic attributes of the wetland. The form for Part I requires the identification of land use, the basin and watershed, hydrologic connectivity to other surface waters and wetlands, and geographic connectivity to uplands. Practitioners must determine if there is any designated use or restrictive classification of the surface water and any other important protection classes such as aquatic preserve. The wetland size must be defined as well as a brief description of the natural plant community. Also on a landscape level, significant nearby features should be listed. These could be anything from airports and industry to state parks, agricultural withdraws, and consumptive use well fields. The relative uniqueness of the feature compared to the regional landscape should also be described; this can be related to expected flora and fauna or the type of wetland being evaluated. Historic uses or permit history should also be examined.

PART II – Quantification of Assessment Area (impact or mitigation)
(See Sections 62-345.500 and .600, F.A.C.)

Site/Project Name	Application Number	Assessment Area Name or Number
Impact or Mitigation	Assessment conducted by:	Assessment date:

Scoring Guidance	Optimal (10)	Moderate(7)	Minimal (4)	Not Present (0)
The scoring of each indicator is based on what would be suitable for the type of wetland or surface water assessed	Condition is optimal and fully supports wetland/surface water functions	Condition is less than optimal, but sufficient to maintain most wetland/surface waterfunctions	Minimal level of support of wetland/surface water functions	Condition is insufficient to provide wetland/surface water functions

.500(6)(a) Location and Landscape Support w/o pres or current / with	
.500(6)(b)Water Environment (n/a for uplands) w/o pres or current / with	
.500(6)(c)Community structure 1. Vegetation and/or 2. Benthic Community w/o pres or current / with	

Score = sum of above scores/30 (if uplands, divide by 20) current or w/o pres / with

If preservation as mitigation,
Preservation adjustment factor =
Adjusted mitigation delta =

For impact assessment areas
FL = delta x acres =

Delta = [with-current]

If mitigation
Time lag (t-factor) =
Risk factor =

For mitigation assessment areas
RFG = delta/(t-factor x risk) =

Form 62-345.900(2), F.A.C. [effective date 02-04-2004]

FIG. 4.3.5.2 UMAM Part II quantification of assessment area.

Scientific literature, field guides, technical reports, and professional expertise can help to describe ecological components of the area of interest. Explicit wetland functions associated with each type of natural community to be evaluated should be listed. Examples of wetland function include specific habitat support for native fauna and physical and biological function such as water storage and nutrient removal. Anticipated wildlife utilization as well as potential listed species are also described. This anticipated function and wildlife utilization are based on reference condition and geographic variation. Part I of the form continues with qualitative observations in the field. The assessor must describe direct evidence of wildlife utilization. Additionally, any unanticipated but relevant observations are to be recorded in order to augment the office analysis conducted prior to visiting the site.

UMAM Indices

The quantitative scoring of the wetland assessment area is based on three categories of wetland function: location and landscape support, water environment, and community structure. The rule has descriptive guidelines to consider when evaluating function per category. The score assigned per category is between 0 and 10. The value represents a percentage of optimal function (100% or the number 10) that a reference wetland would exhibit.

Location and Landscape Support

The category location and landscape support scores wetland function through physical and ecological connectivity to support expected fauna. The rule lists descriptions of function indicators to guide potential scores. To receive a score of 10, the full range of habitat needed to fulfill the life history requirements of expected wildlife should be present in or surrounding the evaluated system. Invasive species should not be present in the landscape. Physical obstructions, hydrologic impediments, or gaps in habitat should not be present as barriers to wildlife movement. Habitat downstream should not be limited from positive upstream interaction. Adjacent land uses should not be incongruous with protection of expected wildlife. Finally, the practitioner must determine if other critically dependent habitats would be impacted if the assessment area were altered.

Water Environment

The water environment category is intended to be evaluated in terms of wetland function support for expected wildlife. Water quality and hydrodynamic evidence should be appropriate for the type of community being evaluated. This category accounts for potential impediments and disturbances in the landscape outside the direct assessment area. The assessor must consider appropriate hydrodynamics for depth, duration, frequency, flow, and seasonality of the wetland hydrology as well as elements of water quality. To receive an optimal score of 10, water levels and evidence of flow patterns should be appropriate in the context of seasonal weather patterns and climatic variability. Water level indicators should be consistent for the expected hydrology. Hydric soil attributes should be consistent with hydrologic indicators. Evaluators should look for signs of excessive drying, inappropriate fire history, soil deposition, and hydrologic stress. Vegetation or benthic zonation should be appropriate for that community. Evidence of wetland-dependent fauna should be present. Species present should not indicate degraded water quality or an altered hydrologic regime. If available, existing water quality data should be reviewed.

Community Structure

Community Structure is evaluated by the presence of appropriate macrophyte composition and cover or, when absent, the submerged benthic community existing in open water. Vegetation species presences as primary producers are foundational in community support of wetland dependent fauna. Cover of species in terms of foraging, nesting, and refugia for fish and wildlife is evaluated in terms of what is appropriate and judged as healthy. Evaluators should look for appropriate complexity, both structurally and compositionally, uneven age class distribution, and regeneration of desirable species. Succession in terms of natural impacts, human activities, or management and stochastic events effect on structure and composition should be considered. Invasive species should not be present.

Calculating Mitigation Requirements

In evaluating each scoring category, the assessor should consider the ecological concepts presented as guidelines or other relevant scientific principles and then consider how the assessment area relates to reference condition and decide on a score that reflects a percentage of optimal function. To determine current condition of a wetland assessment area, sum the three category scores and divide that value by 30. The result will be a number between 0 and 1. Likewise, if evaluating a supporting upland, sum scores and divide by 20 for a result between 0 and 1. Calculating the wetland score between 0 and 1 allows for easier comparison of overall condition to a reference wetland as a percentage of optimal condition, where 1 represents the highest condition.

Time Lag

After evaluating wetland condition, further calculations are made that influence the score if mitigation is required to offset impacts. If the wetland evaluation is for a proposed impact site that requires mitigation and the mitigation has not already occurred, then a factor for time lag must be applied to the score. Time lag will be specific to the wetland type and the type of impact that is to occur. Time lag accounts for the function lost at the impact site associated with biological, chemical, and physical processes. The evaluator will apply reasonable scientific judgment to the amount of time in years it will take for a mitigation area to achieve the level of function being lost at the impact site. A table of factors (T) is provided in the rule that correlates with the time lag in years.

Risk

In a mitigation scenario, there is a level of inherent uncertainty in whether mitigation can achieve the desired level of lost wetland function. The level of risk increases with the amount of time needed to achieve the lost function. Risk values are assigned in quarter increments between 1 (no risk, the mitigation is likely to succeed) and 3 (successful restoration of function is unlikely). To help evaluate risk, practitioners should consider landscape support and potential factors that could impede success. Hydrologic complexity, if a system is dependent on engineering or has a naturally less predictable hydrology, could be considered risky. Vegetative community composition and proximity to exotic vegetation could also increase risk to successful mitigation. The potential for impacts to water quality or other impacts that could degrade the mitigation area should also be considered when evaluating risk.

Mitigation Determination

The three categories of wetland function are evaluated under the context of impact or mitigation scenarios. "Without mitigation, with mitigation, and with impact" are intended as likely scenarios based on understood intentions for the site being evaluated. "Without mitigation" would be evaluated as current condition with the likelihood the wetland will maintain function without protection or preservation plans. If the wetland is proposed for preservation under a mitigation plan, function is evaluated "with mitigation," which is the persistence of wetland function under the proposed protection. The wetland can also be evaluated "with impact:" what is the loss of function if the intended impact that will be mitigated is to occur? The scores may represent no change from current condition or increases or decreases in wetland function based on actions or inactions related to mitigation scenarios. Depending on which scenario is being considered, a formula is applied in Part II of the UMAM forms using the category scores to determine functional loss or relative functional gain. The functional gain or loss results are then used to determine mitigation credit or area needed to offset wetland impacts.

VERIFICATION, VALIDATION, AND CALIBRATION EFFORTS

In recent years, state regulators requested input from the professional and scientific community for feedback in applying the rule. Regulators held public meetings and workshops as well as created a survey to elicit comments. Primarily practitioners found the rule useful but sought more guidance in interpreting the rule and evaluating the location and landscape support category specifically (Florida DEP, 2014). Rule makers created a series of online tools to aid in the interpretation of the rule and to help frame the ecological basis for assessing wetland function. The state created four online vignettes to aid practitioners in the understanding of landscape ecology theory and to help in interpretation of those concepts for understanding wetland function. The vignettes and other YouTube videos describe GIS-based desktop evaluations for ecological concepts such as patch size, fragmentation, and edge effect while also providing definitions of common terms. Outside of initial rule making and training sessions, no formal published verification, validation, and calibration efforts have occurred or have been announced to the public.

TIME CONSIDERATION

UMAM has been through continual but periodic modifications since its establishment, with the most recent adoption of changes to 62-345.300 Assessment Method Overview and Guidance and 62-345.900 Forms (Repealed) in June 2016. 62-345.200 Definitions remain from the original rule-making in February 2004, with changes to 62-345.100 Intent and Scope in April 2005 and 62-345.400 Qualitative Characterization—Part I, 62-345.500 Assessment and Scoring—Part II, and 62-345.600 Time Lag, Risk, and Mitigation Determination in September 2007. Several of these sections are under rule revision development.

Completion of UMAM does not follow a specific time frame for office or field effort. Under Part I, the assessor is advised to gather sufficient information to identify functions through "aerial photographs, topographic maps, geographic information system data and maps, site visits, scientific articles, journals, other professional reports, field verification when needed, and reasonable scientific judgment. For artificial systems, such as borrow pits, ditches, and canals, and for altered systems, refer to the native community type it most closely resembles." Likewise, guidance for specific time requirements is absent for completion of Part II. In practice, the time necessary to complete a UMAM evaluation varies widely based on site conditions, complexity of impacts, assessor's background and familiarity with the system, and other such factors.

LESSONS LEARNED

The development and vision for UMAM began some 20 years ago, and UMAM is currently the standard rule used in assessing compensatory mitigation requirements to offset permitted impacts to Florida wetlands and surface waters. As the UMAM is part of the state administrative code, changes can only be made through a formal rule change process. Since its adoption, some studies and user comments have come forth identifying discrepancies in applying the method and asking for clarification. These issues are described in this section (Lessons Learned) followed by how these issues are currently being addressed with rule modification in "Prospects for the Future" section.

UMAM was developed to provide a flexible and consistent evaluation of mitigation efforts for wetlands and surface waters across system types and geographic regions within Florida, but a lack of statewide unbiased and scientifically sound research evaluating mitigation outcomes prohibits evaluating whether this goal has been met. Two studies offer partial evidence on the suitability of UMAM to determine effectiveness of mitigation outcomes. Beever et al. (2013) compared UMAM, wetland rapid assessment procedure (WRAP), and hydrogeomorphic method (HGM) scores within coastal wetlands in Charlotte Harbor and concluded that the use of UMAM resulted in low mitigation ratios and net loss of wetlands in regards to both function and area. Likewise, earlier Reiss et al. (2007) determined that UMAM and individual Part II categories were not significantly correlated with WRAP, HGM, macrophyte, and macroinvertebrate Florida Wetland Condition Index (FWCI) and Landscape Development Intensity (LDI) index scores for wetlands in mitigation banks. Both studies found poor correlation of UMAM and other assessment methods.

Florida DEP (2013b) identified four obstacles to UMAM: cognitive biases, nonstandardized field protocols, insufficient reference data, and training. Regarding cognitive bias (or assessor bias), in an uncontrolled experiment, 196 UMAM workshop participants were administered a two-question UMAM survey. While all surveys had identical information for the Part II quantification of assessment area location and landscape support explanation, the surveys were assigned two different "staff scores." Half the surveys had been assigned a score of 1 (just above the lowest possible score of 0) and the other half were assigned a score of 10 (optimal) for the same exact survey data, just the scores were different. Discussion within groups led to consensus of changing the scores to a median of 3 and median 7, respectively, based on the given assigned "staff score." The outcome was a four-point spread in median scores for the same category and with the same descriptive data, perhaps due to the initial bias of the fictitious "staff score." While Alvarez (as stated in SJRWMD, 2012) had previously cautioned that UMAM evaluation must start in the same "place" in describing the target ecological community with a focus on the historic ecological community, in the UMAM workshop survey exercise even having identical descriptive data prevented disparate quantification of Part II location and landscape support.

Nonstandardized field protocols were identified as a second obstacle to UMAM (Florida DEP, 2013b). In research studies, often a limited field team is involved in carrying out the sampling. For example, in the study by Reiss et al. (2007), the evaluations were limited to three individuals, and in the study by Gaucherand et al. (2015) a single evaluator carried out all the sampling efforts. In practice, UMAM is completed by a variety of professionals, working for a variety of companies or agencies, and with a varied professional background. Limitations in consistently applying UMAM come from observer bias and professional experience as UMAM relies on best professional judgment. Assessors applying the rule in a variety of systems across Florida would need an understanding of a highly heterogeneous and complex landscape. Also, interpretation of wetland function can be misjudged if seasonal and temporal variation and ecological response to stochastic events are not adequately understood.

A third obstacle to UMAM, as identified by Florida DEP (2013b), is insufficient reference data. Gaucherand et al. (2015) described UMAM as using a "culturally unaltered" reference standard, which suggests that even what we might classify as reference today would fail to meet the UMAM standard of not impacted by human actions. While in theory a culturally unaltered reference standard may promote description of the most pristine conditions, that none such references occur across the Florida landscape poses a perceptible challenge in adequately characterizing Part I, which is used as the basis of assigning Part II. Further, determination of loss of wetland function at a wetland in an urban setting facing development may underrepresent the function being performed in the urban environment, where the importance of increased functional capacity in nutrient retention or flood storage capacity may far exceed the expected level of function of the same wetland community in a reference landscape. In this example, the calculated current condition for the urban wetland may minimize the actual realized current function by comparing the urban wetland to a relic functional expectation. Of course, in the opposite scenario, some functions lowered in the urban environment, such as habitat support for wildlife, may be far diminished in the urban wetland. The unknown is how these changes in realized function balance in the calculations of mitigation. Are functions interchangeable, or which functions are more important in the no net loss framework of wetland management? Owing to this type of site bias, the UMAM may not be sensitive to evaluating "working wetlands" situated in developed landscapes.

The final obstacle Florida DEP (2013b) identified in regards to UMAM is training. Because so much of the UMAM focuses on best professional judgment but a standardized approach to completing Part I and gathering data to assess Part II is lacking, training could help safeguard consistency in assessor scoring. For example, as Reiss et al. (2007) point out, guidance on the spatial extent to investigate Part II's location and landscape support is lacking, leading to uncertainty in how far out from the assessment area the assessor should consider. Similarly, Reiss et al. (2007) questioned the appropriate scale for consideration of wildlife movement, hydrologic connectivity, and surrounding land use and land cover because UMAM does not offer a standard scale for consideration. The researchers concluded that the location and landscape support score could vary widely, depending on the spatial extent used in consideration. Florida DEP (2013b) also noted that a standardized framework for field data collection could be considered while maintaining the rapid design of UMAM.

A few other considerations have surfaced in the decade plus time frame of UMAM use in regulatory mitigation. For example, Gaucherand et al. (2015) identified the especially strong correlation between water environment and community structure in Part II, and noted that hydrology alone explained 92% of the variance in the overall UMAM score. While they noted that connections between hydrology and vegetation are expected, they suggested that score components should incorporate unrelated information in order to limit redundancy in the method.

PROSPECTS FOR THE FUTURE

Currently, the state is developing a new draft rule for the UMAM with a public workshop scheduled for July 2017 and intends to further develop tools that will address the four obstacles impeding consistent application of UMAM identified by Florida DEP (2013b). One goal is to reduce the qualitative narrative in the method by using specific species detail and avoiding template descriptions and buzzwords. Some of the suggestions being evaluated in the revision include natural community-specific quantitative values that could categorically bracket qualitative descriptions in the rule. These values could provide more quantitative structure to descriptions in the rule that are currently open to interpretation, words such as "sufficient," "limited," and "minimal."

Another strategy to reduce cognitive bias in UMAM is to provide more guidance on a standard field data collective method. While Florida DEP (2013b) is careful to maintain that the field component of UMAM must continue to meet expectations for a rapid method, there is still the possibility of a more systematic survey of a site where, for a specific community type, there are direct steps for gathering evidence to complete the assessment. There is also intent to force the evaluator to systematically consider specific questions that contribute to condition to direct the evaluation away from picking and choosing what the observer considers the relevant factors.

Reiss et al. (2007) and Florida DEP (2013b) determined that UMAM guidance did not require a direct comparison to reference condition but instead required assessors to describe assessment areas with a classification scheme such as Florida Land Use and Cover Codes. The 2007 report went on to describe the Florida Natural Area Inventories (FNAI) Natural Communities Guide, updated in 2010, as a useful basis for an initial literary reference to describe baseline expectation for composition, structure, and ecosystem processes in a reference system (Reiss et al., 2007). Florida DEP (2013b) also determined the FNAI guide to be an appropriate tool and further described goals for development of regional examples of reference wetlands that can be visited for calibration and validation of UMAM application. There are suggestions for the development of an accessible ecosystem model, funded by an Environmental Protection Agency grant, which will aid in the desktop evaluation of the location and landscape support category of UMAM (Florida DEP, 2013b). Finally, as of 2014 there were plans to develop a relational database of approved UMAM evaluations so that regional community examples, descriptions, and past scores can be researched, though as of early 2017 no public update on this effort has been released. As UMAM has been in place for more than 13 years, sufficient data should exist to evaluate the application of UMAM across mitigation determinations. A central repository of UMAM determinations should be maintained in order to promote scholarly review of the tool and its influence on fairness and transparency in mitigation.

Regulatory staff does not currently have the infrastructure to support an institutional training program for the UMAM (Florida DEP, 2013b). With adequate support, the state preference would be to have geographically distributed local experts to train practitioners. Centralized training of the instructors would be controlled at the state level to calibrate and control application of the rule for consistency. Local trainers could increase the frequency of training intervals, and the regionally identified experts could also aid in the development of UMAM evaluations for geographically distributed reference wetland sites. Even with the call for specific targeted variables, quantitative and categorical criteria for scoring, a process for standardizing estimation through comparison and modification, a reduction in narrative, and an improved reference framework, UMAM would remain a RAM (Florida DEP, 2013a,b).

Immediate updates to the UMAM, as identified for the scheduled July 2017 public workshop, include clarification of assessment standards to existing UMAM scoring guidelines and newly developed criteria specific to freshwater streams and submerged (seagrass and hardbottom) aquatic habitats. As identified in the announcement for the proposed rule change, changes to official UMAM forms will improve efficiency and ease of use, including the elimination of narrative justification with multiple-choice, fill-in-the-blank, and check-the-box style formats. The tentative agenda presents minimal changes in the UMAM process and is unlikely to address the four main obstacles identified by the state in 2013.

REFERENCES

Bardi, E., Brown, M.T., Reiss, K.C., Cohen, M.J., 2007. UMAM Uniform Mitigation Assessment Method Training Manual: Web-Based Training Manual for Chapter 62-345, FAC for Wetlands Permitting [PowerPoint slides]. Retrieved from: http://www.dep.state.fl.us/Water/wetlands/mitigation/umam/toolbox.htm.

Beever III, J.W., Gray, W., Cobb, D., Walker, T., 2013. A watershed analysis of permitted coastal wetland impacts and mitigation assessment methods within the Charlotte Harbor National Estuary Program. Fla. Sci. 76 (2), 310–327.

Bersok, C., 2011. Uniform Mitigation Assessment Method [PowerPoint slides]. Retrieved from: http://www.dep.state.fl.us/Water/wetlands/mitigation/umam/toolbox.htm (Chapter 62-345).

Florida Department of Environmental Protection [DEP], 2013a. 5-Minute Vignette: Ecological function. Retrieved from: https://youtu.be/tlQteSb6FoE (YouTube video).

Florida Department of Environmental Protection [DEP], 2013b. UMAM Workshop #1. Retrieved from: http://www.dep.state.fl.us/water/wetlands/mitigation/umam/rule.htm (YouTube video).

Florida Department of Environmental Protection [DEP], 2014. Uniform Mitigation Assessment Method Survey Summary Report—January 10, 2014. Retrieved from: http://www.dep.state.fl.us/water/wetlands/mitigation/umam/docs/rule-dev/UMAM_Survey_Results.pdf (PDF document).

Gaucherand, S., Schwoertzig, E., Clement, J., Johnson, B., Quétier, F., 2015. The cultural dimensions of freshwater wetland assessments: lessons learned from the application of US rapid assessment methods in France. Environ. Manag. 56, 245–259. https://doi.org/10.1007/s00267-015-0487-z.

Hull, C., G. Lowe, R. Robbins, and C. Bersok. 2011. Florida's Uniform Mitigation Assessment Method. (PowerPoint Slides; Chapter 62-345)

Reiss, K.C., Hernandez, E., Brown, M.T., 2007. An Evaluation of the Effectiveness of Mitigation Banking in Florida: Ecological Success and Compliance with Permit Criteria. HT Odum Center for Wetlands/Florida Department of Environmental Protection, Gainesville, FL/Okeechobee, FL. Retrieved from: http://www.dep.state.fl.us/water/wetlands/mitigation/reference.htm.

St. Johns River Water Management District (SJRWMD), 2012. Mitigation and UMAM. Retrieved from: https://youtu.be/MNrZM981hgc (YouTube video).

St. Johns River Water Management District (SJRWMD), 2013. Environmental Resource Permit Applicant's Handbook, Volume I (General and Environmental). Retrieved from: http://www.sjrwmd.com/handbooks/erphandbook.html (PDF document).

FURTHER READING

Florida Department of Environmental Protection [DEP]. n.d. Uniform Mitigation Assessment Method U.M.A.M. Retrieved from: http://www.dep.state.fl.us/water/wetlands/mitigation/umam/index.htm.

Chapter 4.3.6

New England Wetland Functional Assessment (NEWFA)

Paul Minkin* **and Erica Sachs-Lambert**†
New England District, US Army Corps of Engineers, Concord, MA, United States, †US Environmental Protection Agency, Boston, MA, United States

Chapter Outline

INTRODUCTION

In 2008 under the federal Clean Water Act, the US Army Corps of Engineers (Corps) issued regulations (33 CFR 332) popularly known as the "Mitigation Rule" in which a variety of recommendations and requirements for developing and assessing compensatory wetland mitigation were established. Wetland impacts and wetland compensation were discussed in terms of debits and credits, necessitating a quantitative means for their evaluation. Since 1995, the Corps New England District (NE District) has used a qualitative functional assessment method known as the Highway Methodology Supplement due to its release as a supplement to a regulatory process planning tool (U.S. Army Corps of Engineers, 1999). Its use is for all projects needing wetland assessment and is not restricted in any way to highway projects. In 2013, the director of the NE District's Regulatory Division tasked technical staff with updating the NE District's functional assessment methods, making them more quantitative to better fit with the most recent national guidance and regulations. An interagency team was established to work on this task. The NE District led the effort but was assisted by staff from the US Environmental Protection Agency (USEPA) and the US Fish and Wildlife Service. The team evaluated many existing wetland functional assessment methods to determine if what was needed already existed or could be developed with little modification. When no examined, existing method met the NE District's needs, a hybrid method was developed that incorporated elements from many different wetland assessment methods (Ainslie et al., 1999; Faber-Langendoen et al., 2012b; Hruby et al., 1999a,b; Kutcher, 2011; Brinson et al., 1995; Gilbert et al., 2006; Noble et al., 2011). Draft models were developed for more than a dozen specific functions, falling within three categories of functions: water quality maintenance, hydrologic integrity, and biota support. The water quality maintenance and hydrologic integrity categories contain true functional assessments evaluating the ability of a wetland to perform each function. However, the evaluation of biota support uses a condition assessment to determine the current capability of a wetland to perform functions in the biota support category. Each function is evaluated independently and no overall functional score is determined; however, when appropriate, scores may be grouped within each function category.

CONCEPTUAL FRAMEWORK FOR THE METHOD

The method is intended for use in all types of wetlands throughout the six New England states. Models are developed for each function, but are not separated for different hydrogeomorphic classes. However, hydrogeomorphic data are used as variables in the models. Specific variables are intended to zero out the model result if the wetland type cannot perform a

Wetland and Stream Rapid Assessments. https://doi.org/10.1016/B978-0-12-805091-0.00010-4

specific function due to its hydrogeomorphic characteristics. In such cases, these functions are considered not applicable to the specific wetland type under review and they receive no score. Because this method is a functional assessment rather than a condition assessment or similarity index, there is no formal use of reference sites. Variables are measured and then processed through models to determine the level of function. The assessment area is the overall wetland or portion of wetland that has the same general hydrology, soils, landscape position, parent material, and plant community.

DESCRIPTION OF THE METHOD

There are models for 14 functions and 3 compound variables (Table 4.3.6.1) and data are collected on a series of variables (see Table 4.3.6.2) that are then scored for each model (variables are scored differently for each model as the specific variable may affect different functions differently; see Table 4.3.6.2). Some of the data are from remote sensing and are collected prior to the field data collection. Any such remotely-collected data is then ground-truthed during the fieldwork and additional data on field variables are collected. It is intended that the field component for a typical wetland site will be able to be sampled within approximately half a day by a wetland-knowledgeable individual or small interdisciplinary team of 2–4 people (2–4 hours with a small interdisciplinary team of 3–6 people was common during preliminary field testing).

Variables are plugged into the model for each of the 14 functions and a level of function, scored 0–10, is then calculated for each function for each wetland under evaluation). Each function is evaluated independently and function scores ARE

TABLE 4.3.6.1 Function List for the New England Method

Function List	
1.	**Particulate Retention**: the ability of the wetland to stop and retain sediment, reducing suspended solids.
2.	**Removing Heavy Metals**: the ability of the wetland to bind heavy metals and retain them in place (usually bound to soil particles), thereby removing them from the water column.
3.	**Nutrient Transformation—Phosphorus**: the ability to remove and transform excess nutrient phosphorus from the water column.
4.	**Nutrient Transformation—Nitrogen**: the ability to remove and transform excess nutrient nitrogen from the water column.
5.	**Carbon Sequestration**: the ability to retain more carbon coming into the system than leaving the system.
6.	**Groundwater Recharge**: the ability for a wetland to serve as a groundwater recharge area. Recharge relates to the potential for the wetland to contribute water to an aquifer.
7.	**Surface Water Detention**: the ability to slow and retain surface water flows.
8.	**Streamflow Maintenance**: the ability to provide groundwater discharge at the headwaters of one or more streams.
9.	**Coastal Storm Surge Detention**: the ability to hold and absorb coastal (Atlantic Ocean and Lake Champlain) storm surge within the wetland.
10.	**Bank Stabilization**: the ability to stabilize and protect streambanks from erosion.
11.	**Shoreline Stabilization**: the ability to stabilize and protect shorelines from erosion.
12.	**Production Export**: the ability of the wetland to produce food for other living organisms (including humans).
13.	**Plant Community Integrity**: the ability to maintain a healthy, natural plant community typical of the geomorphic and climatic conditions.
14.	**Wildlife Habitat Integrity**: the ability to maintain a healthy, natural faunal community typical of the geomorphic, vegetative, and climatic conditions.
Complex Variable Models	
15.	**Retentionability**: the ability of the wetland to hold and retain surface water and sediment (determined by using Particulate Retention score as a proxy).
16.	**Velocity Reduction**: the physical process of slowing surface water movement as it comes into contact with surface features of the wetland.
17.	**Groundwater Discharge**: the ability of the wetland to convey groundwater to the surface, generally yielding some form of surface waters.

TABLE 4.3.6.2 Variable List for the New England Method

Variable Name	Description	Model #
Remote Sensing Data		
Size	Total size of overall wetland system	7, 9
Width	Width of AU perpendicular to shoreline (if applicable)	9, 11
HabProx	Proximity to nearest similar habitat	14
Corridor	Number of undisturbed travel corridors	14
BufferPerimeter	Percent of perimeter with vegetated buffer	13, 14
BufferWidth	Average width of vegetated buffer	13, 14
LandCover	Land Cover Index Score for land use within 500 m of AU	14
Overall Site Field Data		
Position	Location of wetland on the landscape	1, 6, 7, 8, (15), (17)
LakeChamp	AU borders Lake Champlain (binary)	9
Shoreline	AU has a Shoreline feature (binary)	9, 11
Channel	AU has a downstream Channel (binary)	8
Bank	AU has a Bank feature (binary)	10
SurficialGeology	Surficial Geology type	6, 8, (17)
SoilText	Soil textures present	2, 3, (17)
OrganicThickness	Thickness of organic layer	3, 5, 6, 8
FlowRestrict	Presence and characteristics of features which restrict flow either at the outlet or internal to the AU	1, 7, 12, (15)
HydroConnect	Hydrologic connections between assessment unit and other aquatic systems	12
Activity	Index of ongoing human activity measured as a combination of activity intensity and proportion of AU affected	13, 14
PStressor	Index of plant stressors measured as a combination of stressor intensity and proportion of AU affected	13
Community Weighted Field Data		
VegCover	Percent of AU that is vegetated weighted by community area	8, 10, 11
TreeCover	Areal cover of trees weighted by community area	4, 5, 8, 12
ShrubCover	Areal cover of shrubs weighted by community area	1, 4, 5, 12, (15)
TotalHerbCover	Aerial cover of herbaceous plants weighted by community area	2, 4, 5, 10, 12
TreeStem	Stem count of trees (per 100 m^2) weighted by community area	
BasalArea	Total basal area (per 100 m^2) weighted by community area	1, 4, 5, 12, (15)
PersisRel	Aerial cover of persistent herbaceous plants (relative to overall vegetative cover) weighted by community area	(16)
NonPersisRel	Aerial cover of nonpersistent herbaceous plants (relative to overall vegetative cover) weighted by community area	(16)
ShrubRel	Aerial cover of shrubs (relative to overall vegetative cover) weighted by community area	(16)
Invasives	Aerial cover of invasive plant species weighted by community area	13

Continued

TABLE 4.3.6.2 Variable List for the New England Method—cont'd

Variable Name	Description	Model #
MicroFeat	Count of microtopographic features (large woody material, boulders, hummocks) weighted by community area	(16)
HabFeat	Presence and diversity of features which enhance wildlife use (number of strata, patches, specialized wildlife features)	14
Complex Variables (Included in Above List)		
Retentionability (15)	Measure of the time that water is held and retained in the wetland (Model 1 is used as a proxy)	2, 3, 4, 5
Velocity Reduction (16)	Measure of structural ability of the wetland to slow surface water movement	1, 7, 9
Groundwater Discharge (17)	Measure of the ability of the wetland to convey groundwater to the surface	8

NOT added together to generate a single score per wetland. A score of 0 for a function means that function is not applicable to that wetland type (e.g., a wetland that does not abut open water would not have shoreline stabilization as one of its functions and is not scored less because of that). Scores for functions that are present range from 1 to 10. Precise numbers used in a functional assessment often give the impression of a greater level of precision and accuracy than the methods are actually capable of producing. Acknowledging this, we note that function scores within a point or two of each other have an essentially similar level of functioning (e.g., scores of 1–3 are all considered low functioning, 4–6 are moderately functioning, and 7–10 are high functioning).

VERIFICATION EFFORTS

As of early 2018, the method is still being finalized; however, field testing of the draft sampling methods took place from late summer through fall of 2015 and again in spring and summer of 2017. These data were then used to test and modify the draft models during winter 2016 and fall 2017. Remote sensing and field data were collected from 20 various wetland sites throughout New England during the first round of field testing. Data collection methods were tested and modified based on the field sampling. These data were then run through the draft models and used to validate and modify them where necessary. The second round of field testing included 23 sites and was used to further refine the methods and models before finalizing the methods for general use.

TIME SPENT IN DEVELOPING/TESTING THE METHOD

The method has been under development for approximately 5 years. During the early stages, there was an attempt to modify existing methods to meet the NE District's specific needs. When it became apparent that this was not going to provide the necessary results, the effort shifted to developing a new method that would provide the required functional assessment and quantitative approach. After 2 years of development, 7 weeks of field testing took place, followed by data and model analysis and modifications. A second round of field testing occurred a year and a half later. The method is currently (early 2018) being finalized.

SAMPLE APPLICATION OF METHOD IN THE FIELD

The method was field tested at 43 sites throughout New England. In addition, some workshops performed field data collection and then ran that data through one or two function models.

TIME SPENT TO APPLY THE METHOD IN THE FIELD

This took 2–4 hours per site with a wetland knowledgeable individual or a small interdisciplinary team of 2–4 people.

HOW WAS/IS THE DATA BEING USED

The primary usage is designed for the Corps of Engineers Regulatory Program to determine existing functions of wetlands proposed to be impacted, impacts to existing functions based on proposed projects, and developing appropriate and adequate compensatory mitigation for any authorized unavoidable impacts. In addition, the method may be used for resource assessment for planning.

WHAT WAS LEARNED

The method is designed to measure the level of various wetland functions within examined wetlands. Other than for some of the biota functions, it is not designed to evaluate conditions of the wetlands.

PROSPECTS FOR THE FUTURE

The method should be completed and released for general use within the year (2018). Subsequent to that, it will be periodically reexamined to fix "bugs" or just generally improve the method based on input from a greater volume of usage.

REFERENCES

Ainslie, W.B., Smith, R.D., Pruitt, B.A., Roberts, T.H., Sparks, E.J., West, L., Godshalk, G.L., Miller, M.V., 1999. A Regional Guidebook for Assessing the Functions of Low Gradient, Riverine Wetlands of Western Kentucky. Technical Report WRP-DE-17, Wetlands Research Program, U.S. Army Engineer Waterways Experiment Station, Vicksburg, MS.

Brinson, M.M., Hauer, F.R., Lee, L.C., Nutter, W.L., Rheinhardt, R.D., Smith, R.D., Whigham, D., 1995. A Guidebook for Application of Hydrogeomorphic Assessments to Riverine Wetlands. Wetlands Research Program Technical Report WRP-DE-1. U.S. Army Engineer Waterways Experiment Station, Vicksburg, MS.

Faber-Langendoen, D., Rocchio, J., Thomas, S., Kost, M., Hedge, C., Nichols, B., Walz, K., Kittel, G., Menard, S., Drake, J., Muldavin, E., 2012b. Assessment of Wetland Ecosystem Condition Across Landscape Regions: A Multi-Metric Approach. Part B. Ecological Integrity Assessment Protocols for Rapid Field Methods (L2). EPA/600/R-12/021b, U.S. Environmental Protection Agency Office of Research and Development, Washington, DC.

Gilbert, M.C., Whited, P.M., Clairain Jr., E.J., Smith, R.D., 2006. A Regional Guidebook for Applying the Hydrogeomorphic Approach to Assessing Wetland Functions of Prairie Potholes. ERDC/EL TR-06-5, U.S. Army Engineer Waterways Experiment Station, Vicksburg, MS.

Hruby, T., Granger, T., Brunner, K., Cooke, S., Dublanica, K., Gersib, R., Reinelt, L., Richter, K., Sheldon, D., Teachout, E., Wald, A., Weinmann, F., 1999a. Methods for Assessing Wetland Functions Volume I: Riverine and Depressional Wetlands in the Lowlands of Western Washington. WA State Department Ecology Publication #99-115.

Hruby, T., Granger, T., Teachout, E., 1999b. Methods for Assessing Wetland Functions. Volume I: Riverine and Depressional Wetlands in the Lowlands of Western Washington. Part 2: Procedures for Collecting Data. Washington State Department Ecology Publication #99-116, Olympia, WA.

Kutcher, T.E., 2011. Rhode Island Rapid Assessment Method User's Guide: RIRAM Version 2.10. Rhode Island Department of Environmental Management Office of Water Resources.

Noble, C.V., Murray, E.O., Klimas, C.V., Ainslie, W., 2011. Regional Guidebook for Applying the Hydrogeomorphic Approach to Assessing the Functions of Headwater Slope Wetlands on the South Carolina Coastal Plain. ERDC/EL TR-11-11, U.S. Army Engineer Research and Development Center, Vicksburg, MS.

U.S. Army Corps of Engineers, 1999. The Highway Methodology Workbook Supplement: Wetland Functions and Values, a Descriptive Approach. NAEEP-360-1-30a, New England District.

FURTHER READING

Brinson, M.M., 1993. A Hydrogeomorphic Classification for Wetlands. Technical Report WRP-DE-4, U.S. Army Engineer Waterways Experiment Station, Vicksburg, MS.

Brinson, M.M., Lugo, A.E., Brown, S., 1981. Primary productivity, decomposition and consumer activity in freshwater wetlands. Annu. Rev. Ecol. Syst. 12, 123–161.

Cowardin, L.M., Carter, V., Golet, F.C., LaRoe, E.T., 1979. Classification of Wetlands and Deepwater Habitats of the United States. FWS/OBS-79/31, USDI Fish and Wildlife Service, Office of Biological Services, Washington, DC. 103 pp.

Faber-Langendoen, D., Hedge, C., Kost, M., Thomas, S., Smart, L., Smyth, R., Drake, J., Menard, S., 2012a. Assessment of Wetland Ecosystem Condition Across Landscape Regions: A Multi-Metric Approach. Part A. Ecological Integrity Assessment Overview and Field Study in Michigan and Indiana. EPA/600/R-12/021a, U.S. Environmental Protection Agency Office of Research and Development, Washington, DC.

Hauer, F.R., Cook, B.J., Gilbert, M.C., Clairain Jr., E.J., Smith, R.D., 2002. A Regional Guidebook for Applying the Hydrogeomorphic Approach to Assessing Wetland Functions of Riverine Floodplains in the Northern Rocky Mountains. ERDC/EL TR-02-21, U.S. Army Engineer Research and Development Center, Vicksburg, MS.

Hruby, T., 2014. Washington State Wetland Rating System for Western Washington: 2014 Update (Publication #14-06-029). Washington Department of Ecology, Olympia, WA.

Klimas, C.V., Murray, E.O., Pagan, J., Langston, H., Foti, T., 2004. A Regional Guidebook for Applying the Hydrogeomorphic Approach to Assessing Wetland Functions of Forested Wetlands in the Delta Region of Arkansas, Lower Mississippi River Alluvial Valley. ERDC/EL TR-04-16, U.S. Army Engineer Research and Development Center, Vicksburg, MS.

Klimas, C.V., Murray, E.O., Pagan, J., Langston, H., Foti, T., 2005. A Regional Guidebook for Applying the Hydrogeomorphic Approach to Assessing Wetland Functions of Forested Wetlands in the West Gulf Coastal Plain Region of Arkansas. ERDC/EL TR-05-12, U.S. Army Engineer Research and Development Center, Vicksburg, MS.

Klimas, C.V., Murray, E.O., Langston, H., Witsell, T., Foti, T., Holbrook, R., 2006. A Regional Guidebook for Conducting Functional Assessments of Wetland and Riparian Forests in the Ouachita Mountains and Crowley's Ridge Regions of Arkansas. ERDC/EL TR-06-14, U.S. Army Engineer Research and Development Center, Vicksburg, MS.

Minnesota Board of Water and Soil Resources, 2010. Minnesota Routine Assessment Method (MnRAM) Evaluating Wetland Function, Version 3.4 (beta). Comprehensive General Guidance.

Murray, E.O., Klimas, C.V., 2013. A Regional Guidebook for Applying the Hydrogeomorphic Approach to Assessing Functions of Forested Wetlands in the Mississippi Alluvial Valley. ERDC/EL TR-13-14, U.S. Army Engineer Research and Development Center, Vicksburg, MS.

NatureServe, 2013. International Ecological Classification Standard: Terrestrial Ecological Classifications. NatureServe Central Databases, Arlington, VA. Data current as of 2 October 2013.

Noble, C.V., Evans, R., McGuire, M., Trott, K., Davis, M., Clairain Jr., E.J., 2002. A Regional Guidebook for Applying the Hydrogeomorphic Approach to Assessing Wetland Functions of Flats Wetlands in the Everglades. Technical Report ERDC/EL TR-02-19, U.S. Army Engineer Research and Development Center, Vicksburg, MS.

Noble, C.V., Wakeley, J.S., Roberts, T.H., Henderson, C., 2007. Regional Guidebook for Applying the Hydrogeomorphic Approach to Assessing the Functions of Headwater Slope Wetlands on the Mississippi and Alabama Coastal Plains. ERDC/EL TR-07-9, U.S. Army Engineer Research and Development Center, Vicksburg, MS.

Noble, C.V., Roberts, T.H., Morgan, K.L., Hill, A.J., Neary, V.S., Cripps, R.W., 2013. Regional Guidebook for Applying the Hydrogeomorphic Approach to Assessing the Functions of Flat and Seasonally Inundated Depression Wetlands on the Highland Rim. ERDC/EL TR-13-12, U.S. Army Engineer Research and Development Center, Vicksburg, MS.

Null, W.S., Skinner, G., Leonard, W., 2000. Wetland Functions Characterization Tool for Linear Projects. Washington State Department of Transportation, Environmental Affairs Office, Olympia.

Rheinhardt, R.D., Rheinhardt, M.G., Brinson, M.M., 2002. A Regional Guidebook for Applying the Hydrogeomorphic Approach to Assessing Wetland Functions of Pine Flats on Mineral Soils in the Atlantic and Gulf Coastal Plains. ERDC/EL TR-02-9, U.S. Army Engineer Research and Development Center, Vicksburg, MS.

Shafer, D.J., Roberts, T.H., Peterson, M.S., Schmid, K., 2007. A Regional Guidebook for Applying the Hydrogeomorphic Approach to Assessing the Functions of Tidal Fringe Wetlands along the Mississippi and Alabama Gulf Coast. ERDC/EL TR-07-2, U.S. Army Engineer Research and Development Center, Vicksburg, MS.

Stutheit, R.G., Gilbert, M.C., Whited, P.M., Lawrence, K.L., 2004. A Regional Guidebook for Applying the Hydrogeomorphic Approach to Assessing Wetland Functions of Rainwater Basin Depressional Wetlands in Nebraska. ERDC/EL TR-04-4, U.S. Army Engineer Research and Development Center, Vicksburg, MS.

Tiner, R.W., 2011. Dichotomous Keys and Mapping Codes for Wetland Landscape Position, Landform, Water Flow Path, and Waterbody Type Descriptors: Version 2.0. U.S. Fish and Wildlife Service, National Wetlands Inventory Program, Northeast Region, Hadley, MA. 51 pp.

U.S. Army Corps of Engineers, 2010. Operational Draft Regional Guidebook for the Functional Assessment of High-Gradient Ephemeral and Intermittent Headwater Streams in Western West Virginia and Eastern Kentucky. ERDC/EL TR-10-11, U.S. Army Engineer Research and Development Center, Vicksburg, MS.

U.S. Fish and Wildlife Service, 1980. Habitat Evaluation Procedures (HEP), Ecological Services Manual 102. U.S. Department of Interior Fish and Wildlife Service, Washington, DC.

Wilder, T.C., Rheinhardt, R.D., Noble, C.V., 2013. A Regional Guidebook for Applying the Hydrogeomorphic Approach to Assessing Wetland Functions of Forested Wetlands in Alluvial Valleys of the Coastal Plain of the Southeastern United States. ERDC/El TR-13-1, U.S. Army Engineer Research and Development Center, Vicksburg, MS.

Chapter 4.3.7

Rapid Assessment of Arid Land Lowland Riverine Wetland Ecosystems: A New Mexico Case Study

Maryann M. McGraw*, Esteban H. Muldavin[†] and Elizabeth R. Milford[†]

**New Mexico Environment Department, Surface Water Quality Bureau, Santa Fe, NM, United States, [†]Natural Heritage New Mexico Division, Museum of Southwestern Biology, Albuquerque, NM, United States*

Chapter Outline

INTRODUCTION

Since 2006, the state of New Mexico has been developing a New Mexico Rapid Assessment Method (NMRAM) for wetlands and riparian areas to meet its objectives for wetlands protection. The method needed to be sufficiently robust to provide essential information for implementing wetlands protection measures, particularly with respect to preventing degradation and impairments to waters of the state (20.6.4 New Mexico Administrative Code (NMAC)), and promoting effective management and protection of the state's wetland resources. Further, under the auspices of the New Mexico Environment Department Surface Water Quality Bureau (SWQB), the goal was not only to improve our knowledge of the ecological condition of New Mexico's wetlands but also to provide a consistent and reproducible method that would foster communication among agencies and organizations to meet the long-term goals for wetlands protection in the state.

Given the predominantly arid environment of New Mexico (it is the third-driest state in the United States), developing NMRAM presented a unique challenge. Unlike that of more mesic conditions, water is not only more limited but it is highly variable across the landscape and through seasons and years. Arid land wetland ecosystems tend to be isolated—many are concentrated along riverine corridors where water collects and where, under natural conditions, highly dynamic and complex fluvial environments support biologically complex ecological communities. This added complexity, in turn, has led to the development of a set of metrics tailored to meet the needs of a wetland rapid assessment in arid lands. To demonstrate this, we focus on our recent development of a lowland riverine NMRAM module within the Gila River Basin, one of the last free-flowing river systems in arid southwestern New Mexico. Our emphasis is on innovative metrics and approaches in an arid-land context that will have applications elsewhere for rapid ecological assessment of lowland riverine wetlands in arid regions.

Wetland and Stream Rapid Assessments. https://doi.org/10.1016/B978-0-12-805091-0.00009-8

CONCEPTUAL FRAMEWORK FOR THE METHOD

The NMRAM focuses on understanding ecological integrity of a wetland complex with measures of ecological condition and, by inference, the functional capacity of a wetland. In other words, if a wetland is in good condition, then it is assumed that the wetland is functioning at reference standard levels.[1] Ecological *integrity* is the "ability of a system to support and maintain a balanced, integrated, adaptive community of organisms having species composition, diversity, and functional organization comparable to the natural habitat of the region" (Fennessy et al., 2007; USEPA, 2010). Ecological *condition* could then be defined as the "ability of a wetland to support and maintain its complexity and capacity of self-organization with respect to species composition, physicochemical characteristics, and functional process as *compared to wetlands of a similar type without human alterations*" (Fennessy et al., 2007, emphasis added). Accordingly, the basic assumption underlying our rapid assessment methodology is that wetland condition will vary along a disturbance gradient and that the resultant state can be evaluated based on a set of landscape-level, geographic information systems-based, and rapidly obtained field-based condition metrics. As with other modules of the NMRAM, we developed both new metrics and modified existing ecological condition metrics from the California Rapid Assessment Method (CRAM; CWMW, 2013) and the Ecological Integrity Assessment (EIA; Faber-Langendoen et al., 2008) (and in some places the hydrogeomorphic (HGM) method of Hauer et al. (2002)) to create a suite of semiquantitative measures that best evaluates the ecological conditions of arid land wetlands and riparian ecosystems across the reference gradient.

Any given rapid assessment is also developed within a specific environmental framework and geographic reference domain. For the overall environmental framework, we followed the HGM classification of Brinson (1993) and defined a lowland riverine subclass of wetlands derived from his riverine class. In broad outline, this subclass refers to wetlands in floodplains associated with relatively large low-gradient rivers that flow through the broad river valleys of the low-elevation arid regions of the state. The purpose of defining a subclass is to constrain the natural variability in wetland types as well as the variability that occurs with latitude, altitude, climate, and geomorphology. The NMRAM also borrows from the ecological systems approach sensu Comer et al. (2003), where wetlands are viewed as a dynamic patch mosaic (DPM) of shifting wetland communities on a changing fluvial geomorphic template that is driven by hydrological processes (Crawford et al., 1993; Hupp and Osterkamp, 1996; Crawford et al., 1999; Latterell et al., 2006; Weisberg et al., 2013). For example, riverine floodplain wetlands are a dynamic patchwork of successional vegetation types of herbaceous, shrubland, and forested wetlands whose development is intertwined with the evolution of fluvial surfaces in response to flooding and channel migration (e.g., a range from small pioneer bars of herbaceous vegetation and shrublands to mature forested wetlands on large terraces). Taken together, these patches make up an ecological system (sensu Comer et al., 2003) to be assessed collectively because their condition and functional capacity are linked to the same environmental drivers, mostly hydrology, in the local landscape. They cannot be viewed in isolation, and the NMRAM metrics have been structured in acknowledgment of this fact.

Specifically, the lowland riverine subclass includes floodplain wetlands of mainstem fifth-order or greater streams (140–1133 m^3/sec 50-year discharge rate) occurring at elevations below 1675 m in broad alluvial valleys where the steam gradient is $<1\%$. It excludes confined valleys that are generally bedrock-controlled and lack significant alluvial floodplains. Streams are by and large perennial, but as is common in arid climates, can have intermittent segments, particularly during droughts. Typically, unlike most temperate region rivers, base flows rise during snowmelt events sporadically during the winter and gradually in the spring following the final snowmelt event. In addition, flows can be flashy during large thunderstorms and fronts that characterize the summer monsoon season of the arid Southwest. In New Mexico, this definition of lowland riverine applies to portions of three out of the five major rivers of the state: the Rio Grande, the Pecos, and the Gila (including the San Francisco River and segments of the Mimbres River) (Fig. 4.3.7.1). Of these, only the Gila, San Francisco, and Mimbres are uncontrolled by major dams in New Mexico and represent some of the last "wild" river segments in the Southwest. As such, they present one of the best available reference standards for developing a lowland riverine rapid assessment method in an arid setting. Accordingly, the development of NMRAM for the lowland reaches of the Gila, San Francisco, and Mimbres watersheds (referred to hereafter as the Gila Reference Domain) was initiated, beginning with a pilot study in 2012. This was followed by two seasons of field work and two years of assessment metric calibration and validation, culminating in the first of its kind rapid assessment for lowland riverine unconfined floodplain wetlands.

1. Reference standard levels or reference standard conditions are "conditions of unimpaired or minimally impaired water bodies characteristic of a water body type in a region" (EPA, 2010).

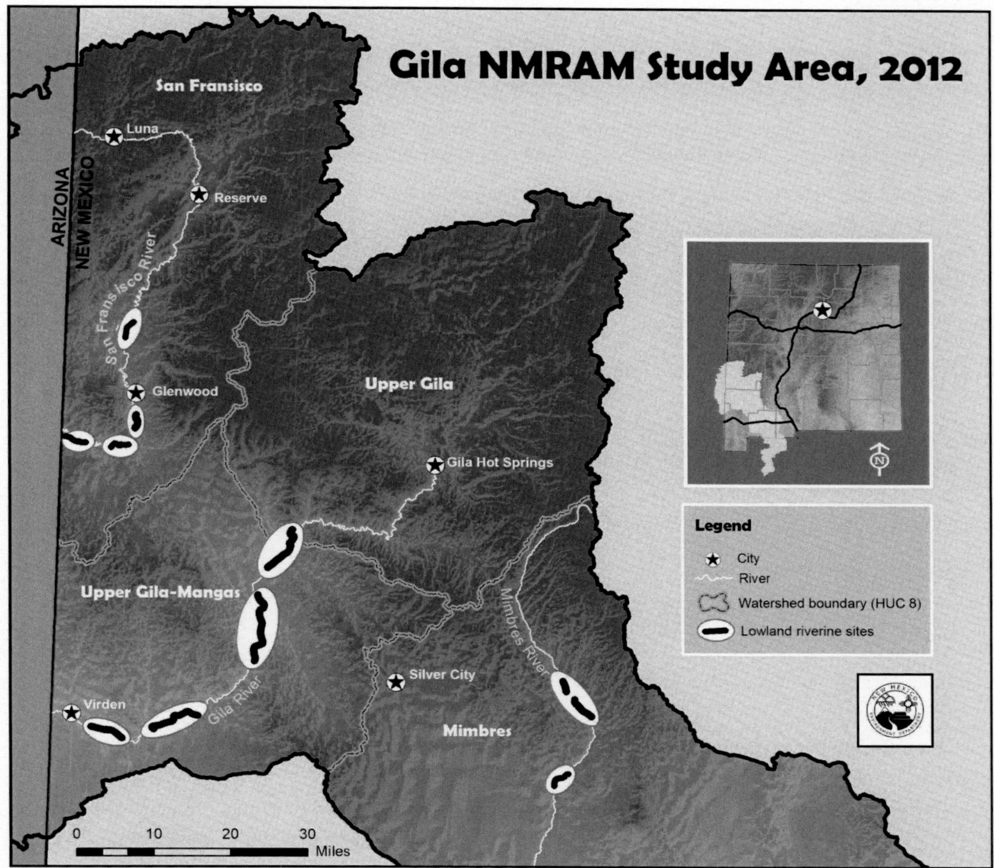

FIG. 4.3.7.1 Gila Reference Domain study area for Lowland Riverine Wetlands NMRAM. Data for the development phase of Lowland Riverine Wetlands NMRAM were collected from 25 sample areas situated within the lowland riverine floodplain sites on this map (wetlands of interest). In addition, high-resolution multispectral images of these floodplain sites were collected to help understand macrotopographic features as well as aerial extent and type of cover classes for our rapid assessment reference set.

DESCRIPTION OF THE METHOD

The NMRAM uses a set of observable and relatively easy to measure landscape and field indicators to express the relative condition of a wetland against a reference disturbance gradient for a given HGM subclass and geographic reference domain. Our objectives were to develop condition metrics that were sensitive to the disturbance gradient and minimally redundant as well as ones that could be used to rank sites by a field team of three in one day, based on a preponderance of evidence of landscape, biotic, and abiotic attributes. In addition, they are designed to compare wetlands equitably across many scales and jurisdictions, and in a variety of project contexts. Specifically, for the Gila Lowland Riverine Module, we developed 13 ecological condition metrics based on testing of several potential metrics and metric designs, and sampling from 25 sites across the disturbance gradient. These are grouped by three major attribute classes: landscape context (four metrics); biotic (five metrics); and abiotic (four metrics) (Table 4.3.7.1). In keeping with the goals of rapid assessment, the landscape context metrics are primarily GIS-based spatial analysis with field validation where possible, that is, Level 1 metrics per the EPA (2006). The biotic and abiotic metrics are semiquantitative, field-based metrics, that is, Level 2 metrics. High-precision Level 3 metrics that require detailed quantitative measurements were avoided except during the development phase as validation data.

Reference sites within the lowland riverine subclass were selected to embody the best available to the most degraded sites along the continuum of conditions reflecting a variety of stressors and stress levels within the Gila Reference Domain. Thirty-five sites that fit the lowland riverine subclass definition were identified and assigned a preliminary ranking based on best professional judgment of the NMRAM development team and local experts. Subsequent data collection at 25 sampling areas (SA) included GIS map evaluations using different land features and land-use map layers; field-based rapid

TABLE 4.3.7.1 NMRAM Lowland Riverine Metrics for the Gila Reference Domain (Index Numbers Refer to Attribute Class and Selected Metric).

Metrics
L1. Buffer Integrity Index
L2. Riparian Corridor Connectivity
L3. Relative Wetland Size
L4. Surrounding Land Use
B1. Relative Native Plant Community Composition
B2. Vegetation Horizontal Patch Structure
B3. Vegetation Vertical Structure
B4. Native Riparian Tree Regeneration
B5. Invasive Exotic Plant Species Cover
A1. Floodplain Hydrologic Connectivity
A2. Physical Patch Complexity
A5. Soil Surface Condition
A6. Channel Mobility

assessment of biotic and abiotic attributes; and completion of stressor checklists grouped by four attributes—land use, vegetation, physical structure, and hydrologic modification stressors.

The 13 metrics are measured within and around one or more SAs within a wetland area of interest (WOI). The WOI is user-defined and determining its boundary is the initial step in the NMRAM process. The determination is made depending on user needs and objectives. The NMRAM requires no specific criteria, but, as a minimum, the "natural rule" is suggested where a lowland riverine wetland should be composed of continuous natural wetland vegetation unbroken by major anthropogenic disturbance patches (e.g., roads, urban development). Further, in keeping with the ecological system approach, the WOI may be a complex of one or more natural vegetation types, but all of them should be part of the same wetland subclass (i.e., lowland riverine) with clear separation from other wetlands or wetland types. Overall, the WOI boundaries should follow the natural-feature patterns of the wetland and avoid major discontinuities or inclusions caused by land use.

For the lowland riverine subclass, SAs are placed along the stream channel on sites that best reflect the hydrological processes and conditions of the local reach within the WOI (e.g., flooding, sediment deposition, scour, and groundwater recharge) and characterize the wetland/riparian vegetation communities that are representative of the subclass. Nonriparian or nonwetland vegetation may occur internally to the SA but should be relatively minor elements. At a minimum, there is one SA per WOI, but for large WOIs, two or more SAs may be required to capture the range of variation. If an SA is constrained by logistical considerations such as ownership and access, some metric scores may be affected.

The size determination of the SA for lowland riverine wetlands is scale-dependent. This is because in riverine ecosystems, the size of vegetation communities and associated fluvial landforms vary with stream discharge (the larger the discharge, the larger the river bars and floodplains and the vegetation stands that grow on them). Thus, among metrics that are scale-dependent, as the SA size goes up, the assessment scores can go up. Therefore, to maintain consistency across sites, an upper limit on SA size needs to be set based on the subclass. Based on our analysis of scores from the Gila reference set, lowland riverine SAs should not exceed approximately 16 ha in overall size while not extending beyond 500 m of river corridor length. Conversely, as SA size goes down from the maximum, scores are likely to decline, but this is considered a measure of lowered ecological integrity and is intrinsic to the assessment scoring. If an SA is on only one bank, it should extend up to the outside edge of the floodplain that is not hydrologically connected (e.g., a high terrace). The break can be either natural or anthropogenic. If the SA is on both sides of a channel, then the width may be split between them; however, on at least one side it should extend to the edge of the active floodplain. For very large floodplains this may not be practical, in which case obtaining a representative sample of the WOI is the primary driver of SA placement and should be foremost in determining how the SA is apportioned across the floodplain.

An assessment of an SA requires a minimum of three people working together on a site—one with training in evaluating the biotic metrics, particularly plant identification of dominant and wetland species; the second with training in physical sciences, particularly hydrologic elements; and the third a technical assistant. All team members should also have training in basic GIS measurements, and as much cross training as possible is desirable. Because lowland riverine wetlands are scaled to the size of the ecosystem pattern, the SAs are relatively large, but the expectation is that a team of three people should still be able to complete the field survey in one day.

The foundations for NMRAM metric protocols are site maps with two different scales—a smaller scale map (1:12,000 or smaller) that is used for navigation to the site and for evaluating the landscape context metrics around an SA, and a larger scale map (1:6000 or larger) that is used for mapping vegetation communities and abiotic attributes within an SA. A set is prepared for each team member prior to going into the field with the WOI and SA boundaries along with the landscape context metric measurements to support their field validation.

Landscape Context Metrics

Four landscape context metrics were developed for the lowland riverine subclass as indicators of ecological condition of the landscape surrounding the SA (Fig. 4.3.7.2). These metrics are based on the concept that significant anthropogenic modification of a landscape and degraded condition around the wetland can influence biotic and abiotic conditions within the wetland itself. The expectation is that impacts immediately adjacent to the SA will have the most effect on wetland conditions and these are measured using the buffer integrity index and riparian corridor connectivity (derived from CWMW, 2013). The effectiveness of buffers is also a function of impinging land uses and this is measured with surrounding land use (derived from Hauer et al., 2002; Faber-Langendoen et al., 2012). As a measure of the reduction of overall functional

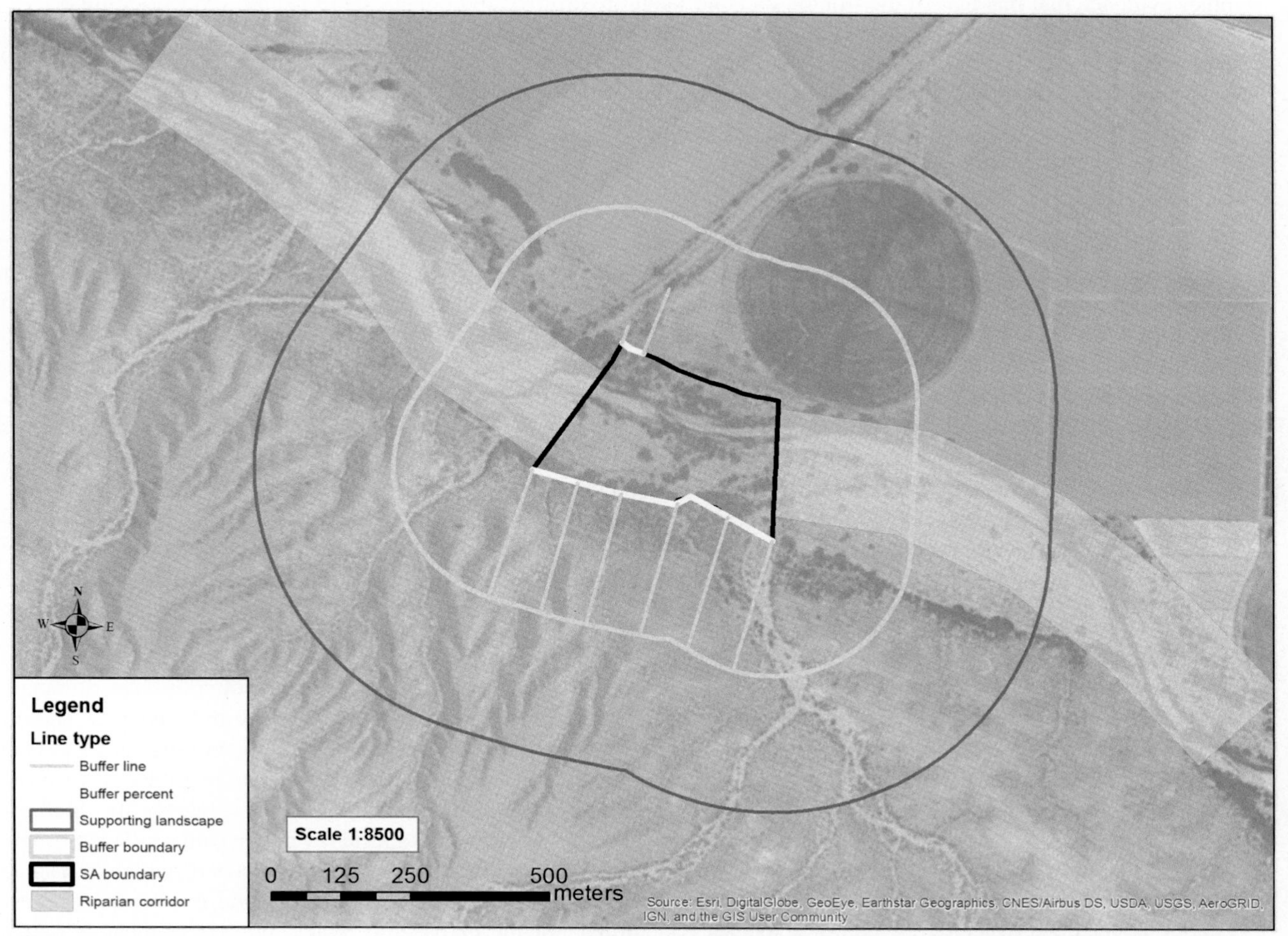

FIG. 4.3.7.2 Landscape context metrics are measured in three zones around an SA: buffer 250m around the SA *(gray contour)*, supporting landscape 500m *(black contour)*, and riparian corridor upstream and downstream 1000m from SA boundaries *(shaded area)*. Buffer percent is measured around the perimeter of the SA *(white line)* and buffer width is measured at eight points extending laterally from the SA boundary *(gray lines)*. Riparian corridor connectivity is evaluated upstream and downstream on both banks. Land use index is evaluated in the supporting landscape.

capacity at a landscape scale, relative wetland size estimates the change in a wetland's size that is due to direct human development (derived from Muldavin et al., 2011).

The landscape context metrics are designed to be measured in GIS using maps and aerial imagery, following these protocols:

- The buffer integrity index is based on the extent of buffer land-cover elements within 250 m of the SA boundary that provides protective services such as reducing pollutant contamination versus nonbuffer land-cover elements that do not. The two submetrics, buffer percent and buffer width, measure the percentage of the perimeter of the SA that is considered a natural or seminatural buffer and the width of the buffer lateral to the SA, respectively.
- Riparian corridor connectivity measures the disruption of natural land connectivity upstream and downstream of the SA with emphasis on detecting intervening obstructions that might inhibit wildlife movement and impact plant populations. The riparian corridor connectivity metric is measured in a zone 1000 m upstream and 1000 m downstream from the SA boundaries along the main channel, and 200 m in width centered on the river-available floodplain (not disconnected by anthropogenic features such as levees), and must include both banks of the river.
- Surrounding land use is based on measurements of the relative extent of land-use elements in an area extending 500 m out from the SA boundary. Each land-use element is weighted for its potential impacts on the SA and a land use index is calculated from the results.
- Relative wetland size is an index of the reduction of the current WOI size relative to its estimated historic extent. The key is determining the lateral extent of the historic floodplain based on photo-interpreted features and historic evidence, then followed by field verification. The assumption is that the valley bottom represents the historic floodplain and the departure of the current active portion of the floodplain is indicated by abandoned terraces and upland vegetation or other evidence that this part of the floodplain is no longer hydrologically connected to natural river flooding activity. This metric assumes that large reductions of area indicate alteration of hydrology or ecosystem processes and may indicate ecological instability, reduced viability, and tendency to lose diversity in the future. As such, relative size is an indicator of potential stress on the remaining extant wetland.

Biotic Metrics

Five biotic metrics are designed to measure key biological attributes within a wetland that reflect ecosystem integrity. Fundamental to ecological health of floodplain riparian and wetland areas is a diverse and dynamic mosaic of vegetation communities that is sustained by natural hydrological processes (Crawford et al., 1993; Muldavin et al., 2017). Such dynamic patch mosaics (DPMs) with their complex compositional and structural vegetation attributes maximize habitat for aquatic and terrestrial wildlife, reflect functional hydrological conditions (Latterell et al., 2006), and enhance overall ecological services. Accordingly, the NMRAM is oriented toward mapping and describing various biotic attributes of the DPM that are indicative of its status. These include vegetation native community composition and diversity; vegetation patch diversity, both across the wetland and with respect to vertical structure; the degree of nonnative invasive species incursion into a wetland; and the presence of native riparian tree regeneration.

Biotic metric measurements are based on the field mapping of vegetation community patches (stands) on the large-scale SA map with an aerial imagery base (Fig. 4.3.7.3). Only polygons of individual patches of homogeneous vegetation >0.25 ha are delineated. Patches smaller than 0.25 ha are considered inclusions in the surrounding patch. Each mapped vegetation community polygon is assigned to a community type during reconnaissance and, in turn, the community types are evaluated with respect to native species composition and their relative abundance.

- Relative native plant community composition is an index of the abundance of native-versus-exotic/invasive-dominated vegetation communities. High native plant species diversity generally indicates overall high biotic diversity, stability of wetland biotic communities, increased wildlife habitat and species diversity, and overall higher resilience and resistance to environmental disturbance. In contrast, high numbers of exotic plant species indicate degraded or disturbed wetlands. Faber-Langendoen et al. (2012) suggest that those ecosystems dominated by native species reflect high ecological integrity. Each polygon is described with respect to the top two dominant species by height strata: tall woody strata composed of trees and shrubs >5 m tall; a short woody strata of trees and shrubs under 5 m; and a herbaceous strata made up of graminoids (grasses and grass-like plants) and forbs. For each of the tall and short woody strata, total strata vegetative canopy cover must exceed 25% before a species is recorded; for the herbaceous strata, total cover must be >10%. The species are recorded in the order of their relative abundance by strata, and a species can appear only once within a community type designation (if a species occurs in two strata, it is assigned to the one in which it is most abundant).

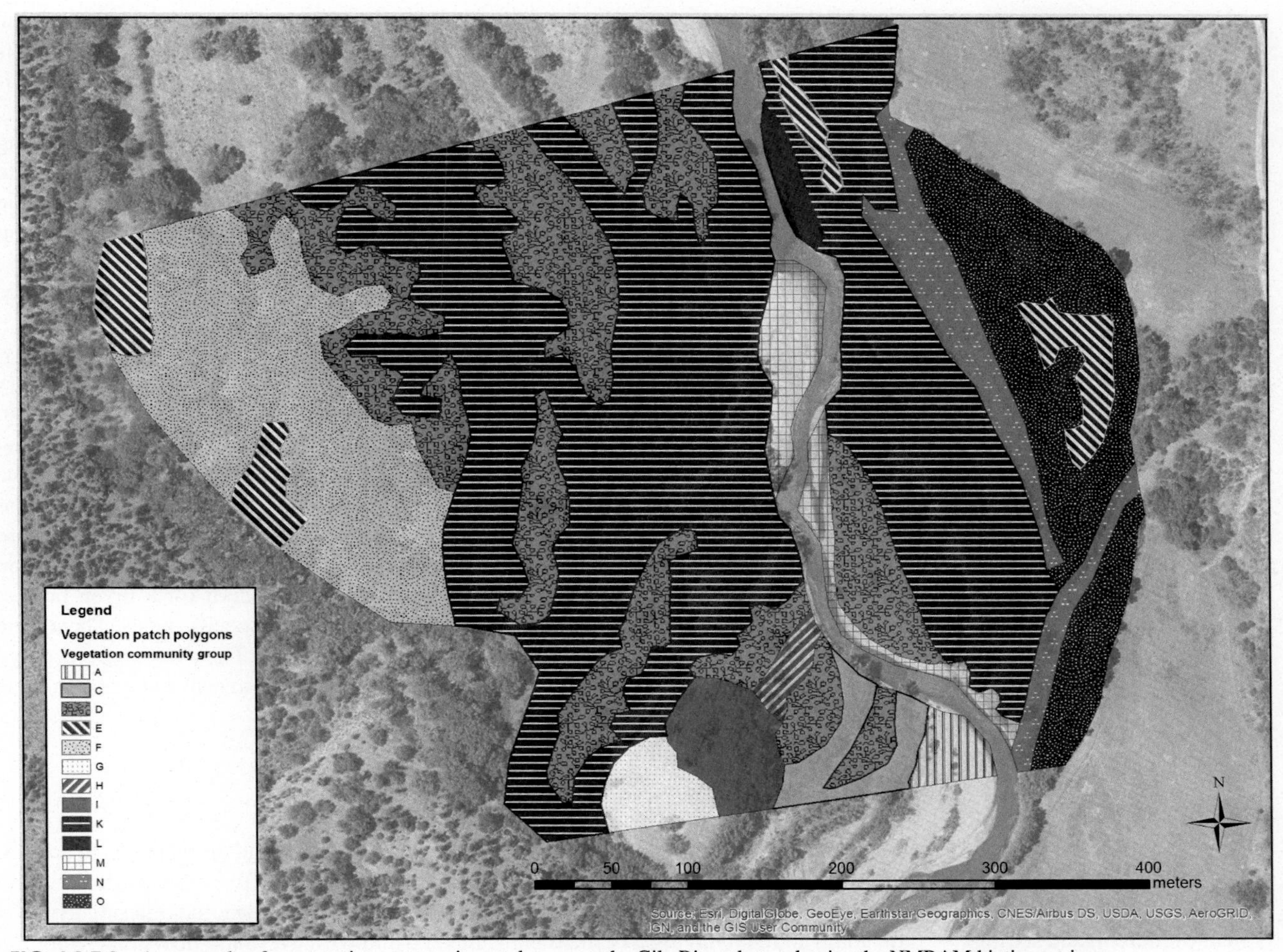

FIG. 4.3.7.3 An example of a vegetation community patch map on the Gila River that underpins the NMRAM biotic metrics.

Based on these basic species data for each polygon and community type assignment, a weighted native plant composition score for the SA is computed and this, in turn, is used to rate relative native plant community composition.

- Vegetation horizontal patch structure is an assessment of diversity of vegetation patches and complexity of the patch pattern (interspersion among vegetation patch types) within an SA. The vegetation horizontal patch structure metric is derived from CRAM horizontal interspersion (CWMW, 2013). It is also related to the concept that complex DPMs reflect functioning riparian and wetland ecosystems (Latterell et al., 2006). Multiple horizontal plant patches across the SA reflect greater ecosystem heterogeneity that generates more diverse habitat structure for wildlife and high biotic diversity in general. A patch mosaic of different vegetation types suggests intact hydrological regimes with associated ecological processes. In contrast, complexes dominated by one community type likely reflect highly altered hydrological regimes or other impacts to ecosystem function. Schematic diagrams of idealized riverine vegetation patterns are matched with the patch patterns in the SA and are assigned a score. A numerical description of the schematics is also provided to further facilitate matching the SA patch pattern.
- Vegetation vertical structure measures the vertical structural complexity of the vegetation canopy layers across the SA, and includes the number of strata and age/size classes. The concept of vegetation vertical structure is derived from CRAM vertical biotic structure (CWMW, 2013) and EIA vegetation structure (Faber-Langendoen et al., 2012). However, the vertical structure class types used here are based on the riparian vegetation structural type classification for the Rio Grande created by Hink and Ohmart (1984) and further elaborated on by Callahan and White (2004). Each mapped polygon patch is assigned one of the seven vertical structure types. The percentage of each vertical structure type is estimated across the SA. The ratings are based on the various combinations of dominant and codominant or subdominant vertical structure types.

- Native riparian tree regeneration assesses the abundance and spatial distribution of riparian tree reproduction across the SA and uses tree regeneration by species native to the SA. This metric is derived from the woody regeneration of Faber-Langendoen et al. (2012) and from Winward (2000) and Burton et al. (2008). Healthy functioning riverine wetlands should consist of a mosaic of woody vegetation stands that include stands of both mature and young regeneration trees. Absence of young trees may indicate ecological dysfunction. Generally, native riparian trees reproduce (seedling recruitment) in patches on disturbed, usually recently flooded moist ground. Because reproduction is closely tied to natural disturbance cycles (Crawford et al., 1993), the presence of numerous patches of differently aged native tree species acts as a surrogate measure for a functional natural disturbance regime, which includes flooding and sediment transport. During the field survey of vegetation community patches, the total percent cover of established native tree seedlings and saplings (under 12.7 cm DBH) is estimated for each polygon. This metric assumes that 5% or greater tree regeneration over the SA provides adequate evidence of mixed generation stands.
- Invasive exotic plant species cover is a measure of the total percent cover of invasive plant species based on the New Mexico list of noxious weeds (NRCS, 2003). Invasive nonnative species can have a significant impact on community diversity and function. High levels of invasive exotic species within a riparian plant community are a direct threat to maintaining wetland function and biodiversity (Stenquist, 2000; Bailey et al., 2001). While the mechanisms underlying the "invasive" character of some species are an active area of research, there are indications that riparian sites that have been altered or significantly impacted by human activity may be more prone to invasion. Invasive exotic species tend to thrive in riparian systems when natural hydrologic and geomorphic functions have been disturbed, particularly where the hydrological regime has been altered and is controlled. Thus, this metric is both a measure of current vegetation condition and an indicator of the status of the hydrological regime. Species of specific concern for a given project or those that are not yet on the New Mexico list of noxious weeds can be included on a project-specific basis. The metric is based on the percentage cover of invasive species in each mapped plant community patch. The average percentage cover of exotic invasive species is estimated for the entire SA, being mindful that when >10% invasive species cover is reached, the SA score is a D (poor).

Abiotic Metrics

Abiotic condition metrics address factors affecting the hydrology, fluvial geomorphic processes, and direct physical anthropogenic disturbances that influence wetland function and condition. Lowland riverine systems offer a challenge for rapid assessment of abiotic attributes because of their size and complexity. Rivers may form single-thread, meandering systems typical of more mesic settings with the main channel having a defined bank and bed and with overbank flows that support a riparian zone. In contrast, in arid regions such as the Gila Reference Domain, rivers may be multithread (braided or anastomosing), caused by deposition and distribution of sediment loads that are naturally high. This is most common where the rivers exit confined valleys into broad river valleys or when seasonal flash flow events occur; rivers drop their sediments in response to floodplain widening, and/or sudden reductions in flow volume, flow velocity, and slope. In the multichannel system, there may be a main low-flow channel with side channels that carry flow during bankfull and higher flow events instead of overbanking. The side channels provide pathways through the riparian zone for inundation and infiltration along with the development of wetlands fed by groundwater pathways. Features of multichannel systems include interlocking channels, longitudinal bars, and high system mobility. They tend to be wide and shallow and their bedload materials are often dominated by noncohesive coarse cobbles, gravel, and sand. Four abiotic metrics were developed tailored to address this geomorphological and physical complexity, which is evaluated with a channel and floodplain survey using indicator checklists and narrative approaches to assessment.

For the channel and floodplain survey, the stream reach is divided into three more or less equal segments (upper, middle, and lower). Each of these segments is surveyed in a lateral traverse from the SA outer boundary to the active channel edge. Important features from the metric checklists plus any feature that is affecting the hydrologic function of the segment, regardless of its inclusion on the checklist, is noted. The indicator checklists are designed to guide surveyors in identifying important parameters and characteristics to apply to ratings tables' narratives. A sketch of major features of the floodplain on the large-scale SA map like that for biotic vegetation communities is encouraged as an aid in filling out the checklist and for later interpretation. In addition, photographs are taken in each survey segment, across the channel to the opposite bank as well as upstream and downstream from the bank edge. Supplemental photographs of features that alter the size of the SA or significantly impact floodplain connectivity are particularly useful.

- Floodplain hydrologic connectivity is an assessment of the ability of water to flow into or out of the wetland and is derived from CRAM (CWMW, 2013). EIA also developed a narrative version for riverine systems based on CRAM

(Faber-Langendoen et al., 2012). HGM has a similar frequency of surface flooding metric (Hauer et al., 2002). The hydrologic connectivity between the river and riverine wetlands formed on its floodplain supports ecologic function and plant and wildlife habitat diversity by promoting exchange of water, sediment, nutrients, and organic carbon. It is evaluated by detecting recent channel and floodplain inundation indicators. That is, evidence of the extent of flood deposits and side-channel wetting that has occurred within the last five years. The assessment is also dependent on the size of a peak flow that has occurred in the last five years—large flows leave more evidence, small flows leave less. When there have only been very small flows in the preceding five years it may be very hard to rate this metric accurately. In such a case, the field team must use its best professional judgment. The worksheets require rating of five key indicators and their extent across the floodplain, as noted during the lateral traverse of each SA segment. These indicators include (1) floodplain inundation deposits, scouring surfaces, and wrack, (2) recent woody debris, (3) indicators of recent flow within side channels, (4) removal or burying of floodplain litter, and (5) removal or burying of side channel litter. A table for estimating the return interval for the peak discharge within the last five years is included, and metric ratings are based on the peak discharge.

- Physical patch complexity is a measure of the physical complexity of the site that contributes to ecological diversity. Physical patch complexity is adapted from CRAM structural patch richness (CWMW, 2013) and HGM macrotopographic complexity (Hauer et al., 2002), but rescaled following EIA physical patch type, which emphasizes condition rather than functional complexity (Faber-Langendoen et al., 2012). Rivers act as conveyor belts of both water and sediment, the movement of which occurs linearly in the direction of flow and horizontally as rivers periodically spill onto the floodplain. Under optimal hydrological conditions, a varied and complex habitat should develop from a variety of fluvial and flood-derived features that can support high biological diversity and create multiple pathways for the operation of ecological processes. But even abandoned floodplains can retain some of the structural elements that support habitat variety. Physical patch complexity is measured as part of each SA segment traverse with check offs of field indicators derived from flooding, flow, the presence of water, and fluvial geomorphic surfaces. The metric rating is based on a narrative description and the number of patch types encountered.
- Soil surface condition reflects anthropogenic soil disturbance impacts within the SA and is derived from Faber-Langendoen et al. (2012), which in turn was based on Mack (2001). This metric evaluates disturbance of the soil and surface substrates that affects biological, physical, and chemical processes that ultimately define broader wetland ecological condition such as plant establishment and vegetation communities. In this capacity, understanding of soil condition, whether natural or modified via land use, is critical to setting restoration goals and developing restoration strategies. Examples of soil surface disturbance include filling and grading, plowing, livestock disturbance, vehicle use, pavement, dredging, and other mechanical disturbances to the surface substrates or soils. Layers of ash or fire pits onsite can change the ability of soils to absorb water (Larsen et al., 2009). Soil disturbance can potentially negatively impact soil nutrient cycling, moisture, chemistry, biodiversity, and structure. This metric is based on a visual assessment of anthropogenic soil disturbance indicators during the SA segment traverses, and an estimate of the percentage of soil disturbance relative to the total area of the SA.
- Channel mobility is an assessment of impediments to a channel to laterally migrate or access side channels, prohibiting the development of a DPM of fluvial landforms on the floodplain that supports wetland and riparian communities. A guiding principal underlying riverine ecosystem health is the maintenance of a dynamic riverscape of shifting ecological communities on a changing fluvial geomorphic template that is driven by hydrological processes (Crawford et al., 1993; Hupp and Osterkamp, 1996; Crawford et al., 1999; Stanford et al., 2004; Latterell et al., 2006; Weisberg et al., 2013). Critical to maintaining this dynamism is ensuring the capacity for lateral channel migration, shifting channel movement across the floodplain, and accessing side channels that lead to the development of new sites for vegetation recruitment and tree regeneration. This metric is assessed at the river end of each SA traverse, looking up and down each side of the river. The metric is rated based on the percentage of channel banks within the SA that is stabilized by artificial elements such as artificial levees, riprap or jetty jacks, and by woody nonnative shrubs and trees. When channels have only minor amounts of armoring, they score higher; when they have little or no open natural banks that can provide opportunity for lateral migration or channel movement during high flows, they are considered impaired.

Metric, Attribute, and SA Scoring

For each SA, there is an *SA rank summary worksheet* where the metric ratings are compiled, weighted by importance and sensitivity, and rolled up for each attribute (landscape context, biotic, and abiotic). Using the attribute scores, the SA is given an overall weighted condition rank. The metric and attribute weighting hierarchy is built into the summary sheet such that individual and attribute category weighted scores can be calculated easily and then rolled up into a final numeric wetland condition score. The digital PDF version of the form *automatically* compiles the scores from the various worksheets and computes a weighted wetland condition score on the SA rank summary worksheet. This ranges from 4.0 (excellent) to 1.0 (poor). The wetland condition score is then given a letter wetland condition rank based on its score (A = Excellent, B = Good, C = Fair, and D = Poor), which is summarized as follows:

- A—(Score >3.25–4.0) wetlands considered in "Excellent" condition have intact wetland functions and processes and diverse wetland/riparian vegetative communities with no exotic invasive vegetation. Their size equals or approaches their historic extent due to intact hydrological conditions. These wetlands are surrounded by a natural protective vegetated buffer and the surrounding riparian corridor is not disconnected by anthropogenic features. These wetlands are largely undisturbed and would be considered the reference standard wetlands for the subclass.
- B—(Score > 2.5–3.25) Wetlands ranked "Good" display some degradation in condition. These wetlands may have disrupted hydrological regimes that somewhat reduce duration, frequency, and extent of floodplain inundation and infiltration, and therefore reduce the current wetland size moderately from its historic size. These wetlands may display a combination of onsite and surrounding anthropogenic disturbances, a reduction of vegetative community and structural diversity, less common native riparian tree regeneration, and the presence (1%–<5%) of exotic invasive vegetation. The overall complexity of floodplain features is reduced. Often, these wetlands would benefit from restoration and would rank as the most likely candidates. Wetlands in good condition may be the best available.
- C—(Score >1.75–2.5) Wetlands ranked in "Fair" condition often exhibit disrupted hydrology that markedly reduces the size of the wetland relative to its historic size. Disruptive buffer and riparian corridor land cover elements are numerous, reducing riparian connectivity and ecosystem processes. Vegetative condition is marked by fewer vegetation community types and lower plant diversity within those communities, often hosting exotic invasive vegetation. Native tree regeneration is uncommon due to lack of floodplain inundation. Land use surrounding the wetland is intense and impacts wetland patterns and processes. These wetlands may have some potential for restoration, depending on the ability to remove stressors and reverse the impacts that are affecting the wetland condition.
- D—(Score ≤ 1.75) Wetlands ranked "Poor" often have a fully disrupted hydrology, poor vegetative composition, and diversity that may be dominated by exotic invasive vegetation and anthropogenic ground disturbance. They may also be extremely small relative to their historic extent. Buffer and riparian corridor connectivity are disrupted with numerous permanent structures, and surrounding land use is intense and may permanently modify adjacent land. These wetlands generally would not be considered candidates for restoration to natural conditions because of permanent modifications and irreversible impacts. However, data from this condition assessment may provide information to restore some of the natural ecological integrity of the wetland.

Stressor Checklists

The NMRAM method includes the identification and evaluation of stressors for each site. Stressor checklists are designed to assess the intensity of stressors that occur within the SA and the buffer area (250 m from the SA boundary). Stressors are anthropogenic disturbances that would be expected to have a negative effect on the condition of the SA. The purpose of the stressor checklists is to provide information that furthers the understanding of the current wetland condition. They are not used in scoring or ranking the condition of the wetland.

Stressor checklists are grouped into four categories; land use, vegetation, hydrologic modifications, and physical structure. Stressor checklists for each of the four categories identify stressors that occur within the SA and the buffer, and rank their intensity as absent, minor (<10%), moderate (10%–50%), or intense (>50%). A final sum of stressors and their intensity within the SA or buffer is calculated for each category. In addition, the final number of stressors and their intensity within the SA or buffer is summarized on the SA rank summary worksheet.

VALIDATION/CALIBRATION EFFORTS UNDERTAKEN WITH THE METHOD

A series of validation studies and sensitivity tests was conducted for lowland riverine metrics using data from 25 reference SAs within the Gila Reference Domain, and representing a continuum of conditions and a variety of stressors. Validation was conducted incrementally over two years and iteratively improved the metric protocols and ratings tables based on field outcomes. Beyond developing and testing protocols, Level 3 data were collected from vegetation plots with full floristic inventories to test aspects of the biotic metrics for their efficacy. Level 3 data on channel and floodplain morphology (Rosgen, 1996) were collected subsequently from five previously sampled SAs to validate channel processes.

Current hydrology studies in the Gila at the same sites as our data collection were referenced as corroborating data (Soles, 2003). Relative sensitivity of each metric was evaluated based on the distribution of scores among the 25 sites, and metrics and ratings were modified as necessary to increase applicability and sensitivity as appropriate. Because the Gila basin is still a predominantly uncontrolled natural river system, there were intrinsic limits to the range of scores possible. Most low-end scores were associated with downriver locations where irrigation drawdowns and partial levee systems can significantly impact the system. Accordingly, we are currently testing metrics and developing supplemental metrics in two fully flow-controlled rivers, the Rio Grande and the Pecos River in central and southern New Mexico, to further understand the lower end functionality and ecosystem integrity of large river systems.

TIME SPENT IN DEVELOPING/TESTING THE METHOD

Overall, four years were spent in development that included convening two meetings and a two-day training exercise on the ground with an advisory committee of wetland specialists, NGOs, state and federal land managers, and state and federal regulators, both from the region and across the state. In the first meeting, we presented our set of potential metrics, received feedback, and integrated comments before field testing. In addition, a pilot study was conducted by the development team to confirm the reference domain and subclass limits, and to address questions regarding the suite of metrics proposed before the data collection teams were sent to the field. A second advisory committee meeting was conducted to present the results of the data collection efforts and the revised metrics while again soliciting input on improvements, particularly on clarifying metric rating rationale. The two-day training exercise was conducted with 15 participants to evaluate how well the metric protocols worked in a typical application; modifications were made accordingly.

SAMPLE APPLICATION OF METHOD IN THE FIELD

Following completion of Version 1.0 of the NMRAM Lowland Riverine Wetlands Field Guide, we had the immediate opportunity to test the method for the New Mexico Department of Game and Fish (NMDGF). Full-scale NMRAMs were conducted on two recent land acquisitions (River Ranch) on the Mimbres River (within the Gila Reference Domain) that had significant riparian corridors to inform the future management of parcels as wildlife habitat areas. The parcels were the WOIs and they were large enough to require three SAs total. SA scores were 3.41 (A Rank) for Upper SA, 3.14 (B Rank) for Middle SA, and 3.44 (A Rank) for Lower SA.

Based on the NMRAM assessment, the River Ranch riparian wetlands overall are currently in excellent to good condition, with both the lower and upper SA rating excellent and the middle SA rating good. The ranch average for both landscape context and biotic metrics was also excellent. The abiotic metrics were rated in the good category. However, the data from some individual metrics point out areas where management is needed to maintain or improve the condition status of the ranch. Both in some of the biotic and abiotic metrics, there are indications of a decline in groundwater and losses in hydrologic connectivity on the ranch. Most measurements showed that these declines were more severe below an irrigation diversion dam that took significant amounts of water from the river during the growing season. In particular, the metric vegetation vertical structure indicated a lack of riparian shrub layers across the lower portion of the ranch, indicating a lowered groundwater table as well as possible removal by livestock in the past. Across the ranch, scores for the metric native riparian tree regeneration were low, with very few young trees observed. This can indicate both a lowered groundwater table, a loss of hydrological connectivity to the floodplain, and/or removal of seedlings by livestock. Finally, the metric hydrologic connectivity indicated a minor to moderate loss in connectivity from the expected, with the lowest connectivity scores coming from the middle SA, just below the irrigation diversion dam. The reports were well received by the agency and are available at https://nhnm.unm.edu/ (River Ranch Riparian Assessment; Milford et al., 2015).

REGULATORY APPLICATIONS

A regulatory module of NMRAM for riverine wetlands is currently under development, refinement, and testing by the US Army Corps of Engineers (USACE), Albuquerque District. The purpose of the regulatory module is to have a stand-alone condition assessment methodology for use in impacts and mitigation area assessments associated with Clean Water Act Section 404 permits issued by USACE. The regulatory module of NMRAM is a modified version from the current riverine versions of NMRAM (both montane and lowland modules) to increase its applicability to the regulatory program in general, and to address application issues including assessment area size related to permit area and impacts within or proposed for WOI and SA. The premise is that the current condition for the WOI applies to the permit area, preproject. NMRAM data are collected using GIS applications surrounding the permit area and on the ground within the permit area, so long as the permit area SA is within a minimum size area as specified in the regulatory module field guide. Using an area smaller than minimum may be reflected in lower scores. In addition, the active channel is included in the SA, similar to lowland riverine NMRAM.

The metrics used for the regulatory module of NMRAM include five each for the landscape, biotic, and abiotic attributes. Two new metrics were developed that measured disturbances internal to the SA; internal riparian corridor connectivity and sample area land use. The internal riparian corridor connectivity measures riparian corridor disruption within the SA. Sample area land use uses the same land use index as that of the surrounding land use metric but scores it for the footprint (area) occupied by the specific land use within the SA. These metrics are particularly useful for determining and scoring project impacts that will occur (e.g., a bridge construction) or already exist (e.g., a bridge widening project) within the SA. A key feature for using the regulatory module of NMRAM is that it can be used for predicting a functional lift based on proposed mitigation, either at the project site or at an alternative site. The metrics have enough scoring spread by attribute to rate and weigh the differences using the mitigation scenario. The metrics themselves along with the stressor checklists provide valuable information for mitigation planning.

NONREGULATORY APPLICATIONS—WETLAND ACTION PLANS

The NMRAM Team is providing guidance and training for watershed groups to develop and prepare "wetlands action plans" as an add-on to their watershed-based plans. A wetlands action plan is developed by established watershed and/or community groups to broaden their planning and resource improvement efforts to include wetlands, riparian, and buffer areas within their watersheds. Watershed groups can use NMRAM to obtain data necessary to effectively prioritize conservation and restoration of wetland/riparian ecosystems in their watershed.

TIME SPENT TO APPLY THE METHOD IN THE FIELD

Large river systems present challenges for rapid application for these primary reasons: increased size of the SA and the increased hydrological and fluvial geomorphic complexity that are driven by larger discharges. Lowland riverine NMRAM data collection on the NMDGF land, described above with three SAs, took three days to complete by three people, not counting travel time. The SAs were sufficiently large per the SA size requirements to avoid as much as possible scale-dependent depression of scores. Although they were on somewhat smaller rivers, they still posed complex hydrological environments that took additional time to assess. Because lowland riverine NMRAM SAs are very large and with tall, dense vegetation, an additional field person is necessary for safety reasons and to expedite the data collection process. Yet, from the NMDGF perspective, the assessments were relatively cost-effective and provided useful data compared to other assessment protocols.

HOW WAS/IS THE DATA BEING USED

A priority of the SWQB Wetlands Program is to develop methods for wetlands assessment that provide essential information for implementing wetlands protection measures. The rapid and accurate classification and assessment of wetlands are critical as part of SWQB's ongoing efforts to promote effective management and protection of the state's wetland resources. In particular, SWQB is developing water quality standards for wetlands that have direct application to wetland types that are threatened or declining. Rapid assessments of wetland condition as well as the identification of wetland stressors provide essential information for wetlands standards development.

The assessment data from the 25 Gila Reference Domain reference SAs have been uploaded to an off-line version of the SWQB's database and a web interface is being developed to make data accessible to the public. The web site will include a map viewer and provide tools for practitioners to upload their assessments as well. When fully realized, this web-based platform for assessment is expected to aid wetland conservation and restoration at an unprecedented level for the state of New Mexico.

WHAT WAS LEARNED

Given the size of lowland riverine ecosystems, significant adjustments to NMRAM metrics were made from our previous module developed for smaller, montane riverine wetlands. The larger, uncontrolled rivers of the Gila Reference Domain have complex fluvial geomorphology, (i.e., multichannel systems predominated over single channels) requiring a different approach to rapid assessment of hydrologic connectivity and overall river function. Large basin landscape-scale stressors are difficult to track compared to the montane riverine wetlands or isolated wetlands.

PROSPECTS FOR THE FUTURE

While we were successful in developing a large-river RAM focused on arid-land hydrology, further testing is needed, particularly in scaling attributes. Our initial scaling analysis on the relationship between discharge volume and resulting fluvial geomorphic surfaces and associated vegetation patterns had limited precision and was confined to a single geologic context. With all else being equal, sediment loads may vary across basins as a function of geologic substrate, climate, and upland condition such that the relationship may not be strictly linear, particularly at very high flows.

The range of scores in the Gila Reference Domain was limited and we suggest that this is because the metrics were developed against a relatively intact reference set in a wild, uncontrolled river system. We suspect scores would be significantly lower in flow-controlled reaches. Accordingly, we are currently working on applying these metrics in the Rio Grande and Pecos River floodplains, two large waterways with significant impoundments and alterations to the hydrological regime. A key question is whether it is appropriate to hold these highly altered river systems to the same scale of ecological condition as wild, uncontrolled systems like that of the Gila Reference Domain, or do they effectively represent a new, highly modified subclass to be evaluated on their own terms? Which approach to adopt could be based on assessment objectives. If conservation of the best remaining riparian wetland landscape is the goal or understanding mitigation values, then the former option of keeping a playing field level by scoring across all large lowland rivers regardless of impacts might be the best option for understanding the cost and benefits of any given conservation action. If restoration or partial restoration of already degraded sites is a goal, then having an assessment method designed and calibrated to restoration objectives might be the better alternative.

Lastly, rapid assessment metrics are by design semiquantitative and further testing of their efficacy with detailed and more intensive studies (Level 3) is a must for growing the science that underpins them. While we have gathered detailed floristic data to support some of our biotic metrics, there is a need for more detailed hydrological measurements followed by modeling of these riverine systems to further correlate outcomes with ecological integrity.

INSTITUTIONAL SUPPORT

Wholehearted support and funding for the development of NMRAM by wetland subclass has been provided by the U.S. Environmental Protection Agency (EPA) Region 6, awarded to SWQB Wetlands Program through Wetlands Program Development Grants. This includes funding for Montane Riverine Wetlands in the Rio Grande and Canadian Watersheds, Playas of the Southern High Plains, and Lowland Riverine Wetlands in the Rio Grande and Pecos River floodplains. Additional funding was provided by Natural Heritage New Mexico, a division of the Museum of Southwestern Biology, University of New Mexico.

Staff of the University of New Mexico Natural Heritage New Mexico and SWQB conduct NMRAM trainings annually. In 2018, training for Lowland Riverine Wetlands NMRAM expanded for the Rio Grande and Pecos River floodplains will be conducted. Please contact the authors for additional information or to attend this training.

REFERENCES

Bailey, J.K., Schweitzer, J.A., Whitham, T.G., 2001. Salt cedar negatively affects biodiversity of aquatic macroinvertebrates. Wetlands 21 (3), 442–447.

Brinson, M.M., 1993. A Hydrogeomorphic Classification for Wetlands (Technical Report WRP-DE-4). Prepared for U.S. Army Corps of Engineers.

Burton, T.A., Smith, S.J., Cowley, E.R., 2008. Monitoring Stream Channels and Riparian Vegetation – Multiple Indicators. Version 5.0, April 2008. Interagency Technical Bulletin. BLM/ID/GI-08/001+1150.

California Wetlands Monitoring Workgroup (CWMW), 2013. California Rapid Assessment Method (CRAM) for Wetlands, Version 6.1. p. 67.

Callahan, D., White, L., 2004. Vegetation Mapping of the Rio Grande Floodplain From Velarde to Elephant Butte. U.S. Bureau of Reclamation, Albuquerque Area Office, Albuquerque, and Technical Service Center, Ecological Planning and Assessment Branch, Denver.

Comer, P., Faber-Langendoen, D., Evans, R., Gawler, S., Josse, C., Kittel, G., Menard, S., Pyne, M., Reid, M., Schulz, K., Snow, K., Teague, J., 2003. Ecological Systems of the United States: A Working Classification of U.S. Terrestrial Systems. NatureServe, Arlington, VA.

Crawford, C.S., Cully, A.C., Leutheuser, R., Sifuentes, M.S., White, L.H., Wilber, J.P., 1993. Middle Rio Grande Ecosystem: Bosque Biological Management Plan. .

Crawford, C.S., Ellis, L.M., Shaw, D., Umbreit, N.E., 1999. Restoration and monitoring in the Middle Rio Grande Bosque: current status of flood pulse related efforts. In: Finch, D.M., Whitney, J.C., Kelly, J.F., Loftin, S.R. (Eds.), Rio Grande Ecosystems: Linking Land, Water, and People. USDA Forest Service, Rocky Mountain Research Station, Ogden, UT, pp. 158–163.

Faber-Langendoen, D., Kudray, G., Nordman, C., Sneddon, L., Vance, L., Byers, E., Rocchio, J., Gawler, S., Kittel, G., Menard, S., Comer, P., Muldavin, E., Schafale, M., Foti, T., Josse, C., Christy, J., 2008. Ecological Performance Standards for Wetland Mitigation: An Approach Based on Ecological Integrity Assessments. NatureServe, Arlington, VA.

Faber-Langendoen, D., Rocchio, J., Thomas, S., Kost, M., Hedge, C., Nichols, B., Walz, K., Kittel, G., Menard, S., Drake, J., Muldavin, E., 2012. Assessment of Wetland Ecosystem Condition Across Landscape Regions: A Multi-metric Approach, Part B, Ecological Integrity Assessment Protocols for Rapid Field Methods (L2), EPA/600/R-12/021b. U.S. Environmental Protection Agency Office of Research and Development, Washington, DC.

Fennessy, M.S., Jacobs, A.D., Kentula, M.E., 2007. An evaluation of rapid methods for assessing the ecological condition of wetlands. Wetlands 27 (3), 543–560.

Hauer, F.R., Cook, B.J., Gilbert, M.C., Clairain Jr., E.J., Smith, R.D., 2002. A Regional Guidebook for Applying the Hydrogeomorphic Approach to Assessing Wetland Functions of Riverine Floodplains in the Northern Rocky Mountains (ERDC/EL TR 02-21). U.S. Army Corps of Engineers, Engineer Research and Development Center, Environmental Laboratory, Vicksburg, MS.

Hink, V.C., Ohmart, R.D., 1984. Middle Rio Grande Biological Survey. U.S. Army Engineer Corps of Engineers, Albuquerque District, Albuquerque, New Mexico (Contract No. DACW47-81-C-0015). Arizona State University, Tempe.

Hupp, C.R., Osterkamp, W.R., 1996. Riparian vegetation and fluvial geomorphic processes. Geomorphology 14 (4), 277–295.

Larsen, I., MacDonald, L.H., Brown, E., Rough, D., Welsh, M.J., Pietraszek, J.H., Libohava, Z., Benavides-Solorio, J.D., Schaffrath, K., 2009. Causes of post-fire runoff and erosion: water repellency, cover or soil heating? Soil Sci. Soc. Am. J. 73 (4), 1393–1407.

Latterell, J.J., Bechtold, J.S., O'Keefe, T.C., Van Pelt, R., Naiman, R.J., 2006. Dynamic patch mosaics and channel movements in an unconfined river valley of the Olympic mountains. Freshw. Biol. 51 (3), 523–544.

Mack, J.J., 2001. Ohio Rapid Assessment Method for Wetlands v. 5.0 User's Manual and Scoring Forms (Ohio EPA Technical Report WET/2001-1). Ohio Environmental Protection Agency, Division of Surface Water, Wetland Ecology Group, Columbus.

Milford, E.R., Muldavin, E.H., Chauvin, Y., 2015. River Ranch Riparian Assessment. University of New Mexico Natural Heritage New Mexico, Albuquerque (Publ. No. 25-GTR-368).

Muldavin, E.H., Bader, B.J., Milford, E.R., McGraw, M.M., Lightfoot, D., Nicholson, B., Larson, G., 2011. New Mexico Rapid Assessment Method: Montane Riverine Wetlands Manual, Version 1.1. New Mexico Environment Department, Surface Water Quality Bureau, Santa Fe, New Mexico 90 p. and appendices.

Muldavin, E., Milford, E., Umbreit, N., Chauvin, Y., 2017. Long-term outcomes of a natural-processes approach to riparian restoration in a large regulated river: the Rio Grande Albuquerque Overbank Project after 16 years. Ecol. Restor. 35 (4), 341–353.

Rosgen, D., 1996. Applied River Morphology. Wildland Hydrology, Pagosa Springs, CO.

Soles, E.S., 2003. Where the River Meets the Ditch: Human and Natural Impacts on the Gila River, New Mexico, 1880–2000 (MS Thesis). Northern Arizona University, Flagstaff.

Stanford, J.A., Lorang, M.S., Hauer, F.R., 2004. In: The shifting habitat mosaic of river ecosystems.29th Congress of the International-Association-of-Theoretical-and-Applied-Limnology, Lahti, Finland, pp. 123–136.

Stenquist, S., 2000. In: Spencer, N.R. (Ed.), Salt cedar integrated weed management and the Endangered Species Act. Proceedings of the X International Symposium on Biological Control of Weeds, 4–14 July 1999. Montana State University, Bozeman, MT, pp. 487–504.

U.S. Department of Agriculture Natural Resources Conservation Service (NRCS), 2003. New Mexico State-Listed Noxious Weeds. https://plants.usda.gov/java/noxious?rptType=State&statefips=35.

U.S. Environmental Protection Agency, 2010. Basics: What are Biocriteria and Bioassessment Data? http://water.epa.gov/scitech/swguidance/standards/criteria/aqlife/biocriteria/basics.cfm. Accessed October 2010.

U.S. Environmental Protection Agency (USEPA), 2006. Application of Elements of a State Water Monitoring and Assessment Program for Wetlands. Wetlands Division, Office of Wetlands, Oceans and Watersheds, U.S. Environmental Protection Agency, Washington, DC.

Weisberg, P.J., Mortenson, S.G., Dilts, T.E., 2013. Gallery forest or herbaceous wetland? The need for multi-target perspectives in riparian restoration planning. Restor. Ecol. 21, 12–16.

Winward, A.H., 2000. Monitoring the Vegetation Resources in Riparian Areas (General Technical Report RMRS-GTR-46). U.S. Department of Agriculture, Forest Service, Rocky Mountain Research Station, Ogden, UT.

FURTHER READING

Faber-Langendoen, D.G., Kittel, K., Schulz, E., Muldavin, M., Reid, C.N., Comer, P., 2008b. Assessing the Condition of Lands Managed by the U.S. Army Corps of Engineers: Level 1 Ecological Integrity Assessment. NatureServe, Arlington, VA.

Muldavin, E.H., Milford, E.R., McGraw, M.M., 2016a. New Mexico Rapid Assessment Method: Montane Riverine Wetlands Field Guide. Version 2.1. New Mexico Environment Department, Surface Water Quality Bureau, Santa Fe, NM.

Muldavin, E.H., Milford, E.R., McGraw, M.M., 2016. New Mexico Rapid Assessment Method: Lowland Riverine Wetlands Field Guide. Version 1.1. New Mexico Environment Department, Surface Water Quality Bureau, Santa Fe, NM.

Chapter 4.3.8

The Development and Implementation of the Ohio Rapid Assessment Method for Wetlands: A Case Study

John J. Mack*, M. Siobhan Fennessy† and Mick Micacchion‡

*Permit and Resource Management Department of Sonoma County, Santa Rosa, CA, United States, †Department of Biology, Kenyon College, Gambier, OH, United States, ‡Midwest Biodiversity Institute, Hilliard, OH, United States

Chapter Outline

INTRODUCTION

The Ohio Rapid Assessment Method for Wetlands Version 5.0 (ORAM) or close adaptations of the method adopted by other states (e.g., Kutcher, 2010; Michigan DNRE, 2010; Kentucky, 2016) and multistate organizations (TVA, 2017) is used in some or all the area of at least eight states (AL, GA, KY, MI, MS, OH, RI, TN, and VA). The method had its genesis in the push by the U.S. EPA in the 1990s for states to develop wetland specific biological criteria and antidegradation policies (U.S. EPA, 1990). For the state of Ohio, strong and consistent encouragement from the Region 5 Office of the U.S. EPA for those states (Illinois, Indiana, Michigan, Minnesota, Ohio, and Wisconsin) was provided to initiate rulemaking to adopt wetland water quality standards (WWQS) into their existing Clean Water Act regulations.

Contemporaneously, the U.S. EPA national wetland program formed the Biological Assessment of Wetlands Workgroup (BAWWG) to bring together wetland scientists and experts in the field of biological assessment and biological criteria to determine if the use of indicator taxa (e.g., fish, invertebrates), which had been successfully used to develop stream biological criteria in the 1970s and 1980s, could be extended to a "new" type of surface water, that is, wetlands (U.S. EPA, 1997). While in principal it appeared that there should be no reason the approach could not be extended to wetlands, there were many questions and concerns that wetlands intrinsically presented issues that might make a biocriteria approach impractical: variability (e.g., hydroperiod), seasonality (e.g., seasonal floods), classification (too many types), stability (e.g., successional changes), lack of good indicator taxa, lack of standardizable sampling methods, etc. (U.S. EPA, 2002a). Further complicating the field was the fact that an alternate approach focusing on the enumeration and quantification of the "functions" and "values" of wetlands had become well entrenched in the lexicon of wetlands protection and assessment. The practical upshot of all this was the award by the U.S. EPA of a series of Wetland Program Development (WPD) Grants to fund state WWQS rulemaking projects and wetland biological assessment research.

By the 1990s, the state of Ohio had already been actively regulating wetlands via the Clean Water Act (CWA) Section 401 certification authority and had a brief but comprehensive in scope rule on the books (adopted in the early 1980s) that provided the basis for a wetland regulatory program and a small wetland-focused staff. In addition, the state of Ohio was a national if not world leader in the development of stream biocriteria (Ohio EPA, 1988a,b, 1989a,b). So, in hindsight, it appears natural that the state would apply for and receive WPD grants that funded a WWQS rulemaking

Wetland and Stream Rapid Assessments. https://doi.org/10.1016/B978-0-12-805091-0.00046-3

process between 1996 and 1998 and wetland research that focused on vascular plants, macroinvertebrates, and amphibians as possible indicator taxa for numeric wetland biocriteria. In reality, hard work and considerable internal advocacy within the Ohio EPA and strong support from U.S. EPA staff at the regional and national levels were what moved this from "something we should do" to "something we are doing."

A key moment occurred on May 1, 1998, when the Ohio EPA adopted narrative wetland water quality standards and a wetland antidegradation rule after a two-year rulemaking process that involved all major stakeholders. The water quality standard part of the rule was only "narrative" because the research into the development of wetland biological criteria was still ongoing; the heart of the rulemaking was found in a creative application of the Clean Water Act's antidegradation policy (Ohio Administrative Code (OAC) Rule 3745-1-54).[1] The wetland antidegradation rule created three explicit categories and one implicit category of wetlands and then applied a tiered level of wetland protection and mitigation to each category based on its overall quality and/or level of functioning, or in a few instances, by its type (Table 4.3.8.1).

TABLE 4.3.8.1 State of Ohio Wetland Antidegradation Categories

Category	OAC Rule Section	Rule Definition	Have Some or All of the Following Characteristics
Category 1	3745-1-54(C)(1)	Wetlands that "...support minimal wildlife habitat, and minimal hydrological and recreational functions," and as wetlands that "...do not provide critical habitat for threatened or endangered species or contain rare, threatened or endangered species."	• Often hydrologically isolated. • Low species diversity and often dominated by disturbance-tolerant native or nonnative species. • Limited potential to achieve beneficial wetland functions and no significant habitat or wildlife use. • Often a predominance of invasive planta, for example, narrow-leaved cattail, Phragmites, purple loosestrife or glossy buckthorn.
Modified Category 2	3745-1-54(C)	"...wetlands that are degraded but have a reasonable potential for reestablishing lost wetland functions."	• An implied category between Category 1 and 2. • Presently of somewhat lower quality than undisturbed Category 2 wetlands. • Able to be restored.
Category 2	3745-1-54(C)(2)	Wetlands that "...support moderate wildlife habitat, or hydrological or recreational functions [and which are...dominated by native species but generally without the presence of, or habitat for, rare, threatened, or endangered species..."	• Generally, do not have rare, threatened, or endangered species or their habitat. • Broad middle category of good quality wetlands. • Functioning, diverse, healthy water resource that has ecological integrity and human value. • Often relatively lacking in human disturbance and can be considered to be naturally of moderate quality. • Others may have been Category 3 wetlands in the past, but have been disturbed down to Category 2 status.
Category 3	3745-1-54(C)(3)	Wetlands that have "...superior habitat, or superior hydrological or recreational functions." They are typified by high levels of diversity, a high proportion of native species, and/or high functional values.	• Contain or provide habitat for threatened or endangered species. • Wetlands with high ecological quality, for example, mature forested wetlands, vernal pools, bogs, fens. • Scarce regionally and/or statewide. • High floral or faunal species diversity. • Lacking in substantial human disturbance or fully recovered from past disturbance. • A wetland may be Category 3 because it exhibits just one or all of superior functions.

1. Delving into the details of this policy at the federal level and the multiple rulemakings and challenges to its implementation in Ohio is beyond the scope of this chapter. Suffice to say, it provided a strong legal basis for advancing a relatively nonprescriptive but comprehensive protection approach to wetlands based on their type and quality. In its briefest characterization, antidegradation can be considered a "no-backsliding" policy such that the "existing" condition of the nation's waters as of Nov. 28, 1975, (40 CFR 131.12(a)) formed the biological/physical/chemical bottom floor from which recovery is measured and no action could be taken or permitted that would "degrade" waters below this "existing" state as of this date.

The broad policy implication of this wetland categorization system was a pragmatic recognition that under state and federal law, some wetlands would always be allowed to be destroyed and replaced elsewhere. If this were to happen, there was a strong legal and ecological preference that low-quality wetlands be "filled' first, that medium-quality wetlands (which include degraded but easily restorable systems) be the broad middle category where regulatory arguments between regulators and the regulated community occur,[2] and impacts to high-quality wetlands be largely avoided absent a strong societal-level interest in the project.

The key legal hook that underlay the development and ultimately widespread acceptance and utilization of ORAM was language in OAC Rule 3745-1-54(B)(2)(a)(ii), which required that wetlands were to be categorized using a "method acceptable to the director" of the Ohio EPA. As discussed below, having or identifying a statutory or regulatory requirement to develop and *require* the use of an assessment method makes the difference between developing something that is put on the shelf versus something that becomes widely used.

Early in the discussions of the regulatory implications for this rule, the regulated community required that a relatively fast and easy method be developed concomitantly with the rulemaking for categorizing wetlands. To accomplish this, a technical workgroup comprised of wetland technical experts from government, academia, and the regulated community was formed to develop a method that would in effect *always* be acceptable to the Ohio EPA. It was decided to investigate a suite of existing methods from other states and regions to determine the range of methods already in use and if a method could be adopted or adapted for use in Ohio.

The various methods were reviewed in the office and tested in the field at multiple study wetlands suggested by the workgroup. The workgroup's attention began to focus seriously on the Western Washington Rapid Assessment Procedure (Washington, 1993), for several reasons: (1) It resulted in a single ordinal score that could be broken into scoring ranges that equated with the proposed antidegradation categories; (2) the patterns of suburban and rural land use and broad weather patterns, especially precipitation in Western Washington, were similar to conditions in Ohio; (3) the types of wetlands found in Western Washington were also equivalent to most types in Ohio; (4) the method focused on relatively easy to observe structural variables that different users could come to agreement on; and (5) it included a set of narrative questions that placed wetlands in higher protection categories regardless of score.

Adaptation of the Western Washington Method quickly moved through a Version 0 to a draft Version 3.0, with Version 3.0 being available for use when the Ohio WWQS rules went into effect on May 1, 1998 (OAC 3745-1-50 through 3745-1-54). Shortly thereafter in 1999, the ORAM name and acronym was settled on and ORAM was finalized for use in Version 4.0 with some minor changes in Version 4.1 (Mack, 2001, Appendices).

Proceeding at the same time as the Ohio WWQS rulemaking and initial ORAM development, Ohio wetland bioassessment research was making substantial progress and had focused on vascular plants, amphibians, and macroinvertebrates as likely indicator taxa groups (e.g., Fennessy et al., 1998a,b). And the federally sponsored BAWWG effort was also meeting regularly throughout this period with extensive collaboration helping to accelerate the participant's individual efforts. The BAWWG effort eventually resulted in the publication of a series of modules focusing on the state of the science and providing guidance on the various aspects of wetland bioassessment development (e.g., U.S. EPA, 2002b). A major hurdle was what can be called the "*x*-axis" problem or the development of replicable and predictive "disturbance gradients" for selecting indicator taxa responsive to human disturbance and for evaluating and ultimately selecting attributes of those taxa that could be usable metrics in the development of composite indices (e.g., index of biotic integrity) (Karr and Chu 1999).

The Ohio wetland bioassessment efforts were working this problem (U.S. EPA, 2002a) and investigating various landscape, chemical, and ordinal ranking disturbance gradients. An obvious choice for investigation was the score derived from ORAM, as it was supposed to broadly rank all types of Ohio wetlands by their overall quality and/or functionality, which was presumably closely associated with the degree of anthropogenic disturbance of the wetland. But, two serious problems had emerged by this time (c.1998): one had to do with the wetlands sampled to date in Ohio's reference wetland dataset; the other had to do with systematic bias in the ORAM score observed by Ohio EPA wetland staff, especially wetlands that were often smaller and isolated within a single vegetation class and with hydroperiods that were seasonally saturated to seasonally inundated.

With regard to the first problem, this was made explicit at the January 1999 BAWWG meeting in Providence, Rhode Island, where Ohio EPA staff presented on their efforts to date and had the opportunity to meet personally with Jim Karr (the arguable founder of the index of biotic integrity approach) to discuss their efforts. After reviewing and discussing numerous

2. For example, disagreements on whether the impact could be avoided altogether, whether unavoidable impacts could be further minimized, and how much mitigation will be required.

scatter plots and box and whisker plots of data from wetlands in Ohio, Karr quickly and simply summed up the problem: you are missing the bottom of the gradient. The wetlands in your data set are all good to excellent and you need to sample the poor and fair systems, then you will see your sites starting to array along a disturbance gradient (Mack, personal communication). This led to a concerted effort in the 1999 field season to locate, obtain access to, and sample highly degraded[3] emergent, shrub, and, especially, forested wetlands.[4] This effort proved successful and directly enabled the development of the first wetland IBI (index of biotic integrity) for the state of Ohio (Mack et al., 2000).

The second problem became clear during a difficult wetland-permitting episode. Without going into the actual details, a large commercial development was proposed in a large remnant of mature forest with multiple embedded vernal pools. The forest as a whole and the vernal pools in particular were very high quality, with actual amphibian use having been determined by direct observation and trapping. However, using the then-current version of ORAM, the sites were solidly categorized as Category 2 and not Category 3 wetlands. In the end, the decision was made that since it was the agency's own assessment protocol that categorized these wetlands into a lower protection category, that is how they would be regulated. This painful (to staff) loss of high-quality wetlands resulted in a close look at the internal mechanics and scoring of ORAM, and it became clear that ORAM questions—and the scoring weights given to those questions—were strongly biased toward large, hemi-marsh type systems with direct hydrologic connections to rivers and lakes.[5]

At the same time, ORAM Version 4.0/4.1 was proving to be, at best, a coarse disturbance gradient surrogate for wetlands. Fortunately, Ohio's stream program had moved across this ground before and had a well-developed, relatively rapid stream habitat index (Rankin, 1989) that focused on easily observed structural characteristics of streams. More importantly, this index had proved to be highly correlated to the quantitative fish and invertebrate IBIs that were the foundation of the state of Ohio's stream biological criteria. The stream habitat index had incorporated an express assessment of the intactness of key stream characteristics that were known to be strongly correlated with stream functions and biological quality and assigned scores based on this qualitative gradient of disturbance:

no observed disturbance.
to
recovered from a past disturbance.
to
in the process of recovering from a disturbance.
to
recent disturbance or no recovery.

It was very easy to conceptualize wetlands in this same manner. There are, practically speaking, three main ways to degrade a wetland: alter its natural hydrology (drain, flood, impound, raise/lower surface or soil inundation/saturation, etc.); alter its vegetation (cut, mow, spray, clear, etc.); or mechanically alter its soils or substrates (plow, fill, scrape, bury in sediment, etc.). By making this disturbance/lack of disturbance evaluation a core aspect of the method, in addition to requiring users to rate a wetland relative to other wetlands of its type or class and retaining some of the function and value elements of the earlier ORAM versions, a robust new method was developed.[6]

DESCRIPTION OF THE METHOD

Version 5.0 of ORAM was officially published on Feb. 1, 2001, although it had existed in draft form since 1999, becoming the de facto method acceptable to the director in the state of Ohio. The user's manual included a short and long form of the field data sheets as well as a narrative rating section and a score categorization worksheet to aid the user in the final step of

3. Somewhat tongue in cheek, these were called good "bad wetlands" (Mack, personal communication).

4. Given that the state of Ohio was predominately forested at the time of European settlement and that c.2000 a large percentage of this forested landscape had been converted to agriculture, the most disturbed type of forested wetland is arguably an emergent marsh in an agricultural landscape. However, highly disturbed forested wetlands that still had a canopy of trees do exist and were located and sampled as part of the Ohio wetland bioassessment efforts.

5. And this was at least partially due to a deeper bias found in many earlier function and value type assessment methods toward the "kidney-on-the-landscape" rationale for why wetlands should be protected and valued.

6. Thus was born ORAM v. 5.0 on the day after the lead author attended Ed Rankin's QHEI Training class. This was one of those rare "aha" moments with which one is occasionally graced. ORAM v. 4.1 was literally snipped into pieces with scissors and pasted on to herbarium paper with elements of QHEI approach written in between these as the key metrics 3, 4a, and 4c and the entire scoring scheme adjusted to a 100-point scale (Mack, personal communication).

determining the appropriate regulatory category of the wetland being assessed (Mack, 2001). Also, a decision was made to publish the numeric scoring breaks in a separate document so that the numeric thresholds between categories could be adjusted, if necessary, as new data was collected (Mack, 2000). Partly in response to the Ohio legislature enshrining version 5.0 in statute for isolated, that is, nonfederal waters of the state, version 5.0 has stayed as is for >15 years, even though a version 5.1 or 5.2 could have been published. The user's manual is now (as of late 2017) somewhat dated and is in need of a thorough updating and refreshing, which could be accomplished without modifying the method per se.

Completion of an ORAM assessment requires five steps: (1) determining the hydrogeomorphic and vegetation class of the wetland being assessed, (2) establishing the scoring boundaries (assessment area) of the wetland, (3) completing the narrative rating, (4) completing the quantitative rating, and (5) completing the score categorization worksheet. Each of these will be discussed below. However, one key requirement implicit in ORAM is how the assessment unit is defined.

All the Ohio wetland assessment methods—Level 1 (landscape), Level 2 (rapid), Level 3 (intensive) (U.S. EPA, 2009)—assess wetlands and not a predetermined assessment area. There is a definite split in approaches to wetland monitoring and assessment between the assessing wetlands versus assessing assessment areas camps. In the end, there are advantages and disadvantages to both approaches and the reason to select one or the other depends on the purposes to be accomplished by the assessment program. In Ohio, the development of wetland assessment protocols was always linked to and even justified by the needs of the 401/404 permitting program and the development of state water quality standards for wetlands. In the permitting program, a boundary is always delineated around something called a wetland and impacts are authorized, or not, to that wetland. To develop an assessment protocol that supports the permitting program, but which could also be used in watershed or state-wide ambient assessment efforts, basically requires you to assess a "wetland" and not an "area," or at least that you can make the leap that your assessment area is representative of the wetland being evaluated for 401/404 permit issuance.

COMPONENTS OF ORAM ASSESSMENT

Determining the Wetland Class

A key aspect of ORAM is a "one-size-fits-all" approach to wetland assessment method such that the user is required to evaluate the wetland being assessed in relation to other wetlands in that region and of the same type and class. This constrains interwetland type variation as well as variation from and intra and interregional differences in land use patterns, soils, glacial surface geology, climax vegetation, etc.

Although not completely fleshed out in 2001, a wetland classification based on dominant vegetation (emergent, forest, shrub plus multiple subtypes) *and* hydrogeomorphic class (depression, slope, riverine, etc.) was formalized in 2004 (Mack, 2004a,b).

In the user's manual and in the annual training provided by Ohio EPA, it is stressed that wetlands being assessed with ORAM *must only* be evaluated in relation to wetlands of the same type and class. Failure to do so can result in improperly setting scoring boundaries and undercategorization or overcategorization.

Establishing the Scoring Boundary

In terms of setting boundaries for the wetland being assessed, that is, defining an assessment area, ORAM uses natural boundaries based on (1) hydrogeomorphic class, and (2) arbitrary but practical rules to divide contiguous linear (usually riverine) systems or other extensive complexes of contiguous wetlands into (hopefully) ecologically coherent units for assessment purposes. The rules outlined for the second part were developed by the authors of the Western Washington Method and adopted almost verbatim by ORAM Version 5.0.

Learning to read the landscape and see the shifts in hydrology, landform, and vegetation that delimited different types of wetlands is obviously a task where experience counts, and, in some instances, this can be accounted as a disadvantage to assessing wetlands versus areas. Most instances where there is strong disagreement in the ORAM score or a wetland's category are attributable to undersetting or oversetting the scoring boundary of the wetland being assessed (Mack, personal observation). A thorough discussion of scoring boundary rules is beyond the purview of this chapter and the reader is referred to the ORAM user's manual, Mack (2004a,b), and the PowerPoint presentation of Ohio EPA's annual ORAM training course for a detailed discussion.

Narrative Rating

Another area where ORAM borrowed heavily from the Western Washington Method was the incorporation of a narrative assessment that automatically placed a wetland in a regulatory category, *regardless of the score on the quantitative rating* (with limited exceptions specified in the score categorization worksheet). The main advantage of the narrative rating is that it provides an additional level of regulatory protection against underassessment for preidentified high-value wetland types.

The narrative rating incorporates categorization language found in the Ohio wetland rules that identified certain types of wetlands (e.g., bogs, fens, old growth forests) or wetland characteristics (e.g., presence of threatened or endangered species) as always being associated with Category 3 wetlands. Other scarce wetland types were identified during the ORAM development process (e.g., remnant wet prairies, hydrologically unmodified Lake Erie coastal marshes). Also, the Ohio wetland rules identified at least one type of automatic Category 1 wetland (acidic, unvegetated, hydrologically isolated ponds on abandoned mine lands).

Except for the automatic Category 1, if the wetland satisfied the narrative criteria specified, two answers were possible: the wetland is categorized as a Category 3 wetland or the rater should evaluate the wetland for possible Category 3 status. An example of the former is narrative rating question 2:

Is the wetland known to contain an individual of or documented occurrences of federal or state-listed threatened or endangered plant or animal species?

If the answer to this question was yes, the wetland was automatically categorized as a Category 3 wetland. An example of the latter is narrative rating question 8B:

Is the wetland a forested wetland with 50% or of upper canopy consisting of deciduous trees with large diameters at breast height (dbh), generally diameters >45 cm (17.7 in) dbh?

If the answer to this question was yes, the rater was alerted to evaluate whether this was a Category 3 wetland, especially during the quantitative rating.

Quantitative Rating

Generally, when people think of ORAM, it is the quantitative rating that they are thinking of. It is the core of the method from a regulatory perspective and the key piece for the Ohio wetland monitoring and assessment program development, as the score was used for the disturbance gradient for wetland index of biotic integrity (IBI) development. Again, a thorough discussion of each metric is beyond the scope of this chapter, but the rationale and highlights for each metric will be briefly discussed.

One overall practical decision for Version 5.0 was to use a simple 100-point scale as the most familiar, intuitive, and transparent: familiar and intuitive because of our shared experiences in grade school and high school; transparent because the weighting of various scoring factors in the method was immediately apparent and not hidden in equations or standardizing calculations.

Metric 1: size. In terms of thinking about functions such as flood detention or habitat for large flocks of migrating waterfowl, size does matter, and all versions of ORAM and the Western Washington Method allocated points based on wetland size. However, ORAM Version 5.0 deemphasized size in the overall point allocation to only 6% of the total score from 10% to 20% in early versions to reduce the bias against naturally small wetland types such as forested vernal pools where bigger was clearly not better in terms of things like ambystomatid salamander habitat. In Ohio, where the state's population of wetlands is strongly skewed toward the size classes of <10 acres, this allocation has worked well with big, high-value wetland complexes in the Upper Cuyahoga River watershed, the ancient Teays River Valley in southeast Ohio, and the Lake Erie coastal marshes being awarded points here. The scoring boundary rules also tend to reduce the size of contiguous complexes, as there are usually multiple hydrogeomorphic classes in a big wetland complex.

Metric 2: upland buffers and surrounding land use. The importance of local upland buffers around most wetland types as well as the negative correlation with intensifying land uses at the 100–1000 m scale is well noted in the literature and demonstrated in multiple studies by Ohio EPA (e.g., Fennessy et al., 2007; Mack, 2006; Mack and Micacchion, 2007). In Metric 2a, scores for buffer width ranged from very narrow (<10 m, 0 pts.) to wide (≥50 m, 7 pts.). ORAM considers land use outside the buffers to be "surrounding land use" and, in Metric 2b, awards points from 7 (very low intensity) to 1 point (high intensity). Up to 14% of the total points could be awarded for having wide buffers and very low surrounding land uses.

Metric 3: hydrology. It is a truism that hydrology is the single most important variable for a wetland, yet it is relatively infrequently monitored using short interval data collection and was not emphasized in early versions of ORAM. Reflecting this, up to 30% of the total points in the quantitative rating are awarded in Metric 3.

This metric also has the first explicit "disturbance" metric in ORAM (Metric 3e), which, along with the substrate (Metric 4a) and habitat alteration (Metric 4c) disturbance metrics, forms the core of ORAM Version 5.0 condition assessment questions.[7] As discussed earlier, the rater is asked to note all possible disturbances to the wetland's hydrology, determine if these alterations were more than trivial, and then assign a score based on where the wetland is on a continuum of hydrologic intactness, from no or no apparent disturbances, recovered from a past disturbance, in the process of recovery, or not recovered. Considerable additional explanatory information is provided in the ORAM long form and user's manual text and approaching this question correctly is a focus of the annual training course. The structure and language of Metric 3e (and 4a and 4c) was borrowed directly from the Ohio stream habitat method (Rankin, 1989).

The remaining Metric 3 questions (3a, source of water, 3b, water depth, 3c, connectivity, 3d, duration of inundation/saturation) have their provenance in more traditional functions and values, the "kidney-on-the-landscape" approach to assessment. Wetlands with deeper water, that are more connected to floodplains or other surface waters, or that have more permanent hydroperiods are able to gain points, while more isolated, precipitation-driven, seasonal wetlands lose points (note that the reverse happens in Metric 6d).[8]

Metric 4: habitat alteration and development. Metric 4a and 4c follow the logic of Metric 3e, except now focusing on alterations to wetland soils and substrates (e.g., plowing, filling, trampling, burying) and wetland habitat (e.g., grazing, mowing, spraying, cutting, clearing). The same logic and continuum of intactness is used as in Metric 3e. Metric 4b evaluates the overall habitat development of the wetland, from poor (1 pt.) to excellent (7 pts.). This question intentionally overlaps with other Metric 4 and 6 questions and is part of the redundancy built into the method. Up to 20% of the quantitative rating points can be awarded in Metric 4.

Metric 5: special wetlands. This metric was included purely to support ORAM as a regulatory categorization tool. It awards up to 10 additional points to wetlands identified in the narrative rating to ensure they score well on the quantitative rating. Except for mature forested wetlands (5 pts.), points for Metric 5 are awarded relatively infrequently, which is not surprising given that these are generally scarce wetland types. In terms of using the ORAM score for condition assessment, in practice, inclusion or exclusion of the Metric 5 has little effect (Mack, unpublished data).

Metric 6: plant communities, interspersion, microtopography. Metric 6 has four submetrics primarily focusing on vegetative structural features and plant community species diversity. Metric 6a notes the presence of vertical (think of the layers of a forested wetland from submersed and floating aquatic species to the forest canopy trees) plant communities and ranks them in terms of community quality. This is a metric where small, seasonal, high-quality forested vernal pools typically gain points vis-a-vis other wetland types that score well on Metric 3a to 3d. Metric 6b is a direct descendant of a metric in the Western Washington Method focusing on the degree of horizontal habitat interspersion of large hemi-marsh complexes. Metric 6c deducts points for high levels of areal cover of a short list of highly invasive plants in Ohio. Finally, Metric 6d scores the presence and quality of small-scale microtopographic features often found in small forested vernal pools (e.g., hummocks, wood debris, amphibian breeding pools) and some emergent wetlands such as fens (e.g., tussocks and hummocks) or beaver ponds (standing dead, coarse woody debris).

Up to 20% of the total points can be awarded in Metric 6. This metric provides some overlap and redundancy with other metrics (e.g., 4b habitat development) and includes features that focus on characteristics typical of smaller (<3 acre) seasonally inundated (forested pools) to saturated (fens, wet meadows) systems.

VALIDATION/CALIBRATION EFFORTS UNDERTAKEN WITH THE METHOD

Validation and calibration occurred in stages as reference wetland datasets from the four main Ohio ecoregions and all wetland types (dominant plant communities such as marsh, forest, wet meadow, etc.) and classes (hydrogeomorpic classes such as depression, riverine, slope, etc.) were acquired and analyzed over the course of a decade (1996–2007) of wetland program development work. The approach developed is illustrated by Fig. 4.3.8.1, taken from the ORAM Score Calibration Report (Mack, 2000) where the Level 2 ORAM score was compared to the Level 3 1996–99 dataset represented by the Ohio Vegetation Index of Biotic Integrity (Mack, 2007). Numeric scoring breaks for the ORAM score were interpolated and then

7. Metric 2 (buffer width, surrounding land use), Metric 4b (habitat development), Metric 6a (vegetation quality), and 6c (invasive plant cover) are also strongly associated with overall wetland condition.

8. A comparison of the points gained or lost based on wetland type in Metric 3 and Metric 6 is a good example of intentional weighting so that small forested vernal pools can compete against big, "kidney-on-the-landscape" wetlands.

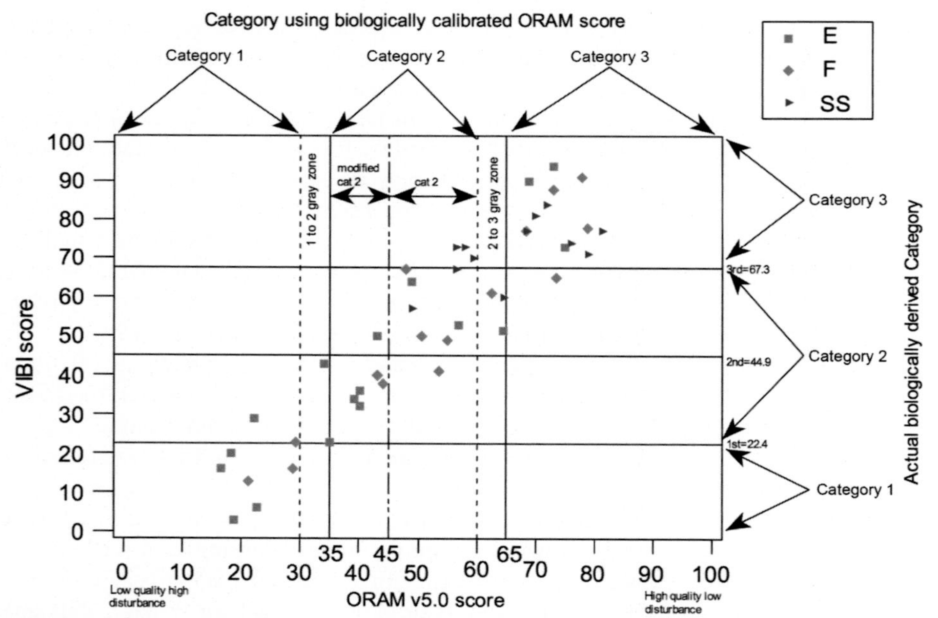

FIG. 4.3.8.1 Initial ORAM score calibration from Mack (2000) comparing ORAM score to Vegetation Index of Biotic Integrity Score and derivation of ORAM score categories.

equated to an aquatic life use category. As new datasets were acquired and analyzed in 2000, 2001, 2004, and 2006 (Mack et al., 2000; Mack, 2001, 2004b, 2007; Mack and Micacchion, 2006), this approach was repeated and reevaluated to determine if, (1) the ORAM score categories needed to be adjusted, or (2) whether separate score categories based on wetland type/class or ecoregion needed to be established. In terms of number one, the conclusion reached was that the score categories continued to work and they have remained in place since originally adopted.

For question number two, the evaluations discussed above provided strong support that ORAM was, at a minimum, relatively insensitive to differences in wetland type or class; however, it was two regional surveys (Fennessy et al., 2007; Durkalec et al., 2009) that unequivocally addressed this question. As discussed above, a fundamental choice in assessment tool development is whether to develop single or multiple tools based on wetland type or class. ORAM specifically took a single score sheet approach with calibrated point totals such that all wetland types can score well or poorly on the same set of questions. The large watershed scale survey undertaken in Fennessy et al. (2007) provided the most comprehensive evaluation of this approach and the design of ORAM Version 5.0. Fennessy et al. (2007) observed no significant (or in most instances even observable) differences in average ORAM scores by condition category for HGM class or plant community (Tables 4.3.8.2a and 4.3.8.2b). In fact, average scores for Category 3 hydrogeomorphic classes (depression, riverine mainstem, riverine headwater, slope, impoundment), that is, reference standard, least-impacted

TABLE 4.3.8.2A Average (Standard Deviation) ORAM Scores by Five HGM Classes for Four Wetland Condition Categories

	ANOVA Results	Depression	Riverine Headwater	Riverine Mainstem	Impoundment	Slope
All sites	*df*=230, *F*=6.11, *P*=.000	49.2(14.4)a	57.0(14.9)b	59.8(12.1)b	58.3(17.7)b	58.5(13.7)b
Category 1	*df*=21, *F*=1.51, *P*=.244	24.1(7.7)	28.3(4.2)	34.0(0.0)	31.0(1.7)	25.0(4.2)
Modified Cat 2	*df*=29, *F*=2.30, *P*=.087	39.8(2.4)	42.1(2.5)	38.5(0.7)	43.0(0.0)	42.3(1.5)
Cat 2	*df*=116, *F*=1.03, *P*=.393	53.5(5.2)	54.8(6.3)	54.9(5.5)	57.1(3.7)	55.9(5.2)
Cat 3	*df*=61, *F*=0.42, *P*=.797	71.0(5.2)	73.1(4.5)	73.3(4.8)	73.0(10.2)	73.6(4.6)

Means without shared letters significantly different after Tukey's multiple comparison test ($P<.05$).
From Fennessy, M.S., Mack, J.J., Deimeke, E., Sullivan, M.T., Bishop, J., Cohen, M., Micacchion, M., Knapp, M., 2007. Assessment of Wetlands in the Cuyahoga River Watershed of Northeast Ohio. Ohio EPA Technical Report WET/2007-4. Ohio Environmental Protection Agency, Division of Surface Water, Wetland Ecology Group, Columbus, OH.

TABLE 4.3.8.2B Average (Standard Deviation) ORAM Scores by Five Plant Community Classes for Four Wetland Condition Categories

	ANOVA Results	Forest Seep	Marsh	Shrub Swamp	Swamp Forest	Wet Meadow
All sites	$df=236$, $F=1.63$, $P=.167$	61.7(14.7)	54.5(16.1)	51.6(14.5)	55.7(11.9)	54.1(14.6)
Category 1	$df=20$, $F=1.59$, $P=.230$	–	26.0(8.1)	23.7(5.8)	33.8(0.8)	28.0(3.0)
Modified Cat 2	$df=31$, $F=1.55$, $P=.225$	–	39.8(2.1)	41.5(2.3)	41.1(2.4)	38.7(3.8)
Cat 2	$df=120$, $F=0.68$, $P=.605$	55.1(6.1)	54.0(5.9)	55.0(5.4)	55.4(5.0)	53.0(4.1)
Cat 3	$df=61$, $F=0.73$, $P=.572$	74.1(5.3)	73.3(6.3)	70.1(4.4)	72.7(4.9)	71.4(4.1)

From Fennessy, M.S., Mack, J.J., Deimeke, E., Sullivan, M.T., Bishop, J., Cohen, M., Micacchion, M., Knapp, M., 2007. Assessment of Wetlands in the Cuyahoga River Watershed of Northeast Ohio. Ohio EPA Technical Report WET/2007-4. Ohio Environmental Protection Agency, Division of Surface Water, Wetland Ecology Group, Columbus, OH.

systems, were basically equal (71.0, 73.3, 73.1, 73.8, and 73.0, respectively) (Table 4.3.8.2a). So, attainable expectations for reference standard wetlands were nearly equal across major HGM classes and similar results were obtained when we compared major plant communities (Table 4.3.8.2b). When all sites were evaluated, we did observe that depressional wetlands had lower ORAM scores than other HGM classes, but this was due to differential patterns of disturbance and not to any inherent bias against depressions by ORAM (Table 4.3.8.2a). In Durkalec et al. (2009), this result was confirmed in a census of all wetlands within the Cleveland Metroparks' 22,000 acre landholdings in five counties of northeast Ohio. Again, no significant, or in most instances even observable, differences were observed when average ORAM scores by condition category for HGM class or plant community were compared and average scores for HGM class and plant community types were nearly identical to those obtained in the Cuyahoga Watershed survey.

TIME SPENT IN DEVELOPING THE METHOD

The primary development process occurred from 1996 to 1999 and the user's manual and training program were developed between June 1999 and February 2001, with periodic updates and revisions to the training program for several years thereafter. Training was recommended although not required and no formal certification program was established in 2001 when ORAM was formally adopted.[9] The Ohio EPA typically scheduled several two-day training classes each year at no cost to participants from 2001 to 2014, with well over 1000 trainees during that time (Mack, personal communication).

TIME SPENT TO APPLY THE METHOD IN THE FIELD

An ORAM assessment, like other methods, would begin with a geographic information systems analysis producing maps showing recent aerial photographs, soil types, hydrography layers, wetland inventory maps, etc., as well as a search of rare species and heritage program databases for previously documented special status species or high-quality plant communities in or near the wetland. Some narrative rating questions must be answered based on this office review. In addition, scoring Metric 1 (size), Metric 2 (buffer width and surrounding land use), Metric 3c (connectivity), and Metric 6c (interspersion) as well as all the disturbance metrics (3e, 4a, 4b) can be aided by a GIS analysis of current and historical GIS information.

Most of the narrative rating questions and all of the quantitative rating questions require a site visit and a fairly thorough walk around and through the wetland. For small wetlands, an experienced user can usually learn what they need to know in 15–30 min, since the delineation boundary is usually also the scoring boundary; for larger wetlands, considerably more time—up to several hours—is usually needed to explore the site, determine the hydrogeomorphic classes present in the complex, set scoring boundaries, and evaluate the potential disturbances and community features. However, once this information is collected, the rating sheets take only a few minutes to fill out for an experienced user.

9. In 2015, Ohio HB64 was signed into law, providing the Ohio EPA rulemaking authority to develop a "water quality certified professional program," which would include certification for categorization of wetlands and presumably mandatory ORAM training, among other things.

HOW WAS/IS THE DATA BEING USED

As discussed above, ORAM has been effectively used in three contexts of an integrated, statewide wetland program:

1. Section 401/404 permit programs.
2. As a disturbance gradient (*x*-axis) for the development of wetland IBIs and numeric wetland biocriteria.
3. Watershed, regional, and statewide ambient wetland condition surveys.

One area where ORAM has not been used is in the context of assessing the quality of mitigation wetlands, although though there have been numerous requests and often substantial pressure to adapt it to this class of wetlands. There are several reasons for this. First, given the quantitative ecological and hydrological data usually required as part of postconstruction mitigation monitoring (data equivalent to Level 3 wetland assessments), there is no need for a Level 2 assessment as this quantitative information is a much better and more accurate evaluation of mitigation wetland quality. Second, a policy decision was made to backload the more intensive data collection into the mitigation wetland process since it had been shown that an effective and protective categorization could be done for most natural wetlands using a rapid method. In effect, the rapid method allowed the upfront assessment required to obtain a permit to be streamlined in terms of the time needed to perform the assessment (and when the applicants were most concerned about slowing down their projects). Since mitigation wetlands had at least five and often 10 years of monitoring, there is no need to triage speed of the assessment procedure. Finally, it is frankly quite easy to game the scores of rapid methods, designed with conservative safeguards to fairly assess natural wetlands, so that poor or mediocre mitigation sites look better than they are.

LESSONS LEARNED

The approach taken in Ohio had several pragmatic assumptions and goals and arguably ORAM has attained them all:

1. *Lumping is better than splitting*. In effect, this was a decision to develop a method that would work for all the state's wetlands, if possible, rather than develop a separate method for each wetland type or class.
2. *Multifunctional*. Here the goal was to develop a method that could fulfill the three main needs of the state's wetland program:
   (a) rapid regulatory categorization tool.
   (b) a reliable predictive wetland disturbance gradient for wetland IBI development,
   (c) usable in the context of watershed, regional, or statewide ambient condition assessment.
3. *Robust to abuse*. As regulatory tool, the method needed to reliably place wetlands into the appropriate protection category even in the face of intentional (due to gaming the score) or unintentional (due to inexperience of the evaluator) misuse. Related to this is an important point that the real goal of the method was to have *low* interuser variability in placing wetlands into *three or four regulatory protection categories*. It was *not* a goal to have low interuser variability in the *actual score or metric subscores*. In fact, score variability of a large, relatively unregulated user community was expected, and redundancy in score allocation was intentionally built into the method. This is to be contrasted with the use of ORAM Version 5.0 as a disturbance gradient for IBI development where the user group is highly trained—and usually highly experienced—in the type and quality of wetlands in the state or region. In this context, low score variability is good and is expected.
4. *Easily categorizes best quality systems*. The core disturbance metrics (2a, 3a, 4a, 4c) receive the highest point allocations by there being a *lack* of observable impacts or disturbances; there is very little room to disagree about this between users. As this is by and large a safe and conservative assumption, that is, that intact systems are of high ecological quality with their associated functions operating at equivalent levels, this type of wetland is quickly placed into the high protection category.[10]
5. *Mistakes result in overcategorization*. This is a good approach in a regulatory setting where, if the method or the use of the method is going to fail, that such mistakes result in over categorization not undercategorization, and hence higher versus lower regulatory protection and mitigation.

That ORAM and its close variants and adaptations are being used across a significant portion of the eastern United States speaks to its effectiveness, ease of use, and future extensibility.

---

10. Similarly, extremely degraded wetlands are relatively easy to categorize. It is the fair to moderate quality systems with moderate to moderately high levels of past disturbance where most disagreement occurs in the use of ORAM and where the knowledge and experience of the user on the type and quality of wetlands typical of the state or region becomes critical.

REFERENCES

Durkalec, M., Weldon, C., Mack, J.J., Bishop, J., 2009. The State of Wetlands in Cleveland Metroparks: Implications for Urban Wetland Conservation and Restoration. Cleveland Metroparks Technical Report 2009/NR-07, Division of Natural Resources, Cleveland Metroparks, Fairview Park, Ohio.

Fennessy, M.S., Gray, M.A., Lopez, R.D., 1998a. An Ecological Assessment of Wetlands Using Reference Sites Volume 1: Final Report, Volume 2: Appendices. Final Report to U.S. Environmental Protection Agency. Wetlands Unit, Division of Surface Water. Grant CD995761-01.

Fennessy, M.S., Geho, R., Elfritz, B., Lopez, R., 1998b. Testing the Floristic Quality Assessment Index as an Indicator of Riparian Wetland Disturbance. Final Report to U.S. Environmental Protection Agency. Wetlands Unit, Division of Surface Water. Grant CD995927.

Fennessy, M.S., Mack, J.J., Deimeke, E., Sullivan, M.T., Bishop, J., Cohen, M., Micacchion, M., Knapp, M., 2007. Assessment of Wetlands in the Cuyahoga River Watershed of Northeast Ohio. Ohio EPA Technical Report WET/2007-4, Ohio Environmental Protection Agency, Division of Surface Water, Wetland Ecology Group, Columbus, OH.

Karr, J.R., Chu, E.W., 1999. Restoring Life in Running Waters, Better Biological Monitoring. Island Press, Washington, DC. 206 p.

Kentucky, D.E.P., 2016. Guidance Manual for Kentucky Rapid Assessment Method (KY-RAM) v. 3.0. Department for Environmental Protection, Division of Water, Frankfort, KY.

Kutcher, T.E., 2010. Rhode Island Rapid Assessment Method v. 2.10 User's Guide. Rhode Island Department of Environmental Management, Office of Water Resources, Providence, RI.

Mack, J.J., 2000. ORAM v. 5.0 Quantitative Score Calibration. Ohio Environmental Protection Agency, Division of Surface Water, 401/Wetland Ecology Unit, Columbus, OH.

Mack, J.J., 2001. Ohio Rapid Assessment Method for Wetlands v. 5.0, User's Manual and Scoring Forms. Ohio EPA Technical Report WET/2001-1, Ohio Environmental Protection Agency, Division of Surface Water, 401/Wetland Ecology Unit, Columbus, OH.

Mack, J.J., 2004a. Integrated Wetland Assessment Program. Part 2: An Ordination and Classification of Wetlands in the Till and Lake Plains and Allegheny Plateau Regions. Ohio EPA Technical Report WET/2004-2, Ohio Environmental Protection Agency, Wetland Ecology Group, Division of Surface Water.

Mack, J.J., 2004b. Integrated Wetland Assessment Program. Part 4: Vegetation Index of Biotic Integrity (VIBI) and Tiered Aquatic Life Uses (TALUs) for Ohio Wetlands. Ohio EPA Technical Report WET/2004-4. Ohio Environmental Protection Agency, Wetland Ecology Group, Division of Surface Water, Columbus, OH.

Mack, J.J., 2006. Landscape as a predictor of wetland condition: an evaluation of the landscape development index (LDI) with a large reference wetland dataset from Ohio. Environ. Monit. Assess. 120, 221–241.

Mack, J.J., 2007. Developing a wetland IBI with statewide application after multiple testing iterations. Ecol. Indic. 7, 864–881.

Mack, J.J., Micacchion, M., 2007. An Ecological and Functional Assessment of Urban Wetlands in Central Ohio. Volume 1: Condition of Urban Wetlands Using Rapid (Level 2) and Intensive (Level 3) Assessment Methods. Ohio EPA Technical Report WET/2007-3A, Ohio Environmental Protection Agency, Wetland Ecology Group, Division of Surface Water, Columbus, OH.

Mack, J.J., Micacchion, M., 2006. Addendum to: Integrated Wetland Assessment Program. Part 4: Vegetation Index of Biotic Integrity for Ohio Wetlands and Part 7: Amphibian Index of Biotic Integrity for Ohio Wetlands. Ohio Environmental Protection Agency, Wetland Ecology Group, Division of Surface Water, Columbus, OH.

Mack, J.J., Micacchion, M., Augusta, L., Sablak, G., 2000. Vegetation Indices of Biotic Integrity for Wetlands and Calibration of the Ohio Rapid Assessment Method for Wetlands v. 5.0. Volume 1: Final Report to U.S. EPA Grant No. CD985276, Interim Report to U.S. EPA Grant No. CD985875, Division of Surface Water, Ohio Environmental Protection Agency, Columbus, OH.

Michigan DNRE, 2010. Michigan Rapid Assessment Method for Wetlands (MiRAM), Version 2.1. Department of Natural Resources and Environment, Lansing, MI.

Ohio Environmental Protection Agency, 1988a. Biological Criteria for the Protection of Aquatic Life: Volume I. The Role of Biological Data in Water Quality Assessment. Ecological Assessment Section, Division of Water Quality Planning and Assessment, Columbus, OH.

Ohio Environmental Protection Agency, 1988b. Biological Criteria for the Protection of Aquatic Life: Volume II. Users Manual for Biological Field Assessment of Ohio Surface Waters. Ecological Assessment Section, Division of Water Quality Planning and Assessment, Columbus, OH.

Ohio Environmental Protection Agency, 1989a. September 30, 1989 Addendum to Biological Criteria for the Protection of Aquatic Life: Volume II, 1988. Ecological Assessment Section, Division of Water Quality Planning and Assessment, Columbus, OH.

Ohio Environmental Protection Agency, 1989b. Biological Criteria for the Protection of Aquatic Life: Volume III. Standardized Biological Field Sampling and Laboratory Methods for Assessing Fish and Macroinvertebrate Communities. Ecological Assessment Section, Division of Water Quality Planning and Assessment, Columbus, OH.

Rankin, E.T., 1989. The Qualitative Habitat Evaluation Index (QHEI): Rationale, Methods, and Application. Ecological Assessment Section, Division of Surface Water, Ohio Environmental Protection Agency, Columbus, OH.

TVA, 2017. Tennessee Valley Authority Rapid Assessment Method for Wetlands. http://sewwrg.rti.org/InformationResources/tabid/60/Default.aspx. Accessed 7 August 2017.

U.S. EPA, 1990. Water Quality Standards for Wetlands, National Guidance. EPA 440-S-90-011, July 1990, U.S. EPA, Office of Water, Washington, DC.

U.S. EPA, 1997. Wetlands: Biological Assessment Methods and Criteria Development Workshop. Proceedings September 18–20, 1996, Boulder, Colorado. U.S. EPA, Office of Water, Washington, DC.

U.S. EPA, 2002a. Methods for Evaluating Wetland Condition: Introduction to Wetland Biological Assessment. EPA 822-R-02-014, March 2002, U.S. EPA, Office of Water, Washington, DC.

U.S. EPA, 2002b. Methods for Evaluating Wetland Condition: Using Vegetation to Assess Environmental Conditions in Wetlands. EPA 822-R-02-020, March 2002, U.S. EPA, Office of Water, Washington, DC.

U.S. EPA, 2009. Core Elements of an Effective State and Tribal Wetlands Program. https://www.ep.gov/sites/production/files/2015-10/documents/2009_03_10_wetlands_initiative_cef_full.pdf.

Washington, D.E., 1993. Washington State Wetlands Rating System, Western Washington, second ed. Washington Department of Ecology. Publication No. 93-74. August 1993.

FURTHER READING

Fennessy, M.S., Jacobs, A.D., Kentula, M.E., 2004. Review of Rapid Methods for Assessing Wetland Condition. U.S. Environmental Protection Agency, Washington, DC. EPA/620/R-04/009.

Mack, J.J., 2009. Development issues with extending plant-based IBIs to forested wetlands in the Midwestern United States. Wetl. Ecol. Manag. 17, 117–130.

Chapter 4.3.9

Ontario Wetland Evaluation System

Regina Varrin and Rebecca Zeran
Ontario Ministry of Natural Resources and Forestry, Peterborough, ON, Canada

Chapter Outline

INTRODUCTION

First published in 1983, the OWES is a science-based system that outlines a process and a set of criteria to define, identify, and assess the functions and values of wetlands and rank them relative to one another. The OWES was developed to support Ontario's land-use planning policy framework to help determine which wetlands should be protected from development. The first edition of the OWES focused on southern Ontario (south and east of the Ontario Shield) and was developed during a time when there was growing international focus on wetland conservation. The system was built around evaluations being undertaken in a field setting (e.g., on-the-ground wetland inventories) and was created well before digital remote sensing technology became established. The OWES underwent a significant revision in 1993 where more extensive criteria for evaluating hydrological functions were included. It was at this time that a second version of the OWES was developed to focus specially on northern parts of Ontario.

The OWES manuals provide two types of guidance: (a) rules and criteria for identification and delineation of wetland boundaries, and (b) criteria for the assessment and scoring of wetland functions and values. This chapter focuses on the method of evaluation (e.g., features and functions assessed) and not on the methods or rules related to wetland mapping and boundary delineation.

Under the OWES, the "provincial significance" of wetlands is assessed based on the value of the wetland in maintaining ecosystem processes and on the benefits that the wetland provides to society. The system was not designed to assess the condition or "health" of the wetland and it was not intended to be an impact assessment (e.g., it does not assess the vulnerability of a wetland to development).

CONCEPTUAL FRAMEWORK FOR THE METHOD

There are two OWES manuals: one for southern Ontario (OMNRF, 2014a) for wetlands in Ecoregions 6 and 7, and one used in northern Ontario (OMNRF, 2014b) for wetlands in Ecoregions 2, 3, 4, and 5 (Fig. 4.3.9.1). There are relatively minor differences between the manuals with the northern manual outlining a slightly different process for assessing hydrological function and including a few additional assessments for species that are present in the north and less common in the south (e.g., black ducks and moose).

The system recognizes both single contiguous wetland areas and wetland complexes where a number of distinct wetland areas, while not being contiguous, may be close enough together to share a functional or ecological linkage (Fig. 4.3.9.2). For example, wetlands may be linked together to form a wetland complex if they are connected by a stream, or if important

Wetland and Stream Rapid Assessments. https://doi.org/10.1016/B978-0-12-805091-0.00047-5

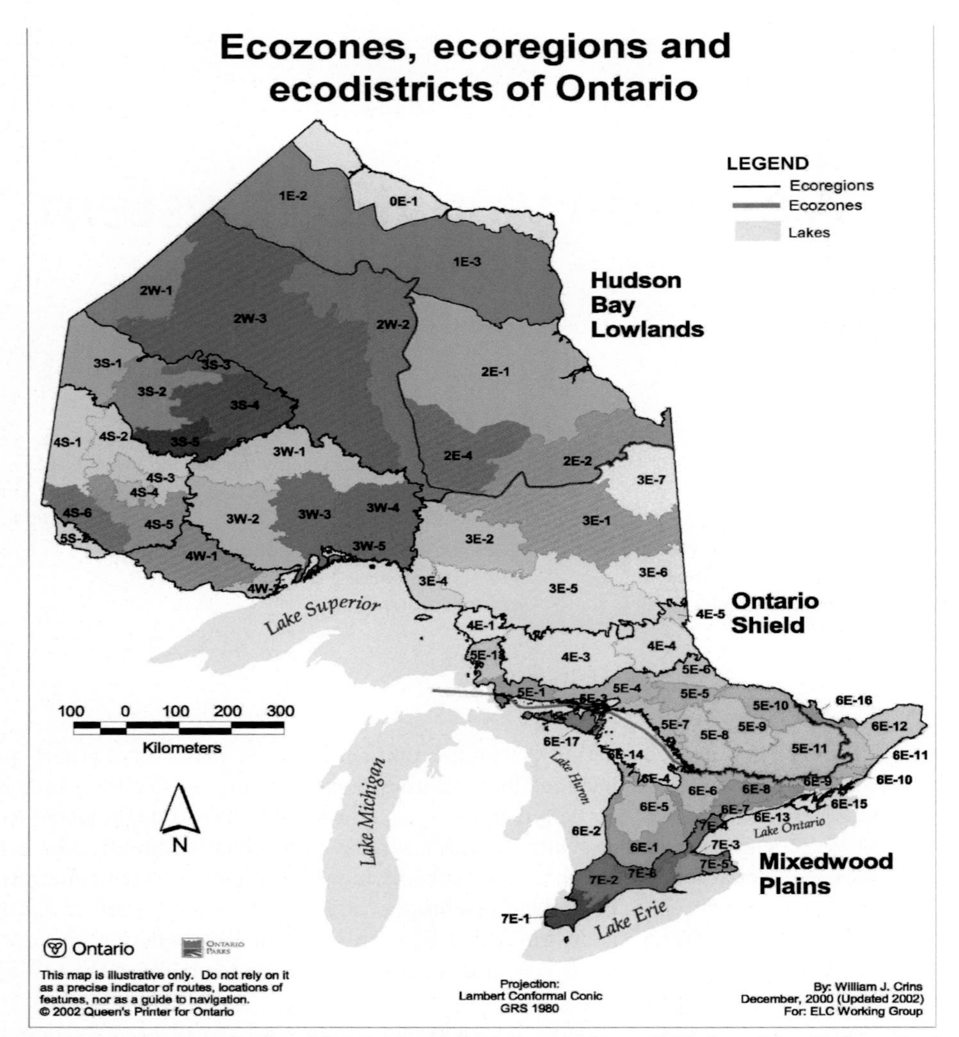

FIG. 4.3.9.1 Ecozones, ecoregions, and ecodistricts of Ontario. An ecodistrict layer is also available through Land Information Ontario. *(From Ontario Ministry of Natural Resources and Forestry (OMNRF), 2014. Ontario Wetland Evaluation System Northern Manual, first ed., Version 1.3. Queen's Printer for Ontario, Toronto, ON. https://www.ontario.ca/page/wetlands-evaluation.)*

species migration occurs between the two wetlands. Once all wetlands that make up a complex are identified, the assessment of wetland function and value using the criteria in the OWES occurs and considers all of the wetland areas in a complex together as if they were one single wetland.

DESCRIPTION OF THE METHOD

Wetland evaluations can be carried out by anyone who has taken an Ontario Ministry of Natural Resources and Forestry (OMNRF)-approved OWES training course. A wetland that has been evaluated using the OWES is known as an "evaluated wetland" and will have a "wetland evaluation file." Wetland evaluation files may be updated from time to time as new information becomes available. All completed evaluations (regardless of who undertakes the evaluation or why the evaluation was undertaken) must be submitted to OMNRF for review and approval.

There are three types of maps that must be prepared as part of a wetland evaluation: (1) the wetland boundary map, depicting all outer boundaries of the wetland and any features within or adjacent to the wetland, (2) the vegetation community map (Fig. 4.3.9.3), depicting all internal vegetation community boundaries, and (3) the catchment basin map, showing the boundary of the wetland's catchment and all other additional wetlands or water bodies within the catchment area. Wetland complexes are identified during the mapping stage of a wetland evaluation. These maps are used during scoring of the some of the wetland's features and functions.

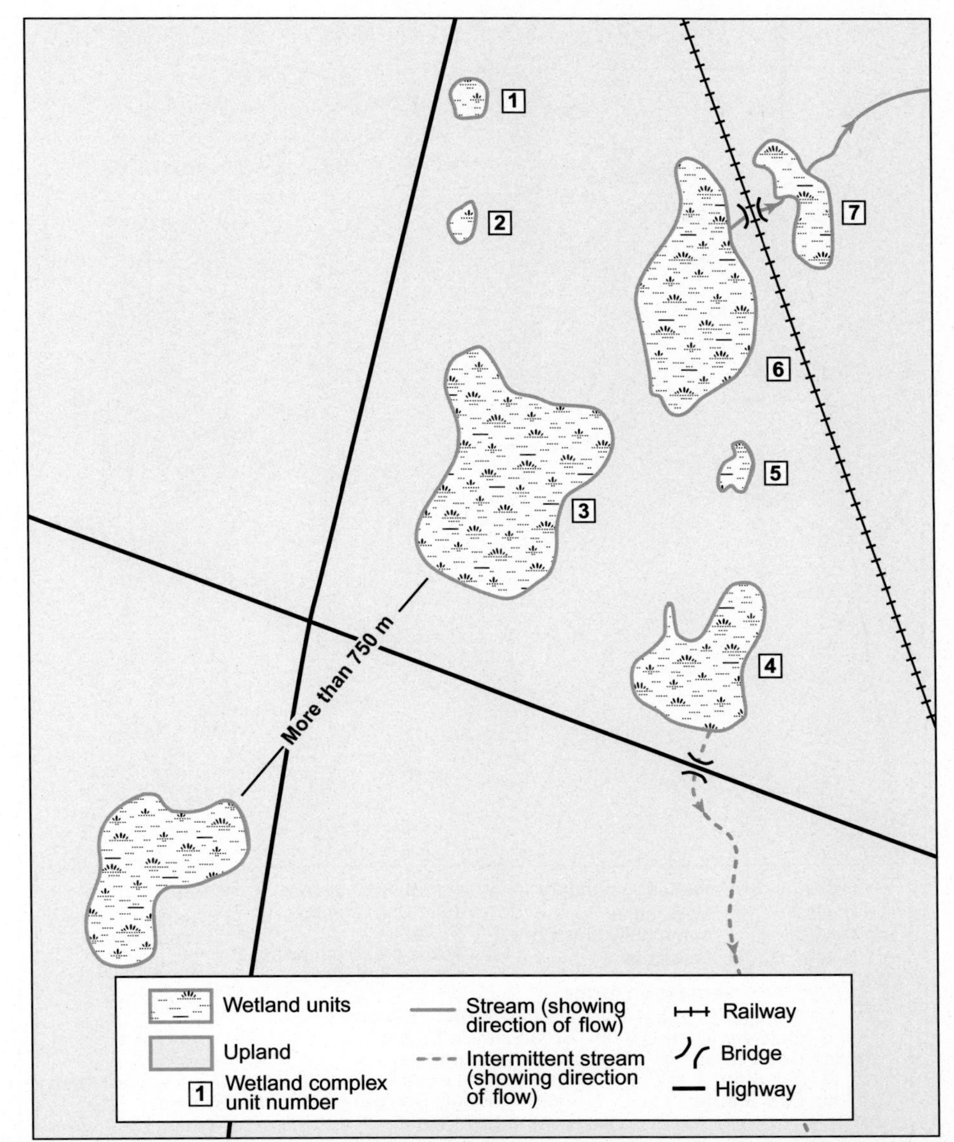

FIG. 4.3.9.2 An example of a wetland complex. Wetlands 1–7 are within 750m of one another and can be evaluated as a complex if they share a function or ecological linkage. The wetland in the bottom left corner of the figure is >750m from the nearest wetland and thus cannot be included in the wetland complex. *(From Ontario Ministry of Natural Resources and Forestry (OMNRF), 2014. Ontario Wetland Evaluation System Southern Manual, third ed., Version 3.3. Queen's Printer for Ontario, Toronto, ON. https://www.ontario.ca/page/wetlands-evaluation; Ontario Ministry of Natural Resources and Forestry (OMNRF), 2014. Ontario Wetland Evaluation System Northern Manual, first ed., Version 1.3. Queen's Printer for Ontario, Toronto, ON. https://www.ontario.ca/page/wetlands-evaluation.)*

The OWES was originally developed as a field-based evaluation system. Over time, technology and information sources have improved so that much of the evaluation can now be undertaken remotely. However, the current system still requires one or more field visits to take place to field-check boundaries and seek out features and functions that cannot be scored using existing information or detected remotely. While there are other approaches for identifying and delineating wetlands in Ontario (e.g., the Ecological Land Classification System), only the mapping rules described in OWES can be used to determine the boundaries of a provincially significant wetland.

Once the wetland (or wetland complex) has been mapped, wetland functions and values are evaluated against almost 50 scored criteria organized under four main categories or components:

- Biological: recognizes that wetlands can differ in terms of productivity and habitat diversity.
- Social: measures some of the direct human uses of wetlands, including economically valuable products (such as wild rice, commercial fish, and furbearers), recreational activities, educational uses, and cultural values.

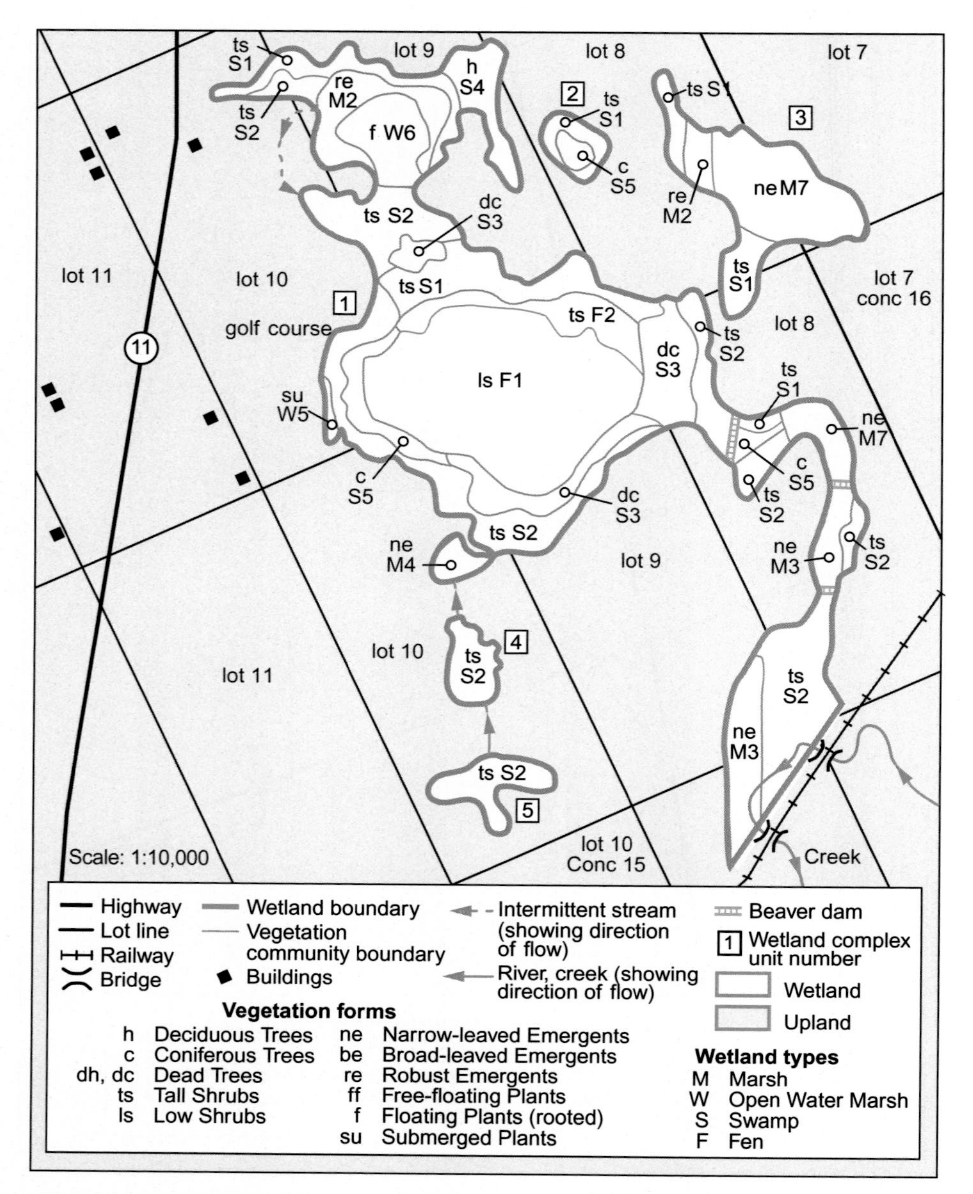

FIG. 4.3.9.3 A vegetation community map illustrating the OWES conventions for mapping wetland type and vegetation form. *(From Ontario Ministry of Natural Resources and Forestry (OMNRF), 2014. Ontario Wetland Evaluation System Southern Manual, third ed., Version 3.3. Queen's Printer for Ontario, Toronto, ON. https://www.ontario.ca/page/wetlands-evaluation; Ontario Ministry of Natural Resources and Forestry (OMNRF), 2014. Ontario Wetland Evaluation System Northern Manual, first ed., Version 1.3. Queen's Printer for Ontario, Toronto, ON. https://www.ontario.ca/page/wetlands-evaluation.).*

- Hydrological: characterizes water-related values of wetlands, such as the reduction of flood peaks, contributions to groundwater recharge and discharge, and improvements to water quality.
- Special features: addresses the geographic rarity of wetlands, the occurrence of species at risk, ecosystem age, and habitat quality for fish and wildlife.

Table 4.3.9.1 provides an overview of the functions and values that are assessed during a wetland evaluation. Because ecosystem processes and functions are interconnected and interdependent, the criteria present in the biological component recognize linkages between productivity, biodiversity, and a wetland's hydrological setting. The social component is intended to evaluate the short-term uses and amenities that wetlands can provide to people. The hydrological component is designed to determine the net hydrological benefit provided by the wetland to the portion of the basin downstream of the

TABLE 4.3.9.1 Functions and Values Assessed During a Wetland Evaluation

Function/Value Assessed	Description
Biological component	
Productivity: Growing degree days (GDD)/soils	Estimates how productive the wetland is based on temperature and substrate; higher temperatures and mineral soils result in higher productivity (max points = 30).
Productivity: Wetland type	Estimates productivity associated with each wetland type (marsh, swamp, bog, fen) present; marshes are considered most productive (max points = 15).
Productivity: Site type	Refers to where the wetland is on the landscape in relation to its water source; different sources of water supply different nutrients and lead to varying productivity; four main site types: isolated (no outflow), palustrine (permanent or intermittent outflow), lacustrine, and riverine (max points = 5).
Biodiversity: Number of wetland types	Uses the number of wetland types (Table 4.3.9.2) present as an indicator that the wetland likely supports greater diversity of species (max points = 30).
Biodiversity: Vegetation communities	Awards more points to wetlands with a higher variety of plant communities (max points = 45).
Biodiversity: Diversity of surrounding habitat	Score is based on the number of different habitats that surround the wetland (more different habitats = higher score) (max points = 7).
Biodiversity: Proximity to other wetlands	Measures connectivity (type and distance) of the wetland to other wetlands (wetlands that are connected via surface water to other wetlands score highest) (max points = 8).
Biodiversity: Interspersion	Measures the number and complexity of ecotones present (i.e., different vegetation communities, wetland types, wetland units), diversity increases with higher amount of ecotones (max points = 30).
Biodiversity: Open water types	Assesses the value of permanently flooded areas to habitat diversity (highest scores for wetlands with 25–75% open water arranged in pattern of small "ponds") (max points = 30).
Size	Assesses how the total size of the wetland relates to its biodiversity; large wetlands tend to score higher, unless they are monocultures (e.g., one species dominates), and thus they can score lower because they tend to contribute less to biodiversity (max points = 50).
Social component	
Economically valuable products	Scores based on: • The area of live trees present (max points = 18 points in south and 14 points in north). • (*North only*) the presence of low bush cranberry (max points = 2). • The presence of wild rice (max points = 6 points in south and 10 points in north). • The presence of baitfish (max points = 12). • The presence of furbearing mammals (max points = 12).
Recreational activities	Scores based on the intensity of use of the wetland (Table 4.3.9.3) for hunting, fishing, or nature enjoyment (max points = 80).
Landscape aesthetics: Distinctness	Evaluates whether the wetland is easily distinguished from the surrounding land (premise is that if the wetland is more recognizable it is of more social value) (max points = 3).
Landscape aesthetics: Absence of human disturbance	Assesses how natural the wetland is based on how much human disturbance may be present (premise is that a more natural wetland is of more value to people) (max points = 7).
Education and public awareness	Scores for 3 aspects: • Educational Uses: how frequently has the wetland been used for formal education purposes (max points = 20)? • Facilities and Programs: does the wetland have any interpretation facilities, nature trails, or public nature programs (max points = 8)? • Research and Studies: has the wetland been the subject of any research or study (max points = 12)?
Proximity to areas of human settlement	Measures how close the wetland is to human populations (the higher the population and the closer the wetland is to that population center, the higher the score), and provides a higher score to

Continued

TABLE 4.3.9.1 Functions and Values Assessed During a Wetland Evaluation—cont'd

Function/Value Assessed	Description
	wetlands closer to settled areas as they are more likely to be identified, viewed, and visited by more people (max points=40).
Ownership	Determines how much of the the wetland is in public or private ownership (wetlands held by a land trust or regulated as parks score highest) (max points=10).
Size	Score assessed based on total size of wetland compared to its value in providing economically valuable products, recreation, and proximity to human settlement (max points=20).
Aboriginal and cultural values	Awards a "bonus" score of 30 points if wetland is of aboriginal or cultural significance (max points=30).
Hydrological component	
Flood attenuation	Estimates the wetland's value in reducing flood peaks in areas downstream, isolated wetlands score full points, coastal wetlands directly adjacent to the Great Lake score no points (max points=100).
Water quality improvement	Estimates the ability of the wetland to improve water quality, based on the rate of flow and position of wetland in the watershed (max points=70 points in south and 90 points in north).
Carbon sink	Uses wetland type and presence of organic soils to estimate amount of peat present that might act as a sink for atmospheric carbon (max points=5 points south, 15 points north).
Shoreline erosion control	Assesses the function of a wetland located along a shoreline to protect against erosion (max points=15).
Groundwater recharge	Occurs when water level in the wetland is higher than that of the surrounding water table and water flows out of the wetland into the surrounding soil, thus recharging the ground; uses wetland site type (Table 4.3.9.4) and soil to indicate a groundwater recharge function (max points=60 points in south and 30 points in north).
Groundwater discharge	Occurs when the water level of the wetland is lower than the surrounding water table and water flows from surrounding lands into the wetland; evaluated by using presence/absence of a number of features that are known to suggest a discharge function (max points=30).
Special features component	
Species rarity	Considers whether the wetland provides habitat for rare species, scores awarded for: • Presence of reproductive habitat for an endangered or threatened species (250 points for each species). • Presence of traditional migration, feeding, or hibernation habitat for an endangered or threatened species (150 points for first species, 75 points each thereafter). • Use of wetland by a provincially significant animal species (species tracked by Natural Heritage Information Center (NHIC)) (first species scores 50 points, sliding scale thereafter). • Presence of a provincially significant plant species (species tracked by NHIC) (first species scores 50 points, sliding scale thereafter). • Presence of a regionally significant bird species (designation based on data from the Ontario Breeding Bird Atlas organized by ecodistrict) (first species scores 20 points, sliding scale thereafter). • Presence of a locally significant species (mostly plants, identified in locally rare species lists) (first species scores 10 points, sliding scale thereafter).
Significant features and habitats	Awards points if the wetland provides habitat for these groups of species and assesses the significance of that habitat: • Colonial waterbirds (nesting or feeding) (max points=50). • Winter cover for wildlife (max points=100). • Waterfowl staging/molting (max points=150). • Waterfowl breeding (max points=150). • Migratory passerine, shorebird, or raptor stopover area (max points=150). • Ungulate habitat (*northern manual only*) (max points=100). • Fish habitat (max points=125).

TABLE 4.3.9.1 Functions and Values Assessed During a Wetland Evaluation—cont'd

Function/Value Assessed	Description
Rarity of wetlands	Considers how rare a wetland is within an ecodistrict (south only) and how rare a wetland type is within an ecodistrict (both manuals) (max points=160 points south, 70 points north).
Black ducks (*northern manual only*)	Awards score based on location of wetland within regions of Black Duck indicated breeding pair density, highest scores given to wetlands in central parts of Ontario (max points=20).
Ecosystem age	Uses wetland type (Table 4.3.9.5) to estimates how long a wetland takes to develop (based on the principle that wetlands needing a long time for natural restoration/development are more valued) (max points=25).
Great lakes coastal wetlands	Awards score based on size of the coastal units of the wetland (max points=75).

wetland. The OWES relies on the assessment of some key hydrological functions using indicators that can be assessed by nonhydrologists. The special features component primarily evaluates biological and ecological features present within the wetland itself. Tables 4.3.9.2–4.3.9.5 provide more detailed examples of the calculations that are required to determine a score for some of the features and functions evaluated by the OWES.

While there are some scored criteria that have no maximum point limit, the total score for each of the four components is capped at 250 points each. The manuals provide direction on how to calculate a score for each function and value evaluated, and how to sum and cap scores for each component. The maximum number of points that any wetland can receive is 1000. Wetlands that score 600 or more points in total and wetlands that score 200 points or more in either the biological or the special features component are considered to be "provincially significant."

TABLE 4.3.9.2 An Example of One of the Calculations for Scoring Part of the Biological Component

Evaluation of Biodiversity Using Number of Wetland Types	Score One Only
One	9 points
Two	13 points
Three	20 points
Four	30 points
Score	(Maximum 30 points)

The OWES recognizes four wetland types (i.e., marsh, swamp, bog, and fen) and includes criteria for classifying them.

TABLE 4.3.9.3 An Example of One of the Calculations for Scoring Part of the Social Component

Evaluation of Recreational Activities	Score One Level for Each of the Three Wetland Uses; Scores are Cumulative		
Intensity of Use	Hunting	Nature Appreciation/Ecosystem Study	Fishing
High	40 points	40 points	40 points
Moderate	20 points	20 points	20 points
Low	8 points	8 points	8 points
Not possible/ No evidence	0 points	0 points	0 points
Score	(Maximum 80 points)		

The manuals provide specific criteria for intensity of use, such as number of hunter days, etc.

TABLE 4.3.9.4 An Example of One of the Calculations for Scoring Part of the Hydrological Component

Evaluation of Groundwater Recharge Using Site Type (Southern Manual)			
(a)	Wetland >50% lacustrine (by area) or located on one of the major rivers; or		0 points
(b)	Wetland not as above. Calculate score as follows: (FA=fractional area, site type/total area of wetland).		
FA	of isolated or palustrine wetland	× 50	
FA	of riverine wetland	× 20	
FA	of lacustrine wetland (not dominant site type)	× 0	
Score			(Maximum 50 points)

The manuals provide direction on how to determine site type and what to consider a major river for the purposes of wetland evaluation. The calculation below is for evaluating groundwater recharge using the southern manual. Where the northern manual applies, the calculation is almost the same, but the maximum score is 20 points and different factors are used for multiplying each site type's fractional area.

TABLE 4.3.9.5 An Example of One of the Calculations for Scoring Part of the Special Features Component

Evaluation of Ecosystem Age (For Each Wetland Type Present Within the Wetland, Multiply Each Fractional Area by Factor Indicated and Add the Results Together)

	Fractional Area (Area of Site Type/Total Area of Wetland)		Score
Bog	FA	× 25 =	
Fen, on deeper soils, floating mats or marl	FA	× 20 =	
Fen, on limestone rock	FA	× 5 =	
Swamp	FA	× 3 =	
Marsh	FA	× 0 =	
Score			(Maximum 25 points)

When the OWES was first created, the 1984 Guidelines for Wetlands Management in Ontario established seven classes of wetland based on the score received during evaluation (OMNR, 1984). Class 1 and 2 were considered to be "provincially significant" while the other classes of wetlands came to be considered "regionally" or "locally" significant. When the OWES manual was revised in 1994, use of wetland classes was retired.

Under the current OWES, an evaluated wetland is either determined to be provincially significant, often called a "provincially significant wetland," or it is not. The OWES no longer classifies wetlands as being "regionally" or "locally" significant. However, to recognize that all wetlands have value, many organizations and planning authorities consider evaluated wetlands not meeting the criteria for provincial significance to be "locally significant."

VALIDATION/CALIBRATION EFFORTS UNDERTAKEN WITH THE METHOD

A statistical analysis of interobserver variability was conducted during field testing of the draft system, and a second analysis was conducted the year after the system was released (OMNR and EC, 1984). Although minor revisions were made to the manuals between the first and second edition, no changes were made to scoring as a result of the statistical analysis (OMNR and EC, 1984).

TIME SPENT IN DEVELOPING/TESTING THE METHOD

Development of the OWES took place from 1980 to 1983 and under the guidance of an interagency steering committee (OMNR and EC, 1983). The process began with a literature review, jurisdictional scan, and assessment of needs of planners and resource managers, followed by testing and comparison of three alternate evaluation systems at 23 well-known wetlands in southern Ontario (Ecologistics, 1981). The resulting preferred system was then tested at 45 sites followed by further improvements in 1981–1982 (OMNR and EC, 1983). A draft was published in 1982 and tested at 110 sites, reviewed by experts, and a statistical analysis was performed on replicated wetland evaluations (OMNR and EC, 1983). The resulting first edition was published in 1983 (OMNR and EC, 1983).

SAMPLE APPLICATION OF METHOD IN THE FIELD

Mapping, names, and status ("provincially significant" or "other") of evaluated wetlands is available to the public through the web-based mapping tool—Make a Map: Natural Heritage Areas (OMNRF, 2017d) available online at http://www.gisapplication.lrc.gov.on.ca/mamnh/Index.html?site=MNR_NHLUPS_NaturalHeritage&viewer=NaturalHeritage&locale=en-US. Mapping and attribute information are also available through Land Information Ontario. Visit Land Information Ontario's Metadata Management Tool (OMNRF, 2017c) at https://www.javacoeapp.lrc.gov.on.ca/geonetwork/srv/en/main.home and search for the "wetland" data class.

TIME SPENT TO APPLY THE METHOD IN THE FIELD

The amount of time it takes for an evaluator to map the wetland and collect the information required for an evaluation depends on the size, complexity, and accessibility of the wetland.

HOW WAS/IS THE DATA BEING USED

Most wetlands are evaluated to inform land-use decisions. For example, the natural heritage policies of the Provincial Policy Statement (OMMAH, 2014), issued under Ontario's Planning Act, protect provincially significant wetlands and coastal wetlands from development and site alteration, depending on their location within the province. Through the Provincial Policy Statement (PPS), development (e.g., the creation of new lots, a change in land use, or the construction of buildings and structures that require an approval under the Planning Act), and site alteration (e.g., grading, excavation, and the placements of fill that change the landform and natural vegetative characteristics of a site) are prohibited within provincially significant wetlands throughout southern and much of central Ontario and provincially significant Great Lakes Coastal Wetlands anywhere in the Great Lakes Basin. The PPS also provides protection for provincially significant wetlands in northern Ontario, unless no negative impacts can be demonstrated. The Natural Heritage Reference Manual provides guidance to planning authorities in implementing the PPS, including identifying negative impacts (e.g., loss of productivity, loss of habitat, etc.), listing sources of information (e.g., where to find wetland mapping of significant wetlands), and examples of mitigation approaches for avoiding negative impacts (e.g., maintaining vegetation buffers, developing and implementing erosion and sediment control plans, etc.) (OMNR, 2010).

The Conservation Land Tax Incentive Program is another way that the government of Ontario uses the results of wetland evaluations to encourage conservation (OMNRF, 2017a). This voluntary program provides tax relief for landowners with provincially significant wetlands on their property. In order to be eligible for the program, the wetlands must be mapped and evaluated using the OWES, and found to be provincially significant. Landowners who enroll in the program receive an exemption from property taxes for that portion of their land that is a provincially significant wetland. To participate in the program, landowners must agree to maintain their land in a manner that contributes to the natural heritage and biodiversity objectives for conserving the land.

WHAT WAS LEARNED

Approximately one-third of Ontario (>35 million hectares) is covered by wetlands. Although the majority of the province's wetlands are found in the Hudson Bay Lowlands, wetland evaluations are not typically conducted in most northern parts of the province. Evaluation efforts have been targeted to areas where historical loss and development pressure have been the greatest, the Mixedwood Plains, where the southern manual applies. More than half the remaining wetlands in the Mixedwood Plains (>515,000 ha) have been evaluated since the mid-1980s and of these, >460,000 ha are identified as provincially significant wetlands (OMNRF, 2017c). In the Ontario Shield, >110,000 ha have been identified as provincially significant since the northern manual was published in 1993 (OMNRF, 2017c).

PROSPECTS FOR THE FUTURE

Ontario's wetland evaluation system has now been in use for >30 years. It is a key pillar in the province's toolkit for wetland conservation. In the recently released *A Wetland Conservation Strategy for Ontario 2017–2030*, the province sets out a vision, targets, and suite of actions that provincial government will undertake by 2030 (OMNRF, 2017b). Improving the evaluation of significant wetlands by reviewing the OWES is identified as a priority action. The review will allow investigation of recent advances in wetland science and mapping technology, increasing clarity where current guidance is limited, and better aligning the system with the current land-use and resource management policies. The review will be informed by more than three decades of experience with wetland evaluation and the wealth of information that has been gathered on Ontario's wetlands over this period.

REFERENCES

Ecologistics, 1981. A Wetland Evaluation System for Southern Ontario. Prepared for the Canada/Ontario Steering Committee on Wetland Evaluation and the Canadian Wildlife Service, Environment Canada.

Ontario Ministry of Municipal Affairs and Housing (OMMAH), 2014. Provincial Policy Statement. Queen's Printer for Ontario, Toronto. http://www.mah.gov.on.ca/Page10679.aspx.

Ontario Ministry of Natural Resources (OMNR), 1984. Guidelines for Wetlands Management in Ontario.

Ontario Ministry of Natural Resources (OMNR), 2010. Natural Heritage Reference Manual for Natural Heritage Policies of the Provincial Policy Statement, 2005, second ed. Queen's Printer for Ontario, Toronto, ON.

Ontario Ministry of Natural Resources and Environment Canada (OMNR and EC), 1983. An Evaluation System for Wetlands of Ontario South of the Precambrian Shield First Edition.

Ontario Ministry of Natural Resources and Environment Canada (OMNR and EC), 1984. An Evaluation System for Wetlands of Ontario South of the Precambrian Shield Second Edition.

Ontario Ministry of Natural Resources and Forestry (OMNRF), 2014a. Ontario Wetland Evaluation System Southern Manual, third ed. Queen's Printer for Ontario, Toronto, ON. Version 3.3, https://www.ontario.ca/page/wetlands-evaluation.

Ontario Ministry of Natural Resources and Forestry (OMNRF), 2014b. Ontario Wetland Evaluation System Northern Manual, first ed. Queen's Printer for Ontario, Toronto, ON. Version 1.3, https://www.ontario.ca/page/wetlands-evaluation.

Ontario Ministry of Natural Resources and Forestry (OMNRF), 2017a. Conservation Land Tax Incentive Program. https://www.ontario.ca/page/conservation-land-tax-incentive-program.

Ontario Ministry of Natural Resources and Forestry (OMNRF), 2017b. A Wetland Conservation Strategy for Ontario 2017–2030. Queen's Printer for Ontario, Toronto, ON. https://www.ontario.ca/page/wetland-conservation-strategy.

Ontario Ministry of Natural Resources and Forestry (OMNRF), 2017c. Land Information Ontario: Wetland Data Class Metadata. https://www.javacoeapp.lrc.gov.on.ca/geonetwork/srv/en/main.home.

Ontario Ministry of Natural Resources and Forestry (OMNRF), 2017d. Make A Map: Natural Heritage Areas. http://www.gisapplication.lrc.gov.on.ca/mamnh/Index.html?site=MNR_NHLUPS_NaturalHeritage&viewer=NaturalHeritage&locale=en-US.

FURTHER READING

Environment Canada, 1984. A Statistical Analysis of "An Evaluation System for Wetlands of Ontario, First Edition (1983)".

Chapter 4.3.10

Case Study—Washington State Rapid Assessment Methods

Thomas Hruby
Washington State Department of Ecology, Lacey, WA, United States

Chapter Outline

INTRODUCTION

In Washington, wetlands are protected and regulated under the state's Clean Water Act, the Shoreline Management Act, the Growth Management Act, and a governor's executive order. In addition, Washington has been delegated authority over wetlands under the federal Clean Water Act. The Washington State Department of Ecology has been the lead agency in the state in developing the tools needed to assess wetlands under these laws. Rapid methods (Level 2, Fennessy et al., 2004) have been an essential part of this toolbox from the beginning. Since the first two methods were introduced in 1991, we have developed 10 Level 2 methods that include 160 calibrated or recalibrated rapid assessment models for individual functions. Table 4.3.10.1 lists the methods and the dates when they were published.

The first methods developed in 1991 focused on characterizing wetlands based on their functions and values. At first it was assumed that assessing wetland functions could also act as surrogates for "biological integrity" and "health." Unfortunately, this hypothesis was tested and found invalid during extensive field calibrations in the late 1990s (Hruby, 2001). The concepts of integrity and health are more difficult to define and assess because they depend on the viewpoint of the individual defining the terms. It was much easier to arrive at a consensus among regulators, potential users, and wetland scientists on the definitions of functions as specific environmental processes than on definitions for biological integrity or ecological health. Thus, subsequent modeling efforts addressed only wetland functions.

A subset of the methods developed, the Wetland Rating Systems, included sections on the societal values a wetland may provide. The rating systems were, and are, being used to meet statutory requirements that specifically call for the preservation of values as well as functions. The scoring and scaling of values were based on consultations with state agencies and the public; efforts were also made to meet the requirements in laws and regulations. Values were not calibrated to data collected in the field. For example, the presence of a threatened or endangered species in a wetland meant that it was categorized as a Category I wetland (the most protective category), regardless of its other functions.

HISTORY OF RAPID ASSESSMENT METHODS IN WASHINGTON

The first rapid assessment methods developed in 1991 (Washington State Wetland Rating Systems for eastern and western Washington) addressed the need to better protect wetlands. The method differentiated wetlands according to specific valuable characteristics and indicators of wetland functions. The method did not rate wetlands as high, medium, or low, but rather placed wetlands into one of four categories based on how much protection they might need. Category I wetlands required the highest level of protection and Category IV the lowest. The variables used to determine the category

Wetland and Stream Rapid Assessments. https://doi.org/10.1016/B978-0-12-805091-0.00048-7

TABLE 4.3.10.1 Rapid Tools Developed by Washington to Categorize and Rate Wetland Functions and Values

1991—Washington State Wetland Rating System—Western Washington (Ecology Publication #91-57) *Not Calibrated*
1991—Washington State Wetland Rating System—Eastern Washington (Ecology Publication #91-058) *Not Calibrated*
1993—Washington State Wetland Rating System—Western Washington (second edition) (Ecology Publication #93-074) *Not Calibrated*
1999—Methods for Assessing Wetland Functions: Riverine and Depressional Wetlands in Western Washington (Ecology Publication #99-115) *Contains 54 individually calibrated models for functions*
2000—Methods for Assessing Wetland Functions: Depressional Wetlands in the Columbia Basin of Eastern Washington (Ecology Publication #00-06-47) *Contains 34 individually calibrated models*
2004—Washington State Wetland Rating System for Eastern Washington—Revised (Ecology Publication #04-06-015) *Contains 12 individually calibrated models*
2004—Washington State Wetland Rating System for Western Washington—Revised (Ecology Publication #04-06-025) *Contains 12 individually calibrated models*
2012—Calculating Credits and Debits for Compensatory Mitigation in Western Washington (Ecology Publication #10-06-011) *Contains 12 individually calibrated models*
2012—Calculating Credits and Debits for Compensatory Mitigation in Eastern Washington (Ecology Publication #10-06-001) *Contains 12 individually calibrated models*
2014—Washington State Wetland Rating System for Eastern Washington: 2014 Update (Ecology Publication #14-06-030) *Contains 12 individually calibrated models*
2014—Washington State Wetland Rating System for Western Washington: 2014 Update (Ecology Publication #14-06-029) *Contains 12 individually calibrated models*
These publications are available in PDF format on the Department of Ecology's web page: http://www.ecy.wa.gov/programs/sea/wetlands/index.html.

included aspects of value, sensitivity to disturbance, rarity, irreplaceability, and functions. Wildlife habitat was the primary function addressed by field indicators because our understanding of other wetland functions was limited at that time.

In the 2 years following the release of this first method, it became apparent that there were textual inconsistencies in the manual, and users had problems in interpreting the descriptions of some indicators. This first method was revised in 1993 to correct some of these issues.

Between 1995 and 1999, the Department of Ecology received several wetland development grants from the US Environmental Protection Agency to develop a set of methods for assessing wetland functions using field indicators. These methods were developed to assess 15 different functions (nine habitat functions, three water quality functions, and three hydrologic functions) in several classes of wetlands using the hydrogeomorphic (HGM) classification (Brinson, 1993). Values and other important characteristics that were in the rating systems were not included. This was a large effort that relied heavily on volunteers to help develop the models, field test them, and collect data at reference sites to help calibrate the models. The department was the fortunate recipient of more than 3000h volunteered by local wetland scientists and specialists. The culmination of these efforts was the publication of regionally calibrated models assessing 15 different wetland functions in two HGM subclasses of riverine wetlands and two subclasses of depressional wetlands in western Washington, and three subclasses of depressional wetlands in the Columbia Basin. The mechanistic models using field indicators were calibrated against separate assessments of each function made by at least six wetland scientists at each site (two scientists with expertise in hydrologic functions, two with expertise in water quality functions, and two in habitat functions). Eight of the reference sites were revisited by the teams after a period of at least 6 months to check on the accuracy of the initial assessments. The methods used and the data collected to establish these assessments of functions are described in Hruby (2001).

The methods for assessing the 15 functions in the different HGM subclasses required collecting data on more than 80 indicators. These methods proved to be too lengthy for most consultants and potential users. Most sites required 1–3 days in the field and at least 1 day in the office to collect the required data. As a result, the methods were rarely used and local jurisdictions ended up adopting the older 1991 or 1993 rating systems rather than the more detailed assessment of functions in establishing their regulations for protecting wetlands.

The knowledge gained in this effort, however, proved to be very useful when it came time to revise the wetland rating system. The 2004 revision of the rating systems (both regional versions), which provided scores and ratings for three groups of functions (wildlife habitat, hydrologic, and water quality), used some of the more important indicators from the models that assessed the 15 wetland functions. The rating systems still included variables representing values, sensitivity to disturbance, rarity, and irreplaceability. Individual models were developed to score the three groups of functions in four HGM classes of wetlands (depressional, riverine, lake-fringe, and slope) in both eastern and western Washington. The 2004 rating system assigned a maximum of 36 points for wildlife habitat and 32 points each for water quality and hydrologic functions, a maximum of 100 points. The original intent was to give each group of functions an equal weight in the overall score, but the presence of some multipliers in the scoring system necessitated slightly unbalanced scoring. Scores for functions were added together because regulators at the local level wanted to regulate wetlands based on just the four categories. Although protecting specific wetland functions would have allowed more meaningful implementation of the policy of no net loss of wetland functions, local planners did not want the added complexity of trying to regulate for each function separately.

Now that the rating system provided a score for three groups of functions, there was a need to develop a rapid assessment tool for calculating how much mitigation is needed to replace wetland functions that are changed by human activities. The department published two methods in 2012 called the Credit/Debit methods for short (Calculating Credits and Debits for Compensatory Mitigation in Wetlands of Eastern/Western Washington—Ecology publications 11-05-015 and 11-06-25). These methods were based on the same three general wetland functions in the rating system.

In the Credit/Debit methods, the calculations and scoring of functions were changed to better reflect the accuracy of the methods as determined by analyses of the data collected in the reference sites. Using the assessment of the level of functions in the reference wetlands in each region and HGM class, it was possible to calculate the difference between the score for a function at a reference site that a model generated and the score developed during the field visits by the experts. The assessment by the experts was used as the independent variable, and the score from the model was treated as the dependent variable. The average deviation for a function in the function assessment methods ranged between 9% and 19% for the 88 models (up to 15 functions in seven different HGM classes. The average deviation by function was calculated from the absolute value of the difference between the score from the model and the independent variable for each site in an HGM subclass. The summary of data for each function at the reference sites is reported in Hruby (2001). Generally, the models scoring the hydrologic functions had the largest deviations and the models for the habitat functions had the lowest deviations from the separate assessments.

However, when the scores from the more simple rating systems (2004 versions) were analyzed, the average deviations for the 24 individual models of function (4 HGM classes × 3 functions × 2 regions) ranged between 14% and 35%. This suggested that scoring a function from 0 to 36, or 0 to 32 as done in the 2004 rating system, was not supported by an analysis of the data. Given the fairly large deviations in the results between the models and the separate assessments, we concluded that the more rapid methods could provide only a qualitative rating of high, medium, or low level of functioning. The presence, or absence, of specific field indicators of a function was now used to assign a qualitative rating of high, medium, or low rather than a score. The procedures for assigning a rating and calibrating the indicators are described in Hruby (2009, 2012).

In the process of developing these tools, we found that both regulators and users of these tools wanted an actual score for a function when making decisions about mitigation, rather than just a rating of high, medium, or low. As a result, the qualitative rating for each function was converted to a score as the final step. Thus, assigning scores of 1, 2, and 3 to the qualitative ratings was a policy decision, and those numbers are not scientifically defensible.

In addition to changing the assessment of functions to a qualitative rating, the methods published in 2012 separated different aspects of wetland functions in the rating. The final qualitative rating was based on the qualitative ratings of three separate aspects of function: (1) the site potential, or the environmental indicators present in the wetland itself that suggest the function is being performed; (2) the landscape potential, or the indicators in the surrounding landscape that support the functions (e.g., presence of wildlife corridors that allow wildlife to use the wetland); and (3) indicators of the value to society that function provides (e.g., presence of damage from flooding downstream of a riverine wetland that provides flood storage).

The new approach to scoring functions was then incorporated in the 2014 updates of the rating system. Many of the indicators and questions used in the 2014 update are the same as those in the 2004 version. The major difference is that the wetland category based on functions is now based on the qualitative ratings rather than a composite score. The 2014 update, however, kept many of the same indicators for values, sensitivity to disturbance, rarity, and irreplaceability.

CONCEPTUAL FRAMEWORK FOR THE METHOD

All of the rapid methods developed to date were designed to be used in the state of Washington, and all reference sites were within the state.

The methods for assessing wetland functions were limited to two geographical regions in the state: the lowlands of western Washington below 3000 ft, and the Columbia Basin. These regions do not match the Level 3 Ecoregions developed by the Environmental Protection Agency (ftp://ftp.epa.gov/wed/ecoregions/us/Eco_Level_III_US.pdf). The resources were not available to develop separate methods for the different HGM classes of wetlands in the Level 3 ecoregions because there are nine Level 3 ecoregions in the state. In western Washington, the models were calibrated for riverine and depressional wetlands below about 3000 ft found in five Level 3 ecoregions. In eastern Washington, the region was limited to depressional wetlands in the following geomorphic settings: (1) channel scablands created by Lake Missoula floods; (2) windblown loess outside the area scoured by Lake Missoula floods; (3) windblown sand dunes within the channel scabland area; (4) glacial kettles or potholes located in Douglas County; and (5) alluvial and basalt terraces, particularly along the Columbia River. This area is smaller than the Level 3 ecoregion called the Columbia Plateau.

The rating systems and Credit/Debit methods, however, were calibrated for all wetlands in eastern and western Washington below approximately 3000 ft. The division of the state into eastern and western regions is based on the definitions of the regions found in state rules for regulating natural resources (Washington Administrative Code 222-16-010).

Between 1996 and 2004, 212 wetlands were selected to act as a reference set for calibrating the methods for assessing functions and the rating systems. These were not randomly selected, but were chosen to represent the range of functions, values, levels of disturbance, and HGM types in wetlands below 3000 ft found throughout the state. The Department of Ecology maintains a file of all field data collected at these sites as well as updates of any changes in the sites that could be observed from recent aerial photographs. As of 2014, 203 of the 212 sites remain. Five were lost to natural processes such as river-bed migration, and four were lost to human activities.

A minimum of 15 sites was chosen for each HGM class in a region (east and west) as reference sites. A statistically random sample was not considered necessary or feasible from a cost standpoint because there was no intention to attempt statistical analyses based on normally distributed data. Furthermore, the indicators used in rapid methods such as these cannot be normalized because they are usually assigned an ordinal number for scoring. Ordinal numbers cannot be used in parametric statistical analyses because they do not represent continuous variables (Stevens, 1946), and even if they do represent a continuous variable, that variable is usually not normally distributed.

The assessment area for all methods developed by the state is defined as the area that falls within the wetland boundary as determined by the US Army Corps of Engineers delineation manual and its supplements. In cases where there may be large contiguous wetlands in valleys or along riverbanks, the methods provide guidance on how to separate individual units for assessing. The guiding principles for separating such wetlands into different units are changes in the water regime or breaks in the cover of wetland plants.

DESCRIPTION OF THE METHOD

The most recent methods (Credit/Debit methods and the 2014 version of the rating systems) require data collected at the site as well as data on land uses within 1 km that can be determined from high resolution aerial photographs, such as found on Google maps. The first step is to map the boundary of the unit being assessed. Generally this boundary is the same as that provided by a wetland delineation.

The second step is to classify the wetland into one of four classes using the HGM classification of Brinson (1993). These are depressional, riverine, lake-fringe, and slope. In Washington we have very few wetlands that could be classified as flats, and these have been combined with the depressional class in the analyses. Classifying a wetland correctly is critical because there are some indicators that are unique to each class, and there may be different scaling factors for indicators that are common to all classes. The methods also provide guidance on how to collect data for a wetland unit that may contain several different HGM classes, such as a slope wetland that grades into a depressional wetland.

At the request of our users, we tested whether a wetland unit could be subdivided into smaller units based on type of vegetation or different land uses. We did numerous tests of this question, and the data showed that both the function assessment methods and the rating systems produced nonsensical results when applied to small areas within a wetland. None of the rapid methods developed by the department are rigorous enough to adequately assess the functions of only a small area within a wetland unit. Such assessments would require monitoring and measuring the actual processes taking place in different parts of a wetland rather than characterizing the structural indicators present.

The third step is to collect the data requested on the forms and to rate three aspects of each function separately as high, medium, or low (the site potential, the landscape potential, and the value to society of that function). Fig. 4.3.10.1 shows part of the cover page of the data forms that summarizes the ratings of the three aspects of functions, and how to calculate the final score and category of a wetland based on those ratings.

1. Category of wetland based on FUNCTIONS

______ **Category I** – Total score = 23–27

______ **Category II** – Total score = 20–22

______ **Category III** – Total score = 16–19

______ **Category IV** – Total score = 9–15

FUNCTION	Improving Water Quality	Hydrologic	Habitat	
	Circle the appropriate ratings			
Site potential	H M L	H M L	H M L	
Landscape potential	H M L	H M L	H M L	
Value	H M L	H M L	H M L	TOTAL
Score based on ratings				

Score for each function based on three ratings
(*order of ratings is not important*)

9 = H,H,H
8 = H,H,M
7 = H,H,L
7 = H,M,M
6 = H,M,L
6 = M,M,M
5 = H,L,L
5 = M,M,L
4 = M,L,L
3 = L,L,L

FIG. 4.3.10.1 Part of the cover page of the data form for the 2014 version of the wetland rating system. It shows how the qualitative rating of the three different functions is changed to a number.

Some questions are best answered by drawing polygons on aerial photos of the site, and by calculating the relative area of these polygons as a percent of total area within the unit or within a polygon that extends 1 km from the wetland edge. Four to nine such maps are required for each site, depending on the HGM class. During the training of more than 1200 users of the methods during the last decade, we have found that estimates of percentage area need to be made using gridded squares or tools based on geographic information systems. Visual estimates of area by our trainees were prone to errors as high as 40%.

CALIBRATION OF THE METHODS

An initial list of indicators identified from a review of the literature was used to develop protocols and data sheets for sampling reference sites. Indicators were divided into three types:

- Those present at the site itself (indicators of site potential).
- Those found in the surrounding landscape (indicators of landscape potential).
- Those that indicate that the function performed is providing some value to society (indicators of value).

The calibration process involved the following steps and was performed before the methods were released:

1. Indicators that could not be readily estimated from aerial photographs or during a brief field visit (<3 h) were deleted. This represents a compromise between the science and the needs of the user. Some important indicators of function were dropped because they could not be adequately characterized within the time allocated, or did not provide reproducible results when used by different environmental scientists, even after a 2-day training session. For example, the percentage of organic matter or clay in wetland soils is an important indicator of chemical processes that improve water quality (Rosenblatt et al., 2001), but these percentages cannot be readily measured in the field. The indicators of organic and clay soils therefore had to be simplified. Users are asked to determine if organic or clay soils are present in the unit based on coarser-scale spatial data from the National Resource Conservation Service (NRCS). If the wetland is not mapped as having organic or clay soils, users are asked to perform one simple "feel" test to determine if the soil can be categorized as clay or organic. In this case, the reproducibility of the data collection among different users was judged to be more important than asking them to estimate the percent of clay or organic matter present. The indicator, however, was important in assessing the function, and dropping it increased the average deviations between the models and the independent variables by 3%–5% in the calibrations.
2. The indicators for site potential were calibrated to the data collected for the Washington State Function Assessment Methods (Hruby et al., 1999) and as described in Hruby (1999, 2009). This involved developing an assessment of how well a reference wetland performs a function (described previously) and then calibrating the scores of the indicators in the models of functions to get the best fit between the models and the independent assessment. This was done for each HGM class or subclass.

3. The indicators for the landscape potential were calibrated by reviewing the literature on landscape indicators and determining what aspect of the indicators has the highest and lowest impact on each of the functions in a wetland. The data for each indicator collected at the reference sites were then sorted based on the values representing the highest level of function to the lowest in the reference wetlands used for calibration.
4. The indicators of value were not calibrated to field data but rather were based on consensus-driven discussions with regulators and experts in state agencies with the responsibility of protecting natural resources. Although statistical methods are being developed for multicriteria decision models (e.g., Ferguson et al., 2007; Fuller et al., 2008), these methods are not yet applicable to a categorization that incorporates values and special characteristics as well as qualitative indicators.
5. The wetland rating systems had one additional step in calibration: selecting the breaks in the scores calculated from the qualitative ratings to determine Category I, II, III, and IV wetlands. This was done using a graphical analysis of the distribution of scores of all reference sites in a region. The goal was to ensure that the updated rating systems did not change the relative numbers of Category I, II, III, and IV wetlands in each region. Every time the rating system has been updated, there has been a concern among some that the Department of Ecology had a hidden agenda to put more wetlands into higher categories that would therefore require more protection. The graphical analysis was used to demonstrate that this was not the case and the data are presented in the 2004 and 2014 versions of the rating systems. Any individual reference wetland may have a higher or lower category, but the overall distribution of categories remained approximately the same between the 1993, 2004, and 2014 versions.

Further details on the approach used to calibrate the rapid assessment methods developed by the Washington State Department of Ecology can be found in Hruby et al. (1999) and Hruby (2001, 2009).

TIME SPENT IN DEVELOPING/TESTING THE METHOD

The methods for assessing 15 wetland functions in seven regional HGM subclasses (published in 1999 and 2000) took 4 years to develop, calibrate, and test. This major effort, however, formed the basis of all future methods and these latter ones were each developed, calibrated, and tested in 12–18 months. We used the same reference sites, where possible, to collect the field data. Initially, the methods for assessing functions only addressed depressional and riverine wetlands in western Washington and depressional wetlands in the Columbia Basin. Since the rating systems and the Credit/Debit Method include wetlands in the lake-fringe and slope classes, we added such sites to our reference set when calibrating these methods.

In addition we examined the variance in results among users, both previously trained and those that had no training, and differences in results that can be attributed to seasonal variations in the indicators. Our results show that variations related to seasonal factors are overshadowed by differences among users, even trained users.

TIME SPENT TO APPLY THE METHOD IN THE FIELD

The recent versions of the rating system and the Credit/Debit Method were designed to be a Level 2 assessment. These assessments should take two people no more than half a day in the field and half a day of preparation and data analysis (Fennessy et al., 2004). So, on average a Level 2 assessment should take 16 h. Washington's rating systems take about 8 h, on average, not including travel time, and the Credit/Debit method takes about 16 h because it involves rating both the impact site and the mitigation site. However, small wetlands may take less time and large ones more.

HOW THE DATA ARE BEING USED

In Washington, the cities and counties are required to protect wetlands as one of five critical areas identified in the state's Growth Management Act (GMA) using the best available science. Because many jurisdictions do not have the staff or resources to develop their own methods to meet the GMA requirements, they adopt the Department of Ecology's methods. Currently, more than 180 cities and counties, out of a total of 320, have adopted the 2014 rating system in their ordinances as one of the primary means to protect the functions and values of wetlands.

PROSPECTS FOR THE FUTURE

The Washington State Wetland rating systems have been an integral part of the state's approach to protecting wetlands. We are currently using the fourth update. With updates occurring in 1993, 2004, and 2014, it can be expected that another update will be needed in 2024 as our knowledge of wetlands increases.

REFERENCES

Brinson, M.M., 1993. A hydrogeomorphic classification for wetlands. US Army Engineer Waterways Experiment Station, Vicksburg, Mississippi. Technical Report WRP-DE-4.

Fennessy, M.S., Jacobs, A.D., Kentula, M.E., 2004. Review of rapid methods for assessing wetland condition. EPA/620/R-04/009.

Ferguson, C.A., Bowman, A.W., Scott, E.M., Carvalho, L., 2007. Model comparison for a complex ecological system. J. Royal Stat. Soc. Ser. A Stat. Soc. 170, 691–711.

Fuller, M.M., Gross, L.J., Duke, S.M., Palmer, M., 2008. Testing the robustness of management decisions to uncertainty: everglades restoration scenarios. Ecol. Appl. 18, 711–723.

Hruby, T., 1999. Assessments of wetland functions: what they are and what they are not. Environ. Manag. 23, 75–85.

Hruby, T., 2001. Testing the basic assumption of the hydrogeomorphic approach to assessing wetland functions. Environ. Manag. 27, 749–761.

Hruby, T., 2009. Developing rapid methods for analyzing upland riparian functions and values. Environ. Manag. 43, 1219–1243.

Hruby, T., 2012. Calculating Credits and Debits for Compensatory Mitigation in Wetlands of Eastern Washington. Washington Department of Ecology, Olympia, WA. Publication #11-06-015.

Hruby, T., Granger, T., Teachout, E., 1999. Methods for Assessing Wetland Functions. Volume I: Riverine and Depressional Wetlands in the Lowlands of Western Washington. Part 2: Procedures for Collecting Data. Washington Department of Ecology, Olympia, WA. Publication #99-116.

Rosenblatt, A.E., Gold, A.J., Stolt, M.H., Groffman, P.M., Kellogg, D.Q., 2001. Identifying riparian sinks for watershed nitrate using soil surveys. J. Environ. Qual. 30, 1596–1604.

Stevens, S.S., 1946. On the theory of scales of measurement. Science 103, 677–680.

Section 4.4

Implementing National-Scale and Regional-Scale Wetland Assessments

Chapter 4.4.1

Creating a Unified Mid-Atlantic Rapid Condition Assessment Protocol for Wetlands

Robert Brooks*, Kirk J. Havens†, Hannah Ingram*, Kory Angstadt†, David Stanhope†, Amy Jacobs‡, Michael Nassry* and Denice Wardrop*

**Department of Geography, Pennsylvania State University, University Park, PA, United States, †College of William and Mary, Virginia Institute of Marine Science, Gloucester Point, VA, United States, ‡The Nature Conservancy, Denton, MD, United States*

Chapter Outline

INTRODUCTION

Monitoring and assessment are critical components of any resource management program where there is a need to evaluate progress and performance over time. The federal Water Pollution Control Act of 1972, Public Law 92-500, or the Clean Water Act (CWA), specifies a need to monitor, compile, analyze, and report on water quality data, broadly defined (CWA§106(e)(1)). Wetlands are included because they are "waters of the United States." Thus, there is both a management imperative and a legal basis to monitor and assess wetlands for multiple purposes (Wardrop et al., 2013). Until relatively recently, the data available to allow a condition assessment of the nation's aquatic resources was not available (Shapiro et al., 2008). The U.S. Environmental Protection Agency (U.S. EPA) and other partners are participating in the National Aquatic Resource Survey (NARS) to assess all the nation's waters. The following waters have been assessed: wadeable streams (2004), lakes (2007), rivers and streams (2008–2009), coastal waters (2010), and wetlands (2011) (http://www.epa.gov/national-aquatic-resource-surveys/nwca). A second round of sampling was completed ending with wetlands in 2016 (Serenbetz, 2016). In addition, the U.S. EPA's Office of Wetlands, Oceans, and Watersheds (OWOW) established a national wetlands monitoring strategy as a means of assisting states and tribes in reaching this goal of protecting and restoring wetlands.

During the 2000s, Pennsylvania, Virginia, and Delaware of the Mid-Atlantic Region (MAR) developed wetland assessment methods designed to support state protection and management programs. As a result of continuing regional collaboration through the Mid-Atlantic Wetlands Work Group (MAWWG, http://mawwg.psu.edu/) and professional and academic exchanges, the methods used by the three states share many common elements without being identical. A three-tiered approach as described by Brooks et al. (2002), and further refined in U.S. EPA's elements letter (U.S. EPA, 2006) was adopted by these states and is designed as a multilevel protocol incorporating a landscape assessment, a rapid field assessment, and an intensive site assessment (Kentula, 2007). Applications of this three-tiered approach for assessing wetland condition can be found in Brooks et al. (2006) and Wardrop et al. (2007). The U.S. EPA (2006); http://www.epa.gov/owow/wetlands/monitor, accessed December 20, 2012) provided guidance to the states and tribes on how to formulate the approach, which was updated in 2012 (U.S. EPA, 2012). A review of North American RAMs

Wetland and Stream Rapid Assessments. https://doi.org/10.1016/B978-0-12-805091-0.00008-6

intended for assessing wetland condition was completed by Fennessy et al. (2007). A more detailed description and chro nology of monitoring and assessment efforts in the MAR region can be found in Wardrop et al. (2013).

Typically, a landscape assessment (Level 1) can be accomplished in the office using only readily available digital data and a geographic information system (GIS) and requires a low level of effort compared to the site assessments. A rapid assessment (Level 2) refines the results of the landscape assessment by incorporating observational indicators of human disturbance to a site into the evaluation of ecological condition; the emphasis is usually on stressors (Adamus and Brandt, 1990; Adamus et al., 2001). An intensive assessment (Level 3) entails detailed data collection on each site and produces the most complete evaluation. Intensive assessments typically apply hydrogeomorphic (HGM) functional models, which require a suite of simple mathematical models to estimate the magnitude at which a wetland performs a suite of ecological functions associated with a specific wetland subclass (Smith and Wakeley, 2001). Alternatively, indices of biotic integrity (IBIs, Karr and Chu, 1999), are also used; these are multimetric indices focused on a specific taxonomic group (e.g., vascular plant communities, Miller et al. 2006; Mack, 2007; Chamberlain and Brooks, 2016; aquatic macroinvertebrates, Burton et al., 1999) that quantitatively assess changes in the structure and composition of those communities likely due to human disturbance. The level of effort appropriate for a monitoring project depends on the resources available and the degree of confidence required in the results, with costs increasing with greater effort but with the benefit of having more confidence in the data.

METHODS

While a tremendous amount of overlap exists in the methods for wetland RAMs in the MAR, conducting a comprehensive regional assessment of wetland conditions necessitated a sampling program capable of providing a synoptic regional data set. It was also essential that this regional assessment protocol be as similar as practical to all the existing individual state protocols. This was to avoid the possible complication of diverging state and regional assessments that might communicate mixed messages, and to make the data obtained in the regional assessment both useful and desirable for each state program. Because a significant amount of cooperative effort was required to design such a unified protocol for the MAR, we used a facilitated discussion at a MAWWG meeting in May 2006 to gain a consensus on how to proceed with an approach to unify the variables and procedures derived from multiple RAMs. Ultimately, we selected three well-developed RAMs from which to construct a Unified Mid-Atlantic Rapid Condition Assessment (UMARCA) protocol for wetlands. We chose RAMs from Pennsylvania (Brooks, 2004; Brooks et al., 2004), Virginia (Virginia DEQ, 2005), and Delaware (DNREC, 2006).

The Level 2 rapid assessment was designed to address three needs. First, it would provide a probabilistic survey of wetland condition for the MAR with the capability of being stratified by states and ecoregions. For classifying wetland types, we used a new classification system with broad applicability across the MAR that blended elements of HGM classification and the existing system used for the National Wetland Inventory (NWI) (Brooks et al., 2009). To test for consistent use of the classification system, field teams underwent training sessions to standardize the terminology and use of the system. These were held at the beginning of each of the two summer field seasons, and after about 1 month of sampling. Field teams completed independent identification of wetland types during a formal quality assurance audit in 2008 to compare their results.

The second need was to generate a landscape profile of wetland types at a variety of geopolitical boundaries. Landscape profiles are critical tools for restoration, management, mitigation, and cumulative impact assessment of naturally occurring wetlands (Bedford, 1996; Gwin et al., 1999). Use of NWI maps alone cannot provide this profile because they do not always include the entire resource and do not typically assign wetland type according to the HGM classification.

The third need addressed by a regional RAM is the calibration of ecosystem service models developed at a landscape scale. These Level 1 models predict stressor levels affecting wetland services based on the composition of surrounding landscapes. The collection of stressor data during the fieldwork can be used to confirm the landscape assessment. Estimation of two wetland function groups (water quality, habitat quality) based on landscape assessment is covered in another report (Hershner et al., 2013), available at http://www.riparia.psu.edu/products and Havens et al. (2018) (this book, Chapter 2.2.8). Thus, the MAR rapid assessment was designed to provide a regional landscape profile of wetland types, various stressor profiles by wetland type and ecoregion, and an assessment of wetland condition.

SELECTION OF WETLAND SITES

Balancing statistical rigor with resource constraints, we sought to assess 400 points located in wetlands throughout the MAR, or about 80 points in each of five ecoregions. Digital versions of the NWI maps provided the best representation

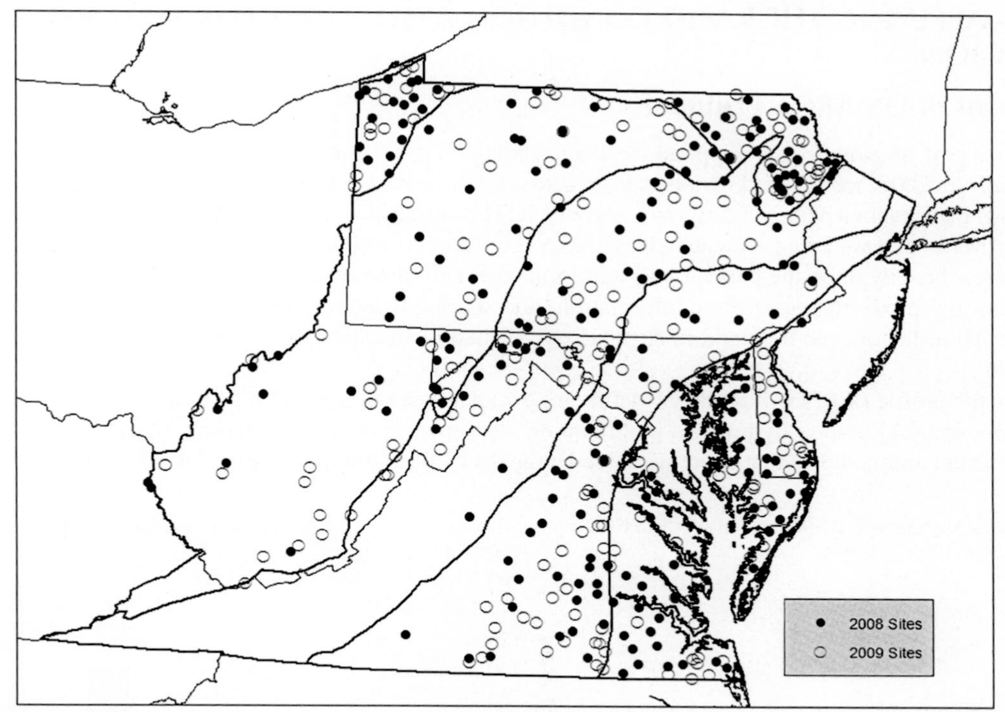

FIG. 4.4.1.1 Wetland sample points ($n = 400$) in 2008 and 2009 in the Mid-Atlantic Region for conducting a Level 2 condition assessment using a rapid assessment protocol.

of wetland occurrence in the MAR (i.e., a frame in statistical terminology). These data were filtered to include only freshwater wetlands. We removed types of no interest, such as farm ponds and quarries. A generalized random tessellation stratified (GRTS) design (Stevens and Olsen, 1999, 2004) was used to draw sample points from the relevant datasets. Briefly, the GRTS design results in a spatially balanced sample with the points ordered so that sequential use of the points as study sites maintains spatial balance. We identified twice as many sites as needed, 800, to serve as an over sample to compensate when access was denied (rarely) or other protocol rules were not met. Using this approach our *target population* was all NWI nontidal vegetated wetlands in the MAR. These were subdivided into *subpopulations* by ecoregions. The *sampling unit*, then, was a bounded NWI wetland about 1 ha in size, around a selected point. To help avoid any interannual bias, the target points representing NWI wetlands were spatially distributed throughout the MAR for both years (Fig. 4.4.1.1).

FIELD SAMPLING

Field sampling was completed by two teams, one operating out of Riparia at Pennsylvania State University (Penn State) and another from the Virginia Institute of Marine Science (VIMS). The sampling territory was divided approximately by ecoregions, with Penn State covering the Allegheny Plateau (glaciated and unglaciated sections) and the Ridge and Valley, and VIMS covering the Piedmont and Coastal Plain (Fig. 4.4.1.1). Beyond a set of standard field gear, each team used map resources, aerial photographs, and a Global Positioning System (GPS) unit to navigate to selected points. Data were recorded primarily on a handheld device programmed to prompt field personnel to record data for all fields of the electronic forms. These data were downloaded to a laptop daily and emailed to colleagues at Penn State for compilation. About 1.0–2.0 h were needed onsite for the rapid assessment.

Following a training session to help standardize field methods, two field teams of 2–3 people per team, one from Penn State and one from VIMS, conducted the sampling throughout the region during the summers of 2008 and 2009. Each of the field sites consisted of a wetland assessment area with a 40 m radius circle, surrounded by a 100 m buffer, which eventually became the sample unit for the National Wetland Condition Assessment (http://www.epa.gov/national-aquatic-resource-surveys/nwca). Our goal was to have a sufficient number of assessed sites in each of the five major ecoregions (80 sites per ecoregion), and where possible, make comparisons across the more common wetlands types (e.g., riverine, depression).

RESULTS—APPLYING THE RAPID CONDITION ASSESSMENT FOR WETLANDS FOR A REGION

Development of UMARCA Protocol

During the first year of sampling, we applied the Pennsylvania, Virginia, and Delaware RAMs to every site. After compiling the data from 2008, we selected variables from across all three RAMs that showed variability in response to stressors across wetlands that spanned a human disturbance gradient. The variables used were derived from at least one of the aforementioned RAMs so that we could recreate a UMARCA score from the 2008 field data. The terminology and text used in the protocol drew heavily upon the publications and forms from all three RAMs. The final protocol contained two sets of factors: nonscoring (providing descriptive information) and scoring (used to compute a quantitative score for each site). The complete list of factors can be found in Appendix A located in an expanded version of the case study at http://www.riparia.psu.edu.

The landscape profile (Fig. 4.4.1.2) shows that riverine wetlands dominate HGM types across the MAR. A simple tally of the most dominant stressors, hydrologic modification, sedimentation, and vegetation alternation, recorded from the wetland assessment areas, shows that ecoregions are being affected differentially (Fig. 4.4.1.3). The qualitative condition

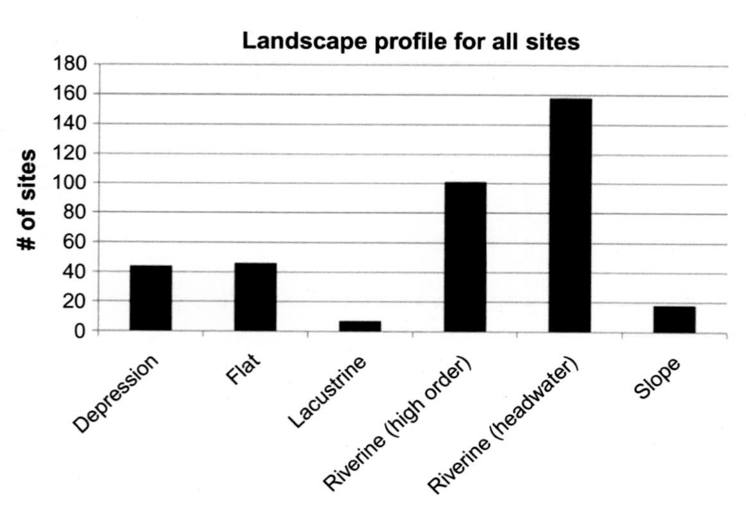

FIG. 4.4.1.2 Landscape profile of 400 wetland sites by HGM subclass for the Mid-Atlantic Region.

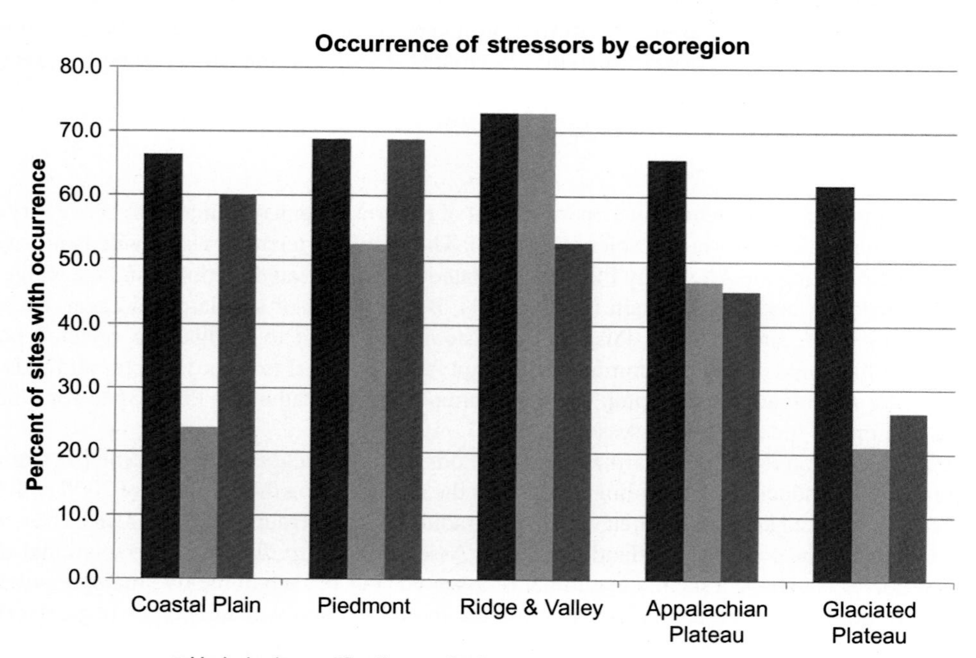

FIG. 4.4.1.3 Occurrence of three major stressors in wetland sites of the MAR.

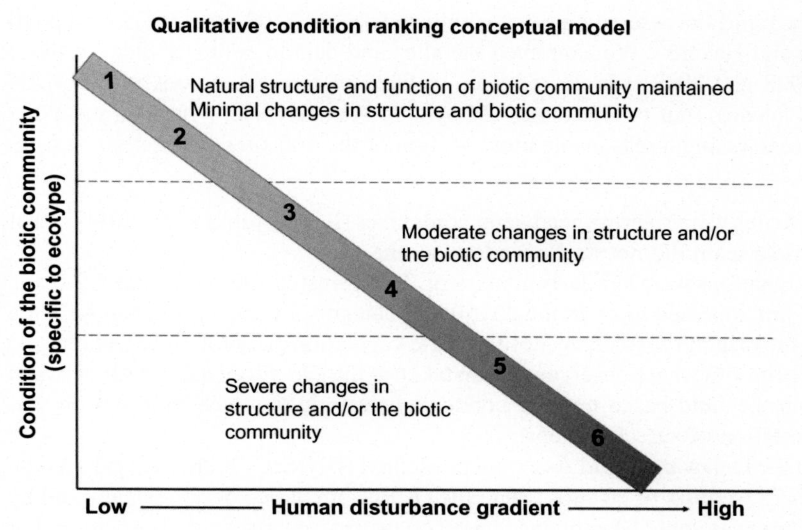

FIG. 4.4.1.4 Diagram of narrative criteria for qualitative ranking of condition (DENREC, 2006, after Davies and Jackson, 2006).

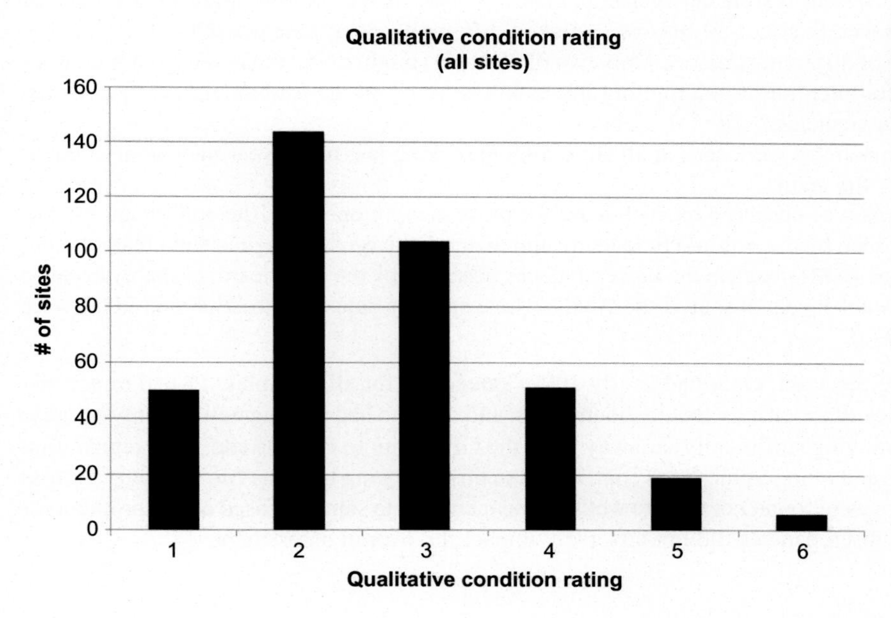

FIG. 4.4.1.5 Condition of wetlands in the MAR based on a QCR alone, using best professional judgment (1 indicates sites of the highest ecological integrity, 6 the lowest).

rating (QCR) score (Fig. 4.4.1.4) showed that most wetland sites were ranked as 2, the lower of the high integrity categories, with the next highest tally ranked as 3, the highest of the moderate category (Fig. 4.4.1.5). More detailed results of this study, including the ecosystem assessment based on landscape data, are posted at http://www.riparia.psu.edu/products.

Quality Assurance

We conducted quality assurance audits twice during the process of creating the protocol, one to assess consistency of measures between two field teams (2008), and one to assess the quality of data obtained by field teams versus more experienced scientists (2010).

Audit for consistency of measures. After about 1 month of sampling in 2008, the two field teams and project directors met for several days to assess field sites near Richmond, Virginia. We were joined by a quality assurance manager from U.S. EPA, Dr. Robert Ozretich. The two field teams consistently identified the HGM subclasses of the sites and tallied the stressors in a similar manner. Other indicators were consistently assessed and recorded, yielding a set of data that could be scored for the entire MAR, incorporating both years of data across all ecoregions.

Audit to assess data quality. A QA audit of the rapid site assessments was conducted on 20 of 400 sites (5%) in fall 2010 by two experienced wetland scientists (Brooks and Havens). We examined the sites and data to compare them to those collected by field teams in the summers of 2008 and 2009. We visited sites in three of five ecoregions in the MAR (Piedmont, Ridge and Valley, and Allegheny Plateau), four of five states (Maryland, Pennsylvania, Virginia, and West Virginia), and a variety of wetland types, thus reassessing a reasonable cross section of the 400 sites originally assessed. The results were as follows:

- *HGM subclass classification*: The majority of sites were riverine headwater complexes (R3c)(Brooks et al., 2011); 19 of 20 (95%) were in agreement. The original field team did not find a wetland at one site.
- *Vegetation community*: Community type designations were similarly consistent. Two forested sites were listed as being >50 years of age by the original field team, but appeared to be in the 25–50 year category by the audit team; both are considered to be mature forest. At two sites, the original field team chose a single vegetation class for a site whereas the audit team chose to list two classes. Much depends on where observations are taken at the site. Close agreement between original fieldwork and the audit suggest that the field-based training conducted early during each field season was useful, and contributed to successful interpretation of site conditions.
- *Quality condition rating*: A QCR came from the Delaware Rapid Assessment Method (DNREC, 2006), and is based on best professional judgment (Fig. 4.4.1.4). There were six of 20 sites where the QCR score varied between teams, all by one unit on the 1–6 unit QCR scale. Most differences could be explained based on the season or year of observation. For two sites where questions remained, differences were between high rankings (1 or 2); one site was upgraded and one site was downgraded by the audit team. Overall, six of 20 sites were ranked differently; four were considered valid in the original sampling due to the presence of different stressors. Only two of 20 (10%) might affect the results, but both were in the upper rankings (1 or 2). Careful attention to interpreting the rankings based on the assessment area versus the buffer would likely reduce these discrepancies.
- *Invasive species*: Generally, the species listed were seen at all sites, although being late in the year the extent of cover was more difficult to estimate during the audit.
- *Stressors list*: A similar set of stressors was observed by each team for most sites. In one case, the audit team did not observe a significant area of garbage seen by the original field team, and in another case, the original field team did not observe an adjacent acid mine drainage (AMD) area in the buffer that was observed by the audit team, probably because higher vegetation heights or lower water levels obscured the AMD. Thus, stressor data in both 2008 and 2009 were accurately recorded during the study.

Overall, for the 20 sites evaluated during the audit, eight had nearly 100% agreement for all variables, 11 had minor discrepancies explained by seasonality or lack of specificity for site boundaries, and one site had poor agreement only because the original team missed the site by not moving sufficiently far away from the GPS point to find the site. Line registration problems between the aerial photographs and topographic maps contributed to this site being listed as "no wetland." Across the 400 sites sampled, only four (1%) sites were found not to have wetlands near enough to sample, based on the protocol for moving points. We believe these minor discrepancies did not adversely impact the overall assessment.

UMARCA Scores

The UMARCA score was derived from synthesizing three state RAMs, and was consistently applied across a major geographic region of the United States, the MAR. Details about protocol variables can be found in Appendix A located at http://www.riparia.psu.edu/products. A QCR based on best professional judgment (DNREC, 2006) and made during field visits showed the majority of wetlands were rated between high and moderate condition (Fig. 4.4.1.5). Alternatively, we wanted to explore developing a scoring system that incorporated more metrics derived from the field measurements. This process was delayed while seeking an appropriate scoring method, but an approach was selected comparable to one used by Herlihy et al. (2008) and Van Sickle and Paulsen (2008) for a national streams assessment.

Initially, we sought a biological metric that was responsive to the human disturbance gradient that spanned the representative sample of sites across the MAR: invasive plant species cover. Plant community composition has displayed correlations to deteriorating site quality as an early indicator of stressors and other impacts of anthropogenic disturbance (Mack, 2007; Chamberlain and Brooks, 2016). Community composition is a reasonable choice to determine site quality because degradation of freshwater wetlands does not result in wetlands moving to a different location in the landscape, changing HGM classification, or switching primary water sources. Additionally, impacts of offsite stressors on freshwater wetlands in the MAR rarely result in the complete disappearance of wetland resources, rather the degradation is expressed through impacts to the structure and function of the wetland ecosystem resulting in a decreased ability to provide ecosystem services. The field assessment categorized invasive plant cover in the site assessment area (AA) into four categories of invasive cover: 0%–5%, 5%–20%, 20%–50%, and >50%. These four site assessment categories were then condensed to three quality assessment categories:

0%–5% invasive cover—High quality site (optimal)
5%–20% invasive cover—Suboptimal site (marginal)
>20% invasive cover—Low quality site (poor)

Due to missing data in some sites, 374 of 400 sites were evaluated using invasive cover, with 72% scoring high, and 17% and 11% falling into the suboptimal and low site quality categories, respectively. This distribution varied by ecoregion with the Coastal Plain having the highest percentage of high quality sites (92%), and the Ridge and Valley ecoregion having the lowest percentage of high quality sites (44%). See figures in the expanded case study at http://www.riparia.psu.edu/products.

Although the 374 sites distributed across five ecoregions is a large sample size for this type of field data, a potential issue arises when using these data to create a scoring system—the small number of low sites in some significant categories. For example, only 2 of 80 Coastal Plain sites are categorized as low ecological condition, and only 4 are suboptimal. Only 6 of 76 sites in the Glaciated Plateau are low, 9 are suboptimal. This trend is evident but less pronounced in other ecoregions, except for the Ridge and Valley. The lack of low quality sites caused difficulty in creating a scoring gradient that accurately predicts low ecological condition across MAR sites.

As an alternative to invasive plant cover, we explored data on site stressors (Fig. 4.4.1.3). In total, 76 stressors were evaluated (present or not present), with 41 of those evaluated in the AA and 35 in the buffer area. They were characterized into five broad categories: hydrologic modification (32), sedimentation (14), vegetation alteration (16), eutrophication (5), and contaminant/toxicity (9).

We followed the approach used by Van Sickle and Paulsen (2008) for scoring the condition of wadeable streams, using terminology and analytics from Van Sickle et al. (2006) for a similar study on condition of streams in the coterminous United States; relative extent and relative risk. Relative extent is the proportion of regional wetlands in low (poor) condition for a given stressor. Relative extent was calculated for all sites, AAs, and buffers as the number of sites with stressors present, thus evaluating the proportion of sites with individual or stressor categories present. Hydrologic modification and sedimentation were the only two stressor categories present at >20% of sites, vegetation alteration was present at about 13% of sites, and eutrophication and contaminant/toxicity stressors were present at <5% of sites. This pattern held constant for the AA while vegetation alteration increased to >20% for buffers.

Relative risk measures the strength of association between the condition classes of a biological response indicator and a single stressor. Relative risk was calculated to determine the link between stressors and poor site quality measured by invasive plant coverage. Relative risk values >1 indicate a link between the stressor and biological condition, which is likely degraded if the stressor is present. The only stressor category exceeding 1 was sedimentation (~1.4), with relative risk for hydrologic modification equaling 1. No stressor group exceeded a relative risk value of 1 in the AA, but sedimentation and hydrologic modification both exceeded 1 for buffers. Relative risk calculations for individual stressors indicated a link between eroding banks and overall site condition, both at a study-wide scale, and for individual ecoregions. Hydrologic modifications (stormwater inputs and grading) had a relative risk over 1 for the Alleghany Plateau, and mowing and brush cutting had an elevated relative risk in the Piedmont.

Calculations recommended for relative risk (Van Sickle and Paulsen, 2008) could not be performed in a robust manner due to the low number of poor quality sites defined by invasive plant coverage, so a different measure of ecological condition was selected. Because a QCR was recorded for each site, we decided to use that metric of overall site disturbance. With the QCR, sites with minimal disturbance (or high integrity) receiving a score of 1 or 2, sites moderately disturbed (or moderate integrity) were scored 3 or 4, and sites with high disturbance (or low integrity) were scored 5 or 6 (Fig. 4.4.1.4). No sites in this dataset received a score of 6, but 51 sites were scored as 4 and 25 sites as 5. These two lower tiers of the available dataset were grouped to form a set of sites with low ecological condition. These sites were used to determine relative extent of stressors and relative risk of individual stressors in low scoring sites.

For all sites, filling and grading in the buffer, eroding banks/slopes in the buffer, and mowing the buffer all exceeded the relative risk threshold of 1.0. A further expansion into ecoregion-specific relative risk values included additional stressors exceeding the relative risk threshold such as stormwater inputs in the buffer and AA, gravel/dirt road in the buffer, eroding banks in the AA, mowing in the AA, and brush cutting in the buffer and AA.

A simple count of the 10 stressors exceeding the relative risk threshold for any ecoregion yielded a pattern of increasing stressor presence associated with decreasing ecological condition as defined by QCR (Fig. 4.4.1.6). Sites scoring 1 ($n = 50$) or 2 ($n = 144$) by QCR had a median stressor count of 0 and 1, respectively, when focusing on only these 10 stressors. Sites

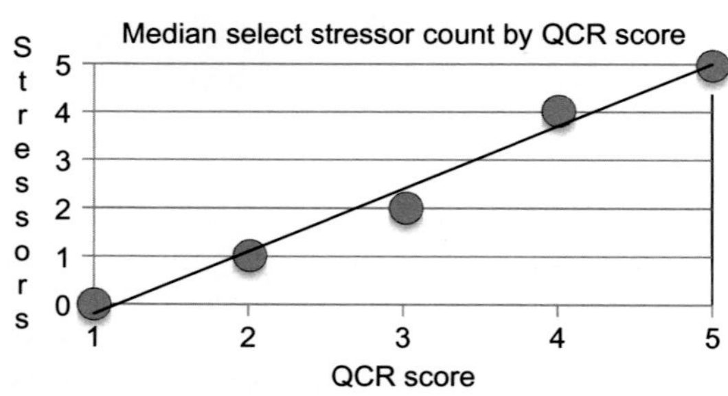

FIG. 4.4.1.6 Median number of stressors for qualitative condition rating (QCR) categories.

TABLE 4.4.1.1 Ten Stressors Used to Score Relative Risk for a Condition Assessment of MAR Wetlands
Filling/grading (buffer)
Stormwater inputs (AA)
Stormwater inputs (buffer)
Gravel/dirt road (buffer)
Eroding banks/slopes (AA)
Eroding banks/slopes (buffer)
Mowing (AA)
Mowing (buffer)
Brush cutting (AA)
Brush cutting (buffer)

scoring 3 ($n=104$), 4 ($n=51$), and 5 ($n=25$) by QCR ratings had median stressor total values of 2, 4, and 5, respectively (Fig. 4.4.1.6). Further refinement using equations that include additional stressors identified in statistical analyses, such as a correlation and regression tree (CART) analysis, might provide a more accurate estimation, but the stressors listed in Table 4.4.1.1 appear to be those most associated with decreased ecological condition at the wetland sites evaluated in this study.

DISCUSSION

By combining relevant parameters from three state wetland RAMs, the UMARCA protocol and scoring system were developed. The protocol was used to collect a consistent dataset across a spatially balanced sample of nearly 400 points. About 80 wetland sites were located in each of the five major ecoregions within the MAR. Quality assurance audits demonstrated that preseason training may have improved the quality and consistency of the data collected. Despite having two teams collect data during 2 years, the data from both teams and both years appeared to be consistent, and therefore suitable to use in a region-wide wetlands condition assessment.

The study showed that riverine wetlands are the dominant HGM subclasses for freshwater wetlands in the MAR. The dominant stressor categories for wetlands in the region, in descending order of abundance, were hydrologic modification, sedimentation, and vegetation alteration. The Ridge and Valley and Piedmont ecoregions have the most stressors in wetlands, with fewer to the east in the Coastal Plain and toward the west in the Allegheny Plateau.

The UMARCA scoring system showed that the number and proportion of stressors varied with the QCR scores with about 52% (194 of 374) of the wetlands in the region ranked in the top two highest categories of condition, 28% (104 of 374) of wetlands were in a moderate condition category, and the remaining 20% (76 of 374) were considered to be in low categories of condition (Fig. 4.4.1.6). We believe these rankings are representative of the population of nontidal freshwater wetlands in the MAR based on a spatially balanced sample of 400 wetlands assessed using rapid field-based methods. It will be up to the individual states as to whether they should adopt the UMARCA versus their individual methods.

REFERENCES

Adamus, P.R., Brandt, K., 1990. Impacts on Quality of Inland Wetlands of the United States: A Survey of Indicators, Techniques, and Application of Community-Level Biomonitoring Data. EPA/600/3-90/073, US Environmental Protection Agency Environmental Research Laboratory, Corvallis, OR.

Adamus, P.R., Danielson, T.J., Gonyaw, A., 2001. Indicators for Monitoring Biological Integrity of Inland, Freshwater Wetlands: A Survey of North American Technical Literature (1990–2000). EPA 843-R-0, U.S. Environmental Protection Agency, Office of Water, Washington, DC.

Bedford, B.L., 1996. The need to define hydrologic equivalence at the landscape scale for freshwater wetland mitigation. Ecol. Appl. 6, 57–68.

Brooks, R.P. (Ed.), 2004. Monitoring and Assessing Pennsylvania Wetlands. Final Report for Cooperative Agreement No. X-827157-01, Between Penn State Cooperative Wetlands Center, Pennsylvania State University, University Park, PA and U.S. Environmental Protection Agency, Office of Wetlands, Oceans, and Watersheds, Washington, DC. Rep. No. 2004-3. Penn State Cooperative Wetlands Center, University Park, PA.

Brooks, R.P., Wardrop, D.H., Cole, C.A., Reisinger, K.R., 2002. In: Tiner, R.W. (Ed.), Using reference wetlands for integrating wetland inventory, assessment, and restoration for watersheds. Watershed-Based Wetland Planning and Evaluation. A Collection of Papers From the Wetland Millennium Event, 6–12 August 2000, Quebec City, Quebec, Canada. Distributed by Assoc. State Wetland Managers, Inc., Berne, NY, pp. 9–15. 141 pp.

Brooks, R.P., Wardrop, D.H., Bishop, J.A., 2004. Assessing wetland condition on a watershed basis in the Mid-Atlantic region using synoptic land-cover maps. Environ. Monit. Assess. 94, 9–22.

Brooks, R.P., Wardrop, D.H., Cole, C.A., 2006. Inventorying and monitoring wetland condition and restoration potential on a watershed basis with examples from the Spring Creek watershed, Pennsylvania, USA. Environ. Manag. 38, 673–687.

Brooks, R.P., McKenney-Easterling, M., Brinson, M., Rheinhardt, R., Havens, K., O'Brian, D., Bishop, J., Rubbo, J., Armstrong, B., Hite, J., 2009. A Stream-Wetland-Riparian (SWR) index for assessing condition of aquatic ecosystems in small watersheds along the Atlantic slope of the eastern U.S. Environ. Monit. Assess. 150, 101–117.

Brooks, R.P., Brinson, M.M., Havens, K.J., Hershner, C.S., Rheinhardt, R.D., Wardrop, D.H., Whigham, D.F., Jacobs, A.D., Rubbo, J.M., 2011. Proposed hydrogeomorphic classification for wetlands of the Mid-Atlantic Region, USA. Wetlands 31 (2), 207–219.

Burton, T.M., Uzarski, D.G., Gathman, J.P., Genet, J.A., Keas, B.E., Stricker, C.A., 1999. Development of a preliminary invertebrate index of biotic integrity for Lake Huron coastal wetlands. Wetlands 19 (4), 869–882.

Chamberlain, S.J., Brooks, R.P., 2016. Testing a rapid floristic quality index on headwater wetlands in Central Pennsylvania, USA. Ecol. Indic. 60, 1142–1149.

Davies, S.P., Jackson, S.J., 2006. The biological condition gradient: a descriptive model for interpreting change in aquatic ecosystems. Ecol. Appl. 16, 1251–1266.

Delaware Department of Natural Resources and Environmental Control. 2006. Delaware Rapid Assessment Procedure Version 3.0, Dover, DE. 30 pp. Available from: http://www.mawwg.psu.edu. Version 6.0 available at http://www.dnrec.delaware.gov (accessed 20.12.12).

Fennessy, M.S., Jacobs, A.D., Kentula, M.E., 2007. An evaluation of rapid methods for assessing the ecological condition of wetlands. Wetlands 27, 543–560.

Gwin, S.E., Kentula, M.E., Shaffer, P.W., 1999. Evaluating the effects of wetland regulation through hydrogeomorphic classification and landscape profiles. Wetlands 19, 477–489.

Havens, K.J., Hershner, C., Rudnicky, T., Stanhope, D., Schatt, D., Angstadt, K., Henicheck, M., Davis, D., Bilkovic, D.M., 2018. Virginia wetland condition assessment tool (WeCAT): a model for management. In: Dorney, J., Tiner, R., Adamus, P., Savage, R. (Eds.), Wetland and Stream Rapid Assessments: Development, Validation, and Application. Elsevier.

Herlihy, A.T., Paulsen, S.G., Van Sickle, J., Stoddard, J.L., Hawkins, C.P., Yuan, L.L., 2008. Striving for consistency in a national assessment: the challenges of applying a reference-condition approach at a continental scale. J. N. Am. Benthol. Soc. 27 (4), 860–877.

Hershner, C., Havens, K., Rudnicky, T., Schatt, D., Wardrop, D., Brooks, R., 2013. Assessment of the potential ecosystem services of nontidal wetlands in the Mid-Atlantic Region. In: Mid-Atlantic State Regional Wetlands Assessment Final Report to U.S. Environmental Protection Agency by Pennsylvania State University and Virginia Institute of Marine Science, pp. 27–46. + appendices A, B, C in Chapter 3, 121 pp.

Karr, J.R., Chu, E.W., 1999. Restoring Life in Running Waters: Better Biological Monitoring. Island Press, Washington, DC. 206 pp.

Kentula, M.E., 2007. Monitoring wetlands at the watershed scale. Wetlands 27, 412–415.

Mack, J.J., 2007. Developing a wetland IBI with statewide application after multiple testing iterations. Ecol. Indic. 7, 864–881.

Miller, S.J., Wardrop, D.W., Mahaney, W.M., Brooks, R.P., 2006. A plant-based index of biological integrity (IBI) for headwater wetlands in central Pennsylvania. Ecol. Indic. 6, 290–312.

Serenbetz, G., 2016. National Wetlands Condition Assessment 2011–016: lessons learned and moving forward. Natl. Wetl. Newslett. 38 (3), 16–19.

Shapiro, M.H., Holdsworth, S.M., Paulsen, S.G., 2008. The need to assess the condition of aquatic resources in the US. J. N. Am. Benthol. Soc. 27 (4), 808–811.

Smith, R.D., Wakeley, J.S., 2001. Hydrogeomorphic Approach to Assessing Wetland Functions: Guidelines for Developing Regional Guidebooks—Chapter 4 Developing Assessment Models. ERDC/EL TR-01-30, U.S. Army Engineer Research and Development Center, Vicksburg, MS.

Stevens Jr., D.L., Olsen, A.R., 1999. Spatially restricted surveys over time for aquatic resources. J. Agric. Biol. Environ. Stat. 4, 415–428.

Stevens Jr., D.L., Olsen, A.R., 2004. Spatially-balanced sampling of natural resources in the presence of frame imperfections. J. Am. Stat. Assoc. 99, 262–278.

U.S. Environmental Protection Agency, 2006. 2006–2011 EPA Strategic Plan: Charting our Course. EPA-190-R-06-001, U.S. Environmental Protection Agency, Office of Planning, Analysis, and Accountability, Washington, DC.

U.S. Environmental Protection Agency, 2012. Core Elements of an Effective State and Tribal Wetlands Program Framework. http://water.epa.gov/grants_funding/wetlands/cefintro.cfm. Accessed 20 December 2012.

Van Sickle, J., Paulsen, S.G., 2008. Assessing the attributable risks, relative risks, and regional extent of aquatic stressors. J. N. Am. Benthol. Soc. 27 (4), 920–931.

Van Sickle, J., Stoddard, J.L., Paulsen, S.G., Olsen, A.R., 2006. Using relative risk to compare the effects of aquatic stressors at a regional scale. Environ. Manag. 38, 1020–1030.

Virginia Department of Environmental Quality, 2005. Commonwealth of Virginia's Wetland Monitoring and Assessment Strategy. Office of Wetland and Water Protection, Richmond, VA. 14 pp. + app.

Wardrop, D.H., Kentula, M.E., Stevens Jr., D.L., Jensen, S.F., Brooks, R.P., 2007. Assessment of wetland condition: an example from the Upper Juniata Watershed in Pennsylvania, USA. Wetlands 27 (3), 416–431.

Wardrop, D.H., Kentula, M., Brooks, R.P., Fennessy, S., Chamberlain, S., Havens, K., Hershner, C., 2013. Monitoring and assessment of wetlands: concepts, case studies, and lessons learned. In: Brooks, R.P., Wardrop, D.H. (Eds.), Mid-Atlantic Freshwater Wetlands: Advances in Wetlands Science, Management, Policy, and Practice. Springer, NY.

Chapter 4.4.2

A Rapid Assessment Method for the Continental United States: USA-RAM

M. Siobhan Fennessy* and Josh Collins†

*Department of Biology, Kenyon College, Gambier, OH, United States, †San Francisco Estuary Institute and Aquatic Science Center, Richmond, CA, United States

Chapter Outline

INTRODUCTION

The increasing pressure that human activities are having on wetland ecosystems (Brinson and Malvarez, 2002; Kentula et al., 2004) has generated considerable interest in developing methods designed to assess the wetland condition. As discussed elsewhere in this book, rapid assessment methods (RAMs) have benefits such as requiring less time in the field and less taxonomic expertise than more quantitative methods, leading to cost savings and potentially larger sample sizes. They often address the need expressed by many wetland managers for a single (often numerical) score for the overall condition, quality, or health of a wetland. For these reasons, RAMs have a key role in the implementation of wetland monitoring and assessment programs and the effective management of the resource (Fennessy et al., 2007).

The USA-Rapid Assessment Method (USA-RAM) was developed as an integral component of the suite of methods used in the 2011 National Wetland Condition Assessment (NWCA) from the U.S. Environmental Protection Agency (U.S. EPA). The three primary objectives of the NWCA were to: (1) report the ecological condition of the nation's wetlands, (2) build state and tribal capacity for wetland monitoring and assessment, and (3) advance the science of wetland assessment. USA-RAM helps meet the first objective by providing relatively less expensive, semiquantitative measures of overall wetland health that complement the more quantitative and expensive NWCA methods for assessing particular aspects of wetland condition or stress. It helps meet the second objective by serving as a RAM template for consideration by states and tribes that wish to adapt it to assess the ecological condition of wetlands they manage. To help meet the third objective, USA-RAM provides data that can support an exploration of the statistical relationships between stress and condition of wetland areas as mediated by their buffers (U.S. EPA, 2016).

CONCEPTUAL FRAMEWORK FOR THE METHOD

USA-RAM is designed for use across the conterminous 48 states. An iterative process of field trials and revisions was conducted over the course of two field seasons, based on the following set of guiding principles or tenets.

(1) The overall condition of a wetland is its capacity or potential to provide its full suite of functions and services, relative to reference sites. It describes the extent to which a given site departs from reference condition by measuring variables that correlate with the integrated combination of the chemical, physical, and biological processes that maintain the system over time.

Wetland and Stream Rapid Assessments. https://doi.org/10.1016/B978-0-12-805091-0.00007-4

(2) The overall condition of a wetland can be assessed partly in terms of the complexity of its form and structure relative to reference sites. For any wetland class, the condition of a wetland area improves as stressors in the wetland and its buffer decrease. USA-RAM was not modified for different wetland types and therefore naturally less complex wetlands tended to get lower scores.

(3) The overall stress on a wetland is the sum total and extent of human-caused (anthropogenic) activities and events that are likely to cause wetland form and structure to diverge from reference condition for that wetland class.

(4) The overall stress on a wetland can be assessed as the combined measures of the abundance, diversity, duration, and magnitude of common stressors evident within a wetland area and its buffer. As the number and extent of stressors accumulates, wetland overall condition declines relative to that of less-impacted reference sites, regardless of wetland type or vegetation community composition.

(5) USA-RAM scores should be responsive to stressors across the full range of form, structure, and stress for all wetland classes and regions, and different practitioners should independently achieve consistent scores for the same locations.

(6) Measures of condition and stress are assessed separately within each wetland area and within its surrounding buffer.

USA-RAM is applied to a 0.5 ha wetland assessment area (AA) and its buffer zone. The buffer zone is defined as the area within 100 m distance from the perimeter of the AA. Ultimately, metrics that assess structure and stressors within a wetland area determine its overall ecological condition, as mediated by its buffer. Thus, the effects of a stressor that originate outside a wetland area are diminished as the stress passes through the buffer, lessening its impact (Fig. 4.4.2.1). In the NWCA application, USA-RAM was used to assess the wetland resource across the sample frame, and not to assess individual wetlands (i.e., wetlands much larger than one AA). USA-RAM can be used to assess large wetlands by assessing multiple randomly located AAs.

DESCRIPTION OF THE METHOD

USA-RAM is organized around four attributes of condition and stress: buffer, hydrology, physical structure, and biological structure (Table 4.4.2.1). Each attribute is assessed using two metrics, except for the hydrology attribute, which is only assessed in terms of its stressors. Metrics of hydrological condition were not developed directly for three reasons: (1) because all aspects of wetland condition are affected by hydrology, its condition is reflected in the other attributes; (2) a survey of how hydrology is treated in other RAMs revealed that it is usually assessed as the amount of departure from natural hydrological conditions due to stress, such that it could be well represented by stressor indicators; and (3) early efforts to develop metrics of hydrological condition indicated that the natural variability of hydrology across wetland classes and regions of the United States was too great to be reasonably represented by a single version of USA-RAM. An assessment of hydrological stressors is critical, however, to account for human activities that alter hydrology and to be better able to interpret the results of the condition assessment. Scores were calculated separately for stressors and condition of each AA and its associated buffer zone.

Each metric consists of a checklist of visible indicators of field conditions, based on reference sites. Narrative descriptions are provided for each indicator, allowing rapid scoring in the field. The data for each metric are used to develop metric scores for the AAs and their buffer zones. The metric scores are unweighted and combined into an overall ecological condition score for each AA (the USA-RAM score).

Stressors are an important component of an assessment because of their effect on condition. Knowledge of the stressors present in and around a wetland is valuable in determining how condition might be improved through management actions.

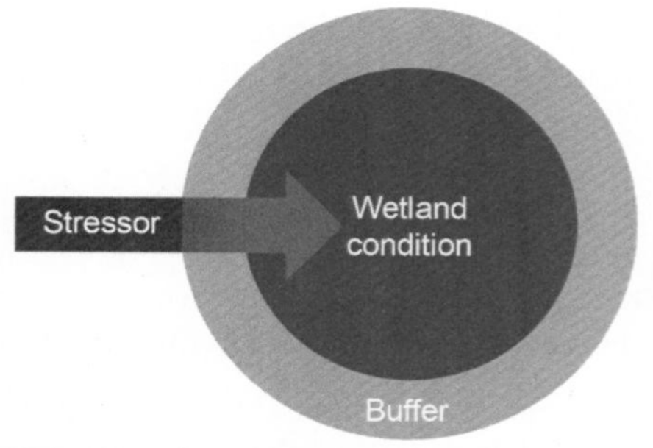

FIG. 4.4.2.1 Conceptual diagram showing the relationship between stressors, buffers, and condition. The effect of a stressor that originates outside a wetland is diminished as it passes through the buffer area that adjoins it.

TABLE 4.4.2.1 USA-RAM Attributes and Metrics of Wetland Condition and Stress

Attributes	Condition Metrics	Stress Metrics
Buffer	Percent of AA having buffer	Stressors to the buffer zone
	Buffer width	
Hydrology	None	Alterations to hydroperiod
		Stressors to water quality
Physical structure	Topographic complexity	Habitat/substrate alterations
	Patch mosaic complexity	
Biological structure	Vertical complexity	Percent cover of invasive plants
	Plant community complexity	Stressors to vegetation

TABLE 4.4.2.2 Guidelines for Assessing Stressor Severity

Description of Stressor Prevalence	Stressor Severity Score
Less than one-third of the buffer or AA is influenced by the stressor	1 (not severe)
Between one-third and two-thirds of the buffer or AA is influenced by the stressor	2 (moderately severe)
More than two-thirds of the buffer or AA is influenced by the stressor	3 (severe)

All stressor metrics are scored based on the number of stressors that are observed (i.e., visibly evident at the time of the assessment) as well as a ranking of their severity. The severity of a stressor is characterized based on the portion of the zone or AA that is obviously influenced by the stressor, as indicated in Table 4.4.2.2.

Assessment Using USA-RAM

There are three metrics designed to evaluate the extent and condition of the buffer zone, four for the condition of the AA, and five to assess stressors impacting the AA (Fig. 4.4.2.2).

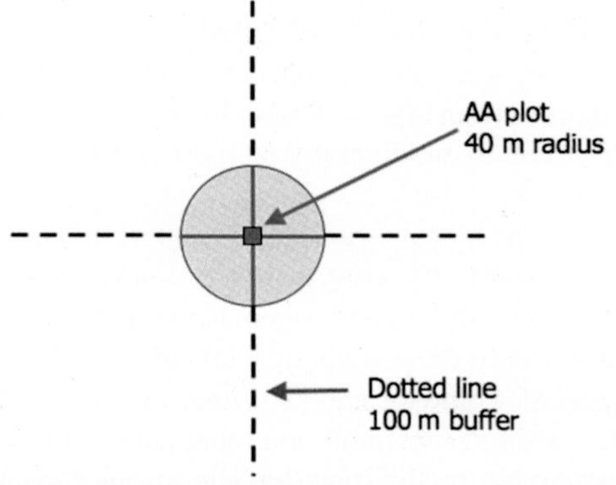

FIG. 4.4.2.2 The buffer zone is the area 100 m distance from the perimeter of the AA. Transects were assessed for land use types and stressor indicators. *(Adapted from U.S. Environmental Protection Agency, 2016. National Wetland Condition Assessment: A Collaborative Survey of the Nation's Wetlands. EPA Publication 843-R-15-005. http://water.epa.gov/type/wetlands/assessment/survey/index.cfm.)*

Section A: Assessment of the Buffer Surrounding the AA

Metric 1: Percent of AA having buffer. The land area adjacent to the AA only qualifies as buffer if it consists of natural land cover types that are capable of "buffering" the AA by protecting it from stress originating in the landscape outside the buffer. This metric tallies the percent of the AA perimeter that adjoins a qualifying buffer land cover.

Metric 2: Buffer width. The ability of an area to buffer a wetland from external stressors depends on the width of the buffer that is present. Minimum effective buffer widths can vary depending on the type of stressors present. A width of 100 m has become a common definition for the sake of assessment in many programs, and land use in the buffer has been found to be correlated with wetland condition (Lopez and Fennessy, 2002). To complete this metric, the average distance that buffer land uses (i.e., natural land covers) extends from the AA perimeter to a nonbuffer land use (human land use types) is calculated, using aerial imagery, by measuring it along eight transects evenly spaced along the AA boundary. The results of the aerial imagery are ground-checked to ensure their accuracy.

Metric 3: Stressor to the buffer zone. This metric tabulates and characterizes the types and severity of stressors that occur in the buffer zone that can act to reduce the effectiveness of the buffer in protecting the AA from human activity in the surrounding landscape. For the sake of this metric, the buffer zone is considered to be the entire 100 m area around the AA, regardless of land use.

Section B: Assessment of Wetland Condition in the AA

Metric 4: Topographic complexity. Topographic relief naturally develops as a result of variations in sediment production or deposition, erosion or oxidation of sediments, sediment "mounding" by vegetation, tree falls, varying hydroperiods, wildlife activity, and some types of human disturbance (e.g., skidder tracks). Increases in both *micro-* and *macro-relief* represent increases in the surface area of a wetland leading to increases in some biological and geochemical processes at the sediment-water or sediment-air interface. It can also represent an increase in biological diversity through an increase in habitat heterogeneity.

Metric 5: Patch mosaic complexity. This metric assesses the horizontal structural complexity of the AA (as viewed from above), a characteristic that is sometimes referred to as *interspersion*. When viewed from above, most wetlands are mosaics of different patches of substrate or plant cover. The complexity of the mosaic consists of the diversity of the component patches and the degree to which they are interspersed. Within a given wetland class, the diversity of natural processes within a wetland mosaic is expected to increase with its overall complexity.

Metric 6: Vertical complexity. The vertical structure of the plant community is defined by the number of plant strata present. Different strata provide different physical and ecological services. For instance, tall vegetation tends to be more efficient at intercepting and holding rainwater, serving as a source of allochthonous inputs and moderating air temperature. The assumption is that more strata provide a greater amount of niche space and broader range of habitat condition as well as more kinds and higher levels of material and energy transformations for the wetland as a whole.

Metric 7: Plant community complexity. This metric evaluates the diversity of plant species that dominate the plant strata. Because different species tend to have different growth patterns and morphometry, an increase in species diversity within a stratum tends to increase its internal architectural complexity. Within a wetland class, the diversity of natural processes and animal species within a wetland is expected to increase with the number and abundance of different plant species.

Section C: Assessment of Stress in the AA

Metric 8: Stressors to water quality. Hydrology has been called the "master variable" that determines the structure, function, and ecosystem services provided by wetlands. Human activities that degrade water quality include discharge from point sources and watershed activities that result in high sediment loads, nutrient runoff, mine drainage, excess salts, etc. As stressors accumulate at a site, functions such as biodiversity support and biogeochemical cycling are compromised, and downstream aquatic systems can become impaired.

Metric 9: Stressors to the hydroperiod. The hydroperiod, or the pattern of water level change over time, affects wetland vegetation community composition and productivity, controls the provision of spawning and nursery grounds for fish and amphibians, and affects migratory waterfowl habitat and biogeochemical processes. Functions such as floodwater storage and flood peak reduction are reflected in the hydroperiods of wetlands.

Metric 10: Stressors to habitat/substrate. Some human activities such as grading, cattle grazing, off-road vehicle use, and vegetation control can severely alter wetland substrates and other parameters of wetland habitats. Some urban wetlands are subject to dumping of yard debris and other trash. Substrate alterations can cause changes in soil quality and drainage that subsequently alter wetland plant communities. Alterations of wetland substrates can lead to invasions by nonnative vegetation.

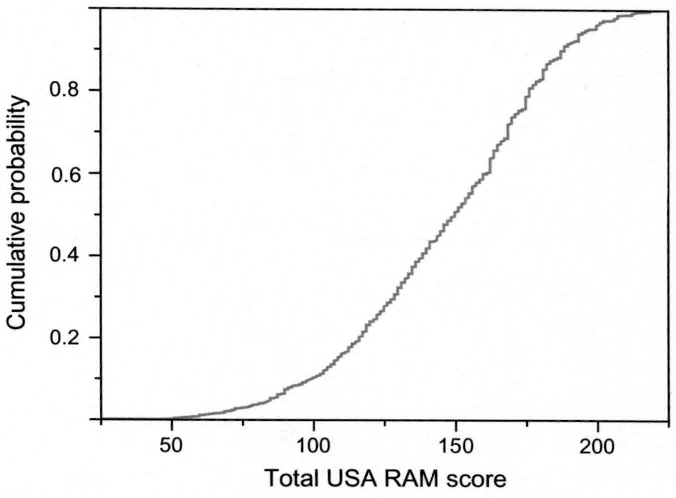

FIG. 4.4.2.3 Cumulative frequency distribution of USA-RAM scores. The possible range of scores is 0–225. While sites at the top end of the condition gradient appear well represented, sites at the low end of the range (<50) are lacking.

Metric 11: The cover of invasive species. Wetland plants are particularly useful as indicators because they are an easily observed, universal component of wetland ecosystems that integrate other aspects of wetland condition or stress. The occurrence (as measured by cover) of invasive species provides a clear and robust signal of human disturbance.

Metric 12: Stressors to the vegetation community. This metric accounts for human activities that directly alter the plant community in the AA. As stressors increase, important wetland services such as biodiversity support and water quality improvement may be affected. Common stressors might include mowing within the AA, excess herbivory, or various management practices to suppress the risk of wildfires.

The cumulative distribution frequency (CDF) of USA-RAM scores for all wetland AAs sampled in the NWCA shows scores ranging from 46 to 220 (the maximum possible range is 0–225) (Fig. 4.4.2.3). While the high end of the range is well represented, the lower end is not; very few AAs had index values <50, and 95% of sites had scores >85. It should be noted that the CDF is based on the number of AAs rather than wetland area. Highly disturbed sites tend to be small and fragmented (Lopez and Fennessy, 2002; Fennessy et al., 2007) and the site selection process in the NWCA seems to have favored larger wetlands, which may explain the lack of lower scoring sites (U.S. EPA, 2016). Furthermore, most sites selected for the survey were much larger than the standard AA, such that the AAs tended to have extensive buffers; 92% of all AAs received the highest score for this metric (i.e., the percent of AA perimeter adjoining a buffer land cover), while only 2% of the AAs were assigned to the lowest scoring category.

Verification of USA-RAM

The USA-RAM was verified after the NWCA fieldwork was completed. While the method was field tested and calibrated across independent users during its development, the fact that it was designed for use in wetlands of all NWCA classes across the conterminous United States meant that collecting the amount of field data necessary to verify or calibrate it before the national survey was not feasible. Therefore, the USA-RAM was evaluated post-hoc based on its ability to distinguish between the least disturbed AAs and most disturbed AAs for each of the 10 NWCA reporting groups, as defined by the NWCA (U.S. EPA, 2016). For each reporting group, the efficacy of USA-RAM was high, as indicated in Fig. 4.4.2.4. For example, for the Central Plains Palustrine Woody sites (CPL-PW), the interquartile range (i.e., between the 25th and 75th percentiles) for the least disturbed sites is well above the range for the most disturbed sites. Palustrine herbaceous wetlands in the Interior Plains (IPL-PH) showed the least difference in scores, indicating a narrow range of overall ecological condition for this group. This ecoregion is one of the most modified by human activities, and herbaceous wetlands are subject to some of the greatest numbers of stressors.

Time Spent in Developing/Testing the Method

The effort to develop USA-RAM began approximately 3 years before the 2011 NWCA field data collection. The USA-RAM lineage stems from inspiration from the Western Washington Wetland Assessment Method (Hruby et al., 1993), which served as the basis for what became the Ohio RAM (Fennessy et al., 1998; Mack, 2001), which in turn was a

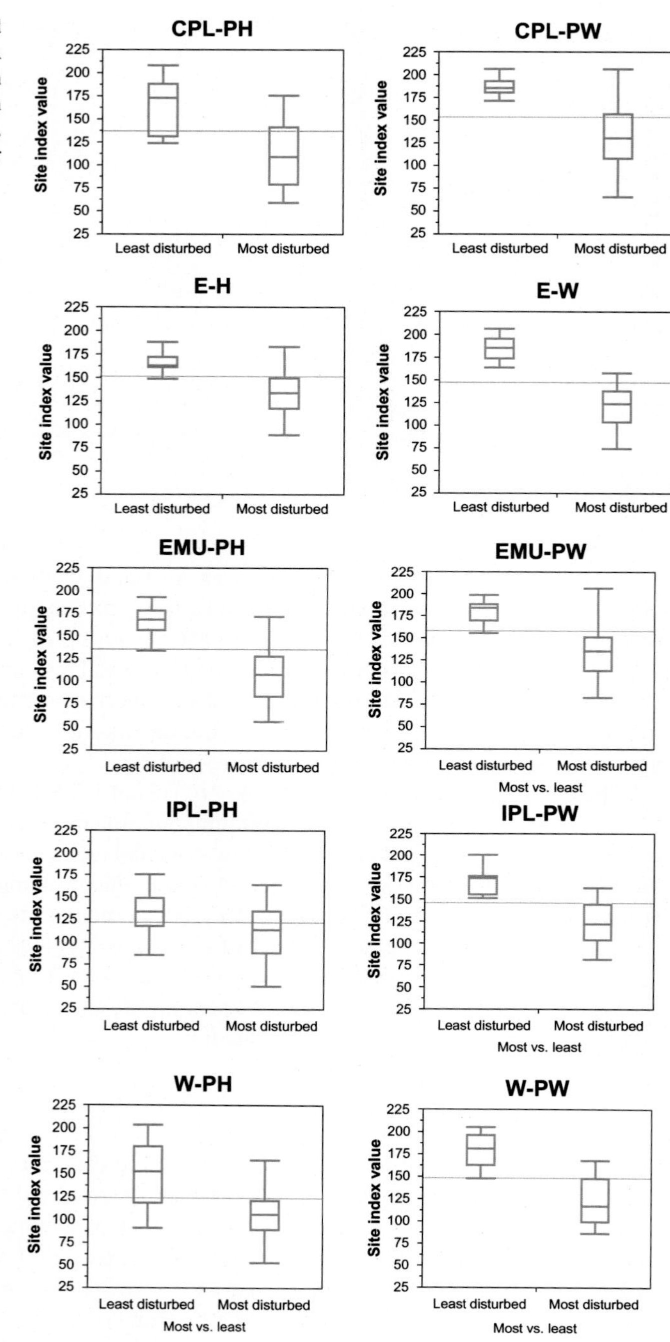

FIG. 4.4.2.4 Box-plots of the USA-RAM scores for the least disturbed and most disturbed AAs (as categorized independently by the NCWA analysis team) for the 10 NWCA reporting groups. Differences in mean scores are highly significant (*t*-tests: $P < .001$) except for IPL-PH with $P < .002$). Note: CPL, Central Plains; E-H, Estuarine herbaceous; E-W, Estuarine woody; EMU, Eastern Mountains Upper Midwest; IPL, Interior Plains; W, Western; PH, Palustrine herbaceous; PW, Palustrine woody.

foundation for what became the California RAM (Collins et al., 2007). The experience gained in these efforts informed the development of USA-RAM, which went through several rounds of development including metric development, field testing, and subsequent metric refinement.

Sample Application of Method in the Field

USA-RAM was used in the 2011 NWCA assessment as part of the effort to: (1) report the ecological condition of the nation's wetlands, (2) build state and tribal capacity for wetland monitoring and assessment, and (3) advance the science of wetland assessment (U.S. EPA, 2016).

Time Spent to Apply the Method in the Field

Like many methods, USA-RAM has an office and field component. Once on site, and depending on the complexity of the site and its buffer, a team of two can complete the method within 3 h.

What Was Learned

USA-RAM demonstrates that it is possible to develop a robust rapid assessment method for use in a diversity of wetland types across broad geographical regions. USA-RAM was designed to assess the full range of form, structure, and stress for all wetland classes and regions in the 2011 NWCA, and was applied consistently by 2011 NWCA field crews, based on quality control procedures.

The sample frame used for surveys of condition across a broad geographic scope can influence the results. In this case, use of the US Fish and Wildlife service status and trends plots to help define the sample frame may have resulted in relatively few smaller wetlands being sampled, resulting in fewer low scores, particularly for the buffer metrics (U.S. EPA, 2016). Similar probability surveys conducted for estuarine wetlands in California have yielded different cumulative frequency distributions than are evident for California from the NWCA (Collins et al., 2007). These data also enabled analyses of the interactions among stress, buffer, and condition (Sutula et al., 2008; Pearce et al., 2016).

Prospects for the Future

Several states have expressed interest in adopting USA-RAM for their own use. For example, Puerto Rico has adapted it as the Antilles Rapid Assessment Method for use in their wetland regulatory program.

REFERENCES

Brinson, M.M., Malvarez, A., 2002. Temperate freshwater wetlands: types, status and threats. Environ. Conserv. 29, 115–133.

Collins, J.N., Stein, E.D., Sutula, M., Clark, R., Fetscher, A.E., Grenier, L., Grosso, C., Wiskind, A., 2007. California Rapid Assessment Method (CRAM) for Wetlands, v. 5.0. California Water Quality Monitoring Council, Sacramento, CA, 151 pp.

Fennessy, M.S., Gray, M.A., Lopez, R.D., 1998. An Ecological Assessment of Wetlands Using Reference Sites. vols. 1 and 2. Ohio Environmental Protection Agency Technical Bulletin. Division of Surface Water, Wetlands Ecology Unit, Columbus, OH. www.epa.state.oh.us/dsw/401/.

Fennessy, M.S., Jacobs, A., Kentula, M., 2007. An evaluation of rapid methods for assessing the ecological condition of wetlands. Wetlands 27, 504–521.

Hruby, T., McMillan, A., Toshach, S., 1993. Washington State Wetlands Rating Systems, second ed. Pub #93–74. Washington Department of Ecology, Lacey, WA.

Kentula, M.E., Gwin, S.E., Pierson, S.M., 2004. Tracking changes in wetlands with urbanization: sixteen years of experience in Portland, Oregon, USA. Wetlands 24, 734–743.

Lopez, R., Fennessy, M.S., 2002. Testing the floristic quality assessment index as an indicator of wetland condition along gradients of human influence. Ecol. Appl. 12, 487–497.

Mack, J.J., 2001. Ohio Rapid Assessment Method for Wetlands, Manual for Using Version 5.0. Ohio EPA Technical Bulletin Wetland/2001-1, Ohio Environmental Protection Agency, Division of Surface Water, Columbus, OH.

Pearce, S., Lowe, S., Collins, J.N., 2016. Relationship Between Wetland Condition, Stress, and Buffer. Report to the California Department of Transportation (CalTrans), http://www.sfei.org/sites/default/files/biblio_files/Task%204%20CRAM%20Stress%20Index.pdf.

Sutula, M., Collins, J.N., Clark, R., Roberts, C., Stein, E., Grosso, C., Wiskind, A., Solek, C., May, M., O'Connor, K., Fetscher, E., Grenier, J.L., Pearce, S., Robinson, A., Clark, C., Rey, K.L., Morrisette, S., Eicher, A., Pasquinelli, R., Ritter, K., 2008. California's Wetland Demonstration Program Pilot. A Final Project Report to the California Resources Agency.

U.S. Environmental Protection Agency, 2016. National Wetland Condition Assessment: A Collaborative Survey of the Nation's Wetlands. EPA Publication 843-R-15-005, http://water.epa.gov/type/wetlands/assessment/survey/index.cfm.

Section 4.5

Other Methods

Chapter 4.5.1

Nearshore Assessment Tool for Alaska: Southeast (NATAK-SE Version 1.0)

Paul Adamus* **and Patricia Harris**[†]

Oregon State University, Corvallis, OR, United States, [†]Marine Biologist (retired), Juneau, AK, United States

Chapter Outline

NATAK-SE is intended to provide a preliminary indication of the relative diversity and importance of the segment as habitat for several biological resources and ecological functions at the scale of a shore segment mapped by the NOAA ShoreZone Program (Harper and Morris, 2014), based mainly on differences in substrate type (rock, sand, mud, etc.) and wave exposure. Shore segments differ in length. The scores and ratings that NATAK-SE assigns to a shore segment aim to represent the segment's relative level of biodiversity and importance as habitat for specific groups. This is based on automated comparison with reference data collected previously using both the rapid and biosurvey components of NATAK-SE at 47 shore segments throughout the region. NATAK-SE is intended to provide a consistent platform for summarizing and applying existing natural resource information. It also provides an approach for collecting and processing new data that is practical and relatively rapid.

NATAK-SE consists of data forms and a spreadsheet. It has a *rapid* component and an optional *biosurvey* component. Both can be completed during a one-day visit to a shore segment (segments are mainly <0.5 km in length). Only the rapid protocol should be used if the user lacks skills at basic identification of seaweeds and intertidal macroinvertebrates. Use of the biosurvey protocol, which adds to the information provided by the rapid protocol, is recommended for persons who do have those identification skills.

During a single visit around the time of daily low tide, NATAK-SE users answer 16 questions ("indicators") based mainly on field observations. In addition, 13 questions are answered using online resources prior to the visit, and additional data pertaining to 29 indicators are pasted into the spreadsheet from a reference table. NATAK-SE then automatically generates scores on a scale of 0 (lowest relative capacity or function) to 10 (highest) that reflect seven attributes of the shore segment (Table 4.5.1.1). No overall score is computed for the assessed shore segment. Shore segments that differ with regard to substrate type, salinity, wave exposure, human disturbances, and other factors can be compared.

The scores and supporting documentation can be used to help evaluate applications for new structures (e.g., piers, bulkheads, bridges) within all or part of a shore segment, or can be used with other tools to help prioritize conservation or restoration opportunities. The NATAK-SE scores could potentially be used to help calculate credits and debits for a particular shore segment, if necessary to compensate for proposed impacts. This being a very recent tool, it has not yet been formally adopted by agencies or used widely. In its present form, it is applicable only to Southeast Alaska. However, it could serve as a template for the development and calibration of RAMs with similar purposes in other coastal regions.

GENERAL PROCEDURE—RAPID COMPONENT

(1) Determine the identification number of the target shore segment by consulting an online map created for this project.
(2) Download and open a database file, available online, that contains attributes derived from geographic information systems (GIS) of every shore segment in the region. Copy the data for the target segment and paste it into the calculator spreadsheet.

Wetland and Stream Rapid Assessments. https://doi.org/10.1016/B978-0-12-805091-0.00006-2

TABLE 4.5.1.1 Attributes of Shore Segments Assessed by NATAK-SE

Resource or Function	Definition
Subsidy Function	Conditions that indicate the capacity of a shore segment and its associated watershed to produce and/or export carbon and associated micronutrients to subtidal waters on a net annual basis.
Food Web Diversity	Conditions that support large number of intertidal macroinvertebrate and seaweed species per unit area.
Focal Fish	Conditions that support relatively large number of focal fish species (salmonids, herring, etc.).
Sea and Shore Birds	Conditions that support relatively high density (concentrations) or large number of bird species that regularly use shoreline habitats: geese, gulls, shorebirds, cranes, some ducks, loons, grebes, cormorants, and alcids.
Pinnipeds	Conditions that support relatively high density (concentrations) of seals or sea lions.
Buffer Wildlife	For songbirds and raptors that regularly use shoreline habitats (shoreline forests and meadows that are within 100ft landward of annual HHW): Conditions that support relatively high density and productivity, large number of taxa, and/or taxa that contribute the most to regional avifauna due to their restricted distribution. Also, conditions that support relatively high density of deer and/or bears at any season.
Filter Function	Conditions that indicate the capacity of a shore segment to detain sediment and/or process some pollutants before those enter subtidal waters.

(3) Make additional measurements of the target segment using Google Earth.
(4) Check local tide tables and plan to arrive at the target segment at least 1 h before the daytime low tide. In addition to a clipboard and the usual field gear, bring the standard data forms and, if possible, a tool to measure salinity, for example, a refractometer.
(5) Answer the questions on the data forms and enter those responses in the calculator spreadsheet.

Scores resulting from only the rapid assessment are automatically computed and shown in a worksheet table contained in the calculator.

GENERAL PROCEDURE—BIOSURVEY COMPONENT

The biosurvey protocol requires at least two people surveying three transects that run perpendicular to the shore, beginning at marine water (where it exists at low tide) and extending upgradient to the annual high water line, usually represented by the tree line, driftwood line, or a road. The survey on each transect requires: (a) estimating seaweed cover in three quadrats, (b) enumerating mussels, clams, periwinkles, and limpets in subquadrats, (c) estimating worm density by excavating, if possible, three shovels of sediment near each quadrat, and (d) recording all seaweed and macroinvertebrate taxa encountered within approximately 1 m on either side of the transect. More detailed instructions are provided in the NATAK manual (Adamus and Harris, 2016).

After completing all three transects, users fill out the parts of the data form that deal with intertidal conditions. Then, close to the time of high tide, users walk through as much of the upland that is within 100 horizontal feet from (inland of) the HHW or tree line and can be safely accessed, and answer the remaining data form questions. After all data are entered in the calculator spreadsheet, scores that reflect inputs from both the rapid component and the biosurvey components are automatically computed and shown in a worksheet table contained in the calculator (Tables 4.5.1.2 and 4.5.1.3).

DEVELOPMENT HISTORY

The work was funded by the Coastal Impact Assistance Program, United States Fish and Wildlife Service, through the Alaska Department of Commerce, Community, and Economic Development, Grant #10-CIAP-0009, "Habitat Mapping and Analysis Project." The authors created the rudiments of the NATAK-SE calculator from their scientific knowledge and a review of the regional literature. A structured workshop of 20 topic experts was then hosted to further identify and discuss important indicators of the region's resources that might be assessed during a single site visit. Next, in order to calibrate a draft

TABLE 4.5.1.2 Field Indicators That Contribute to a NATAK-SE Assessment of Attributes in Table 4.5.1.1
Percent of Intertidal Flooded by High Tide (most days)
Seaweed Cover: Percent of Intertidal
Canopy Kelp and Seagrasses: Percent of Segment Length
Tide Pools
Trees Fallen in Water
Cloudy Water
Human Use Indicators
Bulkheads, Seawalls, and Levees (Shoreline Armoring)
Artificial Muting of Tidal Prism
Potential Disturbance of Wildlife by Boats
Woody Diameter Classes
Alder and Sweetgale Cover
Berry Producers
Wildlife Sign
Salinity

TABLE 4.5.1.3 Indicators From Secondary Sources That Contribute to a NATAK-SE Assessment of Attributes in Table 4.5.1.1
Predominant Coastal Class
Scarcity of This Coastal Class in Region
Subregion
Exposure Class
Slope of Intertidal
Number of Biobands (main)
Number of Biobands (all)
Seagrass
Canopy Kelp
Marsh
Mussel Bioband
Eulachon Spawning
Herring Spawning
Distance to Anadromous Stream
Seal/SeaLion/SeaOtter Concentration
Seabird Density: Summer
Seabird Density: Winter
Deer Wintering Suitability

Continued

TABLE 4.5.1.3 Indicators From Secondary Sources That Contribute to a NATAK-SE Assessment of Attributes in Table 4.5.1.1—cont'd

Bear Habitat Suitability
Number of Mapped Cover Types in Buffer
Salmon Habitat (score for all species), Watershed Scale
Salmon Habitat (score for species with best habitat), Watershed Scale
Estuarine Predominance, Watershed Scale
Marbled Murrelet Nesting Habitat, Watershed Scale
Older Growth Riparian Forest, Watershed Scale
Cumulative Index of Human Activity (TNC index)
Watershed Carbon Output, Spring
Watershed Carbon Output, Fall
Karst Geology
Width of Intertidal Zone: Maximum
Width of Vegetated Intertidal (marsh only) Maximum
Width of Wooded Buffer: Minimum
Percent Slope of Contributing Area
Distance to Nearest Mapped Stream
Distance to Nearest Lake/Pond
Distance to Nearest Town
Distance to Nearest Residence or Busy Access Point
Distance to Nearest Segment of Same Coastal Class
Number of Different Coastal Classes within 1 mile
Number of Segments of Same Coastal Class within 1 mile
Number of Eagle Nests within 2 miles
Notable Bird Concentrations
IBA (Important Bird Area)

of the tool to actual conditions in the region, a series of reference shore segments was identified and assessed. To inform the selection process, existing spatial data with resource themes potentially relevant to this tool were queried using GIS by the Juneau Office of the Nature Conservancy. The resulting Excel database quantifies >50 attributes of each of Southeast Alaska's 88,677 shore segments. Using that database, the author implemented a statistical procedure (k-means cluster analysis) to help objectively select 50 geographically distributed shore segments, one segment per cluster. This procedure has been used by other scientists to select sample sites elsewhere in Alaska (Hoffman et al., 2013) and defines statistical clusters based on similarity, using characteristics compiled initially with a spatial data query. The k-means data processor was instructed to select 50 clusters because we determined that no more than 50 shore segments could be visited and assessed during the single summer available for the field calibration effort.

To create the calibration database, the authors and assistants visited one shore segment per cluster and assessed it using both the rapid component and the optional biosurvey component. The spreadsheet calculator then uses those data to automatically normalize the raw scores it calculates for the seven shore segment attributes. An accompanying report (Adamus, 2016) summarizes the calibration data (algal and macroinvertebrate richness and composition) and describes correlations with environmental factors.

REFERENCES

Adamus, P.R., 2016. Technical Data Report for the Nearshore Assessment Tool for Alaska: Southeast. Prepared for Southeast Alaska Land Trust, Juneau, AK.

Adamus, P.R., Harris, P., 2016. Manual for the Nearshore Assessment Tool for Alaska: Southeast (NATAK-SE Version 1.0). Report and Spreadsheet Calculator Prepared for Southeast Alaska Land Trust, Juneau, AK.

Harper, J.R., Morris, M.C., 2014. Alaska ShoreZone, Coastal Habitat Mapping Protocol. Bureau of Ocean Energy Management, Anchorage, AK.

Hoffman, F.M., Kumar, J., Mills, R.T., Hargrove, W.W., 2013. Representativeness-based sampling network design for the State of Alaska. Landsc. Ecol. 28, 1567–1586.

Chapter 4.5.2

Floristic Quality Index and Forested Floristic Quality Index: Assessment Tools for Restoration Projects and Monitoring Sites in Coastal Louisiana

Kari F. Cretini*, William B. Wood†, Jenneke M. Visser‡, Ken W. Krauss§, Leigh Anne Sharp†, Gregory D. Steyer¶, Gary P. Shaffer‖ and Sarai C. Piazza#

**Cherokee Nation Technologies, Wetland and Aquatic Research Center, Lafayette, LA, United States, †Coastal Protection and Restoration Authority of Louisiana, Lafayette Field Office, Lafayette, LA, United States, ‡School of Geosciences, University of Louisiana at Lafayette, Lafayette, LA, United States, §U.S. Geological Survey, Wetland and Aquatic Research Center, Lafayette, LA, United States, ¶U.S. Geological Survey, Southeast Region, Baton Rouge, LA, United States, ‖Department of Biological Sciences, Southeastern Louisiana University, Hammond, LA, United States, #U.S. Geological Survey, Wetland and Aquatic Research Center, Baton Rouge, LA, United States*

Chapter Outline

INTRODUCTION

In 2003, the Coastwide Reference Monitoring System (CRMS) program was established in coastal Louisiana marshes and swamps to assess the effectiveness of individual coastal restoration projects and the cumulative effects of multiple projects at regional and coastwide scales (Steyer et al., 2003). In order to make these assessments, analytical teams were assembled for each of the primary data types sampled under the CRMS program, including vegetation, hydrology, landscape, and soils. These teams consisted of scientists and support staff from the US Geological Survey and other federal agencies, the Coastal Protection and Restoration Authority of Louisiana, and university academics. Each team was responsible for developing or identifying parameters, indices, or tools that can be used to assess coastal wetlands at various scales. The CRMS Vegetation Analytical Team has developed a Floristic Quality Index (FQI) for coastal Louisiana to determine the quality of a wetland based on the composition and abundance of its herbaceous plant species (Cretini et al., 2012). The team has also developed a Forested Floristic Quality Index (FFQI) that uses basal area by species to assess the quality and quantity of the overstory at forested wetland sites in Louisiana (Wood et al., 2017). Together these indices can provide an estimate of wetland vegetation health in coastal Louisiana marshes and swamps.

The FQI has been developed and used for several regions throughout the United States to provide an objective assessment of the vegetation quality or biological integrity of wetland plant communities. The FQI was first developed as a weighted average of the native plant species at a site (Swink and Wilhelm, 1979). It is based on a coefficient of conservatism (CC) score that is scaled from 0 to 10 and is applied to each plant species in a local flora. The score reflects a species' tolerance to disturbance and specificity to a particular habitat type. Species adapted to disturbed

Wetland and Stream Rapid Assessments. https://doi.org/10.1016/B978-0-12-805091-0.00049-9

areas are often not habitat specific and, as such, have a low CC score. In contrast, habitat-specific species are generally not tolerant to disturbances and, as such, have a high CC score. A group of experts on local plants agrees upon and assigns CC scores.

The FFQI, which is similar to the FQI, was developed to evaluate ecosystem structural changes among forested wetland sites. The FFQI will be used to (1) evaluate forested wetland sites on a continuum from severely degraded to healthy, (2) assist in defining areas where forested wetland restoration is needed, and (3) determine the effectiveness of future restoration projects aiming to return degraded forested wetlands to healthy ecosystems. While the FQI is based on the percent cover of emergent herbaceous species, the FFQI uses this emergent herbaceous layer data in conjunction with the basal area at a species level and canopy cover. As such, the FFQI is a natural extension of the FQI and can be used in conjunction with the FQI of the understory herbaceous community in forested wetland systems, as there is typically an inverse relation between tree and herbaceous layer vegetation dominance in Louisiana's coastally restricted forested wetlands that represents natural succession (Conner and Day, 1992a; Shaffer et al., 2009; Nyman, 2014). As environmentally driven temporal shifts occur in the ecosystem, the FFQI contains valuable information that depicts a trajectory in system function. Generally, coastal flooded forested wetlands have transitioned to shrub-scrub; fresh, floating, and intermediate marshes; and open water. Conversely, in a few select locations, such as the Atchafalaya River Delta, the natural deltaic cycle causes the reversal of this trend. In this emerging deltaic environment, the succession of fresh marsh is transitioning into young forested wetlands populated by low value pioneer and disturbance woody species, leading to the development of fledgling swamps (Johnson et al., 1985; Shaffer et al., 1992). These two contrasting successional trajectories occurring within the same coastal system and same monitoring network highlight the need for a multivariable and index approach to site and restoration assessment.

METHODS

Coefficients of Conservatism

A comprehensive list of wetland plant species occurring in coastal Louisiana was first compiled from previous work by the authors and then distributed to a group of 40 Louisiana coastal vegetation experts. The panel of experts then assigned CC scores to each species by using the descriptions in Table 4.5.2.1. A group of eight local experts from the larger group met and (1) combined the scores, (2) resolved inconsistencies in scoring from the larger group, (3) amended the original list of species to include additional species, and (4) by consensus assigned CC scores to the additional species. Groups of plants—including floating or submerged aquatics and nonrooting parasitic plants—are not routinely assigned percent-cover values within coastal Louisiana monitoring projects and programs (e.g., CRMS; Folse et al., 2014). Therefore, species within these groups were excluded from the analysis and were not assigned CC scores. A total of 849 plants were assigned CC scores (Cretini et al., 2012).

TABLE 4.5.2.1 Criteria for Assigning Coefficient of Conservatism (CC) Scores to Plant Species for Coastal Louisiana

CC Score	Louisiana Description
0	Nonnative plant species
1–3	Plants that are opportunistic users of disturbed sites
4–6	Plants that occur primarily in less vigorous coastal wetland communities
7–8	Plants that are common in vigorous coastal wetland communities
9–10	Plants that are dominants in vigorous coastal wetland communities

Nonnative status according to USDA PLANTS Database (USDA, 2008). Vigorous implies that a coastal wetland community is composed generally of native species and that it is minimally influenced by disturbance.

Modified from Swink, F., Wilhelm, G.S., 1979. Plants of the Chicago Region, third ed., Revised and Expanded Edition with Keys. The Morton Arboretum, Lisle, IL; Swink, F., Wilhelm, G.S., 1994. Plants of the Chicago Region, fourth ed. The Morton Arboretum, Lisle, IL; Andreas, B.K., Lichvar, R.W., 1995. Floristic Index for Assessment Standards: A Case Study for Northern Ohio (Wetlands Research Program Technical Report WRP-DE-8). U.S. Army Corps of Engineers, Waterways Experiment Station, Vicksburg, MS.

Herbaceous FQI

The FQI was modified from the standard equation developed by Swink and Wilhelm (1979, 1994; see Eq. (1) in Fig. 4.5.2.1) for coastal Louisiana herbaceous vegetation, (1) to include percent cover (see Folse et al., 2014 for sampling protocol); (2) to include all plant species (native and nonnative); and (3) to scale the score from 0 to 100 (Cretini et al., 2012). The modified FQI (see Eqs. (2), (3) in Fig. 4.5.2.1) was applied to the herbaceous vegetation data collected annually from monitoring stations within the CRMS sites (Fig. 4.5.2.2). These stations are distributed within swamp forests and marshes (fresh, intermediate, brackish, and saline) in coastal Louisiana.

Eq. (2) is used when $TOTAL\ COVER_t$ is 100 or less. By using Eq. (2), a lower FQI score will be calculated in cases where the area including the monitoring station consists of species found in vigorous wetlands (i.e., CC score is 7–10), but the amount of cover is low because of environmental stressors such as drought or prolonged flooding. Eq. (3) is used when the percent cover ($TOTAL\ COVER_t$) in the area including the monitoring station exceeds 100 because of overlapping canopies (i.e., carpet, herbaceous, shrub, tree layers). After computation, FQI scores are scaled from 0 to 100. FQI scores are

$$FQI_{std} = \left(\frac{\sum (CC_i)}{\sqrt{N_{native\ species}}} \right) \quad \text{Eq. (1)}$$

$$FQI_{mod\,t} = \left(\frac{\sum (COVER_{it} \times CC_i)}{100} \right) \times 10 \quad \text{Eq. (2)}$$

$$FQI_{mod\,t} = \left(\frac{\sum (COVER_{it} \times CC_i)}{\sum (TOTAL\ COVER_t)} \right) \times 10 \quad \text{Eq. (3)}$$

CC_i is the coefficient of conservatism for species i;

$N_{native\ species}$ is the total number of native species within a monitoring station;

$COVER_{it}$ is the percent cover for a given species i at a monitoring station at a given time t; and

$TOTAL\ COVER_t$ is the cumulative percent species cover within a monitoring station.

FIG. 4.5.2.1 Equations used to calculate the Floristic Quality Index (FQI). Eq. (1) is the standard equation developed by Swink and Wilhelm (1979). Eqs. (2), (3) are the modified FQI developed for coastal Louisiana wetlands. Eq. (2) is used when percent cover ($TOTAL\ COVER_t$) is 100 or less. Eq. (3) is used when $TOTAL\ COVER_t$ in a monitoring station exceeds 100 because of overlapping canopies.

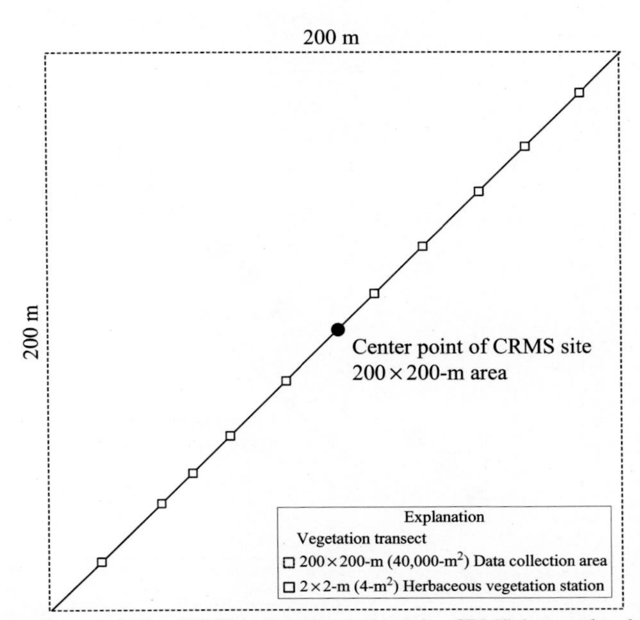

FIG. 4.5.2.2 A typical vegetation station layout within a CRMS herbaceous wetland site. CRMS data used to develop the Floristic Quality Index (FQI) for the CRMS network were collected from each of the 10, 2 × 2 m vegetation stations. The FQI is calculated for each station and then averaged to determine a site FQI. *(Recreated from Folse, T.M., Sharp, L.A., West, J.L., Hymel, M.K., Troutman, J.P., McGinnis, T., Weifenbach, D., Boshart, W.M., Rodrigue, L.B., Richardi, C.C., Wood, W.B., Miller, C.M., 2014. A Standard Operating Procedures Manual for the Coastwide Reference Monitoring System-Wetlands: Methods for Site Establishment, Data Collection, and Quality Assurance/Quality Control. Louisiana Coastal Protection and Restoration Authority, Office of Coastal Protection and Restoration, Baton Rouge, LA, 228 p.)*

calculated annually for herbaceous vegetation monitoring stations (Fig. 4.5.2.2). FQI scores for other spatial scales (e.g., CRMS site, restoration project, hydrologic basin) are calculated by averaging modified FQI scores of monitoring stations that occur within the desired geographic boundary.

Forested FQI

A base FFQI score (see Eq. (1) in Fig. 4.5.2.3) was developed for coastal Louisiana forested wetlands (Wood et al., 2017) using, (1) CC scores provided by a panel of experts (Cretini et al., 2012), and (2) species-specific basal area within CRMS forested monitoring stations (Fig. 4.5.2.4). Based on a literature review, a basal area of 80 m^2/ha was considered to be the total possible basal area occurring in healthy Louisiana coastal forested wetlands (Table 4.5.2.2). Where the sum of basal areas by species at an overstory station within a CRMS site (Fig. 4.5.2.4) at time t is less than or equal to 80 m^2/ha, the base FFQI (Eq. (1) in Fig. 4.5.2.3) was calculated. This equation allows for stations containing sparse cover with vigorous (high CC score) trees and stations containing dense cover with less desirable (low CC score) species to have a similarly low index value. Highly dense stands (i.e., basal areas greater than 80 m^2/ha) were lacking from the CRMS data and as such an equation was not developed for these stands.

The diameter at breast height of all trees larger than 5 cm within the three overstory stations was measured to calculate the species-specific basal area (m^2/ha). The basal area for each tree species is collected a minimum of every 3 years. The basal area was multiplied by the species-specific CC score ranging from 0 to 10. The calculated species-specific base FFQI value was then summed across the station and divided by 80 m^2/ha, which was found to be greater than the maximum basal

FIG. 4.5.2.3 Equations used to calculate the Forested Floristic Quality Index (FFQI). Eq. (1) is the base FFQI developed in Wood et al., 2017 and based on a literature review of maximum basal area reported for coastal Louisiana forested wetlands. Eq. (2) is the modified FFQI that incorporates annual changes in herbaceous percent cover and canopy cover at the CRMS site.

$$BaseFFQI_t = \left(\frac{\sum(BASAL\ AREA_{it} * CC_i)}{80}\right) * 10 \quad \text{Eq. (1)}$$

$$FFQI_t = CCM * BaseFFQI_t * ISM \quad \text{Eq. (2)}$$

$BASAL\ AREA_{it}$ is the sum of basal area for species i at an overstory station within a CRMS site at time t;

CC_i is the coefficient of conservatism for species i;

$CCM = 0.75 + 0.005 * Percent\ Canopy\ Cover$;

$ISM = 0.75 + 0.0025 * Percent\ Net\ Indicator\ Species$.

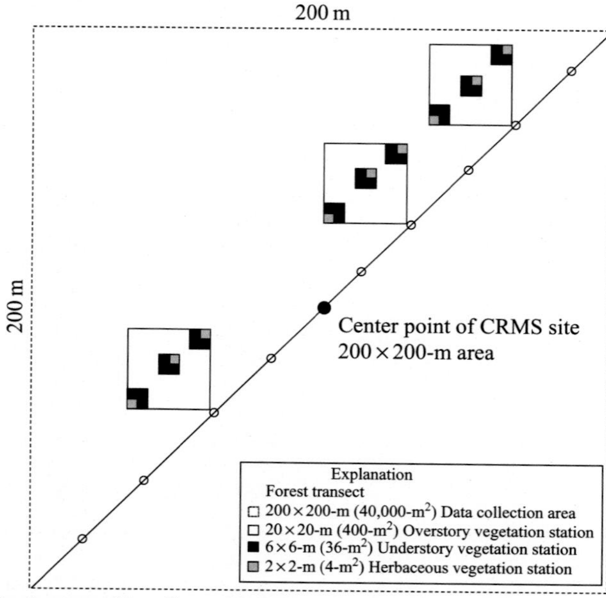

FIG. 4.5.2.4 A typical vegetation station layout within a Coastwide Reference Monitoring System (CRMS) forested wetland site. CRMS data used to develop the Forested Floristic Quality Index (FFQI) for the CRMS network were collected from overstory vegetation stations at each of the 20 × 20 m stations and the 2 × 2 m herbaceous layer vegetation stations.

TABLE 4.5.2.2 Literature Review of Maximum Basal Area Reported for Coastal Louisiana Forested Wetlands

Maximum Basal Area (m^2/ha)	Source
81.6	Conner and Brody (1989)
71.6	Conner et al. (1986)
68*	Shaffer et al. (2009)
65*	Hesse et al. (1998)
54.4	Conner et al. (2002)
70.6	Krauss et al. (2009)
45*	Visser and Sasser (1995)
38.3	Conner and Day (1992b)
35*	Conner et al. (1981)

Values with *asterisk* estimated from publication figures.

area among the CRMS forested wetland sites in Louisiana. This value was multiplied by 10 to scale the final product from 0 to 100. The CRMS station level base FFQI (Eq. (1) in Fig. 4.5.2.3) scores were then modified on the basis of both the station-specific canopy cover and the herbaceous layer vegetation present at the forest floor to incorporate habitat characteristics that quickly respond to stimuli on an annual basis. The canopy cover modifier (CCM) and a herbaceous layer indicator species modifier (ISM) were given equal weight to affect the final FFQI value at each CRMS station (Eq. (2) in Fig. 4.5.2.3).

Canopy cover was recorded annually via a spherical densiometer on a scale from 0.0 to 100.0 at the center of each 20×20 m station. This measurement was recorded in duplicate by separate personnel and not finalized until separate readings were within 20% of one another (Folse et al., 2014). The CCM ranged linearly from fully closed canopy multiplying the base FFQI by 1.25 times to an open absent canopy reducing the station base FFQI by as much as 0.25 (Fig. 4.5.2.5).

$$CCM = 0.75 + 0.005^{*}\ Percent\ Canopy\ Cover$$

The live herbaceous layer vegetation was measured inside three 2×2 m plots per station embedded diagonally in the larger 20×20 m overstory plot (Fig. 4.5.2.4). Species-specific percent cover, total cover, height of the dominant species, and

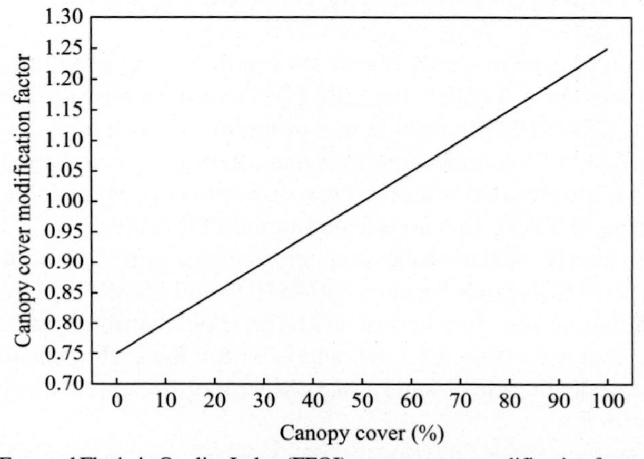

FIG. 4.5.2.5 The relation between the Forested Floristic Quality Index (FFQI) canopy cover modification factor and canopy cover. A modification factor of 1.25 was applied to the station level FFQI for fully closed canopies (100% canopy cover value), where station level FFQI for open canopies was reduced by a modification factor of 0.25.

percent cover of the nonvegetated surface were recorded in each 2×2 m plot (Folse et al., 2014). Species-specific percent covers were estimated by no fewer than two observers agreeing to the nearest whole number from 0 to 100. These data were used in concert with a list of indicator species (Table 4.5.2.3) developed by a panel of Louisiana wetland forest experts and species-specific salinity scores developed for a marsh type assignment algorithm (Visser et al., 2002). This indicator species list consists of two main groups: (1) positive indicator species often seen in association with healthy swamp ecosystems, and (2) negative indicator species common to floating, intermediate, brackish, and saline marshes and disturbed forested wetlands (Table 4.5.2.3). The individual species within a single 2×2 m plot were categorized into positive, negative, or inert percentages of indicator type present and then averaged at the plot level and then again at the 20×20 m station level to spatially correspond to the other constituents of the FFQI. The indicator species cover theoretically ranged from 100% positive to 100% negative with all possible intermediate states.

The ISM was given the same weight as the CCM with 100% positive indicator species modifying the base FFQI by 1.25, whereas a 100% cover of negative indicator species would reduce the base FFQI by as much as 25% (Fig. 4.5.2.6). After the base FFQI was modified via the CCM and the ISM, the new FFQI scores were rescaled by the maximum station level value, thereby creating a 0–100 FFQI scale.

$$ISM = 0.75 + 0.0025^{*}\, Percent\, Net\, Indicator\, Species$$

This approach allows the FFQI to change annually even as basal area and species composition of the overstory plots remain fixed for 3 years, by using the yearly collected canopy cover and herbaceous layer indicator species data.

DISSEMINATION

Herbaceous FQI

FQI scores are calculated annually for each of the 334 CRMS marsh sites and the 58 swamp sites. Scores for each year are displayed graphically, along with species composition data, on the CRMS website (http://www.lacoast.gov/crms2/Home.aspx). These graphical displays can be done for one site at one time (year), for one site across time, for a site compared to scores of the same marsh type and within the same hydrologic basin, and as a report card. These scores can also be displayed spatially with a mapping feature available on the website. This versatility allows managers to view the data from various extents and to compare sites and/or restoration projects. The FQI is an important component of a suite of indices and tools that allows informed management decisions to be made based on observations from the monitoring network compared across various spatial and temporal scales.

The CRMS site-level FQI scores during the 2016 survey year ranged from 0.00 to 92.00. The range of scores displayed on a map of the Louisiana coast and classified by quartiles gives a visual representation of scores across the coast (Fig. 4.5.2.7). CRMS sites with FQI scores of 38.00 and lower are within the 25th quartile and are regarded as poor (triangles), scores greater than 38.00 and less than or equal to 71.00 are regarded as fair (squares), and scores greater than 71.00 are regarded as sites with healthy vegetation (circles). It appears that each quartile comprises sites spread across the coast without regard to basin or marsh type.

The CRMS site FQI scores can be used to assess vegetation condition over a temporal scale. For example, Figs. 4.5.2.8 and 4.5.2.9 show annual FQI scores for a specific site (CRMS1024 and CRMS0178, respectively) along with the annual species composition at the sites. CRMS1024 is located in a saline marsh with high cover values of *Spartina alterniflora* (Fig. 4.5.2.8) and as such has a high FQI score because *Spartina alterniflora* is a dominant species in coastal marshes. By comparison, CRMS0178, which is also located in a saline marsh, has lower cover values of *Spartina alterniflora* and other species with lower CC scores (Fig. 4.5.2.9); this is reflected in the FQI score.

The CRMS site FQI score can also be used to assess vegetation condition across various spatial scales. Figs. 4.5.2.10 and 4.5.2.11, for example, show the 2016 FQI scores for sites CRMS1024 and CRMS0178, respectively, in comparison to other sites in the same marsh type (saline), in the same hydrologic basin (Pontchartrain), and all other CRMS sites. CRMS0124 scores well when compared to other sites (Fig. 4.5.2.10) and is within the 75th quartile of FQI scores while CRMS0178 scores lower than other saline marsh site scores and most sites within the same hydrologic basin (Fig. 4.5.2.11), as it is between the 25th and 75th quartiles.

TABLE 4.5.2.3 Indicator Taxa That Modify Base Forested Floristic Quality Index (FFQI) Scores

Positive Indicator Taxa	Negative Indicator Taxa
Acer rubrum var. *drummondii* (Drummond's maple)	Saltwater intrusion
Boehmeria cylindrica (smallspike false nettle)	*Distichlis spicata* (saltgrass)
Crinum americanum (seven sisters)	*Iva frutescens* (Jesuit's bark)
Fraxinus caroliniana (Carolina ash)	*Panicum repens* (torpedo grass)
Fraxinus profunda (pumpkin ash)	*Schoenoplectus americanus* (chairmaker's bulrush)
Hymenocallis occidentalis var. *occidentalis* (Northern spiderlily)	*Bolboschoenus robustus* (sturdy bulrush)
Iris virginica (Virginia iris)	*Spartina alterniflora* (smooth cordgrass)
Juncus effusus (common rush)	*Spartina patens* (saltmeadow cordgrass)
Nyssa aquatica (water tupelo)	*Symphyotrichum subulatum* (eastern annual saltmarsh aster)
Nyssa biflora (swamp tupelo)	*Symphyotrichum tenuifolium* (perennial saltmarsh aster)
Phanopyrum gymnocarpon (savanna-panicgrass)	All other species with a salinity score of 2.75 and above (Visser et al., 2002)
Quercus sp. (oak)	Flood stress
Sagittaria latifolia (broadleaf arrowhead)	*Bacopa monnieri* (herb of grace)
Saururus cernuus (lizard's tail)	*Bidens laevis* (smooth beggartick)
Tradescantia ohiensis (bluejacket)	*Habenaria repens* (waterspider bog orchid)
Taxodium distichum (bald cypress)	*Hydrocotyle ranunculoides* (floating marshpennywort)
	Ludwigia peploides (floating primrose-willow)
	Pontederia cordata (pickerelweed)
	Sacciolepis striata (American cupscale)
	Schoenoplectus californicus (California bulrush)
	Typha L. (cattail)
	Zizaniopsis miliacea (giant cutgrass)
	Disturbance/transition to marsh
	Amaranthus australis (southern amaranth)
	Cephalanthus occidentalis (common buttonbush)
	Eleocharis sp. (spikerush)
	Ludwigia sp. (primrose-willow)
	Salix nigra (black willow)
	Sesbania drummondii (poisonbean)
	Sesbania herbacea (bigpod sesbania)
	Triadica sebifera (Chinese tallow)

Positive indicator taxa are often observed in association with healthy forested wetland ecosystems. Negative indicator taxa are common in floating, intermediate, brackish, and saline marsh types and in disturbed forested wetlands.

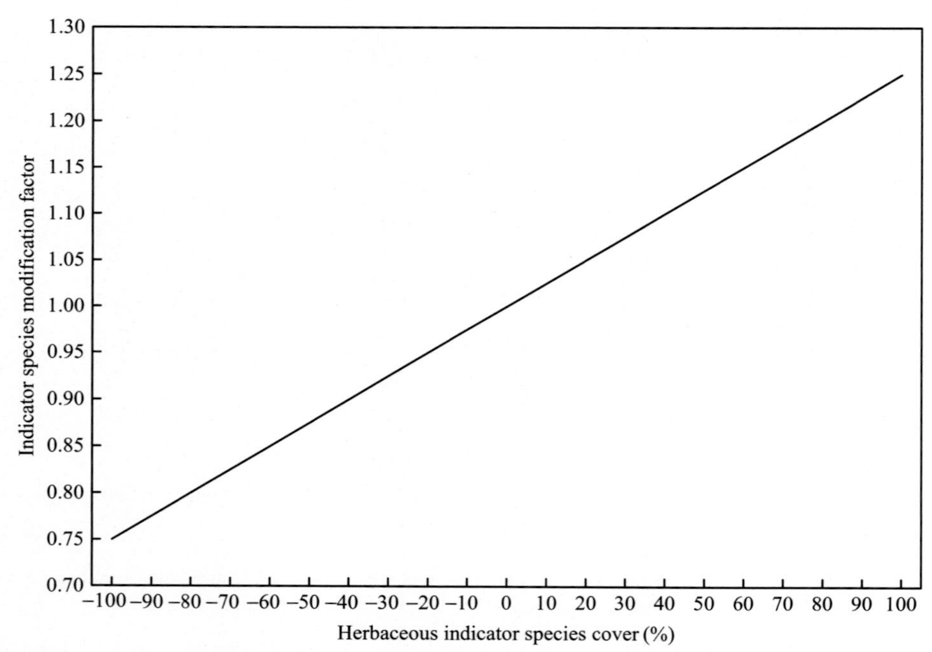

FIG. 4.5.2.6 The relation between the herbaceous indicator species cover and the Forested Floristic Quality Index (FFQI) canopy cover modification factor. The canopy cover modifier with 100% positive indicator species modified the base FFQI by a modification factor of 1.25, and 100% cover of negative indicator species reduced the base FFQI by a modification factor of 0.25.

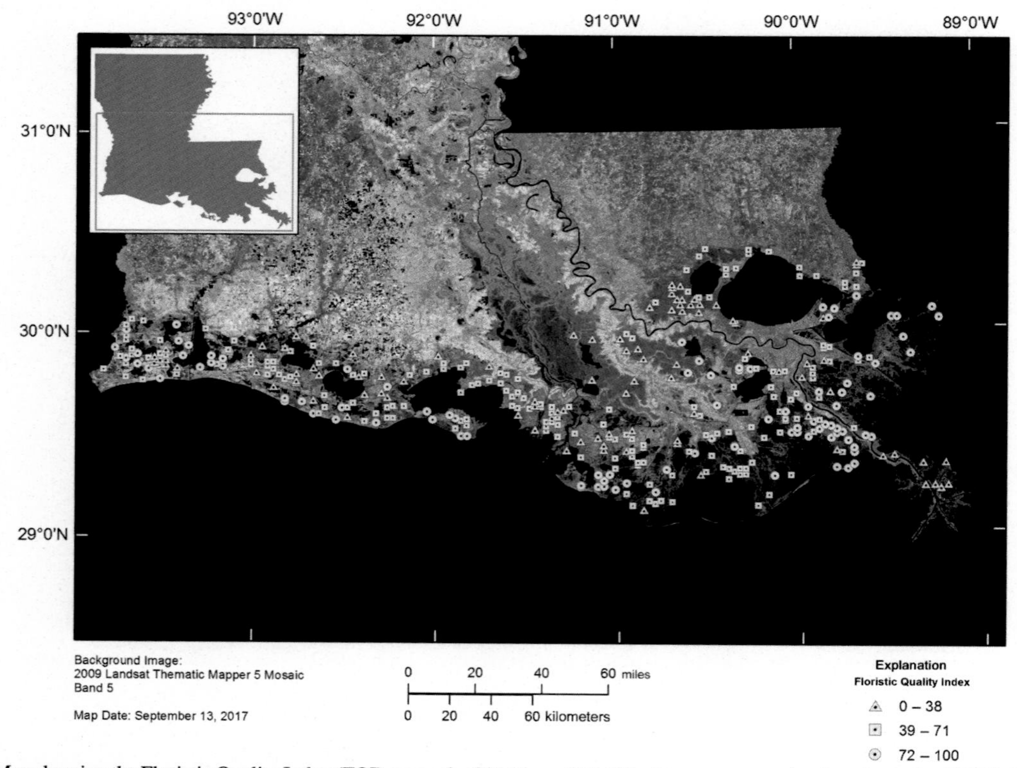

FIG. 4.5.2.7 Map showing the Floristic Quality Index (FQI) scores for 2016 from CRMS herbaceous vegetation data (includes CRMS sites in both marsh and swamp) separated into the 25th and 75th quartiles. Site FQI scores in the 25th quartile, designated as triangles, are below 39.00. Site FQI scores between the 25th and 75th quartiles are designated as squares and site FQI scores in the 75th quartile are designated as circles (FQI scores greater than 71.00).

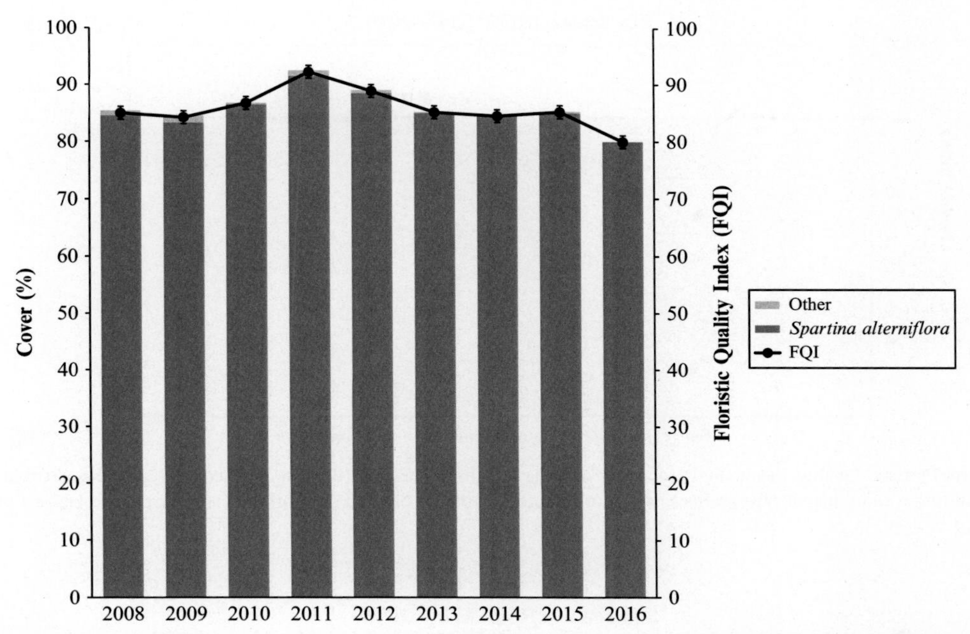

FIG. 4.5.2.8 Floristic Quality Index (FQI) scores for saline marsh site CRMS1024 by year shown with the percent cover values of the species present at the site.

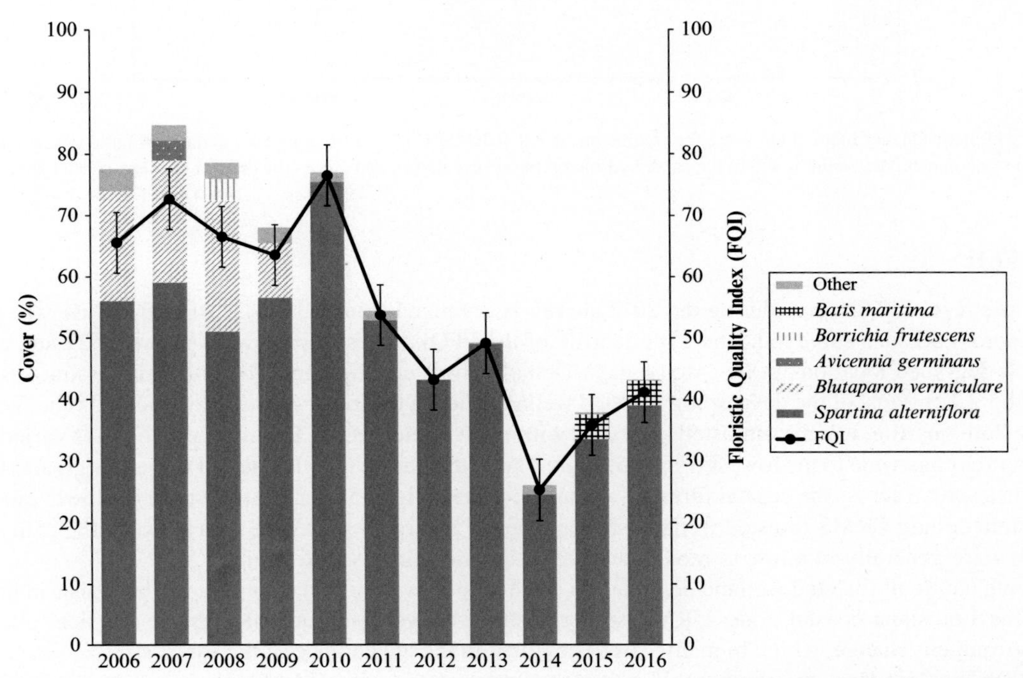

FIG. 4.5.2.9 Floristic Quality Index (FQI) scores for saline marsh site CRMS0178 by year shown with the percent cover values of the species present at the site.

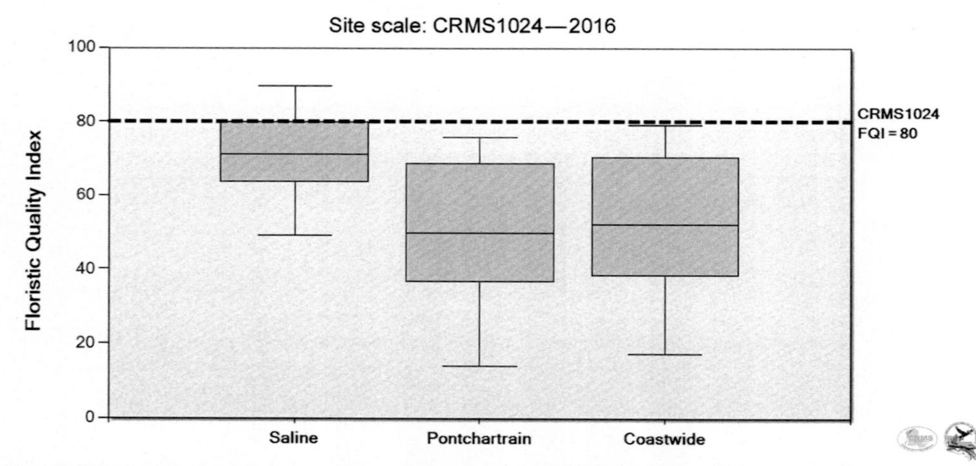

FIG. 4.5.2.10 The Floristic Quality Index (FQI) score for a saline marsh site (CRMS1024) in 2016 compared to the distribution of scores for all coastwide sites within the same marsh type (saline), within the same hydrologic basin (Pontchartrain), and across the entire Louisiana coastal zone (coastwide).

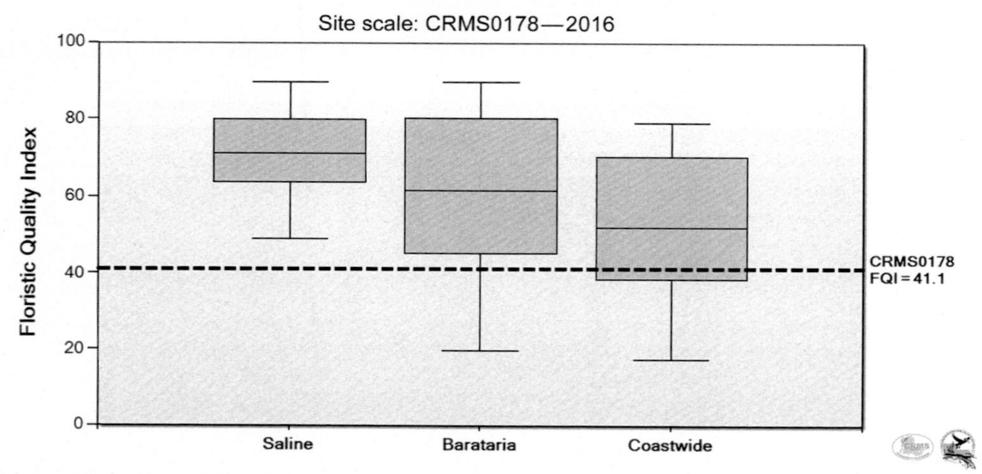

FIG. 4.5.2.11 Floristic Quality Index (FQI) score for a saline marsh site (CRMS0178) in 2016 compared to the distribution of scores for all coastwide sites within the same marsh type (saline), within the same hydrologic basin (Barataria), and across the entire Louisiana coastal zone (coastwide).

Forested FQI

The CRMS site level FFQI scores during the 2015 survey year ranged from 0.39 to 78.01 with an average score of 33.02 (SE = 3.05) across all sites surveyed. The 25th quartile of the FFQI scores ($n = 14$) was below 11.27 and contained both emergent and degraded locations in the Atchafalaya Delta/Teche-Vermilion and Pontchartrain Basins, respectively (triangles, Fig. 4.5.2.12). Most of the developing forested wetland sites in the lower Atchafalaya Delta/Teche-Vermilion Basin scored in the 25th quartile, which contrasted with sites within the Pontchartrain Basin, where the sites varied from some of the highest scores coastwide to the lowest (Fig. 4.5.2.12). The 75th quartile ($n = 14$) had FFQI scores greater than 52.08 and comprised sites spread across the coastal forested wetlands (circles, Fig. 4.5.2.12). Sites within the lower quartile represent both poorly functioning CRMS forested wetland sites and newly emergent sites. The sites between the 25th and 75th quartiles ($n = 29$) were generally in a low to moderate state of deterioration (squares, Fig. 4.5.2.12).

The varying nature of forested wetland degradation coastwide is evident in the continuum that exists in the FFQI scores throughout the Louisiana coastal zone. CRMS sites transition between the quartiles as the physical characteristics of the local environment change, which in many cases is evident in the appearance of the forest structure at each CRMS site (Fig. 4.5.2.13). The 25th through 50th quartile grouping illustrates the general lack of a complete overstory canopy, typically with ample herbaceous layer vegetation (Fig. 4.5.2.13B). The more densely wooded 75th quartile maintains

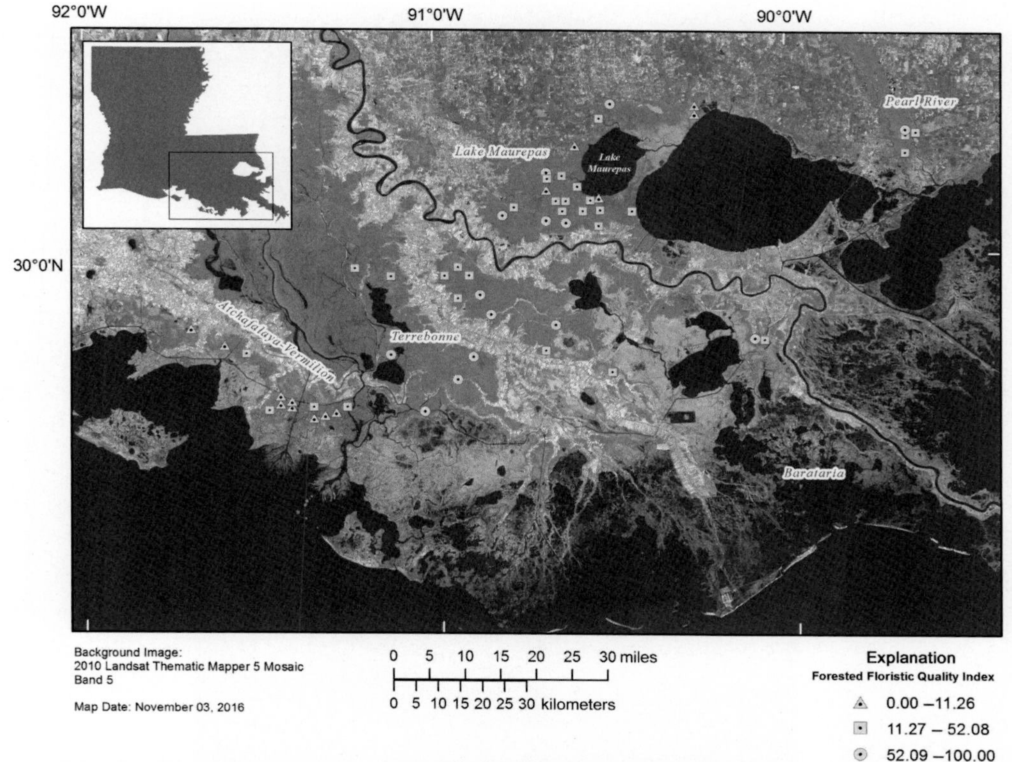

FIG. 4.5.2.12 Map showing the Forested Floristic Quality Index (FFQI) scores for 2015 from CRMS forested wetland sites separated into the 25th and 75th quartiles. Most sites in the lower Atchafalaya-Vermilion Basin scored in the 25th percentile because of low basal area and presence of early successional species. Contrast this with the Lake Maurepas Basin, where the sites varied from some of the highest scores coastwide to the lowest as saltwater intrusion, nutrient limitation, and impoundments near Lake Maurepas degrade the forest structure.

noticeably larger canopy structure, reducing the amount of growing season sunlight at the forest floor (Fig. 4.5.2.13C). The more open 25th quartile has a savanna-like visual aspect that generally contains relic individual overstory trees and isolated clusters of overstory trees in areas of increased elevations, often along waterways (Fig. 4.5.2.13A).

The site-level FFQI can be computed annually as canopy cover and indicator species data are collected by using the previously collected (every 3 years) basal area by species, which is typically very stable on an annual basis. The average annual percent cover of positive and negative indicator species in the forested plots can display trends in the understory community composition as it relates to forest persistence. These individual CRMS sites can be tracked annually through time as data are collected to determine what specific vegetative attribute is generating change in the site FFQI score, whether basal area and species composition are static (Figs. 4.5.2.14–4.5.2.16). However, as the Louisiana coastal zone does contain active delta formation and deterioration, the FFQI alone cannot differentiate between newly forming forests and severely degrading habitat.

FQI vs. FFQI

A simple linear regression shows the inverse relation between the FFQI and the corresponding herbaceous-only FQI at the same CRMS forested wetland sites (Fig. 4.5.2.17). A degraded forested site, CRMS6209 (Fig. 4.5.2.13A), had an FFQI score of 0.39 in 2015 and an FQI score of 55.82. CRMS0089 (Fig. 4.5.2.13B) had an FFQI score of 30.41 in 2015 and an FQI score of 37.93. In 2015 the densely wooded forested site, CRMS5672 (Fig. 4.5.2.13C), had FFQI and FQI scores of 53.16 and 4.43, respectively.

The FFQI and FQI are inversely correlated to one another when a continuum of forested wetland degradation is assessed, as basal area and stable late stage successional woody species are replaced with more disturbance-oriented shrubs

FIG. 4.5.2.13 Coastwide Reference Monitoring System (CRMS) forested wetland sites and their associated 2015 Forested Floristic Quality (FFQI) scores. (A) Site CRMS6209 and (B) site CRMS0089 are located in the Lake Maurepas Basin and scored 0.39 (25th quartile) and 30.41 (25th–75th quartile), respectively. (C) Site CRMS5672 is located within the Barataria Basin and received an FFQI score of 53.16 (75th quartile).

and multiple herbaceous vegetation communities that change depending on the cause of habitat transition and light availability. There are also many subtle relations between the two indices on a specific scale. For example, an upward trend in FQI coupled with a stable FFQI score over time may indicate that the location is losing sapling and midstory stems while maintaining the overstory stress, thereby beginning the transition toward more herbaceous vegetation. Alternatively, sites with low but increasing FFQI scores may be indicative of a succession from a marsh or shrub-scrub community toward a forested wetland, either through natural processes or as a consequence of restoration activities. A steady decrease in FQI and stable FFQI scores could suggest that herbaceous vegetation quality in an area is declining, either through natural succession toward a mature forested community or through disturbance and/or restoration activities meant to encourage tree colonization and growth.

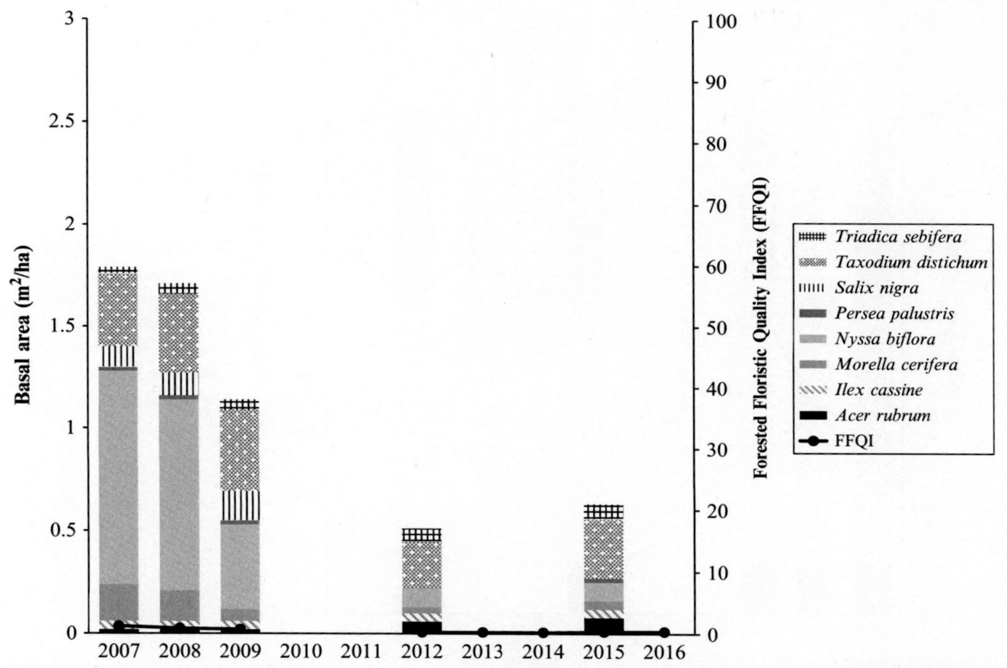

FIG. 4.5.2.14 Forested Floristic Quality Index (FFQI) scores for forested site CRMS6209 by year shown with the basal area (m^2/ha) of the overstory species present at the site.

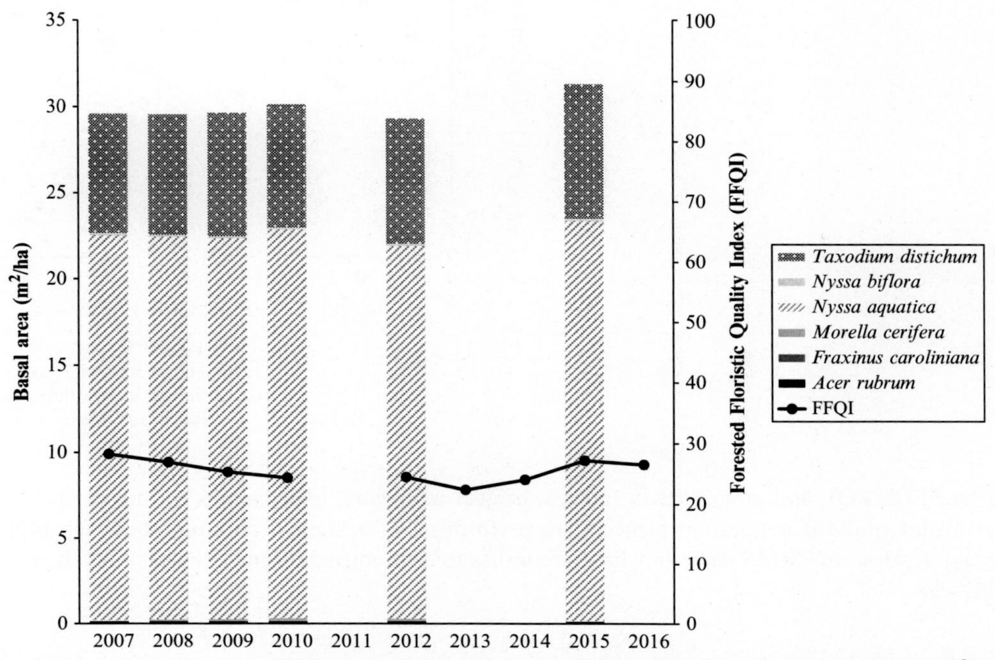

FIG. 4.5.2.15 Forested Floristic Quality Index (FFQI) scores for forested site CRMS0089 by year shown with the basal area (m^2/ha) of the overstory species present at the site.

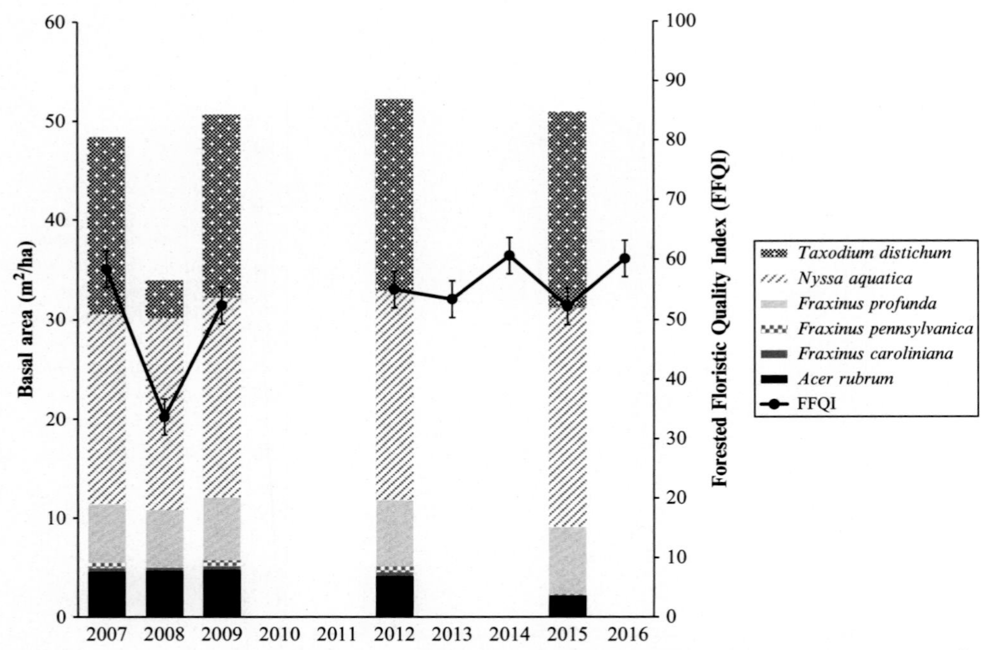

FIG. 4.5.2.16 Forested Floristic Quality Index (FFQI) scores for forested site CRMS5672 by year shown with the basal area (m^2/ha) of the overstory species present at the site.

FIG. 4.5.2.17 The relation between the Forested Floristic Quality Index (FFQI) score and the Floristic Quality Index (FQI) score. A significant negative correlation exists between the two vegetation indices: as the FFQI decreases, the FQI increases, thereby depicting a transitional habitat.

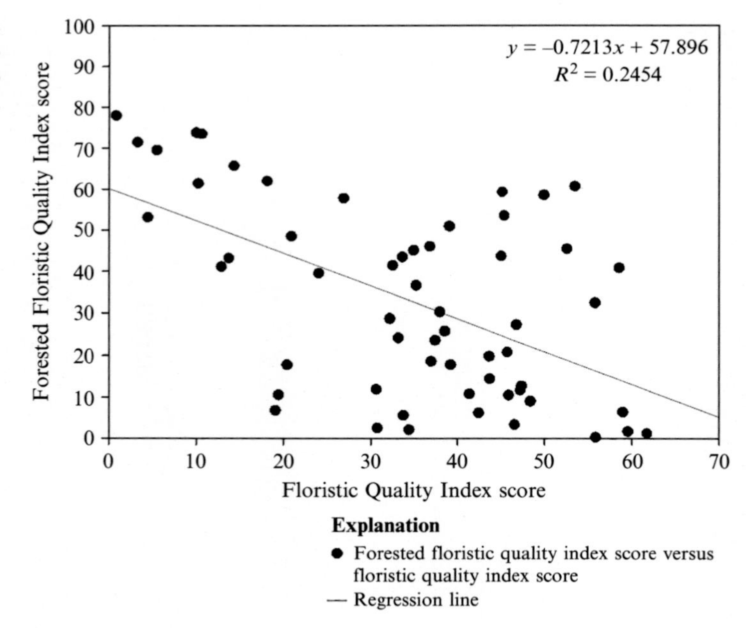

By using the FFQI, FQI, and other CRMS indices, project managers, team members, and the public will be able to evaluate areas to determine if restoration projects are performing as designed. Although both the FQI and FFQI were developed for applications to CRMS data, they have the utility to be modified for many coastal marsh and forested wetland systems worldwide.

REFERENCES

Conner, W.H., Brody, M., 1989. Rising water levels and the future of southeastern Louisiana swamp forests. Estuaries 12, 318–323.

Conner, W.H., Day Jr., J.W., 1992a. Water level variability and litterfall productivity of forested freshwater wetlands in Louisiana. Am. Midl. Nat. 128, 237–245.

Conner, W.H., Day Jr., J.W., 1992b. Diameter growth of *Taxodium distichum* (L.) Rich. and *Nyssa aquatic* L. from 1979-1985 in four Louisiana swamp stands. Am. Midl. Nat. 127, 290–299.

Conner, W.H., Gosselink, J.G., Parrondo, R.T., 1981. Comparison of the vegetation of three Louisiana swamp sites with different flooding regimes. Am. J. Bot. 68, 320–331.

Conner, W.H., Toliver, J.R., Sklar, F.H., 1986. Natural regeneration of baldcypress (*Taxodium distichum* (L.) Rich.) in a Louisiana swamp. For. Ecol. Manag. 14, 305–317.

Conner, W.H., Mihalia, I., Wolfe, J., 2002. Tree community structure and changes from 1987 to 1999 in three Louisiana and three South Carolina forested wetlands. Wetlands 22, 58–70.

Cretini, K.F., Visser, J.M., Krauss, K.W., Steyer, G.D., 2012. Development and use of a floristic quality index for coastal Louisiana marshes. Environ. Monit. Assess. 184, 2389–2403.

Folse, T.M., Sharp, L.A., West, J.L., Hymel, M.K., Troutman, J.P., McGinnis, T., Weifenbach, D., Boshart, W.M., Rodrigue, L.B., Richardi, C.C., Wood, W.B., Miller, C.M., 2014. A Standard Operating Procedures Manual for the Coastwide Reference Monitoring System-Wetlands: Methods for Site Establishment, Data Collection, and Quality Assurance/Quality Control. Louisiana Coastal Protection and Restoration Authority, Office of Coastal Protection and Restoration, Baton Rouge, LA. 228 p.

Hesse, I.D., Day Jr., J.W., Doyle, T.W., 1998. Long-term growth enhancement of baldcypress (*Taxodium distichum*) from municipal wastewater application. Environ. Manag. 22, 119–127.

Johnson, W.B., Sasser, C.E., Gosselink, J.G., 1985. Succession of vegetation in an evolving river delta, Atchafalaya Bay, Louisiana. J. Ecol. 73, 973–986.

Krauss, K.W., Duberstein, J.A., Doyle, T.W., Conner, W.H., Day, R.H., Inabinette, L.W., Whitbeck, J.L., 2009. Site condition, structure, and growth of baldcypress along tidal/non-tidal salinity gradients. Wetlands 29, 505–519.

Nyman, J.A., 2014. Integrated successional ecology and the delta lobe cycle in wetland research and restoration. Estuar. Coasts 37, 1490–1505.

Shaffer, G.P., Sasser, C.E., Gosselink, J.G., Rejmanek, M., 1992. Vegetation dynamics in the emerging Atchafalaya Delta, Louisiana, USA. J. Ecol. 80, 677–687.

Shaffer, G.P., Wood, W.B., Hoeppner, S.S., Perkins, T.E., Zoller, J., Kandalepas, D., 2009. Degeneration of baldcypress-water tupelo swamp to marsh and open water in southeastern Louisiana, USA—an irreversible trajectory? J. Coast. Res. 54, 152–165.

Steyer, G.D., Sasser, C.E., Visser, J.M., Swensen, E.M., Nyman, J.A., Raynie, R.C., 2003. A proposed coast-wide reference monitoring system for evaluating wetland restoration trajectories in Louisiana. Environ. Monit. Assess. 81, 107–117.

Swink, F., Wilhelm, G.S., 1979. Plants of the Chicago Region, third ed., Revised and Expanded Edition with Keys The Morton Arboretum, Lisle, IL.

Swink, F., Wilhelm, G.S., 1994. Plants of the Chicago Region, Fourth ed. The Morton Arboretum, Lisle, IL.

United States department of Agriculture (USDA), Natural Resource Conservation Service, 2008. The PLANTS Database. National Plant Data Center, Baton Rouge, LA. http://plants.usda.gov.

Visser, J.M., Sasser, C.E., 1995. Changes in tree species composition, structure and growth in bald cypress-water tupelo swamp forest, 1980–1990. For. Ecol. Manag. 72, 119–129.

Visser, J.M., Sasser, C.E., Chabreck, R.H., Linscombe, R.G., 2002. The impact of a severe drought on the vegetation of a subtropical estuary. Estuaries 25, 1184–1195.

Wood, W.B., Shaffer, G.P., Visser, J.M., Krauss, K.W., Piazza, S.C., Sharp, L.A., Cretini, K.F., 2017. Forested Floristic Quality Index: An Assessment Tool for Forested Wetland Habitats Using the Quality and Quantity of Woody Vegetation at Coastwide Reference Monitoring System (CRMS) Vegetation Monitoring Stations. (U.S. Geological Survey Open-File Report 2017-1002). 15 p.

FURTHER READING

Andreas, B.K., Lichvar, R.W., 1995. Floristic Index for Assessment Standards: A Case Study for Northern Ohio. U.S. Army Corps of Engineers, Waterways Experiment Station, Vicksburg, MS (Wetlands Research Program Technical Report WRP-DE-8).

Cretini, K.F., Visser, J.M., Krauss, K.W., Steyer, G.D., 2011. CRMS Vegetation Analytical Team Framework—Methods for Collection, Development, and Use of Vegetation Response Variables. (U.S. Geological Survey Open-File Report 2011-1097). 60 p.

Chapter 4.5.3

Ecological Assessment and Rehabilitation Prioritization for Improving Springs Ecosystem Stewardship

Kyle Paffett*, **Lawrence E. Stevens**[†] **and Abraham E. Springer***

**Northern Arizona University, Flagstaff, AZ, United States,* [†]*Museum of Northern Arizona, Flagstaff, AZ, United States*

Chapter Outline

INTRODUCTION

As described elsewhere in this book, adaptive ecosystem management generally requires a secure administrative context; stakeholder involvement and concurrence on clearly defined goals and objectives; inventory-based understanding of the ecological integrity of the ecosystems under consideration, including existing resource conditions and the processes affecting them; assessment of resource conditions, associated values, and risks; planning and implementation of management actions; and monitoring those actions, with feedback to improve effectiveness in relation to program goals (Haynes et al., 1996; Millennium Ecosystem Assessment, 2005; Proctor and Drechsler, 2006; Suter, 2006). Critical to this management formula is the quality of the inventory information and the efficiency and credibility of assessment valuation for prioritization of stewardship actions. Also, consideration of multiple ecological criteria coupled with risk assessment is needed for a sound, structured decision analysis and implementation (Gregory et al., 2006; Gamper and Turcanu, 2007; Gregory et al., 2012). This process is most effective when managers incorporate goals and inventory information to elucidate potential opportunities and trade-offs. However, adaptive management of large ecosystems, such as rangelands, forests, rivers, lakes, and oceans typically involves many stakeholders, and integration of scientific advice into appropriate management actions is often fraught with challenges related to incomplete understanding of basic issues, priorities, and risks; undisclosed agendas and competing interests among stakeholders and scientists; and socioeconomic trade-offs (Gregory et al., 2006, 2012). In contrast, inventory, assessment, and management of smaller ecosystems such as springs often can be undertaken quickly and efficiently (Davis et al., 2011). We suggest that prioritization of ecosystem stewardship requires two steps: (1) credible quantitative scientific inventory and assessment, and (2) integration of that information into the context of the management culture. Here we test this two-step approach using data from US Forest Service springs ecosystems across the southern Colorado Plateau in Arizona.

Springs are surface-linked, groundwater-dependent headwater ecosystems where groundwater emerges and often flows from the Earth's surface (Meinzer, 1923; Springer and Stevens, 2009). Springs often are associated with wetland and stream-riparian ecosystems in both arid and humid environments. Groundwater provides the baseflow for many, if not most, streams and springs water drives much of the hydrology of wetlands. Springs provide unique and essential habitat

Wetland and Stream Rapid Assessments. https://doi.org/10.1016/B978-0-12-805091-0.00051-7

for wetland-specific and many upland species. Springs are usually small, well-defined ecosystems, encompassing <0.01% of the US land area, but are abundant, varying in density from at least 0.01 to 0.2 per km^2 among the coterminous United States (Stevens and Meretsky, 2008). Springs support many federally listed species in the United States as well as thousands of rare, springs-dependent species, and springs often have considerable cultural, historical, and economic significance (Kreamer et al., 2015). However, springs are vulnerable to many human impacts, including groundwater pumping and pollution, and surface-based impacts, particularly flow diversion and livestock grazing practices (Kodrick-Brown and Brown, 2007; Grand Canyon Wildlands Council, Inc., 2002; Anderson et al., 2003; Scott et al., 2004; Unmack and Minckley, 2008). Degradation and loss of springs ecosystems is a global environmental crisis, a challenge that requires improved understanding, tools, and stewardship attention to groundwater and surface water management (Stevens and Meretsky, 2008; Morrison et al., 2013; Hershler et al., 2014). Indigenous tribes, the US Forest Service, several US states, and other organizations have begun to improve inventory information on springs distribution, ecological status, and management potential of springs (e.g., USDA Forest Service, 2012; Springs Stewardship Institute, 2016). Davis et al. (2011) reported that springs ecosystem rehabilitation efforts often were highly successful, and that if the supporting aquifer was functioning, point-source conservation actions can effectively protect critical biological and cultural resources, values, and processes, thereby improving the ecological functioning of large landscapes. Nonetheless, the gap between science and management still prevents or retards progress in springs stewardship.

In this study, we examined and tested application of coupling the Museum of Northern Arizona Springs Stewardship Institute (SSI, 2012) rapid, quantitative springs ecosystem assessment protocol (SEAP) with steward-based definition of stewardship priority criteria (SPC) to improve springs ecosystem rehabilitation planning and implementation. We used field site visit data from 200 randomly selected springs across the semiarid landscapes of Coconino and Kaibab National Forests in northern Arizona during 2012 and 2013 (Fig. 4.5.3.1). Field site visits consisted of single-visit inventories using

SEAP

Spring Name Delinator Date 2017-06-20 Page 8 of 9 Obs JDC

Information Source

Aquifer/WQ	Cond	Risk
Spring dewatered (Y/N)		
Aquifer functionality	6	2
Spring discharge		2
Flow naturalness	6	2
Flow persistance	6	2
Water quality	6	2
Algal and periphyton cover	6	1
Geomorphology		
Site obliterated (Y/N)		
Geomorphic functionality	6	1
Runout channel Geometry	6	1
Soil integrity	6	1
Geomorphic diversity		1
Natural physical disturbance	5	2

Habitat	Cond	Risk
Isolation		2
Habitat patch size		1
Microhabitat quality	6	2
Native plant ecological role	5	2
Trophic dynamics	5	2
Score		
Biotic Integrity		
Native plant richness/diversity	5	2
Native faunal diversity	5	2
Sensitive plant richness	5	2
Sensitive faunal richness	5	2
Nonnative plant rarity	5	2
Nonnative faunal rarity	5	2
Native plant demography	5	2
Native faunal demography	5	2

Human Influence	Cond	Risk
Surface water quality	6	2
Flow regulation	6	2
Road/trail/railroad	5	1
Fencing	6	1
Construction	6	0
Herbivory	5	2
Recreational	5	2
Adjacent conditions	5	2
Fire influence	5	2

Administrative Context	Cond	Risk
Information quality/quantity		
Cultural significance		
Historical significance		
Recreational significance		
Economic value		
Conformance to mgmt plan		
Scientific/educational value		
Environmental compliance		
Legal status		

Notes:

Recommendations: No management action is indicated at this time, although 5 year monitoring would be recommended. The site is subject to surface runoff, & likely influenced by the road upslope. Otherwise it is in very good condition.

Entered by Melissa Carrillo-Galaviz Date 06/27/2017 Checked by Date

FIG. 4.5.3.1 Springs Stewardship Institute (2012) springs ecosystem assessment protocol (SEAP) field sheet.

SSI protocols (Stevens et al., 2012, updated 2016) and SEAP (Stevens et al., 2012). We undertook the second phase of integrated management prioritization by convening a workshop with US Forest Service natural resource managers. We presented inventory and assessment data to the stewards and had them describe criteria that influenced decision-making in relation to management actions. The workshop participants ranked and weighted their SPC variables. We then used inventory and SEAP data to develop quantitative SPC, and conducted further discussions with the forest managers about the administrative context of each springs ecosystem, including water rights information. We archived those data into SSI's Springs Online database (Ledbetter et al., 2014). In concert, these elements were used to develop and prioritize springs stewardship recommendations for the participating forests. This interdisciplinary coupled SEAP and SPC assessment and prioritization approach provided clear, quantitative linkage from science to stewardship through collaborative engagement of the stewards. Although this study was conducted in the arid Southwest, our results are likely broadly applicable to springs stewardship and management of many other types of ecosystems, including springs in arid landscapes throughout the world.

METHODS

Springs Ecosystem Assessment

Several ecological integrity assessment protocols have been proposed for springs ecosystems, including those proposed by Thompson et al. (2002) for White Sands Proving Grounds, Sada and Pohlmann (2006) for the National Park Service in Mojave and Chihuahuan Desert lands, Stevens et al. (2012) for Colorado Plateau National Parks, the Bureau of Land Management's (Prichard et al., 1998a, b, 2003) proper functioning condition for lentic and lotic stream systems, the USDA Forest Service (2012) for groundwater-dependent fens, and the Springs Stewardship Institute's springs ecosystem assessment protocol (SSI, 2012, 2016). These approaches use varying amounts of inventory information, with many similar response variables among these protocols, most except the latter method employing qualitative ("yes-no," "true-false") answers to management assessment questions. Unfortunately, such qualitative information does not readily provide accurate reporting of site conditions, nor does it facilitate prioritization of ecosystem stewardship options.

SSI's SEAP (SSI, 2012, 2016) was designed to facilitate conversations with springs stewards through quantitative comparison of the ecological integrity of, and risks to, springs, either for individual springs or among many springs in large, topographically and geomorphically diverse landscapes (e.g., Springer et al., 2014; Ledbetter et al., 2016; Fig. 4.5.3.1). Inventory and rapid SEAP assessment are conducted on individual springs ecosystems by a team of experts in geography, ecohydrology, and ecosystem assessment. SEAP scoring involves ranking answers to 42 questions within six categories based on quantitative measurements, calculations derived from the SSI Springs Online inventory database (Ledbetter et al., 2014), expert opinions from the inventory team, and information provided by the springs stewards themselves. SEAP information categories include aquifer and water quality, geomorphology, habitat, biology, human influences, and administrative (management) context. The first four of those categories include natural resource variables, each with five to eight subcategory variables, and the human impacts category contains nine variables. All subcategory variables are scored for condition from 0 (low condition) to 6 (pristine condition), and for risk from 0 (no risk) to 6 (high risk). Individual variable scores can be analyzed separately, by category, or averaged to create an overall natural resource condition score. Human risk subcategory and category scores are generally inversely related to the natural resources condition score and to the restoration potential of the site. The administrative context category is designed to focus discussion between the springs stewards and the expert team to clarify desired management conditions and futures. Scoring criteria are defined for each variable's condition and risk score (SSI, 2016; SpringsData.org).

The SEAP approach also was designed for versatility. SEAP scoring can be conducted at three levels: (1) completion of condition and risk scoring of 42 variables among six categories of information; (2) rapid condition and risk scoring for each of those six categories; or (3) very rapid assignment of natural resource and anthropogenic impacts condition and risk scores for the whole site. Such dimensionality makes the SEAP process attractive for managers with limited time, money, and staffing, but scoring by (2) or (3) above benefits from a sound understanding of the ecological status of the springs being considered. However, SEAP scores are not sufficient in and of themselves to clearly guide management. Nonetheless, SEAP scores can be used individually or in combination with other variables to calculate steward-defined prioritization criteria through multicriteria analysis within and among springs. Springs inventory information is sometimes highly sensitive due to the presence of archeological sites, other culturally sensitive issues, endangered species, and additional steward concerns. Therefore, access to data within Springs Online is password protected by the steward, and access permission is required to ensure the security of intellectual property.

Study Area

We conducted this study on two large US Forest Service land units in Northern Arizona: Coconino National Forest (CNF) and Kaibab National Forest (KNF), both of which occur on the Colorado Plateau physiographic province (Fig. 4.5.3.2). CNF occupies 7371 km^2 (1,821,495 ac) and lies in the north-central portion of Arizona (USDA Forest Service, 2013), extending from the San Francisco Peaks (elevation 2850 m) south to the Mogollon Rim and the Verde River Valley. CNF elevations range from 792 to 3850 m, and springs in CNF occur from 938 to 3475 m on the San Francisco Peaks. Most of CNF is dominated by ponderosa pine (*Pinus ponderosa*), with lower elevations dominated by desert shrub vegetation and highest elevations by alpine tundra. Annual precipitation varies by elevation, ranging from 330 mm on the floor of the Verde Valley to >660 mm/year at Happy Jack Ranger Station in central southern CNF (Western Regional Climate Center, 2014). Flagstaff is the largest community within CNF.

KNF occupies 6474 km^2 and is situated west-northwest of CNF (U.S. Forest Service, 2014). KNF is divided into three ranger districts: the Williams District in the south, the Tusayan District in the central area, and the North Kaibab District north of Grand Canyon. KNF elevations range from 914 to 3175 m. The lowest KNF elevations are vegetated with high-elevation desert shrub species while plateau elevations are covered by ponderosa pine and Gambel's oak (*Quercus gambelii*), and highest elevations support mixed conifer forests. The lowest elevation springs in the KNF lie at elevations near 1115 m and the highest springs lie at or above 2664 m. The town of Williams is the largest community in KNF.

Springs on the southern Colorado Plateau generally discharge either from deep regional karstic aquifers in relatively flat-lying but fractured lower to middle Paleozoic sedimentary rock layers, or from perched aquifers in younger volcanic units on top of Phanerozoic sedimentary sequences (Springer et al., 2016; Tobin et al., 2017). Springs generally occur where aquifer edges are exposed through incision along escarpments, or in channels. Perched aquifers tend to be localized and occur where there is either the expression of contacts and/or volcanic topography. Ledbetter et al. (2016) estimate that total discharge from KNF springs is <100 L/s while discharge from deeper aquifers underlying the forests approaches 10^4 L/s. Sedimentary rock units in the region contain few minerals that are capable of dissolving and influencing the water quality. Groundwater discharge from most of the region's springs is potable, except where deep regional faults mix crustal water with the meteoric water (Crossey et al., 2009; Crossey and Karlstrom, 2012). Application of road salt as a winter ice suppressant and the discharge of municipally treated wastewater pose significant but localized impacts on groundwater quality, but thus far the region has sustained relatively few groundwater quality impacts from mining, agriculture, or industrial activity (Bills et al., 2007).

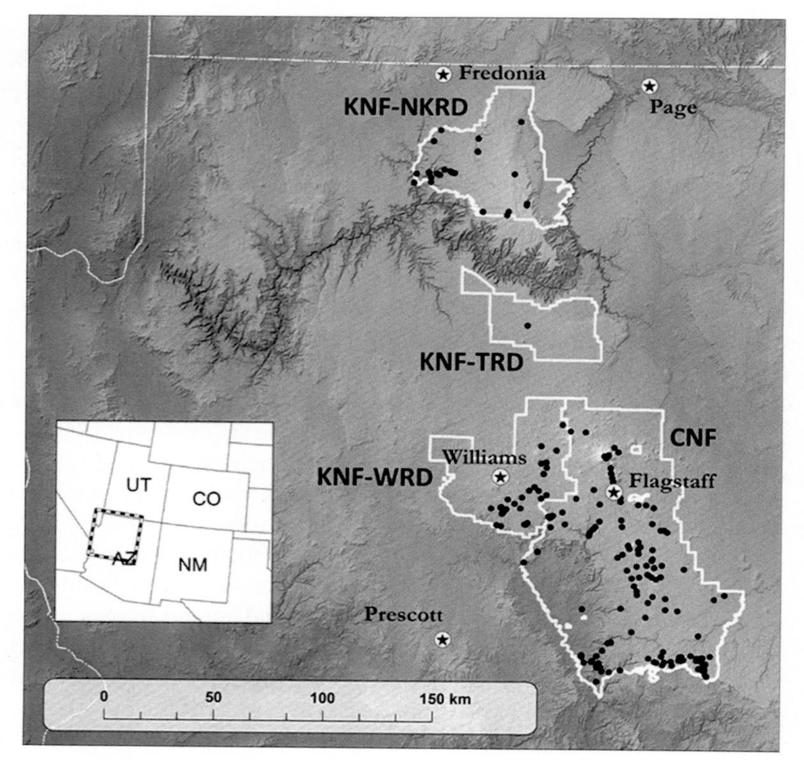

FIG. 4.5.3.2 Map of Coconino National Forest (CNF) and Kaibab National Forest North Kaibab Ranger District (KNF-NKRD), Tusayan Ranger District (NKF-TRD), and Williams Ranger District (KNF-WRD), and the springs selected for this study in northern Arizona.

Study Site Selection

The SSI Springs Online database (2016) maintains information on the reported and inventoried springs of western North America. Springs distribution data for the two forests studied here were compiled from the Arizona State Land Office, the National Hydrography Database, the US Forest Service, the National Park Service, the Grand Canyon Trust, the Grand Canyon Wildlands Council, and databases from individual researchers (http://SpringStewardshipInstitute.org). Springs in the SSI database have varying levels of verification as to their existence and the accuracy of location.

The spatial distribution of CNF and KNF springs was analyzed using a combination of ArcGIS ArcMap (ESRI, 2010) and the SSI's (2012) springs geodatabase. A total of 535 springs were reported in the two forests in the SSI database, from which we randomly selected a subset of 200 (Fig. 4.5.3.2). A random selection of study sites was generated with the Matlab random number generator (The MathWorks, Inc., 2012). The selection of the 200 study springs was stratified by the percentage of springs reported in each national forest: CNF contains 70% of the reported springs in both forests and therefore 70% of the springs selected for assessment were chosen from CNF, with the remaining sites randomly selected from KNF.

Field Data Collection

Site visits were conducted during the summers of 2012 and 2013, during or immediately after the peak of the growing season. We verified the locations of 153 of the 200 selected springs. The unverified remainder of springs had been mismapped or flowed so rarely that no evidence of spring flow, geomorphology, or biota were present. Thus, we inventoried, assessed, and prioritized stewardship potential for a total of 153 randomly selected springs for this study.

The inventory team used an abbreviated SSI springs Level II inventory protocol (Stevens et al., 2012, revised 2016) to note, measure, and compile the following information at each springs ecosystem visited: springs type, georeferencing data, site photography, solar radiation budget using a Solar Pathfinder (Solar Pathfinder, Inc., Linden, Tennessee), aquifer and structural geology, geomorphology, flow, water quality (temperature, specific conductivity, dissolved oxygen, pH, oxidation reduction potential, and total dissolved solids), plant species present, and hydrologic data, and drew a sketch map indicating site configuration and locations where measurements were made. Comments and management recommendations for each site were recorded on the field data sheets. Field data sheets were returned to the laboratory for entry into the SSI database. Specimens of unrecognized plants were collected and returned to the laboratory for identification. Plant specimens are curated at the NAU Deaver Herbarium and the MNA McDougall Herbarium, both in Flagstaff.

Following a site inventory, the team scored the ecological integrity and risk to the springs ecosystem using SSI's SEAP criteria (Table 4.5.3.1; SpringStewardshipInstitute.org/SEAP). Springs with moderate condition scores (3–5) and modest negative impacts (2–4) were considered higher priority for stewardship because they were more likely to respond positively

TABLE 4.5.3.1 Springs Ecosystem Assessment Protocol (SEAP) Variable Categories and Subcategory Variables

SEAP Category	Associated Variables
1. Aquifer/groundwater quality	A/WQ (1) Aquifer functionality; (2) springs discharge*; (3) flow naturalness; (4) flow persistence; (5) aquifer water quality; (6) algal/periphyton cover
2. Geomorphology	GEOM (1) Geomorphic functionality; (2) runout channel geomorphology; (3) soils integrity; (4) geomorphic diversity*; (5) natural physical disturbance regime
3. Habitat	HAB (1) Isolation*; (2) habitat patch area*; (3) microhabitat quality; (4) native plant ecological role; (5) trophic dynamics
4. Biota	BIO (1–2) Native plant and faunal richness, diversity; (3–4) sensitive plant and faunal richness; (5–6) nonnative plant and faunal rarity; (7–8) native plant and faunal demography
5. Freedom from human impacts	FFHI (1) Surface water quality; (2) flow regulation; (3) road, trail, railroads; (4) fencing; (5) construction (nonflow-related); (6) herbivory; (7) recreation; (8) adjacent conditions; (9) fire
6. Administrative context	AC (1) Extent of information; (2) cultural significance; (3) historical significance; (4) recreational significance; (5) economic value; (6) conformance to management plan; (7) science/education value; (8) status of environmental compliance; (9) legal status

SEAP provides individual subcategory, category, natural resource (categories 1–4), and anthropogenic impacts and risks scores (category 5). Individual subcategory variables are scored for both condition (value) and risk from 0 (low) to 6 (high), and condition and risk scores are averaged to provide category scores. Variables are scored based on site visit data and inventory team expert opinions. *Asterisks* indicate a measured variable or one calculated from inventory data. Scoring criteria are provided in SSI (2012, 2016).

and efficiently to stewardship actions. Springs with extensive negative anthropogenic impacts were considered to be of intermediate priority due to increased costs of stewardship. Springs with high ecological integrity (condition ≥5) and low risk (<2) were considered unlikely to require immediate attention, except perhaps for occasional monitoring, and therefore received low priority ranking.

The SEAP provides relatively coarse, rapid prioritization to support consideration of additional management issues, including project urgency, costs, efficiency, and potential for success in relation to agency goals and mission. The US Forest Service clearly defines its management goals in its motto "Caring for the Land and Serving People": a long-term, multiple resource use management policy designed to maintain long-term productivity (USDA Forest Service, 2002, 2012, 2014). Recently completed (KNF—USDA Forest Service, 2014) or pending (CNF) forest plans emphasize goals of improving stewardship of springs, and the Forest Service has initiated a nationally applicable springs inventory protocol and information management system to improve understanding, planning, and management of groundwater-dependent ecosystems (USDA Forest Service, 2012).

To investigate these additional prioritization considerations, we polled CNF and KNF managers about criteria they used to evaluate potential stewardship actions. We presented the results to them at an agency workshop where SPC variables were ranked in order of managerial importance. Forest managers identified selected 10 issues related to springs management, and their prioritization of those issues provided a weighting scale for each factor (Table 4.5.3.2). We then developed scoring calculations for those SPC variables using springs inventory data and SEAP subcategory scores, which were applied with the weighting factor to generate a unique prioritization score for each springs ecosystem (Paffett, 2014). The lessons learned in undertaking springs rehabilitation on high-priority sites were expected to inform efforts applied to lower-priority, generally more remote sites.

Analyses

We used multiple linear regression (SAS, 2013) to determine relationships between springs conditions and variables that might influence the SEAP and SPC scores. Pairwise analysis between SPCs and natural resource conditions of the springs were regarded as significant if the correlation coefficient (R) exceeded 0.8 or had a P value below 0.05. All 10 weighted SPC values were then summed to generate a total site score to provide a prioritized list of springs in relation to stewardship potential on the two forests. The total site score was calculated as:

TABLE 4.5.3.2 Stewardship Prioritization Criteria and Weighting Values Derived From Discussions With US National Forest Workshop Participants, Coconino and Kaibab National Forests, Arizona

Stewardship Prioritization Criteria	Comments/Questions	Weighting Factor
1. Ease of restoration	How readily can the needed actions be accomplished?	1
2. Water rights	USFS water rights increase agency interest	0.9
3. Presence of federally listed species	Sensitive species are high management priority	0.8
4. Return spring to natural sphere of discharge	What is the original sphere of discharge?	0.7
5. Eradicating or absence of exotic species	What nonnative species exist and how difficult is control?	0.6
6. Location of springs in priority watershed	USFS has greater interest in priority watersheds	0.5
7. Culturally or historically sensitive springs	What indigenous, historical, and socioeconomic resources exist?	0.4
8. Exclusion of ungulates from springs source	How difficult will it be to exclude wild and domestic ungulates?	0.3
9. Increasing accessibility for native animals	Will the action(s) facilitate wildlife habitat use?	0.2
10. Proximity to municipalities	Springs nearer municipalities may signal groundwater management issues	0.1

$$S_i = \Sigma((R_{SPC1i} \cdot W_{SPC1}) + (R_{SPC2i} \cdot W_{SPC2}) + \cdots + (R_{SPC10i} \cdot W_{SPC10})) \quad (4.5.3.1)$$

where S_i is the total weighted site score for spring i; $R_{SCP1i...10i}$ are the 10 SPC prioritization criteria identified by the USFS, each of which was calculated from a unique combination of inventory, geographic information systems, and SEAP condition or risk scores for that springs ecosystem, and which were then ranked from 1 (lowest priority) to 153 (highest priority) among this sample of 153 springs; and $W_{SPC1...10}$ are the weighting values for each of the 10 SPC defined by the Forest Service managers, which ranged from 1.0 down (highest) to 0.1 (lowest) in increments of 0.1. The larger the S_i value, the greater the potential likelihood of rehabilitation success. The 50 most highly ranked springs were reviewed for stewardship recommendations and planning, and that list and the rationale for stewardship was provided to the appropriate forest.

RESULTS

Springs Ecosystem Conditions

A total of 200 reported springs were initially randomly selected from the CNF and KNF for the springs inventory and assessment. Selected springs were located in all Ranger Districts of both National Forests. Field site visits subsequently revealed that at least 76.5% of the reported springs were still functioning springs ecosystems in the two forests; however, of the 200 randomly selected springs that we visited, a total of 47 (23.5%) sites either revealed no evidence of springs flow, geomorphology, biota, or anthropogenic activity within 50 m, or had been mismapped. Inaccurate georeferencing is a common problem in springs inventories (e.g., Stevens and Meretsky, 2008; Ledbetter et al., 2012, 2016). Thus, our overall sample size was reduced to 153 springs.

Of the 153 springs used in this analysis, a total of 43% were hillslope springs, 28% were rheocrenes, 11% were helocrenes, 11% were natural or anthropogenic limnocrenes, and 7% were other types (*sensu* Springer and Stevens, 2009). Overall, the following relative abundance of springs types was observed:

$$\text{Hillslope} \gg \text{Rheocrene} \gg \text{Helocrene} = \text{Limnocrene} > \text{Hanging Garden} > \text{Cave} > \text{Exposure} = \text{Mound} = \text{Gushet}$$

Discharge varied widely among the verified springs, from unmeasurably small seepage at sites like Black Spring to >1200 L/s at Fossil Springs, one of Arizona's largest springs and one that is continuously gauged by the US Geological Survey. Low elevation springs and those emerging from Permian Toroweap and Coconino Sandstone Formations had elevated specific conductance. Springs water temperature was negatively related to elevation and aquifer depth. Springs discharging from shallow perched aquifers sustained reduced seasonal temperature variations while springs from deeper regional aquifers had steadier temperatures and flow (Rice, 2007; Ledbetter et al., 2016). According to the Arizona Department of Agriculture's endangered species list, we found five species of "salvage-protected native plants," including three orchid species (*Habernaria viridis*, *Spiranthes romanzoffiana*, *Platanthera sparsiflora*), *Maianthemum racemosum* (Lilliaceae), and *Veratrum californicum* (Melianthaceae; V. Markgraf, written communication). We did not conduct searches for rare invertebrate species but found several leopard frog populations (*Lithobates* spp.), which are species of management concern, at springs on the south side of the Colorado River and springs in the Verde Valley.

Springs Ecosystem Assessment

The springs visited during this study ranged widely in ecological integrity (condition; Figs. 4.5.3.3 and 4.5.3.4). A total of 118 (77.1%) of the 153 randomly selected springs were in degraded ecological condition (natural condition scores ≤ 3), including 60 (39.2%) that were in poor to obliterated condition (natural condition score ≤ 2). Those in poor condition were typically named and lay in close proximity to roads. For example, Van Deren Spring had a composite natural resource condition score of 3.1 (somewhat degraded condition) and a human influences risk score of 2.8 (moderate risk; Fig. 4.5.3.4). A slight improvement of ecological integrity of that springs ecosystem, coupled with a slight reduction of risk, could enhance its ecological sustainability. Approximately one-third of the springs in CNF and the southern KNF are naturally ephemeral, and many of those selected were dry at the time of the site visit.

We noted great variation in the SEAP condition scores between assessed springs due to differing levels of development and management. Springs often had been developed as a source of water for either agriculture (particularly livestock watering), domestic, or industrial use. The impacts of these uses on the springs depended on the extent of development and the level of use. Higher use of springs by livestock, wild ungulates, or humans often was negatively related to springs ecological integrity.

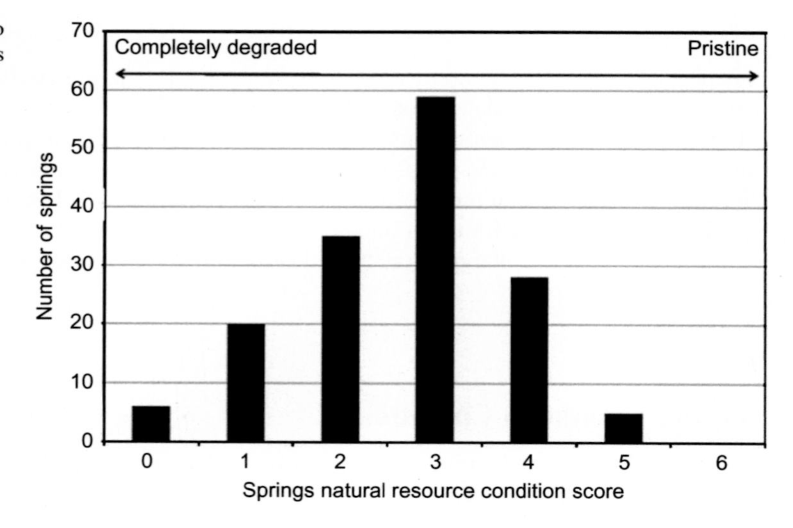

FIG. 4.5.3.3 Frequency of springs ecosystems with low (0) to high (6) natural resource condition scores among those springs selected for analysis in this study.

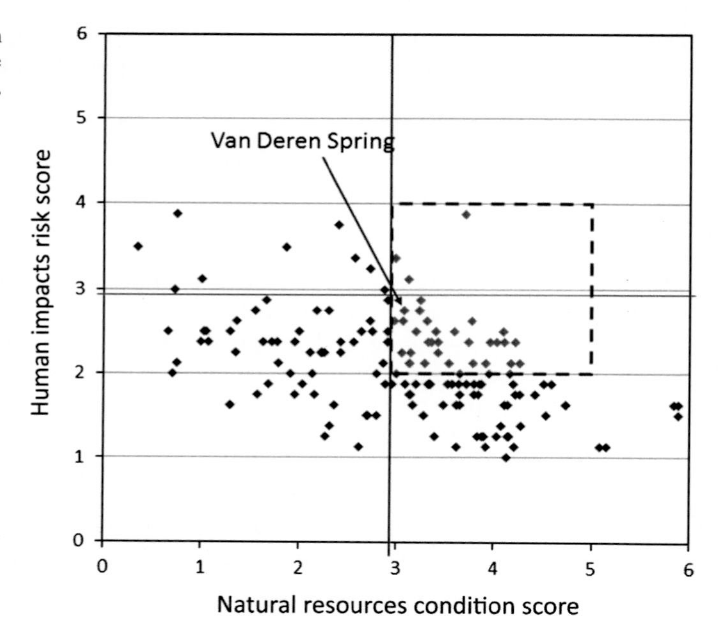

FIG. 4.5.3.4 Natural resource condition score in relation to human influences risk score for 153 springs assessed with the SEAP in the Coconino and Kaibab National Forests in northern Arizona, 2012–2013.

Prioritization

SEAP analysis provided general identification of sites that might readily benefit from stewardship actions, based on natural resource conditions, anthropogenic risk, and management criteria. From that analysis, we identified springs that were in reasonably good ecological condition but had moderate levels of risk. Thirty-seven of the 153 springs in this study were considered likely to respond positively to management attention with a minimum of rehabilitation expense. These included stream channel (rheocrene) and wet fen (helocrene; low-gradient groundwater-fed wet meadow) springs that had been subjected to various levels of development. For example, Jones Springs was a moderately steep wet meadow spring that had sustained some development, but had a relatively high SEAP natural resource condition score of 4.3.

We calculated the SEAP natural resource condition score against the human impacts risk score for all springs, revealing the typical negative relationship between natural resource condition and risk from human influences (Fig. 4.5.3.3). Natural resource condition scores between 3 and 5 and anthropogenic risk scores between 2 and 4 were sites at which stewardship benefit-cost ratios may be maximized. The SEAP condition-risk graphical approach identified 37 springs (24%) that may warrant stewardship attention.

Inventory Variable and SEAP and SPC Scores

Pearson correlation analyses indicated generally limited statistical relationships among SPC variables, except for SPC-8 (exclusion of ungulates) and SPC 9 (increased accessibility for native wildlife). Those criteria were calculated in an identical manner, but to ensure inclusion of management concerns, both scores were included in the final weighted total SPC site score. Positive relationships existed for both SPC 8 and SPC-9 with SPC-5 (ease of nonnative species eradication; $R=0.72$ for both), for both SPC-8 and -9 with SPC-4 ($R=0.40$ for both), between SPC-5 and SPC-4—(geomorphic rehabilitation; $R=0.43$), and between SPC-1 (ease of stewardship) and SPC-10 (proximity to municipalities; $R=0.31$). The generally low and nonsignificant correlation coefficients among these pairwise comparisons indicated that variables used to generate SPC scores were acceptably independent for the response modeling we conducted.

Several relationships between inventory SEAP variables and natural resource condition scores were used to calculate SPC scores, but not all were equally robust. Both proximity to roads and the extent of flow regulation were negatively related to SEAP natural resource condition ($P=0.04$ and 0.01, respectively). However, neither the presence of containment structures ($P=0.96$), nor the elevation difference between the springs and nearest road ($P=0.49$) was statistically related to springs condition. Low correlation coefficients among these variables and natural resource condition scores ($R^2<0.37$) indicated that other variables may negatively affect springs ecological integrity in addition to the accessibility and development of the springs source.

Several of the management criteria influenced ranking of the top priority springs selected for potential stewardship attention. For example, SPC-3 (presence of listed species) was a criterion for most of the higher-ranking springs, and 45 (29%) of the highest ranking 50 springs had elevated ranking due to the large amount of critical habitat designated for the endangered Mexican Spotted Owl (*Strix occidentalis lucida*) in both forests. However, only 14 (9%) of the 153 assessed springs were located in the designated priority watersheds (SPC-6). Conflicting impacts on prioritization among SPC indicate the need for more detailed discussion about criteria choices and interactions within the stakeholder group.

DISCUSSION

While scientists can provide background information on management options and can provide monitoring information to inform future management, society and its managers need to request such advice and information, and decide how to use such information effectively and efficiently (e.g., Schmidt et al., 1998). The coupled SEAP and agency-based springs ecosystem assessment and stewardship prioritization approach undertaken in our study sought to facilitate this process. It differs from other approaches to ecosystem assessment in several respects. Few other efforts quantitatively couple scientific and management valuation to provide guidance to stewards and reciprocally to the scientific community on issues and potential actions. Furthermore, quantified prioritization provides a clear path from science to rationale and decision-making, and stands in marked contrast to evaluation protocols based on, for example, check boxes or nominal data. This is not to say that our approach is perfect. The agency-based priorities and weighting are likely to vary among different stewardship groups. Nonetheless, discussions that arise from stakeholder involvement clarify agency valuation of issues and priorities that reciprocally inform scientists about what information and approaches are most relevant to the decision-makers.

Many stream-riparian ecosystem assessment and management approaches have been proposed (reviewed in Stromberg et al., 2004; Stevens et al., 2005), but such protocols are not necessarily applicable to springs. Springs channel geomorphology and biota often are not reliant on surface flooding disturbance and habitat renewal (Stevens et al., 2005; Griffiths et al., 2008). Ecologically appropriate springs stewardship involves assessment of aquifer integrity and site history, which may be poorly understood and must incorporate information on springs type and function in the surrounding region (Springer and Stevens, 2009; e.g., Burke et al., 2015). Springs ecosystem ecology in general has remained poorly studied, in part because of underrecognized influences of springs type, and the tendency of springs ecosystems to be highly individualistic in terms of ecosystem characteristics and processes, assemblage composition, and responses to anthropogenic impacts. Nonetheless, resource management and conservation can be relatively easily accomplished if assessment provides clear, prioritized guidance and a sound basis for geomorphological and species restoration as well as monitoring, as long as the supporting aquifer is relatively intact.

The goal of this project was to evaluate the effectiveness of a two-stage assessment process for prioritizing stewardship actions using the SSI SEAP rapid assessment methods (SSI, 2012, 2016) and steward-defined weighted ranking method. We tested this approach using inventory data from a large, randomly selected group of springs on CNF and KNF. Such a combination integrated science and management perspectives to ensure that springs managers had the necessary information and understanding to accomplish springs ecosystem rehabilitation. The forest managers chose key attributes needed for springs management, and those issues were then quantified using springs inventory and SEAP data.

The quantitative and expert-based assessment approach of the SEAP and the weighted ranking of steward-based priorities incorporated multiple lines of evidence to provide coupled scientific and managerial stewardship prioritization. Graphing SEAP natural resource condition scores against the anthropogenic risk scores for a large array of springs in the landscape (Fig. 4.5.3.4) allows for rapid identification of a suite of springs potentially warranting management attention. We report that 77% of the springs in these two national forests are ecologically degraded, a finding that is consonant with other springs assessment studies (e.g., Ledbetter et al., 2012 in Nevada; Springer et al., 2014 in Alberta), and which indicate the need for improved assessment and prioritization efforts.

The springs considered for stewardship here had slightly to moderately altered ecological integrity and risk, and therefore were likely to benefit from increased management attention. Springs that were in highly degraded (condition <2) require more intensive stewardship and investment than typically are available, while springs with high condition scores (>5) likely were ecologically intact. Ecological enhancement of highly degraded and at-risk springs or of largely intact springs is generally likely to be costly. Middle scores for condition (3–5) and risk (2–4) are more likely to have higher benefit-cost ratios, and therefore are better candidates for stewardship attention. However, the weighted steward-based metrics applied here were needed to clarify priorities.

We recognize that some springs ecosystems may be of particular management concern (e.g., those supporting endangered species, or those that have high cultural or economic value), even though they are relatively highly degraded. Such sites may warrant enhanced rehabilitation attention, although those costs may be large. For example, the presence of individual endangered and endemic plants, invertebrates, and fish was the motivation for costly rehabilitation of several springs in Ash Meadows, Nevada (Scoppettone et al., 2005; Scoppettone, 2013). Cultural values are often strong drivers for management decision-making in large landscapes, particularly Native American reservations. Our methods emphasized natural resource values more than cultural values, and a more robust quantification of cultural values would be needed to more fully incorporate those values into management prioritization. Identification and stewardship assessment and prioritization of additional sites valued for ecological goods and services and/or cultural values is likely to emerge from scientist-manager discussions, and should be weighted appropriately to achieve agency goals.

Quantification of SPC required combinations of inventory, GIS, and SEAP assessment data through the stewardship workshops and subsequent discussions. The integration of field data with GIS and examination of preexisting data was the most rigorous and informative approach because it incorporated a broad array of indicators (multicriteria assessment) and directly engaged springs stewards in the evaluation process. The 50 highest-prioritized springs were selected for further consideration of stewardship action by the resource managers.

Our analysis identified and refined a different subset of springs for stewardship than did the US Forest Service management criteria method (USDA Forest Service, 2012). Our and the US Forest Service approaches both selected a total of 84 springs of the 153 considered in our study, but only 13 springs were selected in common by SEAP, SPC, and the U.S. Forest prioritization approaches, whereas an average of 27 springs were selected using two of the prioritization approaches, and on average 44 springs were selected using just one of the methods.

Other methods of springs assessment we reviewed provided detail about management requirements through nonquantitative check-box scoring (e.g., Prichard et al., 2003; USDA Forest Service, 2012), but nominal or yes/no, true/false, or presence/absence ordinal information does not readily assist managers to prioritize potential actions among sites. The SEAP approach incorporates field inventory data gathered during the site visits as well as quantitative data compiled through GIS and automated calculations in the Springs Online database.

While the process of science-based, collaborative decision-making was employed to clarify stewardship prioritization, several inconsistencies were identified in the agency-based SPC approach. For example, ease of access (SPC-1) and proximity to municipalities (SPC-10) resulted in selection of springs lying near roads and settlements, even though such proximity increased the likelihood of anthropogenic alteration and colonization by nonnative plant species (SPF-4), and livestock use intensity (SPC-8). In addition, development of springs near roads or communities (SPC-1 and SPC-10) might be more strongly desired by the public but often is more difficult to restore. Resolving such inconsistencies requires follow-up discussion and re-evaluation of importance ranking with the stewards.

The scope of this project limited inventory efforts at each springs ecosystem to basic information on geographic, geologic, geomorphic, botanical, and site conditions and risks. More intensive inventories and monitoring data would provide more robust information and potentially better stewardship guidance. Such information is recommended to improve the utility of these methods. Additional focused inventory and assessment of individual, high-priority springs should include detailed monitoring of flow and water quality variation, habitat quality, and the distribution of aquatic and wetland biota. Both forests are proceeding with more rigorous inventories (e.g., Ledbetter et al., 2016) and revisiting this approach is likely to yield additional refinement to stewardship planning.

MANAGEMENT IMPLICATIONS

Springer et al. (2014), Ledbetter et al. (2016), other regional surveys of springs, and this study revealed that springs in western North America support rare and sometimes endangered species, but often exist in highly degraded ecological conditions. The primary impacts on springs in rural and wild lands in southwestern North America is not necessarily attributable to groundwater overdraft, but is often due to underinformed local flow regulation, livestock and wildlife watering, and recreational management practices. While groundwater overdraft and pollution are common stressors on aquifers and springs in developed landscapes elsewhere, springs rehabilitation options to reduce livestock and recreation management impacts can be relatively easily adjusted to ensure the sustainability of both natural ecological function and anthropogenic use.

Our analysis was conducted on a large, randomly selected suite of springs in two large National Forests. However, not all springs of importance to forest managers were included in the analysis. To ensure that critical springs are not neglected in assessment and prioritization, inventories should be conducted not only on a suitable sample size of randomly selected springs, but also on those of particular management concern within the landscape. This approach was used by Springer et al. (2014) in southern Alberta, and greatly enriched management consideration of valuation and prioritization. Coupled expert and agency-based stewardship prioritization is greatly strengthened with high-quality inventory information; however, our approach also should work in an exploratory fashion in landscapes where springs distribution is poorly known, where springs types and ecological integrity have not been determined, and where the status of springs-dependent species is unknown. The SEAP and agency-based SPC approach also can be used as a monitoring template for treated sites.

Informed by SEAP data and the results of this study, springs ecosystem rehabilitation planning and implementation have been adopted in forest plan development (USDA Forest Service and Coconino National Forest, 2014; USDA Forest Service, 2014). Kaibab National Forest used the SEAP and manager-based discussion prioritization process to select two springs for a pilot rehabilitation project. Castle and Big Springs on the North Kaibab Ranger District were selected for rehabilitation planning and implementation, based on use of the SEAP and discussion among forest managers, the allotment holder, and two Native American tribes. The SEAP also was subsequently used to monitor rehabilitation success at those two springs, revealing improved ecological condition and reduced risk in 1- and 2-year site revisits. Thus, the results from this study are helping guide and refine planning and management of forest springs, and the approach warrants further consideration for stewards interested in adaptive assessment, rehabilitation planning, and monitoring of springs ecosystems.

ACKNOWLEDGMENTS

This project was graciously funded by the Nina Mason Pulliam Charitable Trust, and we thank their staff deeply for that support.Northern Arizona University (project 1002969) and the Museum of Northern Arizona Springs Stewardship Institute provided project administration, and we thank ESRI for supporting our GIS analyses. Data collection and analysis was accomplished by volunteers from the Grand Canyon Trust and the Grand Canyon Wildlands Council, Inc., as well as by employees of Coconino and Kaibab National Forests. This project would not have been possible without the analytic assistance of the MNA-SSI staff, particularly Jeri Ledbetter and Jeff Jenness, whom we thank deeply. We also thank Krista Sparks for coordination of field and laboratory volunteers, and for information management. We thank the many volunteers who contributed to gathering and entering data, particularly Vera Markgraf for her botanical expertise. We thank the editors and anonymous reviewers of the original draft for insightful comments that improved the text.

REFERENCES

Anderson, D., Springer, A., Kennedy, J., Odem, W., DeWald, L., 2003. Verde River Headwaters Restoration Demonstration Project: Final Report. Arizona Department of Water Resources Water Protection Fund. Grant No. 98-059.

Bills, D.J., Flynn, M.E., Monroe, S.A., 2007. Hydrogeology of the Coconino Plateau and adjacent areas. U.S. Geological Survey, Coconino and Yavapai Counties, AZ. Scientific Investigations Report 2005–5222.

Burke, K.J., Harcksen, K.A., Stevens, L.E., Andress, R.J., Johnson, R.J., 2015. Collaborative rehabilitation of pakoon springs in grand canyon-parashant national monument, Arizona. In: Huenneke, L.F., van Riper III, C., Hayes-Gilpin, K.A. (Eds.), The Colorado Plateau VI: Science and Management at the Landscape Scale. University of Arizona Press, Tucson, pp. 312–330.

Crossey, L.J., Karlstrom, K.E., 2012. Travertines and travertine spring in eastern Grand Canyon: what they tell us about groundwater, paleoclimate, and incision of Grand Canyon. In: Timmons, J.M., Karlstrom, K.E. (Eds.), Grand Canyon Geology: Two Billion Years of Earth's History. Geological Society of America Special Paper 489, pp. 131–143.

Crossey, L.J., Karlstrom, K.E., Springer, A.E., Newell, D., Hilton, D.R., Fischer, T., 2009. Degassing of mantle-derived CO_2 and He from springs in the southern Colorado Plateau region—neotectonic connections and implications for groundwater system. Geol. Soc. Am. Bull. 121, 1034–1053. https://doi.org/10.1130/B26394.1.

Davis, C.J., A.E. Springer, and L.E. Stevens. 2011. Have arid land springs restoration projects been effective in restoring hydrology, geomorphology, and invertebrate and plant species composition comparable to natural springs with minimal anthropogenic disturbance? Collaboration for Environmental Evidence, Systematic Review CEE 10-002, 72 p.

ESRI, 2010. ArcGIS: ArcMap Desktop: Release 10. Environmental Systems Research Institute, Redlands, CA.

Gamper, D., Turcanu, C., 2007. On the governmental use of multi-criteria analysis. Ecol. Econ. 62, 298–307.

Grand Canyon Wildlands Council, Inc., 2002. A hydrological and biological inventory of springs, seeps and ponds of the Arizona Strip: Final Report. Arizona Water Protection Fund, Phoenix.

Gregory, R., Failing, L., Ohlson, D., McDaniels, T.L., 2006. Some pitfalls of an overemphasis on science in environmental risk management decisions. J. Risk Res. 9, 717–735.

Gregory, R., Failing, L., Harstone, M., Long, G., McDaniels, T., Ohlson, D., 2012. Structured Decision-Making: A Practical Guide to Environmental Management Choices. Wiley-Blackwell, West Sussex.

Griffiths, R.E., Anderson, D.E., Springer, A.E., 2008. The morphology and hydrology of small spring-dominated channels. Geomorphology. https://doi.org/10.1016/j.geomorph.2008.05.038.

Haynes, R.W., Graham, R.T., Quigley, T.M. (Eds.), 1996. A framework for ecosystem management in the Interior Columbia Basin including portions of the Klamath and Great Basins. US Department of Agriculture Forest Service Pacific Northwest Research Station, Portland. General Technical Report PNW-GTR-374.

Hershler, R., Liu, H.-S., Howard, J., 2014. Springsnails: a new conservation focus in western North America. Bioscience 64, 693–700.

Kodrick-Brown, A., Brown, J.H., 2007. Native fishes, exotic mammals, and the conservation of desert springs. Front. Ecol. Environ. 5, 549–553.

Kreamer, D.K., Stevens, L.E., Ledbetter, J.D., 2015. Groundwater dependent ecosystems—science, challenges, and policy. In: Adelana, S.M. (Ed.), Groundwater. Nova Science Publishers, Inc., Hauppauge, ISBN 978-1-63321-759-1, pp. 205–230.

Ledbetter, J.D., Stevens, L.E., Springer, A.E., Brandt, B., 2014. Springs Online—Springs and Springs-Dependent Species Database. Ver. 1.0. Museum of Northern Arizona Springs Stewardship Institute, Flagstaff, AZ. Available from: springsdata.org (Accessed 23 May 2015).

Ledbetter, J.D., Stevens, L.E., Hendrie, M., Leonard, A., 2016. Ralston, B.E. (Ed.), Ecological inventory and assessment of springs ecosystems in Kaibab National Forest, northern Arizona. Proceedings of the 12th Biennial Conference of Research on the Colorado Plateau. U.S. Geological Survey, pp. 25–40. Scientific Investigations Report 2015-5180.

Meinzer, O.E., 1923. Outline of ground-water hydrology, with definitions. U.S. Geological Survey Water Supply Paper 494, Washington, DC.

Millennium Ecosystem Assessment, 2005. Ecosystems and Human Well-Being: Synthesis. Island Press, Washington, DC. Available from http://www.millenniumassessment.org/documents/document.356.aspx.pdf (Accessed 1 July 2016).

Morrison, R.R., Stone, M.C., Sada, D.W., 2013. Environmental response of a desert springbrook to incremental discharge reductions, Death Valley National Park, California, USA. J. Arid Environ. 99, 5–13.

Paffett, K.P., 2014. Analysis of Springs Assessment Data for Stewardship in the Coconino and Kaibab National Forests, Northern Arizona. Northern Arizona University, Flagstaff. Master's of Science Thesis.

Prichard, D., Bridges, C., Krapf, R., Leonard, S., Hagenbuck, W., 1998a. Riparian area management: process for assessing proper functioning condition for lentic riparian-wetland areas. Technical Reference 1737-11, Bureau of Land Management (45 pp).

Prichard, D., Anderson, J., Corell, C., Fogg, J., Gebhardt, K., Krapf, R., Leonard, S., Mitchell, B., Staats, J., 1998b. Riparian area management: a user guide to assessing proper functioning condition and the supporting science for lotic areas. TR 1737-15, Bureau of Land Management. BLM/RS/ST-98/001+1737.

Prichard, D., Berg, F., Hagenbuck, W., Krapf, R., Leinard, R., Leonard, S., Manning, M., Noble, C., Staats, J., 2003. Riparian area management: a user guide to assessing proper functioning condition and the supporting science for lentic areas. Technical Reference 1737-16. BLM/RS/ST-99/001+1737 +REV03, Bureau of Land Management, National Applied Resource Science Center, CO.

Proctor, W., Drechsler, M., 2006. Deliberative multicriteria evaluation. Eviron. Plann. C. Gov. Policy 24, 169–190. Available from: http://waterwiki.net/index.php?title=Deliberative_Multi-Criteria_Evaluation#Tool_Overview (Accessed 1 July 2016).

Rice, S.E., 2007. Springs as Indicators of Drought: Physical and Geochemical Analyses in the Middle Verde River Watershed, Arizona. M.S. thesis, Northern Arizona University, Flagstaff, AZ.

SAS, 2013. JMP Pro 10. Cary, NC.

Schmidt, J.C., Webb, R.H., Marzolf, R.G., Valdez, R.A., Stevens, L.E., 1998. Science and values in river restoration in the Grand Canyon. BioScience 48, 735–747.

Scoppettone, G.G., 2013. Information to support monitoring and habitat restoration on Ash Meadows National Wildlife Refuge. U.S. Geological Survey Open-file Report 2013-1022.

Scoppettone, G.G., Rissler, P., Gourley, C., Martinez, C., 2005. Habitat restoration as a means of controlling non-native fish in a Mojave Desert oasis. Restor. Ecol. 13, 247–256.

Scott, T.M., Means, G.H., Meegan, R.P., Upchurch, S.B., Copeland, R.E., Jones, J., Roberts, T., Willet, A., 2004. Springs of Florida. Florida Geological Survey Bulletin 66, Tallahassee.

Springer, A.E., Stevens, L.E., 2009. Spheres of discharge of springs. Hydrogeol. J. 17, 83–93.

Springer, A.E., Stevens, L.E., Ledbetter, J.D., Schaller, E.M., Gill, K.M., Rood, S.B., 2014. Ecohydrology and stewardship of Alberta springs ecosystems. Ecohydrology 8, 896–910. https://doi.org/10.1002/eco.1596.

Springer, A.E., Schaller, E.M., Junghans, K.M., 2016. Local vs regional groundwater flow from stable isotopes at Western North America springs. Groundwater. https://doi.org/10.1111/gwat1242.

Springs Stewardship Institute (SSI), 2012. Springs Ecosystem Assessment Protocol Scoring Criteria. revised 2016, http://www.springstewardship.org.

Springs Stewardship Institute, 2016. Springs ecosystem inventory protocols. Springs Stewardship Institute, Museum of Northern Arizona, Flagstaff, Arizona. Online at: http://docs.springstewardship.org/PDF/ProtocolsBook.pdf (Accessed 1 April 2018).

Stevens, L.E., Meretsky, V.J., 2008. Aridland Springs in North America: Ecology and Conservation. University of Arizona Press, Tucson.

Stevens, L.E., Stacey, P.B., Jones, A., Duff, D., Gourley, C., Caitlin, J.C., 2005. A protocol for rapid assessment of southwestern stream-riparian ecosystems. In: van Riper III, C., Mattson, D.J. (Eds.), Fifth Conference on Research on the Colorado Plateau. University of Arizona Press, Tucson, pp. 397–420.

Stevens, L.E., Springer, A.E., Ledbetter, J.D., 2012. Springs Inventory Protocols. Museum of Northern Arizona Springs Stewardship Institute, Flagstaff. Available from: Springstewardshipinstitute.org (Accessed 23 May 2015 revised 2016).

Stromberg, J., Briggs, M., Gourley, C., Scott, M., Shafroth, P., Stevens, L., 2004. Human alterations of riparian ecosystems. In: Baker Jr., M.B., Ffolliott, P.F., DeBano, L., Neary, D.G. (Eds.), Riparian Areas of the Southwestern United States: Hydrology, Ecology, and Management. Lewis Publishers, Boca Raton, pp. 99–126.

Suter II, G.W., 2006. Ecological Risk Assessment, second ed. CDC Press, Boca Raton.

The MathWorks, Inc., 2012. MATLAB and Statistics Toolbox Release. The MathWorks, Inc., Natick, MA.

Thompson, B.C., Matusik-Rowan, P.L., Boykin, K.G., 2002. Prioritizing conservation potential of arid-land montane natural springs and associated riparian areas. J. Arid Environ. 50, 527–547.

Tobin, B.W., Springer, A.E., Kreamer, D.K., Schenk, E., 2017. Review: the distribution, flow, and quality of Grand Canyon springs, Arizona (USA). Hydrogeol. J. 26 (3), 721–732. https://doi.org/10.1007/s10040-017-1688-8.

Unmack, P.J., Minckley, W.L., 2008. The demise of desert springs. In: Stevens, L.E., Meretsky, V.J. (Eds.), Aridland Springs in North America: Ecology and Conservation. University of Arizona Press, Tucson, pp. 11–34.

U.S. Forest Service, 2014. Land and Resource Management Plan for the Kaibab National Forest: Coconino. Yavapai, and Mojave Counties, Arizona. U.S. Department of Agriculture Forest Service, Southwestern Region MB-R3-07-17, Washington, DC.

USDA Forest Service, 2002. The Process Predicament: How Statutory, Regulatory, and Administrative Factors Affect National Forest Management. Available from www.fs.fed.us/publications.html (Accessed 15 April 2005).

USDA Forest Service, 2012. Groundwater-dependent ecosystems: level II inventory field guide. U.S Department of Agriculture, Forest Service. General Technical Report WO-86b, 124 pp.

USDA Forest Service, 2013. Coconino National Forest. http://www.fs.usda.gov/main/coconino/about-forest.

USDA Forest Service, 2014. Land and Resource Management Plan for the Kaibab National Forest—Coconino, Yavapai, and Mojave Counties, AZ. USDA Forest Service.

USDA Forest Service, Coconino National Forest, 2014. S_R03_COC.Road. Flagstaff, Arizona.

Western Regional Climate Center, 2014. Cooperative Climatological Data Summaries. Retrieved 6 June 2012, from, http://www.wrcc.dri.edu/summary/Climsmaz.html.

FURTHER READING

Arizona Department of Water Resources, 2012. GIS Download Page: Surface Water Filings Active. Retrieved April 15, 2012, http://www.azwater.gov/azdwr/gis/.

Blasch, K.W., Hoffman, J.P., Graser, L.F., Bryson, J.R., Flint, A.L., 2006. Hydrogeology of the upper and middle Verde River watersheds, central Arizona. U.S. Geological Survey. Scientific Investigations Report 2005-5198.

Chimmer, R.A., Lemly, J.M., Cooper, D.J., 2010. Mountain fen distribution, types and restoration priorities, San Juan Mountains, Colorado, USA. Wetlands 30, 763–771.

Farley, G.H., Ellis, L.M., Stuart, J.N., Scott, N.J., 1994. Avian species richness in different aged stands of riparian forest along the Middle Rio Grande, New Mexico. Conserv. Biol. 8, 1098–1108.

Fraser, J., Martinez, C., 2002. Restoring a desert oasis. Endangered Species Bull. 27, 18–19. Available online at: http://www.fws.gov/endangered/bulletin/2002/03-06/18-19.pdf (Accessed 23 May 2015).

Minckley, W.L., Deacon, J.E., 1991. Battle Against Extinction: Native Fish Management in the American West. University of Arizona Press, Tucson.

Parker, J.T.C., Steinkampf, W.C., Flynn, M.E., 2004. Hydrogeology of the Mogollon Highlands, central Arizona. U.S. Geological Survey. Scientific Investigations Report 2004-5294.

Springer, A.E., Stevens, L.E., Anderson, D.E., Parnell, R.A., Kreamer, D.K., Levin, L., Flora, S., 2008. A comprehensive springs classification system: integrating geomorphic, hydrogeochemical, and ecological criteria. In: Stevens, L.E., Meretsky, V.J. (Eds.), Aridland Springs in North America: Ecology and Conservation. University of Arizona Press, Tucson, pp. 49–75.

U.S. Fish and Wildlife Service, 2005. Critical Habitat Portal. Retrieved 29 October 2013, http://ecos.fws.gov/crithab/.

USDA Forest Service, 2011. Watershed condition classification technical guide. U.S. Department of Agriculture, Forest Service. FS-978. 41 p.

USDA Forest Service, 2013. Beale's Wagon Road Historic Trail. Kaibab National Forest Williams Ranger District RG-R3-07-5, Williams. Available from: http://www.fs.usda.gov/Internet/FSE_DOCUMENTS/stelprdb5434056.pdf (Accessed 23 May 2015).

White, R.D., Withers, N., Kanim, K., Stubbs, T.K., Hanlon, J., 1995. U.S. Fish and Wildlife Service, Region I, report on conservation actions undertaken during 1993 for federally listed and candidate fishes and other aquatic species in California, Idaho, Nevada, and Oregon. Proc. Desert Fishes Council 24 (1994), 5–9.

Section 5

Non-North American Methods

Chapter 5.1

Introduction and Overview—John Dorney

John Dorney
Moffatt and Nichol, Raleigh, NC, United States

Most of the early RAMs for wetlands were published in North America, beginning in the 1960s, and focused on waterfowl habitat; stream RAMs date from the late 1990s (see Chapter 1.0). In the United States, the enactment of the National Environmental Policy Act and Clean Water Act created a need for more accurate and comprehensive RAMs. More recently, various stream-related regulations (especially riparian buffer rules) have created a need for stream RAMs.

Most of the more recent RAMs in North America have targeted wetlands or streams in a specific state (or province). However, creating and calibrating a RAM that is acceptably accurate for areas as large as an entire nation [such as the United States RAM (Chapter 4.4.2, EPA 2016)] or a large region [such as the Mid-Atlantic RAM (Chapter 4.4.1)] is far more challenging. These geographically broader RAMs often must address a wider variety of wetlands and streams as well as climatic, elevational, geologic variability, and different historic land uses.

A key underpinning to the development of these wetland and stream RAMs is a long, in-depth history of intensive wetland and stream research that is locally based and therefore can be relied upon to develop, verify, and calibrate these RAMs. In most countries outside North America, less research has been published that is pertinent to rapidly estimate wetland and stream functions or ecological uplift. In many nations, this paucity of information hinders development of reliable regionally appropriate RAMs. Those RAMs that are developed must be based largely on the judgment of subject experts, knowledge shared by indigenous peoples, and/or inference from North American research or research from other nearby areas. One of the broader lessons from the French work reported in this section (also see Gaucherand et al., 2015) is that methods developed in North America may not be readily applicable to landscapes with a longer history of human settlement (such as Europe or Asia) because identification of "old-growth" or "pristine" condition is problematic in these areas. In these landscapes, as Gaucherand et al. (2015) point out, a novel view of reference standard may be needed, especially for vegetation and landscape-related metrics.

The following chapters describe wetland RAMs in these countries outside North America-France, New Zealand, Nepal, Jamaica, Costa Rica, and South Africa. These particular RAMs are included for no reason other than the interest and willingness of qualified persons in these countries to prepare a description for this book. Note that some of these RAMs are intended to address only wetland integrity while others intend to address wetland functions or ecosystem services (see Section 3 for an explanation of this critical distinction). In addition, although separate chapters are not included in this book, we summarize other published non North American RAMs of which we are aware, as follows:

Australia: Australia began developing RAMs in the late 1990s (Spencer et al., 1998) with a method developed and tested for southeastern Australia. This method used four main measures (with 13 specific metrics)—soil, fringing vegetation, aquatic vegetation, and water quality. It was intended to assess the integrity of permanent floodplain wetlands in the Darling Basin of Australia. The results of the method were calibrated using a long-term monitoring data set. More recent work was reported by the Victoria Department of Sustainability and Environment (2005). The work of Spencer et al. (1998) has evolved in the intervening decades into the Index of Wetland Condition (IWC) used in Victoria (Papas and Morris, 2014) for conducting a rapid (defined as taking no more than a few hours) assessment of all naturally occurring, nonflowing wetlands without a marine hydrological influence. The main factors of the IWC are catchment size, physical form, water regime, soils, water properties, and biota.

Iran: The Caspian Rapid Assessment Method (CRAM/Iran, Pour et al., 2015) was developed to assess the quality of wetlands and develop management strategies for wetlands along the southern fringe of the Caspian Sea. The authors examined 16 methods from the United States and assessed their usefulness through a survey of local government officials to reach consensus on the importance of criteria to use in the method. CRAM/Iran uses six main criteria and 12 subcriteria, many of which are based on the Ohio Rapid Assessment Method (Chapter 4.3.8, c, ii). Those main criteria and subcriteria are: (1) wetland size, (2) buffer zone and surrounding land uses (average buffer width and density of surrounding land uses),

Wetland and Stream Rapid Assessments. https://doi.org/10.1016/B978-0-12-805091-0.00050-5

(3) hydrology (water sources, connectivity, maximum water depth, changes in natural hydrologic regime), (4) habitat changes (soil disturbance, habitat changes), (5) wetlands of special concern, and (6) plant species and dispersion (wetland vegetation communities, horizontal stratification, coverage of invasive plants, and microtopography). A series of wetlands was then ranked in terms of ecological condition and rehabilitation potential.

China: A nationwide effort is underway using rapid field-based methods to develop a large-scale meta-assessment method (linking field data from multiple sites to land use cover data) by wetland type to evaluate wetland functionality and ecosystem services at the landscape scale. China completed its first national ecosystem services assessment in 2015 where scientists coupled land-cover data, field-based data, and biophysical models to rapidly evaluate changes in ecosystem area, ecosystem services, and biodiversity from 2000 to 2010 (Ouyang, et al., 2016). Chinese scientists and policymakers are working to refine these methods to establish a national wetland ecosystem services monitoring program. They are testing a multitude of rapid assessment techniques across various sites, representing a diversity of wetland types to determine credible and practical protocols. Work has been completed for lakes and marshlands, and scientists are currently testing approaches on coastal wetlands (Wong, et al., 2016). The goal of the large-scale meta-assessment methodology is to be able to assess wetland ecosystem services rapidly across China using GIS data, grounded in empirical relationships from field-based data (C. Wong, personal communication, June 29, 2017). This work follows up on intensive work in China using the ecosystem services approach linked to wetland policy (Wong et al., 2015; Jiang et al., 2015; Ouyang et al., 2016).

India: An effort has begun to standardize ways of rapidly assessing the biodiversity and ecosystem services of wetlands (Gopal, 2015). These include hydrological, biological, and cultural services. Biological services address a wetland's capacity to support macrophytes, microphytes, zooplankton, benthic macroinvertebrates, waterfowl, herptofauna, and fish. In general, Gopal suggests the following general process—first, collection of baseline information on the wetland, second, identify and prioritize stakeholders (especially citizens who live in or on the wetland), and third, conduct an inventory of wetland services.

Spain: A rapid assessment method (QBR index) for determining riparian habitat quality has been developed and tested for three river basins in Catalonia, NE Spain (Munne et al., 2002). This evaluation is completed in the field using a two-sided form by an observer familiar with local tree and shrub species in about 10–20 min. The index ranges from 0 to 100 and is the sum of four scores: total riparian vegetation cover, cover structure, cover quality, and channel alterations. The authors suggest that the QBR could be used in other geographic areas with modifications to reflect local conditions. A recent development is an APP that can readily used by citizens to assess both the hydrological and biological status (using aquatic macroinvertebrates) of streams (N. Prat, personal communication, March 8, 2017). The QBR index has been studied along with other indices in Cantabria, Northern Spain (Barquin et al., 2011), and found to be a useful method to determine riparian habitat quality.

REFERENCES

Barquin, J., Fernandez, D., Alvarez-Babria, M., Penas, F., 2011. Riparian quality and habitat heterogeneity assessment in Cantabrian rivers. Limnetica 30 (2), 329–346.

Gaucherand, S., Schwoertzig, E., Clement, J.-C., Johnson, B., Quetier, F., 2015. The cultural dimensions of freshwater wetland assessments: lessons learned from the application of US rapid assessment methods in France. Environ. Manag. 56, 245–259.

Gopal, B. (Ed.), 2015. Guidelines for Rapid Assessment of Biodiversity and Ecosystem Services of Wetlands, Version 1.0. Asia-Pacific Network for Global Change Research (APN-GCR)/National Institute of Ecology, Kobe/Delhi. 134 pp.

Jiang, B., Wong, C.P., Chen, Y., Cui, L., Ouyang, S., 2015. Advancing wetland policies using ecosystem services—China's way out. Wetlands 35, 983–995.

Munne, A., Prat, N., Sola, C., Bonada, N., Rieradevall, M., 2002. A simple field method for assessing the ecological quality of riparian habitat in rivers and streams: QBR index. Aquat. Conserv. Mar. Freshwat. Ecosyst. 13, 147–163.

Papas, P., Morris, K., 2014. An Assessment of Quality Assurance and Quality Control Measures for the Index of Wetland Condition. Arthur Rylah Institute for Environmental Research Technical Report Series No. 254, Department of Environment and Primary Industries, Heidelberg, VIC. Available from: https://www.ari.vic.gov.au/__data/assets/pdf_file/0022/36517/ARI-Technical-Report-254-An-assessment-of-QAQC-measures-for-the-Index-of-Wetland-Condition.pdf. Accessed 20 July 2017.

Ouyang, Z., Zheng, H., Xiao, Y., Polasky, S., Liu, J., Xu, W., Wang, Q., Zhang, L., Xiao, Y., Rao, E., Jiang, L., Lu, F., Wang, X., Yang, G., Gong, S., Wu, B., Zeng, Y., Yang, W., Daily, G.C., 2016. Improvements in ecosystem services from investments in natural capital. Science 352, 1455–1459.

Pour, S.K., Monavari, S.M., Riazi, B., Korasani, N., 2015. Caspian rapid assessment method: a localized procedure for assessment of wetlands at the southern fringe of the Caspian Sea. Environ. Monit. Assess. 187 (420), 1–12.

Spencer, C., Robertson, A.I., Curtis, A., 1998. Development and testing of a rapid appraisal wetland condition index in South-Eastern Australia. J. Environ. Manag. 54, 143–159.

Victoria Department of Sustainability and Environment, 2005. Index of Wetland Condition. Victoria Government, Department of Sustainability and Environment, Melbourne. Available from: https://wetlandinfo.ehp.qld.gov.au/wetlands/resources/tools/assessment-search-tool/19/. Accessed 20 July 2017.

Wong, C.P., Jiang, B., Linzig, A.P., Lee, K.N., Ouyang, Z., 2015. Linking ecosystem characteristics to final ecosystem services for public policy. Ecol. Lett. 18, 108–118.

Wong, C.P., Jiang, B., Lee, K.N., Ouyang, Z., Cui, L., Ma, D., 2016. In: Implementing ecosystem services for inclusive green growth transitions. Green Growth Knowledge Platform Fourth Annual Conference, Jeju, Korea (Conference Paper).

Chapter 5.2

Wetland Assessment in France—Development, Validation, and Application of a New Method Based on Functions

Guillaume Gayet*, Florence Baptist[†], Pierre Caessteker[‡], Jean-Christophe Clément[§], Maxime Fossey*, Juliette Gaillard*, Stéphanie Gaucherand[¶], Francis Isselin-Nondedeu[‖], Claire Poinsot[†] and Fabien Quétier[†]

**Muséum national d'Histoire naturelle, UMS Patrimoine Naturel, Paris, France, [†]Biotope, Mèze, France, [‡]Agence Française pour la Biodiversité, Direction de la Recherche, de l'Expertise et du Développement des Compétences, Vincennes, France, [§]Centre, Alpin de Recherche sur les Réseaux Trophiques et les Écosystèmes Limniques-UMR INRA-USMB, Le Bourget du Lac, France, [¶]Institut national de recherche en sciences et technologies pour l'environnement et l'agriculture, Unité de Recherche sur les Ecosystèmes Montagnards, Saint Martin d'Hères, France, [‖]Ecole Polytechnique de l'Université François Rabelais, UMR 7324, CNRS CITERES, Tours, France*

Chapter Outline

INTRODUCTION—WETLAND IMPACT MITIGATION IN FRANCE

In France, developments that impact wetlands are regulated through a dedicated permitting regime anchored in the European Union's Water Framework Directive (WFD), recognizing the contribution of wetlands to achieving the Directive's goal of "good ecological status" of waters. Permit applicants must apply the mitigation hierarchy, which involves taking steps to avoid and reduce harm to wetlands and offsetting any residual impacts. The mitigation hierarchy was introduced in French environmental law in 1976, but its profile was recently raised through a series of legal and procedural changes at both the national (see "*Loi n° 2016-1087 du 8 août 2016 pour la reconquête de la biodiversité, de la nature et des paysages*") and European levels (Quétier et al., 2014). For wetlands, the most recent River Basin Management Plans (RBMP) established under the WFD now require planned developments that impair wetlands to include offset measures that restore, rehabilitate, or create wetlands that are equivalent on a functional level. Although some RBMPs allow applicants to use an area ratio (e.g., 1:1, 2:1) to determine the amount of offset that is required, for many developers and decision-makers, demonstrating that a project will not lead to a "net loss" of wetland functions is a recent challenge.

In this context, and until recently, there was no widely used approach or method that could demonstrate with tangible and verifiable evidence that offsets for wetland impacts meet regulatory requirements. Without such a method, projects would be exposed to considerable (technical and financial) uncertainty and increasing legal challenges. It would also enable the effectiveness of the permitting regime to be better evaluated in terms of its contributions to the policy objectives. This is the aim of rapid assessment methods (RAM) (e.g., Uniform Mitigation Assessment Method Florida, Ohio Rapid Assessment Method, California Rapid Assessment Method, and others described in Chapter 2 of this book) developed in the United States and Canada for a variety of purposes, including streamlining mitigation for wetland impact. In this chapter, we describe a new RAM that was recently adopted by French authorities for assessing losses and gains of wetland functions for use in applying the mitigation hierarchy to development projects.

Wetland and Stream Rapid Assessments. https://doi.org/10.1016/B978-0-12-805091-0.00053-0

Existing RAMs from the United States could not readily fill France's need for an operational wetland assessment method. As demonstrated by Gaucherand et al. (2015), some underlying assumptions are not met in France. Specifically, the reference states against which wetland losses and gains are measured and that guide restoration objectives are not the same in France as in the United States and Canada. It is common in North America to aim for a more "pristine" state that is considered as having a higher level of functional performance. In Europe, restoration of specific functions is generally pursued (Davidsson et al., 2000) without reference to a pristine state, although when some types of "natural" habitats are impacted by planned development, offsets generally aim to restore similar habitats (Wende et al., 2018). To our knowledge, no wetland RAM had been developed in Europe (Carletti et al., 2004), although there is a growing diversity of approaches and methods used in applying the mitigation hierarchy in Europe (Wende et al., 2018) and elsewhere (Quétier and Lavorel, 2011; Bezombes et al., 2017).

In 2012, to enable developers and decision-makers in applying the new "no net loss" requirement, the French Biodiversity Agency (AFB, formerly the National Office for Water and Aquatic Environments) initiated the development of a dedicated method. The intention was to build a method that would help harmonize the terms and associated definitions used in applying the mitigation hierarchy (e.g., what is a function?) and identify and prioritize wetland features in relation to their functional performance (which functions?) to improve the design and implementation of planned developments and their offset programs. Another goal for the AFB was to make it easier for government agencies to control the effectiveness of the offset programs and to strengthen their permits when these are challenged in court.

In essence, the method named "*méthode nationale d'évaluation des fonctions des zones humides*" (i.e., National Method to Assess Wetland Functions) was designed to answer the following question: are the functional losses on the impacted site compensated by the functional gains on the offset site, after offset implementation?

The first version of the method was developed for continental wetlands in accordance with the expectations of the AFB. It does not apply to tidal wetlands, river beds, and areas of open water or to French overseas regions and territories.

APPROACH USED TO DEVELOP THE METHOD

The method was developed through an iterative process of design, review, and testing, which we describe below. It was also developed collaboratively, with public, private, scientific, and technical partners, in order to: (1) provide operational definitions of functions, (2) make the method consistent with French regulations and guidance on applying the mitigation hierarchy, and offsetting impacts in particular, and (3) ensure a robust scientific basis, which was considered critical for the method's suitability and scientific credibility. The method's development benefitted from the experience of developing and using RAMs in North America (see Bartoldus, 1999; Carletti et al., 2004; Cole, 2006; Fennessy et al., 2004; Sutula et al., 2006). Most practical requirements to develop RAMs from Sutula et al. (2006) were followed.

Like most RAMs, the method is not a scientific protocol. Rather, it is a cost and time-effective tool that provides basic information on the level of functional performance of a wetland for technical staff with a tight schedule. The actual functional performance of the wetland is not quantitatively measured. Instead, the level of functional performance of the wetland is assessed semi-quantitatively, based on structural features (parameters) that were hypothesized and tested to ensure they are indicative of function levels.

Definition and Identification of Target Functions

In France, wetlands are defined by Ministerial Order DEVO0813942A of 2008 on the basis of physical (hydromorphic soils or water table depth) and/or biological (composition of plant communities) indicators. This means that wetlands are not restricted to areas of "natural" or "pristine" wetland vegetation, but also include many degraded and former wetlands (including cropland). There are not, however, any legal definitions in France of wetland functions. Many definitions have been proposed, and confusion with "services" is common (Lamarque et al., 2011). Functions are defined here as physical, chemical, and biological processes at work in wetlands, with no consideration for the resulting benefits or damages (i.e., services and disservices) and values for society. This definition was inspired from the ones proposed by Maltby et al. (1996) or Smith et al. (1995) and values and services are clearly outside the scope of the method.

Ten functions were selected, in three categories:

- *Hydrological functions*: run-off reduction, groundwater recharge, and sediment retention;
- *Biogeochemical functions*: denitrification, nitrogen assimilation by plants, phosphorus adsorption and precipitation, orthophosphates assimilation by plants, and carbon sequestration;
- *Complete species life cycle functions*: habitat support and habitat connectivity.

Assessing the presence of threatened or endangered species and addressing likely impacts to these is outside the scope of the method, and is subject to a different permitting process in France.

These functions were selected by considering representativeness, relevance, feasibility and the need for a robust scientific basis. Functions such as sustaining summer (dry season) stream flows and water levels, or pesticide retention, among others, are not included in the current version of the method. Work is underway to include them in the future as the method is updated with feedback from implementation and more research and testing.

Scientific and Technical Considerations

Given the practical context in which wetland offsets are designed, sized, and implemented in France, the method had to meet several conditions:

- *No specialized knowledge is needed*: the method is targeted at a technical audience with basic knowledge in ecology, soil science and geographic information systems (GIS);
- *Pragmatism*: the method must be simple to use;
- *Fast to implement*, that is, less than 1 day to assess a wetland site under 5 ha (including field and office work), as planned development projects in France mostly impact small wetland areas;
- *Objectivity*: the method is fact-based, using measurable indicators;
- *Independence to phenology*, as assessments and monitoring must be conducted year-round;
- *Data availability*: assessments are based on information that is readily available across France.

Given these conditions, the method was developed by identifying parameters (and their limitations) that could be used to reflect function levels (performance) for at least one of the 10 selected functions. This was done through a review of published scientific studies to identify relevant parameters and their limits, before calibrating related indicators (see Gayet et al., 2016a for details). The method includes tools for handling the collected data and synthesizing results for users.

Consistency With Applicable Regulations

The method was designed to conform to applicable regulations and guidance on the mitigation hierarchy. Its current version is directly relevant to four key principles laid out in French law: equivalence, geographical proximity, efficiency, and ecological additionality (Table 5.2.1).

Field Testing

An iterative process was used to develop the method: two prototypes were developed and tested in 2014 and 2015, before Version 1 of the method was published in 2016. Overall, the method has been tested by tens of users on ca. 220 sites across France (Fig. 5.2.1).

Test sites were selected to cover: (i) various sizes of wetlands, with a focus on sites smaller than 5 ha, (ii) various levels of initial degradation, including quasi "pristine" wetlands (e.g., riverine forests, peat bogs), intensely used wetlands (e.g., grazed riverine meadows, drained riverine forests), and highly modified areas (e.g., cropland on hydromorphic soils that are sometimes legally qualified as wetlands under French regulations), and (iii) various ecological contexts and wetland types: vegetation types, hydrogeomorphological characteristics, elevation (from sea level to ca. 2550 m a.s.l.), surrounding anthropogenic pressures (e.g., forested, agricultural, and urban landscapes), etc.

A few sites ($n=19$) with ongoing intensive monitoring were also included in the tests. This demonstrated the consistency between the results given by the method and the results given by repeated quantitative measurements of wetland states and functions (Jaymond et al., 2015) and aided in calibration and validation.

Such testing was also essential to verify the reproducibility of the assessments (consistency across users), and the time required to conduct them. Testing revealed that the method is largely applicable year-round (with some exceptions such as during summer drought, freezing or snowy conditions, or important waterlogging of the soil resulting from recent rain or flooding). Field tests also played a role in the choice of parameters measured, related indicators, and checking that parameters could be assessed by the target audience.

TABLE 5.2.1 Compensation Principles Laid Down in the Current French Code of Environment and Their Consideration by the "méthode nationale d'évaluation des fonctions des zones humides" (Version 1 Published in 2016: ✔; Version 2 in Prep.: ✓)

Principles	Short Description	Verified by the Method
Proportionality	Expected details in a permit application must be appropriate considering the risks for the impacted site and the magnitude of the impacts. The quality of the baseline survey, impact assessment, and set of avoidance, reduction, and offset measures submitted by the permit applicant must be consistent with the magnitude of the predictable impacts of the planned development.	
Equivalence	Offset measures must target the same ecosystem components and features as those destroyed or altered (species, habitats, functions).	✔
	Sizing of offsets must be consistent with the magnitude of the predictable residual negative impacts.	✔
Geographical and temporal proximity	Offsets must be implemented close to the impacted site, in an area with similar physical and anthropogenic characteristics.	✔
	The offset measures must be effective quickly to prevent irreversible harm.	✓
Feasibility	Offsets are in kind and must deliver measurable outcomes on the ground. Financial compensation is not an appropriate offset.	✓
	The efficiency of the proposed ecological engineering/restoration techniques must be demonstrated and be technically feasible.	✓
Efficiency, sustainability	Offsets must reach their objectives.	✔
	Offsets are monitored and must be completed on time.	
	Available resources (e.g., land, funding, human resources) and ecological goals must be clearly stated, precise, and verifiable.	
Additionality	Offsets must generate environmental "gains" equivalent or greater than "losses" beyond public commitments for nature protection and restoration; it consolidates them without replacing them.	✔
	A given offset cannot be used to address residual impacts from different development projects, at the same time or over the time, and cannot be used to implement already existing commitments by developers or other third parties.	
Coherence	Offsets required under various regulations must be conciliated (e.g., wetlands, protected species, Natura 2000).	

Adapted from Onema, 2015. Pour une conception et une réalisation des IOTA de moindre impact environnemental—Modalités d'expertise, préconisations techniques et retours d'expériences—Tome 5: expertise des mesures de compensation écologique. Collection 'Guides et protocoles'. Onema, 76 pp. (in French)

Dissemination and Uptake

To ensure the method reached its target audience, three key publications were made available upon its publication (in 2016). A *general report* presents theoretical, scientific, and technical foundations with all references used to develop the method (see Gayet et al., 2016a). A *handbook* presents a summary of the general report, a detailed notice to implement the method, and instructions for interpreting results (see Gayet et al., 2016b). The notice is crucial: it provides detailed instructions and answers to every question in the assessment. It ensures measurement objectivity and reproducibility. Finally, a *spreadsheet* associated with the handbook allows users to enter assessment data. Tables and figures are automatically produced that give users rapid feedback on their assessment (these are illustrated below). These documents are available on http://www.onema.fr/node/3981. In addition, a section with frequently asked questions is regularly updated on the method's website, together with tutorial videos and case-study examples.

Developing a user-friendly set of tools is not enough. A key step in developing the method was close engagement with decision-makers who largely drive which methods developers (and their consultants) choose. An ad hoc oversight committee was put in place by AFB to generate buy-in for the method. Although AFB, which commissioned the method, is

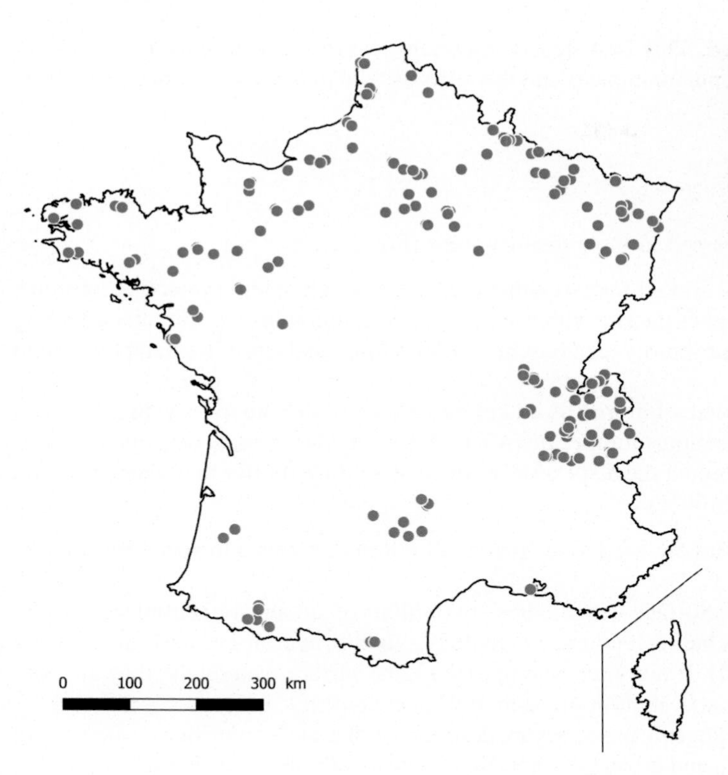

FIG. 5.2.1 Distribution of field tests with prototypes developed in 2014 and 2015. Each dot indicates a site where one prototype has been tested. *(Adapted from Gayet, G., Baptist, F., Baraille, L., Caessteker, P., Clément, J.-C., Gaillard, J., Gaucherand, S., Isselin-Nondedeu, F., Poinsot, C., Quétier, F., Touroult, J., Barnaud, G., 2016. Guide de la méthode nationale d'évaluation des fonctions des zones humides - version 1.0. Onema, collection Guides et protocoles (in French).)*

involved in the permitting process, getting official endorsement from the Ministry of Environment was important in getting the method accepted. The Ministry was in the committee and issued a formal letter to all permitting agencies encouraging them to use of the method whenever possible.

Official endorsement generated momentum for some developers, local governments, and public agencies to look to the method to improve wetland management in the context of development projects, plans, and programs. In this context, the partners involved in developing the method are actively providing training with a dedicated training package. Training is not necessary to apply the method, but it can make its use easier and more consistent.

DESCRIPTION OF THE METHOD

The method is a two-step process. The first step—called *Context diagnosis*—is focused on the overall characteristics of the site being assessed and its surrounding environment. This allows the user to check that impacted and offset sites are comparable. The second step—called *Functional diagnosis*—assesses the site's functional performance, which is hereafter called "functional capacity." Comparisons between losses of functional performance on the impacted site and gains of functional performance on the offset site allow decision-makers to draw conclusions on whether no net loss is likely to be achieved.

Step 1: Context Diagnosis

The first step of the method is the context diagnosis, which describes the geomorphological, hydrological, and ecological characteristics of the assessed wetlands on both the impacted and offset sites. It ensures that the offset program is both "on site" and "in kind," according to the definition of Brinson and Rheinhardt (1996). This means that offsets are close to the impacted site (e.g., same watershed at the local level) and restore similar wetlands to those impacted. If that is not the case ("off site" and/or "out of kind"), site properties are considered too different to compare their features in a relevant manner and the method cannot be applied.

This context diagnosis is necessary to ensure that the principles of geographical proximity and equivalence are met (see Table 5.2.1): the offset site must be in an area with similar characteristics to the impacted site, which means that both sites and their surroundings must be assessed and compared. This first step is important to avoid a widespread outcome of wetland mitigation and offsetting regimes where development impacts and their offsets tend to occur in distinct watersheds (Bendor et al., 2009).

The Five Criteria of the Context Diagnosis

The context diagnosis uses five criteria, which are assessed in three distinct areas (Fig. 5.2.2):

- The site itself, which is the wetland area that was delimited for assessment (e.g., the development project's footprint);
- The "contributory area" of the site, which likely supplies the site with water, sediments, nutrients, etc., directly affecting hydrological and biogeochemical functions. The contributory area is delimited by a brief analysis of topography around the site, as proposed by Maltby (2009), and
- The "landscape" of the site. The landscape is delimited arbitrarily as a 1 km buffer around the site to prevent users from considering different areas as relevant "landscape" settings for the site. A small area enables users to describe the landscape rapidly, but this means it does not take into account the dispersion abilities of wildlife (buffer may seem too wide or narrow for species with low or high dispersal abilities).

Existing and accessible national databases and typologies are used to document all criteria to reduce the need for users to look for relevant data sources.

Criteria 1: belonging to a given surface waterbody. Surface waterbodies are portions of streams or coastal waters that are considered homogenous based on several factors, including topography, geology, climate, hydrology, and climate. They have been delimited during implementation of the WFD. If two sites belong to the same surface waterbody, they are considered to have a similar topographic, geological, climatic, hydrologic, and anthropic context.

Criteria 2: upland anthropogenic pressures. Anthropogenic pressures are described in the site's contributory area, using national databases. They include agricultural, industrial, and urban pressures that are generally the major sources of hydrological (e.g., sediments) and biogeochemical (e.g., N and P) inputs to wetlands. In lowland riverine or lacustrine sites where no differences are likely to be found on anthropogenic pressures between the impacted and offset sites, a description of anthropogenic pressures in the contributory area is not necessary as the goal of Step 1 is to compare these two sites.

Criteria 3: landscape composition and structure. Landscape composition and structure are described based on aerial photographs and using level 1 of the EUNIS typology of Davies et al. (2004). EUNIS is a habitat typology widely used in Europe.

Criteria 4: hydrogeomorphologic system (site-level). The geomorphological configuration, the dominant source of water, and the hydrodynamics of the impacted and offset sites are identified to ensure they share the same hydrological functioning. The method relies on the hydrogeomorphologic classification of Brinson (1993a,b). It distinguishes riverine, depressional, slope, lacustrine, and wet flat wetlands.

Criteria 5: habitat composition and structure (site-level). Composition and structure of habitats in the impacted site (before impact) and in the offset site (after offsets) must be similar to ensure that functions are comparable. Habitats are described at EUNIS level 3. Level 3 was chosen because it is accurate to describe habitats without requiring specialized skills in botany or phytosociology.

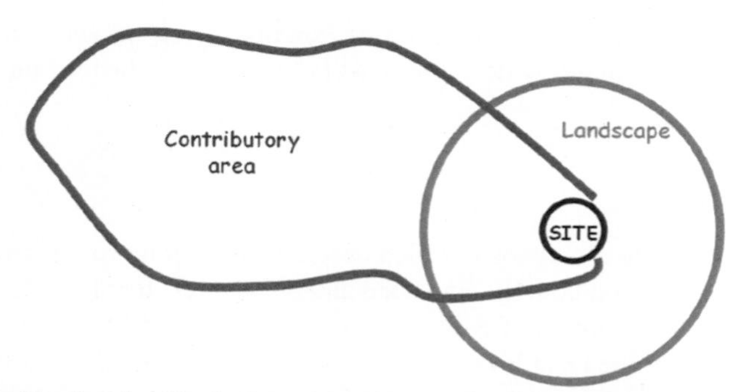

FIG. 5.2.2 The three major areas considered to describe the site and its setting in the context diagnosis (step 1). *(Adapted from Gayet, G., Baptist, F., Baraille, L., Caessteker, P., Clément, J.-C., Gaillard, J., Gaucherand, S., Isselin-Nondedeu, F., Poinsot, C., Quétier, F., Touroult, J., Barnaud, G., 2016. Méthode nationale d'évaluation des fonctions des zones humides. Fondements théoriques, scientifiques et techniques. Onema, MNHN 310 pp. (in French).)*

Interpretation

At the end of the context diagnosis step, it is easy to check if the five criteria are similar for the impacted and offset sites. An exact match is required for criteria 1 and 4 while only strong similarity is required for criteria 2, 3, and 5. Some French RBMP rules may be more restrictive (i.e., the impacted and offset sites must be geographically close, and belonging to the same surface waterbody may not be enough). Others are more lenient regarding criteria 1.

The five criteria of Step 1 are not independent. In practice, sites that belong to the same surface waterbody (criteria 1) and with the same hydrogeomorphologic system (criteria 4) commonly have similar contributory areas (criteria 2) and landscapes (criteria 3). Therefore, it is generally not too challenging to find a matching offset site.

For criteria 5, it remains difficult to achieve very specific habitat types through restoration that precisely match those on the impacted site before impact (see Hobbs et al., 2009). Thus, flexibility is required when analyzing the similarity between habitats, as proposed by Keenleyside et al. (2013). Furthermore, if habitats on the impacted site (before impact) are very degraded (this usually comes with very low functional performance), instead of restoring a similarly degraded habitat on the offset site, it is possible for stakeholders to determine a more appropriate trade up to a more "natural" habitat type with higher functional performance.

Illustration of Step 1 Results

Fig. 5.2.3 provides a (hypothetical) example of how results from Step 1 are presented for an offset program. In this case, three potential offset sites were assessed but only offset sites 1 and 2 meet the on site and in-kind requirements (geographical proximity and equivalency principles—Table 5.2.1). Offset site 3 is not suitable because its context diagnosis shows it is too different from the impacted site (criteria 1, 2, 4, and 5). At this stage, choosing between offset sites 1 and 2 would reflect a preference for a mosaic of three or two habitats, to restore reedbeds rather than sedge meadows. Step 2 in the method looks at the functional performance of the two sites, and the net outcome of using one or the other to offset the loss of functions on the impacted site.

	Impacted site before impact (1.6 ha)		Expected offset outcomes on proposed offset sites: Option #1 (2.88 ha)	Option #2 (3.5 ha)	Option #3 (2.7 ha)
In site environment					
Criteria 1: belonging to a given surface waterbody*	FRAR38 Noye	=	FRAR38 Noye	FRAR38 Noye	FRAR06 Arve
Criteria 2: upland anthropogenic pressures	33,123 ha. Crops : 76%, meadows : 3.7% urban : 0.8%	≃	32,993 ha. Crops : 75.5%, meadows : 4.2%, urban : 0.9%	35,654 ha. Crops : 69%, meadows : 6%, urban : 0.6%	272 ha. Crops : 0.8%, meadows : 0.7%, urban : 0.1%
Criteria 3: landscape composition and structure**	361 ha. Cultivated habitats: 45%, woodlands and forests: 20%, mires, bogs and fens: 10%	≃	389 ha. Cultivated habitats: 51%, woodlands and forests: 20%	421 ha. Cultivated habitats: 37%, woodlands and forests: 18%, mires, bogs and fens: 11%	388 ha. Cultivated habitats: 40%, mires, bogs and fens: 19%, woodlands and forests: 15%
In site					
Criteria 4: hydrogeomorphologic system	Riverine	=	Riverine	Riverine	Wet flat
Criteria 5: habitat composition and structure***	D5.1 Reedbeds normally without free-standing water: 40% G1.C Highly artificial broadleaved deciduous forestry plantations: 60%	≃	D5.1 Reedbeds normally without free-standing water: 40% G1.1 Riparian and gallery woodland, with dominant *Alnus, Betula, Populus* or *Salix*: 40% E3.4 Moist or wet eutrophic and mesotrophic grassland: 20%	D5.2 Beds of large sedges normally without free-standing water : 30% G1.1 Riparian and gallery woodland, with dominant *Alnus, Betula, Populus* or *Salix* : 70%	F9.2 *Salix* carr and fen scrub : 80% G1.B Nonriverine *Alnus* woodland : 20%

* Surface waterbody international codes. ** Short description of EUNIS level 1 units (only those ≥ 10% are shown). *** EUNIS level 3 habitat codes and names after Davies et al. (2004).

FIG. 5.2.3 Context diagnosis (step 1) of the impacted site (before impact) and three proposed offset sites (with expected offset outcomes when offsets are still being planned but not yet carried out). A spreadsheet allows users to compile data and a table is automatically filled to compare sites. The sign = indicates criteria that must be equal between impacted site and offset site. The sign ≃ indicates criteria that must be similar. See full text for exceptions to these requirements.

Step 2: Functional Diagnosis

The second step of the method assesses a wetland's functional performance, which is necessary to demonstrate whether no net loss of functions is achievable or achieved by comparing the impacted site before and after impact with the offset site before and after offset measures. Through this functional diagnosis, users can ensure that the principles of ecological additionality, equivalence, and efficiency are respected (see Table 5.2.1).

Parameters: Field and Desktop Information to Calculate Indicators

National databases and field data are used to gather physical, chemical, or biological information (named parameters), which are expected to indicate functional performance. This definition of parameters is similar to that of Bartoldus (1994) for parameters, Maltby (2009) for controlling variables, or Adamus et al. (1991) for predictors. Chosen parameters are scientifically founded (i.e., general relationship between its measurement and functional performance) and its measurement must be objective, inexpensive, and feasible year-round. It must also discriminate sites within a given ecological context and be responsive to impacts (degradation) and offsets alike (restoration). Parameters were chosen so that their measurement in the field could be done with no prior knowledge of the assessed site (e.g., times series such as flooding frequency or nutrient inputs are not included). Field parameters are based on vegetation cover, drainage systems, erosion signs, soil, and habitats (Table 5.2.2). In total, 22 parameters (GIS and field measurements) were selected.

TABLE 5.2.2 Parameters Harvested to Fill Indicators and Assess Functional Performance Diagnosis (Step 2)

Functions	Parameters Harvested to Fill Indicators
Run-off reduction	*Vegetation cover*: type of plant cover *Drainage systems*: low depth ditch, ditch, deep ditch
Groundwater recharge	*Drainage systems*: low depth ditch, ditch, deep ditch, underground drain *Soil*: surface texture [0–30 cm], deep texture [30–120 cm]
Sediment retention	*Vegetation cover*: permanent cover, type of plant cover *Drainage systems*: low depth ditch, ditch, deep ditch, ditch vegetalization *Erosion*: gully, bank vegetalization *Soil*: humic matter, surface texture [0–30 cm]
Denitrification	*Vegetation cover*: permanent cover, type of plant cover *Drainage systems*: low depth ditch, ditch, deep ditch, ditch vegetalization, underground drain *Erosion*: gully, bank vegetalization *Soil*: humic matter, buried humic matter, surface texture [0–30 cm], deep texture [30–120 cm], waterlogging
Nitrogen assimilation by plants	*Vegetation cover*: permanent cover, type of plant cover *Drainage systems*: low depth ditch, ditch, deep ditch, ditch vegetalization, underground drain *Erosion*: gully, bank vegetalization *Soil*: humic matter, buried humic matter
Phosphorus adsorption-precipitation	*Vegetation cover*: permanent cover, type of plant cover *Drainage systems*: low depth ditch, ditch, deep ditch, ditch vegetalization, underground drain *Erosion*: gully, bank vegetalization *Soil*: pH
Orthophosphates assimilation by plants	*Vegetation cover*: permanent cover, type of plant cover *Drainage systems*: low depth ditch, ditch, deep ditch, ditch vegetalization, underground drain *Erosion*: gully, bank vegetalization *Soil*: pH
Carbon sequestration	*Vegetation cover*: type of plant cover *Soil*: humic matter, buried humic matter, histic horizons, buried histic horizon, waterlogging
Habitat support	*Habitats*: habitat number, habitat repartition, edge, habitat artificialization, invasive plant cover
Habitat connectivity	*Habitats*: distance to the similar nearest habitat unit, similarity with landscape

A Set of Indicators to Assess Functional Performance

Many RAMs are based on a reference wetland (or hypothetical state) against which functions are assessed, using a composite score or index indicative of an ideal state of a given wetland type. The French method does not use such a reference. Instead, functions are assessed separately with a set of simple indicators derived from parameters, and no composite indicators are used. This is because interpreting composite indices is challenging, and there are often hidden assumptions in how indicators are combined (see Cole, 2006; Girardin et al., 1999 in Bockstaller and Girardin, 2003, and Chapters 3.6 and 3.7 of this book).

The 22 parameters account for 32 indicators. There are more indicators than parameters because one measurement on a single parameter can (i) be used for several indicators (e.g., habitat richness or equitability are derived from the parameter "habitat"), (ii) or indicate favorable conditions for the performance of one function and unfavorable conditions for another.

Because of trade-offs between functions (see Adamus et al., 1991 for information about complex relations between functions), a high value for one indicator may automatically mean a low value for another. For example, a given pH value can indicate favorable conditions for phosphorus adsorption/precipitation but an unfavorable condition for phosphorus plant uptake. In such a case, the measurement on the single parameter generates an indicator per function. Indicator values are not independent. Also, the redundancy of indicators (multicollinearity) and correlation between the method's indicators and those used in intensive surveys and monitoring was tested to address these issues (Jaymond et al., 2015).

It is important to note that some indicators are specific to riverine wetlands (they are not considered for other wetlands) and that some indicators may not always be applicable, e.g., some soil indicators when soil sampling cannot be done (rocky, too wet, frozen, etc.).

Functional Capacity: An Area × Quality Type Metric

The capacity of the assessed wetland to perform a given function depends on a set of characteristics and the wetland's area (Celada and Bogliani, 1993; Forman and Godron, 1981; Fustec and Lefeuvre, 2000; Hilty et al., 2006; Hooftman et al., 2003; Houlahan et al., 2006; Oertli et al., 2002; Peintinger et al., 2003; Woltemade, 2000). In this method, the value of every indicator is assessed first per unit area, and comprised between 0 (min) and 1 (max). This is labeled CAP_{REL} for "relative functional capacity." An "absolute functional capacity," called CAP_{ABS}, is then calculated for each indicator by multiplying CAP_{REL} of a given indicator by the site's area. The resulting value for any indicator is therefore comprised between 0 (function is absent) and the area of the assessed wetland:

Site CAP_{ABS} for the indicator	=	Site CAP_{REL} for the indicator	×	Site area (ha)

The higher the per unit area value of a given indicator (i.e., close to 1) and the larger the size of a site, the higher the absolute functional capacity will be, and thus the higher the expected level of performance of the wetland for this function after this indicator. Note that when a planned development impact is such that the impacted site completely loses its wetland character (i.e., wetland area is then 0 ha after impact), then the value of all indicators is 0 after impact. Fig. 5.2.4 illustrates the process through which indicators were calibrated.

No Net Loss of Functions?

Comparisons between losses of functional capacities on the impacted site and gains in functional capacity on the offset site allow users to draw conclusions on whether no net loss is likely to be achieved. In this method, this comparison is done separately for each indicator (Fig. 5.2.5), rather than for a single composite indicator. Although this may achieve more complex results (no single total score or ranking), it enables the most at-risk functions to be identified and thus specifically managed on both the impacted and offset sites. This limits possible bias from comparing processes that aren't readily comparable (e.g., loss of soil texture compared with gain of habitat richness)—as discussed in Quétier and Lavorel (2011) and Maseyk et al. (2016). This is an important element of Step 2 of the method, and distinguishes it from many other RAMs.

It is generally accepted that simply equating losses and gains isn't enough to claim no net loss, for example, to account for uncertainties and/or delays in offset outcomes (Quétier and Lavorel, 2011; Bull et al., 2016). Instead, multipliers or area-based ratios are often introduced to ensure that functional gains exceed function losses, and that the equivalency and ecological additionality principles are respected (see Table 5.2.1). This generally means that offset areas have to be increased relative to those that would be calculated simply on the basis of functional indicators (e.g., Pickett et al., 2013). Ideally, such a ratio should be supported by scientific and technical evidence, such as observed time lags in restoration (Moilanen et al.,

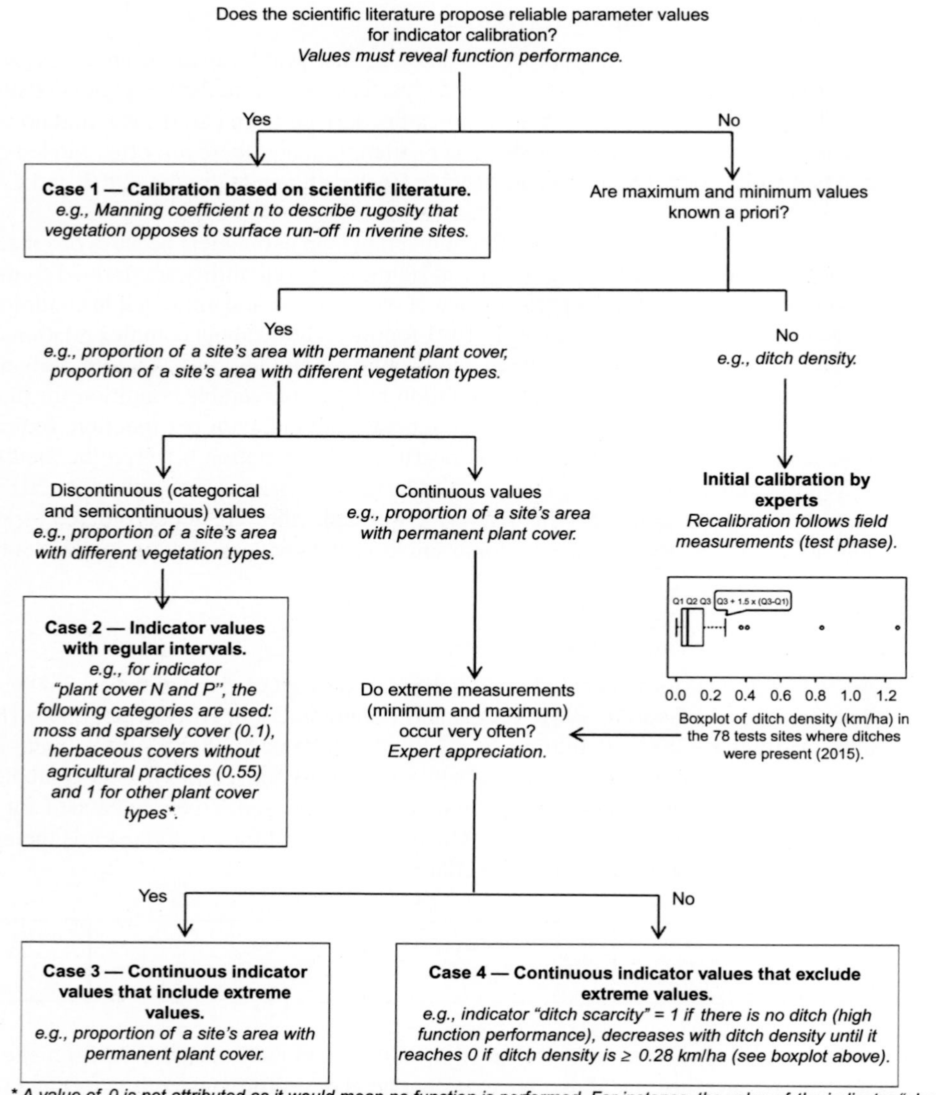

* *A value of 0 is not attributed as it would mean no function is performed. For instance, the value of the indicator "plant cover N and P" in a site where herbaceous cover (0.55) occupies half of the site and the other half is sparsely covered (0.1), will be 0.325.*

FIG. 5.2.4 Flowchart for indicator calibration. Deviation to the flowchart is possible, but it must be clearly mentioned and justified when it occurs. *(Adapted from Gayet, G., Baptist, F., Baraille, L., Caessteker, P., Clément, J.-C., Gaillard, J., Gaucherand, S., Isselin-Nondedeu, F., Poinsot, C., Quétier, F., Touroult, J., Barnaud, G., 2016. Méthode nationale d'évaluation des fonctions des zones humides. Fondements théoriques, scientifiques et techniques. Onema, MNHN 310 pp. (in French).)*

2009; Laitila et al., 2014; Bull et al., 2016) but this was outside the scope of Version 1 of the method. It was agreed, however, that the area of offset sites should never be lower than that of the corresponding impacted wetland sites.

Illustration of Step 2 Results

In conducting the Step 2 assessment, users fill a template spreadsheet for all assessed sites (impacted and effective or candidate offset sites). Another table is automatically filled to allow an easy comparison of functional diagnosis results before/after impact, to ensure impacts are sufficiently offset. Fig. 5.2.6 provides a (hypothetical) example of the output for each indicator from Step 2. It is the same offset program as that presented in "Illustration of Step 1 Results" section (Fig. 5.2.3), but only offset sites 1 and 2, whose context diagnosis were close to the one of the impacted site (Step 1), are shown. Here, an arbitrary ratio of 1.5:1 was required, that is, functional gains should be at least 1.5 times greater than losses, on the basis that impacts were irreversible on c.40% of the impacted site's area because restoring reedbeds and grasslands is likely to be fast but woodland restoration takes much longer.

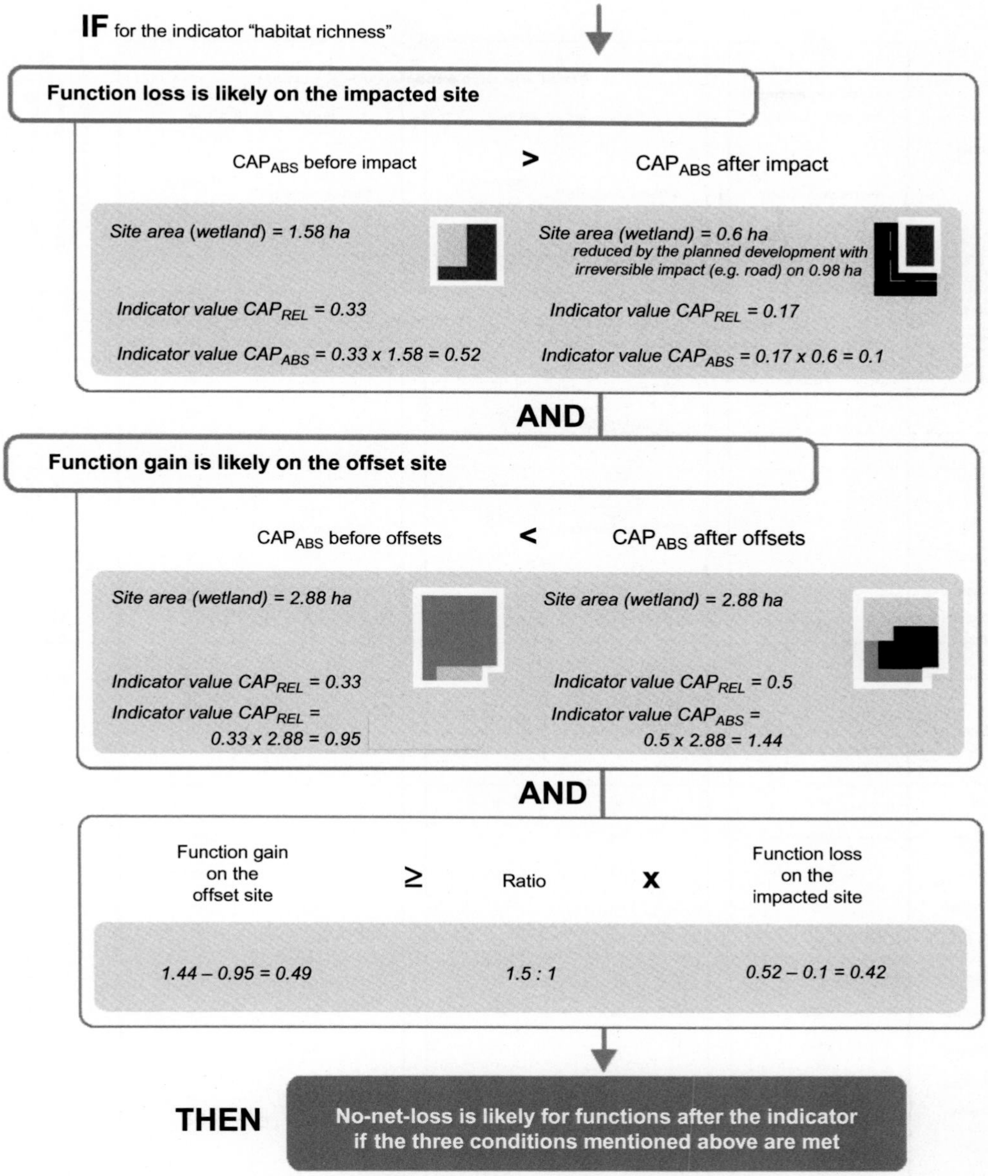

IF for the indicator "habitat richness"

Function loss is likely on the impacted site

CAP_{ABS} before impact > CAP_{ABS} after impact

Site area (wetland) = 1.58 ha
Indicator value CAP_{REL} = 0.33
Indicator value CAP_{ABS} = 0.33 x 1.58 = 0.52

Site area (wetland) = 0.6 ha
reduced by the planned development with irreversible impact (e.g. road) on 0.98 ha
Indicator value CAP_{REL} = 0.17
Indicator value CAP_{ABS} = 0.17 x 0.6 = 0.1

AND

Function gain is likely on the offset site

CAP_{ABS} before offsets < CAP_{ABS} after offsets

Site area (wetland) = 2.88 ha
Indicator value CAP_{REL} = 0.33
Indicator value CAP_{REL} = 0.33 x 2.88 = 0.95

Site area (wetland) = 2.88 ha
Indicator value CAP_{REL} = 0.5
Indicator value CAP_{ABS} = 0.5 x 2.88 = 1.44

AND

Function gain on the offset site	≥	Ratio	x	Function loss on the impacted site
1.44 – 0.95 = 0.49		*1.5 : 1*		*0.52 – 0.1 = 0.42*

THEN **No-net-loss is likely for functions after the indicator if the three conditions mentioned above are met**

FIG. 5.2.5 Summary of the comparison between an impacted site and a proposed offset site for "habitat richness" (a single indicator). Here, gains (0.49) are marginally higher than losses (0.42). If an additional multiplier or "ratio" is introduced to account for, for example, uncertainties and time lags (e.g., 1.5:1), then those gains may be considered insufficient. *(Adapted from Gayet, G., Baptist, F., Baraille, L., Caessteker, P., Clément, J.-C., Gaillard, J., Gaucherand, S., Isselin-Nondedeu, F., Poinsot, C., Quétier, F., Touroult, J., Barnaud, G., 2016. Méthode nationale d'évaluation des fonctions des zones humides. Fondements théoriques, scientifiques et techniques. Onema, MNHN 310 pp. (in French).)*

Offset site 1 had been drained by ditches, cultivated with crops, and heavily fertilized. The proposed offset involves filling ditches, scraping the topsoil to promote reedbeds by getting closer to the water table, making artificial grasslands and intensive monocultures more diversified (e.g., with hay transfer), and planting trees to recover woodlands. Expected outcomes may generate gains for 11 indicators out of 32 ("YES" in Fig. 5.2.6). Among these 11 indicators, three indicators are likely to meet the no net loss requirement. This promotes functional gain and no net loss on hydrological, biogeochemical, and accomplishment of species life cycle functions that are key issues in this region (see Fig. 5.2.3, e.g., heavy agricultural pressures on uplands and forested landscape). Ecological additionality, equivalency, and efficiency principles laid down in the French Code of Environment (Table 5.2.1) are considered to be respected after these three indicators.

Offset site 2 was not considered an appropriate offset. It is composed of beds of large sedges and woodlands with no major direct disturbances. Developers have proposed filling a few ditches and preserving the site in its present condition.

Indicator name	Impacted site with expected impact (1.6 ha)	Offset site with expected offset outcomes			
		Option #1 (2.88 ha)		Option #2 (3.5 ha)	
	Function loss?	Function gain?	No-net-loss?	Function gain?	No-net-loss?
Vegetation cover					
Permanent cover	YES	YES (1.6 time the loss)	YES	no	no
Plant cover N and P	YES	no	no	no	no
Plant cover C	YES	YES (0.8 time the loss)	no	no	no
Rugosity	YES	YES (1 time the loss)	no	no	no
Drainage systems					
Low depth ditch scarcity	YES	YES (0.4 time the loss)	no	no	no
Ditch scarcity	YES	YES (1.5 time the loss)	YES	YES (0.7 time the loss)	no
Deep ditch scarcity	YES	no	no	no	no
Ditch vegetalisation	YES	no	no	no	no
Underground drain scarcity	YES	no	no	no	no
Erosion					
Gully erosion scarcity	YES	no	no	no	no
Bank vegetalization	no information	no information	no information	no	no
Soil					
Neutral pH	YES	no	no	no	no
Acid or basic pH	YES	no	no	no	no
Humic matter	YES	no	no	no	no
Buried humic matter	no	no	no	no	no
Histic horizons	no	no	no	no	no
Buried histic horizon	no	no	no	no	no
Surface erodability	YES	no	no	no	no
Surface denitrication	YES	no	no	no	no
Deep denitrification	YES	no	no	no	no
Surface hydraulic conductivity	YES	no	no	no	no
Deep hydraulic conductivity	YES	no	no	no	no
Water-logging	YES	no	no	no	no
Habitats					
Habitat richness (EUNIS level 1)	YES	YES (1.1 time the loss)	no	no	no
Habitat equitability (EUNIS level 1)	YES	YES (1 time the loss)	no	no	no
Habitat proximity	YES	no	no	no	no
Similarity with landscape	YES	no	no	no	no
Habitat richness (EUNIS level 3)	YES	YES (1.1 time the loss)	no	no	no
Habitat equitability (EUNIS level 3)	YES	YES (1 time the loss)	no	no	no
Edge scarcity	YES	no	no	no	no
Habitat artificialization scarcity	YES	YES (1.8 time the loss)	YES	no	no
Invasive plant scarcity	YES	YES (0.4 time the loss)	no	no	no

Related functions: Run-off reduction; Groundwater recharge; Sediment retention; Denitrification; N assimilation; P adsorption; P assimilation; Carbon sequestration; Habitat support; Habitat connection

FIG. 5.2.6 Functional diagnosis (Step 2) results for a fictive impacted site and two proposed offset sites (with expected offset outcomes when offsets are still being planned but not yet carried out). A spreadsheet allows users to compile data and a table is automatically filled to analyze no net loss. Sites are the same than on Figs. 5.2.3 and 5.2.6. Here, no net loss is considered achievable only when gains represent at least "1.5 times the loss" to account for uncertainties and time lags (ratio of 1.5:1). "No information" means the indicator was not documented (e.g., no river bank for "bank vegetalization"), which is the case for this fictive example.

Fig. 5.2.6 shows that gains would only be expected for a single indicator, which does not meet the no net loss requirement. Some stakeholders, however, expressed interest in conserving the site, believing this to be an acceptable way to address residual impacts from the developer's project. Because it is rigorously analytical and transparent, the method is helpful in explaining the benefits of a functional loss-gain approach and why site 2 is, in fact, inappropriate.

Although offset site 1 is a much better offset site than site 2, it may be too small. For instance, for "rugosity," offset site 1 would have to be 0.5 time larger to generate a functional gain of at least 1.5 times the loss, and meet the no net loss goal with this ratio.

Other tables are automatically filled for a detailed analysis of indicators and reasons why no net loss is reached or not but are not presented here.

CONCLUSIONS AND LESSONS LEARNED

The method developed by AFB helps developers ensure their offset programs are compliant by (i) identifying priority functions in and around the offset site(s) (analysis of contributory areas, landscape, surface waterbody sites), (ii) verifying that no net loss is achievable or achieved for the indicators that are relevant to these functions, and (iii) ensuring that offsets are focused on addressing the major disturbances to the offset site (e.g., ditch or drainage networks).

As the (hypothetical) example above illustrates, offsets cannot achieve gains on all indicators at the same time, not least because of complex interactions between functions (positive, negative, and not necessarily linear, with thresholds), as discussed by Adamus et al. (1991). By providing a standardized and transparent tool with which to assess gains and losses, the method highlights both the benefits and the limits of an offset program, and how difficult it is to replace all impaired functions, thereby providing a valuable incentive to avoid and minimize residual impacts from development.

The method produces standardized results that guide decision-makers on whether to accept a project and its residual impacts on wetlands, and on the design and acceptability of proposed offsets to address these impacts. It doesn't provide the actual decision, which necessarily involves case-by-case appreciations of trade-offs with other concerns and constraints, including biodiversity (protected species and habitats), other environmental issues (e.g., noise, dust), social issues, and cost. In this, the method is aligned with the "Weight of Evidence Approach" that intends to improve environmental decisions (Burton et al., 2002; Keenleyside et al., 2013). Each criteria (Step 1—context diagnosis) and indicator (Step 2—functional diagnosis), and how it responds to the decisions being considered offers a useful line of evidence to decision-makers.

Feedback from over a year of use in France indicates that no net loss, with varying ratios depending on offset programs, can generally be achieved on a maximum of eight indicators (out of 32). No net loss of functions is generally easier to reach for a significant number of indicators when (i) the impacted site is highly impaired (low functional performance), (ii) impacts are temporary or reversible in the short term, (iii) the impacted site is small compared to the offset site, and (iv) the offset site is very degraded and offers plenty of options for gains. High potential for gains, however, may come with increased uncertainty.

The successful uptake of the method, since it was published, results, in part, from the way it was developed. First, the multipartner team involved in its development brought together scientists from different disciplines, consultants, and field staff who are confronted with operational challenges and work with developers, and people from public agencies that are involved in reviewing permit applications. Every effort was made to ensure the method was embedded in the regulatory landscape. Most of the key concepts, criteria, and indicators in the method are familiar to those involved in wetland management in France. This includes concepts from the Water Framework Directive such as surface waterbodies, and existing tools and datasets such as the EUNIS habitat typology. This has made the method readily applicable by its target audience of environmental consultants and permitting authorities. The method's endorsement by permitting authorities played an important role in generating buy-in and incentivizing developers to use it.

Although it was designed to assess compliance with regulatory requirements, the method can also be used in other ways, such as supporting the design of wetland assessment, management, and restoration outside the specific context of impact mitigation and offsetting. The method can also be used to compare, rank, and prioritize wetlands within a given landscape or jurisdiction in the context of watershed-level planning and management. This, of course, makes it relevant to strategic environmental assessments (as discussed in Vaissière et al., 2016).

LIMITS AND PERSPECTIVES FOR IMPROVEMENT

As with other RAMs, the method has a number of built-in limitations (not all of which are highlighted here). It was designed to answer a specific question in implementing the mitigation hierarchy and does not replace thorough assessments of actual ecological processes by specialists with dedicated instrumentation when this is practical. Plenty of protocols and methods exist to more precisely assess wetland functional performance, but generally these are not readily applicable in the context of mitigation and offsetting (e.g., for Europe, the functional assessment procedure of Maltby, 2009).

An important limitation is that the method only assesses 10 functions out of many (known and unknown) functions that may be impaired by planned development projects. This should be kept in mind when making decisions on whether to allow development to go forward or on how to offset residual impacts. In addition, the method does not consider ecosystem services and social or economic values tied or derived from wetlands. It also ignores biodiversity conservation goals and can't be used to assess a site's overall ecological/conservation status (e.g., for sites within the European Natura 2000 network—Maciejewski et al., 2016).

Finally, functional losses and gains are analyzed on the site with no consideration for external impacts (indirect and cumulative impact of planned developments). All these aspects need consideration during impact assessment and the design of appropriate offset programs.

To address some of these limitations, an updated version (2) of the method is already under preparation. It will include tidal wetlands and additional functions (e.g., summer stream flow support) as well as address additional principles of French regulations and guidance on the mitigation hierarchy (e.g., feasibility, temporal proximity—Table 5.2.1). An improved user interface (e.g., automated GIS module) is also expected. It is expected that such updates will be regular, as with many RAMs, to ensure the method remains fit for its purpose. Maintaining a community of practice that can provide timely and constructive feedback on the method will be an important aspect of its continuous improvement over time.

ACKNOWLEDGMENTS

The method was developed by the following institutions: the French Muséum national d'Histoire naturelle, Agence Française de la Biodiversité, Biotope, Centre d'études et d'expertise sur les risques, l'environnement, la mobilité et l'aménagement, Forum des Marais Atlantiques, Institut national de recherche en sciences et technologies pour l'environnement et l'agriculture, Université François Rabelais de Tours, Université Grenoble Alpes, Université Savoie Mont-Blanc.

The authors and these institutions are grateful for the financial support provided by AFB and the Conseil Départemental de l'Isère, and for the endorsement by the French Ministry of Environment. Additional support is acknowledged from the European Union's Seventh Framework Program (FP7/2007-2013) under grant agreement 308393 "OPERAs." Many thanks are due to the numerous people who took part in field tests (AFB Directions InterRégionales of Compiègne, Metz, Rennes and Toulouse), and those who are now using the method and providing useful feedback for its subsequent improvements.

REFERENCES

Adamus, P.R., Stockwell, L.T., Clairain Jr., E.J., Morrow, M.E., Rozas, L.P., 1991. Wetlands Evaluation Technique (WET). Volume 1: Literature Review and Evaluation Rationale. DTIC Document.

Bartoldus, C.C., 1994. EPW: a procedure for the functional assessment of planned wetlands. Water Air Soil Pollut. 77, 533–541.

Bartoldus, C.C., 1999. A Comprehensive Review of Wetland Assessment Procedures: A Guide for Wetland Practitioners. Environmental Concern Inc., St. Michaels, MD. ISBN 188322604X 9781883226046.

Bendor, T., Sholtes, J., Doyle, M.W., 2009. Landscape characteristics of a stream and wetland mitigation banking program. Ecol. Appl. 19, 2078–2092.

Bezombes, L., Gaucherand, S., Kerbiriou, C., Reinert, M.E., Spiegelberger, T., 2017. Ecological equivalence assessment methods: what trade-offs between operationality, scientific basis and comprehensiveness? Environ. Manag. 60 (2), 216–230.

Bockstaller, C., Girardin, P., 2003. How to validate environmental indicators. Agric. Syst. 76, 639–653.

Brinson, M.M., 1993a. A Hydrogeomorphic Classification for Wetlands. DTIC Document, USACE, Waterways Experiment Station Report WRP-DE-4.

Brinson, M.M., 1993b. Changes in the functioning of wetlands along environmental gradients. Wetlands 13, 65–74.

Brinson, M.M., Rheinhardt, R., 1996. The role of reference wetlands in functional assessment and mitigation. Ecol. Appl. 6, 69–76.

Bull, J.W., Lloyd, S.P., Strange, N., 2016. Implementation gap between the theory and practice of biodiversity offset multipliers. Conserv. Lett. 10 (6), 656–666.

Burton, G.A., Chapman, P.M., Smith, E.P., 2002. Weight-of-evidence approaches for assessing ecosystem impairment. Hum. Ecol. Risk. Assess. 8, 1657–1673.

Carletti, A., Leo, G.A.D., Ferrari, I., 2004. A critical review of representative wetland rapid assessment methods in North America. Aquat. Conserv. Mar. Freshwat. Ecosyst. 14, S103–S113.

Celada, C., Bogliani, G., 1993. Breeding bird communities in fragmented wetlands. Ital. J. Zool. 60, 73–80.

Cole, C.A., 2006. HGM and wetland functional assessment: six degrees of separation from the data? Ecol. Indic. 6, 485–493.

Davidsson, T., Kiehl, K., Hoffmann, C.C., 2000. Guidelines for monitoring of wetland functioning. EcoSys 8, 5–50.

Davies, C.E., Moss, D., Hill, M.O., 2004. EUNIS Habitat Classification Revised 2004. Report to: European Environment Agency-European Topic Centre on Nature Protection and Biodiversity, pp. 127–143.

Fennessy, M.S., Jacobs, A.D., Kentula, M.E., 2004. Review of Rapid Methods for Assessing Wetland Condition. US Environmental Protection Agency, Washington, DC. EPA/620/R-04/009.

Forman, R.T., Godron, M., 1981. Patches and structural components for a landscape ecology. Bioscience 31, 733–740.

Fustec, E., Lefeuvre, J.-C., 2000. Les fonctions des zones humides: des acquis et des lacunes. In: Dunod, (Ed.), Fonctions et Valeurs Des Zones Humides, pp. 17–38. Paris (in French).

Gaucherand, S., Schwoertzig, E., Clement, J.-C., Johnson, B., Quétier, F., 2015. The cultural dimensions of freshwater wetland assessments: lessons learned from the application of US rapid assessment methods in France. Environ. Manag. 56 (1), 245–259.

Gayet, G., Baptist, F., Baraille, L., Caessteker, P., Clément, J.-C., Gaillard, J., Gaucherand, S., Isselin-Nondedeu, F., Poinsot, C., Quétier, F., Touroult, J., Barnaud, G., 2016a. Méthode nationale d'évaluation des fonctions des zones humides. Fondements théoriques, scientifiques et techniques. Onema, MNHN, p. 310 (in French).

Gayet, G., Baptist, F., Baraille, L., Caessteker, P., Clément, J.-C., Gaillard, J., Gaucherand, S., Isselin-Nondedeu, F., Poinsot, C., Quétier, F., Touroult, J., Barnaud, G., 2016b. Guide de la méthode nationale d'évaluation des fonctions des zones humides - version 1.0. Onema, collection Guides et protocoles (in French).

Girardin, P., Bockstaller, C., Ven der Werf, H., 1999. Indicators: tools to evaluate the environmental impacts of farming systems. J. Sustain. Agric. 13, 5–21 (In Bockstaller and Girardin 2003).

Hilty, J.A., Lidicker Jr., W.Z., Merenlender, A., 2006. Corridor Ecology: The Science and Practice of Linking Landscapes for Biodiversity Conservation. Island Press, San Diego.

Hobbs, R.J., Higgs, E., Harris, J.A., 2009. Novel ecosystems: implications for conservation and restoration. Trends Ecol. Evol. 24, 599–605.

Hooftman, D.A., Van Kleunen, M., Diemer, M., 2003. Effects of habitat fragmentation on the fitness of two common wetland species, *Carex davalliana* and *Succisa pratensis*. Oecologia 134, 350–359.

Houlahan, J.E., Keddy, P.A., Makkay, K., Findlay, C.S., 2006. The effects of adjacent land use on wetland species richness and community composition. Wetlands 26, 79–96.

Jaymond, D., Buelhoff, K., Gaucherand, S., Gayet, G., 2015. Etude d'une méthode d'évaluation rapide des zones humides—La validation et l'application d'une méthode d'évaluation rapide des fonctions des zones humides. Irstea, Grenoble (in French).

Keenleyside, K.A., Dudley, N., Cairns, S., Hall, C.M., Stolton, S., 2013. Restauration Écologique Pour les Aires Protégées: Principes, Lignes Directrices et Bonnes Pratiques. UICN, Gland (in French).

Laitila, J., Moilanen, A., Pouzols, F.M., 2014. A method for calculating minimum biodiversity offset multipliers accounting for time discounting, additionality and permanence. Methods Ecol. Evol. 5 (11), 1247–1254.

Lamarque, P., Quétier, F., Lavorel, S., 2011. The diversity of the ecosystem services concept: implications for quantifying the value of biodiversity to society. C. R. Biol. 334 (5–6), 441–449.

Maciejewski, L., Lepareur, F., Viry, D., Bensettiti, F., Puissauve, R., Touroult, J., 2016. État de conservation des habitats: propositions de définitions et de concepts pour l'évaluation à l'échelle d'un site Natura 2000. Revue d'Ecologie (Terre et Vie) 71 (1), 3–20 (in French).

Maltby, E., 2009. Functional Assessment of Wetlands. Towards Evaluation of Ecosystem Services. Woodhead Publishing, Abington, Cambridge.

Maltby, E., Hogan, D.V., McInnes, R.J., 1996. Functional Analysis of European Wetland Ecosystems: Improving the Science Base for the Development of Procedures of Functional Analysis. The Function of River Marginal Wetland Ecosystems. Phase 1 (FAEWE). Office for Official Publications of the European Communities.

Maseyk, F.J.F., Barea, L.P., Stephens, R.T.T., Possingham, H.P., Dutson, G., Maron, M., 2016. A disaggregated biodiversity offset accounting model to improve estimation of ecological equivalency and no net loss. Biol. Conserv. 204, 322–332.

Moilanen, A., Van Teeffelen, A.J., Ben-Haim, Y., Ferrier, S., 2009. How much compensation is enough? A framework for incorporating uncertainty and time discounting when calculating offset ratios for impacted habitat. Restor. Ecol. 17 (4), 470–478.

Oertli, B., Joye, D.A., Castella, E., Juge, R., Cambin, D., Lachavanne, J.-B., 2002. Does size matter? The relationship between pond area and biodiversity. Biol. Conserv. 104, 59–70.

Peintinger, M., Bergamini, A., Schmid, B., 2003. Species-area relationships and nestedness of four taxonomic groups in fragmented wetlands. Basic Appl. Ecol. 4, 385–394.

Pickett, E.J., Stockwell, M.P., Bower, D.S., Garnham, J.I., Pollard, C.J., Clulow, J., Mahony, M.J., 2013. Achieving no net loss in habitat offset of a threatened frog required high offset ratio and intensive monitoring. Biol. Conserv. 157, 156–162.

Quétier, F., Lavorel, S., 2011. Assessing ecological equivalence in biodiversity offset schemes: key issues and solutions. Biol. Conserv. 144, 2991–2999.

Quétier, F., Regnery, B., Levrel, H., 2014. No net loss of biodiversity or paper offsets? A critical review of the French no net loss policy. Environ. Sci. Pol. 38, 120–131.

Smith, R.D., Ammann, A., Bartoldus, C., Brinson, M.M., 1995. An Approach for Assessing Wetland Functions Using Hydrogeomorphic Classification, Reference Wetlands, and Functional Indices. (DTIC Document).

Sutula, M.A., Stein, E.D., Collins, J.N., Fetscher, A.E., Clark, R., 2006. A practical guide for the development of a wetland assessment method: the California experience 1. J. Am. Water Resour. Assoc. 42, 157–175.

Vaissière, A.C., Bierry, A., Quétier, F., 2016. Mieux compenser les impacts sur les zones humides: modélisation de différentes approches dans la région de Grenoble. Sci. Eaux Territ. 21, 64–69 (in French).

Wende, W., Tucker, G., Quétier, F., Rayment, M., Darbi, M. (Eds.), 2018. Biodiversity Offsets—European Perspectives on no Net Loss of Biodiversity and Ecosystem Services. Springer, Cham.

Woltemade, C.J., 2000. Ability of restored wetlands to reduce nitrogen and phosphorus concentrations in agricultural drainage water. J. Soil Water Conserv. 55, 303–309.

FURTHER READING

Onema, 2015. Pour une conception et une réalisation des IOTA de moindre impact environnemental—Modalités d'expertise, préconisations techniques et retours d'expériences—Tome 5: expertise des mesures de compensation écologique. Collection 'Guides et protocoles'. Onema. 76 pp. (in French).

Chapter 5.3

Monitoring Wetland Condition in New Zealand

Beverley Clarkson* **and Brian Sorrell**†
*Landcare Research, Hamilton, New Zealand, †Aarhus University, Aarhus, Denmark

Chapter Outline

INTRODUCTION

As signatory to the Ramsar Convention on Wetlands and the Convention on Biological Diversity, New Zealand has international obligations to monitor the health and condition of wetlands. Government agencies, in particular the Department of Conservation and the Ministry for the Environment, share responsibilities for meeting the requirements of the conventions. The Resource Management Act (1991) is the principal national legislation governing the use of natural and physical resources and environments through land-use planning in New Zealand. It identifies the protection and management of wetlands as matters of national importance. Local authorities (regional, district, and city councils[1]) have responsibility for maintaining native biodiversity and ecosystems, with more recent policy and legislation, for example, a 2003 amendment to the Resource Management Act, providing statutory direction to prevent damage and degradation of wetlands (Myers et al., 2013). While most of the larger nationally significant wetlands are on public land and managed by the Department of Conservation, the majority of wetlands, which contribute to the nation's full diversity, are in private ownership. These are subject to council policies and rules through regional and district plans, which are a key mechanism for implementing national legislative direction.

New Zealand has lost 90% of its historic wetlands (Ausseil et al., 2008), and the remaining wetlands are under increasing pressure from both direct and indirect human activities. Monitoring wetland condition is essential for meeting international obligations as well as for assessing the effectiveness of national and regional policies and regulations with respect to protecting wetlands. To assist in state of the environment reporting for wetland ecosystems, a national system for reporting on changes in ecological condition of terrestrial wetlands was developed. The focus was on palustrine (freshwater) wetland systems, particularly swamp, marsh, fen, and bog, which represent the main functional types present in New Zealand (Johnson and Gerbeaux, 2004). This chapter outlines the methodology, indicator context, and assessment; how to apply the approach both in a single wetland and at multiwetland scales; and how to use it to answer a range of monitoring questions. It also provides a case study on changes in wetland condition over time, indicates how the data collected in a national wetland database are being used for other projects, and summarizes subsequent refinements to the basic methodology.

1. Regional councils are responsible for regional environmental matters, and include city and district councils within their boundaries. Unitary authorities are city or district councils that also perform the functions of a regional council. New Zealand has 11 regional councils and six unitary authorities.

Wetland and Stream Rapid Assessments. https://doi.org/10.1016/B978-0-12-805091-0.00054-2

CONCEPTUAL FRAMEWORK

During the development of environmental indicators for freshwater, as part of the national Environmental Performance Indicators Program run by the Ministry for the Environment (Ward and Lambie, 1999), a lack of information on indicators for monitoring wetlands became evident. In addition, as indicators may vary according to wetland type, there was a need for a wetland classification system that was relevant to the New Zealand context after the introduction of the Resource Management Act (1991). As a result, a project on "Coordinated monitoring of New Zealand wetlands" was funded from 1998 to 2002 under the Sustainable Management Fund administered by the Ministry for the Environment. The aims of the project were to develop a nationally consistent methodology for mapping and monitoring New Zealand wetlands and provide practical tools and guidelines for managers to enable them to assess and report on wetlands. The project was funded in stages. Phase One began in 1998, and involved developing a hierarchical classification system for New Zealand wetlands (Johnson and Gerbeaux, 2004) and a method for monitoring changes in spatial extent (Ward and Lambie, 1999). These were important prerequisites for Phase Two, namely developing a set of cultural indicators for wetland condition (Harmsworth, 2002) and producing a handbook of science-based indicators of wetland condition (Clarkson et al., 2004b).

Stakeholders were closely involved as collaborators and partners throughout all phases of the project. The process involved workshops, hui (meetings with Māori partners), field trials, input from working and steering groups, technical advice, feedback loops, and sharing of information. This was essential in order for the tools and guidelines to be useful for effective, coordinated wetland inventory, monitoring, and management. The field trials and discussions with a wide range of partners around New Zealand were particularly important for development of the science-based indicators.

METHODS FOR ASSESSING ECOLOGICAL CONDITION

Overview

The ecological condition of a wetland is assessed and compared against an assumed prehuman natural or reference state. Although Polynesian Māori first arrived in New Zealand around 1280 CE (Wilmshurst et al., 2008), the major transformative changes in wetlands (to pastoral agriculture) did not occur until after 1840, when European settlement began in earnest. Five indicators of condition are scored to reflect the extent and impact of the modification; a high degree of modification equates to a low score. The indicators relate to the major threats and stressors known to damage New Zealand wetlands and are based on changes in:

1. Hydrological integrity.
2. Physicochemical parameters (soils/nutrients).
3. Ecosystem intactness.
4. Browsing, predation, and harvesting regimes (a measure of introduced animal and human impacts).
5. Dominance of native plants ("weed-free" measure).

The sum of the indicator scores provides the wetland condition index (WCI), which reflects the current ecological state (condition) of the wetland.

The indicators are scored at both a broad wetland-wide scale and a more detailed plot scale to account for differences in scale and monitoring requirements, and to underpin scores with quantitative scientific data. The plot-based approach is the technique advocated for monitoring United States wetlands by Tiner (1999a) and is the foundation for the Protected Natural Areas Program in New Zealand (Myers et al., 1987). This approach initiates a process in which detailed field reconnaissance, ground-truthing, establishment of representative plots, and collection of data, together with integrating existing information, are necessary steps to facilitate informed assessment and scoring of indicators.

Using the Method

The selection of wetlands for monitoring depends on the specific requirements of the monitoring project. At regional or subregional scales, the wetlands should be representative examples of the natural ecological diversity of wetland ecosystems in the project area. This can be based on various environmental frameworks or filters (see Myers et al., 1987) and should reflect former wetland type and extent, altitudinal and geographical range, and vegetation structure and composition. Maps of both former and present-day wetland extent, class, and vegetation type are prerequisites for the robust selection of a representative subset of wetlands.

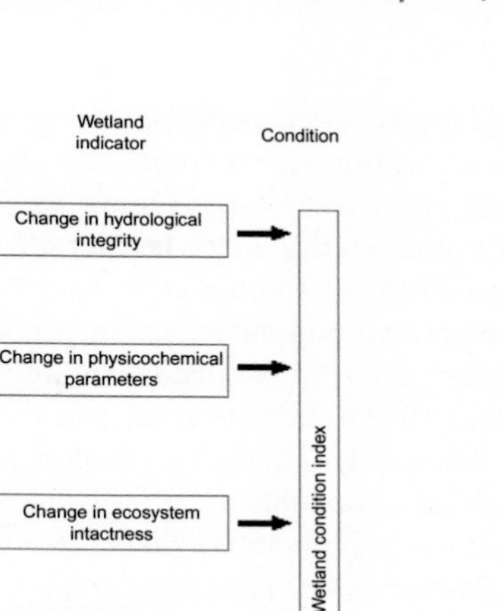
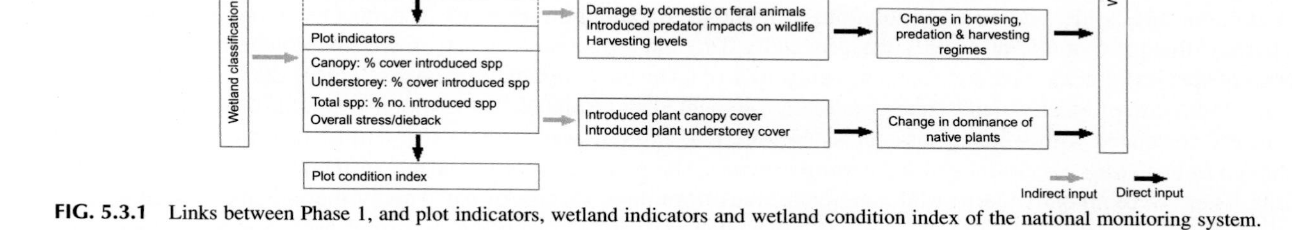

FIG. 5.3.1 Links between Phase 1, and plot indicators, wetland indicators and wetland condition index of the national monitoring system.

The links between the different layers of the wider wetland monitoring project and the wetland and plot information are summarized in Fig. 5.3.1. Phase 1 involves classifying the wetland (hydrosystem, class, etc.) and producing a detailed map using a geographic information system (GIS) of the vegetation types and habitats present. This provides information on the number and areal extent of vegetation types at an appropriate scale. Permanent plots, representative of the vegetation types, are established with a minimum of one plot per major vegetation type. These plots yield quantitative data that are used to assess the nature of the vegetation within the plot, and provide a baseline to detect future changes in physical, chemical, and biological parameters. Each wetland indicator comprises 2–4 indicator components, which are scored according to a 0–5 scale of semiquantitative descriptors following field reconnaissance of the whole wetland, consideration of historical and other information, and interpretation of the plot data. Wetland indicator scores are determined by averaging the scores of the associated indicator components, and then tallying them to give a total WCI out of the maximum achievable of 25.

Wetland pressures, that is, external factors that threaten future wetland condition, are assessed as a separate score. These include modifications to catchment (watershed) hydrology, proportion of catchment in introduced (nonnative) vegetation, and key undesirable species in the catchment; they are also scored on a 0–5 scale. Wetlands that score highly are most at risk of being degraded.

More intensive monitoring may sometimes be required in specific wetlands, for example, assessing threatened species populations, effects of drainage, numbers of introduced predators, and rates of weed invasion. The basic monitoring regime can be extended and modified by increasing the number of permanent plots, sampling across gradients of interest, and adding further elements, as outlined in Clarkson et al. (2004b).

Indicators

Change in hydrological integrity. Hydrology is probably the most important environmental determinant of wetland structure and function (Mitsch and Gosselink, 2000). As such, any modification to natural hydrological regimes will impact wetland processes, and therefore condition. Many hydrological parameters of a palustrine wetland can be determined relatively easily from continuous measurements of water levels, for example, hydroperiod, frequency and duration of flooding, and water depth. These are useful in interpreting whether significant changes to the water budget have been caused by natural climatic events or result from human modifications. As hydrology is difficult to score meaningfully from single site visits, surrogates are also used, guided by visual clues and guidelines for assessing the degree of modification (Clarkson et al., 2004b). These are based on the impact of manmade structures, for example, drains and stop banks (levees), likely to interfere with hydrological integrity, and the presence and abundance of dryland (upland) plants. In recent years, New Zealand wetland plant species have been classified according to their wetland indicator status (obligate wetland OBL, facultative wetland FACW, facultative FAC, facultative upland FACU, Upland UPL; Clarkson, 2014a) following the US wetland delineation system (Environmental Laboratory, 1987; Tiner, 1999b), to distinguish between "dryland" and "wetland" plants.

Change in physicochemical parameters. The physicochemical parameters under this indicator—sedimentation, nutrient enrichment, and fire—most commonly affect New Zealand wetlands. Sediment input is usually accompanied by nutrient enrichment, particularly for wetlands in lowland agricultural environments, but also includes inorganic sediment particles that reduce water clarity. In the last decade, increasing agricultural land-use intensification, especially dairying, throughout New Zealand has put additional pressures on nutrient regimes of wetlands. Anthropogenic inputs of N and P can alter the nature of nutrient limitation in wetlands, potentially causing changes in the structure, composition, and productivity of plant communities. Fires in wetlands are mostly human-induced and temporarily increase nutrient levels (Wilbur and Christensen, 1983), degrade peat, and remove vegetation and other biota. A fourth physicochemical parameter is the von Post index (Clymo, 1983), which applies only to peatland systems, providing a simple and useful measure of peat decomposition or health.

Change in ecosystem intactness. This indicator relates to the loss in area of the original wetland and barriers to connectivity. Wetlands that have been greatly reduced are likely to have modified disturbance-recovery cycles, habitats below minimum area thresholds, and some populations vulnerable to extinction. Barriers to connectivity include both aquatic (e.g., barriers that prevent migration of fauna, especially fish) and terrestrial (e.g., excessive isolation detrimentally affects dispersal of species). Because New Zealand has only 10% of its historic wetland area (Ausseil et al., 2008, 2011), wetlands, particularly those in productive landscapes, are often fragmented and isolated. The ecological condition has been shown to be inversely correlated with wetland loss across swamp, fen, and bog wetland types ($P < .001$; Clarkson et al., 2015).

Change in browsing, predation and harvesting regimes. The prehuman New Zealand vertebrate fauna was dominated by birds. There were no native terrestrial mammals, apart from three species of bat. This bird dominance ended with the arrival of humans (particularly Europeans) and associated introduced browsing and predatory mammals, facilitating changes in community structure, and species composition and abundance. Domestic stock, for example cattle, can cause severe damage to soil and plants from trampling and browsing; however, they are relatively easy to exclude with the appropriate use of fences. Other introduced animals, for example, mustelids, possums, and rats, damage both flora (by browsing, granivory, and substrate disturbance) and fauna (by predation and displacement). They are much harder to control or eradicate as they are small, wary, and usually nocturnal. Because of the crucial role these creatures play in driving species extinctions and biodiversity decline in New Zealand, extensive introduced predator control programs are now a typical feature of management of native ecosystems, including wetlands.

Change in dominance of native plants. This indicator is assessed by determining the degree to which native plants have been displaced by introduced plants at both the canopy and understory levels. Introduced plants are one of the major threats to wetland condition in New Zealand as they can modify wetland structure, composition, and function, in many cases in otherwise intact systems. In particular, fast-growing deciduous trees such as willow (*Salix cinerea* and *S. fragilis*), which have no native ecological equivalent, can invade and dominate swamps and fens characterized by low-growing vegetation. Recent invasions often have a nonnative canopy overtopping the former native plant community, and restoration is relatively straightforward. However, over time, the understory usually also becomes dominated by introduced species that are better adapted to the changed light, nutrients, and other regimes. These are more difficult to manage and restore.

METHOD DEVELOPMENT AND VALIDATION

A science team comprising wetland ecologists from around New Zealand developed the method over a period of 2 years. The process involved an initial workshop with stakeholders to scope criteria needed to develop condition indicators that would be practical and useful to wetland managers. A draft set of indicators was developed and then tested in 15 wetlands throughout New Zealand, selected on the basis of representativeness, existing information, significance, threats, and land-use pressures. The field trials were undertaken in collaboration with wetland managers, iwi (Māori tribes), landowners, and other interested parties. Indicators were refined and improved during the course of the field trials, follow-up meetings, and project steering group input, and then presented and evaluated at a second stakeholder workshop. Further refinements were incorporated into the final set of indicators, which were considered appropriate for satisfying the reporting requirements and incorporated in a web-based handbook (Clarkson et al., 2004b).

Uptake of the monitoring system by end users has been facilitated by ongoing training workshops, presentations, field visits, audits, and ongoing feedback and advice. We have found that consistency improves with experience as the wetland field ecologists become more familiar with the system. As with any scoring or ranking system, the robustness of the WCI depends on the quality and comprehensiveness of the underpinning data, and their fitness for purpose. Wetlands with reliable information may be assessed with more confidence than those with incomplete data. We adopted a pragmatic approach to assess wetland condition using rapid survey techniques (Myers et al., 1987) in which field visits to individual wetlands, including establishing permanent plots, were usually completed within 1 day.

As part of the validation process, the WCI scores were compared with an ecological integrity index developed from GIS indicators (Ausseil et al., 2008, 2011). This incorporates six measures of anthropogenic pressures known to impact wetland ecological integrity and which could be measured consistently using GIS indicators at a national level: naturalness of the upper catchment cover, artificial impervious cover, nutrient enrichment, introduced fish, woody weeds, and drainage. The WCI assesses actual change (state) compared with predicted change from the GIS ecological integrity index. Analysis showed they were highly correlated ($P < .001$; Clarkson et al., 2013).

CASE STUDY: LAKE MARATOTO PEATLAND

The method was applied to assess changes in peatland vegetation and condition at Lake Maratoto, North Island, based on 20 years of monitoring permanent plots established following a fire in 1993 (Table 5.3.1 and Fig. 5.3.2). Although the heath shrub-dominated vegetation recovered rapidly in the early years postfire, there were concerns that the peatland was starting to degrade. Since 2001, water tables had declined significantly ($\bar{x}_{2001} = -0.44$ m; $\bar{x}_{2013} = -0.63$ m, $n = 5$, $P < .05$), and discernible increases in weeds, forest species, and other "dryland" plants had occurred (Clarkson, 2014b). The WCI_{2013} of 11.96/25 was very low compared with other peatland sites in the region and nationally. However, there is potential to increase condition though management and restoration measures focused on restoring hydrology and native plant communities (and ultimately peat-forming species), in association with pest and weed control programs.

USE OF DATA

The data can be used to analyze change at different scales and within different layers of the system, from the separate indicators or components to the WCI, and from individual plot/wetland through regional to national scales. The WCI and indicator/components are left as unweighted scores to allow transparency for comparison between monitoring visits, and in recognition of inconsistency of the nature and extent of impacts across different wetlands. For example, willow invasion is a major threat to swamps and most fens but not bogs. Any weighting can be added later if required.

Some examples of how the monitoring system can be used include (Clarkson et al., 2004b):

- The effectiveness of a fencing/stock exclusion education program assessed by using the indicator component "damage by domestic or other introduced animals" compared at $time = 1$ (preprogram) and $time = 2$ (postprogram).
- The trend in wetland condition for a region summarized by a pie chart showing what proportion of wetlands is deteriorating, improving, or remaining steady. This may be analyzed by wetland area or by wetland class, geographic distribution, altitude, or other suitable filters.
- Priorities for wetland management determined using both WCI (state) and pressure indicators. For example, wetlands with a high WCI and a high wetland pressure score would be obvious candidates for targeting resources or further monitoring.
- Using the WCI for setting restoration goals in restoration projects, for example, a 50% increase in WCI (from 8 to 12/25) by year five through restoration of a water supply, fencing the wetland, and planting native species.

At least 12 agencies (10 regional/unitary councils out of a total of 17, one industry partner, and the Department of Conservation) have used the national wetland monitoring system for state of the environment reporting and to assist in wetland significance assessment. Two more councils are in the early stages of establishing a wetland monitoring network, namely undertaking an inventory of historic and present-day wetland information, as the basis for selecting a set of priority wetlands. Agencies that have better-developed monitoring systems tend to be larger, have a greater funding base, and employ more wetland science capability.

The monitoring data, which are entered into the Landcare Research Wetland Database (as at Feb. 2, 2017, $n = 204$ wetlands) are valuable for multiple purposes. They have been used to characterize New Zealand wetlands, for example, soil carbon stocks (Ausseil et al., 2015), soil and plant nutrient status (Appendix II in Clarkson et al., 2004a,b), and vegetation-environment patterns (Clarkson et al., 2004a) as well as to inform wetland management, for example, environmental limits for maintaining wetland ecological integrity (Clarkson et al., 2015). They provide baselines for wetland condition (compared with historic levels) as a tool for assessing environmental pressures and determining wetland priorities. For example, wetlands in developed agricultural catchments have significantly lower WCI than those in indigenous-dominated catchments ($n = 72$, $P < .001$; Clarkson et al., 2013), and should be priorities for monitoring under the current trend of increasing agricultural land-use intensification.

TABLE 5.3.1 Case Study: Wetland Record Sheet and Plot Sheet for Lake Maratoto Peatland

Wetland Record Sheet

Wetland name:	Lake Maratoto	Date:	20 Nov 2014
Region:	Waikato	GPS:	E1802428 N5804356
Altitude:	55 m a.s.l.	No. of plots sampled:	5
Classification: I System	**IA Subsystem**	**II Wetland Class**	**IIA Wetland Form**
Palustrine	Permanent	Bog	Raised bog

Field Team: BRC, CHW, SB, DT

Indicator	Indicator Components	Specify and Comment	Score 0–5[a]	Mean Score
Change in hydrological integrity	Impact of manmade structures	Very high. Drainage ditches on boundaries. Lake has weir but water levels were and continue to be lowered	1	1.33
	Water table depth	Significantly lower than originally, particularly since 1993. Very low in summer	1	
	Dryland plant invasion	High—forest and other non-wetland species invading. PI significantly lower since 1993	2	
Change in physicochemical parameters	Fire damage	None recently. Last fire in 1993 was extensive	5	3.00
	Degree of sedimentation/ erosion	Local—from race along N edge of lake, and farmed rolling catchment to E, S and N	4	
	Nutrient levels	Significantly higher TP, TN TK, BD, than in 1993. Much higher than "intact" bogs	2	
	Von Post index	Very high—highly decomposed peat	1	
Change in ecosystem intactness	Loss in area of original wetland	Extensive. Probably sole remnant of 6400 ha raised bog drained for farmland	0.1	0.55
	Connectivity barriers	Virtually all connections to other wetlands and water bodies lost except for adjacent Lake Maratoto, a peat lake	1	
Change in browsing, predation & harvesting regimes	Damage by domestic or other introduced animals	Fenced but occasional stock access to Māori fishing village site (pastured). Other browsers, e.g., possums likely present in high numbers	3	3.33
	Introduced predator impacts on wildlife	Pest animals have ready access from surrounding farmland and likely present in high numbers, e.g., stoats, rats	3	
	Harvesting levels	None observed. Māori formerly caught tuna (eels)	4	
Change in dominance of native plants	Introduced plant canopy cover	Abundant on margins and tracks. Starting to invade as mānuka canopy ages and collapses. Weed control program commenced	4	3.75
	Introduced plant understorey cover	Abundant on margins and tracks. Shade-tolerant blackberry invading mānuka understorey. Weed control program commenced	3.5	
Total wetland condition index/25				11.96

Main vegetation types:
Mānuka scrub and shrubland (dominant). Minor areas of grey willow treeland (although now mainly cleared), gorse/blackberry scrub, rank pasture (on old pā site), *Eleocharis sphacelata* reedland (peat lake edge), *Machaerina* sedgeland

Native fauna:
Fantail, Australasian harrier, pukeko, lake aquatic fauna (see separate reports)

Other comments: protected as a Queen Elizabeth II National Trust Open Space Covenant, comprising 46 ha, including 16 ha Lake Maratoto. Remnant surrounded by pasture. Land use: dairying. Peatland vegetation recovered within 5 years of 1993 fire.

Nutrient levels significantly higher than other relatively intact restiad bogs in region, e.g., peat N, P, K (volumetric), and bulk density, and mānuka foliage nutrients, particularly P.

Peatland remnant has dried out considerably since 2001 (29 piezometers have been installed in and around the wetland). Lake levels need to be raised (e.g., by adjusting the lake drain weir) to increase long term viability of the covenant. Pest and weed control programs also recommended, along with re-establishment of extirpated peat-forming species, *Empodisma robustum* and *Sporadanthus ferrugineus* (Restionaceae), once water levels in the peatland have been raised.

The wetland pressure score (below) is high indicating significant external pressures and stresses on the wetland, which, without management intervention, can lead to further degradation in condition.

Pressure	Score[b]	Specify and Comment
Modifications to catchment hydrology	5	Ongoing lowering of lake levels, drainage via boundary ditches, regular droughts in summer
Water quality within the catchment	3	Farm run-off directly into lake
Animal access	3	Currently fenced but easily accessible from surrounding farms
Key undesirable species	3	Grey willow, blackberry, *Osmunda regalis*, privet in catchment and also in peatland
% catchment in introduced vegetation	5	Peatland within agricultural landscape
Other land-use threats	3	Ongoing land clearance, e.g., significant vegetation clearance for pasture on E and S of covenant
Total wetland pressure index/30	22	

Wetland Plot Sheet

Wetland name: Lake Maratoto	Date: 20 Nov 2013	Plot no: 5
Plot size: 2 m × 2 m	Altitude: 55 m a.s.l.	GPS: E1802198 N5804439
Field leader: BRC	Structure: shrubland	Composition: mānuka

Canopy (Bird's Eye View)			Subcanopy			Groundcover		
Species[c] (or Substrate)	**%**	**H**	**Species**	**%**	**H**	**Species**	**%**	**H**
Leptospermum scoparium	45	6.5	*Rubus fruticosus**	25	1.1	*Nertera scapanioides*	0.5	0.05
			Pteridium esculentum	8	1.2	*Pseudopanax crassifolius*	0.5	0.15
			Hypolepis distans	55	0.9	*Chiloglottis cornuta*	0.5	0.03
			Asplenium polyodon	4	0.6			
			Histiopteris incisa	0.5	0.7			
			Cordyline australis	0.5	0.35			

Additional species in vicinity in same vegetation type: *Microsorum pustulatum* (epiphyte), *Stellaria media**				
Comments: Some fallen dead stems of mānuka (*Leptospermum scoparium*) in canopy and understorey—many old moribund mānuka shrubs throughout wetland.				
Prevalence index = 3.2939 (PI_{2001} = 3.0100) indicating (from the vegetation composition) the plot has probably become drier.				
Water level at nearby logger = −0.82 m. Peat depth c.6 m.				
Indicator (Use Plot Data Only)	**%**	**Score 0–5**[d]	**Specify & Comment**	
Canopy: % cover introduced species	0	5	Tall mānuka canopy. Also dead stems	
Understorey: % cover introduced spp[e]	26	3	Blackberry has established since 2001	
Total species: % number introduced spp	10	4	One introduced species—blackberry	
Total species: overall stress/dieback	NA	4	Some collapse of tall mānuka stems	
Total plot condition index/20	NA	16		
Field Measurements				
Water table cm	Not accessible	Water conductivity uS (if present)	Not accessible	
Water pH (if present)	Not accessible	von Post peat decomposition index	H8	
Soil Core Laboratory Analysis (2 Soil Core Subsamples)				
Water content % dry weight	168	Total C %	50.4	
Bulk Density T/m^3	0.18	Total N %	2.3	
pH	3.65	Total P %	0.034	
Conductivity dS/m	0.33	Total K %	0.037	
Foliage Laboratory Analysis (Leaf/Culm Sample of Dominant Canopy Species)				
Species	%N	%P	%K	%C
Leptospermum scoparium	2.08	0.13	0.51	53

[a]*Assign degree of modification as follows: 5 = v. low/none, 4 = low, 3 = medium, 2 = high, 1 = v. high, 0 = extreme.* [b]*Assign pressure scores as follows: 5 = very high, 4 = high, 3 = medium, 2 = low, 1 = very low, 0 = none.* [c]*% = % cover: total canopy % cover = 100%; H = maximum height in m; indicate introduced species by **. [d]*5 = 0%: none, 4 = 1%–24%: very low, 3 = 25–49%; low, 2 = 50%–75%: medium, 1 = 76%–99%: high, 0 = 100%; very high.* [e]*Add subcanopy and groundcover % cover for introduced species.*

FIG. 5.3.2 Lake Maratoto peatland showing locations of permanent plots (P). The 1993 fire burned the entire western tract.

PROSPECTS FOR THE FUTURE

The monitoring system has also evolved to account for subsequent changes in reporting requirements, as new legislation, policies, and plans are developed to protect wetland biodiversity. Regional (and unitary) councils now have additional responsibilities to protect wetlands on private land, and monitoring indicators have been refined slightly (Clarkson and Bartlam, 2017) to capture additional monitoring components recommended for terrestrial ecosystems (Lee and Allen, 2011). The Prevalence Index (PI: Wentworth et al., 1988; U.S. Army Corps of Engineers, 2010) has also been incorporated, which can be used as a surrogate for the "wetness"—or more correctly, "dryness"—of a plot based on the vegetation. The PI is a weighted average value derived from the abundance and wetland indicator status for each plant species within a permanent plot. As plants integrate and reflect the environmental conditions at a site, significant changes in hydrological regimes will be apparent in changes in vegetation composition and cover. For example, influxes of FACU and upland UPL species may be promoted by the lowering of the water table following drain construction, and will result in increases in PI values. In the case study presented (Lake Maratoto peatland; Table 5.3.1), the PI increased significantly between 2001 and 2013 (PI_{2001} mean = 3.1028, PI_{2013} = 3.4574, $n = 5$, $P < .01$), reflecting drier conditions at the site.

The wetland monitoring system has also been adapted to produce a less technical wetland assessment monitoring kit (WETMAK; Denyer and Peters, 2015) for community groups wishing to undertake their own monitoring. WETMAK is an online resource presented as separate modules aimed at assisting the groups in assessing the impact of their restoration work. Other monitoring programs have been developed in recent years and some are being applied to wetlands. For example, the Department of Conservation Biodiversity and Reporting Systems (http://www.doc.govt.nz/our-work/monitoring-and-reporting-system/) aim to provide consistent, comprehensive information about biodiversity across public conservation lands, and potentially across New Zealand. This system is part of an ongoing program with the aim to develop a cohesive approach to managing biodiversity across all New Zealand's land and water ecosystems. It is more detailed, costly, and time consuming than the WCI approach; however, it will contribute significantly to our knowledge of wetland condition and trends. At a global scale, wetland monitoring systems and approaches will continue to evolve over time with new knowledge of functional processes and their measurement, and as new threats to wetlands emerge. Sharing findings with the wider wetland community is essential for the development and adaptation of indicators to ensure they are suitable for decision-making and advancing wetland management outcomes.

REFERENCES

Ausseil, A.-G.E., Gerbeaux, P., Chadderton, W.L., Stephens, R.R.T., Brown, D., Leathwick, J.R., 2008. Wetland Ecosystems of National Importance for Biodiversity. Landcare Research Contract Report LC0708/158.

Ausseil, A.-G.E., Chadderton, W.L., Gerbeaux, P., Stephens, R.T.T., Leathwick, J.R., 2011. Applying systematic conservation planning principles to palustrine and inland saline wetlands of New Zealand. Freshw. Biol. 56, 142–161.

Ausseil, A.-G.E., Jamali, H., Clarkson, B.R., Golubiewski, N.E., 2015. Characterising soil carbon stocks of New Zealand wetlands. Wetl. Ecol. Manag. 23, 947–961.

Clarkson, B.R., 2014a. A vegetation tool for wetland delineation in New Zealand. Landcare Research Contract Report LC1793.

Clarkson, B.R., 2014b. Lake Maratoto Peatland: 20 Years of Vegetation Change (1993–2013). Landcare Research Contract Report LC1809 for Waikato Regional Council.

Clarkson, B.R., Bartlam, S., 2017. State of the Environment Monitoring of Hawke's Bay Wetlands: Tukituki Catchment. Landcare Research Contract Report LC2713.

Clarkson, B.R., Schipper, L.A., Lehmann, A., 2004a. Vegetation and peat characteristics in the development of lowland restiad peat bogs, North Island, New Zealand. Wetlands 24, 133–151.

Clarkson, B.R., Sorrell, B.K., Reeves, P.N., Champion, P.D., Partridge, T.R., Clarkson, B.D., 2004b. Handbook for Monitoring Wetland Condition. Coordinated Monitoring of New Zealand Wetlands. A Ministry for the Environment Sustainable Management Fund Project. 74 pp. http://www.landcareresearch.co.nz/publications/researchpubs/handbook_wetland_condition.pdf (accessed 30.03.17).

Clarkson, B.R., Ausseil, A.-G.E., Gerbeaux, P., 2013. Wetland ecosystem services. In: Dymond, J.R. (Ed.), Ecosystem Services in New Zealand—Conditions and Trends. Manaaki Whenua Press, Lincoln, pp. 192–202.

Clarkson, B.R., Overton, J.M., Robertson, H.A., Ausseil, A.-G.E., 2015. Towards Quantitative Limits to Maintain the Ecological Integrity of Freshwater Wetlands: Interim Report. Landcare Research Report LC1933.

Clymo, R.S., 1983. Peat. In: Gore, A.J.P. (Ed.), Ecosystems of the World 4A Mires: Swamp, Bog, Fen and Moor. Elsevier Scientific, Amsterdam, pp. 159–224.

Denyer, K., Peters, M., 2015. WETMAK—Wetlands Monitoring and Assessment Kit. New Zealand Landcare Trust. http://www.landcare.org.nz/wetmak (accessed 30.03.17).

Environmental Laboratory, 1987. Corps of Engineers Wetlands Delineation Manual Technical Report Y-87-1. US Army Engineer Waterways Experiment Station, Vicksburg, MS. 100 pp. + appendices.

Harmsworth, G., 2002. Coordinated Monitoring of New Zealand Wetlands, Phase 2, Goal 2: Maori environmental performance indicators for wetland condition and trend. A Ministry for the Environment Sustainable Management Fund Project. Landcare Research Contract Report LC 0102/099.

Johnson, P.N., Gerbeaux, P., 2004. Wetland Types in New Zealand. Department of Conservation, Wellington. 184 pp.

Lee, W.G., Allen, R.B., 2011. Recommended Monitoring Framework for Regional Councils Assessing Biodiversity Outcomes in Terrestrial Ecosystems. Landcare Research Contract Report LC144.

Mitsch, W.J., Gosselink, J.G., 2000. Wetlands, third ed. John Wiley & Sons, New York.

Myers, S.C., Park, G.N., Overmars, F.B., 1987. The New Zealand Protected Natural Areas Programme: A Guidebook for the Rapid Ecological Survey of Natural Areas. New Zealand Biological Resources Centre Publication No.6, Department of Conservation, Wellington.

Myers, S.C., Clarkson, B.R., Reeves, P.N., Clarkson, B.D., 2013. Wetland Management in New Zealand—are current approaches and policies sustaining wetland ecosystems in agricultural landscapes? Ecol. Eng. 56, 101–120.

Tiner, R.W., 1999a. Wetland Indicators: A Guide to Wetland Formation, Identification, Delineation, Classification, and Mapping. Lewis Publishers, CRC Press, Boca Raton, FL.

Tiner, R.W., 1999b. Wetland Monitoring Guidelines. Operational Draft. US Fish and Wildlife Service, Hadley, MA. 78 pp.

U.S. Army Corps of Engineers, 2010. Regional Supplement to the Corps of Engineers Wetland Delineation Manual: Western Mountains, Valleys, and Coast Region (Version 2.0). ERDC/EL TR-08-28U.S. Army Engineer Research and Development Center, Environmental Laboratory, Vicksburg, MS. 152 pp.

Ward, J.C., Lambie, J.S., 1999. Monitoring Changes in Wetland Extent: An Environmental Performance Indicator for Wetlands. Coordinated Monitoring of New Zealand Wetlands. A Ministry for the Environment SMF Project. Lincoln Environmental, Lincoln University, Canterbury.

Wentworth, T.P., Thompson, G.P., Kologiski, R.I., 1988. Designation of wetlands by weighted averages of vegetation data: preliminary evaluation. Water Resour. Bull. 24 (2), 389–396.

Wilbur, R.B., Christensen, N.L., 1983. Effects of fire on nutrient availability in a North Carolina coastal plain pocosin. Am. Midl. Nat. 110, 54–61.

Wilmshurst, J.M., Anderson, A.J., Higham, T.F.G., Worthy, T.H., 2008. Dating the late prehistoric dispersal of Polynesian rats to New Zealand using the commensal Pacific rat. Proc. Natl. Acad. Sci. U. S. A. 105, 7676–7680.

Chapter 5.4

Rapid Bioassessment for the Himalayan Rivers

Subodh Sharma
Aquatic Ecology Centre, Kathmandu University, Kathmandu, Nepal

Chapter Outline

INTRODUCTION

Nepal's economy is mainly based on agriculture, which provides a livelihood to more than 70% of the population. Due to a rapid increase in population, the demand of land for agriculture is constantly increasing. As a result, two major developments can subsequently be observed: extension into marginal lands and intensifying agriculture in other areas. With both developments, increased pressure has been placed on water resources. While water quantity has always been a major concern in the hills of Nepal, recently water quality has also become a new challenge and concern in wide areas of both rural and urban Nepal. This is mainly connected with increased fertilizer and pesticide applications in rural areas and water abstraction and organic pollution in the urban areas. In the Kathmandu Valley and in some urban areas of the plains (*Tarai*), increasing population has fostered increased urbanization. This leads to increased garbage, new industries, and more and more pollutants finding their way into rivers. The pollution level in some of Nepal's rivers that pass through urbanized areas, such as the Bagmati river, has become critical. These rivers still are used for drinking water, agricultural purposes, and industrial use. The concern about the degradation of water quality is widespread in the public sector. It was in the late 1970s and early 1980s that people began to document the issue of water pollution in the rivers of Nepal, based on some preliminary research. It was in the 1990s that it became clear that the conditions of some rivers in the country were getting out of hand. This development begs for close investigation. The Bagmati Mega Clean-up Campaign was launched on May 18, 2013, focusing on cleaning the Bagmati River with the people's participation. According to the High Powered Committee for Integrated Development of Bagmati Civilization (HPCIDBC), more than 7000 metric tons of solid waste has been extracted from the Bagmati River since the campaign was launched.

Sharma and Moog (1996) applied 12 types of indices and score methods in use worldwide to assess the water quality of Nepalese rivers. Sharma (1998a, b) reported on the river water quality of the Western rivers of Nepal. Sharma (1999) investigated the water quality status of the Eastern rivers of Nepal. Moog and Sharma (1999) proposed the Nepalese biotic score (NEPBIOS) as the most appropriate biotic score method for assessing the water quality of surface waters in Nepal. The effect of land use and altitude has also been studied on the macroinvertebrate communities of Nepal (Rundle et al., 1993; Suren, 1994), organic pollution (Sharma, 1999; Feld et al., 2010), river classification strategies (Tachamo Shah, 2011), and impacts of climate change on water quality (Tachamo Shah et al., 2015).

There exists an adequate database of water chemistry in most of the countries in the Himalayas. Essentially all monitoring programs in the region are focused on the physicochemical analysis of water samples. The knowledge of chemical monitoring is immense. On the other hand, there is virtually no information on biomonitoring, except in India and Pakistan (Habermann and Reisenberger, 2000), and that is to a very limited extent. In India, the Central Pollution Control Board, Delhi was reported to be actively involved in not only the physical and chemical characterization of rivers but also the

Wetland and Stream Rapid Assessments. https://doi.org/10.1016/B978-0-12-805091-0.00052-9

biological monitoring of 126 rivers on a regular basis. Habermann and Reisenberger (2000) have tested the efficiency of, in addition to NEPBIOS, five other indices developed in Europe for monitoring the water quality of rivers in Pakistan. It was stated clearly that the reason for limiting their study to the family level was the lack of adequate literature describing the aquatic fauna of Pakistan. The trend of generating chemical data is prevalent in South Asia and a good technical workforce is also available for carrying out chemical analyses, but the matter of data handling and interpretation remains questionable.

SAMPLING TECHNIQUES

As indicated in the conceptual framework (Fig. 5.4.1), the first step in river quality assessment considering macroinvertebrates as indicators is sampling, which is performed by qualitative techniques that are not restricted to a certain mesh size or specific sampling gear type. Sampling is conducted qualitatively from downstream by using all possible net sizes, including hand-picking. The sampling of macroinvertebrates for the application of methods such as rapid field bioscreening (RFB) or NEPBIOS is based on the experience that the sampling should be continued until the collector is assured that no more new

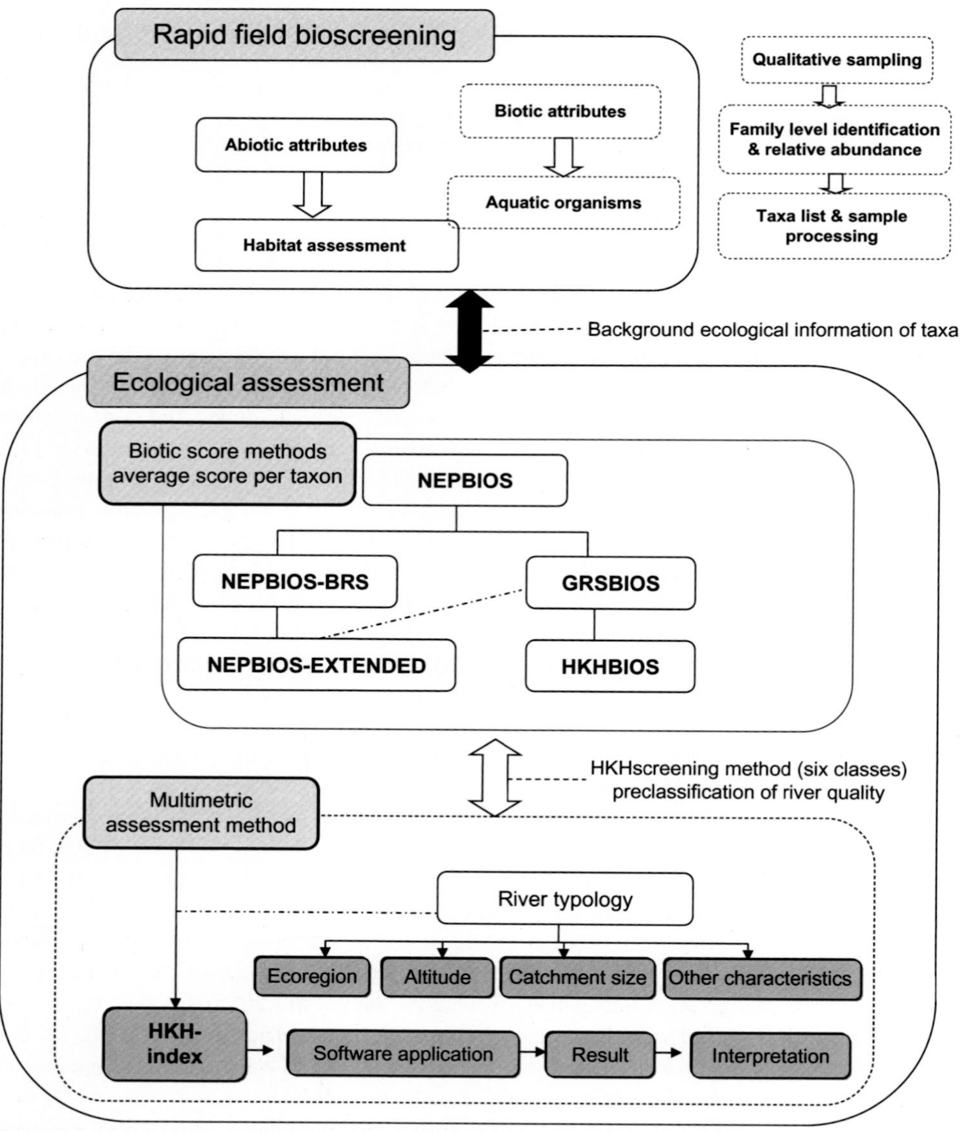

FIG. 5.4.1 Methods in use for the assessment of Himalayan rivers, their origin, and development procedures for application to wadeable mountain and lowland streams.

taxa are likely to occur at the site under investigation. In the case of ecological assessment, however, multiple habitat sampling (MHS) (Barbour et al., 2006) is recommended as a standard technique for sampling macroinvertebrates. The sample reach or station under consideration should extend to a 100m distance, having no major tributaries in the area. It should be conducted at least 100m upstream of any road or bridge crossing to minimize the effects on stream velocity, depth, and overall habitat. Sampling is conducted by jabbing the D-frame net into productive and stable habitats 20 times (in Nepal, where habitats are less heterogenous, a 10 jab approach was found to be equally effective). A single jab consists of forcefully thrusting the net into a productive habitat for a linear distance of 1 m. Different types of habitat should be sampled in rough proportion to their frequency within reach.

In large and deep rivers, which are nonwadeable, an artificial substrate is placed for some duration to allow the macroinvertebrates to colonize. Assessment of large and deep rivers considering colonization samples in artificial substrates for macroinvertebrates has not been practiced so far in the Himalayan rivers. The RFB method is found equally effective in assessing the water quality of smaller streams, which are located close to the spring sources. But the species diversity in such stretches of the streams is distinctly reduced, making it not comparable with other sections of the same streams.

DESCRIPTION OF THE ASSESSMENT METHODS

Altogether, five biotic indices and one multimetric assessment method have been applied for Himalayan river water quality assessment besides RFB, which is considered as the basis for all the biotic index methods for preclassification of the sites to be studied. The five biotic indices methods include; NEPBIOS (Sharma, 1996, Sharma and Moog, 2005), NEPBIOS-BRS (Pradhan, 1998), GRSBIOS (Nesemann, 2006), NEPBIOS-EXTENDED (Sharma et al., 2009), and HKHBIOS (Ofenböck et al., 2010). In the first decade of the 21st century, a 3-year research project funded by the European Union (Contract number: INCO-CT-2005-003659) allowed the combination of knowledge from Asian and European scientists hailing from Bangladesh, Bhutan, India, Nepal, Pakistan, Austria, the Czech Republic, and Germany. The aim of this research focused on the development of an assessment system to evaluate the ecological status of rivers in the Hindu Kush-Himalayan region (ASSESS-HKH). The ASSESS-HKH research work primarily intended to achieve applicable tools for river assessment. This led to the development and validation of a three-tier methodology to detect hot spots in rivers of the HKH region. The three methods provide three levels of increasing discriminatory power to detect the effects of environmental impacts on the benthic invertebrate assemblages. The RFB was modified to HKH screening, which allows a rapid field assessment of mountain and lowland rivers (Hartmann et al., 2010). The HKHBIOS adapts the biotic score concept for the HKH region, which includes taxonomic families as well as the species and genus level for selected indicators (Ofenböck et al., 2010). The HKH index represents the most advanced ecological river evaluation method and is based on the multimetric index procedure (Korte et al., 2010). The usefulness of using macroinvertebrates as bioindicators for assessing the water quality status in the southwest streams of Bhutan is described by Kinzang (2016) in his PhD dissertation. However, there is still a need to use existing biotic scores or biotic index methods to assess stream health in Bhutan, leading toward the development of a more effective method.

A description of sampling site features was recorded in a site protocol by Sharma (1996). Determination of benthic macroinvertebrates was done to the highest possible taxonomic precision level. Preclassification of water quality was by experts' judgment, taking into consideration parameters such as turbidity, color, odor, foaming, algal composition and cover, suspended solids, wastes, anaerobic oxidation, and microbiological and biological characteristics; details can be obtained from Moog and Sharma (1999). Water quality was classified into four main and three transitional classes (Table 5.4.1), the definitions of which are given in Sharma and Moog (1996).

Rapid Field Bioscreening (RFB) (Moog, 1991)

RFB is a simple yet effective method for preclassifying the ecological status of rivers. The method is based on the assessment of sites in the field based on a protocol named decision support table (DST), which compares the result of that rapid assessment with the narrative description of water quality classes. Each appropriate observation of the listed criteria has to be scored in the DST. For the classification of water quality classes, the entries of each column are summed up. The majority of points per column "water quality class" are decisive for the preclassification of the organic pollution status of the river under study. This method has later evolved into HKH screening. The HKH screening allows a rapid field assessment of mountain and lowland rivers separately (Hartmann et al., 2010).

TABLE 5.4.1 Classification of River Water Quality Classes

River Quality Color Code and Class Category[a]		Description	River Quality Color Code and Class Category[b]		Description
Blue	I	No to very slightly polluted	Blue	I	Not polluted
Blue-green	I-II	Little pollution			
Green	II	Moderate pollution	Green	II	Slightly polluted
Green-yellow	II-III	Critical pollution			
Yellow	III	Heavy pollution	Yellow	III	Moderately polluted
Yellow-red	III-IV	Very heavy pollution	Orange	IV	Heavily polluted
Red	IV	Extreme pollution	Red	V	Extremely polluted

[a]*Moog and Sharma (2001).*
[b]*Shrestha et al., 2008*

Nepalese Biotic Score (NEPBIOS-Original) (Sharma, 1996)

This is a biological method of assessing river water quality, which takes into account the presence and absence of families of macroinvertebrates or higher taxa. This method involves the determination of an average score per taxon (ASPT) of the indicator species present in a sample of macroinvertebrates; more detail can be obtained from Sharma (1996). The NEPBIOS-Original list of benthic macroinvertebrates consists of 92 (82 family level indicators, 8 genus level, and 2 species level) indicators.

Hundreds of biological methods exist worldwide for water quality assessment. The efficiency of 10 such methods was tested in assessing the water quality of Nepalese rivers (Sharma and Moog, 1996). It was finally recommended that the newly adopted NEPBIOS/ASPT method should be applied to assess the water quality of Nepalese rivers. Pradhan (1998). in her attempt to apply NEPBIOS for the Bagmati river system (BRS), has also formulated NEPBIOS-BRS (also called NEPBIOS-BBS) and contributed an expansion of the list of NEPBIOS. Scholz (2001) applied the NEPBIOS and Wassmann/Xylander methods for biological water quality assessment of the river Seti and its tributaries in the Pokhara Valley in Nepal. In order to provide the most suitable assessment system for Nepal, the applicability of both the original NEPBIOS and the NEPBIOS-BRS was tested (Sharma and Moog, 2002). To increase the comparability of NEPBIOS within different ecoregions of Nepal, Sharma and Moog (2002) suggested a deviation from a class-defined system to a reference-based evaluation scheme (Table 5.4.2).

Applying a reference-based evaluation system in ecoregional frameworks offers valuable tools for inventorying and assessing environmental resources, for setting resource management goals, and for developing biological criteria and water quality standards (Hughes, 1989; Griffith et al., 1994).

TABLE 5.4.2 NEPBIOS/ASPT Transformation Scale Proposed for Midland and Lowland Nepal Separately

NEPBIOS/ASPT Original Scale	NEPBIOS/ASPT for Midland	NEPBIOS/ASPT for Lowland	Water Quality Class
8.00–10.00	7.50–10.00	6.50–10.00	I
7.00–7.99	6.51–7.49	6.00–6.49	I–II
5.50–6.99	5.51–6.50	5.00–5.99	II
4.00–5.49	4.51–5.50	4.00–4.99	II–III
2.50–3.99	3.51–4.50	2.50–3.99	III
1.01–2.49	2.01–3.50	1.01–2.49	III–IV
1	1.00–2.00	1	IV

Source: Sharma S., Moog O., 2002. A reference based "Nepalese Biotic Score" and its application in the Midland Hills and Lowland Plains for river water quality assessment and management. In: Proceedings of the Second International Conference on Plants and Environmental Pollution, February 4–9, 2002, NBRI, Lucknow, India.

Nepalese Biotic Score for the Bagmati River System (NEPBIOS-BRS) (Pradhan, 1998)

NEPBIOS-BRS, when applied in assessing the water quality of the BRS, increased the R^2 value to 0.83 from 0.78 in the case of NEPBIOS-Original. This precision is due to modifications in the scores that include nine additional families added to its list, and the adjustment is done in the values of tolerant species belonging to the family Baetidae, Chironomidae, and Salifidae with identification to the higher taxonomic level. Altogether, 71 families were listed and 54 families were given a family level score while 12 families were given genus and species levels. These new scores were named NEPBIOS-BRS, for which the scores were initially proposed. In NEPBIOS-Original, altogether 103 families were listed and 82 families were scored. The scores given were family level-based except for the genera and species belonging to families Baetidae, Heptageniidae, Leptophlebiidae, Hydraenidae, and Perlidae. While comparing the scores between NEPBIOS-BRS and NEPBIOS-Original, out of 45 scored families of NEPBIOS, 17 families got the same score while the rest of the families got different scores. Twelve families of NEPBIOS-BRS got one score less than NEPBIOS-Original. The NEPBIOS-BRS has received one score higher than that of NEPBIOS-Original to eight families. Habermann and Reisenberger (2000) have also compared the scores allotted by NEPBIOS-Original and its extension to NEPBIOS-BRS.

Ganga River System Biotic Score (GRSBIOS) (Nesemann, 2006)

The Ganga river system biotic score, abbreviated as GRSBIOS, includes 420 Taxa (Family level 145, Subfamily level 3, Genera level 24, Species level 223, Unidentified 25) genus and species level. Scores are assigned according to ecological preference of each taxon. It is updated with all the new information of macroinvertebrates of the last decade (Nesemann, 2006). All noninsecta are classified according to investigator's personal knowledge based on ecological habitats of macroinvertebrates. GRSBIOS-ASPT is calculated by dividing the sum of individual GRSBIOS with the number of scored taxa and the resulting numerical value is compared to its transformation table (Sharma and Moog, 2005) for determination of river quality classes of running water.

NEPBIOS-EXTENDED (Sharma et al., 2009)

Data from 216 sites published in Sharma (1996) and Pradhan (1998) were the bases for this analysis. The NEPBIOS-Original list of benthic macroinvertebrates with 92 (82 family level indicators, 8 genus level, and 2 species level) indicators is extended to 116 indicators (87 family level, 15 genus level, and 14 species level). No new data were collected for this study. Available data were used based on all earlier studies, most of which are unpublished. Macroinvertebrates collected from Nepal from 1993 to 1999 with saprobic water quality classes known are categorized as from Himalayas, Midland hills, and *Tarai*. Description of seven water quality classes is based on professional experiences and judgments.

HKHBIOTIC SCORE (HKHBIOS) (Ofenböck et al., 2010)

The HKH Biotic Score (HKHBIOS) is a score-based assessment method and adopts the biotic score concept for the HKH region, and also includes (besides taxonomic families) the species and genus level for selected indicators (Ofenböck et al., 2010). The HKH index represents the most sophisticated ecological river evaluation method available and is based on the multimetric index procedure (Korte et al., 2010). The indicator species of benthic macroinvertebrates are ranked using a 10-point scoring system. The scores reflect the sensitivity of taxa to organic pollution and chemical pollution as well as on hydromorphological deficits. Taxa with high scores (e.g., score of 10) indicate high sensitivity to environmental stressors and taxa with low scores (e.g., score of 1) indicate high tolerance to stressors (Ofenböck et al., 2010). To accomplish this, a total of 199 taxa were scored for the HKHBIOS. The final HKH taxa scoring list comprises two taxa on class level, 139 taxa on family level, four taxa on genus/subfamily level, and 51 genera and three taxa on species level. The method was designed for the stream or river quality assessment of both mountainous and lowland HKH-based ecoregions. A total of 186 taxa were assigned scores for highland streams or rivers while scores are available for 155 taxa for lowland streams or rivers (Ofenböck et al., 2010). The final HKHBIOS is generated as a weighted mean of scores. The resulting value (i.e., HKH biotic score) can directly be compared to the corresponding river quality class. As the score-based assessment method is mainly based on higher taxonomy groups, it can be easily applied by any trained aquatic biologist. It also claims to be more applicable to a wider geographical range and various substrate types (Ofenböck et al., 2010). The HKHBIOS is the expansion of the score-based method used for assessing the Nepalese rivers (e.g., NEPBIOS; Sharma and Moog, 1996).

The HKH-Index (Korte et al., 2010)

The HKH-Index is the multimetric assessment method for detailed analysis of biological data on the basis of the best available taxonomic resolution. This method integrates the effects of a wide range of environmental stressors such as organic pollution, eutrophication, hydromorphological deterioration, and acidification on the river ecosystems (Korte et al., 2010; Shrestha et al., 2008). In this method, metrics such as the richness measures (e.g., number of family, number of EPT taxa), composition measures (e.g., percent of EPT, percent of Diptera taxa), diversity measures (e.g., the Shannon Weiner Index) or functional measures (e.g., functional feeding groups) are determined and calculated from the mean value using a set of core metrics that are derived from a given sampling site. This analysis yields the value of HKH-Index. To accomplish this, the measured metrics are transformed into unit-less scores with values between 0 and 1. Finally, the multimetric HKH-Index is calculated by averaging the scores and each multimetric HKH-Index could be directly assigned to a river quality class for interpretation. However, it is said that this method only allows a rough estimation of ecological quality class and therefore it is recommended that the method be calibrated in concordance with results of HKH screening and HKHBIOS (Hartmann and Moog, 2008) for its application to assess the quality of Himalayan rivers.

DISCUSSION

If a site fulfills the criteria set up for the respective stream type, a preclassification is necessary to assign the site to a quality class between high (reference) and bad. While selecting the sampling sites to be assessed, any overlap due to different stressors must be avoided. A five-class scheme for preclassification of the impact of organic pollution in the field is based on the detailed description of the saprobic water quality classes followed by the application of the results obtained from a rapid field DST (Moog, 1991). The DST is a protocol that helps in preclassification of sites that are under the influence of organic pollution based on an assessment of sensory features, periphyton, bacteria, and biotic diversity of the macroinvertebrates. Sensory features include turbidity, color, foam, odor, and wastes disposed along the study area. Occurrence of ferrosulfide reduction in mud and stony habitats is also observed and the intensity assessed in preclassification. Bacteria and sewage fungi occur in massive quantities on stones or macrophytes as well as benthic macroinvertebrates. The extent or magnitude of periphyton and filamentous blue-green algae is a good indicator of nutrient input into the system. Benthic macroinvertebrates are identified mostly to family level in assessing the water quality of small to moderate river types. Comparison of the results is based on summing up the score given on each matrix under a certain column. Usually the assessment of the quality of rivers/streams is based on a standardized timing period of 1 year, and the taxonomic level of identification is based on the ability of the researchers to identify these bioindicators infield. However, with a week-long intensive training on identification of benthic macroinvertebrates and the use of a DST, researchers can use this method more efficiently.

REFERENCES

Barbour, M.T., Stribling, J.B., Verdonschot, P.F.M., 2006. The multihabitat approach of USEPA's rapid bioassessment protocols: benthic macroinvertebrates. Limnetica 25 (3), 839–850.

Feld, C.K., Tangelder, M., Klomp, M.J., Sharma, S., 2010. Comparison of river quality indices to detect the impact of organic pollution and water abstraction in Hindu Kush-Himalayan Rivers of Nepal. J. Wetl. Ecol. 2091-0363. 4, 112–127. Open access at www.nepjol.info/index.php/jowe. Wetland Friends of Nepal at www.wetlandfriends.co.cc.

Griffith, G.E., Omernik, J.M., Wilton, T.F., Pierson, S.M., 1994. Ecoregions and subregions of Iowa: a framework for water quality assessment and management. J. Iowa Acad. Sci. 10 (1), 5–13.

Habermann, B., Reisenberger, M., 2000. Water Quality Assessment in Malam Jabba Valley, N.W.F.P. Pakistan. Department of Hydrobiology, Fisheries and Aquaculture, University of Agricultural Sciences, Vienna.

Hartmann, A., Moog, O., 2008. In: Moog, O., Hering, D., Sharma, S., Stubauer, I., Korte, T. (Eds.), Development of a field screening methodology to evaluate the ecological status of the streams in the HKH region.ASSESSHKH: Proceedings of the Scientific Conference "Rivers in the Hindu Kush-Himalaya—Ecology & Environmental Assessment", pp. 17–24.

Hartmann, A., Moog, O., Stubauer, I., 2010. HKHscreening: a field bio-assessment to evaluate the ecological status of streams in the Hindu Kush-Himalayan region. Hydrobiologia 651 (1), 25–37.

Hughes, R.M., 1989. Ecoregional biological criteria. In: Water Quality Standards for 21st Century, pp. 147–151.

Kinzang, D., 2016. An assessment of the use of macroinvertebrates as bioindicators to assess stream health in Bhutan. A PhD thesis submitted to the Science and Engineering faculty of Queensland University of Technology in partial fulfillment of the degree of Doctor of Philosophy, 278 pp.

Korte, T., Baki, A.B.M., Ofenböck, T., Moog, O., Sharma, S., Hering, D., 2010. Assessing river ecological quality using benthic macroinvertebrates in the Hindu Kush-Himalayan region. Hydrobiologia 651 (1), 59–76.

Moog, O., 1991. Biologische Parameter zum Bewerten der Gewässergüte von Fließgewässern. Landschaftswasserbau 11, 235–266. TU Wien.

Moog, O., Sharma, S., 1999. In: Reporting nepalese biotic score.Proceedings of the International Conference on Agriculture and Biodiversity, Kathmandu, Nepal.

Moog, O., Sharma, S., 2001. Nepalese Biotic Score for water quality assessment. In: Environment and Agriculture as Proceedings of the International Conference on Environment and Agriculture (Nov. 1–3, 1998). Ecological Society (ECOS), Kathmandu, Nepal, pp. 503–506.

Nesemann, H.F., 2006. Macroinvertebrate non-insects' fauna and their role in bio-monitoring of the Ganga River System. A dissertation submitted in partial fulfillment of the requirements for the Master of Science by research in Environmental Science at Kathmandu University (KU), Dhulikhel, Nepal.

Ofenböck, T., Moog, O., Sharma, S., Korte, T., 2010. Development of the HKHBIOS: a new biotic score to assess the river quality in the Hindu Kush-Himalaya. Hydrobiologia 651 (1), 39–58.

Pradhan, B., 1998. Water Quality Assessment of the Bagmati River and Its Tributaries, Kathmandu Valley, Nepal. BOKU Univ. Department of Hydrobiology, Vienna.

Rundle, S.D., Jenkins, A., Ormerod, S.J., 1993. Macroinvertebrate communities in streams in the Himalaya, Nepal. Freshw. Biol. 30, 169–180.

Scholz, C., 2001. Organic pollution of Seti River in Pokhara region, Nepal. Diploma thesis submitted to the faculty of hospital and clinical engineering, environmental and biotechnology, University of Applied Sciences, Giessen, Germany.

Sharma, S., Moog, O., 2002. In: A reference based "Nepalese Biotic Score" and its application in the Midland Hills and Lowland Plains for river water quality assessment and management. Proceedings of the Second International Conference on Plants and Environmental Pollution, February 4–9, 2002. NBRI, Lucknow.

Sharma, S., Moog, O., 2005. In: A reference based "Nepalese Biotic Score" and its application in the Midland Hills and Lowland Plains for river water quality assessment and management.Proc. Internat. Conf. on Plants & Environmental Pollution (ICPEP-2), Lucknow, India, pp. 1–11.

Sharma, S., 1996. Biological assessment of water quality in the rivers of Nepal. A Ph.D. dissertation at the University of Agricultural Sciences, Vienna, Austria.

Sharma, S., 1998a. Pollution Level of Nepalese Water Resources. Nepal-Nature's Paradise, Majupurias, pp. 504–526.

Sharma, S., 1998b. Water Quality Assessment of the Western Rivers of Nepal—A Biological Approach. Environment, HMG, Ministry of Population and Environment, Kathmandu.

Sharma, S., 1999. Water quality assessment of the Saptakosi river and its tributaries in Nepal: a biological approach. Nepal J. Sci. Technol. 1, 103–114. RONAST.

Sharma, S., Moog, O., 1996. In: The applicability of Biotic indices and scores in water quality assessment of Nepalese rivers.Proceedings of the Ecohydrology Conference on High Mountain Areas, March 23–26, 1996, Kathmandu, Nepal, pp. 641–657.

Sharma, S., Nesemann, H.F., Pradhan, B., 2009. In: Application of Nepalese biotic score and its extension for river water quality management in the Central Himalaya.International Conference on Environment, Energy and Water in Nepal: recent researches and direction for future (31 March–01 April 2009), Kathmandu, Nepal. Organized by IGES, Japan.

Shrestha, M., Pradhan, B., Shah, D.N., Tachamo, R.D., Sharma, S., Moog, O., 2008. In: Moog, O., Hering, D., Sharma, S., Stubauer, I., Korte, T. (Eds.), Water quality mapping of the Bagmati river basin, Kathmandu Valley.ASSESS-HKH: Proceedings of the Scientific Conference "Rivers in the Hindu Kush-Himalaya – Ecology & Environmental Assessment", pp. 189–197 (202 pp.).

Suren, A.M., 1994. Macroinvertebrate communities of streams in western Nepal: effects of altitude and land use. Freshw. Biol. 32, 323–336.

Tachamo Shah, R.D., 2011. Stream Order based stream typology for Indrawati River System, Nepal: An approach towards river management. Lambert Academic Publishing, Germany. ISBN: 978-3-8454-0085-3.

Tachamo Shah, R.D., Sharma, S., Haase, P., Jähnig, S.C., Pauls, S.U., 2015. The climate sensitive zone along an altitudinal gradient in central Himalayan rivers: a useful concept to monitor climate change impacts in mountain regions. Clim. Change 0165-0009. 130(4).

Chapter 5.5

Rapid Assessment Methods Developed for the Mangrove Forests of the Great Morass, St. Thomas, Eastern Jamaica

Ainsley Henry*, Dale Webber† and Mona Webber‡

*University of the West Indies (Mona), Kingston, Jamaica, †School for Graduate Studies and Research, University of the West Indies, Kingston, Jamaica, ‡Centre for Marine Sciences, Department of Life Sciences, University of the West Indies, Kingston, Jamaica

Chapter Outline

INTRODUCTION

Wetlands are arguably among the most productive ecosystems worldwide and are estimated to cover approximately 6% of Earth's total land area (Matthews and Fung, 1987). The 2005 UN-Millennium ecosystem assessment indicates that wetlands are among the most threatened ecosystems: 84% of Ramsar-listed wetlands have recorded ecological change and approximately 50% of wetlands worldwide have been estimated to have disappeared since 1990. The values and functions of wetlands are diverse and include providing nurseries for many juvenile coral reef fish. For instance, the biomass of several commercially important species is more than doubled when adult habitat is connected to mangroves (Mumby et al., 2004).

Mangroves constitute an important tropical/subtropical wetland type with species that provide important social and economic resources, yielding goods and services of high value. They provide a range of supporting, regulating, and provisioning ecosystem services (Webber et al., 2016). They are highly productive and create a wide variety of niches (Granek et al., 2009), serving as a habitat for a wide variety of migratory and sedentary species (supporting and provisioning services). They are important as biological filters and sinks for pollutants; they are also considered to be carbon sinks (Donato et al., 2011; Alongi, 2012, 2014). Also, they are valued for their role in carbon sequestration and burial (regulating services). Additionally, their strategic coastal location allows mangroves to protect the land from hurricane impacts and storms surges (Barbier et al., 2008; Koch et al., 2009) as well as to protect marine ecosystems from terrestrial sediments.

The St. Thomas Great Morass is the easternmost wetland area in Jamaica, covering an area of approximately 1660 ha/ 16.6 km^2 (Scott and Carbonell, 1986). The Forestry Department classifies the Great Morass as a mangrove forest, thereby clearly identifying the dominant feature of this wetland. The St. Thomas Great Morass is one of Jamaica's largest remaining wetlands; due to its isolation, it still has areas that are relatively undisturbed.

Mangrove wetlands are estimated to cover <2% of Jamaica's total surface (NEPA, 2011). They have declined significantly over the years due mainly to human conversion because of development pressure and agriculture. Data on individual wetlands exist but there is little documentation of how anthropogenic pressures have affected the health and status of remaining wetland ecosystems or how mangrove cover has changed over time. The Forestry Department, using satellite imagery from the LANDSAT TM scaled at 1:100,000, has generated the most accurate estimate of Jamaican mangroves as

Wetland and Stream Rapid Assessments. https://doi.org/10.1016/B978-0-12-805091-0.00005-0

totaling 9751 ha (Evelyn and Camirand, 2000). However, there is no authoritative catalogue of Jamaica's mangrove and coastal wetlands (NEPA, 2011). Furthermore, mangrove areas in Jamaica have not been adequately or consistently described (using standard methods) nor functionally assessed. Therefore, proper data related to the mangrove forests of Jamaica are sparse.

This research aims to present a standard approach to mangrove assessment that was used to provide evidence to support the designation of the Great Morass as a Ramsar site, or a wetland of international importance. One of the requirements of the Ramsar Convention is that parties do assessments of wetland resources within their jurisdictions and that they determine the status and threats associated with these areas. Furthermore, the Great Morass represents a significant portion of the intertidal/terrestrial wetland area of Jamaica and it has not been the subject of extensive research. The methods applied will aim to provide information on the status, uses, threats, and biodiversity of one of the largest wetland areas in Jamaica, the Great Morass of St. Thomas.

METHODS

Due to the size of the area, it was determined that the best approach to a comprehensive but rapid assessment would be to utilize both remote sensing/desktop analysis and fieldwork. The extent of the wetland was therefore calculated by aerial and satellite imagery and disputed areas were then ground-truthed by walking along the fringes of the area with a handheld global positioning satellite (GPS) system. The ground-truthing was particularly important for areas that had cloud cover in the images and for those areas in which the distinction was not very clear. Once the boundaries of the entire site had been established, a systematic process of data collection with the aid of belt transects was used to examine the physicochemical and floristic parameters at random sites throughout the area. This process was facilitated by the demarcation of the areas into four stands of vegetation that appeared to represent different features and were delimited by physical discontinuities such as roads, waterways, and footpaths. Every effort was made to ensure that the sampling sites selected were distributed across the entire area. As such, more transects (16) were located in the largest stand (Stand III) while the remaining three stands had four transects in each.

Remote Sensing and GIS

Google Earth-based images (2003–07) were used for the remote sensing of the study area, especially the mangroves, as outlined by Dahdoug-Guebas et al. (2005). These images were compared with the 1991–92 colored aerial photo series and the 1:50,000 maps georeferenced in the JAD 2001 projection from the Survey Department of Jamaica. Care was taken to determine where the interface between terrestrial and mangrove species existed; this interface was confirmed during field surveys. Some of the areas slated for vegetation transects were predetermined during desktop studies based on the shape and size of the wetland area and/or distance from previously surveyed points or the observation of potential entry points. Others were selected during the field exercise and were facilitated by using preexisting trails, rivers, and terrestrial intrusions.

The Magellan Platinum GPS handheld units, which are enabled with the wide area augmentation system (WAAS) that allows for accuracy up to 5 m, were used extensively in the survey. Land surveys were conducted by penetrating the periphery of the wetland wherever possible with GPS points being collected at both the start and end of each transect. This area also has several "rivers" and the mouth of these also offered points of penetration by means of canoes into the wetland to evaluate the homogeneity of the vegetation.

Over the course of the investigation, 28 belt transects were established, with each being 10 m long by 2 m wide. Each transect took approximately 1½ h to sample due in part to the ruggedness of the terrain.

Field Methods

The belt transect method employed is a projection of the plot-less line intercept method, appropriate for sampling dense shrub-dominated vegetation. At each site, the survey consisted of two belt transects with each pair of transects separated by 10 m. Transects were taken from the edge of the wetland inward with measurements taken at each meter; the bearing of each transect was also recorded. Photographs were taken of unidentified flora and fauna as well as of distinctive landscapes and transitions between vegetation types. Water quality, soil temperature, species identification, percentage cover, diameter at breast height (DBH), tree height, bare ground percentage, number of seeds and seedlings, and avifauna were recorded on specially designed data sheets.

The in situ water quality measurements temperature (°C), salinity (ppt), dissolved oxygen (mg L^{-1}), and pH were taken at 2 m intervals along the transect using a Yellow Springs Instruments (YSI) model 556 multiprobe meter, wherever there was surface water associated with the transect. Where there was no surface water, only soil temperature was recorded at 5 m intervals. In addition, whole water samples were collected at 10 points where it was observed that water was either flowing into or out of the wetland system. The whole water samples were processed for labile nutrients (nitrate-N and orthophosphate).

All floral species observed within the belt transect were recorded and identified. Where in situ identification was not possible, notes were made and the species photographed for later identification with the aid of Adams (1972). Additional species identification for the general morass area (external to the transects) was done by recording all flora and fauna observed while traversing each stand before and between transect surveys. This was designed to provide an indication of the general biodiversity (or taxonomic richness) of the area.

Estimates of the mean percentage canopy cover were made along each meter of each transect using two methods. First, the rope used to measure the 2 m width was evaluated and the amount of sunlight hitting the rope estimated (line intercept). Second, the tree crowns were viewed from below and the gaps estimated as a percentage (crown closure). Both figures were usually within 5% of each other. Cover has been recognized to be of ecological significance as a measure of plant distribution, based on the observation that it gives a very good measure of plant biomass, which is the first and second order criterion for structural classification in some classification schemes (Mueller-Dombois and Ellenberg, 1974). Mean DBH for each meter distance along the transects was determined at approximately 1 m from the ground using a *Haglof* caliper and recorded in centimeters. Tree height was taken as the linear vertical distance between the ground and the tip of the tree crown (Cintron and Schaeffer-Novelli, 1984). Mean tree height was measured using a clinometer in the field, a 10 m rope, or a 3 m pole depending on the terrain and conditions. The percentage of bare ground was estimated by visual examination and by determining how much was covered by aboveground vegetation, not including specially adapted roots. The number of seedlings (rooted seeds and newly germinated plants) was determined by a visual count within each meter square of the belt transect.

All faunal species observed were identified where possible. Where identification was not possible, photographs were taken and notes made of the observation. The assessment was not designed to exhaustively census the mobile fauna within the morass. Despite this limitation, observations were made of vertebrate and invertebrate fauna including butterflies, moths, reptiles, insects, crabs, and fish. A total of 43 points were examined for avifauna using the Fixed Radius Point Count Census Method (Bibby et al., 1998) and the Observance of Pool Areas method, providing detailed bird count information throughout the Great Morass. These bird points were close to the transect sites. Avifaunal surveys were done over a 2-day period.

RESULTS

Spatial Extent

An analysis of the images and the resultant areas of the polygons created using Arc Map resulted in the determination that the 2007 wetland extent estimate of the Great Morass is approximately 1536.884 ha (Figs. 5.5.1 and 5.5.2). This estimate is less than the estimate used by Scott and Carbonell (1986) by 123.116 ha, but based on the correlation between the estimates from the time periods and the methods, is likely to be an accurate estimate of the total area of the Great Morass (Table 5.5.1). The significant change in size recorded in Stand III is explained by evidence of at least one 15 m wide band of mangroves in Stand III dying because of the inland retreat of a sand dune after the passage of a hurricane.

Physicochemical

Generally, water quality for wetland systems is complex and does not always conform to expectations (Campbell et al., 2008). Water quality data for the Great Morass confirmed this observation, as many parameters did not conform to a coastal ecosystem or literature-derived expectations. According to Scott and Carbonell (1986), the salinity range for the Great Morass is expected to be between 25 and 35 ppt. The readings taken during the assessment, however, ranged between 2.5 and 26.8 ppt. Dissolved oxygen, pH, and temperature values were generally wide-ranging and did not provide conclusive descriptors of the wetland nor qualification for Ramsar site designation.

The whole water samples collected from the Great Morass during the dry season yielded results ranging from 0.76 to 7.04 m gL^{-1} for nitrates with mean values lowest in Stand III (0.76 $mg L^{-1}$) and highest in Stand II (3.37 $mg L^{-1}$). Values for orthophosphate ranged between <0.02 and 0.18 $mg L^{-1}$ with lowest mean values (<0.2 $mg L^{-1}$) in Stands II and III and highest mean values (0.09 $mg L^{-1}$) in Stand IV. While nitrate maximum values were just within national ambient freshwater standards (7.5 $mg L^{-1}$), phosphate values were well within the national ambient freshwater standards (0.8 $mg L^{-1}$).

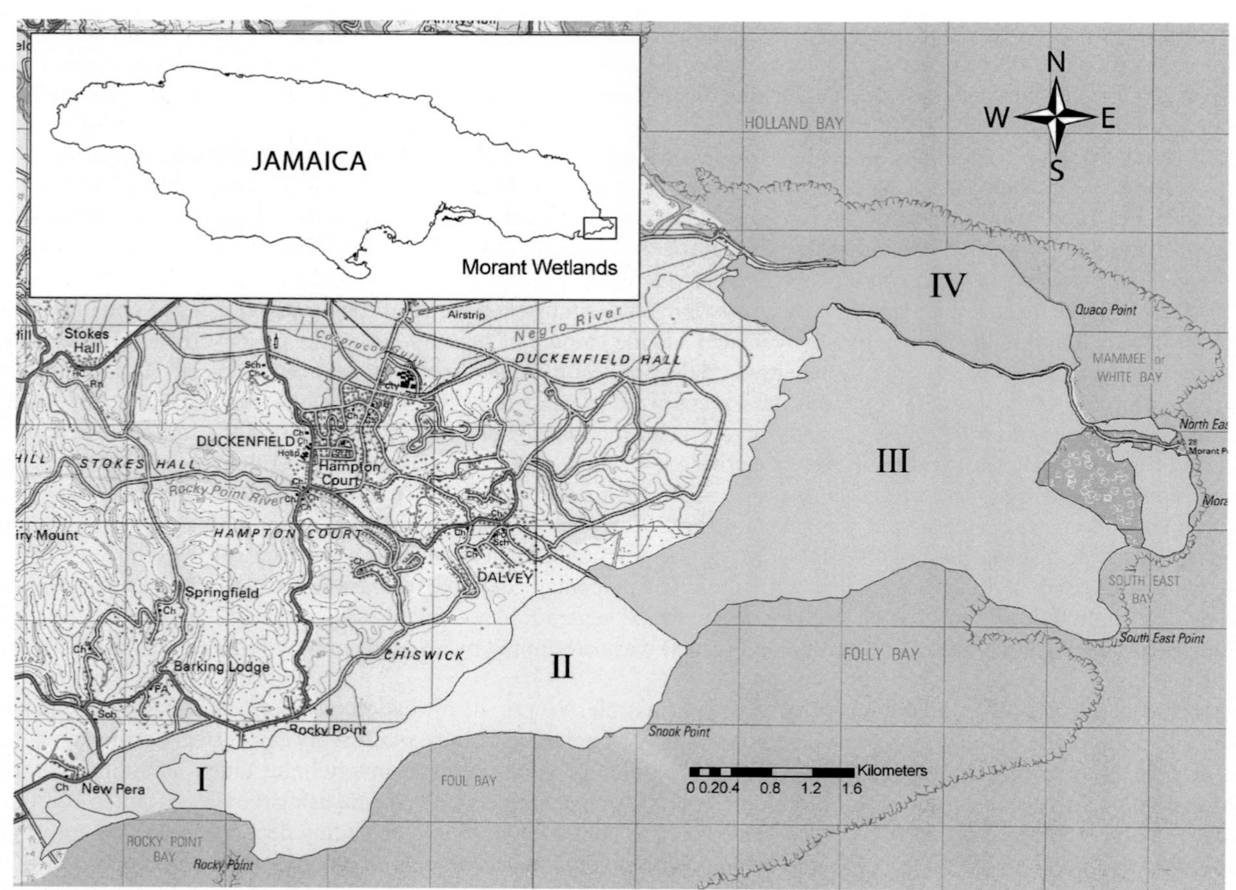

FIG. 5.5.1 Map of Jamaica (inset) indicating Morant Wetlands study area and enlarged map of study area using Arc Map to generate polygons by stand over a 1:50,000 Jamaica Survey Department Base Map (1991–92).

FIG. 5.5.2 Map of study area using Arc Map to generate polygons by stand on the Google satellite imagery (2003–07).

TABLE 5.5.1 Differences in Area (Hectares) and by Stand Between the Two Methods and Number of Transects Used in the Analysis of Each Stand

Stands	1:50,000 Survey Map	Google Earth	# of Transects
I	56.286	73.299	4
II	427.153	420.932	4
III	959.711	781.028	16
IV	254.519	261.625	4
Total	1697.669	1536.884	28

Soil temperatures were similar across Stands II–IV. However, Stand I, which was the most disturbed area, had much higher mean and ranges. Stand I mean temperatures ranged from 25 to 32°C while Stand II ranged from 23 to 28°C. Soil temperature in Stand III ranged from 21 to 28.5°C while for Stand IV, the soil temperature in areas measured ranged from 23.5 to 25°C.

Floristics

The range of plant characteristics (floristics) measured is presented below, according to stands and grouped on different plots for convenience. Thus, percentage cover, DBH, and height by species are presented in Fig. 5.5.3. The number of species in each stand versus the numbers corrected by the number of transects are given in Figs. 5.5.4 and 5.5.5. Finally, the number of mangrove species by stand (Fig. 5.5.6) and the number of seedlings across stands are presented. While it may

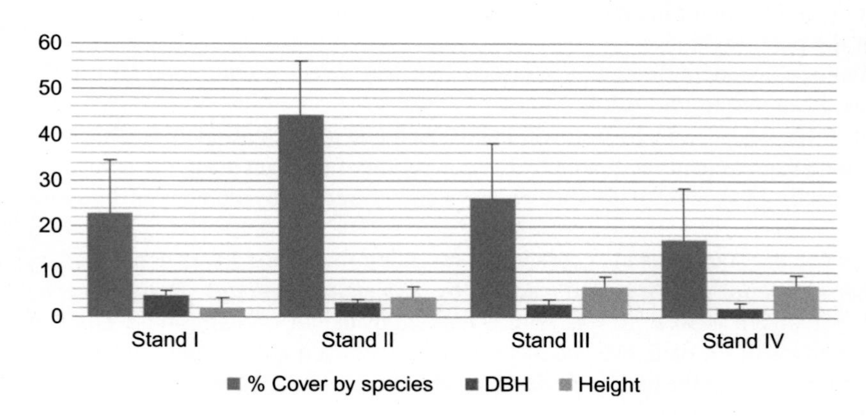

FIG. 5.5.3 Variation of physical characteristics of vegetation across stands.

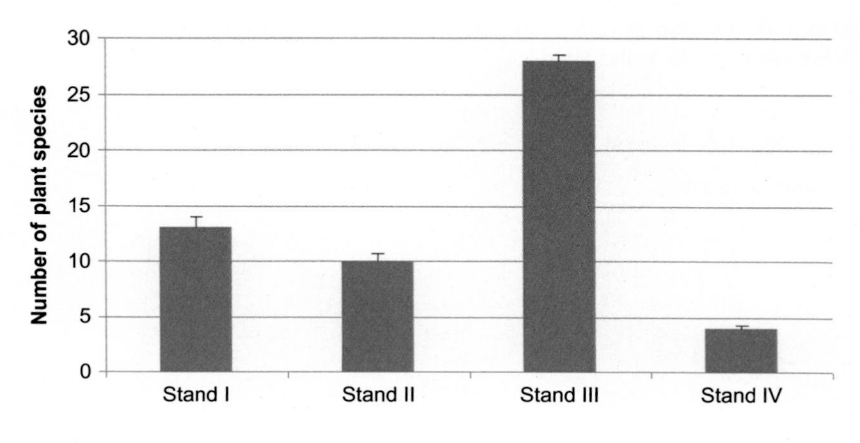

FIG. 5.5.4 Total number of mangrove species identified within each stand.

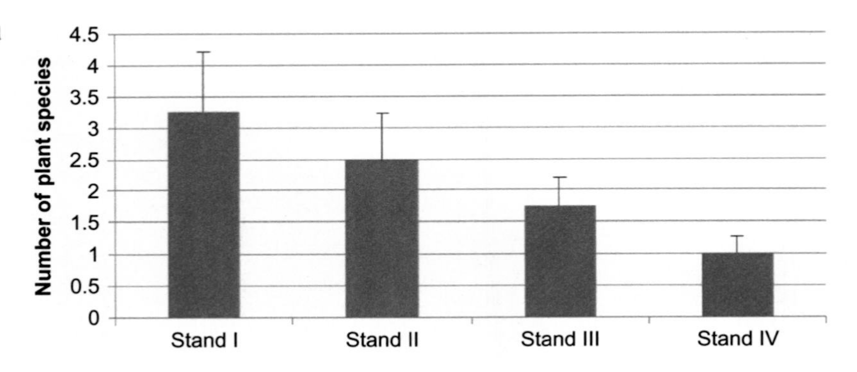

FIG. 5.5.5 Number of mangrove species identified per stand corrected by the number of transects used.

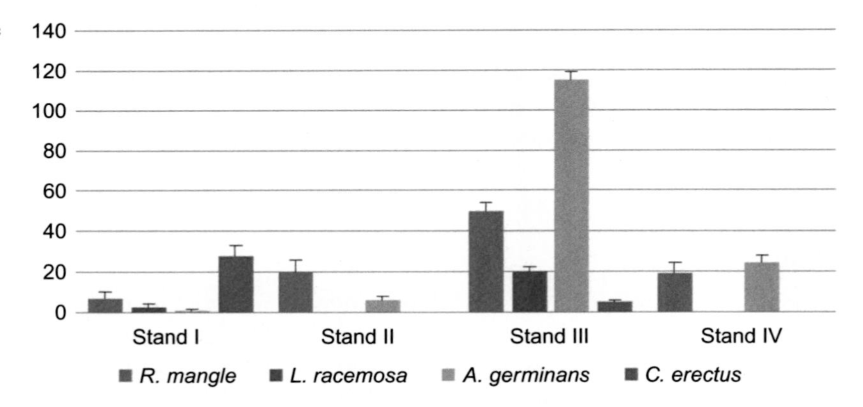

FIG. 5.5.6 The number and types of mangrove species identified per stand.

be important to assess each stand in greater detail by examining individual transects, in an REA of such a large area, it is important to give a good general picture of the plant features.

A comparison of the total number of plant species identified on all transects by stand revealed that Stand III had the greatest number and Stand IV the lowest, as illustrated in Fig. 5.5.4.

While the differences in the number of plant species per mangrove forest stand was found to be significant, as illustrated by the nonoverlapping error bars, a more detailed assessment facilitated by the correction of the data using the number of transects in each stand showed interesting results. The number of species per stand corrected by number of transects actually decreased from Stands I to IV (Fig. 5.5.5).

The number and type of mangroves observed per stand was also compared, revealing that Stand I was dominated by *Conocarpus erectus*, Stand II by *Rhizophora mangle*, and Stand III and IV by *Avicenia germinans*. Dominance by *A. germinans* in Stand III was significant in number as well as size with a maximum height of 18 m and a DBH of 20 cm recorded for one tree. A graphical comparison of the data revealed that the number of each species identified in each stand was significantly different from each other except in the case of *R. mangle*, which was similar between Stands II and IV (Fig. 5.5.6).

The mean number of mangrove seeds and seedlings associated with the transects were extremely variable across the area (Table 5.5.2), with the absolute range being 0 in some quadrats to 1200 seeds and 200 seedlings, respectively, in others. The largest numbers m^{-2} (mean and absolute) were recorded in Stand III.

TABLE 5.5.2 Mean Number of Seeds and Seedlings Per Transect Sampled in Each Stand

Stands	Mean Number of Seeds	Mean Number of Seedlings
I	11.8	0.0
II	0.0	0.0
III	36.1	1.8
IV	17.9	18.7

Fauna

During the six-month survey of the Great Morass, 82 species of fauna were observed and identified. However, based on the purpose for which the assessment was being done, the vertebrates of this group, avifauna, were given greatest emphasis. Furthermore, the importance of the avifauna to the overall biodiversity of a wetland required a more comprehensive evaluation.

A total of 46 species of birds were identified during the survey with 30% being migrants, 61% residents, and 9% long-staying migrants. Of the 46 species identified, 12 were water birds, nnie were endemic, 14 were migrant only, 19 were resident only, and four were long-staying migrants. Water birds accounted for 43.5% of the total observed and between 25% and 58% of these were migrants. The Mangrove cuckoo, which is a typical bird of a mangrove forest, was observed; there were also many birds' nests seen throughout the surveys. Of the total bird population observed, 61%–70% were resident birds, of which 61% were resident only and 20% were endemic. It is also significant to note that two species of migrant ducks, the Blue Winged Teal (*Anas discors*) and Northern Shoveler (*Anas clypeata*), were identified during reconnaissance of the area.

As with other wetland areas in Jamaica, the Great Morass is also home to several species of reptiles, the largest of these being the protected American crocodile (*Crocodylus acutus*). It has been reported that a population of at least 22 resident crocodiles, the majority of which are adults, exist, supported by the Plantain Garden River and the ponds throughout the wetland (R. Nelson, NEPA, personal communication 2007). The area has also been reported to be a habitat for at least nine other species of reptiles, including five endemics (Natural Resources Conservation Authority (NRCA), 1999). Some of these were observed and photographed during the assessments.

Despite the reported adult crocodile population, there were, however, only three sightings throughout the assessment. It should be noted that all the crocodiles observed were fairly large adults measuring approximately 1.5–1.9 m. One crocodile was observed in the grass bordering the wetland along the road that forms the border between Stands III and IV. Crocodiles were also observed in the Negro River on two occasions during the study of the site. This river was also observed to be periodically heavily impacted by wastewater from the Duckenfield sugar factory. While not exhaustive, Table 5.5.3 provides a list of the other vertebrate species either reported or observed during the assessments.

Observed Anthropogenic Impacts

It is important to note activities associated with wetlands that can directly or indirectly impact the area. Especially in the case of the Great Morass, which covers such a large area, a wide variety of uses were being made of the area. These include but are likely not limited to supporting free range (free roaming) cattle, pigs and goats; subsistence agriculture; providing a base of operations for artisanal fisheries, bird shooting as evidenced by the presence of shotgun shells; charcoal production (coal kilns) using mangrove wood; extraction of wood for fence posts (sugar company); providing access to recreational areas (the adjacent beach); and subsistence fishing (in rivers and canals).

TABLE 5.5.3 Other Vertebrate Species of the Great Morass

Scientific Name	Common name	Status	Comments
Albula vulpses	Bonefish	Common	Observed
Anolis grahamicrocod	Graham's Anole	Common	Observed
Anolis lineatopus	Jamaican Brown Anole	Common	Observed
Awaouf tajasica	Giant River Goby	Unknown	Observed
Centropomus sp.	Snook	Common	Observed
Crocodylus acutus	American Crocodile	Protected	Observed
Dermochelys coriacea	Leatherback Turtle	Critically endangered, protected	Reported associate (nests on beaches)
Eretmochelys imbricata	Hawksbill Turtle	Critically endangered, protected	Reported associate (nests on beaches)
Mugil sp.	Mullet	Common	Observed

The Great Morass provides a livelihood for many individuals and supports a significant fishery resource within the fresh, brackish, and saline ponds and waterways that are present throughout the wetland. Frequent observations of residents returning from successful fishing activities are evidence of the resource. While the true extent of this resource could not be determined without detailed investigation, it should be noted that based on interviews with the residents, for some this is a daily activity.

The Great Morass is also a source of wood for the charcoal manufacturing sector with numerous kilns, both old and new, encountered adjacent to areas in which transects were established. The distribution of the charcoal kilns, some of which were mapped, seems to indicate that the coal-burning activities are centered near existing residential communities at both Rocky Point and the sugar factory. Where there were no kilns observed in some areas, chopped trees and accumulated wood signified the intense reaping occurring within the coastal forests. Further stresses on the coastal forests are exerted by natural disasters such as hurricanes and tropical storms. This was observed toward the eastern limits of the Great Morass, where a significant portion of the coastal forest was destroyed by a storm surge associated with Hurricane Ivan in 2004.

DISCUSSION

The REA was conducted primarily because there was a paucity of scientific information on the Great Morass. Baseline data were urgently needed for a relatively large area of wetland. The analysis of extent and lack of change over time are acceptable deliverables of the imagery methods. This evaluation also provided information on the level of threat and the homogeneity of the wetlands as well as facilitated comparison between this area and other Jamaican wetlands. Analysis of wetland extent, condition, and fauna (especially birds) was needed to provide a basis for the designation of the morass as a wetland of international importance, or a Ramsar site. An analysis on the status and species distribution in the wetland indicated that the areal extent is stable and the wetland is still distributed across the range that it occupied 30 years ago. It is, however, clear that there are several issues affecting the resource.

The inconclusive nature of the water quality indices in mangroves is often the result of chemical changes due to tannins and other substances, high organic matter, and tidal and intrusion-related fluctuations in salinity, all of which may make physicochemical parameters in mangrove areas unique and difficult to interpret (Campbell et al., 2008). High variability in nutrient values—although these values were within national standards—make this index inconclusive for mangrove wetlands evaluation. However, the high variability does help to highlight differences between stands.

The floristic assessment for designated stands, following on the imagery and derived from the ground-truthing and survey (percentage cover, DBH, and species identification), are valuable indices of status and biodiversity within the stands of the Great Morass. A simple (surrogate) index such as tree height can be used as an index of age within a species/stand. Roth (1992) and McCoy et al. (1996) used this parameter to suggest that the mangroves in the Great Morass are predominantly mature trees (because of their height). Furthermore, tree diameter (DBH) can be seen in a similar manner where it is recognized that 2.5 cm DBH $\approx$ >5 years (Proffitt and Devlin, 2005), there are several mangroves in excess of 40 years of age at the site. It is of interest that the largest mean DBH by stand was in Stand I, which represented the most disturbed area, while the maximum DBH of an individual was in Stand III, the most complex forest community.

The number of species data demonstrates that there is a significant difference across all stands. Stands I–III had the highest numbers of species; this was to be expected due to the edge effect created by the relative levels of disturbance and the sizes of these stands. Stand III having the highest recorded number of species can be explained both by its relative size and the large number of transects used to analyze that area. Data corrected to account for the number of transects demonstrated that the number of species decreased. However, the difference between stands remained significant. The number of mangrove species and their relative occurrence revealed significant differences in the variation between relative numbers by species across stands. Also, there were significant differences in terms of the occurrence of species by stand.

In the case of Stand II, there is strong evidence to suggest that the ecological character is changing as the incidence of charcoal burning close to the local fishing village was the second highest recorded for all four stands. It is expected that Stand II, the second largest area, would also have significant absorptive capacity, hence helping to filter runoff from land before it reaches the marine environment. The size of the stand also makes it a likely habitat for a significant quantity of flora and fauna. With these factors in mind, this stand has been designated as a high priority for protection/conservation.

Stand III is currently the largest of the four and houses a significant number of ponds, beaches, and limestone islands. These factors suggest that there are diverse habitats available within the area. Also of significance is that of all the stands, the lowest concentrations of nutrients were recorded at Stand II. It is important that further studies include seasonal bird counts and an assessment of the limestone islands that intersperse this area. The diversity of habitats and, potentially, species within this area is a significant factor indicating the potential for this stand to be sustainably used in ecotourism.

Stand III has therefore been designated as the highest priority for conservation and further detailed studies. Due to the relatively mild levels of disturbances, it has also been designated a medium priority for restoration and rehabilitation.

As with Stand II, Stand IV was the subject of intense harvesting for charcoal production and is also known to be exploited for fence posts. A total of 43 charcoal kilns were recorded during this survey and the majority (32) were observed within this stand. Further, the majority of the kilns observed were either quite recent or active at the time of observation. It is also noteworthy that there seemed to be a demonstrated preference for black mangrove as there was evidence of selective cutting of that species.

With >13 anthropogenic and natural impacts identified and a relatively low incidence of seeds and seedlings, the resilience and regenerative capacity of this area seems low. These factors, therefore, serve to increase the levels of threat incident on the different parts of the wetland ecosystem and are therefore compelling arguments that the Great Morass is a threatened wetland.

SUMMARY

The wetlands of the Great Morass provide habitat for a range of species, including freshwater fish, the giant river goby (*Awaouf tajasica*) crayfish, the endangered American crocodile (*C. acutus*), and potentially the West Indian Whistling Duck (*Dendrocygna arborea*). The endangered marine turtles Hawksbill turtle (*Eretmochelys imbricata*) and Leatherback turtle (*Dermochelys coriacea*) are also known to nest on the beaches of the area. Bird counts were conducted at 43 points in the area during the dry season. As such, these would not have included many species that are likely to use the area during the wet season. This also indicates that they provide staging and wintering areas for a wide variety of migrant birds, including ducks previously identified as well as shorebirds and warblers such as *Arenaria interpes* (Ruddy Turnstone) and *Dendroica petechia* (Yellow Warbler). The best time for conducting the bird count is in the morning from sunrise until about 10 a.m. in the lowlands. It is recognized that as the day continues it gets hotter and the ability to detect birds decreases due to lack of movement (Wunderle, 1994). Furthermore, the change in behavior of birds during the breeding and nonbreeding seasons affects detection. The assessment for the Great Morass was done in the nonbreeding season when birds are less vocal (Wunderle, 1994), which may result in an underestimate.

Comparisons between stands clearly indicated significant differences. This therefore means that using the stands as discrete units, the area could be zoned according to the proposed activities in each. The low diversity and proximity to anthropogenic impacts makes Stand I the least attractive. It has therefore been assigned the lowest priority for conservation, restoration, rehabilitation, and further detailed studies in characterizing a location for conservation by being named a Ramsar site.

The Ramsar convention has criteria, named and listed as 1–9, for designation of sites as internationally important. Of those nine criteria, the Great Morass meets at least three. This is significant because designation as a Ramsar site only requires that a minimum of one criterion be met. The specific criteria that the Great Morass satisfies are as follows:

> **CRITERION NUMBER 1**. The Great Morass area is within ecoregion 236—Western Tropical Atlantic Greater Antillean Marine (WWF) and contains wetland types that are representative and near natural for this region, including mangrove forests, karst areas, and permanent and seasonal pools/ponds.
>
> **CRITERION NUMBER 2**. The area is important as it contains the threatened ecological community of the mangroves and functions as a critical habitat for animals listed by the IUCN Red List (2004) as vulnerable and/or endangered. These include the *C. acutus* (American crocodile—VU), *E. imbricata* (Hawksbill turtle—CR), and potentially *D. arborea* (West Indian Whistling Duck—CR).
>
> **CRITERION NUMBER 3**. The site serves as a refuge for many animals at many stages in their lifecycle and during adverse weather conditions. In addition, it also provides habitat for winter migrants and shorebirds.

The data collected using the methodology outlined here allowed for the completion of a Ramsar information sheet (RIS) for the Great Morass. It is therefore reasonable to conclude that a rapid assessment using the inexpensive methods outlined could form the basis for the submission of a completed RIS when designating sites in Jamaica. Similarly, this rapid assessment method could be used in other Caribbean locations as well as elsewhere where there is interest in a Ramsar designation or in determining the quality of a particular mangrove site. Moreover, it is recommended that this REA method should prove invaluable in Jamaica where the National Environment and Planning Agency (NEPA) is constantly challenged by the demand for coastal development and the balance sought in environmental conservation. Not only would the Mangrove REA provide the decision-makers with adequate information, it would also standardize the approach of the development proponents and their consultants, making submissions consistent and comparable across the island.

This research has therefore presented a standard approach to mangrove wetland assessments by combining known methods with innovative methods. The methods combined were aerial photography with ground-truthing, water quality, forest floristics, and faunal assessment. These methods, when used together, provided evidence to support the designation of the Great Morass as a Ramsar site or a wetland of international importance using three given criteria.

REFERENCES

Adams, C.D., 1972. Flowering Plants of Jamaica. University of the West Indies, Mona, Jamaica.

Alongi, D.M., 2014. Carbon cycling and storage in mangrove forests. Annu. Rev. Mar. Sci. 6, 195–219.

Alongi, D.M., 2012. Carbon sequestration in mangrove forests, a review. Carbon Manage. 3, 313–322.

Barbier, E.B., Koch, E.W., Silliman, B.R., Hacker, S.D., Wolanski, E., Primavera, J., Reed, D.J., 2008. Coastal ecosystem-based management with non-linear ecological functions and values. Science 319, 321–323.

Bibby, C., Jones, M., Marsden, S., 1998. Expedition Field Technique Bird Surveys. London Expedition Advisory Centre, London.

Campbell, P.E., Manning, J.E., Webber, M.K., Webber, D.F., 2008. Planktonic communities as indicators of water quality in mangrove lagoons; a Jamaican case study. Transit. Waters Bull. 3, 39–63.

Cintron, G., Schaeffer-Novelli, Y., 1984. Methods for studying mangrove structure. In: Sneadaker, S.C., Sneadaker, J.G. (Eds.), The Mangrove Ecosystem: Research Methods. UNESCO, Paris, pp. 3–17.

Dahdoug-Guebas, F., Van Hiel, E., Chan, J.C.-W., Jayatissa, P., Koedam, N., 2005. Qualitative distinction of congeneric and introgressive mangrove species in a mixed patchy forest assemblage using high spatial resolution remotely sensed imagery (IKONOS). Syst. Biodivers. 2, 113–119.

Donato, D.C., Kauffman, J.B., Murdiyarso, D., Kurnianto, S., Stidham, M., Kanninen, M., 2011. Mangroves among the most carbon-rich forests in the tropics. Nat. Geosci. 4, 293–297.

Evelyn, O.B., Camirand, R., 2000. Forest cover and deforestation in Jamaica: An analysis of forest cover estimate over time. In: Based on Remote Sensing Developed by the Forestry Department/Trees for Tomorrow Project LANDSAT TM Scale 1:100,000. Forestry Department Jamaica.

Granek, E.F., Compton, J.E., Phillips, D.L., 2009. Mangrove-exported nutrient incorporation by sessile coral reef invertebrates. Ecosystems 12, 462–472.

Koch, E.W., Barbier, E.D., Silliman, B.R., Reed, D.J., Perillo, G.M.E., Hacker, S.D., Granek, E.F., Primavera, J.H., Muthiga, N., Polasky, S., Halpern, B.S., Kennedy, C.J., Wolanski, E., Kappel, C.V., 2009. Non-linearity in ecosystem services: temporal and spatial variability in coastal protection. Front. Ecol. Environ. 7, 29–37.

Matthews, E., Fung, I., 1987. Methane emission from natural wetlands: global distribution, area, and environmental characteristics of sources. Glob. Biogeochem. Cycles 1, 61–86. https://doi.org/10.1029/GB001i001p00061.

McCoy, E.D., Mushinsky, H.R., Johnson, D., Meshaka Jr., W.E., 1996. Mangrove damage caused by Hurricane Andrew on the southwestern coast of Florida. Bull. Mar. Sci. 59, 1–8.

Mueller-Dombois, D., Ellenberg, H., 1974. Aims and Methods of Vegetation Ecology. John Wiley and Sons, New York.

Mumby, P.J., Edwards, A.J., Arias-Gonzales, J.E., Lindeman, K.C., Blackwell, P.G., Gall, A., Gorczynska, M.I., Harborne, A.R., Pescod, C.L., Renken, H., Wabnitz, C.C., Llewellyn, G., 2004. Mangroves enhance the biomass of coral reef fish communities in the Caribbean. Nature 427 (6974), 533–536.

National Environment and Planning Agency, 2011. State of the Environment Report 2010. NEPA.

Natural Resources Conservation Authority (NRCA), 1999. Policy for Jamaica's System of Protected Areas. Government of Jamaica with USAID-DEMO project support.

Proffitt, C.E., Devlin, D.J., 2005. Long-term growth and succession in restored and natural mangrove forests in South-Western Florida. Wetl. Ecol. Manag. 13, 531–551.

Roth, L.C., 1992. Hurricanes and mangrove regeneration: effects of Hurricane Joan, October 1988, on the vegetation of Isla del Venado, Bluefields, Nicaragua. Biotropica 24, 375–384.

Scott, D.A., Carbonell, M., 1986. A Directory of Neotropical Wetlands. International Union for Conservation of Nature and Natural Resources.

Webber, M., Calumpong, H., Ferreira, B., Granek, E., Green, S., Ruwa, R., Soares, M., 2016. Mangroves. In: First Global Integrated Marine Assessment (First World Ocean Assessment. United Nations (Chapter 48).

Wunderle Jr., J.M., 1994. Census methods for Caribbean land birds. General Technical Report S0-98, Forest Service, Southern Forest Experiment Station, US Department of Agriculture, New Orleans.

FURTHER READING

Asprey, G.F., Robbins, R.G., 1953. The vegetation of Jamaica. Ecol. Monogr. 23, 359–412.

National Environment and Planning Agency, 1997. Policy for Jamaica's System of Protected Areas. Government of Jamaica with USAID-DEMO Project Support.

Ramsar, (n.d.) http://www.ramsar.org/ris/key_ris_types.htm (accessed 2008; site now discontinued).

Ramsar Convention Bureau, 1977. The Ramsar Convention Manual: A Guide to the Convention on Wetlands (Ramsar, Iran, 1971), second ed. Ramsar Convention Bureau, Gland.

Chapter 5.6

Rehabilitation of Wetlands in the Tempisque River Lower Basin: Mata Redonda National Wildlife Refuge as a Case Study

Juan Bravo*, José C. Leal†, Miriam Miranda* and Marcela Gutiérrez‡

*Wetlands Project, Heredia, Costa Rica, †National System of Conservation Areas, San José, Costa Rica, ‡National University of Costa Rica, Heredia, Costa Rica

Chapter Outline

INTRODUCTION

In Costa Rica, the low basin of the Tempisque River (CBT) has an approximate surface area of 129,100 ha containing 743 wetland units—marshes and lakes—representing 22,697 ha of this type of ecosystem in that territory (SINAC, 2016). CBT wetlands were recognized in 1991 by the Ramsar Convention as protected wetlands of international importance. As a consequence, the Costa Rican government committed itself to managing them in a sustainable way to ensure their conservation and rational use.

This article describes the rehabilitation work carried out in the state-owned lands of the Mata Redonda National Wildlife Refuge (RNVSMR). In 1994, the wetland located at RNVSMR was established as a public trough to favor small local cattle ranchers. The Mata Redonda lagoon has an area of 136 ha and covers 0.1% of the CBT (Fig. 5.6.1).

The RNVSMR has important resources that have been overexploited. This situation has compromised its ecological balance, directly affecting the welfare of local populations. The inhabitants, who live inside or on the periphery of the wetland areas, confront complex social problems derived mainly from the nonrational use of the resources offered by the ecosystem and the impact of climate change on the CBT.

The degradation of CBT wetlands is due to natural and anthropogenic causes. Agroindustrial crops (cane and rice), extensive livestock farming, and the illegal practice of controlled agricultural burning have led to excessive sediment and nutrient transport to the CBRT ecosystems and to the watersheds of the San Lázaro and Charco rivers. These decade-long practices have clogged marshes, natural drainages, intertidal water flows, and water mirrors. This condition has favored propagation of invasive floristic species to water mirrors and natural drainage (*Mimosa pigra* and *Ipomea carnea*, among others). In 2011, 70% of the wetland area was covered by invasive species.[1]

The RNVSMR is a marsh wetland whose territory is of vital importance as a feeding and breeding site for more than 60 species of resident and migratory waterbirds. Some of these birds, including the Jaribu stork (*Jabiru mycteria*), are in danger of extinction and represent one of the key attractions of the tourist development of the area.

By 2014, 100% of the wetland ecosystems of the RNVSMR were lost or transformed. Water mirrors lost their ecological function due to the clogging of their base level, with sediment thicknesses between 0.35 and 80 cm (Bravo, 2014).

1. The life cycle of invasive species allows them to contribute with organic matter to the filling of water mirrors and drainage.

Wetland and Stream Rapid Assessments. https://doi.org/10.1016/B978-0-12-805091-0.00055-4

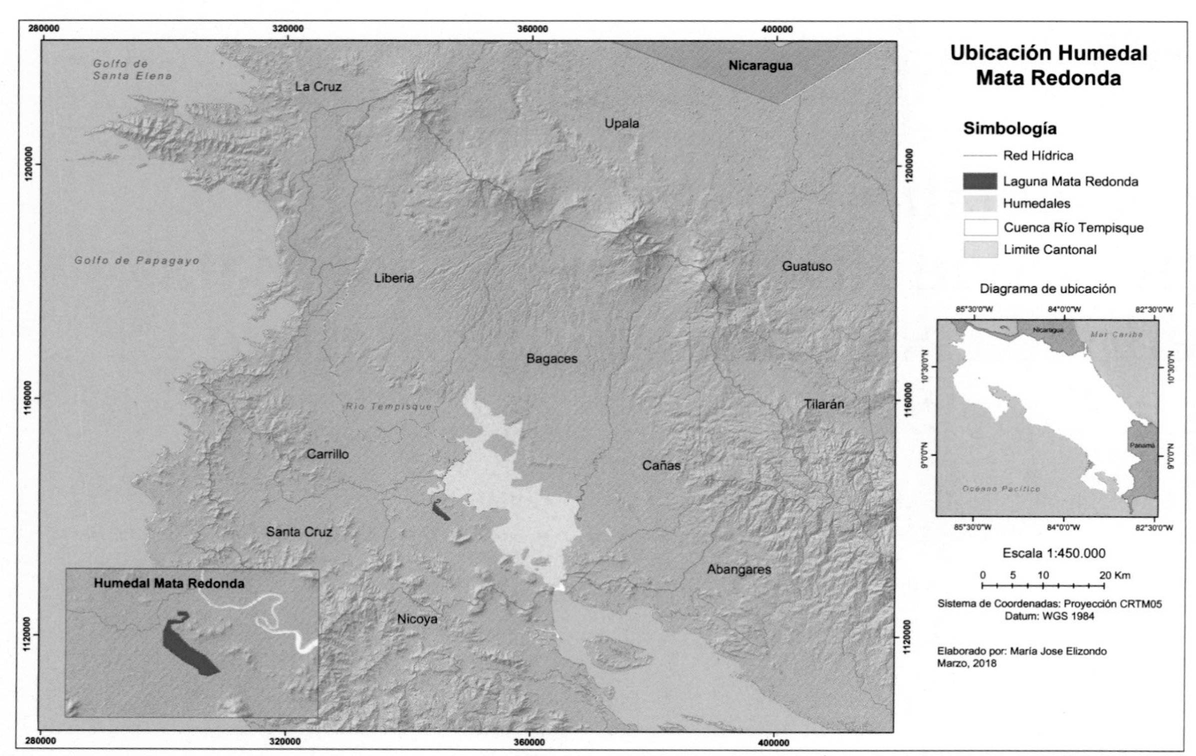

FIG. 5.6.1 Location of wetland in Mata Redonda Lagoon. *(Source: SINAC (Sistema Nacional de Áreas de Conservación), 2016. Sistematización del proceso: intervención ecológica en humedales de la Cuenca Baja del Tempisque, Área de Conservación Tempisque. Informe Final San José, Costa Rica.)*

Consequently, the ichthyofauna and avifauna disappeared while the public trough lost its role, hitting the local economy heavily. In addition, climatic variability, expressed in extreme seasonal changes with long periods of drought, impacted the ecosystem and the livelihoods of surrounding communities.

Taking into account the basin deterioration and negative consequences on the inhabitants, the Costa Rican government, through the Tempisque Conservation Area (ACT), began the process of preparing the general management plan for RNVSMR in 2011. This is the official response for participatory management of this protected wilderness area. This plan emphasizes restoring water mirrors and water systems as well as promoting rational use and protection of the basin with the participation of various local and institutional groups.

In 2015, in order to follow up on the management initiative led by the ACT and the Association Costa Rica Forever, the "Wetlands Project" financed by Global Environmental Facilites (GEF) and implemented by the System National Conservation Areas (SINAC)-United Nations Development Program (UNDP) joined the initiative.

CONCEPTUAL FRAMEWORK

Restoration actions of the RNVSMR are based on Resolution VIII.16, 2002 of the Ramsar Convention, which states that a wetland restoration project must meet two criteria: social participation (at various stages of the process) and a rehabilitation proposal (reflected in a document that includes site, goals, objectives, and performance criteria).

The proposal developed in RNVSMR is based on the Ramsar guidelines for wetland restoration (2010). This qualitative method is based on local reality marked by the overuse of natural resources and the absence of scientific data. The method is based on the experience and historical knowledge of the various actors to reconstruct the landscape of previous years when the ecosystem was in balance and offered the public satisfaction. This proposal includes the historical observation of wetland conditions and involves various public and private organizations with a presence in the area. In this methodology, the experience of the historical inhabitants of the wetland is the source of knowledge and guides restoration efforts. Table 5.6.1 presents the key elements of this methodology.

TABLE 5.6.1 Ramsar Guidelines for Wetland Restoration

(1) Community involvement and/or potential involvement
(2) Declaration of goals, objectives, and achievement indicators
(3) Clear description of the site
(4) Constant monitoring of objectives and degree of compliance with indicators
(5) Inventory degraded and/or destroyed wetlands, including functions impacted
(6) Recognize possible project limitations (environmental, social, and economic)

Sources: Principles and guidelines for wetland restoration (RAMSAR, 2010. Directrices para la evaluación ecológica rápida de la biodiversidad de las zonas costeras, marinas y de aguas continentales. Recuperado 10 de marzo de 2016 http://www.ramsar.org/sites/default/files/documents/library/lib_rtr01_s.pdf).

According to the 2010 Ramsar Convention, Bravo et al. (2013), and De Groot et al. (2007), wetland rehabilitation impacts human health and well-being, reduces risks caused by extreme hydrometeorological phenomena, impacts food security, and represents a tool for mitigation and adaptation to climate change.

WETLAND REHABILITATION RAPID ASSESSMENT METHODOLOGY

The methodology for wetland rehabilitation was developed, focusing on intervening in a wetland located on the Lower Tempisque Basin's (CBT) right bank, specifically wetlands that form the marsh system of Mata Redonda.

The objective of this initiative was to improve the water flow of the heavily clogged natural drainage as well as the water mirror of the wetland, with the purpose of restoring ecological functions and rational use. This is because the small farmers of the surrounding communities—Rosario and Puerto Humo—depend on this ecosystem for their subsistence.

The methodological proposal for rehabilitation was defined based on the Ramsar Convention guidelines for wetland restoration as well as the local historical experience of nearby communities. About 250 ha of water mirrors were rehabilitated. In addition, invasive species were controlled on 100 ha while 1.7 km of the secondary channels of the Charco River were rehabilitated.[2]

The development of the proposal was carried out in four phases:

(1) *Cabinet Phase*: a rapid ecological assessment (REA)[3] was carried out as the first stage due to the urgent need for taking action based on reliable information for the restoration of the ecological functions of the wetland. The REA focused on overall biodiversity and therefore gathered information about the ecosystem. The data obtained gave information about species, types of habitat, species richness, abundances, relative sizes of populations, distribution and areas, cultural importance, and biodiversity. It also collected data on geography, geology, climate, and habitat. The REA was based on criteria for the conservation of wetlands where integral management, conservation of the ecosystem, and conservation of the environment were essential. This methodology was based on previous studies that showed progressive deterioration of the RNVSMR wetland ecosystem (Echeverría et al., 1998; Alpízar et al., 1998). In addition, the knowledge and experiences of the area's residents in dealing with water management for agricultural and livestock purposes, taking into account the qualities and water conditions of the area, were taken into consideration. The indirect influence of tides and topographical conditions was accounted for as well.

This methodology provided the best fit for current conditions for this ecological intervention as current surveys of the environment were missing. The evaluation carried out a survey of primary and secondary sources on the state of the biodiversity of the Mata Redonda National Wildlife Refuge. This wild area contains sets of natural communities such as mangrove, evergreen forest, paddocks, charrales, and herbaceous marshes. The REA included information on mollusks, fish, reptiles, birds, and mammals. The main findings for the flora and fauna (mollusks, fish, reptiles, birds, and mammals) are presented below.

Findings in flora: A total of 38 floristic species were identified in the estuary system. The shrub and tree species developed over a narrow semiflooded fringe of the wetland. The species identified were mangrove *Languncularia racemosa*, *Conocarpus erecta*, *Pithecolobium dulce*, *Avicennia germinans*, and *American Ximenia*.

In the periphery of the wetland, there are wooded fringes consisting of *Coccoloba caracasana*, *Coccoloba uvifera*, *Enterolobium cyclocarpum*, and *Samanea saman* as well as the spiny shrub species *Acacia cornígera*, *Acacia balmesiana*,

2. This was achieved by extracting sediments and weeds that filled up the water channels.
3. The method of analysis used in the evaluation of this component is the one proposed by the Ramsar Convention (2010).

and *M. pigra*. These formations provide ecological niches for various species, especially birds and mammals. A total of 100% of the refuge area is inhabited by floating, perennial, and deep-rooted aquatic plants.

Findings in fauna: Only two mollusk species were identified: *Nephronais tempisquemsis* (endemic to the area) and *Pomacea costaricana*.

A total of 22 species of fish belonging to eight families of ecological and economic value were recognized. These include *Symbranchus marmoratus*, *Cichlasoma* spp., *Ariopsis seemanni*, and *Oreochromis niloticus* (not native).

Nine reptiles were recorded from seven families: *Crocodylus acutus, Iguana iguana, Ctenosaura similis, Ameiba festiva, Ameiba quadrilineata, Boa constricto, Crotalus durisus, Cnemidophorus deppeii,* and *Kinosternon scorpoide.*

Sixty species of resident and migratory waterfowl birds were found, including *Ana discors, Platalea ajaja,* and *Mycteria americana.* The protection of bird populations was one of the priority objectives for the creation of the refuge.

In relation to the mammals, three terrestrial species (*Nasua narica*, *Procyon lotor*, and *Potos flavus*) and eight species of bats, belonging to four families, were counted.

The complex nature and diversity of this wetland ecosystem means that there is no single rapid assessment method (RAM) that can be applied to this area. The REA represented an adequate tool to understand the preliminary state of the habitats inside the wetland and, with that, to take immediate action to improve its ecological balance and to collaborate with the sustenance of the local communities.

In addition, a biophysical analysis[4] of the wetland was carried out as well as an analysis of the land uses in the San Lázaro and Charco watersheds (these influence the hydrological dynamics of the Mata Redonda wetland). The main activities carried out at this stage include:

- REA that provided key information on the environmental conditions of the RNVSMR.
- Preparation of the proposal for the rehabilitation of ecological functions, based on the results of the REA. The intervention was proposed and expected products were defined. Several indicators of implementation were determined: (1) hectares of *M. pigra* removed, (2) hectares of water mirror area recovered, and (3) kilometers of the main channel and secondary channels that are now sediment-free.

(2) *Awareness-raising phase*: At this stage, various groups organized at the local level were involved to determine rehabilitation actions that would be carried out with the objective of improving environmental conditions of wetland and surrounding areas. The organizational network of local farmers and other groups of independent neighbors (wetland users) affected by the deterioration of the ecosystem was used.

Citizen participation was a key element in the development of the proposal. Moreover, coordination activities were developed to prepare the implementation phase, including organization and execution of the bidding process established in the proposal as well as the adjustment of the proposed rehabilitation to the Ramsar Convention and Costa Rican legislation.

(3) *Phase of implementation of the rehabilitation*: In this phase three activities stand out:

- Construction of gates: in May 2015, a dam was built to retain water from tidal brackish flows, especially during the extreme dry season (February–May). Previously, geomorphological studies were carried out in order to identify and evaluate the optimal location for the construction of the site. This gate was built in two modules with dimensions of 2.40×2.20 m that allow the management of incoming and outgoing water levels of the wetland (21.52 m^3/s) in their maximum capacity. It was necessary to adapt the floor's gate to stabilize the base of the substrate, so it was deepened with machinery to a depth of 8 m to reach stable substrate. In addition, the bases for fixing the infrastructure of the gates were prepared. Resistance tests of the material used were then completed as a final step. The construction of the slats that worked as gates was done in a specialized workshop as iron sheets with dimensions of 2.40 by 2.20 m with 35 mm of thickness with an angular frame of 4 in. reinforced with tubes of 2 in. in width and ¼ inch in thickness. Parallel to the construction of the gates, the margins were stabilized in the first 50 m near the infrastructure site using 120 m^3 of rocky material 30 cm in diameter and gravel, using heavy machinery. For the assembly and anchoring of the sheets, four pins with thread welded to the infrastructure were installed. The gate works manually, rises to let the tides flow, and is lowered to retain it.
- Rehabilitation of water mirrors: Based on the photogrammetric survey and taking advantage of the dry climatic conditions typical of the region, heavy machinery was used to restore three mirrors of water inside the wetland.

4. The biophysical analysis was performed as part of the general management plan of the Mata Redonda National Wildlife Refuge and includes biogeographic aspects, hydroclimatic conditions, wetland hydrology, and soil analysis of the wetland.

They cover an area of 1.2 ha with a depth of between 35 and 70 cm. The substrate extracted from the water mirrors served to create promontories or islets of low elevation (between 1.5 and 1.8 cm) suitable for the rest of the associated fauna, especially reptiles and birds, as well as spaces for bird watching and/or ecological monitoring.
- Control of invasive floristic species. The control of the *M. pigra* was carried out through collection actions that also allowed the removal of sediments along the natural channel. *Mimosapigra*, weeds, and sediment were removed from the mouth of the wetland to the site of construction of the floodgate (1780 m). This activity requires monthly maintenance work for at least 5 years with the objective of ensuring the proper functioning of the authorized drains. In addition, it was also reforested with native species along the margins of the canals to stop erosion and improve this habitat for various species.

(4) *Monitoring phase*: As a last step, a plan was developed to carry out ecological monitoring as part of the monitoring of the restoration of the wetland. For this purpose, the following elements are specifically considered: diversity of floristic communities (aquatic and semiaquatic), drainage, water mirrors, waterfowl, hydrobiological fauna, and water quality. The goal of this activity is to have an updated baseline of RVSMR natural resources in 5 years that will contribute to the development of a database containing the ecological basis of the wetlands of the lower Tempisque basin.

LESSONS LEARNED

The work carried out in the RNVSMR is an interinstitutional effort that required commitment of all institutions and organizations involved. This initiative is the first experience in Costa Rica's Protected Wild Areas to restore wetland ecosystems within the framework of mitigation and adaptation to the impacts of climate. The success of the work will depend to a large extent on the follow-up process that is done. Therefore, we highlight these lessons learned from this project to implement for similar projects:

- Prepare a plan by the ACT-SINAC to ensure the follow up of the restoration actions carried out in the intervened areas in a continuous process.
- Encourage establishment of alliances through cooperation agreements with local organizations that contribute to the strengthening of restoration processes in the intervened areas.
- Implement by the ACT-SINAC the execution of ecological monitoring in the intervened areas.
- Establish physical limits with reference to the lands that are the natural heritage of the state.
- Expand restoration of the el Charco drainage to 5 m in width to avoid sedimentation and the invasive effects of plants that obstruct the channel.
- Construct traps in sites identified for sediment capture to avoid obstruction of the channel and interior clogging of the lagoon.
- Monitor the restoration process inside the Mata Redonda lagoon.
- Strengthen the involvement of local organizations through the development of training processes on issues related to wetland management, environmental education, good agricultural practices, and ecotourism initiatives.
- The activities contemplated in the terms of reference were not executed according to the proposed schedule, due to the need to work in a dry period. Therefore, it was necessary to work simultaneously on diagnoses and execution of work. This greatly delayed the work in the first year in terms of products and expenditure.
- Planning TDR should consider relevant aspects such as the seasons (dry or rainy) and the need for continuous actions without interruptions.

REFERENCES

Alpízar, E., Bolaños, R., Bravo, J., Canessa, G., Echeverría, J., 1998. Plan de acción para la cuenca del Río Tempisque. Diagnóstico funcional Vol. II: Zonas de vida, biodiversidad, áreas protegidas y humedales. Centro Científico Tropical.

Bravo, J., 2014. Propuesta para el Desarrollo de un Plan de intervención ecológica de los humedales de la cuenca baja del Rio Tempisque. Informe final de consultoria, Guanacaste.

Bravo, J., Criado, J., Marín, M., 2013. Conservación, uso sostenible de la biodiversidad y el mantenimiento de servicios eco-sistémicos de los humedales protegidos de importancia internacional. Informe final consultoría en Ciencias Naturales, San José.

De Groot, R.S., Stuip, M.A.M., Finlayson, C.M., Davidson, N., 2007. Valoración de humedales: Lineamientos para valorar los beneficios derivados de los servicios de los ecosistemas de humedales. Informe Técnico de Ramsar núm. 3/núm. 27 de la serie de publicaciones técnicas del CDB. Secretaría de la Convención de Ramsar, Gland (Suiza), y Secretaría del Convenio sobre la Diversidad Biológica, Montreal (Canadá), ISBN: 2-940073-31-7.

Echeverría, J., Aguilar, G., Arias, D., Bolaños, R., Bravo, J., Burgos, J., Canessa, G., Cervantes, S., Losilla, M., Mata, A., 1998. Plan de acción para la cuenca del río Tempisque. Zonificación ambiental y plan de acción, vol. VI. Centro Científico Tropical, San José, Costa Rica.

RAMSAR, 2010. Directrices para la evaluación ecológica rápida de la biodiversidad de las zonas costeras, marinas y de aguas continentales. Recuperado 10 de marzo de 2016, http://www.ramsar.org/sites/default/files/documents/library/lib_rtr01_s.pdf.

SINAC (Sistema Nacional de Áreas de Conservación), 2016. Sistematización del proceso: intervención ecológica en humedales de la Cuenca Baja del Tempisque, Área de Conservación Tempisque. Informe Final San José, San José, Costa Rica.

FURTHER READING

Área de Conservación Tempisque, 2018a. Humedal Palustrino Corral de Piedra: características e información. [En línea]. Recuperado 5 de marzo 2017 de: http://www.actempisque.org/hcorraldepiedra.htm.

Área de Conservación Tempisque, 2018b. Refugio Nacional de Vida Silvestre Mata Redonda: características e información. [En línea]. Recuperado 23 de enero 2017 de: http://www.actempisque.org/rnvsmataredonda.htm.

Barbier, E., Acreman, M., Knowler, D., 1997. Valoración económica de los humedales: Guía para decisores y planificadores. [En línea]. Recuperado 5 de mayo 2017 de: http://ramsar.rgis.ch/pdf/lib/lib_valuation_s.pdf.

Convención de Ramsar, 2006. Manual de la Convención de Ramsar: Guía a la Convención sobre los Humedales (Ramsar, Irán, 1971), fourth ed. Secretaría de la Convención de Ramsar, Gland (Suiza).

Costa Rica, FUNDACA, 2015a. Propuesta para el desarrollo de un plan de intervención ecológica de los humedales de la cuenca baja del rio tempisque, I versión. Informe Final, San José Costa Rica.

Costa Rica, FUNDACA, 2015b. Propuesta para el desarrollo de un plan de intervención ecológica de los humedales de la cuenca baja del rio tempisque, II versión. Informe Final, San José, Costa Rica.

Hernández, R., Fernández, C., Baptista, P., 2000. Metodología de la Investigación. Edic. MCGraw-Hill, México.

Hulme, M., 2016. Toward Integrated Historical Climate Research: The Example of Atmospheric Circulation Reconstructions Over the Earth. [En línea]. Recuperado 20 de marzo 2017 de: https://www.researchgate.net/publication/291075416.

Instituto Nacional de Biodiversidad (INBIO) (n.d.). Informe Final: Aves acuáticas del Refugio de Vida Silvestre Laguna Mata Redonda, Costa Rica. [En línea]. Recuperado 5 de febrero 2017 de http://www.inbio.ac.cr/es/estudios/PDF/Informe_AvesMataRedonda.pdf.

Porta, L., Silva, M., n.d. La investigación cualitativa, El Análisis de Contenido en la investigación educativa.

RAMSAR, 1999. Una propuesta metodológica participativa para la protección y conservación de humedales: Avance del Proyecto: "Refugio de Vida Silvestre 'Laguna' Mata Redonda, Guanacaste, Costa Rica" [En línea] Recuperado 3 de marzo 2016 de http://www.ramsar.org/es/nuevas/refugio-de-vida-silvestre-laguna-mata-redonda-guanacaste-costa-rica.

Secretaría de la Convención de Ramsar, 2010a. Informe Técnico de Ramsar número. p. 1.

Secretaría de la Convención de Ramsar, 2010b. Inventario, evaluación y monitoreo: Marco Integrado para el inventario, la evaluación y el monitoreo de humedales, fourth ed. Manuales Ramsar para el uso racional de los humedales, vol. 13. [En línea] Recuperado 8 de marzo 2017 de http://www.ramsar.org/sites/default/files/documents/pdf/lib/hbk4-13sp.pdf.

UNED-SINAC-CEMEDE-UNA, 2013. Plan General de Manejo del Refugio Nacional de Vida Silvestre Mata Redonda. Herramienta de Manejo Adaptativo y Planificación Estratégica. Universidad Estatal a Distancia. Sistema Nacional de Áreas de Conservación, Área de Conservación Tempisque. Centro Mesoamericano para el.

Chapter 5.7

WET-Health, a Method for Rapidly Assessing the Ecological Condition of Wetlands in Southern Africa

Donovan C. Kotze*, Douglas M. Macfarlane† and Dean J. Ollis‡

**University of KwaZulu-Natal, Pietermartizburg, South Africa, †Eco-Pulse Environmental Consulting Services, Hilton, South Africa, ‡Freshwater Research Centre, Cape Town, South Africa*

Chapter Outline

INTRODUCTION

The need to assess the ecological condition of wetlands in South Africa continues to grow in the face of increasing human impacts on these ecosystems. Decision-makers require simple, user-friendly and cost-effective tools to facilitate ecologically sound decisions, but the challenge is to develop methods that balance ease and speed of use with the complexity and detail that is captured (De Leo and Levin, 1997; Kotze et al., 2012). In the United States and Canada, as discussed in this book, a number of tools have been developed using biotic indicators of wetland health, such as the multimetric indices of biotic integrity (IBIs) developed by the United States Environmental Protection Agency (EPA) (2002) and others (e.g., Mack, 2007), and the methods more recently developed for the National Wetland Condition Assessment (EPA, 2011). However, South Africa is still a long way from using biotic indices that account for its considerable diversity of wetland types and biota, which have been the subject of limited field studies (Kotze et al., 2012). In addition, it can be time consuming to identify and describe the biota associated with wetlands, and a high level of specialist input is required. Biotic indices also do not necessarily indicate the nature of the human stressors that have contributed to the decline in ecological condition and, therefore, often provide little insight into management requirements (Kotze et al., 2012).

Due to the challenges of developing robust biotic indices for wetland ecosystems, the approach that has been followed in South Africa to date has been to focus on the development of tools based primarily on the assessment of human stressors to wetland "health." WET-Health was therefore developed as a rapid method for assessing the ecological condition of all palustrine wetland types in South Africa based on the impacts of human stressors on hydrogeomorphic processes and vegetation (Macfarlane et al., 2007). There are two other methods based on a similar approach that are also used in South Africa, namely the rather dated present ecological status (PES) method for floodplain and other palustrine wetlands (Duthie, 1999) and the more recently developed Wetland Index of Habitat Integrity for floodplain and channeled valley-bottom wetlands (DWAF, 2007).

The primary purpose of WET-Health is to assess the ecological condition (health) of wetlands, together with assisting in the diagnosis of problems impacting wetland health, which can all be used to inform management/rehabilitation decisions. The WET-Health method arose out of the legislative need for a robust method to assess the ecological condition of wetlands in South Africa as well as for planning and assessing wetland rehabilitation, in particular rehabilitation undertaken through the national Working for Wetlands program. Development of the method was also encouraged and supported by the

Wetland and Stream Rapid Assessments. https://doi.org/10.1016/B978-0-12-805091-0.00056-6

business sector, in particular the forestry plantation sector through their acknowledged need to assess and report on the ecological condition of wetlands within their landholdings.

WET-Health has been applied to a diverse range of applications, including:

- Assessing wetlands for state of the environment reporting and environmental impact assessments (EIAs).
- Assessing the present ecological state in ecological reserve determination studies.
- Rehabilitation planning to identify important problems requiring rehabilitation and to establish potential gains that would be achieved in terms of ecological condition.
- As part of the planning for biodiversity offsets to assist in determining wetland gains and losses for offset accounting.
- Building an understanding of how wetlands function, for example, through training courses.
- Assist in assessing the sustainability of land-use practices in wetlands and informing policy.

CONCEPTUAL FRAMEWORK FOR THE METHOD

The ecological condition of a wetland can be degraded by a variety of stressors, and indicators of wetland condition can be based either on the response of the wetland to the stressors (e.g., decreased abundance of sensitive native species) or on the stressors themselves (Fennessy et al., 2007). Wetland assessment methods may use only stressor indicators, only biotic-response indicators, or a mixture of both types of indicators. Given the lack of baseline reference wetland studies in South Africa, WET-Health focuses mainly on stressor indicators but uses some response indicators, notably those of wetland vegetation (Kotze et al., 2012).

Indicators are aggregated in structured algorithms to derive an ecological condition score. As in the case of the HGM method (Brinson and Rheinhardt, 1996), the algorithms are not simulation models but are designed to generate an index that reflects the extent of departure from an unimpacted condition. The indicators are combined in a way that represents the authors' understanding of their relative importance at the time of developing the method. The rationale behind the selection of each of the indicators is provided, together with the rationale for combining the scores of the different indicators into a single score (Kotze et al., 2012). Although the relationship of the method's indices and indicators to the underlying processes have not been systematically validated in a quantitative sense, the assumptions behind the method are provided. Therefore WET-Health is open to progressive refinement where specific assumptions are found to be inadequate or incorrect (Kotze et al., 2012).

Although important elements of the HGM approach were used in the development of WET-Health, it is not based upon sets of reference wetlands nor is it a tool for assessing wetland functions, as in the case of the HGM approach (Wardrop et al., 2007). WET-Health also does not predefine an ideal state for the particular HGM type and region within which the wetland occurs. Instead, it relies on the assessor being familiar with the characteristics of wetlands in a particular region (Kotze et al., 2012). WET-Health does, however, respond to the need for a general and rapid approach to assessing wetland health that is tailored according to the wetland's hydrogeomorphological type and its climatic setting (Kotze et al., 2012).

DESCRIPTION OF THE METHOD

WET-Health is a method designed to assess the health or integrity of a wetland by scoring the perceived deviation from a theoretical reference condition where the reference condition is defined as an unimpacted condition in which ecosystems show little or no influence of human actions (Anderson, 1991).

For the purposes of assessment, wetland health is assessed across three separate modules:

Hydrology is defined in this context as the distribution and movement of water through a wetland and its sediments. This module focuses on (i) changes in water inputs that result from human alterations to the catchment, which affect water inflow quantity and pattern, and (ii) modifications within the wetland itself that alter the water distribution and retention patterns of the wetland (e.g., artificial drainage channels).

Geomorphology is defined in this context as the distribution and retention patterns of sediment within the wetland. This module focuses on evaluating current geomorphological health through the presence of indicators of excessive sediment inputs and/or erosional losses for clastic (minerogenic) and organic sediment (peat).

Vegetation is defined in this context as the structural and compositional state of the vegetation within a wetland. This module evaluates changes in vegetation composition and structure as a consequence of current and historic onsite transformation and/or disturbance.

The method uses a stressor-based approach for those human activities that do not produce clearly visible responses in wetland structure and function. The impact of irrigation or afforestation in the catchment, for example, produces invisible impacts on water inputs. This is the approach used in the hydrological assessment. A response-based approach is used for activities that produce clearly visible responses in wetland structure and function, such as the presence of erosion gullies or nonnative plant species. This approach is mainly used in the assessment of geomorphological and vegetation health, but also considers stressors. Application of each of the three modules follows a broadly similar approach. Prior to assessment, the wetland is divided into hydrogeomorphic (HGM) units and their associated catchments. These are analyzed separately for hydrological, geomorphological, and vegetation health based on an assessment of the extent and intensity of various impacts within an HGM unit, which are then combined and translated into an overall health score for each module.

The extent of impact is measured as the proportion of a wetland and/or its catchment that is affected by an activity, and is expressed as a percentage. The intensity of impact is estimated by evaluating the degree of alteration that results from a given activity. Scores range from 0 (no impact) to 10 (critical impact). The magnitude of impact for individual activities is the product of extent and intensity. For example, if the extent of cultivation in a wetland was 40% and the intensity of impact on wetland vegetation within this area was 8 (out of a maximum of 10), then the magnitude of impact would be $40/100 \times 8 = 3.2$. The magnitude of individual activities in each HGM unit is combined in a structured and transparent way to calculate the overall impact of all activities on each component of wetland health. The scores are then used to place the wetland into one of several Present Ecological State health categories that are consistently applied across all freshwater assessments in South Africa. These categories include: A: Natural (0–0.9), B: Largely natural (1–1.9), C: Moderately modified (2–3.9), D: Largely modified (4–5.9), E: Severely modified (6–7.9), and F: Critically modified (8–10). While not specifically encouraged, individual scores from each component can be combined into an overall health score by weighting the component scores according to a predefined formula ((Hydrology score) $\times$ 3 + (Geomorphology score) $\times$ 2 + (Vegetation score) $\times$ 2) $\div$ 7).

Using a combination of threat and/or vulnerability, an assessment is also made in each module on the likely trajectory of change within the wetland based on the following categories: large improvement (↑↑), slight improvement (↑), remains the same (→), slight decline (↓), and large decline (↓↓). The overall health of a wetland is then presented for each module by jointly presenting the Present Ecological State category and the likely trajectory of change, for example, C↓.

WET-Health provides the user with two levels of detail for conducting the assessment: Level 1, which is largely desktop-based, and Level 2, which involves a more detailed assessment of field indicators. Level 1 assessments are designed for situations where many wetlands need be assessed at a relatively low resolution, for example, in order to broadly establish the ecological condition of wetlands in a catchment. Such assessments are primarily desktop-based, but a subsample of these assessed wetlands should be visited in the field. The Level 2 assessment is typically applied where higher resolution and/or confidence assessments are required of individual wetlands.

TIME SPENT TO APPLY THE METHOD IN THE FIELD

Depending on the complexity of the HGM unit and the number of different human disturbance subunits in the wetland, a Level 2 assessment usually requires from 2 to 6 h for a single trained assessor to carry out the fieldwork, and an additional 3–8 h of preparation and assessment in the office per assessment. Level 1 assessments require less time to complete because fewer indicators need to be assessed. Field time at an individual wetland can often be completed in 1–4 h while time in the office is typically reduced to 2–4 h for Level 1 assessments.

VERIFICATION/VALIDATION EFFORTS UNDERTAKEN WITH THE METHOD

The method has been subject to very limited true validation. However, some testing has been undertaken of the repeatability of WET-Health when applied by different operators, relative to other methods. Ollis and Malan (2014) report that in a test of WET-Health Level 1, independent operators (with environmental training and some experience with wetland assessment) scored relatively closely for systems that were minimally impacted, but some widely divergent scores were assigned to wetlands subject to moderate to moderate-high levels of transformation. There was also a relatively high degree of inconsistency between the "gut-feel" ecological categories recorded by assessors and the categories obtained by the same assessor using WET-Health. Similar results were reported for another rapid assessment tool, the Wetland Index of Habitat Integrity (DWAF, 2007), which was developed for floodplain and valley-bottom wetlands (Ollis and Malan, 2014). In a formal test of the robustness of WET-Health Level 2 by Botes (2008) and Bodman (2011), independent operators with environmental training scored closer than in the study of Ollis and Malan (2014). Nonetheless, there was still some divergence, and a clear need was identified for refinements to improve the repeatability of the method.

TABLE 5.7.1 The Impact Scores and Hectare Equivalents Recorded for Both Wetland Systems Pre- and Postrehabilitation

	Killarney		Kruisfontein	
	Prerehabilitation (2005)	Postrehabilitation (2011)	Prerehabilitation (2005)	Postrehabilitation (2012)
Hydrology	3.0	1.0	9	7.5
Geomorphology	1.6	0.5	2.5	1.8
Vegetation	3.6	2.1	8.2	6.7
Overall impact score	2.8	1.2	6.9	5.6
Hectare equivalents[a]	30.7	37.5	8.2	11.5
Hectare equivalents reinstated[b]	6.8		3.3	

[a] *Hectare equivalents were determined as follows: Total wetland area × (10 − Overall impact score)/10. For Killarney the total wetland area was 42.4 ha and for Kruisfontein it was 26.4 ha.*
[b] *Hectare equivalents reinstated were determined as follows: (postrehabilitation hectare equivalents) − (prerehabilitation hectare equivalents).*
From Cowden, C., Kotze, D.C., Ellery, W.N., Sieben, E.J.I., 2014. Assessment of the long-term response to rehabilitation of two wetlands in KwaZulu-Natal, South Africa. Afr. J. Aquat. Sci. 39, 237–247.

SAMPLE APPLICATION OF METHOD IN THE FIELD

WET-Health assessments of the Killarney and Kruisfontein wetlands located in KwaZulu-Natal (South Africa) were undertaken in 2005 prior to rehabilitation intervention, and again in 2011 and 2012, following rehabilitation of the wetlands. At both sites, the rehabilitation interventions were designed primarily to neutralize the effects of artificial drainage channels in the wetland. Results of these assessments are summarized in Table 5.7.1 to illustrate the application of the method. Improvements in hydrological, geomorphological, and vegetation integrity were recorded in both wetlands but to a lesser extent for the Kruisfontein wetland (Table 5.7.1), particularly with respect to vegetation, which in 2012 was still largely dominated by pioneer species. Cowden et al. (2014) attributed this to an invasive nonnative hydric species, *Paspalum dilitatum*, which appeared to be providing little opportunity for colonization of the site by indigenous hydric species. When translated into "hectare equivalents," the effective contribution of the rehabilitation intervention at Killarney was twice that at Kruisfontein (Table 5.7.1).

HOW THE DATA ARE BEING USED

In South Africa, wetland assessment and reporting is a standard requirement for any developments with a potential impact on wetlands. Therefore, wetland assessments are routinely applied by wetland specialists to inform environmental authorizations as part of the development application process. WET-Health is central to such assessments as it helps regulating authorities to understand the present ecological condition of wetlands that could be impacted by development activities.

WET-Health assessments have also been used extensively to inform planning of wetland rehabilitation projects undertaken by private developers, NGOs, and the national Working for Wetlands Program. Over the last 11 years, WET-Health (initially as a prototype and more recently in its current form) has been used to plan rehabilitation for more than 400 wetlands across South Africa. Unfortunately, to date Working for Wetlands has undertaken little evaluation of the ecological outcomes of their rehabilitation. However, a few wetlands have been assessed pre- and postrehabilitation using WET-Health (e.g., Kotze and Ellery, 2009; Cowden et al., 2014) (see Table 5.7.1). These evaluations have improved understanding in terms of feasible improvements in ecological condition to be expected for different types and magnitudes of impact addressed with different rehabilitation interventions.

WET-Health has been used in state of environment reporting, for example, reporting on the health of the priority wetlands of KwaZulu-Natal province (Macfarlane et al. (2012) and reporting on the health of the priority wetlands within the landholdings of Mondi, a major plantation forestry company in South Africa (Walters et al., 2011). WET-Health has also been applied in assessments of the sustainability of wetland cultivation, for example, by Kotze (2011) in selected wetlands in Malawi and by Marambanyika et al. (2017) in selected wetlands in Zimbabwe.

WHAT WAS LEARNED

WET-Health has significant heuristic value in helping to develop an understanding of the key drivers that affect wetland health. This is particularly important in rehabilitation planning as it allows interventions to be planned with due consideration to the causes of wetland degradation, rather than simply focusing on visible impacts. As can be seen from Table 5.7.1, the WET-Health results can also be used to provide a semiquantitative currency for assessing ecological outcomes of rehabilitation interventions. This provides a useful means of comparing different sites and contexts, which is helpful in terms of prioritizing among different rehabilitation options during the planning phase of wetland rehabilitation and later in the summative evaluation of project outcomes. Such calculations have also proved very useful for wetland offset planning, where an understanding of gains need to be costed and compared against a range of potential offset sites.

Developing two levels of the tool (Levels 1 and 2) to cater for different user requirements has also been an important learning point. While the different levels have certainly improved the uptake of the tool, in practice, users often opt for the simplest level even if it is not necessarily best suited for their intended purpose.

Applying the method to state of the environment assessments that involve the assessment of numerous and often extensive wetland systems has highlighted the opportunity to better integrate land-cover datasets into desktop assessments. This has prompted the authors to start investigating how best to integrate readily available land-cover data into revisions of the methodology.

WET-Health results have also contributed to revealing specific vulnerabilities of wetlands and have contributed toward informing policy in other African countries. For example, a study of wetlands in a selected district in Malawi revealed that the geomorphological vulnerability and postdisturbance recovery of vegetation varied greatly across individual wetland systems (Kotze, 2011). In an example from Zimbabwe, WET-Health assessments provided useful information on the impact of different cultivation practices on wetlands that can be used to inform the development of a national wetland policy, which is currently missing in Zimbabwe (Marambanyika et al., 2017).

PROSPECTS FOR THE FUTURE

A Water Research Commission project is currently underway to comprehensively revise the method, taking into account the issues revealed in the repeatability testing, feedback from users, and learning from a diversity of applications. Key focal areas to be addressed in this revision include:

- Forcing users to be more explicit about their understanding of the perceived natural reference state before embarking on an assessment.
- Better tailoring the method to assess the full suite of wetland types occurring in South Africa.
- Catering more effectively for different levels of rapid assessment, from purely desktop-based assessments covering broad (regional to national) spatial scales to more detailed, site-specific, field-based assessments.
- Integrating water quality as a new component of the method.
- Improving the repeatability of the assessment across users by reducing ambiguity and providing more explicit guidance.

WET-Health has been a focus of training courses in wetland assessment for consultants and government officials since 2009, notably the training course run annually by Rhodes University. It is envisaged that the above revisions will be incorporated into ongoing training on WET-Health.

Initial application of the method in other countries suggests that the method can be reliably applied to other contexts, and can be particularly useful where no locally developed tools are available. Clearly, opportunities exist to refine the WET-Health framework to better adjust for the specific climatic and hydrogeomorphic types present in other countries. Such investments would be considerably more cost-effective than trying to develop a new method or to tailor an existing method that requires in-depth knowledge of reference wetlands, which is often lacking, particularly in developing countries.

Website link: http://www.wrc.org.za/Pages/DisplayItem.aspx?ItemID=9025&FromURL=%2FPages%2FDefault.aspx%3F.

REFERENCES

Anderson, J.E., 1991. A conceptual framework for evaluating and quantifying naturalness. Conserv. Biol. 5, 347–352.

Bodman, L., 2011. An Investigation on the Variation of Scores from Users of the WET-Health Tool. Honours Thesis, University of KwaZulu-Natal, Pietermaritzburg.

Botes, W., 2008. Measuring the Success of Individual Wetland Rehabilitation Projects in South Africa. MSc Thesis, University of KwaZulu-Natal, Durban.

Brinson, M.M., Rheinhardt, R., 1996. The role of reference wetlands in functional assessment and mitigation. Ecol. Appl. 6, 69–76.

Cowden, C., Kotze, D.C., Ellery, W.N., Sieben, E.J.I., 2014. Assessment of the long-term response to rehabilitation of two wetlands in KwaZulu-Natal, South Africa. Afr. J. Aquat. Sci. 39, 237–247.

De Leo, G.A., Levin, S., 1997. The multifaceted aspects of ecosystem integrity. Conserv. Ecol. 1 (1), 3. http://www.consecol.org/vol1/iss1/art3.

Duthie, A., 1999. IER (floodplain wetlands): present ecological status (PES) method. In: Appendix W4: 1–6 in DWAF: Resource Directed Measures for Protection of Water Resources, Volume 4—Wetland Ecosystems. Pretoria, Department of Water Affairs and Forestry.

DWAF, 2007. Manual for the Assessment of a Wetland Index of Habitat Integrity for South African Floodplain and Channelled Valley Bottom Wetland Types. Resource Quality Services, Department of Water Affairs and Forestry, Pretoria. RQS Report No. N/0000/00/WEI/0407.

Fennessy, M.S., Jacobs, A.D., Kentula, M.E., 2007. An evaluation of rapid methods for assessing the ecological condition of wetlands. Wetlands 27, 543–560.

Kotze, D.C., 2011. The application of a framework for assessing ecological condition and sustainability of use to three wetlands in Malawi. Wetl. Ecol. Manag. 19, 507–520.

Kotze, D.C., Ellery, W.N., 2009. WET-Outcome Evaluate: An Evaluation of the Rehabilitation Outcomes at Six Rehabilitation Sites in South Africa. Water Research Commission, Pretoria. WRC Report TT 343/09.

Kotze, D.C., Ellery, W.N., Macfarlane, D.M., Jewitt, G.P.W., 2012. A rapid assessment method for coupling anthropogenic stressors and wetland ecological condition. Ecol. Indic. 13, 284–293.

Macfarlane, D.M., Kotze, D.C., Ellery, W.N., Walters, D., Koopman, V., Goodman, P., Goge, M., 2007. WET-Health: A Technique for Rapidly Assessing Wetland Health. Water Research Commission, Pretoria. WRC Report No. TT 340/09.

Macfarlane, D.M., Walters, D., Cowden, C., 2012. A Wetland Health Assessment of KwaZulu-Natal's Priority Wetlands. Unpublished Report, Ezemvelo KZN Wildlife, Pietermaritzburg.

Mack, J.J., 2007. Developing a wetland IBI with statewide application after multiple testing iterations. Ecol. Indic. 7, 864–881.

Marambanyika, T., Beckedahl, H., Ngetar, N.S., 2017. Assessing the environmental sustainability of cultivation systems in wetlands using the WET-Health framework in Zimbabwe. Phys. Geogr. 38, 62–82.

Ollis, D.J., Malan, H.L., 2014. Development of Decision-Support Tools for Assessment of Wetland Present Ecological Status (PES). Volume 1: Review of Available Methods for the Assessment of the Ecological Condition of Wetlands in South Africa. Water Research Commission, Pretoria. WRC Report No. TT 608/14.

U.S. EPA, 2002. Methods for Evaluating Wetland Condition: Introduction to Wetland Biological Assessment. Office of Water, U.S. Environmental Protection Agency, Washington, DC. USEPA Report EPA-822-R-02-014.

U.S. EPA, 2011. National Wetland Condition Assessment: Field Operations Manual. United States Environmental Protection Agency, Washington, DC. USEPA Report EPA-843-R-10-001.

Walters, D., Kotze, D.C., Job, N., 2011. Mondi State of the Wetlands Report: A Health and Ecosystem Services Assessment of a Selection of Priority Wetlands Across Mondi Landholdings. Unpublished Report, WWF Mondi Wetlands Programme, Irene, Pretoria.

Wardrop, D.H., Kentula, M.E., Jensen, S.F., Stevens, D.L., Hychka, K.C., 2007. Assessment of wetlands in the upper Juniata watershed in Pennsylvania, USA using the hydrogeomorphic approach. Wetlands 27, 432–445.

Summary and Conclusions

Paul Adamus*, John Dorney†, Ralph W. Tiner‡ and Rick Savage§

**Oregon State University, Corvallis, OR, United States, †Moffatt and Nichol, Raleigh, NC, United States, ‡Institute for Wetland & Environmental Education & Research, Leverett, MA, United States, §Carolina Wetlands Association, Raleigh, NC, United States*

Rapid assessment methods (RAMs) are standardized procedures that generate a score, index, or rating for a specified site (individual wetland, stream, watershed, etc.) and/or for its individual ecosystem services (functions, values) or other attributes (e.g., condition, sensitivity), based mainly on ground-level observations and/or by using aerial imagery/GIS as described in this book. The observational nature of RAM procedures contrasts with procedures based mainly on ground-level measurements, such as kilograms of nutrients removed by a wetland or the diversity of the aquatic community in a stream. If a site visit is needed, RAMs commonly require just one such visit lasting less than 1 day. Most RAMs do not require a comprehensive knowledge of plant, insect, or soil taxonomy.

This book will help government scientists or others when they are tasked with developing a wetland or stream RAM that is optimized for a particular region, modifying and calibrating an existing RAM from another region so that it is optimal for the user's region, or calibrating and verifying an existing RAM. The book also will help RAM users make better use of wetland and stream RAMs as they gain a deeper understanding of how RAMs have been developed, tested, and applied. Section and chapter introductions describe many of the lessons we, as editors, have learned from developing RAMs in various regions. Additional insights on how wetland and stream RAMs have been developed and applied can be gained by reading this book's case histories.

RAM development has largely been spurred by a desire for systematic tools to help evaluate impacts to wetlands and streams from development, and to achieve appropriate mitigation based on both the quality and quantity of the resource. RAMs provide key information to those responsible for making decisions about development or conservation of wetlands and streams. RAMs characterize individual stream or wetland sites and may compare them with others. They do so for evaluating project impacts, designing appropriate mitigation, and/or developing strategies that promote conservation. RAMs provide vital information for decision making, but most jurisdictions do not require that RAM results be the sole determinant of a decision about a wetland or stream reach.

Many RAMs are localized or regional adaptations of a general template developed previously, such as the hydrogeomorphic (HGM) assessment framework (Smith et al., 2013), the WESP template (Adamus, 2016, Chapter 4.3.2), the Ohio Rapid Assessment Method (Chapter 4.3.8), or the LLWW and NWI-Plus templates for landscape-level assessments (Tiner, 2003; case studies in Section 2.0). RAMs for assessing wetlands are more common than those for streams. Many RAMs are being refined and adapted for use in new geographic areas as well as in response to expanding scientific knowledge, technology, software capabilities, and spatial data availability. In the United States, the impetus for developing wetland and stream RAMs was the Clean Water Act of 1972—the keystone for wetland regulation across the country. Most of the published RAMs have been sponsored by North American states and provinces and are intended to enhance the incorporation of scientific knowledge and existing spatial data into regulatory decisions affecting wetlands and streams.

Nearly all RAMs provide one or more scores and/or ratings for an individual site or a larger landscape by using models to process a user's standardized data inputs. Most often, the models are mathematical formulas that the RAM developer has constructed based partly on published science and partly on correlations thought to exist between various conditions of several indicator variables and an endpoint (wetland or stream functions, values, and/or ecological condition). Some models consist of narrative criteria rather than formulas, and some RAMs do not include explicit models but rather leave it to the user's judgment to connect the user's observations of indicator variables to a rating or score intended to represent the endpoint. For only a very few RAMs have the sponsors attempted to validate the presumed correlations between indicators and endpoints by making detailed measurements of the endpoints and comparing those measurements with model outputs to determine if sites are ranked similarly. More often, RAM models are simply verified by comparing the ranking of a series of sites based on model outputs with a ranking based on the independent opinions of local experts on each function or status condition. The repeatability (consistency) of RAM outputs among independent users visiting the same series of sites has apparently been tested in only a few instances.

As reflected in the structure of this book, wetland and stream RAMs are commonly developed as either landscape-level or field-level methodologies. Also, based on what they aim to assess, they are often characterized as either function-assessment RAMs (which may also assess the values of those wetland or stream functions as related partly to ecosystem services) or condition-assessment RAMs (which address specific values such as wetland or stream health, quality, status, or integrity). Several RAMs incorporate combinations of these approaches and themes. Condition-assessment RAMs cannot consistently predict all or most functions, and function-assessment RAMs cannot reliably predict the condition of wetlands or streams as it is conventionally defined.

Landscape-level wetland RAMs use remotely sensed imagery, existing spatial data layers, and GIS combined with existing knowledge of wetlands to compile input data for models that largely resemble those used in field-level RAMs, and similarly produce scores or ratings believed to reflect wetland or stream functions and/or condition. At present, landscape-level stream RAMs are rare to nonexistent. Landscape-level RAMs are mainly employed when a goal is to assess all wetland or stream sites across a large area (e.g., watershed, river basin, ecoregion) because cost and time requirements to visit and assess all those sites would be prohibitive, even if all sites were accessible for examination. This type of RAM also is an attractive tool for assessing stream segments or wetlands where field-based RAMs have not been developed, or where use of field-level RAMs is difficult or impossible because sites are too remote or hazardous to access, are too large to assess meaningfully during a single 1-day visit, or are inaccessible due to resistance from land owners or managers.

Landscape analysis is a first-step, basic assessment derived from interpretation of imagery and maps and existing knowledge of the variety of wetlands that occur across a geographic region. Such analyses can be done any place with suitable imagery where wetlands or streams can be detected and where the relationship between interpreted variables (such as dominant vegetation, hydrology, landscape position, and connectivity) and wetland function or condition is reasonably understood. As mentioned, landscape analyses provide a preliminary assessment as they are limited by source data (aerial imagery and/or GIS data/maps) for a number of reasons, including: (1) not all wetlands and small streams are readily identified on aerial imagery, (2) scale issues (e.g., limitations to the smallest size of wetland or stream that can be identified remotely and displayed on a map), (3) misidentification of an upland as a wetland and inclusions of upland in wetland units and vice versa, (4) misclassification of wetland type, and (5) registration issues (i.e., positional accuracy). Moreover, many important indicator variables (e.g., plant community composition, soil texture, downed wood, microtopography, high water marks on vegetation, concealed outlets, water quality parameters, aquatic macroinvertebrate communities) can be reliably detected only during a site visit and so are lacking in landscape-level RAM models. Consequently, landscape-level RAMs are usually intended for gaining a broad perspective on wetland and stream functions and condition across a watershed or larger geographic area (e.g., analyzing cumulative impacts of development on wetland functions), developing conservation strategies, general planning purposes, and educating the public on expected functions and condition, rather than for making regulatory or management decisions about individual wetlands or project areas.

The accuracy of landscape-level assessments can be improved by conducting field investigations to verify wetland and stream classifications and the interpretations of wetland and stream functions or condition. In Alberta, the rank order of a series of wetland sites scored by a field-level RAM was compared with the rank order of the same sites scored independently by a landscape-level RAM, and little concordance was found in the scores or ratings of several functions. An example of the mutually reinforcing effect of landscape-level and field-based analyses is described by work done on geographically isolated wetlands in North and South Carolina (RTI International et al., 2011). The goal of this work was to derive a statistically valid estimate of the extent and condition of isolated wetlands in an eight-county study area in these two states. First, a GIS map was prepared based on a wide variety of GIS layers. Then, a stratified random sample of sites was chosen and field work conducted to determine if the sites were isolated wetlands and, if so, to evaluate their condition using NC WAM (Chapter 4.3.1). Based on this field evaluation, it was determined that the GIS map correctly identified wetlands 69% of the time but only correctly identified isolated wetlands 22% of the time. This high rate of misidentification of isolated wetlands had two main causes: (1) many small ditches were present in the field but not discernable from GIS layers, which then made the sites not isolated, and (2) many sites had been falsely identified as wetlands but were either upland or ponds. However, because the field sample was statistically developed, the authors were able to use data from the GIS map and the NC WAM results to make definitive conclusions about the extent of isolated wetlands in the study area as well as their condition. The combined use of a landscape-level model followed by statistically valid field-level investigations provides a good conceptual model to take advantage of the strength of both these approaches and to minimize the weaknesses of both these approaches.

Landscape analysis is often part of field-based RAMs because one cannot see everything from the ground or take time to traverse the entire project area. In addition, landscape-level analysis is often used to address broader watershed or wetland buffer issues that influence the functioning or condition of a wetland or stream. In general, results from the field-level RAM are preferred over landscape-level assessments because they incorporate many more indicators than landscape-level assessments and involve close-up inspection and verification. However, because ground-level field visits are often physically hazardous, are limited to a 1-day visit, and time-consuming for multiple wetlands, use of remote sensing and GIS analysis can be more accurate for predicting some functions of very large sites or those with poor access. In addition, calibrating a field-based RAM prior to its being made available for public use requires significant time and effort.

Of the dozens of field-based RAMs developed and calibrated over the past four decades, only a relatively small number are still used regularly. Why is that? A primary determinant of sustained use is the degree to which sponsors *require* the use of a particular RAM legally or as official agency policy, or at least unofficially encourage its use (Arnold, 2012, 2014, 2015). Degree of use also depends on the sponsor's outreach efforts, lack of other RAMs that have been calibrated to the specific region and/or wetland type being assessed, the RAM's ease of use, the availability of low-cost training for the RAM's potential users, and the ability of sponsors to invest in continued refinement of the RAM in response to the availability of improved spatial data and imagery, new scientific findings, wider peer review and verification, expanded repeatability testing, and validation by comparison with more intensive measures of function and condition. RAM sponsors must decide how frequently to release revisions and updates so as to avoid creating a "moving target" that can frustrate users if a site's rating changes over time as a result of a sponsor's well-intended revisions of the RAM.

As agency budgets and availability of qualified staff decrease, RAM developers and their sponsors will likely find themselves pressured to make their RAM shorter, faster, and simpler. Can this be done while still achieving sufficient accuracy and achieving a desired application outcome such as avoidance or minimization of impacts? Often the requests for RAM streamlining originate from the regulated community out of a desire to minimize project costs by shortening the time required to do an assessment. In some instances limited streamlining is possible without compromising the scientific integrity, sensitivity, repeatability, or accuracy of the RAM, but regulatory agencies with input from scientists must decide the extent to which this is possible.

While most RAMs have been developed by regulatory or natural resource agencies for specific uses in specific areas, the scientific assumptions their models employ to assess what seems to be the same thing (e.g., a particular function) sometimes differ among RAMs. For example, some wetland RAMs assign highest scores for water storage function to river-associated sites whereas others assign the highest to depressional (closed basin, isolated) wetlands. This could be due to differing definitions of the water storage function (especially whether a site's area is included in the score calculation), a confusion of functions with values, a misunderstanding of physics and wetland science, or regional differences in how wetlands function. Better-documented rationales and broader consensus need to be sought concerning each of the hundreds of correlations assumed between various functions and their indicator variables.

Support is warranted for the development or regional adaptation of templates for assessing wetlands in states, provinces, and nations that currently lack wetland or stream RAMs. Even more pressing is the need for developing or adapting RAMs applicable to the functions of stream and riparian segments outside the Mid-Atlantic and Pacific Northwest portions of North America, and RAMs for assessing functions of aquatic habitats other than wetlands and streams. For many RAMs, more effort should be made to incorporate knowledge of indigenous peoples when identifying and modeling suitable indicators of functions, values, and site condition. The potential for specifying the use of emerging technologies (e.g., drones, sensors, software, and modeling techniques) in RAMs should be explored fully with a goal of producing more consistent and accurate results at less cost. Better scientific information is needed with regard to how to quantify, quickly and accurately, the short and long term as well as cumulative effects of various types of stressors on site functions using only one-visit field observations and conventionally available imagery.

As described in Section 3.0, most wetland and stream RAMs focus either on assessing site functions or assessing site health (condition or quality), and these two are not synonymous. Additional research is warranted to determine under what conditions (which functions, regions, wetland types, and metrics) the RAM-generated predictions of wetland health match RAM-generated predictions of each wetland function, and vice versa. This may increase confidence that the results of using one or the other type of RAM adequately represent what is assessed by the other.

The term "tested" is widely used by RAM authors hoping to inspire confidence among users, yet a review of existing RAMs suggests it can mean anything from casual examination of RAM results from a few wetlands by its author to correlation with results from other RAMs to rigorous comparison of results with direct, long-term measurements of function across a large enough series of sites verified to statistically encompass the variation of conditions present in a region. A need exists for measuring the repeatability of many RAMs (and, especially, using the resulting confidence intervals around the scores to ensure the correct use of the scores for interpreting functional "lift" or impact over time). As budgets and time allow, more effort should especially be devoted to verifying and validating parts of the many RAMs currently in use as well as measuring the repeatability of RAM scores and ratings among independent users assessing the same sites.

Guidance is needed from statisticians for appropriate ways of determining the adequate number and type of sites for field calibrating a RAM, and for evaluating under what conditions a RAM can be validly extended into regions where it was not field calibrated. With such guidance, the adequacy of the number of sites where existing RAMs have been calibrated should be examined to determine if more calibration sites need to be added in each RAM's focal region. Adding more calibration sites within any region will allow existing RAMs to be calibrated more precisely by allowing the RAM's scores to be normalized by specific wetland types or settings. Thus, a rating of "high," for example, could represent a high level of a particular function or condition relative to other sites *of that wetland type*, or relative only to (say) other wetlands in an urban setting rather than relative to *all* wetlands within the RAM's region regardless of type or setting, which is often the case with existing RAMs.

What then is the future of RAMs? Based on the work that the authors have done and the chapters included in this book, we foresee several trends and make related recommendations:

(1) The work to validate RAMs will continue and, we recommend, should be accelerated. We recommend that government agencies (and other funding sources) recognize the importance of this effort and adequately fund this work. In the case of the United States, we acknowledge the importance of the U.S. EPA's Wetland Program Development Grant Program and suggest that it explicitly list validation of both function and condition RAMs as a funding priority.
(2) RAMs for streams and other ecosystems will be more common in the future. In the United States, this seems like a logical result of the joint mitigation rule of the U.S. Army Corps of Engineers and U.S. EPA (US Environmental Protection Agency and US Army Corps of Engineers, 2008). We suggest that these agencies prioritize the development and testing of regional RAMs. In addition, the administrators of these agencies should prioritize implementation of the joint mitigation rule by their staffs.
(3) Despite a growing need, RAMs in developing countries will largely continue to be based on professional judgment due to scarcity of scientific data, spatial data, and skilled professionals. Improving wetland inventories for application of landscape-level approaches should be the first step in wetland assessment for large geographic regions, providing the broad overview of wetlands and their functions for strategic conservation planning. We suggest that international organizations (such as the United Nations and the Ramsar Convention) make development and testing of RAMs in developing countries a high priority. A significant role may be played by regional organizations such as the Organization of American States for RAM development in Latin America because RAMs are apparently very sparse in this region. As a parallel effort, we suggest that scientific organizations (such as the Society of Wetland Scientists and the Society of Freshwater Scientists) make outreach and assistance to these countries to develop and test RAMs a priority.
(4) RAMs that require "relatively undisturbed" reference sites to calibrate their scores will need to consider other ways to establish reference condition in highly developed landscapes, such as many in Europe, India, and China. Given contemporary environmental conditions, the most practical reference sites may just have to be those that seem to be subjected to the fewest current stressors, rather than pristine sites (Stoddard et al., 2006).
(5) The outputs of RAMs will be more explicitly incorporated into regulatory programs as the value of the results of these RAMs becomes more apparent to the regulated community, agencies, and the general public. To accelerate this process in the United States, we suggest that organizations such as the Association of State Wetland Managers and agencies such as the U.S. EPA and the U.S. Army Corps of Engineers expand their current efforts and funding for explaining the utility of wetland and stream RAMs to the general public. More funding for updating and enhancing the U.S. Fish and Wildlife Service's National Wetlands Inventory (NWI) will make NWI data more useful for landscape-level wetland function assessments across the country.

REFERENCES

Adamus, P.R., 2016. Manual for the Wetland Ecosystem Services Protocol (WESP). Version 1.3. Internet: people.oregonstate.edu/~adamusp/WESP.

Arnold, G., 2012. Assessing Wetland Assessment: Understanding State Bureaucratic Use and Adoption of Rapid Wetland Assessment Tools. PhD dissertation, Indiana University.

Arnold, G., 2014. Policy learning and science policy innovation adoption by street-level bureaucrats. J. Publ. Policy 34 (03), 389–414.

Arnold, G., 2015. When cooperative federalism isn't: how US federal interagency contradictions impede effective wetland management. Publius 45 (2), 244–269.

RTI International, NC Department of Environment and Natural Resources, South Carolina Department of Health and Environmental Control, the University of South Carolina, 2011. Assessing Geographically Isolated Wetlands in North and South Carolina—The Southeast Isolated Wetlands Assessment (SEIWA). Final Report. http://www.northinlet.sc.edu/training/media/2011/06142011isolatedwetlands/resources/seiwa_final_report.pdf. Accessed 30 October 2017.

Smith, R.D., Noble, C.V., Berkowitz, J.F., 2013. Hydrogeomorphic (HGM) approach to assessing wetland functions: guidelines for developing guidebooks (Version 2). ERDC/EL-TR-13-11, Environmental Lab, Engineer Research and Development Center, Vicksburg, MS.

Stoddard, J.L., Larsen, D.P., Hawkins, C.P., Johnson, P.K., Norris, R.H., 2006. Setting expectations for the ecological condition of streams: the concept of reference condition. Ecol. Appl. 16 (4), 1267–1276.

Tiner, R.W., 2003. Correlating Enhanced National Wetlands Inventory Data With Functions for Watershed Assessments: A Rationale for Northeastern U. S. Wetlands. U.S. Fish and Wildlife Service, National Wetlands Inventory Program, Hadley, MA.

US Environmental Protection Agency, US Army Corps of Engineers, 2008. Compensatory mitigation for losses of aquatic resources. Final Rule. April 10, 2008. Fed. Regist. 73 (70), 19594–19705. https://www.epa.gov/sites/production/files/2015-03/documents/2008_04_10_wetlands_wetlands_mitigation_final_rule_4_10_08.pdf. Accessed 30 October 2017.

Index

Note: Page numbers followed by *f* indicate figures and *t* indicate tables.

V

W

Printed in the United States
By Bookmasters